Dynamic Geometry on Time Scales

Dynamic Geometry on Time Scales

Svetlin G. Georgiev

CRC Press is an imprint of the
Taylor & Francis Group, an **informa** business

First edition published 2022
by CRC Press
6000 Broken Sound Parkway NW, Suite 300, Boca Raton, FL 33487-2742

and by CRC Press
2 Park Square, Milton Park, Abingdon, Oxon OX14 4RN

CRC Press is an imprint of Taylor & Francis Group, LLC

Library of Congress Cataloging-in-Publication Data
A catalog record has been requested for this book

ISBN: 978-1-032-07078-0 (hbk)
ISBN: 978-1-032-07080-3 (pbk)
ISBN: 978-1-003-20526-5 (ebk)

DOI: 10.1201/9781003205265

Typeset in Palatino
by Newgen Publishing UK

Dedicated to Professor Marcel Berger

Contents

Preface xi

1. Curves in $\mathbb{R}^n$ 1
- 1.1 Frenet Curves in $\mathbb{R}^n$ 1
- 1.2 Analytical Representations of Curves 14
  - 1.2.1 Plane Curves 15
    - 1.2.1.1 Parametric Representation 15
    - 1.2.1.2 Explicit Representation 16
    - 1.2.1.3 Implicit Representation 16
  - 1.2.2 Space Curves 18
    - 1.2.2.1 Parametric Representation 18
    - 1.2.2.2 Explicit Representation 18
    - 1.2.2.3 Implicit Representation 19
- 1.3 The Tangent Line and the Normal Plane. The Normal of a Plane Curve 21
  - 1.3.1 Parametric Representation 23
  - 1.3.2 Explicit Representation 28
  - 1.3.3 Implicit Representation 29
- 1.4 Envelopes of Plane Curves 31
- 1.5 The Osculating Plane 35
- 1.6 The Curvature of a Curve 40
- 1.7 The Frenet Frame. Frenet Formulae 41
- 1.8 Osculating Circles. Evolutes 44
- 1.9 Oriented Space Curves. The Frenet Frame of an Oriented Space Curve 49
- 1.10 General Helices 49
- 1.11 Bertrand Curves 51
- 1.12 The Behaviour of the Frenet Frame under a Rigid Motion 55
- 1.13 The Uniqueness Theorem 58
- 1.14 The Existence Theorem 59
- 1.15 Advanced Practical Problems 61
- 1.16 Notes and References 63

2. General Theory of Surfaces 65
- 2.1 Preliminaries 65
- 2.2 Parameterized Surfaces 69

2.3 Representations of Surfaces 73
2.3.1 Parametric Representation 73
2.3.2 Implicit Representation 73
2.4 Tangent Vector Spaces. Normals to a Surface 77
2.5 Orientations of a Surface 94
2.6 Advanced Practical Problems 95
2.7 Notes and References 96

3. First Fundamental Forms 97
3.1 Differentiable Maps on a Surface 97
3.2 The Differential of a Smooth Map Between Surfaces 99
3.3 Spherical Maps. Shape Operators 105
3.4 First Fundamental Forms 123
3.5 Lengths of a Curve on a Surface 131
3.6 Angles between Two Curves on a Surface 139
3.7 Areas of Parameterized Surfaces 142
3.8 Advanced Practical Problems 147
3.9 Notes and References 148

4. The Second Fundamental Forms 151
4.1 Definitions 151
4.2 Normal Curvatures. Variant of the Meusnier Theorem 157
4.3 Asymptotic Directions. Asymptotic Lines on a Surface 183
4.4 Principal Directions. Gaussian Curvatures. Mean Curvatures 187
4.5 The Joachimstahl Theorem 209
4.6 The Deteremination of the Lines of Curvatures 213
4.7 The Computation of the Curvatures of a Surface 223
4.8 Advanced Practical Problems 253
4.9 Notes and References 255

5. The Fundamental Equations of a Surface 257
5.1 The Differentiation Rules. The Christoffel Coefficients 257
5.2 The Weingarten Coefficients 265
5.3 The Gauss and Godazzi-Mainardi Equations for a Surface 289
5.4 Darboux Frames 301
5.5 The Geodesic Torsions 304
5.6 The Geodesic Curvatures. The Geodesic Lines 306
5.7 Advanced Practical Problems 308
5.8 Notes and References 309

6. Minimal Surfaces 311
6.1 Statement of the Variational Problem 311
6.2 First and Second Variation 312
6.3 Euler's Condition 316

6.4 Minimal Surfaces 320
6.5 Advanced Practical Problems 324
6.6 Notes and References 325

7. The Delta Nature Connection 327
7.1 The Directional Derivative 327
7.2 Tangent Spaces. Vector Fields 342
7.3 Delta Covariant Differentiation. The Delta Nature Connection 343
7.4 The Lie Brackets 347
7.5 The Algebra of Dynamic Forms 354
7.6 Exterior Differentiations 359
7.7 Advanced Practical Problems 366
7.8 Notes and References 368

Appendix A: Implicit Function Theorem 369

References 377

Index 379

Preface

The theory of time scales was introduced by Stefan Hilger in his PhD thesis [5] in 1988 (supervised by Bernd Aulbach) in order to unify continuous and discrete analysis and to extend the continuous and discrete theories to cases "in between". This book presents an introduction to the theory of curves and surfaces on time scales. The book is primarily intended for senior undergraduate students and beginning graduate students of engineering and science courses. Students in mathematical and physical sciences will find many sections of direct relevance. This book contains seven chapters, and each chapter consists of results with their proofs, numerous examples, and exercises with solutions. Each chapter concludes with a section featuring advanced practical problems with solutions, followed by a section on notes and references, explaining its context within existing literature.

In Chapter 1 we introduce the concept for Frenet curves in $\mathbb{R}^n$. They are given different representations for planes and space curves. They are deduced from equations of tangent lines and normal planes for space curves as well as the equations of the normal line to a plane curve. The osculating plane and the curvature of a space line are investigated. Conditions are given when a plane curve is an envelope for a family of plane curves. The Frenet frame is and the main Frenet equations are deduced. As applications, general helices and Bertrand curves are defined and we deduce some of their properties. The behaviour of the Frenet frame under a rigid motion is investigated and the existence and uniqueness theorem is proved. In Chapter 2 surfaces are introduced and the local and global parameterization of surfaces. Tangent planes, σ_1-tangent planes, σ_2-tangent planes, normals, σ_1-normals and σ_2-normals to a surface are defined and their equations are deduced. In Chapter 3 differentiable maps on surfaces, the differential of a smooth map between surfaces are introduced and some of their properties are deduced. Spherical maps, σ_1- and σ_2-spherical maps, the shape operator, and σ_1- and σ_2-shape operators are defined. It is proven that the shape operators are not symmetric operators when one of the time scales does not coincide with $\mathbb{R}$. In this chapter the first fundamental form and the first σ_1- and σ_2-fundamental forms of a surface are defined. The σ_1- and σ_2 length of a curve on a surface and the area of a parameterized surface are introduced as applications. Chapter 4 is devoted to the second fundamental forms for oriented surfaces. Some cases of the Meusnier theorem are deduced. Normal curvatures and asymptotic directions for oriented surfaces, asymptotic curves on surfaces, principal directions, principal curvatures, Gauss curvatures and mean curvatures of oriented surfaces are introduced. The points on oriented surfaces are classificed. In this chapter some variants of the Joachimstahl Theorem are formulated and proved. In

Chapter 5 the main differentiation rules using the first and second fundamental forms of a surface are introduced. The Christoffel and Weingarten coefficients are defined and using them the Gauss and Godazzi-Mainardi equations are deduced. In Chapter 6 the first and second variations are investigated. They are given necessary and sufficient conditions for an extremum of a function. The Euler equations are deduced. In the chapter minimal surfaces are defined and some of their properties are deduced. In this chapter the directional derivative and some of its properties are deduced. Using the directional derivative the tangent space on time scales and covariant differentiation on time scales is introduced and the delta nature connection on time scales is defined. In Chapter 7 the Lie bracket is defined and some of its properties are given. The definition for k-dynamics forms on time scales and the $\Delta_{\sigma_{ij}}$-exterior differentiation of k-dynamic forms is given. In the appendix the implicit function theorem is proven.

The aim of this book is to present a clear and well-organized treatment of the concept behind the development of mathematics as well as solution methods. The text material of this book is presented in a readable and mathematically solid format.

Paris

Svetlin Georgiev
January 2021

1

Curves in $\mathbb{R}^n$

Suppose that $\mathbb{T}$ is a time scale with forward jump operator and delta differentiation operator σ and Δ, respectively.

1.1 Frenet Curves in $\mathbb{R}^n$

Let $a, b \in \mathbb{T}$, $a < b$.

Definition 1.1 A time scale curve (shortly curve) in $\mathbb{R}^n$ is an rd-continuous function $f : [a,b] \to \mathbb{R}^n$. The function f that defines the curve is called a parametrization of the curve and the curve is a parametric curve.

Example 1.1 Let

$$\mathbb{T} = \left\{0, \frac{1}{3}, \frac{1}{2}, \frac{5}{6}, 1, \frac{9}{8}, \frac{11}{8}, \frac{7}{4}, 2, \frac{7}{3}, 3, \frac{13}{4}, \frac{7}{2}, \frac{15}{4}, 4\right\}.$$

Suppose that $f : \mathbb{T} \to \mathbb{R}$ is defined by

$$f(t) = (t^2, t^2 + t^4), \quad t \in \mathbb{T}.$$

Take $a = 0$, $b = 3$. Then $f : [a,b] \to \mathbb{R}$ is a time scale curve and its graph is shown in Figure 1.1.

Definition 1.2 A regular parameterized time scale curve (shortly regular parameterized curve) is a function $f : [a,b] \to \mathbb{R}^n$ such that $f \in \mathscr{C}^1_{rd}([a,b])$ and $f^\Delta(t) \neq (0, \ldots, 0)$ for any $t \in [a,b]$.

Definition 1.3 Let $f : [a,b] \to \mathbb{R}$ be a parameterized curve such that $f \in \mathscr{C}^1_{rd}([a,b])$. The vector $f^\Delta(t_0)$, $t_0 \in [a,b]$, is called the tangent vector to f at t_0 and the line spanned by this vector through $f(t_0)$ is called the tangent to f at this point.

Example 1.2 Let $\mathbb{T} = \mathbb{N}_0^2 \bigcup \{0\}$, $f : \mathbb{T} \to \mathbb{R}^2$ be defined by

$$f(t) = \left(t^2, t^3 + t^2 + 1\right), \quad t \in \mathbb{T}.$$

Take $a = 0$ and $b = 9$. Set

$$f_1(t) = t^2,$$

DOI: 10.1201/9781003205265-1

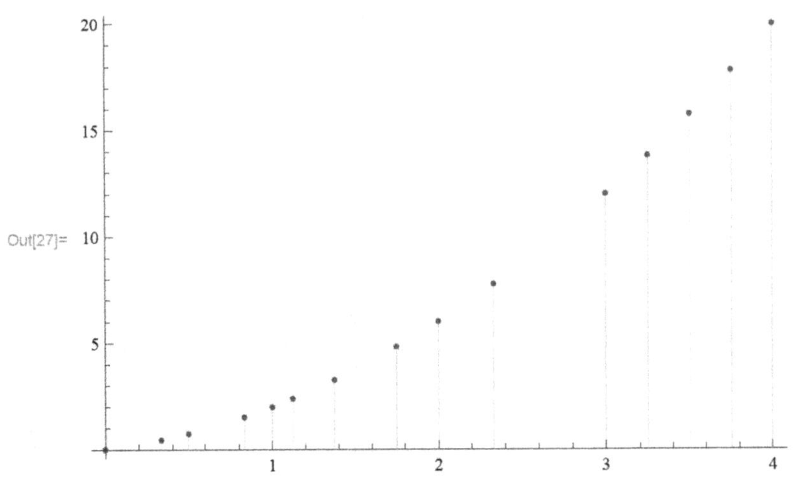

FIGURE 1.1

$$f_2(t)=t^3+t^2+1,\quad t\in\mathbb{T}.$$

Then

$$\sigma(t)=(\sqrt{t}+1)^2,\quad t\in\mathbb{T},$$

and

$$\begin{aligned}
f_1^{\Delta}(t)&=\sigma(t)+t\\
&=(\sqrt{t}+1)^2+t\\
&=2t+2\sqrt{t}+1,\\
f_2^{\Delta}(t)&=(\sigma(t))^2+t\sigma(t)+t^2+\sigma(t)+t\\
&=(\sqrt{t}+1)^4+t(\sqrt{t}+1)^2+t^2+(\sqrt{t}+1)^2+t\\
&=(2t+2\sqrt{t}+1)^2+t(2t+2\sqrt{t}+1)+t^2+(t+2\sqrt{t}+1)+t\\
&=4t^2+4t+1+8t\sqrt{t}+4t+4\sqrt{t}+2t^2\\
&\quad+2t\sqrt{t}+t+t^2+2t+2\sqrt{t}+2\\
&=7t^2+10t\sqrt{t}+11t+6\sqrt{t}+2,\quad t\in\mathbb{T}.
\end{aligned}$$

Hence,

$$f^{\Delta}(t)=(2t+2\sqrt{t}+1,7t^2+10t\sqrt{t}+11t+6\sqrt{t}+2),\quad t\in\mathbb{T},$$

and $f^{\Delta}(t)\neq(0,0)$ for any $t\in[0,9]$. Therefore $f:[0,9]\to\mathbb{R}^2$ is a regular parameterized curve.

Example 1.3 Let $\mathbb{T}=\mathbb{R}$ and f be defined as in Example 1.2.

Then

$$f'(t)=(2t,3t^2+2t),\quad t\in[0,9].$$

Observe that $f'(0) = (0,0)$ and $f : [0,9] \to \mathbb{R}^2$ is not a regular parameterized curve.

Exercise 1.1 Let $\mathbb{T} = \left(\frac{1}{3}\right)^{\mathbb{N}_0} \bigcup\{1\}\bigcup\left\{\frac{4}{3},\frac{5}{3},7,8\right\}, f : \left[\frac{1}{27},\frac{5}{3}\right] \to \mathbb{R}^3$ be defined by

$$f(t) = \left(e_1(t,1)+t^2,\quad \cos_2(t,1)+t, \sin_3(t,1)+t+t^2\right),\quad t \in \left[\frac{1}{27},\frac{5}{3}\right].$$

Check if $f : \left[\frac{1}{27},\frac{5}{3}\right] \to \mathbb{R}^3$ is a regular parameterized curve.

Definition 1.4 Let $\widetilde{\mathbb{T}}$ be a time scale with forward jump operator and delta differentiation operator $\widetilde{\sigma}$ and $\widetilde{\Delta}$, respectively, and $[\alpha,\beta] \subset \widetilde{\mathbb{T}}, f : [a,b] \to \mathbb{R}^n$ be a regular parameterized curve and $\phi : [\alpha,\beta] \to [a,b]$ be a bijection, $\phi \in \mathscr{C}^1_{rd}([\alpha,\beta])$, $\phi^{\widetilde{\Delta}} > 0$ on $[\alpha,\beta]$, and $\widetilde{\sigma}(\phi(s)) = \phi(\sigma(s))$ for $s \in [\alpha,\beta]$. Then $f \circ \phi$ and f are said to be positively equivalent. The four $(f,\phi,\mathbb{T},\widetilde{\mathbb{T}})$ will be called the admissible parameterizing four.

Example 1.4 Let $\mathbb{T} = \widetilde{\mathbb{T}} = 2^{\mathbb{N}_0}$ and

$$\alpha = 81,\quad \beta = 6561,\quad a = 3,\quad b = 9.$$

Consider $\phi : [81,6561] \to [3,9]$, defined by

$$\phi(t) = \sqrt[4]{t},\quad t \in [81,6561],$$

and $f : [3,9] \to \mathbb{R}^3$, defined by

$$f(t) = (t^2 + t^4, e_1(t,1), 1 + t^2),\quad t \in [3,9].$$

We have that $\sigma(t) = 2t$, $t \in \mathbb{T}$.

Set

$$\begin{aligned} f_1(t) &= t^2 + t^4,\\ f_2(t) &= e_1(t,1),\\ f_3(t) &= 1 + t^2,\quad t \in [3,9]. \end{aligned}$$

Then

$$\begin{aligned} f_1^{\Delta}(t) &= (\sigma(t))^3 + t(\sigma(t))^2 + t^2\sigma(t) + t^3 + \sigma(t) + t\\ &= 8t^3 + 4t^3 + 2t^3 + t^3 + 2t + t\\ &= 15t^3 + 3t,\\ f_2^{\Delta}(t) &= e_1(t,1),\\ f_3^{\Delta}(t) &= \sigma(t) + t \end{aligned}$$

$$= 2t + t$$
$$= 3t, \quad t \in [3,9].$$

Thus,

$$f^{\Delta}(t) = (15t^3 + 3t, e_1(t,1), 3t)$$
$$\neq (0,0,0), \quad t \in [3,9].$$

Therefore $f : [3,9] \to \mathbb{R}^3$ is a regular parameterized curve.

Next, $\phi : [81,6561] \to [3,9]$ is a bijection and

$$\begin{aligned}
\phi^{\Delta}(t) &= \frac{\sqrt[4]{\sigma(t)} - \sqrt[4]{t}}{\sigma(t) - t} \\
&= \frac{\sqrt[4]{\sigma(t)} - \sqrt[4]{t}}{(\sqrt[4]{\sigma(t)} - \sqrt[4]{t})(\sqrt[4]{\sigma(t)} + \sqrt[4]{t})(\sqrt{\sigma(t)} + \sqrt{t})} \\
&= \frac{1}{(\sqrt[4]{\sigma(t)} + \sqrt[4]{t})(\sqrt{\sigma(t)} + \sqrt{t})} \\
&= \frac{1}{(\sqrt[4]{2t} + \sqrt[4]{t})(\sqrt{2t} + \sqrt{t})} \\
&= \frac{1}{t^{\frac{3}{4}}(\sqrt[4]{2} + 1)(\sqrt{2} + 1)} \\
&> 0, \quad t \in [81,6561].
\end{aligned}$$

Moreover,

$$\begin{aligned}
f(\phi(t)) &= ((\phi(t))^2 + (\phi(t))^4, e_1(\phi(t),1), 1 + (\phi(t))^2) \\
&= \left(\sqrt{t} + t, e_1\left(\sqrt[4]{t},1\right), 1 + \sqrt{t}\right), \quad t \in [81,6561].
\end{aligned}$$

We have that $f \circ \phi$ and f are equivalent.

Exercise 1.2 Let $\tilde{\mathbb{T}} = \mathbb{T} = \mathbb{Z}$, $\alpha = 0$, $\beta = 15$, $a = 2$, $b = 17$,
$\phi : [0,15] \to [2,17]$ is defined by $\phi(t) = t + 2$, $t \in [0,15]$,
$f : [2,17] \to \mathbb{R}^4$ is defined by

$$f(t) = \left(t, t + t^2 + t^3, \frac{1+t}{1+3t}, t \right), \quad t \in [2,17].$$

Prove that

1. $f : [2,17] \to \mathbb{R}^4$ is a regular parameterized curve.
2. $f \circ \phi$ and f are equivalent.

Example 1.5 Let $\widetilde{\mathbb{T}} = \mathbb{N}_0$, $\mathbb{T} = \mathbb{N}_0^2$, $\phi : \widetilde{\mathbb{T}} \to \mathbb{T}$ and $f : \mathbb{T} \to \mathbb{R}^3$ are defined by

$$\phi(t) = t^2, \quad t \in \widetilde{\mathbb{T}},$$

$$f(t) = \left(1+t, \frac{1+2t}{1+3t}, t\right), \quad t \in \mathbb{T},$$

respectively.

Here

$$\widetilde{\sigma}(t) = t+1, \quad t \in \widetilde{\mathbb{T}},$$

$$\sigma(t) = (\sqrt{t}+1)^2, \quad t \in \mathbb{T}.$$

Set

$$f_1(t) = 1+t,$$

$$f_2(t) = \frac{1+2t}{1+3t},$$

$$f_3(t) = t, \quad t \in \mathbb{T}.$$

Then

$$\begin{aligned}
f_1^{\Delta}(t) &= 1, \\
f_2^{\Delta}(t) &= \frac{2(1+3t)-3(1+2t)}{(1+3t)(1+3\sigma(t))} \\
&= \frac{2+6t-3-6t}{(1+3t)(1+3(\sqrt{t}+1)^2)} \\
&= -\frac{1}{(1+3t)(4+3t+6\sqrt{t})}, \\
f_3^{\Delta}(t) &= 1, \quad t \in \mathbb{T}.
\end{aligned}$$

Therefore

$$\begin{aligned}
f^{\Delta}(t) &= \left(1, -\frac{1}{(1+3t)(4+3t+6\sqrt{t})}, 1\right) \\
&\neq (0,0,0), \quad t \in \mathbb{T}.
\end{aligned}$$

Thus, $f : \mathbb{T} \to \mathbb{R}^3$ is a regular parameterized curve.

Next,

$$\begin{aligned}
\phi^{\widetilde{\Delta}}(t) &= \widetilde{\sigma}(t) + t \\
&= t+1+t \\
&= 2t+1 \\
&> 0, \quad t \in \widetilde{\mathbb{T}}.
\end{aligned}$$

Consequently f and $f \circ \phi$ are equivalent.

Exercise 1.3 Let $\widetilde{\mathbb{T}} = \mathbb{N}_0^3$, $\mathbb{T} = \mathbb{N}_0$, $\phi : \widetilde{\mathbb{T}} \to \mathbb{T}$ and $f : \mathbb{T} \to \mathbb{R}^4$ be defined by

$$\phi(t) = \sqrt[3]{t}, \quad t \in \widetilde{\mathbb{T}},$$
$$f(t) = \left(\frac{1+t}{1+2t}, \cos_2(t,1), \sin_3(t,1), t^2\right), \quad t \in \mathbb{T},$$

respectively.

Prove that

1. f is a regular parameterized curve.
2. f and $f \circ \phi$ are equivalent.

Definition 1.5 Let $f : [a,b] \to \mathbb{R}^n$ be a regular parameterized curve and

$$f(t) = (f_1(t), \ldots, f_n(t)), \quad t \in [a,b].$$

The arc length parameter $L_f(t,a)$, $t \in [a,b]$, is defined as follows

$$\begin{aligned} L_f(t,a) &= \int_a^t \|f^\Delta(s)\| \Delta s \\ &= \int_a^t \sqrt{(f_1^\Delta(s))^2 + \cdots + (f_n^\Delta(s))^2}\, \Delta s. \end{aligned}$$

The total length of f is defined to be the number $L_f(b,a)$.

Example 1.6 Let $\mathbb{T} = 2^{\mathbb{N}_0} \bigcup \{0\}$, $a = 0$, $b = 4$, $f : [0,2] \to \mathbb{R}^3$ be defined by

$$f(t) = (t^2, 1+t+t^2, t^3), \quad t \in [0,2].$$

Here

$$\begin{aligned} f_1(t) &= t^2, \\ f_2(t) &= 1+t+t^2, \\ f_3(t) &= t^3, \quad t \in [0,2]. \end{aligned}$$

We have $\sigma(0) = 1$ and $\sigma(1) = 2$.

Then

$$\begin{aligned} f_1^\Delta(t) &= \sigma(t) + t, \\ f_2^\Delta(t) &= 1 + \sigma(t) + t, \\ f_3^\Delta(t) &= (\sigma(t))^2 + t\sigma(t) + t^2, \quad t \in [0,2], \\ f_1^\Delta(0) &= \sigma(0) \\ &= 1, \end{aligned}$$

$$\begin{aligned}
f_2^{\Delta}(0) &= 1+\sigma(0)\\
&= 1+1\\
&= 2,\\
f_3^{\Delta}(0) &= (\sigma(0))^2\\
&= 1,\\
f_1^{\Delta}(1) &= \sigma(1)+1\\
&= 2+1\\
&= 3,\\
f_2^{\Delta}(1) &= 1+\sigma(1)+1\\
&= 2+2\\
&= 4,\\
f_3^{\Delta}(1) &= (\sigma(1))^2+\sigma(1)+1\\
&= 4+2+1\\
&= 7.
\end{aligned}$$

Hence,

$$\begin{aligned}
L_f(2,0) &= \int_0^2 \sqrt{\left(f_1^{\Delta}(s)\right)^2+\left(f_2^{\Delta}(s)\right)^2+\left(f_3^{\Delta}(s)\right)^2}\,\Delta s\\
&= \mu(0)\sqrt{\left(f_1^{\Delta}(0)\right)^2+\left(f_2^{\Delta}(0)\right)^2+\left(f_3^{\Delta}(0)\right)^2}\\
&\quad +\mu(1)\sqrt{\left(f_1^{\Delta}(1)\right)^2+\left(f_2^{\Delta}(1)\right)^2+\left(f_3^{\Delta}(1)\right)^2}\\
&= \sqrt{1+4+1}+\sqrt{9+16+49}\\
&= \sqrt{6}+\sqrt{74}.
\end{aligned}$$

Exercise 1.4 Let

$$\mathbb{T}=\left\{0,\frac{1}{4},\frac{7}{12},1,\frac{7}{3},\frac{5}{2},\frac{5}{3},2,3,4\right\},\quad f:[0,2]\to\mathbb{R}^4 \text{ be defined by}$$

$$f(t)=\left(\frac{1+t}{1+2t},1+t^2,1+t+t^2,t^2\right),\quad t\in[0,2].$$

Find $L_f(2,0)$.

Theorem 1.1 *Let* $f:[a,b]\to\mathbb{R}^n$ *be a regular parameterized curve,* $[\alpha,\beta]$ *be a time scale with forward jump operator and delta differentiation operator* $\widetilde{\sigma}$ *and* $\widetilde{\Delta}$*, respectively,* $L_f(\cdot,a):[a,b]\to[\alpha,\beta]$ *be a bijection,* $\widetilde{\sigma}(L_f(t,a))=L_f(\sigma(t),a)$*,*

$L_f^{-1}(\tilde{\sigma}(s),\alpha) = \sigma(L_f^{-1}(s,\alpha))$, $s = L_f(t,a)$, $t \in [a,b]$. *Then* f *and* $f\left(L_f^{-1}(\cdot,a)\right)$ *are equivalent and*

$$f^{\Delta}(t) = \left(f\left(L_f^{-1}(\cdot,\alpha)\right)\right)^{\tilde{\Delta}}(s)\|f^{\Delta}(t)\|, \quad t \in [a,b], \quad s = L_f(t,a).$$

Proof. We have

$$L_f(t,a) = \int_a^t \|f^{\Delta}(\tau)\|\,\Delta\tau, \quad t \in [a,b],$$

and

$$\begin{aligned} L_f^{\Delta}(t,a) &= \|f^{\Delta}(t)\| \\ &> 0, \quad t \in [a,b]. \end{aligned}$$

Next,

$$\begin{aligned} \left(L_f^{-1}(\cdot,\alpha)\right)^{\tilde{\Delta}}(s) &= \lim_{s_1 \to s} \frac{L_f^{-1}(\tilde{\sigma}(s),\alpha) - L_f^{-1}(s_1,\alpha)}{\tilde{\sigma}(s) - s_1} \\ &= \lim_{s_1 \to s, t_1 \to t} \frac{\sigma\left(L_f^{-1}(s,\alpha)\right) - L_f^{-1}(s_1,\alpha)}{\tilde{\sigma}(L_f(t,a)) - L_f(t_1,a)} \\ &= \lim_{t_1 \to t} \frac{\sigma(t) - t_1}{L_f(\sigma(t),a) - L_f(t_1,a)} \\ &= \frac{1}{\lim_{t_1 \to t} \frac{L_f(\sigma(t),a) - L_f(t_1,a)}{\sigma(t) - t_1}} \\ &= \frac{1}{L_f^{\Delta}(t,a)} \\ &= \frac{1}{\|f^{\Delta}(t)\|} \\ &> 0, \quad s \in [\alpha,\beta]. \end{aligned}$$

Therefore

$$\left(L_f^{-1}(\cdot,\alpha)\right)^{\tilde{\Delta}}(s) > 0, \quad s \in [\alpha,\beta],$$

and $f\left(L_f^{-1}(\cdot,\alpha)\right)$ and f are equivalent.

Moreover,

$$\begin{aligned} f^{\Delta}(t) &= \lim_{t_1 \to t} \frac{f(\sigma(t)) - f(t_1)}{\sigma(t) - t_1} \\ &= \lim_{s_1 \to s} \frac{f\left(\sigma\left(L_f^{-1}(s,\alpha)\right)\right) - f\left(L_f^{-1}(s_1,\alpha)\right)}{\sigma\left(L_f^{-1}(s,\alpha)\right) - L_f^{-1}(s_1,\alpha)} \end{aligned}$$

$$
\begin{aligned}
&= \lim_{s_1 \to s} \frac{f\left(L_f^{-1}(\widetilde{\sigma}(s),\alpha)\right) - f\left(L_f^{-1}(s_1,\alpha)\right)}{L_f^{-1}(\widetilde{\sigma}(s),\alpha) - L_f^{-1}(s_1,\alpha)} \\
&= \lim_{s_1 \to s} \left(\frac{f\left(L_f^{-1}(\widetilde{\sigma}(s),\alpha)\right) - f\left(L_f^{-1}(s_1,\alpha)\right)}{\widetilde{\sigma}(s) - s_1} \right. \\
&\qquad \left. \times \frac{1}{\frac{L_f^{-1}(\widetilde{\sigma}(s),\alpha) - L_f^{-1}(s_1,\alpha)}{\widetilde{\sigma}(s) - s_1}} \right) \\
&= \left(f\left(L_f^{-1}(\cdot,\alpha)\right)\right)^{\widetilde{\Delta}}(s) \frac{1}{\left(L_f^{-1}(\cdot,\alpha)\right)^{\widetilde{\Delta}}(s)} \\
&= \left(f\left(L_f^{-1}(\cdot,\alpha)\right)\right)^{\widetilde{\Delta}}(s) \|f^{\Delta}(t)\|,
\end{aligned}
$$

$s \in [\alpha,\beta]$, $t = L_f^{-1}(s,\alpha)$. This completes the proof.

Definition 1.6 Let $f : [a,b] \to \mathbb{R}^n$ be a regular parameterized curve such that $f, L_f, L_f^{-1}, [a,b], [\alpha,\beta]$ satisfy all conditions of Theorem 1.1. Then we say that f has the arc length property (shortly (ALP)). In this case, if we denote

$$
g(s) = f\left(L_f^{-1}(s,\alpha)\right), \quad s \in [\alpha,\beta],
$$

then

$$
\begin{aligned}
f^{\Delta}(t) &= g^{\widetilde{\Delta}}(s) s^{\Delta}(t) \\
&= g^{\widetilde{\Delta}}(s) \|f^{\Delta}(t)\|, \quad s \in [\alpha,\beta], \quad t \in [a,b].
\end{aligned}
$$

Example 1.7 Let $\mathbb{T} = 2^{\mathbb{N}_0}$, $a = 1$, $b = 8$, $f : [1,8] \to \mathbb{R}^3$ be defined by

$$
\begin{aligned}
f(t) &= (f_1(t), f_2(t), f_3(t)) \\
&= \left(\frac{1}{7}t^3, \frac{\sqrt{2}}{15}t^4, \frac{1}{31}t^5\right), \quad t \in [1,8],
\end{aligned}
$$

$$
\begin{aligned}
[\alpha,\beta] &= \{0, 2, 42, 1130\}, \\
L_f^{-1}(1130,0) &= 8, \\
L_f^{-1}(42,0) &= 4, \\
L_f^{-1}(2,0) &= 2, \\
L_f^{-1}(0,0) &= 1.
\end{aligned}
$$

We have $\sigma(t) = 2t$, $t \in [1,8]$,

and

$$\begin{aligned}
f_1^{\Delta}(t) &= \frac{1}{7}((\sigma(t))^2 + t\sigma(t) + t^2) \\
&= \frac{1}{7}(4t^2 + 2t^2 + t^2) \\
&= t^2, \\
f_2^{\Delta}(t) &= \frac{\sqrt{2}}{15}((\sigma(t))^3 + t(\sigma(t))^2 + t^2\sigma(t) + t^3) \\
&= \frac{\sqrt{2}}{15}(8t^3 + 4t^3 + 2t^3 + t^3) \\
&= \sqrt{2}t^3, \\
f_3^{\Delta}(t) &= \frac{1}{31}\left((\sigma(t))^4 + t(\sigma(t))^3 + t^2(\sigma(t))^2 + t^3\sigma(t) + t^4\right) \\
&= \frac{1}{31}(16t^4 + 8t^4 + 4t^4 + 2t^4 + t^4) \\
&= t^4, \\
f^{\Delta}(t) &= t^2(1, \sqrt{2}t, t^2), \\
\|f^{\Delta}(t)\| &= \sqrt{(f_1^{\Delta}(t))^2 + (f_2^{\Delta}(t))^2 + (f_3^{\Delta}(t))^2} \\
&= \sqrt{t^4 + 2t^6 + t^8} \\
&= \sqrt{t^4(1 + 2t^2 + t^4)} \\
&= \sqrt{t^4(1 + t^2)^2} \\
&= t^2(1 + t^2), \quad t \in [1, 8].
\end{aligned}$$

Next,

$$\begin{aligned}
\widetilde{\sigma}(0) &= 2, \\
\widetilde{\sigma}(2) &= 42, \\
\widetilde{\sigma}(42) &= 1130, \\
\widetilde{\sigma}(1130) &= 1130, \\
\left(f\left(L_f^{-1}(\cdot, 0)\right)\right)^{\widetilde{\Delta}}(0) &= \frac{f\left(L_f^{-1}(\widetilde{\sigma}(0), 0)\right) - f\left(L_f^{-1}(0, 0)\right)}{\widetilde{\sigma}(0) - 0} \\
&= \frac{f\left(L_f^{-1}(2, 0)\right) - f\left(L_f^{-1}(0, 0)\right)}{2} \\
&= \frac{1}{2}\left(\left(\frac{8}{7}, \frac{16\sqrt{2}}{15}, \frac{32}{31}\right) - \left(\frac{1}{7}, \frac{\sqrt{2}}{15}, \frac{1}{31}\right)\right)
\end{aligned}$$

$$
\begin{aligned}
&= \frac{1}{2}(1, \sqrt{2}, 1), \\
\|f^{\Delta}(1)\| &= 2, \\
f^{\Delta}(1) &= (1, \sqrt{2}, 1).
\end{aligned}
$$

Therefore

$$
f^{\Delta}(1) = \left(f\left(L_f^{-1}(\cdot, 0)\right)\right)^{\tilde{\Delta}}(0)\|f^{\Delta}(1)\|.
$$

Moreover,

$$
\begin{aligned}
f^{\Delta}(2) &= (4, 8\sqrt{2}, 16), \\
f^{\Delta}(4) &= (16, 64\sqrt{2}, 256), \\
f^{\Delta}(8) &= 64(1, 8\sqrt{2}, 64), \\
\|f^{\Delta}(2)\| &= 4\sqrt{1+8+16} \\
&= 4\sqrt{25} \\
&= 20, \\
\|f^{\Delta}(4)\| &= 16\sqrt{1+32+256} \\
&= 16\sqrt{289} \\
&= 272, \\
\|f^{\Delta}(8)\| &= 64\sqrt{1+128+4096} \\
&= 64\sqrt{4225} \\
&= 64 \cdot 65 \\
&= 4160.
\end{aligned}
$$

Consequently

$$
\begin{aligned}
\left(f\left(L_f^{-1}(\cdot, 0)\right)\right)^{\tilde{\Delta}}(2) &= \frac{f^{\Delta}(2)}{\|f^{\Delta}(2)\|} \\
&= \frac{1}{20}(4, 8\sqrt{2}, 16) \\
&= \frac{1}{5}(1, 2\sqrt{2}, 4), \\
\left(f\left(L_f^{-1}(\cdot, 0)\right)\right)^{\tilde{\Delta}}(42) &= \frac{f^{\Delta}(4)}{\|f^{\Delta}(4)\|} \\
&= \frac{1}{272}(16, 64\sqrt{2}, 256) \\
&= \frac{1}{17}(1, 4\sqrt{2}, 16),
\end{aligned}
$$

$$\left(f\left(L_f^{-1}(\cdot,0)\right)\right)^{\tilde{\Delta}}(1130) = \frac{f^{\Delta}(8)}{\|f^{\Delta}(8)\|}$$
$$= \frac{64}{4160}(1,8\sqrt{2},64)$$
$$= \frac{1}{65}(1,8\sqrt{2},64).$$

Suppose that $f:[a,b]\to\mathbb{R}^n$ has (ALP).

Then

$$f^{\Delta}(t) = g^{\tilde{\Delta}}(s)\|f^{\Delta}(t)\|,$$
$$\|g^{\tilde{\Delta}}(s)\| = 1.$$

Then

$$g^{\tilde{\Delta}}(s)\cdot g^{\tilde{\Delta}}(s) = 1$$

and hence,

$$0 = g^{\tilde{\Delta}^2}(s)\cdot g^{\tilde{\Delta}}(s) + g^{\tilde{\Delta}\tilde{\sigma}}(s)\cdot g^{\tilde{\Delta}^2}(s)$$
$$= g^{\tilde{\Delta}^2}(s)\cdot\left(g^{\tilde{\Delta}}(s) + g^{\tilde{\Delta}\tilde{\sigma}}(s)\right), \quad s\in[\alpha,\beta].$$

Thus,

$$g^{\tilde{\Delta}^2} \perp g^{\tilde{\Delta}} + g^{\tilde{\Delta}\tilde{\sigma}}.$$

Example 1.8 Let $\mathbb{T}$, a, b, f, α, β be as in Example 1.7.

Then

$$g^{\tilde{\Delta}}(\tilde{\sigma}(0)) = g^{\tilde{\Delta}}(2)$$
$$= \frac{1}{5}(1,2\sqrt{2},4),$$
$$g^{\tilde{\Delta}^2}(0) = \frac{1}{\tilde{\sigma}(0)-0}\left(g^{\tilde{\Delta}\tilde{\sigma}}(0) - g^{\tilde{\Delta}}(0)\right)$$
$$= \frac{1}{2}\left(\frac{1}{5}(1,2\sqrt{2},4) - \frac{1}{2}(1,\sqrt{2},1)\right)$$
$$= \frac{1}{2}\left(\frac{1}{5}-\frac{1}{2}, \frac{2\sqrt{2}}{5}-\frac{\sqrt{2}}{2}, \frac{4}{5}-\frac{1}{2}\right)$$
$$= \frac{1}{2}\left(-\frac{3}{10}, -\frac{\sqrt{2}}{10}, \frac{3}{10}\right)$$
$$= \left(-\frac{3}{20}, -\frac{\sqrt{2}}{20}, \frac{3}{20}\right),$$
$$g^{\tilde{\Delta}}(0) + g^{\tilde{\Delta}\tilde{\sigma}}(0) = g^{\tilde{\Delta}}(0) + g^{\tilde{\Delta}}(2)$$

$$= \left(\frac{1}{2}, \frac{\sqrt{2}}{2}, \frac{1}{2}\right) + \left(\frac{1}{5}, \frac{2\sqrt{2}}{5}, \frac{4}{5}\right)$$

$$= \left(\frac{1}{2} + \frac{1}{5}, \frac{\sqrt{2}}{2} + \frac{2\sqrt{2}}{5}, \frac{1}{2} + \frac{4}{5}\right)$$

$$= \left(\frac{7}{10}, \frac{9\sqrt{2}}{10}, \frac{13}{10}\right).$$

Thus,

$$g^{\tilde{\Delta}^2}(0) \cdot \left(g^{\tilde{\Delta}}(0) + g^{\tilde{\Delta}\tilde{\sigma}}(0)\right) = -\frac{21}{200} - \frac{18}{200} + \frac{39}{200} = 0.$$

Consequently

$$g^{\tilde{\Delta}^2}(0) \perp g^{\tilde{\Delta}}(0) + g^{\tilde{\Delta}\tilde{\sigma}}(0).$$

Definition 1.7 A curve $f : [a,b] \to \mathbb{R}^n$ that has (ALP) and

$$g^{\tilde{\Delta}}(s), \quad g^{\tilde{\Delta}^2}(s), \quad \ldots, g^{\tilde{\Delta}^{n-1}}(s)$$

exist and are rd-continuous and are linearly independent at any point $s \in [\alpha, \beta]$ is said to be a Frenet curve.

Example 1.9 Let f be the curve in Example 1.7. We consider it as a curve with respect to its arc length on $[0,2] \subset [\alpha, \beta]$.

By Example 1.8, we have

$$g^{\tilde{\Delta}^2}(0) = \left(-\frac{3}{20}, -\frac{\sqrt{2}}{20}, \frac{3}{20}\right).$$

Also,

$$g^{\tilde{\Delta}^2}(2) = \frac{g^{\tilde{\Delta}}(\tilde{\sigma}(2)) - g^{\tilde{\Delta}}(2)}{\tilde{\sigma}(2) - 2}$$

$$= \frac{g^{\tilde{\Delta}}(42) - g^{\tilde{\Delta}}(2)}{42 - 2}$$

$$= \frac{1}{40}\left(\frac{1}{17}(1, 4\sqrt{2}, 16) - \frac{1}{5}(1, 2\sqrt{2}, 4)\right)$$

$$= \frac{1}{40}\left(\frac{1}{17} - \frac{1}{5}, \frac{4\sqrt{2}}{17} - \frac{2\sqrt{2}}{5}, \frac{16}{17} - \frac{4}{5}\right)$$

$$= \frac{1}{40}\left(-\frac{12}{85}, \frac{(20 - 34)\sqrt{2}}{85}, \frac{80 - 68}{85}\right)$$

$$= \frac{1}{40}\left(-\frac{12}{85}, \frac{6\sqrt{2}}{85}, \frac{12}{85}\right)$$

$$= \left(-\frac{3}{850}, \frac{3\sqrt{2}}{1700}, \frac{3}{850}\right)$$

Thus,

$$g^{\tilde{\Delta}}(0) = \left(\frac{1}{2}, \frac{\sqrt{2}}{2}, \frac{1}{2}\right),$$

$$g^{\tilde{\Delta}^2}(0) = \left(-\frac{3}{20}, -\frac{\sqrt{2}}{20}, \frac{3}{20}\right)$$

are linearly independent, and the vectors

$$g^{\tilde{\Delta}}(2) = \left(\frac{1}{5}, \frac{2\sqrt{2}}{5}, \frac{4}{5}\right),$$

$$g^{\tilde{\Delta}^2}(2) = \left(-\frac{3}{850}, \frac{3\sqrt{2}}{1700}, \frac{3}{850}\right)$$

are linearly independent. Therefore the curve f is a Frenet curve on $[0,2]$.

Exercise 1.5 Let $\mathbb{T} = 2\mathbb{Z}, f : [0,10] \to \mathbb{R}^4$ is defined by

$$f(t) = \left(1+t, 1+t^2, \frac{1+t}{1+t^2}, 1+t^4\right), \quad t \in [0,10].$$

1. Find $f \circ L_f^{-1}$.
2. Prove that f and $f \circ L_f^{-1}$ are equivalent.
3. Prove that f is a Frenet curve.

1.2 Analytical Representations of Curves

Let $I \subset \mathbb{T}$.

Definition 1.8 A subset $M \subset \mathbb{R}^n$ is called a regular curve or a 1-dimensional smooth manifold of $\mathbb{R}^n$ if for each point $t_0 \in M$ there is a regular parameterized curve $f : I \to \mathbb{R}^n$ whose support $f(I)$ is an open neighbourhood in M of the point t_0, i.e., is a set of the form $M \cap U$, where U is a neighbourhood of t_0 in $\mathbb{R}^n$, while the map $f : I \to f(I)$ is a homeomorphism with respect to the topology of subspace of $f(I)$. A parameterized curve with these properties is called a local parametrization of the curve M arround the point t_0. If for a curve M there is a local parametrization which is global, i.e., $f(I) = M$, the curve is called simple.

1.2.1 Plane Curves

Definition 1.9 A regular curve $M \subset \mathbb{R}^3$ is called plane if it is contained in a plane π. We shall usually assume that the plane π coincides with the coordinate plane xOy.

1.2.1.1 *Parametric Representation*

We choose an arbitrary local parametrization $(I, f(t)) = (f_1(t), f_2(t), f_3(t))$, of the curve. Then the support $f(t)$ of this local parametrization is an open subset of the curve. For a global parametrization of a simple curve, $f(I)$ is the entire curve. Thus, any point t_0 of the curve has an open neighbourhood which is the support of the parameterized curve

$$\begin{aligned} x &= f_1(t) \\ y &= f_2(t). \end{aligned} \tag{1.1}$$

Definition 1.10 The equations (1.1) are called the parametric equations of the curve in the neighbourhood of the point t_0. Usually, unless the curve is simple, we cannot use the same set of equations to describe the points of an entire curve.

Example 1.10 Let $\mathbb{T} = \mathbb{Z}, I = [-1, 10]$. Then

$$f_1(t) = t + 1$$
$$f_2(t) = \frac{1+t^2}{1+t^4}, \quad t \in I,$$

is a parametric representation of a plane curve.

Example 1.11 Let $\mathbb{T} = 2\mathbb{Z}, I = [0, 4]$,

$$f_1(0) = 1, \quad f_1(2) = 3, \quad f_1(4) = 5,$$
$$f_2(0) = 3, \quad f_2(2) = -1, \quad f_2(4) = 0,$$

is a parametric representation of a plane curve.

Example 1.12 Let $\mathbb{T} = 3\mathbb{Z}, I = [0, 81]$.

Then

$$f_1(0) = 0 \quad f_1(t) = t + 1, \quad t \in [1, 27], \quad f_1(81) = 0$$
$$f_2(0) = 1, \quad f_2(t) = 3t + 1, \quad t \in [1, 27], \quad f_2(81) = 0$$

is a parametric representation of a plane curve.

1.2.1.2 *Explicit Representation*

Suppose that I is an open interval in $\mathbb{T}$ and $f : I \to \mathbb{R}$ is a smooth function. Then its graph is

$$C = \{(t, f(t)) : t \in I\} \tag{1.2}$$

is a simple curve, which has the global representation

$$x = t$$
$$y = f(t), \quad t \in I.$$

Definition 1.11 The equation

$$y = f(x)$$

is called the explicit equation of the curve (1.2). Sometimes for the explicit representation of a plane curve it is, also, used the term nonparametric form.

Example 1.13 Let $\mathbb{T} = 2^{\mathbb{N}_0}$, $I = (1, 128)$.

Then

$$y = \frac{x+1}{1+x+x^2}, \quad x \in I,$$

is an explicit representation of a plane curve.

Example 1.14 Let $\mathbb{T} = \mathbb{N}_0$, $I = (0, 15)$.

Then

$$y = x + 1 + \sinh_1(x, 1), \quad x \in I,$$

is an explicit representation of a plane curve.

Example 1.15 Let $\mathbb{T} = 4\mathbb{Z}$, $I = (0, 24)$.

Then

$$y = \frac{1}{1+x^2} + \cosh_1(x, 0), \quad x \in I,$$

is an explicit equation of a plane curve.

1.2.1.3 *Implicit Representation*

Let $\mathbb{T}_1$ and $\mathbb{T}_2$ be time scales with forward jump operators and delta differentiation operators σ_1, σ_2 and Δ_1, Δ_2, respectively. Suppose that $D \subset \mathbb{T}_1 \times \mathbb{T}_2$. Let $F : D \to \mathbb{R}$ be a smooth function and

$$C = \{(x, y) \in D : F(x, y) = 0\}$$

be the 0-level set of the function F. In the general case, C is not a regular curve. Nevertheless, if at the point $(x_0, y_0) \in C$ the vector gradient

$$\operatorname{grad} F(x_0, y_0) = \left(\frac{\Delta}{\Delta x} F(x_0, y_0), \frac{\Delta}{\Delta y} F(x_0, y_0) \right)$$

is not vanishing, then there exists an open neighbourhood U of the point (x_0,y_0) and a smooth function $y = f(x)$ defined on an open neighbourhood $I \subset \mathbb{T}_1$ of the point x_0 such that

$$C \bigcap U = \{(x,f(x)) : x \in I\}.$$

If grad $F \neq 0$ in all points of C, then C is a regular. curve.

Example 1.16 Let $\mathbb{T}_1 = \mathbb{Z}$, $\mathbb{T}_2 = 2^{\mathbb{N}_0}$, $F : \mathbb{T}_1 \times \mathbb{T}_2 \to \mathbb{R}$, $F(x,y) = x^2 - y^2$, $(x,y) \in \mathbb{T}_1 \times \mathbb{T}_2$.

We have

$$C = \{(x,y) \in \mathbb{T}_1 \times \mathbb{T}_2 : x^2 - y^2 = 0\}$$

and

$$\begin{aligned} \sigma_1(x) &= x+1, \quad x \in \mathbb{T}_1, \\ \sigma_2(y) &= 2y, \quad y \in \mathbb{T}_2, \\ \frac{\Delta}{\Delta x}F(x,y) &= \sigma_1(x)+x \\ &= x+1+x \\ &= 2x+1, \\ \frac{\Delta}{\Delta y}F(x,y) &= -\sigma_2(y)-y \\ &= -2y-y \\ &= -3y, \quad (x,y) \in \mathbb{T}_1 \times \mathbb{T}_2. \end{aligned}$$

Hence,

$$\begin{aligned} \text{grad}F(x,y) &= \left(\frac{\Delta}{\Delta x}F(x,y), \frac{\Delta}{\Delta y}F(x,y)\right) \\ &= (2x+1,-3y) \\ &\neq (0,0), \quad (x,y) \in \mathbb{T}_1 \times \mathbb{T}_2. \end{aligned}$$

Thus, C is a regular curve.

Remark 1.1 Note that the condition for nonsingularity of the grad$F$ is only a sufficient condition for the equation $F(x,y) = 0$ to represent a curve. If grad$F(x_0,y_0) = 0$ for some $(x_0,y_0) \in D$, then we cannot claim that the equation represent a curve in the neighbourhood of that point and the opposite.

Example 1.17 Let $\mathbb{T}_1 = \mathbb{T}_2 = \mathbb{R}$ and $F : \mathbb{R}^2 \to \mathbb{R}$ be given by

$$F(x,y) = (x-y)^2, \quad (x,y) \in \mathbb{R}^2.$$

We have

$$\text{grad}F(x,y) = 2(x-y,y-x), \quad (x,y) \in \mathbb{R}^2.$$

Hence,

$$\text{grad}F(x,y)=0 \quad \text{for} \quad x=y, \quad x,y\in\mathbb{R}.$$

If we denote

$$C=\{(x,y)\in\mathbb{R}^2 : F(x,y)=0\},$$

then $\text{grad}F(x,y)=0$ at any point $(x,y)\in C$. But, clearly, C is a curve.

1.2.2 Space Curves

1.2.2.1 Parametric Representation

As in the case of plane curves, with local parametrization

$$\begin{aligned} x &= f_1(t) \\ y &= f_2(t) \\ z &= f_3(t), \quad t\in I, \end{aligned}$$

we can represent either the entire curve, or only a neighbourhood of one of its points.

Example 1.18 Let $\mathbb{T}=\mathbb{Z}$.

Then

$$\begin{aligned} x &= t+1 \\ y &= t^2+1 \\ z &= \frac{1}{1+t+t^2}, \quad t\in\mathbb{T}, \end{aligned}$$

is a parametric representation of a space curve.

Example 1.19 Let $\mathbb{T}=2\mathbb{Z}$, $I=[0,10]$.

Then

$$\begin{aligned} &f_1(0)=0, \quad f_1(t)=t+1, \quad t\in[2,10], \\ &f_2(0)=1, \quad f_2(t)=\frac{1}{1+t^2}, \quad t\in[2,10], \\ &f_3(0)=4, \quad f_3(t)=e_1(t,1), \quad t\in[2,10], \end{aligned}$$

is a parametric representation of a space curve.

1.2.2.2 Explicit Representation

Let $f,g: I\to\mathbb{R}$ be two smooth functions on an open interval $I\subset\mathbb{T}$.

Then the set

$$C=\{(x,f(x),g(x))\subset\mathbb{R}^3 : x\in I\}$$

is a simple curve with a global representation given by

$$x = t$$
$$y = f(t)$$
$$z = g(t), \quad t \in I.$$

Definition 1.12 The equations

$$y = f(x)$$
$$z = g(x), \quad x \in I,$$

are called the explicit equations of the curve.

Example 1.20 Let $\mathbb{T} = \mathbb{Z}, I = [0,20]$.

Then

$$y = x + 1$$
$$z = x^2 + x + 1, \quad x \in I,$$

is an explicit representation of a space curve.

Example 1.21 Let $\mathbb{T} = 2^{\mathbb{N}_0}, I = [1,64]$.

Then

$$y = e_1(x,1) + x$$
$$z = x + 1, \quad x \in I,$$

is an explicit representation of a space curve.

1.2.2.3 *Implicit Representation*

Suppose that $\mathbb{T}_1$, $\mathbb{T}_2$ and $\mathbb{T}_3$ are time scales with forward jump operators and delta differentiation operators σ_1, σ_2, σ_3 and Δ_1, Δ_2, Δ_3, respectively. Let $D \subset \mathbb{T}_1 \times \mathbb{T}_2 \times \mathbb{T}_3$ and $F, G : D \to \mathbb{R}$ be smooth functions. Consider the set

$$C = \{(x,y,z) \in D : F(x,y,z) = 0, \quad G(x,y,z) = 0\},$$

i.e., the set of solutions of the system

$$F(x,y,z) = 0$$
$$G(x,y,z) = 0.$$

In the general case, the set C is not a regular curve. Nevertheless, if $a = (x_0, y_0, z_0) \in C$ and

$$\operatorname{rank} \begin{pmatrix} \frac{\Delta_1}{\Delta_1 x} F(a) & \frac{\Delta_2}{\Delta_2 y} F(a) & \frac{\Delta_3}{\Delta_3 z} F(a) \\ \frac{\Delta_1}{\Delta_1 x} G(a) & \frac{\Delta_2}{\Delta_2 y} G(a) & \frac{\Delta_3}{\Delta_3 z} G(a) \end{pmatrix} = 2,$$

then there is an open neighbourhood $U \subset D$ of the point a such that $C \bigcap U$ is a curve. If the rank of the matrix

$$\begin{pmatrix} \frac{\Delta_1}{\Delta_1 x}F & \frac{\Delta_2}{\Delta_2 y}F & \frac{\Delta_3}{\Delta_3 z}F \\ \frac{\Delta_1}{\Delta_1 x}G & \frac{\Delta_2}{\Delta_2 y}G & \frac{\Delta_3}{\Delta_3 z}G \end{pmatrix}$$

is equal to two, then C is a curve.

Example 1.22 Let $\mathbb{T}_1 = \mathbb{T}_2 = \mathbb{T}_3 = \mathbb{N}$,

$$F(x,y,z) = x^2 + 3y^2 + z^2 - 100,$$
$$G(x,y,z) = x - y - z, \quad (x,y,z) \in \mathbb{T}_1 \times \mathbb{T}_2 \times \mathbb{T}_3.$$

Consider the system

$$x^2 + 3y^2 + z^2 = 100$$
$$x - y - z = 0.$$

We have

$$x = y + z$$
$$100 = 4y^2 + 2z^2 + 2yz, \quad (x,y,z) \in \mathbb{T}_1 \times \mathbb{T}_2 \times \mathbb{T}_3,$$

and

$$\sigma_1(x) = x + 1, \quad x \in \mathbb{T}_1,$$
$$\sigma_2(y) = y + 1, \quad y \in \mathbb{T}_2,$$
$$\sigma_3(z) = z + 1, \quad z \in \mathbb{T}_3.$$

Hence,

$$\begin{aligned} \frac{\Delta_1}{\Delta_1 x}F(x,y,z) &= \sigma_1(x) + x \\ &= x + 1 + x \\ &= 2x + 1, \\ \frac{\Delta_2}{\Delta_2 y}F(x,y,z) &= 3(\sigma_2(y) + y) \\ &= 3(y + 1 + y) \\ &= 6y + 3, \\ \frac{\Delta_3}{\Delta_3 z}F(x,y,z) &= \sigma_3(z) + z \\ &= z + 1 + z \\ &= 2z + 1, \end{aligned}$$

$$\frac{\Delta_1}{\Delta_1 x}G(x,y,z) = 1,$$
$$\frac{\Delta_2}{\Delta_2 y}G(x,y,z) = -1,$$
$$\frac{\Delta_3}{\Delta_3 z}G(x,y,z) = -1.$$

From here,

$$\begin{pmatrix} \frac{\Delta_1}{\Delta_1 x}F(x,y,z) & \frac{\Delta_2}{\Delta_2 y}F(x,y,z) & \frac{\Delta_3}{\Delta_3 z}F(x,y,z) \\ \frac{\Delta_1}{\Delta_1 x}G(x,y,z) & \frac{\Delta_2}{\Delta_2 y}G(x,y,z) & \frac{\Delta_3}{\Delta_3 z}G(x,y,z) \end{pmatrix} = \operatorname{rank}\begin{pmatrix} 2x+1 & 6y+3 & 2z+1 \\ 1 & -1 & -1 \end{pmatrix}$$
$$= 2, \quad (x,y,z) \in \mathbb{T}_1 \times \mathbb{T}_2 \times \mathbb{T}_3.$$

Therefore

$$C = \{(x,y,z) : x^2 + 3y^2 + z^2 - 100 = 0, \quad x - y - z = 0\}$$

is a curve.

Exercise 1.6 Let $\mathbb{T}_1 = \mathbb{Z}$, $\mathbb{T}_2 = 2\mathbb{Z}$, $\mathbb{T}_3 = 2^{\mathbb{N}_0}$,

$$F(x,y,z) = x - y - z,$$
$$G(x,y,z) = x^3 + y^3 + z^3 - 10000, \quad (x,y,z) \in \mathbb{T}_1 \times \mathbb{T}_2 \times \mathbb{T}_3.$$

Check if

$$C = \{(x,y,z) \in \mathbb{T}_1 \times \mathbb{T}_2 \times \mathbb{T}_3 : F(x,y,z) = 0, \quad G(x,y,z) = 0\}$$

is a curve.

1.3 The Tangent Line and the Normal Plane. The Normal of a Plane Curve

Let $[a,b] \subset \mathbb{T}$ and $f : [a,b] \to \mathbb{R}^n$ be a regular parameterized curve. If $(f, \phi, \mathbb{T}, \widetilde{\mathbb{T}})$ is an admissible parameterized four, then the tangent vectors of f and $f \circ \phi$ are collinear. Let $t_0 \in [a,b]$.

Definition 1.13 Let $P_0 = (f_1(t_0), \ldots, f_n(t_0))$ be a point on the curve f and L be a line through P_0^σ, where

$$P_0^\sigma = (f_1(\sigma(t_0)), \ldots, f_n(\sigma(t_0))).$$

Take on f any point P. Denote by d the distance of the point P from the point P_0^σ and by δ the distance of P from the line L. If $\frac{\delta}{d} \to 0$ as $P \to P_0$, $P \neq P_0^\sigma$, then we say that L is the tangent line to the curve f at the point P_0.

Let f have the tangent line L at the point P_0. Then PP_0^σ converges to L as $P \to P_0$, $P \neq P_0^\sigma$. Conversely, if PP_0^σ converges to some line as $P \to P_0$, $P \neq P_0^\sigma$, then this limiting line will be the tangent line at P_0. Really, if α is the angle between lines L and PP_0^σ, then

$$\frac{\delta}{d} = \sin\alpha \to 0.$$

Theorem 1.2 *Every regular curve f has at any point $P_0 = (f_1(t_0), \ldots, f_n(t_0))$ the tangent line that has the vector $f^\Delta(t_0)$ as its direction vector.*

Proof. Suppose that the curve f has the tangent line L at the point P_0. With l we will denote a unit vector on the line L. Then the distance of the point P from the point P_0^σ is

$$d = \|f(t) - f(\sigma(t_0))\|.$$

The distance of the point P from the line L is

$$\delta = \|(f(t) - f(\sigma(t_0))) \times l\|.$$

Then, by the definition of the tangent line, we get

$$\begin{aligned}\frac{\delta}{d} &= \frac{\|(f(t) - f(\sigma(t_0))) \times l\|}{\|f(t) - f(\sigma(t_0))\|} \\ &\to 0, \quad \text{as} \quad t \to t_0, \quad t \neq \sigma(t_0).\end{aligned}$$

On the other hand,

$$\begin{aligned}\frac{\|(f(t) - f(\sigma(t_0))) \times l\|}{\|f(t) - f(\sigma(t_0))\|} &= \frac{\left\|\frac{(f(t) - f(\sigma(t_0)))}{t - \sigma(t_0)} \times l\right\|}{\left\|\frac{f(t) - f(\sigma(t_0))}{t - \sigma(t_0)}\right\|} \\ &\to \frac{\left\|f^\Delta(t_0) \times l\right\|}{\left\|f^\Delta(t_0)\right\|}, \quad \text{as} \quad t \to t_0.\end{aligned}$$

Therefore

$$\frac{\left\|f^\Delta(t_0) \times l\right\|}{\left\|f^\Delta(t_0)\right\|} = 0.$$

Since f is regular, we get

$$\left\|f^\Delta(t_0) \times l\right\| = 0.$$

Thus, $f^\Delta(t_0)$ and l are collinear. Conversely, let L be a line through the point P_0^σ and have the vector $f^\Delta(t_0)$ as its direction vector.

Then

$$\begin{aligned}\frac{\delta}{d} &= \frac{\left\|(f(t)-f(\sigma(t_0)))\times\frac{f^{\Delta}(t_0)}{\|f^{\Delta}(t_0)\|}\right\|}{\|f(t)-f(\sigma(t_0))\|}\\ &= \frac{\left\|\frac{(f(t)-f(\sigma(t_0)))}{t-\sigma(t_0)}\times\frac{f^{\Delta}(t_0)}{\|f^{\Delta}(t_0)\|}\right\|}{\left\|\frac{f(t)-f(\sigma(t_0))}{t-\sigma(t_0)}\right\|}\\ &\to \frac{\left\|f^{\Delta}(t_0)\times\frac{f^{\Delta}(t_0)}{\|f^{\Delta}(t_0)\|}\right\|}{\|f^{\Delta}(t_0)\|}\\ &= 0, \quad \text{as} \quad t\to t_0.\end{aligned}$$

This completes the proof.

The vectorial equations of the tangent line read as follows

$$F(\tau)=f(t_0)+\tau f^{\Delta}(t_0), \tag{1.3}$$

where τ is a parameter.

Definition 1.14 The normal plane at the point $f(t_0)$ to the curve $f=f(t)$ is the plane which passes through $f(t_0)$ and is perpendicular to the tangent line to the curve at $f(t_0)$. If $f=f(t)$ is a plane parameterized curve, then the normal to the curve at the point $f(t_0)$ will be the straight line through $f(t_0)$, which is perpendicular to the tangent line to the curve at the point $f(t_0)$.

The vectorial equations of the normal plane (line) are given as follows

$$(F(\tau)-f(t_0))\cdot f^{\Delta}(t_0)=0, \tag{1.4}$$

where τ is a parameter.

1.3.1 Parametric Representation

We start with the vectorial equation (1.3) and we obtain the parametric equations of the tangent line for space curves

$$\begin{aligned}X(\tau)&=f_1(t_0)+\tau f_1^{\Delta}(t_0)\\ Y(\tau)&=f_2(t_0)+\tau f_2^{\Delta}(t_0)\\ Z(\tau)&=f_3(t_0)+\tau f_3^{\Delta}(t_0)\end{aligned}$$

and for the plane curve

$$\begin{aligned}X(\tau)&=f_1(t_0)+\tau f_1^{\Delta}(t_0)\\ Y(\tau)&=f_2(t_0)+\tau f_2^{\Delta}(t_0).\end{aligned}$$

If we eliminate the parameter τ, we get the canonical equations of the tangent line for space and plane curves

$$\frac{X-f_1(t_0)}{f_1^\Delta(t_0)}=\frac{Y-f_2(t_0)}{f_2^\Delta(t_0)}=\frac{Z-f_3(t_0)}{f_3^\Delta(t_0)}$$

and

$$\frac{X-f_1(t_0)}{f_1^\Delta(t_0)}=\frac{Y-f_2(t_0)}{f_2^\Delta(t_0)},$$

respectively.

Example 1.23 Let $\mathbb{T}=2^{\mathbb{N}_0}$. Consider the parameterized curve

$$\begin{aligned}x&=t+1\\ y&=t^2+t-1,\quad t\in\mathbb{T}.\end{aligned}$$

We have

$$\begin{aligned}\sigma(t)&=2t,\\ f_1(t)&=t+1,\\ f_2(t)&=t^2+t-1,\quad t\in\mathbb{T}.\end{aligned}$$

Then

$$\begin{aligned}f_1^\Delta(t)&=1,\\ f_2^\Delta(t)&=\sigma(t)+t+1\\ &=2t+t+1\\ &=3t+1,\quad t\in\mathbb{T}.\end{aligned}$$

Therefore the canonical equations of the tangent line at any point to the curve being considered are given by

$$X-t-1=\frac{Y-t^2-t+1}{3t+1},\quad t\in\mathbb{T}.$$

Example 1.24 Let $\mathbb{T}=2\mathbb{Z}$ and $I=[0,100]$. Consider the parameterized curve

$$\begin{aligned}x&=\frac{1}{1+t^2}\\ y&=t^3\\ z&=1+t^2,\quad t\in I.\end{aligned}$$

Here

$$\begin{aligned}\sigma(t)&=t+2,\\ f_1(t)&=\frac{1}{1+t^2},\end{aligned}$$

$$f_2(t) = t^3,$$
$$f_3(t) = 1 + t^2, \quad t \in I.$$

Then

$$\begin{aligned}
f_1^{\Delta}(t) &= -\frac{\sigma(t) + t}{(1+t^2)(1+(\sigma(t))^2)} \\
&= -\frac{t+2+t}{(1+t^2)(1+(t+2)^2)} \\
&= -\frac{2(t+1)}{(1+t^2)(t^2+4t+5)}, \\
f_2^{\Delta}(t) &= (\sigma(t))^2 + t\sigma(t) + t^2 \\
&= (t+2)^2 + t(t+2) + t^2 \\
&= t^2 + 4t + 4 + t^2 + 2t + t^2 \\
&= 3t^2 + 6t + 4, \\
f_3^{\Delta}(t) &= \sigma(t) + t \\
&= t + 2 + t \\
&= 2(t+1), \quad t \in \mathbb{T}.
\end{aligned}$$

Hence, the canonical equations of the tangent line at any point on the curve being considered are given by

$$\frac{X - \frac{1}{1+t^2}}{-\frac{2(t+1)}{(1+t^2)(t^2+4t+5)}} = \frac{Y - t^3}{3t^2 + 6t + 4} = \frac{Z - 1 - t^2}{2(t+1)}, \quad t \in \mathbb{T}.$$

Exercise 1.7 Let $\mathbb{T} = 3^{\mathbb{N}_0}$. Find the canonical equations of the tangent line at any point on the parameterized curve

$$x = \frac{1}{1+t^2} + t + 1$$
$$y = \frac{1}{1+t^4} + t^2 + t - 1, \quad t \in \mathbb{T}.$$

Exercise 1.8 Let $\mathbb{T} = 2^{\mathbb{N}_0}$. Find the canonical equations of the tangent line at any point on the parameterized curve

$$x = e_1(t,1) + \sin_2(t,2) + \frac{1}{1+t}$$
$$y = 1 + t + \frac{1}{1+t^2}$$
$$z = \frac{1}{1+t^4}, \quad t \in \mathbb{T}.$$

For the equation of the normal line at the point t_0 to a plane curve we use the equation (1.4). We get

$$(X-f_1(t_0),Y-f_2(t_0))\cdot\left(f_1^{\Delta}(t_0),f_2^{\Delta}(t_0)\right)=0$$

or

$$(X-f_1(t_0))f_1^{\Delta}(t_0)+(Y-f_2(t_0))f_2^{\Delta}(t_0)=0.$$

In the case of a space curve, we get the equation

$$(X-f_1(t_0),Y-f_2(t_0),Z-f_3(t_0))\cdot\left(f_1^{\Delta}(t_0),f_2^{\Delta}(t_0),f_3^{\Delta}(t_0)\right)=0,$$

whereupon

$$(X-f_1(t_0))f_1^{\Delta}(t_0)+(Y-f_2(t_0))f_2^{\Delta}(t_0)+(Z-f_3(t_0))f_3^{\Delta}(t_0)=0.$$

Example 1.25 Let $\mathbb{T}=\mathbb{Z}$. Consider the plane curve

$$\begin{aligned} x &= t+1,\\ y &= \frac{1+t}{1+t^2}, \quad t\in\mathbb{T}. \end{aligned}$$

Here

$$\begin{aligned} \sigma(t) &= t+1,\\ f_1(t) &= t+1,\\ f_2(t) &= \frac{1+t}{1+t^2}, \quad t\in\mathbb{T}. \end{aligned}$$

We have

$$\begin{aligned} f_1^{\Delta}(t) &= 1,\\ f_2^{\Delta}(t) &= \frac{1+t^2-(1+t)(\sigma(t)+t)}{(1+t^2)(1+(\sigma(t))^2)}\\ &= \frac{1+t^2-(1+t)(t+1+t)}{(1+t^2)(1+(1+t)^2)}\\ &= \frac{1+t^2-(1+t)(2t+1)}{(1+t^2)(2+2t+t^2)}\\ &= \frac{1+t^2-2t-1-2t^2-t}{(1+t^2)(2+2t+t^2)}\\ &= -\frac{3t+t^2}{(1+t^2)(2+2t+t^2)}, \quad t\in\mathbb{T}. \end{aligned}$$

Hence, the equation of the normal line at any point on the plane curve being considered is

$$X-t-1-\left(Y-\frac{1+t}{1+t^2}\right)\frac{3t+t^2}{(1+t^2)(2+2t+t^2)}=0, \quad t\in\mathbb{T}.$$

Example 1.26 Let $\mathbb{T} = 2^{\mathbb{N}_0}$. Consider the space curve

$$\begin{aligned} x &= e_1(t,1) + t \\ y &= 1 + t^2 \\ z &= t^3, \quad t \in \mathbb{T}. \end{aligned}$$

Here

$$\begin{aligned} \sigma(t) &= 2t, \\ f_1(t) &= e_1(t,1) + t, \\ f_2(t) &= 1 + t^2, \\ f_3(t) &= t^3, \quad t \in \mathbb{T}. \end{aligned}$$

Then

$$\begin{aligned} f_1^\Delta(t) &= e_1(t,1) + 1, \\ f_2^\Delta(t) &= \sigma(t) + t \\ &= 2t + t \\ &= 3t, \\ f_3^\Delta(t) &= (\sigma(t))^2 + t\sigma(t) + t^2 \\ &= (2t)^2 + 2t^2 + t^2 \\ &= 4t^2 + 3t^2 \\ &= 7t^2, \quad t \in \mathbb{T}. \end{aligned}$$

Then the equation of the normal plane at any point on the space curve being considered is

$$(X - e_1(t,1) - t)(e_1(t,1) + 1) + 3(Y - 1 - t^2)t + 7(Z - t^3)t^2 = 0, \quad t \in \mathbb{T}.$$

Exercise 1.9 Let $\mathbb{T} = 2\mathbb{Z}$. Find the equation of the normal line at any point on the following plane curve

$$\begin{aligned} x &= \frac{\sin_1(t,1) + t^2}{1 + t^4} + t \\ y &= 1 + t^3, \quad t \in \mathbb{T}. \end{aligned}$$

Exercise 1.10 Let $\mathbb{T} = \mathbb{Z}$. Find the equation of the normal plane at any point on the following space curve

$$\begin{aligned} x &= \frac{1 + t}{1 + t^4} - 3t^2 - 7t + 1 \\ y &= \frac{1}{1 + t^2} + t^2 + 1 \\ z &= \frac{1}{1 + t^2 + t^4} + t, \quad t \in \mathbb{T}. \end{aligned}$$

1.3.2 Explicit Representation

For a plane curve, given by the equation

$$y = f(x)$$

we have the following parametric representation

$$x = t$$
$$y = f(t).$$

Then the equation of the tangent line at the point t_0 is given by

$$X - t_0 = \frac{Y - f(t_0)}{f^{\Delta}(t_0)}$$

and the equation of the normal line at the point t_0 is as follows

$$X - t_0 + (Y - f(t_0))f^{\Delta}(t_0) = 0.$$

Example 1.27 Let $\mathbb{T} = 2\mathbb{Z}$. Consider the plane curve

$$y = \frac{1}{1+x^2} + x^3, \quad x \in \mathbb{T}.$$

Here

$$\sigma(x) = x + 2,$$
$$f(x) = \frac{1}{1+x^2} + x^3, \quad x \in \mathbb{T}.$$

Then

$$\begin{aligned} f^{\Delta}(x) &= -\frac{\sigma(x) + x}{(1+x^2)(1+(\sigma(x))^2)} + (\sigma(x))^2 + x\sigma(x) + x^2 \\ &= -\frac{x+2+x}{(1+x^2)(1+(x+2)^2)} + (x+2)^2 + x(x+2) + x^2 \\ &= -\frac{2(x+1)}{(1+x^2)(5+4x+x^2)} + x^2 + 4x + 4 + x^2 + 2x + x^2 \\ &= -\frac{2(x+1)}{(1+x^2)(5+4x+x^2)} + 3x^2 + 6x + 4, \quad x \in \mathbb{T}. \end{aligned}$$

Therefore the equation of the tangent line at any point on the plane curve being considered is given by

$$X - t = -\frac{Y - \frac{1}{1+t^2} - t^3}{\frac{2(t+1)}{(1+t^2)(5+4t+t^2)}}, \quad t \in \mathbb{T},$$

and the equation of the normal line at any point on the plane curve being considered is as follows

$$X - t - 2\left(Y - \frac{1}{1+t^2} - t^3\right)\frac{1+t}{(1+t^2)(5+4t+t^2)} = 0, \quad t \in \mathbb{T}.$$

Exercise 1.11 Let $\mathbb{T} = 2^{\mathbb{N}_0}$. Find the equations of the tangent line and the normal line for the following plane curve

$$y = \frac{1+x+x^2}{1+x^4} - x - e_1(x,1), \quad x \in \mathbb{T}.$$

Now, we consider a space curve given by the equations

$$y = f_1(x)$$
$$z = f_2(x).$$

We can represent it in parametric form

$$x = t$$
$$y = f_1(t)$$
$$z = f_2(t).$$

Then the equations of the tangent line at the point t_0 are given by

$$X - t_0 = \frac{Y - f_1(t_0)}{f_1^{\Delta}(t_0)} = \frac{Z - f_2(t_0)}{f_2^{\Delta}(t_0)}$$

and the equation of the normal plane at the point t_0 on the considered space curve is as follows

$$X - t_0 + (Y - f_1(t_0))f_1^{\Delta}(t_0) + (Z - f_2(t_0))f_2^{\Delta}(t_0) = 0.$$

Exercise 1.12 Let $\mathbb{T} = 3\mathbb{Z}$. Find the equations of the tangent line and the equation of the normal plane at any point on the following space curve

$$y = 1 + x^2$$
$$z = \frac{1+x}{1+x^4} + x, \quad x \in \mathbb{T}.$$

1.3.3 Implicit Representation

Consider a space curve given by the implicit equations

$$\begin{aligned} F(x,y,z) &= 0 \\ G(x,y,z) &= 0, \end{aligned} \tag{1.5}$$

where $x, y \in \mathbb{T}$. Suppose that $f, g : \mathbb{T} \to \mathbb{R}$ are given functions such that

$$f(\mathbb{T}) = \mathbb{T}_1, \quad g(\mathbb{T}) = \mathbb{T}_2,$$

where $\mathbb{T}_1$ and $\mathbb{T}_2$ are time scales with forward jump operators and delta differentiation operators σ_1, σ_2 and Δ_1, Δ_2, respectively, and

$$f(\sigma(x)) = \sigma_1(f(x)), \quad g(\sigma(x)) = \sigma_2(g(x)), \quad x \in \mathbb{T},$$

f and g are Δ-differentiable. Let (x_0, y_0, z_0) be an arbitrary point and F, G be σ-completely Δ-differentiable functions. We differentiate each equation of the system (1.5) with respect to x and we find

$$F_x^{\Delta}(x,f(x),g(x)) + F_y^{\Delta_1}(\sigma(x),f(x),g(x))f^{\Delta}(x) + F_z^{\Delta_2}(\sigma(x),f(\sigma(x)),g(x))g^{\Delta}(x) = 0$$
$$G_x^{\Delta}(x,f(x),g(x)) + G_y^{\Delta_1}(\sigma(x),f(x),g(x))f^{\Delta}(x) + G_z^{\Delta_3}(\sigma(x),f(\sigma(x)),g(x))g^{\Delta}(x) = 0$$

or

$$F_y^{\Delta_1}(\sigma(x),f(x),g(x))f^{\Delta}(x) + F_z^{\Delta_2}(\sigma(x),f(\sigma(x)),g(x))g^{\Delta}(x) = -F_x^{\Delta}(x,f(x),g(x))$$
$$G_y^{\Delta_1}(\sigma(x),f(x),g(x))f^{\Delta}(x) + G_z^{\Delta_3}(\sigma(x),f(\sigma(x)),g(x))g^{\Delta}(x) = -G_x^{\Delta}(x,f(x),g(x)).$$

Suppose that

$$\begin{aligned} D(x_0,y_0) &= F_y^{\Delta_1}(\sigma(x_0),f(x_0),g(x_0))G_z^{\Delta_2}(\sigma(x_0),f(\sigma(x_0)),g(x_0)) \\ &\quad - G_y^{\Delta_1}(\sigma(x_0),f(x_0),g(x_0))F_z^{\Delta_2}(\sigma(x_0),f(\sigma(x_0)),g(x_0)) \\ &\neq 0. \end{aligned}$$

Then

$$\begin{aligned} f^{\Delta}(x_0) &= -\frac{1}{D(x_0,y_0)}\Big(F_x^{\Delta}(x_0,f(x_0),g(x_0))G_z^{\Delta_2}(\sigma(x_0),f(\sigma(x_0)),g(x_0)) \\ &\quad - F_z^{\Delta_2}(\sigma(x_0),f(\sigma(x_0)),g(x_0))G_x^{\Delta}(x_0,f(x_0),g(x_0))\Big), \\ g^{\Delta}(x_0) &= \frac{1}{D(x_0,y_0)}\Big(F_x^{\Delta}(x_0,f(x_0),g(x_0))G_y^{\Delta_1}(\sigma(x_0),f(x_0),g(x_0)) \\ &\quad - G_x^{\Delta}(x_0,f(x_0),g(x_0))F_y^{\Delta_1}(\sigma(x_0),f(x_0),g(x_0))\Big). \end{aligned}$$

Thus, the equations of the tangent line at the point (x_0, y_0, z_0) to the considered curve are given by

$$X - x_0 = \frac{Y - y_0}{f^{\Delta}(x_0)} = \frac{Z - z_0}{g^{\Delta}(x_0)}$$

and the equation of the normal plane at the point (x_0, y_0, z_0) to the space curve being considered is as follows

$$X - x_0 + (Y - y_0)f^{\Delta}(x_0) + (Z - z_0)g^{\Delta}(x_0) = 0.$$

Now, we consider the plane curve

$$F_1(x,y) = 0. \tag{1.6}$$

Let (x_0, y_0) be an arbitrary point. Assume that

$$y = f_1(x),$$

where $f_1(\mathbb{T}) = \mathbb{T}_3$, $\mathbb{T}_3$ is a time scale with forward jump operator and delta differentiation operator σ_3 and Δ_3, respectively, f_1 is Δ-differentiable and F_1 is σ-completely $f_1(\sigma(x)) = \sigma_3(f_1(x))$, $x \in \mathbb{T}$. We can write the equation (1.6) in the form

$$F_1(x, f_1(x)) = 0.$$

We differentiate the last equation with respect to x and we find

$$F_{1x}^{\Delta}(x, f_1(x)) + F_{1y}^{\Delta_3}(\sigma(x), f_1(x)) f_1^{\Delta}(x) = 0,$$

whereupon

$$f_1^{\Delta}(x) = -\frac{F_{1x}^{\Delta}(\sigma(x), f_1(x))}{F_{1y}^{\Delta_3}(\sigma(x), f_1(x))}.$$

Then the equation of the tangent line at the point (x_0, y_0) is given by

$$X - x_0 = \frac{Y - y_0}{-\frac{F_{1x}^{\Delta}(x_0, f_1(x_0))}{F_{1y}^{\Delta_3}(\sigma(x_0), f_1(x_0))}}$$

and the equation of the normal line at the point (x_0, y_0) is as follows

$$X - x_0 - (Y - y_0)\frac{F_{1x}^{\Delta}(x_0, f_1(x_0))}{F_{1y}^{\Delta_3}(\sigma(x_0), f_1(x_0))} = 0.$$

1.4 Envelopes of Plane Curves

Suppose that $\mathbb{T}$, $\mathbb{T}_1$, $\mathbb{T}_2$ are time scales with forward jump operators and delta differentiation operators σ, σ_1, σ_2 and Δ, Δ_1, Δ_2, respectively.

Let $[a,b] \subseteq \mathbb{T}$, $[c,d] \subseteq \mathbb{T}_1$ and there exists a function $h : [c,d] \to [a,b]$ such that

$$[a,b] = \{h(\lambda) : \lambda \in [c,d]\}, \quad \sigma(h(\lambda)) = h(\sigma_1(\lambda)), \quad \lambda \in [c,d].$$

Also, let

$$f = f(t,\lambda), \quad t \in [a,b], \quad \lambda \in [c,d], \tag{1.7}$$

be a family of plane parameterized curves that are σ-completely delta differentiable on $[a,b] \times [c,d]$.

Definition 1.15 The envelope of the family (1.7) is a parameterized curve $([a,b], g)$ which, at each point, is tangent to a curve from the family.

Theorem 1.3 *The points of the family* (1.7) *are subject to*

$$f = f(t,\lambda), \quad (t,\lambda) \in [a,b] \times [c,d],$$

$$f_t^{\Delta}(t,\lambda) \times f_{\lambda}^{\Delta_1 \sigma}(t,\lambda) = 0, \quad (t,\lambda) \in [a,b] \times [c,d].$$

Proof. Let $([a,b],f)$ be the envelope of the family (1.7) and P be a point of g. Then P is a tangency point between g and a curve of the family corresponding to some value of the parameter λ. Thus, the equation of g will be

$$f_1 = f_1(\lambda), \quad \lambda \in [c,d].$$

Since P is on a curve $f(t,\lambda)$, it verifies

$$f_1 = f(h(\lambda),\lambda), \quad \lambda \in [a,b].$$

Since

$$f_{1\lambda}^{\Delta_1}(h(\lambda),\lambda) = f_t^{\Delta}(h(\lambda),\lambda)h^{\Delta}(\lambda) + f_{\lambda}^{\Delta_1\sigma}(h(\lambda),\lambda), \quad \lambda \in [a,b],$$

we obtain

$$\begin{aligned} 0 &= f_{1\lambda}^{\Delta_1}(h(\lambda),\lambda) \times f_t^{\Delta}(h(\lambda),\lambda) \\ &= \left(f_t^{\Delta}(h(\lambda),\lambda)h^{\Delta}(\lambda) + f_{\lambda}^{\Delta_1\sigma}(h(\lambda),\lambda)\right) \times f_t^{\Delta}(h(\lambda),\lambda) \\ &= f_{\lambda}^{\Delta_1\sigma}(h(\lambda),\lambda) \times f_t^{\Delta}(h(\lambda),\lambda), \quad \lambda \in [c,d]. \end{aligned}$$

This completes the proof.

Example 1.28 Let $\mathbb{T} = 2^{\mathbb{N}_0}$, $\mathbb{T}_1 = 4^{\mathbb{N}_0}$, $a = 1$, $b = 16$, $c = 1$, $d = 256$.

Then

$$h(\lambda) = \sqrt{\lambda}, \quad \lambda \in [c,d],$$

and

$$h([c,d]) = [a,b].$$

Here

$$\begin{aligned} \sigma(t) &= 2t, \quad t \in \mathbb{T}, \\ \sigma_1(t_1) &= 4t_1, \quad t_1 \in \mathbb{T}_1. \end{aligned}$$

Hence,

$$\begin{aligned} h(\sigma_1(\lambda)) &= \sqrt{\sigma_1(\lambda)} \\ &= \sqrt{4\lambda} \\ &= 2\sqrt{\lambda} \\ &= \sigma(h(\lambda)), \quad \lambda \in [c,d]. \end{aligned}$$

Consider the family

$$f(t,\lambda) = (t^2 + \lambda^2, \lambda t), \quad t \in [a,b], \quad \lambda \in [c,d].$$

Then

$$\begin{aligned} f_1(t,\lambda) &= t^2 + \lambda^2, \\ f_2(t,\lambda) &= \lambda t, \end{aligned}$$

$$\begin{aligned}
f_{1t}^{\Delta}(t,\lambda) &= \sigma(t)+t\\
&= 2t+t\\
&= 3t,\\
f_{1\lambda}^{\Delta_1}(t,\lambda) &= \sigma_1(\lambda)+\lambda\\
&= 4\lambda+\lambda\\
&= 5\lambda,\\
f_{1\lambda}^{\Delta_1\sigma}(t,\lambda) &= 5\lambda,\\
f_{2t}^{\Delta}(t,\lambda) &= \lambda,\\
f_{2\lambda}^{\Delta_1}(t,\lambda) &= t,\\
f_{2\lambda}^{\Delta_1\sigma}(t,\lambda) &= \sigma(t)\\
&= 2t, \quad t\in[a,b], \quad \lambda\in[c,d].
\end{aligned}$$

Therefore

$$\begin{aligned}
f_t^{\Delta}(t,\lambda) &= \left(f_{1t}^{\Delta}(t,\lambda), f_{2t}^{\Delta}(t,\lambda)\right)\\
&= (3t,\lambda),\\
f_\lambda^{\Delta_1\sigma}(t,\lambda) &= \left(f_{1\lambda}^{\Delta_1\sigma}(t,\lambda), f_{2\lambda}^{\Delta_1\sigma}(t,\lambda)\right)\\
&= (5\lambda,2t), \quad t\in[a,b], \quad \lambda\in[c,d],
\end{aligned}$$

and

$$\begin{aligned}
f_t^{\Delta}(t,\lambda)\times f_\lambda^{\Delta_1\sigma}(t,\lambda) &= (0,0,6t^2-5\lambda^2)\\
&= 0, \quad t\in[a,b], \quad \lambda\in[c,d],
\end{aligned}$$

if and only if

$$6t^2-5\lambda^2=0 \quad \text{or} \quad t=\sqrt{\frac{5}{6}}\lambda, \quad t\in[a,b], \quad \lambda\in[c,d].$$

Thus,

$$\begin{aligned}
g(t) &= \left(\frac{5}{6}\lambda^2+\lambda^2, \sqrt{\frac{5}{6}}\lambda^2\right)\\
&= \left(\frac{11}{6}\lambda^2, \sqrt{\frac{5}{6}}\lambda^2\right), \quad t\in[a,b], \quad \lambda\in[c,d],
\end{aligned}$$

is the envelope of the family being considered.

Example 1.29 Let $\mathbb{T}=\mathbb{Z}$, $\mathbb{T}_1=2\mathbb{Z}$, $a=0$, $b=5$, $c=0$, $d=10$.

Take

$$h(\lambda)=\frac{\lambda}{2}, \quad \lambda\in[c,d].$$

Then

$$h([c,d]) = [a,b]$$

and

$$\begin{aligned}
\sigma(t) &= t+1, \quad t \in \mathbb{T}, \\
\sigma_1(t_1) &= t_1 + 2, \quad t_1 \in \mathbb{T}_1, \\
h(\sigma_1(\lambda)) &= \frac{\sigma_1(\lambda)}{2} \\
&= \frac{\lambda+2}{2} \\
&= \frac{\lambda}{2} + 1 \\
&= \sigma(h(\lambda)), \quad \lambda \in [c,d].
\end{aligned}$$

Consider the family

$$f(t,\lambda) = (\lambda + 2a_1 c_1(t,0), \lambda + a_1 s_1(t,0)), \quad t \in [a,b], \quad \lambda \in [c,d],$$

where $a_1 \in \mathbb{R}$, $a_1 \neq 0$.

We have

$$\begin{aligned}
f_1(t,\lambda) &= \lambda + 2a_1 c_1(t,0), \\
f_2(t,\lambda) &= \lambda + a_1 s_1(t,0), \quad t \in [a,b], \quad \lambda \in [c,d].
\end{aligned}$$

From here,

$$\begin{aligned}
f_{1t}^{\Delta}(t,\lambda) &= 2a_1(c_1(t,0) - s_1(t,0)), \\
f_{1\lambda}^{\Delta_1}(t,\lambda) &= 1, \\
f_{1\lambda}^{\Delta_1\sigma}(t,\lambda) &= 1, \\
f_{2t}^{\Delta}(t,\lambda) &= a_1(c_1(t,0) + s_1(t,0)), \\
f_{2\lambda}^{\Delta_1}(t,\lambda) &= 1, \\
f_{2\lambda}^{\Delta_1\sigma}(t,\lambda) &= 1, \quad t \in [a,b], \quad \lambda \in [c,d].
\end{aligned}$$

Therefore

$$\begin{aligned}
f_t^{\Delta}(t,\lambda) &= \left(f_{1t}^{\Delta}(t,\lambda), f_{2t}^{\Delta}(t,\lambda)\right) \\
&= (2a_1(c_1(t,0) - s_1(t,0)), a_1(c_1(t,0) + s_1(t,0))), \\
f_{\lambda}^{\Delta_1\sigma}(t,\lambda) &= (1,1), \quad t \in [a,b], \quad \lambda \in [c,d],
\end{aligned}$$

and

$$\begin{aligned}
f_t^{\Delta}(t,\lambda) \times f_{\lambda}^{\Delta_1\sigma}(t,\lambda) &= (0,0,a_1(c_1(t,0) - 3s_1(t,0))) \\
&= 0, \quad t \in [a,b], \quad \lambda \in [c,d],
\end{aligned}$$

if and only if

$$c_1(t,0) = 3s_1(t,0), \quad t \in [a,b].$$

By the equality

$$(c_1(t,0))^2 + (s_1(t,0))^2 = 1, \quad t \in [a,b],$$

we find

$$9(s_1(t,0))^2 + (s_1(t,0))^2 = 1, \quad t \in [a,b],$$

or

$$s_1(t,0) = \pm\frac{1}{\sqrt{10}}, \quad t \in [a,b],$$

and

$$c_1(t,0) = \pm\frac{3}{\sqrt{10}}, \quad t \in [a,b].$$

Therefore

$$g(t) = \left(\lambda \pm \frac{6a_1}{\sqrt{10}}, \lambda \pm \frac{a_1}{\sqrt{10}}\right), \quad t \in [a,b], \quad \lambda \in [c,d],$$

is the envelope of the considered family.

Exercise 1.13 Let $\mathbb{T} = \mathbb{N}_0$, $\mathbb{T}_1 = \mathbb{N}_0^2$, $a = 0$, $b = 5$, $c = 0$, $d = 25$. Find the envelope of the family

$$f(t,\lambda) = (t + \lambda + \lambda t, t^2 - \lambda^2), \quad t \in [a,b], \quad \lambda \in [c,d].$$

1.5 The Osculating Plane

In this section, we suppose that $n = 3, f : \mathbb{T} \to \mathbb{R}^3$ is two times Δ-differentiable.

Definition 1.16 A parameterized curve $f : \mathbb{T} \to \mathbb{R}^3$ is called biregular at the point t_0 if the vectors $f^{\Delta}(t_0)$ and $f^{\Delta^2}(t_0)$ are not colinear, that is

$$f^{\Delta}(t_0) \times f^{\Delta^2}(t_0) \neq 0.$$

The parameterized curve f is said to be biregular, if it is biregular at each point.

We have

$$\begin{aligned} f(t) &= (f_1(t), f_2(t), f_3(t)), \\ f^{\Delta}(t) &= \left(f_1^{\Delta}(t), f_2^{\Delta}(t), f_3^{\Delta}(t)\right), \\ f^{\Delta^2}(t) &= \left(f_1^{\Delta^2}(t), f_2^{\Delta^2}(t), f_3^{\Delta^2}(t)\right), \quad t \in \mathbb{T}. \end{aligned}$$

Then

$$f^{\Delta}(t)\times f^{\Delta^2}(t)=\Big(f_2^{\Delta}(t)f_3^{\Delta^2}(t)-f_2^{\Delta^2}(t)f_3^{\Delta}(t),$$
$$f_1^{\Delta^2}(t)f_3^{\Delta}(t)-f_1^{\Delta}(t)f_3^{\Delta^2}(t),$$
$$f_1^{\Delta}(t)f_2^{\Delta^2}(t)-f_1^{\Delta^2}(t)f_2^{\Delta}(t)\Big),\quad t\in\mathbb{T}.$$

Therefore the curve f is biregular at the point t_0 if

$$f_2^{\Delta}(t_0)f_3^{\Delta^2}(t_0)-f_2^{\Delta^2}(t_0)f_3^{\Delta}(t_0)\neq 0$$

or

$$f_1^{\Delta^2}(t_0)f_3^{\Delta}(t_0)-f_1^{\Delta}(t_0)f_3^{\Delta^2}(t_0)\neq 0,$$

or

$$f_1^{\Delta}(t_0)f_2^{\Delta^2}(t_0)-f_1^{\Delta^2}(t_0)f_2^{\Delta}(t_0)\neq 0.$$

Example 1.30 Let $\mathbb{T}=\mathbb{Z}, f(t)=(t,t+1,t^2+t), t\in\mathbb{T}$.

Here

$$\sigma(t)=t+1,$$
$$f_1(t)=t,$$
$$f_2(t)=t+1,$$
$$f_3(t)=t^2+t,\quad t\in\mathbb{T}.$$

Hence,

$$f_1^{\Delta}(t)=1,$$
$$f_1^{\Delta^2}(t)=0,$$
$$f_2^{\Delta}(t)=1,$$
$$f_2^{\Delta^2}(t)=0,$$
$$f_3^{\Delta}(t)=\sigma(t)+t+1$$
$$=t+1+t+1$$
$$=2(t+1),$$
$$f_3^{\Delta^2}(t)=2.$$

Therefore

$$f^{\Delta}(t)\times f^{\Delta^2}(t)=(1,1,2(t+1))\times(0,0,2)$$
$$=(2,-2,0)$$
$$\neq(0,0,0).$$

Thus, the considered curve is biregular.

Example 1.31 Let $\mathbb{T} = 2^{\mathbb{N}_0}, f : \mathbb{T} \to \mathbb{R}^3$ be given by

$$f(t) = (1+t^2, 1+t, t^3), \quad t \in \mathbb{T}.$$

Here

$$\begin{aligned} \sigma(t) &= 2t, \\ f_1(t) &= 1+t^2, \\ f_2(t) &= 1+t, \\ f_3(t) &= t^3, \quad t \in \mathbb{T}. \end{aligned}$$

Then

$$\begin{aligned} f_1^{\Delta}(t) &= \sigma(t) + t \\ &= 2t + t \\ &= 3t, \\ f_2^{\Delta}(t) &= 1, \\ f_3^{\Delta}(t) &= (\sigma(t))^2 + t\sigma(t) + t^2 \\ &= (2t)^2 + 2t^2 + t^2 \\ &= 7t^2, \\ f_1^{\Delta^2}(t) &= 3, \\ f_2^{\Delta^2}(t) &= 0, \\ f_3^{\Delta^2}(t) &= 7(\sigma(t) + t) \\ &= 7(2t + t) \\ &= 21t, \quad t \in \mathbb{T}. \end{aligned}$$

Therefore

$$\begin{aligned} f^{\Delta}(t) \times f^{\Delta^2}(t) &= (3t, 1, 7t^2) \times (3, 0, 21t) \\ &= (21t, -42t^2, -3) \\ &\neq (0,0,0), \quad t \in \mathbb{T}. \end{aligned}$$

Thus, the considered curve is biregular.

Exercise 1.14 Let $\mathbb{T} = 2\mathbb{Z}$. Check if the curve $f : \mathbb{T} \to \mathbb{R}^3$ given by

$$f(t) = \left(\frac{1+t}{1+t^2}, t, \frac{1-t}{1+t^4} \right), \quad t \in \mathbb{T},$$

is biregular.

Definition 1.17 Let $f : \mathbb{T} \to \mathbb{R}^3$ be a biregular curve at the point $t_0 \in \mathbb{T}$. The osculating plane of the curve at $f(t_0)$ is the plane through $f(t_0)$, parallel to the vectors $f^{\Delta}(t_0)$ and $f^{\Delta^2}(t_0)$, that is the equation of the osculating plane is

$$(F - f(t_0)) \cdot \left(f^{\Delta}(t_0) \times f^{\Delta^2}(t_0)\right) = 0,$$

where $F = (X, Y, Z)$, or

$$\det \begin{pmatrix} X - f_1(t_0) & Y - f_2(t_0) & Z - f_3(t_0) \\ f_1^{\Delta}(t_0) & f_2^{\Delta}(t_0) & f_3^{\Delta}(t_0) \\ f_1^{\Delta^2}(t_0) & f_2^{\Delta^2}(t_0) & f_3^{\Delta^2}(t_0) \end{pmatrix} = 0.$$

Example 1.32 Let $\mathbb{T} = \mathbb{Z}, f : \mathbb{T} \to \mathbb{R}^3$ be given by

$$f(t) = (t^2, 1 + t + t^2, t^2 - t), \quad t \in \mathbb{T}.$$

Here

$$\begin{aligned} \sigma(t) &= t + 1, \\ f_1(t) &= t^2, \\ f_2(t) &= 1 + t + t^2, \\ f_3(t) &= -t + t^2, \quad t \in \mathbb{T}. \end{aligned}$$

We have

$$\begin{aligned} f_1^{\Delta}(t) &= \sigma(t) + t \\ &= t + 1 + t \\ &= 2t + 1, \\ f_2^{\Delta}(t) &= 1 + \sigma(t) + t \\ &= 1 + t + 1 + t \\ &= 2(t + 1), \\ f_3^{\Delta}(t) &= -1 + \sigma(t) + t \\ &= -1 + t + 1 + t \\ &= 2t, \\ f_1^{\Delta^2}(t) &= 2, \\ f_2^{\Delta^2}(t) &= 2, \\ f_3^{\Delta^2}(t) &= 2, \quad t \in \mathbb{T}. \end{aligned}$$

We have

$$\begin{aligned} f^{\Delta}(t) \times f^{\Delta^2}(t) &= (2t + 1, 2t + 2, 2t) \times (2, 2, 2) \\ &= (4t + 4 - 4t, 4t - 4t - 2, 4t + 2 - 4t - 4) \end{aligned}$$

$$= (4,-2,-2)$$
$$\neq (0,0,0).$$

Therefore the considered curve is biregular at any point. Now, we will find the equation of the osculating plane.

We have

$$\begin{aligned}
0 &= \det\begin{pmatrix} X-f_1(t) & Y-f_2(t) & Z-f_3(t) \\ f_1^{\Delta}(t) & f_2^{\Delta}(t) & f_3^{\Delta}(t) \\ f_1^{\Delta^2}(t) & f_2^{\Delta^2}(t) & f_3^{\Delta^2}(t) \end{pmatrix} \\
&= \det\begin{pmatrix} X-t^2 & Y-1-t-t^2 & Z+t-t^2 \\ 2t+1 & 2t+2 & 2t \\ 2 & 2 & 2 \end{pmatrix} \\
&= 2\det\begin{pmatrix} Y-1-t-t^2 & Z+t-t^2 \\ 2t+2 & 2t \end{pmatrix} - 2\det\begin{pmatrix} X-t^2 & Z+t-t^2 \\ 2t+1 & 2t \end{pmatrix} \\
&\quad +2\det\begin{pmatrix} X-t^2 & Y-1-t-t^2 \\ 2t+1 & 2t+2 \end{pmatrix} \\
&= 2\Big(\Big(2t(Y-1-t-t^2)-(2t+2)(Z+t-t^2)\Big) \\
&\quad -\Big(2t(X-t^2)-(2t+1)(Z+t-t^2)\Big) \\
&\quad +\Big((2t+2)(X-t^2)-(2t+1)(Y-1-t-t^2)\Big)\Big) \\
&= 2\Big((-2t+2t+2)X+(2t-2t-1)Y+(-2t-2+2t+1)Z \\
&\quad -2t-2t^2-2t^3-2t^2+2t^3-2t+2t^2+2t^3+2t^2 \\
&\quad -2t^3+t-t^2-2t^3-2t^2+2t+2t^2+2t^3+t+t^2+1\Big) \\
&= 2(2X-Y-Z+1)
\end{aligned}$$

or

$$2X-Y-Z=-1.$$

Exercise 1.15 Let $\mathbb{T}=2\mathbb{Z}$ and $f:\mathbb{T}\to\mathbb{R}^3$ be given by

$$f(t)=\left(\frac{1+t}{1+t^2},1-t+t^2,t^3-t\right),\quad t\in\mathbb{T}.$$

Find the osculating plane.

1.6 The Curvature of a Curve

Let (I,f) be a regular parameterized curve and (J,g) be a naturally parameterized curve and equivalent to it.

Then

$$s(\sigma(t)) = \tilde{\sigma}(s(t)), \quad t \in I, \quad s \in J,$$

and

$$\|g^{\tilde{\Delta}}\| = 1. \tag{1.8}$$

Also,

$$\begin{aligned} f(t) &= g(s(t)), \\ f^{\Delta}(t) &= g^{\tilde{\Delta}}(s(t))s^{\Delta}(t), \end{aligned}$$

whereupon

$$\begin{aligned} \|f^{\Delta}(t)\| &= \|g^{\tilde{\Delta}}(s(t))\| s^{\Delta}(t) \\ &= s^{\Delta}(t), \quad t \in I. \end{aligned}$$

Next,

$$\begin{aligned} f^{\Delta^2}(t) &= g^{\tilde{\Delta}^2}(s(t))s^{\Delta}(t)s^{\Delta\sigma}(t) + g^{\tilde{\Delta}}(s(t))s^{\Delta^2}(t) \\ &= g^{\tilde{\Delta}^2}(s(t))\|f^{\Delta}(t)\|\,\|f^{\Delta}(t)\|^{\sigma} + g^{\tilde{\Delta}}(s(t))s^{\Delta^2}(t), \quad t \in I. \end{aligned}$$

Since

$$\left(s^{\Delta}(t)\right)^2 = \|f^{\Delta}(t)\|^2, \quad t \in I,$$

we get

$$s^{\Delta^2}(t)\left(s^{\Delta}(t) + s^{\Delta\sigma}(t)\right) = f^{\Delta^2}(t) \cdot \left(f^{\Delta}(t) + f^{\Delta\sigma}(t)\right), \quad t \in I,$$

or

$$s^{\Delta^2}(t)\left(\|f^{\Delta}(t)\| + \|f^{\Delta}(t)\|^{\sigma}\right) = f^{\Delta^2}(t) \cdot \left(f^{\Delta}(t) + f^{\Delta\sigma}(t)\right), \quad t \in I,$$

or

$$s^{\Delta^2}(t) = \frac{f^{\Delta^2}(t) \cdot \left(f^{\Delta}(t) + f^{\Delta\sigma}(t)\right)}{\|f^{\Delta}(t)\| + \|f^{\Delta}(t)\|^{\sigma}}, \quad t \in I.$$

Now, we use that

$$g^{\tilde{\Delta}}(s(t)) = \frac{f^{\Delta}(t)}{\|f^{\Delta}(t)\|}, \quad t \in I,$$

and we find

$$\begin{aligned} f^{\Delta^2}(t) &= g^{\tilde{\Delta}^2}(s(t))\|f^{\Delta}(t)\|\,\|f^{\Delta}(t)\|^{\sigma} \\ &\quad + \frac{f^{\Delta}(t)}{\|f^{\Delta}(t)\|}\left(\frac{f^{\Delta^2}(t) \cdot \left(f^{\Delta}(t) + f^{\Delta\sigma}(t)\right)}{\|f^{\Delta}(t)\| + \|f^{\Delta}(t)\|^{\sigma}}\right), \quad t \in I. \end{aligned}$$

Consequently

$$g^{\tilde{\Delta}^2}(s(t)) = \frac{f^{\Delta^2}(t)}{\|f^{\Delta}(t)\|\,\|f^{\Delta}(t)\|^{\sigma}} - \frac{f^{\Delta}(t)}{\|f^{\Delta}(t)\|^2}\left(\frac{f^{\Delta^2}(t)}{\|f^{\Delta}(t)\|^{\sigma}} \cdot \frac{f^{\Delta}(t)+f^{\Delta\sigma}(t)}{\|f^{\Delta}(t)\|+\|f^{\Delta}(t)\|^{\sigma}}\right), \quad t \in I.$$

Definition 1.18 The vector

$$\mathbf{k}(t) = g^{\tilde{\Delta}^2}(s(t)), \quad t \in I,$$

will be called the curvature vector of the parameterized curve $f = f(t)$ at the point $t \in I$, while the norm

$$k(t) = \|g^{\tilde{\Delta}^2}(s(t))\|, \quad t \in I,$$

will be called the curvature of the parameterized curve at the point $t \in I$.

Exercise 1.16 Let $\mathbb{T} = 3^{\mathbb{N}_0}$. Find the curvature of the curve $f : [1,81] \to \mathbb{R}^3$ given by

$$f(t) = \left(\frac{t+1}{t+2}, e_1(t,1)+t^2, \cos_1(t,1)\right), \quad t \in [1,81].$$

Theorem 1.4 *If the curvature of a regular parameterized curve is identically zero, then the support of the curve lies on a straight line.*

Proof. Let $(J, g = g(s))$ be a naturally parameterized curve.

By the hypothesis,

$$g^{\tilde{\Delta}^2}(s) = 0, \quad s \in J,$$

whereupon

$$g(s) = a + bs, \quad s \in J, \quad a, b \in \mathbb{R},$$

that is the support $g(J)$ lies on a straight line. Since two equivalent parameterized curves have the same support, the result holds for non-naturally parameterized curves. This completes the proof.

1.7 The Frenet Frame. Frenet Formulae

Definition 1.19 Let $f : [a,b] \to \mathbb{R}^n$ be a Frenet curve.

If the system $\{e_j\}_{j=1}^n$ satisfies the conditions

1. $\{e_j\}_{j=1}^n$ is an orthonormal basis.

2. For any $k \in \{1,\ldots,n-1\}$ one has

$$\mathrm{Span}\{e_1,e_2,\ldots,e_k\} = \mathrm{Span}\left\{f^{\Delta},f^{\Delta^2},\ldots,f^{\Delta^k}\right\}.$$

3. $f^{\Delta^k} \cdot e_k > 0,\ k \in \{1,\ldots,n\}$,

then the orthonormal basis $\{e_j\}_{j=1}^n$ is called the Frenet n-frame.

For arbitrary $n \in \mathbb{N}$, one can obtain $e_1, e_2, \ldots, e_{n-1}$ from $f^{\Delta}, f^{\Delta^2}, f^{\Delta^3}, \ldots, f^{\Delta^{n-1}}$ by using the Gram-Schmidt orthogonalization process as follows.

Take $e_1 = \dfrac{f^{\Delta}}{\|f^{\Delta}\|}$. Then we will search $\widetilde{e}_2$ in the form

$$\widetilde{e}_2 = f^{\Delta^2} + b_1 e_1,$$

where b_1 is a constant that will be determined by the condition

$$0 = \widetilde{e}_2 \cdot e_1.$$

We have

$$\begin{aligned} 0 &= \widetilde{e}_2 \cdot e_1 \\ &= f^{\Delta^2} \cdot e_1 + b_1 (e_1 \cdot e_1) \\ &= f^{\Delta^2} \cdot e_1 + b_1, \end{aligned}$$

whereupon

$$b_1 = -f^{\Delta^2} \cdot e_1.$$

Then

$$\widetilde{e}_2 = f^{\Delta^2} - \left(f^{\Delta^2} \cdot e_1\right) e_1.$$

Now, we take

$$e_2 = \frac{f^{\Delta^2} - \left(f^{\Delta^2} \cdot e_1\right) e_1}{\|f^{\Delta^2} - \left(f^{\Delta^2} \cdot e_1\right) e_1\|}.$$

Continuing in this manner, we get

$$e_k = \frac{f^{\Delta^k} - \left(f^{\Delta^k} \cdot e_1\right) e_1 - \cdots - \left(f^{\Delta^k} \cdot e_{k-1}\right) e_{k-1}}{\|f^{\Delta^k} - \left(f^{\Delta^k} \cdot e_1\right) e_1 - \cdots - \left(f^{\Delta^k} \cdot e_{k-1}\right) e_{k-1}\|}, \quad k \in \{2,\ldots,n-1\}.$$

Then, we take

$$e_n = \frac{e_1 \times e_2 \times \cdots \times e_{n-1}}{\|e_1 \times e_2 \times \cdots \times e_{n-1}\|}.$$

For $n = 2$ and $n = 3$, we have

$$e_1 = \frac{f^{\Delta}}{\|f^{\Delta}\|},$$

$$\begin{aligned}\widetilde{e}_2 &= f^{\Delta^2} - \frac{1}{\|f^{\Delta}\|}\left(f^{\Delta^2}\cdot f^{\Delta}\right)\frac{f^{\Delta}}{\|f^{\Delta}\|}\\ &= \frac{\|f^{\Delta}\|^2 f^{\Delta^2} - \left(f^{\Delta^2}\cdot f^{\Delta}\right)f^{\Delta}}{\|f^{\Delta}\|^2}.\end{aligned}$$

Then

$$\begin{aligned}e_2 &= \frac{\|f^{\Delta}\|^2 f^{\Delta^2} - \left(f^{\Delta^2}\cdot f^{\Delta}\right)f^{\Delta}}{\|\,\|f^{\Delta}\|^2 f^{\Delta^2} - \left(f^{\Delta^2}\cdot f^{\Delta}\right)f^{\Delta}\|},\\ e_1 \times e_2 &= \frac{\|f^{\Delta}\|^2\left(f^{\Delta}\times f^{\Delta^2}\right)}{\|f^{\Delta}\|\,\|\,\|f^{\Delta}\|^2 f^{\Delta^2} - \left(f^{\Delta^2}\cdot f^{\Delta}\right)f^{\Delta}\|}\\ &= \frac{\|f^{\Delta}\|\left(f^{\Delta}\times f^{\Delta^2}\right)}{\|\,\|f^{\Delta}\|^2 f^{\Delta^2} - \left(f^{\Delta^2}\cdot f^{\Delta}\right)f^{\Delta}\|}\end{aligned}$$

and

$$\begin{aligned}e_3 &= \frac{e_1\times e_2}{\|e_1\times e_2\|}\\ &= \frac{f^{\Delta}\times f^{\Delta^2}}{\|f^{\Delta}\times f^{\Delta^2}\|}.\end{aligned}$$

Definition 1.20 Let $n = 3$. Then e_2 will be called the principal normal of the space curve.

Definition 1.21 Let $n = 3$. Then e_3 will be called the binormal of the space curve.

Remark 1.2 Let $n = 3$. Then the plane determined by the vectors $\{e_1, e_2\}$ is the osculating plane. The plane determined by the vectors $\{e_2, e_3\}$ is the normal plane.

Definition 1.22 Let $n = 3$. Then the plane determined by the vectors $\{e_1, e_3\}$ will be called the rectifying plane.

For arbitrary $n \in \mathbb{N}$, denote

$$\begin{aligned}a_{ij} &= e_i^{\Delta}\cdot e_j, \quad i,j\in\{1,\ldots,n\}, \quad i\le j,\\ a_{ij} &= -e_j^{\Delta}\cdot e_i^{\sigma}, \quad i,j\in\{1,\ldots,n\}, \quad i>j.\end{aligned}$$

Then, using

$$e_i\cdot e_j = 0, \quad i,j\in\{1,\ldots,n\},$$

and

$$e_i^{\Delta}\cdot e_j = -e_i^{\sigma}\cdot e_j^{\Delta}, \quad i,j\in\{1,\ldots,n\}, \tag{1.9}$$

we get

$$e_i^{\Delta} = \sum_{j=1}^{n} a_{ij} e_j, \quad i \in \{1, \ldots, n\}. \tag{1.10}$$

Definition 1.23 The equations (1.10) will be called the Frenet formulae.

Remark 1.3 When $\mathbb{T} \neq \mathbb{R}$, using (1.9), we get in the general case

$$e_i^{\Delta} \cdot e_i \neq 0, \quad i \in \{1, \ldots, n\},$$

and

$$a_{ij} \neq -a_{ji}, \quad i, j \in \{1, \ldots, n\}, \quad i \neq j.$$

Remark 1.4 When $\mathbb{T} = \mathbb{R}$, we get the classical Frenet-Serret formulae.

Below, suppose that $\|f^{\Delta}(t)\| = 1$, $t \in [a, b]$. Let $n = 2$. Then, applying the Taylor formula, we find

$$\begin{aligned} f(t) &= f(t_1) + h_1(t, t_1) f^{\Delta}(t_1) + h_2(t, t_1) f^{\Delta^2}(t_1) + O(h_3(t, t_1)) \\ &= f(t_1) + h_1(t, t_1) e_1(t_1) + h_2(t, t_1) \Big(a_{11}(t_1) e_1(t_1) \\ &\quad + a_{12}(t_1) e_2(t_1) \Big) + O(h_3(t, t_1)) \\ &= f(t_1) + (h_1(t, t_1) + a_{11}(t_1) h_2(t, t_1)) e_1(t_1) \\ &\quad + a_{12}(t_1) h_2(t, t_1) e_2(t_1) + O(h_3(t, t_1)), \quad t, t_1 \in [a, b]. \end{aligned}$$

Now, take $n = 3$. Applying the Taylor formula, we get

$$\begin{aligned} f(t) &= f(t_1) + h_1(t, t_1) f^{\Delta}(t_1) + h_2(t, t_1) f^{\Delta^2}(t_1) + O(h_3(t, t_1)) \\ &= f(t_1) + h_1(t, t_1) e_1(t_1) + h_2(t, t_1) e_1^{\Delta}(t_1) + O(h_3(t, t_1)) \\ &= f(t_1) + h_1(t, t_1) e_1(t_1) \\ &\quad + h_2(t, t_1)(a_{11} e_1(t_1) + a_{12} e_2(t_1) + a_{13} e_3(t_1)) + O(h_3(t, t_1)) \\ &= f(t_1) + (h_1(t, t_1) + a_{11} h_2(t, t_1)) e_1(t_1) \\ &\quad a_{12} h_2(t, t_1) e_2(t_1), \quad t, t_1 \in [a, b]. \end{aligned}$$

1.8 Osculating Circles. Evolutes

Let $n = 2$ and a_{ij}, $i, j \in \{1, 2\}$, be the coefficients in the Frenet formulae for plane curves.

Theorem 1.5 *Let $f : [a,b] \to \mathbb{R}^2$ be a biregular plane curve, $t_0 \in [a,b]$ be such that $a_{21}(t_0) \neq 0$. Then a_{12} is a constant if and only if*

$$g(t) = \int_a^t e_1(s)\Delta s - \frac{1}{a_{21}(t_0)}\left(\int_a^t a_{22}(s)e_2(s)\Delta s - e_2(t)\right), \quad t \in [a,b],$$

is a constant.

Proof. We have that g is a constant on $[a,b]$ if and only if

$$g^{\Delta}(t) = 0, \quad t \in [a,b],$$

or if and only if

$$\begin{aligned}
0 &= e_1(t) - \frac{1}{a_{21}(t_0)}\left(a_{22}(t)e_2(t) - e_2^{\Delta}(t)\right) \\
&= e_1(t) - \frac{1}{a_{21}(t_0)}\left(a_{22}(t)e_2(t) - a_{21}(t)e_1(t) - a_{22}(t)e_2(t)\right) \\
&= e_1(t) + \frac{a_{21}(t)}{a_{21}(t_0)}e_1(t) \\
&= \left(1 + \frac{a_{21}(t)}{a_{21}(t_0)}\right)e_1(t), \quad t \in [a,b],
\end{aligned}$$

if and only if

$$a_{21}(t) = a_{21}(t_0), \quad t \in [a,b].$$

This completes the proof.

Remark 1.5 Suppose that all conditions of Theorem 1.5 hold.

Then

$$\begin{aligned}
\left\| g(t) - \int_a^t e_1(s)\Delta s + \frac{1}{a_{21}(t_0)}\int_a^t a_{22}(s)e_2(s)\Delta s \right\| &= \sqrt{\left(\frac{1}{a_{21}(t_0)}e_2(t), \frac{1}{a_{21}(t_0)}e_2(t)\right)} \\
&= \frac{1}{|a_{21}(t_0)|}.
\end{aligned}$$

Definition 1.24 Let $f : [a,b] \to \mathbb{R}^2$ be a biregular curve with $a_{21}(t_0) \neq 0$ for some $t_0 \in [a,b]$. Then the circle centered at

$$\int_a^{t_0} e_1(s)\Delta s - \frac{1}{a_{21}(t_0)}\int_a^{t_0} a_{22}(s)e_2(s)\Delta s$$

with radius $\dfrac{1}{|a_{21}(t_0)|}$ is called the e_2-osculating circle of f.

The curve

$$t \to \int_a^t e_1(s)\Delta s - \frac{1}{a_{21}(t_0)}\left(\int_a^t a_{22}(s)e_1(s)\Delta s - e_2(t)\right), \quad t \in [a,b],$$

is called the e_2- evolute or the e_2-focal curve of f.

Theorem 1.6 *Let $f:[a,b]\to\mathbb{R}^2$ be a biregular plane curve, $t_0\in[a,b]$ be such that $a_{12}(t_0)\neq 0$. Then a_{12} is a constant if and only if*

$$h(t)=\int_a^t e_2(s)\Delta s-\frac{1}{a_{12}(t_0)}\left(\int_a^t a_{11}(s)e_1(s)\Delta s-e_1(t)\right),\quad t\in[a,b],$$

is a constant.

Proof. Note that h is a constant on $[a,b]$ if and only if

$$h^{\Delta}(t)=0,\quad t\in[a,b],$$

that is, if and only if

$$\begin{aligned}0&=e_2(t)-\frac{1}{a_{12}(t_0)}\left(a_{11}(t)e_1(t)-e_1^{\Delta}(t)\right)\\&=e_2(t)-\frac{1}{a_{12}(t_0)}\left(a_{11}(t)e_1(t)-a_{11}(t)e_1(t)-a_{12}(t)e_2(t)\right)\\&=e_2(t)+\frac{a_{12}(t)}{a_{12}(t_0)}e_2(t)\\&=\left(1+\frac{a_{12}(t)}{a_{12}(t_0)}\right)e_2(t),\quad t\in[a,b],\end{aligned}$$

if and only if

$$a_{12}(t)=a_{12}(t_0),\quad t\in[a,b].$$

This completes the proof.

Remark 1.6 Suppose that all conditions of Theorem 1.6 hold.

Then

$$\begin{aligned}\left\|h(t)-\int_a^t e_2(s)\Delta s+\frac{1}{a_{12}(t_0)}\int_a^t a_{11}(s)e_1(s)\Delta s\right\|&=\sqrt{\left(\frac{1}{a_{12}(t_0)}e_1(t),\frac{1}{a_{12}(t_0)}e_1(t)\right)}\\&=\frac{1}{|a_{12}(t_0)|}.\end{aligned}$$

Definition 1.25 Let $f:[a,b]\to\mathbb{R}^2$ be a biregular curve with $a_{12}(t_0)\neq 0$ for some $t_0\in[a,b]$. Then the circle centered at

$$\int_a^{t_0} e_2(s)\Delta s-\frac{1}{a_{12}(t_0)}\int_a^{t_0} a_{11}(s)e_1(s)\Delta s$$

with radius $\dfrac{1}{|a_{11}(t_0)|}$ is called the e_1-osculating circle of f. The curve

$$t\to\int_a^t e_2(s)\Delta s-\frac{1}{a_{12}(t_0)}\left(\int_a^t a_{11}(s)e_2(s)\Delta s-e_1(t)\right),\quad t\in[a,b],$$

is called the e_1- evolute or the e_1-focal curve of f.

Theorem 1.7 *Let* $f:[a,b]\to\mathbb{R}^2$ *be a biregular curve,* $t_0\in[a,b]$ *be such that* $a_{21}(t_0)\neq 0$, $a_{12}(t_0)\neq 0$. *Then* a_{12} *and* a_{21} *are constants if and only if*

$$m(t)=\int_a^t(e_1(s)+e_2(s))\Delta s-\frac{1}{a_{21}(t_0)}\left(\int_a^t a_{22}(s)e_2(s)\Delta s-e_2(t)\right)$$
$$-\frac{1}{a_{12}(t_0)}\left(\int_a^t a_{11}(s)e_1(s)\Delta s-e_1(t)\right),\quad t\in[a,b],$$

is a constant.

Proof. Note that

$$m(t)=g(t)+h(t),\quad t\in[a,b].$$

Then, by Theorem 1.5 and Theorem 1.6, we get

$$\begin{aligned}m^\Delta(t)&=g^\Delta(t)+h^\Delta(t)\\&=\left(1+\frac{a_{21}(t)}{a_{21}(t_0)}\right)e_1(t)+\left(1+\frac{a_{12}(t)}{a_{12}(t_0)}\right)e_2(t)\\&=0,\quad t\in[a,b],\end{aligned}$$

if and only if

$$a_{21}(t)=a_{21}(t_0),\quad t\in[a,b],$$

and

$$a_{12}(t)=a_{12}(t_0),\quad t\in[a,b].$$

This completes the proof.

Remark 1.7 Suppose that all conditions of Theorem 1.7 hold.

Then

$$\left\|m(t)-\int_a^t(e_1(s)+e_2(s))\Delta s+\frac{1}{a_{21}(t_0)}\int_a^t a_{22}(s)e_2(s)\Delta s+\frac{1}{a_{12}(t_0)}\int_a^t a_{11}(s)e_1(s)\Delta s\right\|$$
$$=\left\|\frac{1}{a_{21}(t_0)}e_2(t)+\frac{1}{a_{12}(t_0)}e_1(t)\right\|$$
$$=\sqrt{\left(\frac{1}{a_{21}(t_0)}e_2(t)+\frac{1}{a_{12}(t_0)}e_1(t),\frac{1}{a_{21}(t_0)}e_2(t)+\frac{1}{a_{12}(t_0)}e_1(t)\right)}$$
$$=\sqrt{\frac{1}{(a_{21}(t_0))^2}+\frac{1}{(a_{12}(t_0))^2}},\quad t\in[a,b].$$

Definition 1.26 Let $f:[a,b]\to\mathbb{R}^2$ be a biregular curve with $a_{12}(t_0)\neq 0$ and $a_{21}(t_0)\neq 0$ for some $t_0\in[a,b]$. Then the circle centred at

$$\int_a^{t_0}(e_1(s)+e_2(s))\Delta s-\frac{1}{a_{21}(t_0)}\int_a^{t_0}a_{22}(s)e_2(s)\Delta s-\frac{1}{a_{12}(t_0)}\int_a^{t_0}a_{11}(s)e_1(s)\Delta s$$

with radius

$$\sqrt{\frac{1}{(a_{12}(t_0))^2}+\frac{1}{(a_{21}(t_0))^2}}$$

is called the osculating circle of f.

The curve

$$t \to \int_a^t (e_1(s)+e_2(s))\Delta s - \frac{1}{a_{21}(t_0)}\left(\int_a^t a_{22}(s)e_2(s)\Delta s - e_2(t)\right)$$
$$-\frac{1}{a_{12}(t_0)}\left(\int_a^t a_{11}(s)e_1(s)\Delta s - e_1(t)\right), \quad t \in [a,b],$$

is called the evolute or the focal curve of f.

Theorem 1.8 *Let $f : [a,b] \to \mathbb{R}^2$ be a biregular curve with $\|f^{\Delta}(t)\| = 1$, $t \in [a,b]$. Then*

$$a_{11}(t) = a_{12}(t) = 0, \quad t \in [a,b],$$

if and only if it lies on a line segment.

Proof. We have

$$f(t) = a_1 + a_2 t, \quad a_1, a_2 \in \mathbb{R}, \quad t \in [a,b],$$

if and only if

$$f^{\Delta^2}(t) = 0, \quad t \in [a,b],$$

i.e., if and only if

$$e_1^{\Delta}(t) = 0, \quad t \in [a,b],$$

if and only if

$$a_{11}(t)e_1(t) + a_{12}(t)e_2(t) = 0, \quad t \in [a,b],$$

if and only if

$$a_{11}(t) = a_{12}(t) = 0, \quad t \in [a,b].$$

This completes the proof.

1.9 Oriented Space Curves. The Frenet Frame of an Oriented Space Curve

Definition 1.27 An orientation of a regular curve $C \subset \mathbb{R}^3$ is a family of local parameterizations $\{(I_\alpha, f_\alpha)\}_{\alpha\in A}$ such that

1. $C = \bigcup_{\alpha\in A} f_\alpha(I_\alpha)$

2. for any connected component $C^b_{\alpha\beta}$ of

$$C_{\alpha\beta} = f_\alpha(I_\alpha) \bigcap f_\beta(I_\beta), \quad \alpha, \beta \in A,$$

the parameterized curves (I^b_α, f^b_α), (I^b_β, f^b_β) with

$$I^b_\alpha = f_\alpha^{-1}(C^b_{\alpha\beta}), \quad f^b_\alpha = f_\alpha \Big|_{I^b_\alpha},$$

$$I^b_\beta = f_\beta^{-1}(C^b_{\alpha\beta}), \quad f^b_\beta = f_\beta \Big|_{I^b_\beta},$$

are positively equivalent.

Definition 1.28 A regular curve $C \subset \mathbb{R}^3$ with an orientation is called an oriented regular curve.

Definition 1.29 A local parameterization (I,f) of an oriented regular curve C is called compatible with the orientation defined by the family $\{(I_\alpha, f_\alpha)\}_{\alpha\in A}$ if on the intersections $f(I) \cap f_\alpha(I_\alpha)$ the parameterized curves (I,f) and (I_α, f_α) are positively oriented.

Definition 1.30 The Frenet frame of an oriented biregular curve C at a point $x \in C$ is the Frenet frame of a biregular parameterized curve $f = f(t)$ at t_0, where $f = f(t)$ is a local parameterization of the curve C, compatible with the orientation such that $f(t_0) = x$.

1.10 General Helices

Definition 1.31 A parameterized curve $([a,b], f)$ in $\mathbb{R}^3$ is called a general helix if its tangents make a constant angle with a fixed direction in the space.

Theorem 1.9 *Let $f : [a,b] \to \mathbb{R}^3$ be a biregular curve with $\|f^\Delta\| = 1$ that is a general helix and*

$$a_{11} = a_{13} = 0, \quad a_{12} > 0.$$

Then

$$a_{21} = c_0 a_{23}$$

for some constant c_0.

Proof. Let c be the versor of the fixed derivative.

Then

$$e_1^\Delta \cdot c = 0$$

and then, using the first equation of the Frenet formulae and $a_{11} = a_{13} = 0$, we find

$$(a_{12}e_2) \cdot c = 0.$$

Therefore

$$e_2 \cdot c = 0 \tag{1.11}$$

and $e_2 \perp c$, and

$$e_3 \cdot c = \sin\alpha_0$$

for some $\alpha_0 \in \left[0, \frac{\pi}{2}\right]$.

By (1.11), we get

$$e_2^{\Delta} \cdot c = 0.$$

Now, using the second equation of the Frenet formulae, we obtain

$$(a_{21}e_1 + a_{22}e_2 + a_{23}e_3) \cdot c = 0$$

or

$$a_{21}(e_1 \cdot c) + a_{22}(e_2 \cdot c) + a_{23}(e_3 \cdot c) = 0,$$
$$a_{21}\cos\alpha_0 + a_{23}\sin\alpha_0 = 0,$$

or

$$a_{21} = -\tan\alpha_0 a_{23}.$$

This completes the proof.

Theorem 1.10 *Let $f : [a,b] \to \mathbb{R}^3$ be a biregular curve with $\|f^{\Delta}\| = 1$ and*

$$a_{11} = -a_{31},$$
$$a_{12} = -a_{32},$$
$$a_{13} = -a_{33}.$$

Then, f is a general helix.

Proof. By the first and third equations of the Frenet formulae, we get

$$e_1^{\Delta} = a_{11}e_1 + a_{12}e_2 + a_{13}e_3$$
$$e_3^{\Delta} = a_{31}e_1 + a_{32}e_2 + a_{33}e_3$$
$$= -a_{11}e_1 - a_{12}e_2 - a_{13}e_3.$$

Thus,

$$e_1^{\Delta} + e_3^{\Delta} = 0$$

and then there is a $c_0 \in \mathbb{R}^3$ so that

$$e_1 + e_3 = c_0.$$

Let

$$c = \frac{e_1 + e_3}{\|e_1 + e_3\|}.$$

Then

$$\begin{aligned}\|e_1+e_3\| &= \sqrt{(e_1+e_3)\cdot(e_1+e_3)}\\ &= \sqrt{2},\\ c &= \frac{e_1+e_3}{\sqrt{2}}\end{aligned}$$

and

$$\begin{aligned}e_1\cdot c &= \left(\frac{e_1+e_3}{\sqrt{2}}\right)\cdot e_1\\ &= \frac{1}{\sqrt{2}}.\end{aligned}$$

Therefore e_1 and c make a constant angle and the curve is a general helix. This completes the proof.

1.11 Bertrand Curves

Definition 1.32 Space curves which have the same family of principal normals will be called Bertrand curves. Obviously, for a Bertrand curve there is only one curve having the same principal normals. We will say that both curves are Bertrand mates or that they are associated, or conjugate Bertrand curves.

Theorem 1.11 *Let* $([a,b],f)$ *and* $([a,b],f^*)$ *be two Bertrand curves such that* $\|f^\Delta\| = \|f^{*\Delta}\| = 1$ *and*

$$a_{11} = a_{13} = a_{22} = a_{31} = a_{33} = 0,$$

where a_{11}, a_{13}, a_{22}, a_{31} *and* a_{33} *are the corresponding coefficients in the Frenet formulae associated with the Frenet frame* $\{e_1,e_2,e_3\}$ *for the curve* f. *Then the angles of the tangents of* f *and* f^* *at the corresponding points is a constant.*

Proof. We have

$$f^* = f + ae_2. \tag{1.12}$$

Let $\{e_1^*,e_2^*,e_3^*\}$ be the Frenet frame associated to the curve f^*. Since f and f^* are Bertrand curves, they have the same principal normals. So,

$$e_2^* = \pm e_2.$$

Now, we differentiate (1.12) and we find

$$\begin{aligned}f^{*\Delta} &= f^\Delta + a^\sigma e_2^\Delta + a^\Delta e_2\\ &= e_1 + a^\sigma(a_{21}e_1 + a_{23}e_3) + a^\Delta e_2\end{aligned}$$

$$= (1 + a^\sigma a_{21}) e_1 + a^\Delta e_2 + a^\sigma a_{23} e_3$$

or

$$e_1^* = (1 + a^\sigma a_{21}) e_1 + a^\Delta e_2 + a^\sigma a_{23} e_3. \tag{1.13}$$

Since e_1^* is a tangent to f^*, we have that $e_1^* \perp e_2^*$ and from here, $e_1^* \perp e_2$. Thus, by (1.13), we find

$$\begin{aligned} 0 &= e_1^* \cdot e_2 \\ &= a^\Delta. \end{aligned}$$

Therefore a is a constant and

$$e_1^* = (1 + aa_{21}) e_1 + aa_{23} e_3.$$

Let

$$w = \angle(e_1, e_1^*).$$

Then

$$e_1 \cdot e_1^* = \cos w$$

and

$$\begin{aligned} e_1^* &= \cos w e_1 + \sin w e_3 \\ e_3^* &= \epsilon(-\sin w e_1 + \cos w e_3), \end{aligned} \tag{1.14}$$

where $\epsilon = \pm 1$. We differentiate (1.14) and using the Pötzsche chain rule, we find

$$\begin{aligned} e_1^{*\Delta} &= -\left(\left(\int_0^1 \sin(w + h\mu w^\Delta) dh\right) w^\Delta\right) e_1 + (\cos w)^\sigma e_1^\Delta \\ &\quad + \left(\left(\int_0^1 \cos(w + h\mu w^\Delta) dh\right) w^\Delta\right) e_3 + (\sin w)^\sigma e_3^\Delta \\ &= -\left(\left(\int_0^1 \sin(w + h\mu w^\Delta) dh\right) w^\Delta\right) e_1 \\ &\quad + \left(\left(\int_0^1 \cos(w + h\mu w^\Delta) dh\right) w^\Delta\right) e_3 \\ &\quad + (a_{12}(\cos w)^\sigma + a_{32}(\sin w)^\sigma) e_2 \end{aligned}$$

or

$$\begin{aligned} e_2^* &= -\left(\left(\int_0^1 \sin(w + h\mu w^\Delta) dh\right) w^\Delta\right) e_1 \\ &\quad + \left(\left(\int_0^1 \cos(w + h\mu w^\Delta) dh\right) w^\Delta\right) e_3 \\ &\quad + (a_{12}(\cos w)^\sigma + a_{32}(\sin w)^\sigma) e_2. \end{aligned}$$

Now, using that

$$e_2^* \perp e_1 \quad \text{and} \quad e_2^* \perp e_3,$$

we get

$$\begin{aligned}\left(\int_0^1 \sin(w+h\mu w^{\Delta})dh\right)w^{\Delta} &= 0\\ \left(\int_0^1 \cos(w+h\mu w^{\Delta})dh\right)w^{\Delta} &= 0.\end{aligned} \tag{1.15}$$

1. Let $\mu \neq 0$. Then (1.15) takes the form

$$\frac{1}{\mu}\left(\cos(w+\mu w^{\Delta})-\cos w\right)=0$$
$$\frac{1}{\mu}\left(\sin(w+\mu w^{\Delta})-\sin w\right)=0$$

or

$$\frac{1}{\mu}\sin\left(\frac{2w+\mu w^{\Delta}}{2}\right)\sin\frac{w}{2}=0$$
$$\frac{1}{\mu}\cos\left(\frac{2w+\mu w^{\Delta}}{2}\right)\sin\frac{w}{2}=0.$$

Then w is a constant.

2. Let $\mu = 0$. Then (1.15) takes the form

$$w^{\Delta}\sin w = 0$$
$$w^{\Delta}\cos w = 0,$$

where upon w is a constant. This completes the proof.

Theorem 1.12 *Let* $([a,b],f)$ *and* $([a,b],f^*)$ *be two Bertrand curves such that* $\|f^{\Delta}\| = \|f^{*\Delta}\| = 1$ *on* $[a,b]$ *and*

$$a_{11} = a_{13} = a_{22} = a_{31} = a_{33} = 0,$$

where $a_{11}, a_{13}, a_{22}, a_{31}$ *and* a_{33} *are the corresponding coefficients in the Frenet formulae associated with the Frenet frame* $\{e_1,e_2,e_3\}$ *for the curve* f. *Then there exist constants* c *and* d *so that*

$$ca_{21} + da_{23} = 1.$$

Proof. Let $\{e_1^*,e_2^*,e_3^*\}$ be the Frenet frame associated to the curve f.
By the proof of Theorem 1.11, we get

$$e_1^* = (1+aa_{21})e_1 + aa_{23}e_3$$
$$e_1^* = \cos w e_1 + \sin w e_2.$$

Hence,

$$1 + aa_{21} = \cos w$$

$$aa_{23} = \sin w$$

and

$$1 + aa_{21} = aa_{23} \cot w,$$

or

$$1 = -aa_{21} + aa_{23} \cot w.$$

Let

$$c = -a,$$
$$d = a \cot w.$$

Then

$$1 = ca_{21} + da_{23}.$$

This completes the proof.

Theorem 1.13 *Let* $([a,b],f)$ *be a biregular curve such that* $\|f^\Delta\| = 1$ *and* a_{12}, a_{21}, a_{23} *and* a_{32} *are constants and*

$$a_{11} = a_{13} = a_{22} = a_{31} = a_{33} = 0,$$

where a_{ij}, $i,j \in \{1,2,3\}$ *are the coefficients in the Frenet formulae associated with the Frenet frame* $\{e_1, e_2, e_3\}$ *for the curve* f. *If* a *is a constant such that*

$$2a_{21} + a\left(a_{21}^2 + a_{23}^2\right) = 0$$
$$(1 + aa_{21})a_{12} + aa_{23}a_{32} = 1$$

or

$$2a_{21} + a\left(a_{21}^2 + a_{23}^2\right) = 0$$
$$(1 + aa_{21})a_{12} + aa_{23}a_{32} = -1,$$

then f *is a Bertrand curve.*

Proof. Let

$$f^* = f + ae_2$$

and $\{e_1^*, e_2^*, e_3^*\}$ is the Frenet frame associated to the curve f.

Then

$$\begin{aligned} f^{*\Delta} &= f^\Delta + ae_2^\Delta \\ &= e_1 + aa_{21}e_1 + aa_{23}e_3 \\ &= (1 + aa_{21})e_1 + aa_{23}e_3 \end{aligned}$$

and

$$\begin{aligned} \|f^{*\Delta}\|^2 &= (1 + aa_{21})^2 + a^2a_{23}^2 \\ &= 1 + 2aa_{21} + a^2a_{21}^2 + a^2a_{23}^2 \end{aligned}$$

$$= 1 + a\left(2a_{21} + a\left(a_{21}^2 + a_{23}^2\right)\right)$$
$$= 1.$$

By the proof of Theorem 1.11, we have

$$e_1^* = (1 + aa_{21})e_1 + aa_{23}e_3,$$

where upon

$$\begin{aligned} e_1^{*\Delta} &= e_2^* \\ &= (1 + aa_{21})e_1^{\Delta} + aa_{23}e_3^{\Delta} \\ &= (1 + aa_{21})a_{12}e_2 + aa_{23}a_{32}e_2 \\ &= ((1 + aa_{21})a_{12} + aa_{23}a_{32})e_2 \\ &= \pm e_2. \end{aligned}$$

Therefore f is a Bertrand curve. This completes the proof.

1.12 The Behaviour of the Frenet Frame under a Rigid Motion

Definition 1.33 A rigid motion of $\mathbb{R}^3$ is a map $\mathscr{D}: \mathbb{R}^3 \to \mathbb{R}^3$,

$$\mathscr{D}(x) = \mathscr{B}x + b,$$

where $x \in \mathbb{R}^3$, $\mathscr{B} \in \mathscr{M}_{3\times 3}(\mathbb{R})$ is an orthogonal matrix, $\mathscr{B}^T\mathscr{B} = I$, with determinant equal to 1. The linear map $B: \mathbb{R}^3 \to \mathbb{R}^3$,

$$B(x) = \mathscr{B}x, \quad x \in \mathbb{R}^3,$$

is called the homogeneous part of the rigid motion.

Definition 1.34 Let $\mathscr{D}: \mathbb{R}^3 \to \mathbb{R}^3$ be a rigid motion with a homogeneous part $B: \mathbb{R}^3 \to \mathbb{R}^3$. The image of $([a,b],f)$ through $\mathscr{D}$ is the parameterized curve

$$([a,b],\tilde{f}) = ([a,b],\mathscr{D}(f)).$$

Theorem 1.14 *Let $\mathscr{D}$ be a rigid motion of $\mathbb{R}^3$ with a homogeneous part B, $([a,b],f)$ be a biregular parameterized curve,*

$$([a,b],\tilde{f}) = ([a,b],\mathscr{D}(f)),$$

$\{e_1(t), e_2(t), e_3(t)\}$ is the Frenet frame associated to the curve f. Then $\{B(e_1(t)), B(e_2(t)), B(e_3(t))\}$ is the Frenet frame associated to the curve $\tilde{f}$ at t.

Proof. Let

$$\begin{aligned} f(t) &= (f_1(t), f_2(t), f_3(t)), \\ \tilde{f}(t) &= (\tilde{f}_1(t), \tilde{f}_2(t), \tilde{f}_3(t)), \\ \mathscr{D}(f(t)) &= \tilde{f}(t), \end{aligned}$$

where

$$\begin{aligned} \tilde{f}_1(t) &= b_{11}f_1(t) + b_{12}f_2(t) + b_{13}f_3(t) + b_1 \\ \tilde{f}_2(t) &= b_{21}f_1(t) + b_{22}f_2(t) + b_{23}f_3(t) + b_2 \\ \tilde{f}_3(t) &= b_{31}f_1(t) + b_{32}f_2(t) + b_{33}f_3(t) + b_3. \end{aligned}$$

Hence,

$$\begin{aligned} \tilde{f}_1^{\Delta}(t) &= b_{11}f_1^{\Delta}(t) + b_{12}f_2^{\Delta}(t) + b_{13}f_3^{\Delta}(t) \\ \tilde{f}_2^{\Delta}(t) &= b_{21}f_1^{\Delta}(t) + b_{22}f_2^{\Delta}(t) + b_{23}f_3^{\Delta}(t) \\ \tilde{f}_3^{\Delta}(t) &= b_{31}f_1^{\Delta}(t) + b_{32}f_2^{\Delta}(t) + b_{33}f_3^{\Delta}(t) \end{aligned}$$

and

$$\begin{aligned} \tilde{f}_1^{\Delta^2}(t) &= b_{11}f_1^{\Delta^2}(t) + b_{12}f_2^{\Delta^2}(t) + b_{13}f_3^{\Delta^2}(t) \\ \tilde{f}_2^{\Delta^2}(t) &= b_{21}f_1^{\Delta^2}(t) + b_{22}f_2^{\Delta^2}(t) + b_{23}f_3^{\Delta^2}(t) \\ \tilde{f}_3^{\Delta^2}(t) &= b_{31}f_1^{\Delta^2}(t) + b_{32}f_2^{\Delta^2}(t) + b_{33}f_3^{\Delta^2}(t), \end{aligned}$$

i.e.,

$$\begin{aligned} \tilde{f}^{\Delta}(t) &= B\left(f^{\Delta}(t)\right), \\ \tilde{f}^{\Delta^2}(t) &= B\left(f^{\Delta^2}(t)\right). \end{aligned}$$

We have

$$\begin{aligned} \|\tilde{f}^{\Delta}(t)\| &= \|B(f^{\Delta}(t))\| \\ &= \sqrt{(\mathscr{B}f^{\Delta}(t)) \cdot (\mathscr{B}f^{\Delta}(t))} \\ &= \sqrt{f^{\Delta}(t) \cdot f^{\Delta}(t)} \\ &= \|f^{\Delta}(t)\|, \\ \tilde{f}^{\Delta}(t) \cdot \tilde{f}^{\Delta^2}(t) &= \left(\mathscr{B}f^{\Delta}(t)\right) \cdot \left(\mathscr{B}f^{\Delta^2}(t)\right) \\ &= f^{\Delta}(t) \cdot f^{\Delta^2}(t), \\ \tilde{e}_1(t) &= \frac{\tilde{f}^{\Delta}(t)}{\|\tilde{f}^{\Delta}(t)\|} \end{aligned}$$

$$
\begin{aligned}
&= \frac{\mathscr{B}f^{\Delta}(t)}{\|f^{\Delta}(t)\|}\\
&= B\left(\frac{f^{\Delta}(t)}{\|f^{\Delta}(t)\|}\right)\\
&= B(e_1(t)),
\end{aligned}
$$

$$
\begin{aligned}
\widetilde{e}_2(t) &= \frac{\|\widetilde{f}^{\Delta}\|^2\widetilde{f}^{\Delta^2} - \left(\widetilde{f}^{\Delta^2}\cdot\widetilde{f}^{\Delta}\right)\widetilde{f}^{\Delta}}{\|\,\|\widetilde{f}^{\Delta}\|^2\widetilde{f}^{\Delta^2} - \left(\widetilde{f}^{\Delta^2}\cdot\widetilde{f}^{\Delta}\right)\widetilde{f}^{\Delta}\|}\\
&= \frac{\|f^{\Delta}\|^2\mathscr{B}f^{\Delta^2} - \left(f^{\Delta^2}\cdot f^{\Delta}\right)\mathscr{B}f^{\Delta}}{\|\,\|f^{\Delta}\|^2\mathscr{B}f^{\Delta^2} - \left(f^{\Delta^2}\cdot f^{\Delta}\right)\mathscr{B}f^{\Delta}\|}\\
&= \frac{B\left(\|f^{\Delta}\|^2f^{\Delta^2} - \left(f^{\Delta^2}\cdot f^{\Delta}\right)f^{\Delta}\right)}{\|B\left(\|f^{\Delta}\|^2f^{\Delta^2} - \left(f^{\Delta^2}\cdot f^{\Delta}\right)f^{\Delta}\right)\|}\\
&= \frac{B\left(\|f^{\Delta}\|^2f^{\Delta^2} - \left(f^{\Delta^2}\cdot f^{\Delta}\right)f^{\Delta}\right)}{\|\,\|f^{\Delta}\|^2f^{\Delta^2} - \left(f^{\Delta^2}\cdot f^{\Delta}\right)f^{\Delta}\|}\\
&= B\left(\frac{\|f^{\Delta}\|^2f^{\Delta^2} - \left(f^{\Delta^2}\cdot f^{\Delta}\right)f^{\Delta}}{\|\,\|f^{\Delta}\|^2f^{\Delta^2} - \left(f^{\Delta^2}\cdot f^{\Delta}\right)f^{\Delta}\|}\right)\\
&= B\left(e_2(t)\right)
\end{aligned}
$$

and

$$
\begin{aligned}
\widetilde{e}_3(t) &= \frac{\widetilde{e}_1(t)\times\widetilde{e}_2(t)}{\|\widetilde{e}_1(t)\times\widetilde{e}_2(t)\|}\\
&= \frac{(\mathscr{B}e_1(t))\times(\mathscr{B}e_2(t))}{\|(\mathscr{B}e_1(t))\times(\mathscr{B}e_2(t))\|}\\
&= \frac{B\left(e_1(t)\times e_2(t)\right)}{\|e_1(t)\times e_2(t)\|}\\
&= B\left(\frac{e_1(t)\times e_2(t)}{\|e_1(t)\times e_2(t)\|}\right)\\
&= B(e_3(t)).
\end{aligned}
$$

This completes the proof.

Corollary 1.1 *Let all conditions of Theorem 1.14 hold.*

Then

$$
\widetilde{a}_{ij} = a_{ij}, \quad i,j \in \{1,\ldots,n\},
$$

where $\widetilde{a}_{ij}$ and a_{ij}, $i,j \in \{1,\ldots,n\}$, are the coefficients in the Frenet formulae for $\widetilde{f}$ and f, respectively.

Proof. We have

$$\begin{aligned}\widetilde{a}_{ij} &= \widetilde{e}_i^{\Delta} \cdot \widetilde{e}_j^{\sigma} \\ &= B\left(e_i^{\Delta}\right) \cdot B(e_j^{\sigma}) \\ &= e_i^{\Delta} \cdot e_j^{\sigma}, \quad i,j \in \{1,\ldots,n\}, \quad i \le j,\end{aligned}$$

and

$$\begin{aligned}\widetilde{a}_{ij} &= -\widetilde{e}_j^{\Delta} \cdot \widetilde{e}_i \\ &= -B\left(e_j^{\Delta}\right) \cdot B(e_i) \\ &= -e_j^{\Delta} \cdot e_i, \quad i,j \in \{1,\ldots,n\}, \quad i > j.\end{aligned}$$

This completes the proof.

1.13 The Uniqueness Theorem

Theorem 1.15 *Let $([a,b],f)$ and $([a,b],\widetilde{f})$ be two biregular parameterized curves and a_{ij}, $\widetilde{a}_{ij}$, $i,j \in \{1,2,3\}$, be the coefficients in the Frenet formulae for f and $\widetilde{f}$, respectively. Suppose that*

$$a_{ij} = \widetilde{a}_{ij}, \quad i,j \in \{1,2,3\},$$

and

$$\|f^{\Delta}(t)\| = \|\widetilde{f}^{\Delta}(t)\|, \quad t \in [a,b].$$

Then there exists a single rigid motion $\mathscr{D}$ of $\mathbb{R}^3$ such that

$$\widetilde{f} = \mathscr{D}(f).$$

Proof. Let $\{e_1,e_2,e_3\}$ and $\{\widetilde{e}_1,\widetilde{e}_2,\widetilde{e}_3\}$ be the Frenet frames corresponding to f and $\widetilde{f}$, respectively. Fix $t_0 \in [a,b]$ arbitrarily.

Set

$$\widetilde{\widetilde{f}}(t) = \mathscr{D}(f(t)), \quad t \in [a,b].$$

Let $\{\widetilde{\widetilde{e}}_1,\widetilde{\widetilde{e}}_2,\widetilde{\widetilde{e}}_3\}$ be the Frenet frame associated to the curve $\widetilde{\widetilde{f}}$ and $\widetilde{\widetilde{a}}_{ij}$, $i,j \in \{1,2,3\}$, be the coefficients in the Frenet formulae for $\widetilde{\widetilde{f}}$. By Theorem 1.14, it follows that

$$\widetilde{\widetilde{e}}_j = e_j, \quad j \in \{1,2,3\}.$$

By Corollary 1.1, we have

$$\widetilde{\widetilde{a}}_{ij} = a_{ij}, \quad i,j \in \{1,2,3\},$$

and

$$\|\widetilde{\widetilde{f}}(t)\| = \|f(t)\|, \quad t \in [a,b].$$

Therefore

$$\{\tilde{\tilde{e}}_1, \tilde{\tilde{e}}_2, \tilde{\tilde{e}}_3\} \quad \text{and} \quad \{\tilde{e}_1, \tilde{e}_2, \tilde{e}_3\}$$

are solutions to the same Frenet equations.
Hence,

$$\tilde{\tilde{e}}_j^{\Delta}(t) - \tilde{e}_j^{\Delta}(t) = 0, \quad j \in \{1,2,3\}, \quad t \in [a,b].$$

Since

$$\tilde{\tilde{e}}_j(t_0) = \tilde{e}_j(t_0), \quad j \in \{1,2,3\},$$

we get

$$\tilde{\tilde{e}}_j(t) = \tilde{e}_j(t), \quad j \in \{1,2,3\}, \quad t \in [a,b].$$

Because $t_0 \in [a,b]$ was arbitrarily chosen, we get the desired result.

1.14 The Existence Theorem

Theorem 1.16 *Let $c_{ij} : [a,b] \to \mathbb{R}^3$, $i,j \in \{1,2,3\}$, be delta differentiable such that $c_{13} = 0$, $c_{12} \geq 0$ on $[a,b]$ and if $C = (c_{ij})$, then*

$$C^T(I + \mu C) + C = 0 \quad on \quad [a,b].$$

Then there is a single parameterized curve $([a,b],f)$ such that $\|f^{\Delta}\| = 1$ on $[a,b]$ and $a_{ij} = c_{ij}$, $i,j \in \{1,2,3\}$, where a_{ij}, $i,j \in \{1,2,3\}$, are the coefficients in the Frenet formulae for f. This curve is uniquely defined, up to a rigid motion of $\mathbb{R}^3$.

Proof. Take $f_0 \in \mathbb{R}^3$, $\|f_0\| = 1$. Let $\{e_{10}, e_{20}, e_{30}\}$ be a direct orthonormal basis of $\mathbb{R}^3$ at the point f_0. Consider the system

$$\begin{aligned} e_1^{\Delta} &= c_{11}e_1 + c_{12}e_2 \\ e_2^{\Delta} &= c_{21}e_1 + c_{22}e_2 + c_{23}e_3 \\ e_3^{\Delta} &= c_{31}e_1 + c_{32}e_2 + c_{33}e_3 \end{aligned} \tag{1.16}$$

with respect to the functions e_1, e_2 and e_3.

Let

$$X(t) = \begin{pmatrix} e_1(t) \\ e_2(t) \\ e_3(t) \end{pmatrix}, \quad t \in [a,b]. \tag{1.17}$$

Then the system (1.16) takes the form

$$X^{\Delta}(t) = C(t)X(t), \quad t \in [a,b].$$

The last system has a single solution subject to

$$X(a) = \begin{pmatrix} e_{10} \\ e_{20} \\ e_{30} \end{pmatrix}.$$

We will show that the vectors $\{e_1(t), e_2(t), e_3(t)\}$, $t \in [a,b]$, form an orthonormal basis. For this purpose, it is enough to show that the matrix $X(t)$, $t \in [a,b]$, is an orthogonal matrix, that is, we will prove that

$$X^T(t)X(t) = I, \quad t \in [a,b]. \tag{1.18}$$

We have

$$\begin{aligned}
(X^T X)^\Delta &= (X^T)^\Delta X^\sigma + X^T X^\Delta \\
&= X^T C^T X^\sigma + X^T C X \\
&= X^T C^T \left(X + \mu X^\Delta\right) + X^T C X \\
&= X^T C^T (X + \mu C X) + X^T C X \\
&= X^T C^T (I + \mu C) X + X^T C X \\
&= X^T \left(C^T (I + \mu C) + C\right) X \\
&= 0.
\end{aligned}$$

Thus,

$$X^T X = \text{const} \quad \text{on} \quad [a,b].$$

On the other hand,

$$\begin{aligned}
X^T(a) &= (e_{10}, e_{20}, e_{30}), \\
X^T(a)X(a) &= I.
\end{aligned}$$

So,

$$X^T X = I \quad \text{on} \quad [a,b].$$

Define

$$f(t) = f(a) + \int_a^t e_1(s)\Delta s, \quad t \in [a,b].$$

Then

$$\begin{aligned}
f^\Delta(t) &= e_1(t), \\
\|f^\Delta(t)\| &= \|e_1(t)\| \\
&= 1, \quad t \in [a,b],
\end{aligned}$$

and

$$\begin{aligned}
f^{\Delta^2}(t) &= c_{11}(t)e_1(t) + c_{12}(t)e_2(t), \\
f^{\Delta^2}(t) \cdot f^\Delta(t) &= c_{11}(t), \quad t \in [a,b],
\end{aligned}$$

and

$$\begin{aligned}
f^{\Delta^2}(t) - \left(f^{\Delta^2}(t) \cdot f^\Delta(t)\right) f^\Delta(t) &= c_{11}(t)e_1(t) + c_{12}(t)e_2(t) - c_{11}(t)e_1(t) \\
&= c_{12}(t)e_2(t),
\end{aligned}$$

$$\frac{f^{\Delta^2}(t)-\left(f^{\Delta^2}(t)\cdot f^{\Delta}(t)\right)f^{\Delta}(t)}{\|f^{\Delta^2}(t)-\left(f^{\Delta^2}(t)\cdot f^{\Delta}(t)\right)f^{\Delta}(t)\|}=\frac{c_{12}(t)e_2(t)}{\|c_{12}(t)e_2(t)\|}$$
$$=e_2(t),\quad t\in[a,b],$$

and hence,

$$e_3(t)=e_1(t)\times e_2(t),\quad t\in[a,b].$$

The uniqueness of f, up to a rigid motion, follows from Theorem 1.15. This completes the proof.

1.15 Advanced Practical Problems

Problem 1.1 Let $\mathbb{T}=2^{\mathbb{N}_0}$ and $f:[1,64]\to\mathbb{R}^4$ be defined by

$$f(t)=\left(1+t,\frac{1+t}{1+t^2},\frac{1-t}{1+t+t^3},t^4\right),\quad t\in[1,64].$$

Check if $f:[1,64]\to\mathbb{R}^4$ is a regular parameterized curve.

Problem 1.2 Let $\tilde{\mathbb{T}}=\mathbb{T}=\mathbb{N}_0$, $\alpha=0$, $\beta=9$, $a=7$, $b=16$, $\phi:[0,9]\to[7,16]$ is defined by $\phi(t)=t+7$, $t\in[0,9]$, $f:[7,16]\to\mathbb{R}^3$ is defined by

$$f(t)=\left(\frac{1+t}{1+4t},1+t+t^2,e_3(t,1)\right),\quad t\in[7,16].$$

Prove that

1. $f:[7,16]\to\mathbb{R}^3$ is a regular parameterized curve.
2. $f\circ\phi$ and f are equivalent.

Problem 1.3 Let $\tilde{\mathbb{T}}=\mathbb{N}_0$, $\mathbb{T}=\mathbb{N}_0^4$, $\phi:\tilde{\mathbb{T}}\to\mathbb{T}$ and $f:\mathbb{T}\to\mathbb{R}^3$ be defined by

$$\phi(t)=t^4,\quad t\in\tilde{\mathbb{T}},$$
$$f(t)=(1+t+t^2,t^3,t^4),\quad t\in\mathbb{T},$$

respectively. Prove that

1. f is a regular parameterized curve.
2. f and $f\circ\phi$ are equivalent.

Problem 1.4 Let

$$\mathbb{T}=\left\{-3,-\frac{8}{3},-\frac{5}{2},-2,-\frac{7}{4},-\frac{5}{3},-1,-\frac{5}{6},-\frac{2}{3},-\frac{1}{2},0,1\right\},$$

$f:\left[-3,-\frac{1}{2}\right]\to\mathbb{R}^3$ be defined by

$$f(t)=\left(1+2t+e_1(t,1),t^3,e_2(t,0)\right),\quad t\in\left[-3,-\frac{1}{2}\right].$$

Find $L_f\left(-\frac{1}{2},-3\right)$.

Problem 1.5 Let $\mathbb{T}=3^{\mathbb{N}_0}, f:[1,81]\to\mathbb{R}^3$ be defined by

$$f(t)=\left(t^2,1+t,t^4\right),\quad t\in[1,81].$$

1. Find $f\circ L_f^{-1}$.
2. Prove that f and $f\circ L_f^{-1}$ are equivalent.
3. Prove that f is a Frenet curve.

Problem 1.6 Let $\mathbb{T}_1=\mathbb{Z}$, $\mathbb{T}_2=2\mathbb{Z}$, $\mathbb{T}_3=3^{\mathbb{N}_0}$,

$$F(x,y,z)=x-y,$$
$$G(x,y,z)=x^2-y^2-z^2,\quad (x,y,z)\in\mathbb{T}_1\times\mathbb{T}-2\times\mathbb{T}_3.$$

Check if

$$C=\{(x,y,z)\in\mathbb{T}_1\times\mathbb{T}_2\times\mathbb{T}_3 : F(x,y,z)=0,\quad G(x,y,z)=0\}$$

is a curve.

Problem 1.7 Let $\mathbb{T}=\mathbb{Z}$ and $I=[0,50]$. Find the canonical equations of the tangent line at any point on the parameterized curve

$$x=\frac{1}{1+t+t^2}+t^3$$
$$y=1+t-t^4,\quad t\in I.$$

Problem 1.8 Let $\mathbb{T}=2\mathbb{Z}$ and $I=[0,100]$. Find the canonical equations of the tangent line at any point on the parameterized curve

$$x=\frac{1+t}{1+t^2}$$
$$y=t^4$$
$$z=t^3,\quad t\in I.$$

Problem 1.9 Let $\mathbb{T}=\mathbb{Z}$. Find the equation of the normal line at any point on the following plane curve

$$x=\frac{1+t}{1+t^2}+e_2(t,3)$$
$$y=\frac{1}{1+t^2}+\sin_2(t,1),\quad t\in\mathbb{T}.$$

Problem 1.10 Let $\mathbb{T} = 3^{\mathbb{N}_0}$. Find the equation of the normal plane at any point of the space curve

$$x = 1 + t + t^2$$
$$y = 1 + t^4 + e_1(t,1)$$
$$z = 1 + t^2 + e_2(t,1), \quad t \in \mathbb{T}.$$

Problem 1.11 Let $\mathbb{T} = 3\mathbb{Z}$. Find the equations of the tangent line and the normal line of the following plane curve

$$y = 1 + x^3 + e_1(x,1) + \sqrt{1 + x^2}, \quad x \in \mathbb{T}.$$

Problem 1.12 Let $\mathbb{T} = 2^{\mathbb{N}_0}$, $\mathbb{T}_2 = 8^{\mathbb{N}_0}$, $a = 1$, $b = 4$, $c = 1$, $d = 256$. Find the envelope of the family

$$f(t,\lambda) = \left(1 + \lambda^2 - t^2, 1 + \lambda t + \lambda \cos_1(t,1)\right), \quad t \in [a,b], \quad \lambda \in [c,d].$$

Problem 1.13 Let $\mathbb{T} = 3^{\mathbb{N}_0}$. Find the equations of the tangent line and the equation of the normal plane at any point of the following space curve

$$y = \frac{1 + x^2}{1 + e_1(x,1) + x^4}$$
$$z = \frac{1 + x^2}{1 + x} + x^3 + 3x + 1, \quad x \in \mathbb{T}.$$

Problem 1.14 Let $\mathbb{T} = h\mathbb{Z}$, $h > 0$. Check if the curve $f : \mathbb{T} \to \mathbb{R}^3$ given by

$$f(t) = \left(1 + t^2, 1 - t^2, 1 + t^4\right), \quad t \in \mathbb{T},$$

is biregular.

Problem 1.15 Let $\mathbb{T} = 2^{\mathbb{N}_0}$ and $f : \mathbb{T} \to \mathbb{R}^3$ be given by

$$f(t) = \left(e_1(t,1), 1 + t^3, \sin_1(t,1)\right), \quad t \in \mathbb{T}.$$

Find the osculating plane.

Problem 1.16 Let $\mathbb{T} = 3^{\mathbb{N}_0}$. Find the Frenet frame for the curve

$$f(t) = \left(1 + t + t^2, \quad \frac{1}{1+t}, \quad 1 + t^2\right), \quad t \in \mathbb{T}.$$

1.16 Notes and References

In this chapter we introduce the concept of Frenet curves in $\mathbb{R}^n$. They are given different representations for plane and space curves. The equations for tangent lines and normal planes for space curves are deduced as well as the equations of the normal line to a plane curve. The osculating plane and the curvature of a space line are investigated. Conditions are given when a plane curve is envelope for a family of plane curves. The Frenet frame is defined and the main Frenet equations are deduced. As applications, general helices and Bertrand curves are defined and some of their properties are deduced. The behaviour of the Frenet frame under a rigid motion is investigated and the existence and uniqueness theorem is proved. Some of the results can be found in [3] and [4].

2

General Theory of Surfaces

Suppose that $\mathbb{T}$, $\mathbb{T}_1$, $\mathbb{T}_2$, $\mathbb{T}_3$, $\mathbb{T}_{(1)}$, $\mathbb{T}_{(2)}$, $\mathbb{T}_{(3)}$ are time scales with forward jump operators and delta differentiation operators σ, σ_1, σ_2, σ_3, $\sigma_{(1)}$, $\sigma_{(2)}$, $\sigma_{(3)}$ and Δ, Δ_1, Δ_2, Δ_3, $\Delta_{(1)}$, $\Delta_{(2)}$ and $\Delta_{(3)}$, respectively. Let $I \subseteq \mathbb{T}$, $U, U_1, W_1 \subseteq \mathbb{T}_1 \times \mathbb{T}_2$, $W_2 \subseteq \mathbb{T}_{(1)} \times \mathbb{T}_{(2)} \times \mathbb{T}_{(3)}$ and $V \subseteq \mathbb{T}_1 \times \mathbb{T}_2 \times \mathbb{T}_3$.

2.1 Preliminaries

In this section we will give some basic definitions needed for the general theory of surfaces.

Definition 2.1 We say that a map $g : W_1 \to W_2$ is a smooth map if it is continuous and its Δ_1 and Δ_2 partial derivatives exist and are continuous on W_1.

Definition 2.2 We say that a map $g : W_2 \to W_1$ is a smooth map if it is continuous and its $\Delta_{(1)}$, $\Delta_{(2)}$ and $\Delta_{(3)}$ partial derivatives exist and are continuous on W_2.

Definition 2.3 We say that a map $g : W_1 \to W_2$ is a time scale homeomorphism (shortly homeomorphism) if it has the following properties.

1. g is a bijection, that is, one-to-one and onto.
2. g is continuous.
3. the inverse map g^{-1} exists and it is continuous.
4. If

$$g = (g_1, g_2, g_3), \quad g^{-1} = (G_1, G_2),$$

then

$$g_j(\sigma_1(t_1), t_2) = g_j(t_1, \sigma_2(t_2)) = \sigma_{(j)}(g_j(t_1, t_2)), \quad j \in \{1,2,3\}, \tag{2.1}$$

$(t_1, t_2) \in W_1$, and

$$\begin{aligned} G_j(\sigma_{(1)}(t_{(1)}), t_{(2)}, t_{(3)}) &= G_j(t_{(1)}, \sigma_{(2)}(t_{(2)}), t_{(3)}) \\ &= G_j(t_{(1)}, t_{(2)}, \sigma_{(3)}(t_{(3)})) \\ &= \sigma_j(G_j(t_{(1)}, t_{(2)}, t_{(3)})), \quad j \in \{1,2\}, \end{aligned} \tag{2.2}$$

$(t_{(1)}, t_{(2)}, t_{(3)}) \in W_2$.

DOI: 10.1201/9781003205265-2

Definition 2.4 A map $g : W_1 \to W_2$ that satisfies (2.1) will be called a $\sigma_1\sigma_2\sigma_{(1)}\sigma_{(2)}\sigma_{(3)}$-map.

Definition 2.5 A map $g : W_2 \to W_1$ that satisfies (2.2) will be called a $\sigma_{(1)}\sigma_{(2)}\sigma_{(3)}\sigma_1\sigma_2$-map.

Definition 2.6 A map $g : W_1 \to W_2$ is called a time scale diffeomorphism (shortly diffeomorphism) if it has the following properties.

1. g is a homemorphism.
2. g and g^{-1} are smooth maps.

Lemma 2.1 *Let $f : W_1 \to W_2$ be a $\sigma_1\sigma_2\sigma_{(1)}\sigma_{(2)}\sigma_{(3)}$-map that is smooth and $g : W_2 \to W_1$ be a $\sigma_{(1)}\sigma_{(2)}\sigma_{(3)}\sigma_1\sigma_2$-map that is smooth. If f is σ_1- or σ_2-completely delta differentiable, then the map*

$$f \circ g : W_2 \to W_2$$

is a smooth map. If g is $\sigma_{(1)}$- or $\sigma_{(2)}$- or $\sigma_{(3)}$-completely delta differentiable, then the map

$$g \circ f : W_1 \to W_1$$

is a smooth map.

Proof. Let

$$f = (f_1, f_2, f_3) \quad \text{and} \quad g = (g_1, g_2).$$

Without loss of generality, suppose that f is σ_1-completely delta differentiable and g is $\sigma_{(1)}$-completely delta differentiable. Take $(t_{(1)}, t_{(2)}, t_{(3)}) \in W_2$ arbitrarily. Then

$$\begin{aligned}
&\lim_{s_{(1)} \to t_{(1)}} \frac{1}{\sigma_{(1)}(s_1) - t_{(1)}} \Big(f_k(g_1(\sigma_{(1)}(s_{(1)}), t_{(2)}, t_{(3)}), g_2(\sigma_{(1)}(s_{(1)}), t_{(2)}, t_{(3)})) \\
&\quad - f_k(g_1(s_{(1)}, t_{(2)}, t_{(3)}), g_2(s_{(1)}, t_{(2)}, t_{(3)})) \Big) \\
&= \lim_{s_{(1)} \to t_{(1)}} \frac{1}{\sigma_{(1)}(s_{(1)}) - t_{(1)}} \Big(f_k(g_1(\sigma_{(1)}(s_{(1)}), t_{(2)}, t_{(3)}), g_2(t_{(1)}, t_{(2)}, t_{(3)})) \\
&\quad - f_k(g_1(t_{(1)}, t_{(2)}, t_{(3)}), g_2(t_{(1)}, t_{(2)}, t_{(3)})) \\
&\quad + f_k(g_1(\sigma_{(1)}(s_{(1)}), t_{(2)}, t_{(3)}), g_2(\sigma_{(1)}(s_{(1)}), t_{(2)}, t_{(3)})) \\
&\quad - f_k(g_1(\sigma_{(1)}(s_{(1)}), t_{(2)}, t_{(3)}), g_2(t_{(1)}, t_{(2)}, t_{(3)})) \Big) \\
&= \lim_{s_{(1)} \to t_{(1)}} \frac{1}{\sigma_{(1)}(s_1) - t_{(1)}} \Bigg(\frac{1}{\sigma_1(g_1(s_1, t_{(2)}, t_{(3)})) - g_1(t_{(1)}, t_{(2)}, t_{(3)})} \\
&\quad \times \Big(f_k(\sigma_1(g_1(s_{(1)}, t_{(2)}, t_{(3)})), g_2(t_{(1)}, t_{(2)}, t_{(3)}))
\end{aligned}$$

$$
\begin{aligned}
&-f_k(g_1(t_{(1)},t_{(2)},t_{(3)}),g_2(t_{(1)},t_{(2)},t_{(3)}))\Big)\\
&\times\Big(g_1(\sigma_{(1)}(s_{(1)}),t_{(2)},t_{(3)})-g_1(t_{(1)},t_{(2)},t_{(3)})\Big)\\
&+\frac{1}{\sigma_2(g_2(s_{(1)},t_{(2)},t_{(3)}))-g_2(t_{(1)},t_{(2)},t_{(3)})}\\
&\times\Big(f_k(\sigma_1(g_1(s_{(1)},t_{(2)},t_{(3)})),\sigma_2(g_2(t_{(1)},t_{(2)},t_{(3)}))\\
&-f_k(\sigma_1(g_1(s_{(1)},t_{(2)},t_{(3)})),g_2(s_{(1)},t_{(2)},t_{(3)}))\Big)\\
&\times\Big(g_2(\sigma_{(1)}(s_{(1)}),t_{(2)},t_{(3)})-g_2(t_{(1)},t_{(2)},t_{(3)})\Big)\Big)\\
=&f_{kt_1}^{\Delta_1}(g_1(t_{(1)},t_{(2)},t_{(3)}),g_2(t_{(1)},t_{(2)},t_{(3)}))\\
&\times g_{1t_{(1)}}^{\Delta_{(1)}}(t_{(1)},t_{(2)},t_{(3)})\\
&+f_{kt_2}^{\Delta_2\sigma_1}(g_1(t_{(1)},t_{(2)},t_{(3)}),g_2(t_{(1)},t_{(2)},t_{(3)}))\\
&\times g_{2t_{(1)}}^{\Delta_{(1)}}(t_{(1)},t_{(2)},t_{(3)}),
\end{aligned}
$$

$k\in\{1,2,3\}$. Therefore there exists

$$(f_k(g_1,g_2))_{t_{(1)}}^{\Delta_{(1)}}(t_{(1)},t_{(2)},t_{(3)}),\quad k\in\{1,2,3\}.$$

As above, one can prove that there exist

$$(f_k(g_1,g_2))_{t_{(2)}}^{\Delta_{(2)}}(t_{(1)},t_{(2)},t_{(3)}),\quad k\in\{1,2,3\},$$

and

$$(f_k(g_1,g_2))_{t_{(3)}}^{\Delta_{(3)}}(t_{(1)},t_{(2)},t_{(3)}),\quad k\in\{1,2,3\}.$$

Next,

$$
\begin{aligned}
&\lim_{s_1\to t_1}\frac{1}{\sigma_1(s_1)-t_1}\Big(g_j(f_1(\sigma_1(s_1),t_2),f_2(\sigma_1(s_1),t_2),f_3(\sigma_1(s_1),t_2))\\
&-g_j(f_1(t_1,t_2),f_2(t_1,t_2),f_3(t_1,t_2))\Big)\\
&=\lim_{s_1\to t_1}\frac{1}{\sigma_1(s_1)-t_1}\Big(g_j(f_1(\sigma_1(s_1),t_2),f_2(t_1,t_2),f_3(t_1,t_2))\\
&-g_j(f_1(t_1,t_2),f_2(t_1,t_2),f_3(t_1,t_2))\\
&+g_j(f_1(\sigma_1(s_1),t_2),f_2(\sigma_1(s_1),t_2),f_3(t_1,t_2))\\
&-g_j(f_1(\sigma_1(s_1),t_2),f_2(t_1,t_2),f_3(t_1,t_2))\\
&+g_j(f_1(\sigma_1(s_1),t_2),f_2(\sigma_1(s_1),t_2),f_3(\sigma_1(s_1),t_2))
\end{aligned}
$$

$$
\begin{aligned}
&-g_j(f_1(\sigma_1(s_1),t_2),f_2(\sigma_1(s_1),t_2),f_3(t_1,t_2))\Big)\\
&=\lim_{s_1\to t_1}\frac{1}{\sigma_1(s_1)-t_1}\left(\frac{1}{\sigma_{(1)}(f_1(s_1,t_2))-f_1(t_1,t_2)}\right.\\
&\times\Big(g_j(\sigma_{(1)}(f_1(s_1,t_2)),f_2(t_1,t_2),f_3(t_1,t_2))\\
&-g_j(f_1(t_1,t_2),f_2(t_1,t_2),f_3(t_1,t_2))\Big)\\
&\times(f_1(\sigma_1(s_1),t_2)-f_1(t_1,t_2))\\
&+\frac{1}{\sigma_{(2)}(f_2(s_1,t_2))-f_2(t_1,t_2)}\\
&\times\Big(g_j(\sigma_{(1)}(f_1(s_1,t_2)),\sigma_{(2)}(f_2(s_1,t_2)),f_3(t_1,t_2))\\
&-g_j(\sigma_{(1)}(f_1(s_1,t_2)),f_2(t_1,t_2),f_3(t_1,t_2))\Big)\\
&\times(f_2(\sigma_1(s_1),t_2)-f_2(t_1,t_2))\\
&+\frac{1}{\sigma_{(3)}(f_3(s_1,t_2))-f_3(t_1,t_2)}\\
&\times\Big(g_j(\sigma_{(1)}(f_1(s_1,t_2)),\sigma_{(2)}(f_2(s_1,t_2)),\sigma_{(3)}(f_3(s_1,t_2)))\\
&-g_j(\sigma_{(1)}(f_1(s_1,t_2)),\sigma_{(2)}(f_2(s_1,t_2)),f_3(t_1,t_2))\Big)\\
&\left.\times(f_3(\sigma_1(s_1),t_2)-f_3(t_1,t_2))\right)\\
&=g_{jt_{(1)}}^{\Delta(1)}(f_1(t_1,t_2),f_2(t_1,t_2),f_3(t_1,t_2))f_{1t_1}^{\Delta_1}(t_1,t_2)\\
&+g_{jt_{(2)}}^{\Delta(2)}(\sigma_{(1)}(f_1(t_1,t_2)),f_2(t_1,t_2),f_3(t_1,t_2))f_{2t_1}^{\Delta_1}(t_1,t_2)\\
&+g_{jt_{(3)}}^{\Delta(3)}(\sigma_{(1)}(f_1(t_1,t_2)),\sigma_{(2)}(f_2(t_1,t_2)),f_3(t_1,t_2))f_{3t_1}^{\Delta_1}(t_1,t_2),
\end{aligned}
$$

$j\in\{1,2\}$. Therefore there exists

$$
(g_j(f_1,f_2,f_3))_{t_1}^{\Delta_1}(t_1,t_2),\quad j\in\{1,2\}.
$$

As above, one can prove that there exists

$$
(g_j(f_1,f_2,f_3))_{t_2}^{\Delta_2}(t_1,t_2),\quad j\in\{1,2\}.
$$

This completes the proof.

2.2 Parameterized Surfaces

Definition 2.7 A smooth map $f : U \to \mathbb{T}_{(1)} \times \mathbb{T}_{(2)} \times \mathbb{T}_{(3)}$,

$$f(t_1,t_2) = (f_1(t_1,t_2), f_2(t_1,t_2), f_3(t_1,t_2)), \quad (t_1,t_2) \in U,$$

is called a σ_1-regular parameterized surface (σ_1-patch) (shortly σ_1-parameterized surface) if f is σ_1-completely delta differentiable and

$$f_{t_1}^{\Delta_1} \times f_{t_2}^{\Delta_2 \sigma_1} \neq 0 \quad \text{on} \quad U. \tag{2.3}$$

The condition (2.3) is called the σ_1-regularity condition. A σ_1-regular parameterized surface is denoted by (U,f) or $(U, f = f(t_1,t_2))$ or $f = f(t_1,t_2)$.

Example 2.1 Let $\mathbb{T}_1 = \mathbb{N}_0$, $\mathbb{T}_2 = 2^{\mathbb{N}_0}$, $\mathbb{T}_{(1)} = \mathbb{T}_{(2)} = \mathbb{T}_{(3)} = \mathbb{R}$ and $f : \mathbb{T}_1 \times \mathbb{T}_2 \to \mathbb{R}^3$ be given by

$$f(t_1,t_2) = (t_1 + t_2^2, -t_1^2 + t_2, t_1^2 + t_2^2), \quad (t_1,t_2) \in \mathbb{T}_1 \times \mathbb{T}_2.$$

Here

$$\begin{aligned}
f_1(t_1,t_2) &= t_1 + t_2^2,\\
f_2(t_1,t_2) &= -t_1^2 + t_2,\\
f_3(t_1,t_2) &= t_1^2 + t_2^2, \quad (t_1,t_2) \in \mathbb{T}_1 \times \mathbb{T}_2,\\
\sigma_1(t_1) &= t_1 + 1, \quad t_1 \in \mathbb{T}_1,\\
\sigma_2(t_2) &= 2t_2, \quad t_2 \in \mathbb{T}_2.
\end{aligned}$$

Then

$$\begin{aligned}
f_{1t_1}^{\Delta_1}(t_1,t_2) &= 1,\\
f_{1t_2}^{\Delta_2}(t_1,t_2) &= \sigma_2(t_2) + t_2\\
&= 2t_2 + t_2\\
&= 3t_2,\\
f_{2t_1}^{\Delta_1}(t_1,t_2) &= -\sigma_1(t_1) - t_1\\
&= -(t_1 + 1) - t_1\\
&= -2t_1 - 1,\\
f_{2t_2}^{\Delta_2}(t_1,t_2) &= 1,\\
f_{3t_1}^{\Delta_1}(t_1,t_2) &= \sigma_1(t_1) + t_1\\
&= t_1 + 1 + t_1\\
&= 2t_1 + 1,\\
f_{3t_2}^{\Delta_2}(t_1,t_2) &= \sigma_2(t_2) + t_2\\
&= 2t_2 + t_2\\
&= 3t_2, \quad (t_1,t_2) \in \mathbb{T}_1 \times \mathbb{T}_2.
\end{aligned}$$

Therefore

$$\begin{aligned} f_{t_1}^{\Delta_1}(t_1,t_2) &= \left(f_{1t_1}^{\Delta_1}(t_1,t_2), f_{2t_1}^{\Delta_1}(t_1,t_2), f_{3t_1}^{\Delta_1}(t_1,t_2)\right) \\ &= (1,-2t_1-1,2t_1+1), \\ f_{t_2}^{\Delta_2}(t_1,t_2) &= f_{t_2}^{\Delta_2\sigma_1}(t_1,t_2) \\ &= \left(f_{1t_2}^{\Delta_2}(t_1,t_2), f_{2t_2}^{\Delta_2}(t_1,t_2), f_{3t_2}^{\Delta_2}(t_1,t_2)\right) \\ &= (3t_2,1,3t_2), \quad (t_1,t_2) \in \mathbb{T}_1 \times \mathbb{T}_2, \end{aligned}$$

and

$$\begin{aligned} f_{t_1}^{\Delta_1}(t_1,t_2) \times f_{t_2}^{\Delta_2\sigma_1}(t_1,t_2) &= (1,-2t_1-1,2t_1+1) \times (3t_2,1,3t_2) \\ &= \Big(-3t_2(2t_1+1)-(2t_1+1), 3t_2(2t_1+1)-3t_2, \\ &\quad 1+3t_2(2t_1+1)\Big) \\ &= \Big(-6t_1t_2-3t_2-2t_1-1, 6t_1t_2+3t_2-3t_2, 1+3t_2+6t_1t_2\Big) \\ &= (-6t_1t_2-2t_1-3t_2-1, 6t_1t_2, 1+3t_2+6t_1t_2), (t_1,t_2) \in \mathbb{T}_1 \times \mathbb{T}_2. \end{aligned}$$

Note that

$$(-6t_1t_2-2t_1-3t_2-1, 6t_1t_2, 1+3t_2+6t_1t_2) \neq (0,0,0), \quad (t_1,t_2) \in \mathbb{T}_1 \times \mathbb{T}_2.$$

Therefore $f : \mathbb{T}_1 \times \mathbb{T}_2 \to \mathbb{R}^3$ is a σ_1-regular parameterized surface.

Definition 2.8 A smooth map $f : U \to \mathbb{T}_{(1)} \times \mathbb{T}_{(2)} \times \mathbb{T}_{(3)}$,

$$f(t_1,t_2) = (f_1(t_1,t_2), f_2(t_1,t_2), f_3(t_1,t_2)), \quad (t_1,t_2) \in U,$$

is called a σ_2-regular parameterized surface (σ_2-patch) (shortly σ_2-parameterized surface) if f is σ_2-completely delta differentiable and

$$f_{t_1}^{\Delta_1\sigma_2} \times f_{t_2}^{\Delta_2} \neq 0 \quad \text{on} \quad U. \tag{2.4}$$

The condition (2.4) is called the σ_2-regularity condition. A σ_2-regular parameterized surface is denoted by (U,f), $(U, f = f(t_1,t_2))$ or $f = f(t_1,t_2)$.

Example 2.2 Let $\mathbb{T}_1 = \mathbb{N}_0^2$, $\mathbb{T}_2 = \mathbb{N}_0$, $\mathbb{T}_{(1)} = \mathbb{T}_{(2)} = \mathbb{T}_{(3)} = \mathbb{R}$ and $f : \mathbb{T}_1 \times \mathbb{T}_2 \to \mathbb{R}^3$ be given by

$$f(t_1,t_2) = \left(t_1^2 + \frac{1}{1+t_2}, t_1, t_2\right), \quad (t_1,t_2) \in \mathbb{T}_1 \times \mathbb{T}_2.$$

Here

$$\begin{aligned} \sigma_1(t_1) &= (\sqrt{t_1}+1)^2, \quad t_1 \in \mathbb{T}_1, \\ \sigma_2(t_2) &= t_2+1, \quad t_2 \in \mathbb{T}_2, \\ f_1(t_1,t_2) &= t_1^2 + \frac{1}{1+t_2}, \end{aligned}$$

$$\begin{aligned} f_2(t_1,t_2) &= t_1, \\ f_3(t_1,t_2) &= t_2, \quad (t_1,t_2) \in \mathbb{T}_1 \times \mathbb{T}_2. \end{aligned}$$

Then

$$\begin{aligned} f_{1t_1}^{\Delta_1}(t_1,t_2) &= \sigma_1(t_1) + t_1 \\ &= (\sqrt{t_1}+1)^2 + t_1 \\ &= t_1 + 2\sqrt{t_1} + 1 + t_1 \\ &= 2t_1 + 2\sqrt{t_1} + 1, \\ f_{1t_2}^{\Delta_2}(t_1,t_2) &= -\frac{1}{(1+t_2)(1+\sigma_2(t_2))} \\ &= -\frac{1}{(1+t_2)(1+t_2+1)} \\ &= -\frac{1}{(1+t_2)(2+t_2)}, \\ f_{2t_1}^{\Delta_1}(t_1,t_2) &= 1, \\ f_{2t_2}^{\Delta_2}(t_1,t_2) &= 0, \\ f_{3t_1}^{\Delta_1}(t_1,t_2) &= 0, \\ f_{3t_2}^{\Delta_2}(t_1,t_2) &= 1, \quad (t_1,t_2) \in \mathbb{T}_1 \times \mathbb{T}_2. \end{aligned}$$

Therefore

$$\begin{aligned} f_{t_1}^{\Delta_1}(t_1,t_2) &= f_{t_1}^{\Delta_1\sigma_2}(t_1,t_2) \\ &= \left(f_{1t_1}^{\Delta_1}(t_1,t_2), f_{2t_1}^{\Delta_1}(t_1,t_2), f_{3t_1}^{\Delta_1}(t_1,t_2)\right) \\ &= (2t_1 + 2\sqrt{t_1} + 1, 1, 0), \\ f_{t_2}^{\Delta_2}(t_1,t_2) &= \left(f_{1t_2}^{\Delta_2}(t_1,t_2), f_{2t_2}^{\Delta_2}(t_1,t_2), f_{3t_2}^{\Delta_2}(t_1,t_2)\right) \\ &= \left(-\frac{1}{(1+t_2)(2+t_2)}, 0, 1\right), \quad (t_1,t_2) \in \mathbb{T}_1 \times \mathbb{T}_2, \end{aligned}$$

and

$$\begin{aligned} f_{t_1}^{\Delta_1}(t_1,t_2) \times f_{t_2}^{\Delta_2}(t_1,t_2) &= (2t_1 + 2\sqrt{t_1} + 1, 1, 0) \\ &\quad \times \left(-\frac{1}{(1+t_2)(2+t_2)}, 0, 1\right) \\ &= \left(1, -2t_1 - 2\sqrt{t_1} - 1, \frac{1}{(1+t_2)(2+t_2)}\right) \\ &\neq 0, \quad (t_1,t_2) \in \mathbb{T}_1 \times \mathbb{T}_2. \end{aligned}$$

Consequently $f : \mathbb{T}_1 \times \mathbb{T}_2 \to \mathbb{R}^3$ is a σ_2-regular parameterized surface.

Definition 2.9 A smooth map $f : U \to \mathbb{T}_{(1)} \times \mathbb{T}_{(2)} \times \mathbb{T}_{(3)}$,

$$f(t_1, t_2) = (f_1(t_1, t_2), f_2(t_1, t_2), f_3(t_1, t_2)), \quad (t_1, t_2) \in U,$$

is called a $\sigma_1^2\sigma_2$-regular parameterized surface ($\sigma_1^2\sigma_2$-patch) (shortly $\sigma_1^2\sigma_2$-parameterized surface) if f is σ_1-completely delta differentiable and

$$f_{t_1}^{\Delta_1\sigma_1\sigma_2} \times f_{t_2}^{\Delta_2\sigma_1^2\sigma_2} \neq 0 \quad \text{on} \quad U. \tag{2.5}$$

The condition (2.5) is called the $\sigma_1^2\sigma_2$-regularity condition. A $\sigma_1^2\sigma_2$-regular parameterized surface is denoted by (U,f), $(U,f = f(t_1,t_2))$ or $f = f(t_1,t_2)$.

Definition 2.10 A smooth map $f : U \to \mathbb{T}_{(1)} \times \mathbb{T}_{(2)} \times \mathbb{T}_{(3)}$,

$$f(t_1, t_2) = (f_1(t_1, t_2), f_2(t_1, t_2), f_3(t_1, t_2)), \quad (t_1, t_2) \in U,$$

is called a $\sigma_1\sigma_2^2$-regular parameterized surface ($\sigma_1\sigma_2^2$-patch) (shortly $\sigma_1\sigma_2^2$-parameterized surface) if f is σ_2-completely delta differentiable and

$$f_{t_1}^{\Delta_1\sigma_1\sigma_2^2} \times f_{t_2}^{\Delta_2\sigma_1\sigma_2} \neq 0 \quad \text{on} \quad U. \tag{2.6}$$

The condition (2.6) is called the $\sigma_1\sigma_2^2$-regularity condition. A $\sigma_1\sigma_2^2$-regular parameterized surface is denoted by (U,f), $(U,f = f(t_1,t_2))$ or $f = f(t_1,t_2)$.

Exercise 2.1 Let $\mathbb{T}_1 = 2^{\mathbb{N}_0}$, $\mathbb{T}_2 = 3^{\mathbb{N}_0}$, $\mathbb{T}_{(1)} = \mathbb{T}_{(2)} = \mathbb{T}_{(3)} = \mathbb{R}$.
Prove that $f : \mathbb{T}_1 \times \mathbb{T}_2 \to \mathbb{R}^3$, given by

$$f(t_1, t_2) = \left(1 + \frac{1}{1+t_1}, \quad t_1^2 + t_2, \quad t_1 t_2 + t_1^2 + t_2^2 + t_1^3\right),$$

$(t_1, t_2) \in \mathbb{T}_1 \times \mathbb{T}_2$, is a $\sigma_1\sigma_2^2$-regular parameterized surface, σ_1-regular parameterized surface and σ_2-regular parameterized surface.

Definition 2.11 The set $f(U) \subset \mathbb{T}_{(1)} \times \mathbb{T}_{(2)} \times \mathbb{T}_{(3)}$ is called the support of the σ_1-regular or σ_2 regular or $\sigma_1^2\sigma_2$-regular or $\sigma_1\sigma_2^2$-regular parameterized surface (U,f).

Example 2.3 Let $\mathbb{T}_1 = \mathbb{T}_2 = \mathbb{T}_{(1)} = \mathbb{T}_{(2)} = \mathbb{T}_{(3)} = \mathbb{R}$ and

$$U = \left\{(t_1, t_2) \in \mathbb{R}^2 : -\frac{\pi}{2} < t_1 < \frac{\pi}{2}, \quad 0 < t_2 < 2\pi\right\}$$

and

$$f(t_1, t_2) = (R\cos t_1 \cos t_2, R\cos t_1 \sin t_2, R\sin t_1).$$

Then the support of the parameterized surface being considered is a sphere of radius R.

Exercise 2.2 Let $\mathbb{T}_1 = \mathbb{T}_2 = \mathbb{T}_{(1)} = \mathbb{T}_{(2)} = \mathbb{T}_{(3)} = \mathbb{R}$. Find the support of the parameterized surface $f : \mathbb{T}_1 \times \mathbb{T}_2 \to \mathbb{R}^3$ given by

$$f(t_1, t_2) = (t_1 + 1, \quad t_2 - t_2, \quad 1 + t_2), \quad (t_1, t_2) \in \mathbb{T}_1 \times \mathbb{T}_2.$$

Definition 2.12 A subset $S \subset \mathbb{T}_{(1)} \times \mathbb{T}_{(2)} \times \mathbb{T}_{(3)}$ is called a σ_1-regular or σ_2-regular or $\sigma_1^2\sigma_2$-regular or $\sigma_1\sigma_2^2$-regular surface if each point $a \in S$ has an open neighborhood W in S such that there is a σ_1-regular or σ_2-regular or $\sigma_1^2\sigma_2$-regular or $\sigma_1\sigma_2^2$-regular parameterized surface (U,f) with $f(U) = W$ and the map $f : U \to W$ is a diffeomorphism. The pair (U,f) is called a local σ_1- or σ_2- or $\sigma_1^2\sigma_2$- or $\sigma_1\sigma_2^2$-parameterization of the surface S around the point a and the support $f(U)$ is called the domain of the parameterization. A surface S that has a global parameterization, that is, a local parameterization (U,f) for which $f(U) = S$, is called a simple surface.

2.3 Representations of Surfaces

2.3.1 Parametric Representation

Definition 2.13 Let S be a surface and (U,f) be a local σ_1- or σ_2- or $\sigma_1^2\sigma_2$- or $\sigma_1\sigma_2^2$- parameterization of S. If

$$f(t_1,t_2) = (f_1(t_1,t_2), f_2(t_1,t_2), f_3(t_1,t_2)), \quad (t_1,t_2) \in U,$$

then the equations

$$\begin{aligned} x &= f_1(t_1,t_2) \\ y &= f_2(t_1,t_2) \\ z &= f_3(t_1,t_2), \quad (t_1,t_2) \in U, \end{aligned}$$

are called the parametric equations of the surface S.

Example 2.4 Let $\mathbb{T}_1 = \mathbb{T}_2 = \mathbb{T}_{(1)} = \mathbb{T}_{(2)} = \mathbb{T}_{(3)} = \mathbb{R}$. The equations

$$\begin{aligned} x &= R\cos t_1 \cos t_2 \\ y &= R\sin t_1 \cos t_2 \\ z &= R\sin t_2, \quad -\frac{\pi}{2} < t_1 < \frac{\pi}{2}, \quad 0 < t_2 < 2\pi, \quad R > 0, \end{aligned}$$

are the parametric equations of a sphere of radius R.

2.3.2 Implicit Representation

Let $F : V \to \mathbb{R}$ be a smooth function. Denote by

$$S = \{(t_1,t_2,t_3) \in V : F(t_1,t_2,t_3) = 0\}$$

the 0-level set of F. Suppose that

$$(F_{t_1}^{\Delta_1}(t_1,t_2,t_3), F_{t_2}^{\Delta_2}(t_1,t_2,t_3), F_{t_3}^{\Delta_3}(t_1,t_2,t_3)) \neq (0,0,0), \quad (t_1,t_2,t_3) \in S.$$

Take $(t_1^0, t_2^0, t_3^0) \in S$ arbitrarily. Without loss of generality, suppose that $F_{t_3}^{\Delta_3}(t_1^0, t_2^0, t_3^0) \neq 0$. Then, by the implicit function theorem (see the of this book), it follows that there exists a neighborhood M of (t_1^0, t_2^0, t_3^0) so that $M \cap S$ is the graph of a continuous function

$$t_3 = f(t_1, t_2).$$

If in addition,

$$\mathbb{T}_3 = f(\mathbb{T}_1, \mathbb{T}_2)$$

and

$$\begin{aligned} f(\sigma_1(t_1), t_2) &= f(t_1, \sigma_2(t_2)) \\ &= \sigma_3(f(t_1, t_2)), \quad (t_1, t_2) \in \mathbb{T}_1 \times \mathbb{T}_2, \end{aligned}$$

and f has continuous Δ_1 and Δ_2 partial derivatives, then there is a local σ_1- or σ_2- or $\sigma_1^2\sigma_2$- or $\sigma_1\sigma_2^2$-parameterization of S around the point (t_1^0, t_2^0, t_3^0). Consequently, S is a surface.

Example 2.5 Let $\mathbb{T}_1 = \mathbb{T}_2 = \mathbb{T}_3 = \mathbb{N}_0$, $\mathbb{T}_{(1)} = \mathbb{T}_{(2)} = \mathbb{T}_{(3)} = \mathbb{R}$ and $F : \mathbb{T}_1 \times \mathbb{T}_2 \times \mathbb{T}_3 \to \mathbb{R}$ is given by

$$F(t_1, t_2, t_3) = t_1 + t_2 - t_3, \quad (t_1, t_2, t_3) \in \mathbb{T}_1 \times \mathbb{T}_2 \times \mathbb{T}_3.$$

We have that the 0-level set of F is as follows

$$0 = t_1 + t_2 - t_3, \quad (t_1, t_2, t_3) \in \mathbb{T}_1 \times \mathbb{T}_2 \times \mathbb{T}_3,$$

or

$$t_3 = t_1 + t_2, \quad (t_1, t_2, t_3) \in \mathbb{T}_1 \times \mathbb{T}_2 \times \mathbb{T}_3.$$

Note that

$$\mathbb{T}_3 = \mathbb{T}_1 + \mathbb{T}_2$$

and

$$\begin{aligned} \sigma_3(t_3) &= t_3 + 1 \\ &= t_1 + 1 + t_2 \\ &= \sigma_1(t_1) + t_2 \\ &= t_1 + \sigma_2(t_2), \quad (t_1, t_2, t_3) \in \mathbb{T}_1 \times \mathbb{T}_2 \times \mathbb{T}_3. \end{aligned}$$

Next,

$$\begin{aligned} F_{t_1}^{\Delta_1}(t_1, t_2, t_3) &= 1, \\ F_{t_2}^{\Delta_2}(t_1, t_2, t_3) &= 1, \\ F_{t_3}^{\Delta_3}(t_1, t_2, t_3) &= -1, \quad (t_1, t_2, t_3) \in \mathbb{T}_1 \times \mathbb{T}_2 \times \mathbb{T}_3, \end{aligned}$$

and

$$\left(F_{t_1}^{\Delta_1}(t_1,t_2,t_3), F_{t_2}^{\Delta_2}(t_1,t_2,t_3), F_{t_3}^{\Delta_3}(t_1,t_2,t_3)\right) \neq (0,0,0)$$

at any $(t_1,t_2,t_3) \in \mathbb{T}_1 \times \mathbb{T}_2 \times \mathbb{T}_3$.

Therefore

$$S = \{(t_1,t_2,t_3) \in \mathbb{T}_1 \times \mathbb{T}_2 \times \mathbb{T}_3 : t_3 = t_1 + t_2\}$$

is a surface. In Figure 2.1 is shown the surface for $t_1, t_2 \in \{1,2,\ldots,10\}$.

Example 2.6 Let $\mathbb{T}_1 = \mathbb{T}_2 = \mathbb{T}_3 = \mathbb{T}_{(1)} = \mathbb{T}_{(2)} = \mathbb{T}_{(3)} = \mathbb{R}$ and

$$F(t_1,t_2,t_3) = \left(\sqrt{t_1^2+t_2^2}-a\right)^2 + t_3^2 - b^2, \quad 0 < b < a.$$

Definition 2.14 The 0-level set

$$T^2 = \{(t_1,t_2,t_3) \in \mathbb{R}^3 : F(t_1,t_2,t_3) = 0\}$$

is called a 2-dimensional torus.

We have

$$F'_{t_1}(t_1,t_2,t_3) = \frac{2t_1}{\sqrt{t_1^2+t_2^2}}\left(\sqrt{t_1^2+t_2^2}-a\right),$$

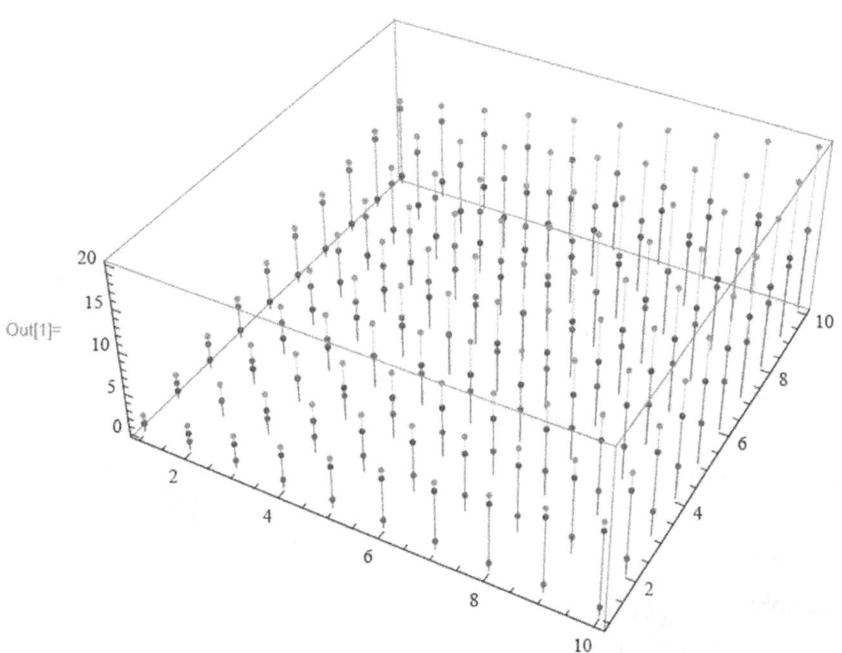

FIGURE 2.1
$\mathbb{T}_1 = \mathbb{T}_2 = \mathbb{N}_0$, $t_3 = t_1 + t_2$, $t_1, t_2 \in \{1,2,\ldots,10\}$.

$$F'_{t_2}(t_1,t_2,t_3) = \frac{2t_2}{\sqrt{t_1^2+t_2^2}}\left(\sqrt{t_1^2+t_2^2}-a\right),$$
$$F'_{t_3}(t_1,t_2,t_3) = 2t_3, \quad (t_1,t_2,t_3) \in \mathbb{R}^3.$$

Hence,

$$\begin{aligned} F'_{t_1}(t_1,t_2,t_3) &= 0 \\ F'_{t_2}(t_1,t_2,t_3) &= 0 \\ F'_{t_3}(t_1,t_2,t_3) &= 0, \quad (t_1,t_2,t_3) \in \mathbb{R}^3, \end{aligned}$$

if and only if

$$\begin{aligned} 0 &= \frac{2t_1}{\sqrt{t_1^2+t_2^2}}\left(\sqrt{t_1^2+t_2^2}-a\right) \\ 0 &= \frac{2t_2}{\sqrt{t_1^2+t_2^2}}\left(\sqrt{t_1^2+t_2^2}-a\right) \\ 0 &= 2t_3, \quad (t_1,t_2,t_3) \in \mathbb{R}^3, \end{aligned}$$

if and only if

$$\begin{array}{rcl} t_1 &=& 0 \\ t_2 &=& 0 \\ t_3 &=& 0, \end{array} \quad \text{or} \quad \begin{array}{rcl} a &=& \sqrt{t_1^2+t_2^2} \\ 0 &=& t_3, \quad (t_1,t_2,t_3) \in \mathbb{R}^3, \end{array} \quad \text{or}$$

$$\begin{array}{rcl} t_1 &=& 0 \\ t_2 &=& \pm a \\ t_3 &=& 0, \end{array} \quad \text{or} \quad \begin{array}{rcl} t_1 &=& \pm a \\ t_2 &=& 0 \\ t_3 &=& 0. \end{array}$$

Note that

$$\begin{aligned} F(0,0,0) &= a^2 - b^2 \\ &\neq 0, \\ F(t_1,t_2,t_3) &= -b^2 \\ &\neq 0 \quad \text{for} \quad \sqrt{t_1^2+t_2^2} = a, \quad t_3 = 0, \\ F(0,\pm a,0) &= -b^2 \\ &\neq 0, \\ F(\pm a,0,0) &= -b^2 \\ &\neq 0. \end{aligned}$$

Therefore

$$\left(F'_{t_1}(t_1,t_2,t_3), \quad F'_{t_2}(t_1,t_2,t_3), \quad F'_{t_3}(t_1,t_2,t_3)\right) \neq (0,0,0),$$

$(t_1,t_2,t_3) \in T^2$. Thus, the 2-dimensional torus is a surface. The 2-dimensional torus for $a = 3$ and $b = 1$ is shown in Figure 2.2.

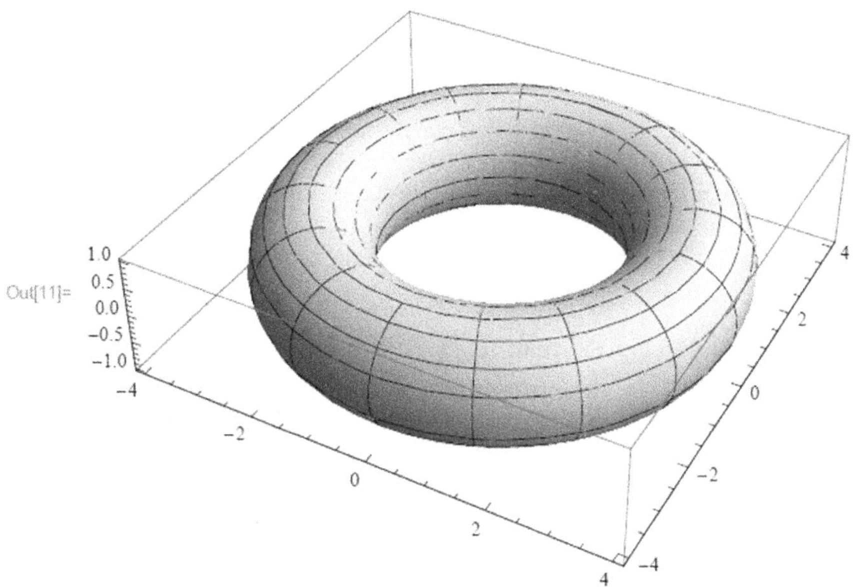

FIGURE 2.2
$\mathbb{T}_1 = \mathbb{T}_2 = \mathbb{T}_3 = \mathbb{R}$, 2-dimensional torus for $a = 3$, $b = 1$..

Definition 2.15 Let S be a surface and (U,f) be a local σ_1- or σ_2- or $\sigma_1^2\sigma_2$- or $\sigma_1\sigma_2^2$- parameterization. Then the map $f^{-1} : W \to U$ is called a σ_1- or σ_2- or $\sigma_1^2\sigma_2$- or $\sigma_1\sigma_2^2$-curvilinear coordinate system on S, respectively, or a σ_1- or σ_2- or $\sigma_1^2\sigma_2$- or $\sigma_1\sigma_2^2$-chart on S.

Definition 2.16 Let (U,f) be a local σ_1- or σ_2- or $\sigma_1^2\sigma_2$- or $\sigma_1\sigma_2^2$-parameterization of a surface S and (I,g) is a smooth parameterized curve such that $g(I) \subset \mathbb{T}_{(1)} \times \mathbb{T}_{(2)} \times \mathbb{T}_{(3)}$. We say that the curve (I,g) lies on S if the support $g(I)$ is included in S and there exist functions $g_1 : I \to \mathbb{T}_1$, $g_2 : I \to \mathbb{T}_2$ that are Δ-differentiable on $\mathbb{T}$ and

$$g(t) = f(g_1(t), g_2(t)), \quad t \in I,$$

and

$$g_1(\sigma(t)) = \sigma_1(g_1(t)), \quad g_2(\sigma(t)) = \sigma_2(g_2(t)), \quad t \in I.$$

2.4 Tangent Vector Spaces. Normals to a Surface

Definition 2.17 Suppose that (U,f) is a local σ_1-parameterization of a surface S,

$$f(t_1,t_2) = (f_1(/t_1,t_2), f_2(t_1,t_2), f_3(t_1,t_2)), \quad (t_1,t_2) \in U.$$

Assume that $(t_1^0,t_2^0)\in U$ and

$$a=\left(f_1(t_1^0,t_2^0),f_2(t_1^0,t_2^0),f_3(t_1^0,t_2^0)\right)\in S.$$

A vector $h\in\mathbb{T}_{(1)}\times\mathbb{T}_{(2)}\times\mathbb{T}_{(3)}$ is called a σ_1-tangent vector to the surface S at the point a if there is a smooth parameterized curve (I,g) on S and a $t_0\in I$ such that $g(t_0)=a$ and $g^\Delta(t_0)=h$. By $T_{a\sigma_1}S$ we will denote the space of all σ_1-tangent vectors to S at a. The vectors $\{f_{t_1}^{\Delta_1}(t_1^0,t_2^0),f_{t_2}^{\Delta_2\sigma_1}(t_1^0,t_2^0)\}$ make up a basis of $T_{a\sigma_1}S$ and it is called the natural basis or coordinate basis of the σ_1-tangent space $T_{a\sigma_1}S$.

Definition 2.18 Suppose that (U,f) is a local σ_2-parameterization of a surface S,

$$f(t_1,t_2)=(f_1(t_1,t_2),f_2(t_1,t_2),f_3(t_1,t_2)),\quad (t_1,t_2)\in U.$$

Assume that $(t_1^0,t_2^0)\in U$ and

$$a=\left(f_1(t_1^0,t_2^0),f_2(t_1^0,t_2^0),f_3(t_1^0,t_2^0)\right)\in S.$$

A vector $h\in\mathbb{T}_{(1)}\times\mathbb{T}_{(2)}\times\mathbb{T}_{(3)}$ is called a σ_2-tangent vector to the surface S at the point a if there is a smooth parameterized curve (I,g) on S and a $t_0\in I$ such that $g(t_0)=a$ and $g^\Delta(t_0)=h$. By $T_{a\sigma_2}S$ we will denote the space of all σ_2-tangent vectors to S at a. The vectors $\{f_{t_1}^{\Delta_1\sigma_2}(t_1^0,t_2^0),f_{t_2}^{\Delta_2}(t_1^0,t_2^0)\}$ make up a basis of $T_{a\sigma_2}S$ and it is called the natural basis or coordinate basis of the σ_2-tangent space $T_{a\sigma_2}S$.

Definition 2.19 Suppose that (U,f) is a local $\sigma_1^2\sigma_2$-parameterization of a surface S,

$$f(t_1,t_2)=(f_1(t_1,t_2),f_2(t_1,t_2),f_3(t_1,t_2)),\quad (t_1,t_2)\in U.$$

Assume that $(t_1^0,t_2^0)\in U$ and

$$a=\left(f_1(t_1^0,t_2^0),f_2(t_1^0,t_2^0),f_3(t_1^0,t_2^0)\right)\in S.$$

A vector $h\in\mathbb{T}_{(1)}\times\mathbb{T}_{(2)}\times\mathbb{T}_{(3)}$ is called a $\sigma_1^2\sigma_2$-tangent vector to the surface S at the point a if there is a smooth parameterized curve (I,g) on S and a $t_0\in I$ such that $g(t_0)=a$ and $g^\Delta(t_0)=h$. By $T_{a\sigma_1^2\sigma_2}S$ we will denote the space of all $\sigma_1^2\sigma_2$-tangent vectors to S at a. The vectors $\{f_{t_1}^{\Delta_1\sigma_1\sigma_2}(t_1^0,t_2^0),f_{t_2}^{\Delta_2\sigma_1^2\sigma_2}(t_1^0,t_2^0)\}$ make up a basis of $T_{a\sigma_1^2\sigma_2}S$ and it is called the natural basis or coordinate basis of the $\sigma_1^2\sigma_2$-tangent space $T_{a\sigma_1^2\sigma_2}S$.

Definition 2.20 Suppose that (U,f) is a local $\sigma_1\sigma_2^2$-parameterization of a surface S,

$$f(t_1,t_2)=(f_1(t_1,t_2),f_2(t_1,t_2),f_3(t_1,t_2)),\quad (t_1,t_2)\in U.$$

Assume that $(t_1^0,t_2^0)\in U$ and

$$a=\left(f_1(t_1^0,t_2^0),f_2(t_1^0,t_2^0),f_3(t_1^0,t_2^0)\right)\in S.$$

A vector $h \in \mathbb{T}_{(1)} \times \mathbb{T}_{(2)} \times \mathbb{T}_{(3)}$ is called a $\sigma_1\sigma_2^2$-tangent vector to the surface S at the point a if there is a smooth parameterized curve (I,g) on S and a $t_0 \in I$ such that $g(t_0) = a$ and $g^{\Delta}(t_0) = h$. By $T_{a\sigma_1\sigma_2^2}S$ we will denote the space of all $\sigma_1\sigma_2^2$-tangent vectors to S at a. The vectors $\{f_{t_1}^{\Delta_1\sigma_1\sigma_2^2}(t_1^0,t_2^0), f_{t_2}^{\Delta_2\sigma_1\sigma_2}(t_1^0,t_2^0)\}$ make up a basis of $T_{a\sigma_1\sigma_2^2}S$ and it is called the natural basis or coordinate basis of the $\sigma_1\sigma_2^2$-tangent space $T_{a\sigma_1\sigma_2^2}S$.

Definition 2.21 The plane parallel to $T_{a\sigma_1}S$ and passing through a is called the σ_1-tangent plane to S at the point a.

The equation of the σ_1-tangent plane at a is given by

$$\det\begin{pmatrix} X - f_1(t_1^0,t_2^0) & Y - f_2(t_1^0,t_2^0) & Z - f_3(t_1^0,t_2^0) \\ f_{1t_1}^{\Delta_1}(t_1^0,t_2^0) & f_{2t_1}^{\Delta_1}(t_1^0,t_2^0) & f_{3t_1}^{\Delta_1}(t_1^0,t_2^0) \\ f_{1t_2}^{\Delta_2\sigma_1}(t_1^0,t_2^0) & f_{2t_2}^{\Delta_2\sigma_1}(t_1^0,t_2^0) & f_{3t_2}^{\Delta_2\sigma_1}(t_1^0,t_2^0) \end{pmatrix} = 0.$$

Definition 2.22 The plane parallel to $T_{a\sigma_2}S$ and passing through a is called the σ_2-tangent plane to S at the point a.

The equation of the σ_2-tangent plane at a is given by

$$\det\begin{pmatrix} X - f_1(t_1^0,t_2^0) & Y - f_2(t_1^0,t_2^0) & Z - f_3(t_1^0,t_2^0) \\ f_{1t_1}^{\Delta_1\sigma_2}(t_1^0,t_2^0) & f_{2t_1}^{\Delta_1\sigma_2}(t_1^0,t_2^0) & f_{3t_1}^{\Delta_1\sigma_2}(t_1^0,t_2^0) \\ f_{1t_2}^{\Delta_2}(t_1^0,t_2^0) & f_{2t_2}^{\Delta_2}(t_1^0,t_2^0) & f_{3t_2}^{\Delta_2}(t_1^0,t_2^0) \end{pmatrix} = 0.$$

Definition 2.23 The plane parallel to $T_{a\sigma_1^2\sigma_2}S$ and passing through a is called the $\sigma_1^2\sigma_2$-tangent plane to S at the point a.

The equation of the $\sigma_1^2\sigma_2$-tangent plane at a is given by

$$\det\begin{pmatrix} X - f_1(t_1^0,t_2^0) & Y - f_2(t_1^0,t_2^0) & Z - f_3(t_1^0,t_2^0) \\ f_{1t_1}^{\Delta_1\sigma_1\sigma_2}(t_1^0,t_2^0) & f_{2t_1}^{\Delta_1\sigma_1\sigma_2}(t_1^0,t_2^0) & f_{3t_1}^{\Delta_1\sigma_1\sigma_2}(t_1^0,t_2^0) \\ f_{1t_2}^{\Delta_2\sigma_1^2\sigma_2}(t_1^0,t_2^0) & f_{2t_2}^{\Delta_2\sigma_1^2\sigma_2}(t_1^0,t_2^0) & f_{3t_2}^{\Delta_2\sigma_1^2\sigma_2} \end{pmatrix} = 0.$$

Definition 2.24 The plane parallel to $T_{a\sigma_1\sigma_2^2}S$ and passing through a is called the $\sigma_1\sigma_2^2$-tangent plane to S at the point a.

The equation of the $\sigma_1\sigma_2^2$-tangent plane at a is given by

$$\det\begin{pmatrix} X - f_1(t_1^0,t_2^0) & Y - f_2(t_1^0,t_2^0) & Z - f_3(t_1^0,t_2^0) \\ f_{1t_1}^{\Delta_1\sigma_1\sigma_2^2}(t_1^0,t_2^0) & f_{2t_1}^{\Delta_1\sigma_1\sigma_2^2}(t_1^0,t_2^0) & f_{3t_1}^{\Delta_1\sigma_1\sigma_2^2}(t_1^0,t_2^0) \\ f_{1t_2}^{\Delta_2\sigma_1\sigma_2}(t_1^0,t_2^0) & f_{2t_2}^{\Delta_2\sigma_1\sigma_2}(t_1^0,t_2^0) & f_{3t_2}^{\Delta_2\sigma_1\sigma_2}(t_1^0,t_2^0) \end{pmatrix} = 0.$$

Example 2.7 Let $\mathbb{T}_1 = 3\mathbb{Z}$, $\mathbb{T}_2 = 2^{\mathbb{N}_0}$, $\mathbb{T}_{(1)} = \mathbb{T}_{(2)} = \mathbb{T}_{(3)} = \mathbb{R}$, and S be a surface with a local parametrization (U,f), where $U = [0,15] \times [1,16]$ and

$$f(t_1,t_2) = \left(t_1^2 + t_2^2, t_1 - t_2^3, t_1t_2\right), \quad (t_1,t_2) \in U.$$

We will find the σ_1-tangent plane and the σ_2-tangent plane at the point $f(0,1)$. We have

$$\begin{aligned}
\sigma_1(t_1) &= t_1+3, \quad t_1 \in \mathbb{T}_1,\\
\sigma_2(t_2) &= 2t_2, \quad t_2 \in \mathbb{T}_2,\\
f_1(t_1,t_2) &= t_1^2+t_2^2,\\
f_2(t_1,t_2) &= t_1-t_2^3,\\
f_3(t_1,t_2) &= t_1t_2, \quad (t_1,t_2) \in U.
\end{aligned}$$

Then

$$\begin{aligned}
f_1(0,1) &= 1,\\
f_2(0,1) &= -1,\\
f_3(0,1) &= 0,\\
f_{1t_1}^{\Delta_1}(t_1,t_2) &= \sigma_1(t_1)+t_1\\
&= t_1+3+t_1\\
&= 2t_1+3,\\
f_{1t_1}^{\Delta_1}(0,1) &= 3,\\
f_{1t_1}^{\Delta_1}(0,\sigma_2(1)) &= f_{1t_1}^{\Delta_1}(0,2)\\
&= 3,\\
f_{2t_1}^{\Delta_1}(t_1,t_2) &= 1,\\
f_{2t_1}^{\Delta_1}(0,1) &= 1,\\
f_{2t_1}^{\Delta_1}(0,\sigma_2(1)) &= f_{2t_1}^{\Delta_1}(0,2)\\
&= 1,\\
f_{3t_1}^{\Delta_1}(t_1,t_2) &= t_2,\\
f_{3t_1}^{\Delta_1}(0,1) &= 1,\\
f_{3t_1}^{\Delta_1}(0,\sigma_2(1)) &= f_{3t_1}^{\Delta_1}(0,2)\\
&= 2,\\
f_{1t_2}^{\Delta_2}(t_1,t_2) &= \sigma_2(t_2)+t_2\\
&= 2t_2+t_2\\
&= 3t_2,\\
f_{1t_2}^{\Delta_2}(0,1) &= 3,\\
f_{1t_2}^{\Delta_2}(\sigma_1(0),1) &= f_{1t_2}^{\Delta_2}(3,1)\\
&= 3,\\
f_{2t_2}^{\Delta_2}(t_1,t_2) &= -(\sigma_2(t_2))^2-t_2\sigma_2(t_2)-t_2^2
\end{aligned}$$

$$\begin{aligned}
&= -(2t_2)^2 - 2t_2^2 - t_2^2 \\
&= -4t_2^2 - 3t_2^2 \\
&= -7t_2^2, \\
f_{2t_2}^{\Delta_2}(0,1) &= -7, \\
f_{2t_2}^{\Delta_2}(\sigma_1(0),1) &= f_{2t_2}^{\Delta_2}(3,1) \\
&= -7, \\
f_{3t_2}^{\Delta_2}(t_1,t_2) &= t_1, \\
f_{3t_2}^{\Delta_2}(0,1) &= 0, \\
f_{3t_2}^{\Delta_2}(\sigma_1(0),1) &= f_{3t_2}^{\Delta_2}(3,1) \\
&= 3.
\end{aligned}$$

Then the σ_1-tangent plane at $(1,-1,0)$ is

$$\begin{aligned}
0 &= \det\begin{pmatrix} X-1 & Y+1 & Z \\ 3 & 1 & 1 \\ 3 & -7 & 3 \end{pmatrix} \\
&= 3(X-1) + 3(Y+1) - 21Z - 3Z + 7(X-1) - 9(Y+1) \\
&= 10(X-1) - 6(Y+1) - 24Z,
\end{aligned}$$

whereupon

$$5X - 5 - 3Y - 3 - 12Z = 0$$

or

$$5X - 3Y - 12Z - 8 = 0.$$

The σ_2-tangent plane at $(1,-1,0)$ is

$$\begin{aligned}
0 &= \det\begin{pmatrix} X-1 & Y+1 & Z \\ 3 & 1 & 2 \\ 3 & -7 & 0 \end{pmatrix} \\
&= 6(Y+1) - 21Z - 3Z + 14(X-1) \\
&= 14(X-1) + 6(Y+1) - 24Z
\end{aligned}$$

or

$$7X - 7 + 3Y + 3 - 12Z = 0,$$

or

$$7X + 3Y - 12Z - 4 = 0.$$

Example 2.8 Let $\mathbb{T}_1 = \mathbb{T}_2 = 3^{\mathbb{N}_0}$, $\mathbb{T}_{(1)} = \mathbb{T}_{(2)} = \mathbb{T}_{(3)} = \mathbb{R}$ and S be a surface with a local parameterization (U,f), where $U = [1,27] \times [1,27]$ and

$$f(t_1,t_2) = \left(t_1^2 + \frac{1}{t_2^2}, \frac{t_1+1}{2+t_2}, t_1 t_2\right), \quad (t_1,t_2) \in U.$$

We will find the σ_1-tangent plane at the point $\left(2, \frac{2}{3}, 1\right)$.

We have

$$\sigma_1(t_1) = 3t_1, \quad t_1 \in \mathbb{T}_1,$$
$$\sigma_2(t_2) = 3t_2, \quad t_2 \in \mathbb{T}_2,$$
$$f_1(t_1,t_2) = t_1^2 + \frac{1}{t_2^2},$$
$$f_2(t_1,t_2) = \frac{1+t_1}{2+t_2},$$
$$f_3(t_1,t_2) = t_1t_2, \quad (t_1,t_2) \in U.$$

Hence,

$$f_1(1,1) = 3,$$
$$f_2(1,1) = \frac{2}{3},$$
$$f_3(1,1) = 1,$$
$$\begin{aligned} f_{1t_1}^{\Delta_1}(t_1,t_2) &= \sigma_1(t_1) + t_1 \\ &= 3t_1 + t_1 \\ &= 4t_1, \end{aligned}$$
$$f_{1t_1}^{\Delta_1}(1,1) = 4,$$
$$f_{2t_1}^{\Delta_1}(t_1,t_2) = \frac{1}{2+t_2},$$
$$f_{2t_1}^{\Delta_1}(1,1) = \frac{1}{3},$$
$$f_{3t_1}^{\Delta_1}(t_1,t_2) = t_2,$$
$$f_{3t_1}^{\Delta_1}(1,1) = 1,$$
$$\begin{aligned} f_{1t_2}^{\Delta_2}(t_1,t_2) &= -\frac{\sigma_2(t_2)+t_2}{t_2^2(\sigma_2(t_2))^2} \\ &= -\frac{3t_2+t_2}{t_2^2(9t_2^2)} \\ &= -\frac{4}{9t_2^2}, \end{aligned}$$
$$f_{1t_2}^{\Delta_2}(1,1) = -\frac{4}{9},$$
$$\begin{aligned} f_{1t_2}^{\Delta_2}(\sigma_1(1),1) &= f_{1t_2}^{\Delta_2}(3,1) \\ &= -\frac{4}{9}, \end{aligned}$$

$$\begin{aligned}
f_{2t_2}^{\Delta_2}(t_1,t_2) &= -\frac{t_1+1}{(2+t_2)(2+\sigma_2(t_2))}\\
&= -\frac{t_1+1}{(2+t_2)(2+3t_2)},\\
f_{2t_2}^{\Delta_2}(1,1) &= -\frac{2}{15},\\
f_{2t_2}^{\Delta_2}(\sigma_1(1),1) &= f_{2t_2}^{\Delta_2}(3,1)\\
&= -\frac{4}{(2+1)(2+3)}\\
&= -\frac{4}{15},\\
f_{3t_2}^{\Delta_2}(t_1,t_2) &= t_2,\\
f_{3t_2}^{\Delta_2}(1,1) &= 1,\\
f_{3t_2}^{\Delta_2}(\sigma_1(1),1) &= f_{3t_2}^{\Delta_2}(3,1)\\
&= 1, \quad (t_1,t_2)\in U.
\end{aligned}$$

Then the σ_1-tangent plane is

$$\begin{aligned}
0 &= \det\begin{pmatrix} X-2 & Y-\frac{2}{3} & Z-1\\ 4 & \frac{1}{3} & 1\\ -\frac{4}{9} & -\frac{4}{15} & 1\end{pmatrix}\\
&= \frac{1}{3}(X-2) - \frac{4}{9}\left(Y-\frac{2}{3}\right) - \frac{16}{15}(Z-1)\\
&\quad + \frac{4}{27}(Z-1) + \frac{4}{15}(X-2) - 4\left(Y-\frac{2}{3}\right)\\
&= \frac{3}{5}(X-2) - \frac{40}{9}\left(Y-\frac{2}{3}\right) - \frac{124}{135}(Z-1)\\
&= \frac{3}{5}X - \frac{40}{9}Y - \frac{124}{135}Z + \frac{400+124-162}{135}\\
&= \frac{3}{5}X - \frac{40}{9}Y - \frac{124}{135}Z + \frac{362}{135}.
\end{aligned}$$

Exercise 2.3 Let $\mathbb{T}_1 = 2^{\mathbb{N}_0}$, $\mathbb{T}_2 = 4^{\mathbb{N}_0}$, S be a surface with a local parameterization (U,f), where $U = [1,8] \times [1,64]$ and

$$f(t_1,t_2) = \left(1+t_1+t_2^2, \frac{1-t_2}{1+t_1^2}, 1+t_1+t_2\right), \quad (t_1,t_2)\in U.$$

Find the σ_1-tangent plane and the σ_2-tangent plane of S at the point $(3,0,3)$.

Definition 2.25 The straight line passing through a point on a surface, perpendicular to the σ_1-tangent plane of the surface at that point, is called the σ_1-normal to the surface at the point being considered.

Definition 2.26 The straight line passing through a point on a surface, perpendicular to the σ_2-tangent plane of the surface at that point, is called the σ_2-normal to the surface at the point being considered.

Definition 2.27 The straight line passing through a point on a surface, perpendicular to the $\sigma_1^2\sigma_2$-tangent plane of the surface at that point, is called the $\sigma_1^2\sigma_2$-normal to the surface at the point being considered

Definition 2.28 The straight line passing through a point on a surface, perpendicular to the $\sigma_1\sigma_2^2$-tangent plane of the surface at that point, is called the $\sigma_1\sigma_2^2$-normal to the surface at the point being considered.

Let (U,f) be a local parameterization of a surface S around the point

$$f(t_1^0,t_2^0)=\left(f_1(t_1^0,t_2^0),f_2(t_1^0,t_2^0),f_3(t_1^0,t_2^0)\right).$$

If f is σ_1-completely delta differentiable, then the σ_1-normal field is given by

$$N_{\sigma_1}(a)=\frac{f_{t_1}^{\Delta_1}(t_1^0,t_2^0)\times f_{t_2}^{\Delta_2\sigma_1}(t_1^0,t_2^0)}{\left\|f_{t_1}^{\Delta_1}(t_1^0,t_2^0)\times f_{t_2}^{\Delta_2\sigma_1}(t_1^0,t_2^0)\right\|}$$

and the equations of the σ_1-normal are as follows

$$\frac{X-f_1(t_1^0,t_2^0)}{\det\begin{pmatrix} f_{2t_1}^{\Delta_1}(t_1^0,t_2^0) & f_{3t_1}^{\Delta_1}(t_1^0,t_2^0)\\ f_{2t_2}^{\Delta_2\sigma_1}(t_1^0,t_2^0) & f_{3t_2}^{\Delta_2\sigma_1}(t_1^0,t_2^0)\end{pmatrix}}=\frac{Y-f_2(t_1^0,t_2^0)}{\det\begin{pmatrix} f_{3t_1}^{\Delta_1}(t_1^0,t_2^0) & f_{1t_1}^{\Delta_1}(t_1^0,t_2^0)\\ f_{3t_2}^{\Delta_2\sigma_1}(t_1^0,t_2^0) & f_{1t_2}^{\Delta_2\sigma_1}(t_1^0,t_2^0)\end{pmatrix}}$$

$$=\frac{Z-f_3(t_1^0,t_2^0)}{\det\begin{pmatrix} f_{1t_1}^{\Delta_1}(t_1^0,t_2^0) & f_{2t_1}^{\Delta_1}(t_1^0,t_2^0)\\ f_{1t_2}^{\Delta_2\sigma_1}(t_1^0,t_2^0) & f_{2t_2}^{\Delta_2\sigma_1}(t_1^0,t_2^0)\end{pmatrix}}$$

If f is σ_2-completely delta differentiable, then the σ_2-normal field is given by

$$N_{\sigma_2}(a)=\frac{f_{t_1}^{\Delta_1\sigma_2}(t_1^0,t_2^0)\times f_{t_2}^{\Delta_2}(t_1^0,t_2^0)}{\left\|f_{t_1}^{\Delta_1\sigma_2}(t_1^0,t_2^0)\times f_{t_2}^{\Delta_2}(t_1^0,t_2^0)\right\|}$$

and the equations of the σ_1-normal are as follows

$$\frac{X-f_1(t_1^0,t_2^0)}{\det\begin{pmatrix} f_{2t_1}^{\Delta_1\sigma_2}(t_1^0,t_2^0) & f_{3t_1}^{\Delta_1\sigma_2}(t_1^0,t_2^0)\\ f_{2t_2}^{\Delta_2}(t_1^0,t_2^0) & f_{3t_2}^{\Delta_2}(t_1^0,t_2^0)\end{pmatrix}}=\frac{Y-f_2(t_1^0,t_2^0)}{\det\begin{pmatrix} f_{3t_1}^{\Delta_1\sigma_2}(t_1^0,t_2^0) & f_{1t_1}^{\Delta_1\sigma_2}(t_1^0,t_2^0)\\ f_{3t_2}^{\Delta_2}(t_1^0,t_2^0) & f_{1t_2}^{\Delta_2}(t_1^0,t_2^0)\end{pmatrix}}$$

$$=\frac{Z-f_3(t_1^0,t_2^0)}{\det\begin{pmatrix} f_{1t_1}^{\Delta_1\sigma_2}(t_1^0,t_2^0) & f_{2t_1}^{\Delta_1\sigma_2}(t_1^0,t_2^0)\\ f_{1t_2}^{\Delta_2}(t_1^0,t_2^0) & f_{2t_2}^{\Delta_2}(t_1^0,t_2^0)\end{pmatrix}}$$

If f is σ_1-completely delta differentiable, then the $\sigma_1^2\sigma_2$-normal field is given by

$$N_{\sigma_1^2\sigma_2}(a) = \frac{f_{t_1}^{\Delta_1\sigma_1\sigma_2}(t_1^0,t_2^0)\times f_{t_2}^{\Delta_2\sigma_1^2\sigma_2}(t_1^0,t_2^0)}{\left\| f_{t_1}^{\Delta_1\sigma_1\sigma_2}(t_1^0,t_2^0)\times f_{t_2}^{\Delta_2\sigma_1^2\sigma_2}(t_1^0,t_2^0)\right\|}$$

and the equations of the $\sigma_1^2\sigma_2$-normal are as follows

$$\begin{aligned}\frac{X-f_1(t_1^0,t_2^0)}{\det\begin{pmatrix} f_{2t_1}^{\Delta_1\sigma_1\sigma_2}(t_1^0,t_2^0) & f_{3t_1}^{\Delta_1\sigma_1\sigma_2}(t_1^0,t_2^0)\\ f_{2t_2}^{\Delta_2\sigma_1^2\sigma_2}(t_1^0,t_2^0) & f_{3t_2}^{\Delta_2\sigma_1^2\sigma_2}(t_1^0,t_2^0)\end{pmatrix}} &= \frac{Y-f_2(t_1^0,t_2^0)}{\det\begin{pmatrix} f_{3t_1}^{\Delta_1\sigma_1\sigma_2}(t_1^0,t_2^0) & f_{1t_1}^{\Delta_1\sigma_1\sigma_2}(t_1^0,t_2^0)\\ f_{3t_2}^{\Delta_2\sigma_1^2\sigma_2}(t_1^0,t_2^0) & f_{1t_2}^{\Delta_2\sigma_1^2\sigma_2}(t_1^0,t_2^0)\end{pmatrix}}\\ &= \frac{Z-f_3(t_1^0,t_2^0)}{\det\begin{pmatrix} f_{1t_1}^{\Delta_1\sigma_1\sigma_2}(t_1^0,t_2^0) & f_{2t_1}^{\Delta_1\sigma_1\sigma_2}(t_1^0,t_2^0)\\ f_{1t_2}^{\Delta_2\sigma_1^2\sigma_2}(t_1^0,t_2^0) & f_{2t_2}^{\Delta_2\sigma_1^2\sigma_2}(t_1^0,t_2^0)\end{pmatrix}}\end{aligned}$$

If f is σ_2-completely delta differentiable, then the $\sigma_1\sigma_2^2$-normal field is given by

$$N_{\sigma_1}(a) = \frac{f_{t_1}^{\Delta_1\sigma_1\sigma_2^2}(t_1^0,t_2^0)\times f_{t_2}^{\Delta_2\sigma_1\sigma_2}(t_1^0,t_2^0)}{\left\| f_{t_1}^{\Delta_1\sigma_1\sigma_2^2}(t_1^0,t_2^0)\times f_{t_2}^{\Delta_2\sigma_1\sigma_2}(t_1^0,t_2^0)\right\|}$$

and the equations of the $\sigma_1\sigma_2^2$-normal are as follows

$$\begin{aligned}\frac{X-f_1(t_1^0,t_2^0)}{\det\begin{pmatrix} f_{2t_1}^{\Delta_1\sigma_1\sigma_2^2}(t_1^0,t_2^0) & f_{3t_1}^{\Delta_1\sigma_1\sigma_2^2}(t_1^0,t_2^0)\\ f_{2t_2}^{\Delta_2\sigma_1\sigma_2}(t_1^0,t_2^0) & f_{3t_2}^{\Delta_2\sigma_1\sigma_2}(t_1^0,t_2^0)\end{pmatrix}} &= \frac{Y-f_2(t_1^0,t_2^0)}{\det\begin{pmatrix} f_{3t_1}^{\Delta_1\sigma_1\sigma_2^2}(t_1^0,t_2^0) & f_{1t_1}^{\Delta_1\sigma_1\sigma_2^2}(t_1^0,t_2^0)\\ f_{3t_2}^{\Delta_2\sigma_1\sigma_2}(t_1^0,t_2^0) & f_{1t_2}^{\Delta_2\sigma_1\sigma_2}(t_1^0,t_2^0)\end{pmatrix}}\\ &= \frac{Z-f_3(t_1^0,t_2^0)}{\det\begin{pmatrix} f_{1t_1}^{\Delta_1\sigma_1\sigma_2^2}(t_1^0,t_2^0) & f_{2t_1}^{\Delta_1\sigma_1\sigma_2^2}(t_1^0,t_2^0)\\ f_{1t_2}^{\Delta_2\sigma_1\sigma_2}(t_1^0,t_2^0) & f_{2t_2}^{\Delta_2\sigma_1\sigma_2}(t_1^0,t_2^0)\end{pmatrix}}\end{aligned}$$

Example 2.9 Let $\mathbb{T}_1 = 2^{\mathbb{N}_0}$, $\mathbb{T}_2 = 3^{\mathbb{N}_0}$, $\mathbb{T}_{(1)} = \mathbb{T}_{(2)} = \mathbb{T}_{(3)} = \mathbb{R}$ and (U,f) be a local parameterization of a surface S, where $U \subset \mathbb{T}_1 \times \mathbb{T}_2$ and

$$f(t_1,t_2) = \left(1+t_1^2+t_2^3, 1+t_1^2t_2, t_1t_2^2\right), \quad (t_1,t_2)\in U.$$

We will find the equations of the σ_1-normal and the equations of the σ_2-normal to S. Here

$$\begin{aligned}\sigma_1(t_1) &= 2t_1, \quad t_1 \in \mathbb{T}_1,\\ \sigma_2(t_2) &= 3t_2, \quad t_2\in\mathbb{T}_2,\\ f_1(t_1,t_2) &= 1+t_1^2+t_2^3,\\ f_2(t_1,t_2) &= 1+t_1^2t_2,\\ f_3(t_1,t_2) &= t_1t_2^2, \quad (t_1,t_2)\in U.\end{aligned}$$

Then

$$\begin{aligned}
f_{1t_1}^{\Delta_1}(t_1,t_2) &= \sigma_1(t_1)+t_1\\
&= 2t_1+t_1\\
&= 3t_1,\\
f_{1t_1}^{\Delta_1\sigma_2}(t_1,t_2) &= 3t_1,\\
f_{2t_1}^{\Delta_1}(t_1,t_2) &= (\sigma_1(t_1)+t_1)t_2\\
&= (2t_1+t_1)t_2\\
&= 3t_1t_2,\\
f_{2t_1}^{\Delta_1\sigma_2}(t_1,t_2) &= 3t_1\sigma_2(t_2)\\
&= 3t_1(3t_2)\\
&= 9t_1t_2,\\
f_{3t_1}^{\Delta_1}(t_1,t_2) &= t_2^2,\\
f_{3t_1}^{\Delta_1\sigma_2}(t_1,t_2) &= (\sigma_2(t_2))^2\\
&= (3t_2)^2\\
&= 9t_2^2,\\
f_{1t_2}^{\Delta_2}(t_1,t_2) &= (\sigma_2(t_2))^2+t_2\sigma_2(t_2)+t_2^2\\
&= (3t_2)^2+t_2(3t_2)+t_2^2\\
&= 9t_2^2+3t_2^2+t_2^2\\
&= 13t_2^2,\\
f_{1t_2}^{\Delta_2\sigma_1}(t_1,t_2) &= 13t_2^2,\\
f_{2t_2}^{\Delta_2}(t_1,t_2) &= t_1^2,\\
f_{2t_2}^{\Delta_2\sigma_1}(t_1,t_2) &= (\sigma_1(t_1))^2\\
&= (2t_1)^2\\
&= 4t_1^2,\\
f_{3t_2}^{\Delta_2}(t_1,t_2) &= t_1(\sigma_2(t_2)+t_2)\\
&= t_1(3t_2+t_2)\\
&= 4t_1t_2,\\
f_{3t_2}^{\Delta_2\sigma_1}(t_1,t_2) &= 4\sigma_1(t_1)t_2\\
&= 4(2t_1)t_2\\
&= 8t_1t_2, \quad (t_1,t_2)\in U.
\end{aligned}$$

Hence,

$$\det\begin{pmatrix} f_{2t_1}^{\Delta_1}(t_1,t_2) & f_{3t_1}^{\Delta_1}(t_1,t_2) \\ f_{2t_2}^{\Delta_2\sigma_1}(t_1,t_2) & f_{3t_2}^{\Delta_2\sigma_1}(t_1,t_2) \end{pmatrix} = \det\begin{pmatrix} 3t_1t_2 & t_2^2 \\ 4t_1^2 & 8t_1t_2 \end{pmatrix}$$
$$= 24t_1^2t_2^2 - 4t_1^2t_2^2$$
$$= 20t_1^2t_2^2,$$

$$\det\begin{pmatrix} f_{3t_1}^{\Delta_1}(t_1,t_2) & f_{1t_1}^{\Delta_1}(t_1,t_2) \\ f_{3t_2}^{\Delta_2\sigma_1}(t_1,t_2) & f_{1t_2}^{\Delta_2\sigma_1}(t_1,t_2) \end{pmatrix} = \det\begin{pmatrix} t_2^2 & 3t_1 \\ 8t_1t_2 & 13t_2^2 \end{pmatrix}$$
$$= 13t_2^4 - 24t_1^2t_2,$$

$$\det\begin{pmatrix} f_{1t_1}^{\Delta_1}(t_1,t_2) & f_{2t_1}^{\Delta_1}(t_1,t_2) \\ f_{1t_2}^{\Delta_2\sigma_1}(t_1,t_2) & f_{2t_2}^{\Delta_2\sigma_1}(t_1,t_2) \end{pmatrix} = \det\begin{pmatrix} 3t_1 & 3t_1t_2 \\ 13t_2^2 & 4t_1^2 \end{pmatrix}$$
$$= 12t_1^3 - 39t_1t_2^3, \quad (t_1,t_2) \in U.$$

Therefore the equations of the σ_1-normal to S are as follows

$$\frac{X-1-t_1^2-t_2^3}{20t_1^2t_2^2} = \frac{Y-1-t_1^2t_2}{13t_2^4-24t_1^2t_2} = \frac{Z-t_1t_2^2}{12t_1^3-39t_1t_2^3}, \quad (t_1,t_2) \in U,$$

provided they exist. Now, we will find the equations of the σ_2-normal. We have

$$\det\begin{pmatrix} f_{2t_1}^{\Delta_1\sigma_2}(t_1,t_2) & f_{3t_1}^{\Delta_1\sigma_2}(t_1,t_2) \\ f_{2t_2}^{\Delta_2}(t_1,t_2) & f_{3t_2}^{\Delta_2}(t_1,t_2) \end{pmatrix} = \det\begin{pmatrix} 9t_1t_2 & 9t_2^2 \\ t_1^2 & 4t_1t_2 \end{pmatrix}$$
$$= 36t_1^2t_2^2 - 9t_1^2t_2^2$$
$$= 27t_1^2t_2^2,$$

$$\det\begin{pmatrix} f_{3t_1}^{\Delta_1\sigma_2}(t_1,t_2) & f_{1t_1}^{\Delta_1\sigma_2}(t_1,t_2) \\ f_{3t_2}^{\Delta_2}(t_1,t_2) & f_{1t_2}^{\Delta_2}(t_1,t_2) \end{pmatrix} = \det\begin{pmatrix} 9t_2^2 & 3t_1 \\ 4t_1t_2 & 13t_2^2 \end{pmatrix}$$
$$= 117t_2^4 - 12t_1^2t_2,$$

$$\det\begin{pmatrix} f_{1t_1}^{\Delta_1\sigma_2}(t_1,t_2) & f_{2t_1}^{\Delta_1\sigma_2}(t_1,t_2) \\ f_{1t_2}^{\Delta_2}(t_1,t_2) & f_{2t_2}^{\Delta_2}(t_1,t_2) \end{pmatrix} = \det\begin{pmatrix} 3t_1 & 9t_1t_2 \\ 13t_2^2 & t_1^2 \end{pmatrix}$$
$$= 3t_1^3 - 117t_1t_2^3, \quad (t_1,t_2) \in U.$$

Consequently, the equations of the σ_2-normal are as follows

$$\frac{X-1-t_1^2-t_2^3}{27t_1^2t_2^2} = \frac{Y-1-t_1^2t_2}{117t_2^4-12t_1^2t_2} = \frac{Z-t_1t_2^2}{3t_1^3-117t_1t_2^3},$$

$(t_1,t_2) \in U$, provided they exist.

Example 2.10 Let $\mathbb{T}_1 = \mathbb{T}_2 = \mathbb{Z}$, $\mathbb{T}_{(1)} = \mathbb{T}_{(2)} = \mathbb{T}_{(3)} = \mathbb{R}$, (U,f) be a local parameterization of a surface S, where $U \subseteq \mathbb{T}_1 \times \mathbb{T}_2$ and

$$f(t_1,t_2) = \left(t_1^2+t_2^2, t_1t_2^2, t_1^2t_2\right), \quad (t_1,t_2) \in U.$$

We will find the equations of the σ_1-normal and the equations of the σ_2-normal to S. Here

$$\begin{aligned}
\sigma_1(t_1) &= t_1+1, \quad t_1 \in \mathbb{T}_1,\\
\sigma_2(t_2) &= t_2+1, \quad t_2 \in \mathbb{T}_2,\\
f_1(t_1,t_2) &= t_1^2+t_2^2,\\
f_2(t_1,t_2) &= t_1t_2^2,\\
f_3(t_1,t_2) &= t_1^2t_2, \quad (t_1,t_2)\in U.
\end{aligned}$$

Then

$$\begin{aligned}
f_{1t_1}^{\Delta_1}(t_1,t_2) &= \sigma_1(t_1)+t_1\\
&= t_1+1+t_1\\
&= 2t_1+1,\\
f_{1t_1}^{\Delta_1\sigma_2}(t_1,t_2) &= 2t_1+1,\\
f_{2t_1}^{\Delta_1}(t_1,t_2) &= t_2^2,\\
f_{2t_1}^{\Delta_1\sigma_2}(t_1,t_2) &= (\sigma_2(t_2))^2\\
&= (t_2+1)^2\\
&= t_2^2+2t_2+1,\\
f_{3t_1}^{\Delta_1}(t_1,t_2) &= (\sigma_1(t_1)+t_1)t_2\\
&= (t_1+1+t_1)t_2\\
&= (2t_1+1)t_2,\\
f_{3t_1}^{\Delta_1\sigma_2}(t_1,t_2) &= (2t_1+1)\sigma_2(t_2)\\
&= (2t_1+1)(t_2+1),\\
f_{1t_2}^{\Delta_2}(t_1,t_2) &= \sigma_2(t_2)+t_2\\
&= t_2+1+t_2\\
&= 2t_2+1,\\
f_{1t_2}^{\Delta_2\sigma_1}(t_1,t_2) &= 2t_2+1,\\
f_{2t_2}^{\Delta_2}(t_1,t_2) &= t_1(\sigma_2(t_2)+t_2)\\
&= t_1(t_2+1+t_2)\\
&= t_1(2t_2+1),\\
f_{2t_2}^{\Delta_2\sigma_1}(t_1,t_2) &= \sigma_1(t_1)(2t_2+1)\\
&= (t_1+1)(2t_2+1),\\
f_{3t_2}^{\Delta_2}(t_1,t_2) &= t_1^2,
\end{aligned}$$

$$f_{3t_2}^{\Delta_2\sigma_1}(t_1,t_2) = (\sigma_1(t_1))^2$$
$$= (t_1+1)^2, \quad (t_1,t_2) \in U.$$

Then

$$\det\begin{pmatrix} f_{2t_1}^{\Delta_1}(t_1,t_2) & f_{3t_1}^{\Delta_1}(t_1,t_2) \\ f_{2t_2}^{\Delta_2\sigma_1}(t_1,t_2) & f_{3t_2}^{\Delta_2\sigma_1}(t_1,t_2) \end{pmatrix} = \det\begin{pmatrix} t_2^2 & (2t_1+1)t_2 \\ (t_1+1)(2t_2+1) & (t_1+1)^2 \end{pmatrix}$$
$$= t_2^2(t_1+1)^2 - t_2(t_1+1)(2t_1+1)(2t_2+1)$$
$$= t_2(t_1+1)(t_2(t_1+1) - (2t_1+1)(2t_2+1))$$
$$= t_2(t_1+1)(t_1t_2+t_2-4t_1t_2-2t_1-2t_2-1)$$
$$= t_2(t_1+1)(-3t_1t_2-2t_1-t_2-1),$$

$$\det\begin{pmatrix} f_{3t_1}^{\Delta_1}(t_1,t_2) & f_{1t_1}^{\Delta_1}(t_1,t_2) \\ f_{3t_2}^{\Delta_2\sigma_1}(t_1,t_2) & f_{1t_2}^{\Delta_2\sigma_1}(t_1,t_2) \end{pmatrix} = \det\begin{pmatrix} (2t_1+1)t_2 & 2t_1+1 \\ (t_1+1)^2 & 2t_2+1 \end{pmatrix}$$
$$= (2t_1+1)t_2(2t_2+1)$$
$$-(2t_1+1)(t_1+1)^2$$
$$= (2t_1+1)(t_2(2t_2+1) - (t_1+1)^2)$$
$$= (2t_1+1)(2t_2^2+t_2-t_1^2-2t_1-1),$$

$$\det\begin{pmatrix} f_{1t_1}^{\Delta_1}(t_1,t_2) & f_{2t_1}^{\Delta_1}(t_1,t_2) \\ f_{1t_2}^{\Delta_2\sigma_1}(t_1,t_2) & f_{2t_2}^{\Delta_2\sigma_1}(t_1,t_2) \end{pmatrix} = \det\begin{pmatrix} 2t_1+1 & t_2^2 \\ 2t_2+1 & (t_1+1)(2t_2+1) \end{pmatrix}$$
$$= (2t_1+1)(t_1+1)(2t_2+1)$$
$$-t_2^2(2t_2+1)$$
$$= (2t_2+1)((2t_1+1)(t_1+1) - t_2^2)$$
$$= (2t_2+1)(2t_1^2+3t_1+1-t_2^2), (t_1,t_2) \in U.$$

Therefore the equations of the σ_1-normal are as follows

$$\frac{X-t_1^2-t_2^2}{t_2(t_1+1)(-3t_1t_2-2t_1-2t_2-1)} = \frac{Y-t_1t_2^2}{(2t_1+1)(2t_2^2+t_2-t_1^2-2t_1-1)}$$
$$= \frac{Z-t_1^2t_2}{(2t_2+1)(2t_1^2+3t_1+1-t_2^2)}, \quad (t_1,t_2) \in U,$$

provided they exist. Now, we will find the equations of the σ_2-normal to S. We have

$$\det\begin{pmatrix} f_{2t_1}^{\Delta_1\sigma_2}(t_1,t_2) & f_{3t_1}^{\Delta_1\sigma_2}(t_1,t_2) \\ f_{2t_2}^{\Delta_2}(t_1,t_2) & f_{3t_2}^{\Delta_2}(t_1,t_2) \end{pmatrix} = \det\begin{pmatrix} (t_2+1)^2 & (2t_1+1)(t_2+1) \\ t_1(2t_2+1) & t_1^2 \end{pmatrix}$$
$$= t_1^2(t_2^2+2t_2+1)$$
$$-(2t_1t_2+t_1)(2t_1t_2+t_2+2t_1+1)$$
$$= t_1^2t_2^2+2t_1^2t_2+t_1^2-4t_1^2t_2^2-2t_1t_2^2$$

$$\begin{aligned}
&-4t_1^2t_2-2t_1t_2-2t_1^2t_2-t_1t_2-2t_1^2-t_1\\
&=-t_1^2-3t_1^2t_2^2-2t_1t_2^2-4t_1^2t_2-3t_1t_2-t_1,
\end{aligned}$$

$$\begin{aligned}
\det\begin{pmatrix} f_{3t_1}^{\Delta_1\sigma_2}(t_1,t_2) & f_{1t_1}^{\Delta_1\sigma_2}(t_1,t_2)\\ f_{3t_2}^{\Delta_2}(t_1,t_2) & f_{1t_2}^{\Delta_2}(t_1,t_2)\end{pmatrix} &= \det\begin{pmatrix}(2t_1+1)(t_2+1) & 2t_1+1\\ t_1^2 & 2t_2+1\end{pmatrix}\\
&=(2t_1+1)(t_2+1)(2t_2+1)\\
&\quad -t_1^2(2t_1+1)\\
&=(2t_1+1)(2t_2^2+3t_2+1)\\
&\quad -t_1^2(2t_1+1)\\
&=4t_1t_2^2+6t_1t_2+2t_1+2t_2^2+3t_2+1\\
&\quad -2t_1^3-t_1^2,
\end{aligned}$$

$$\begin{aligned}
\det\begin{pmatrix} f_{1t_1}^{\Delta_1\sigma_2}(t_1,t_2) & f_{2t_1}^{\Delta_1\sigma_2}(t_1,t_2)\\ f_{1t_2}^{\Delta_2}(t_1,t_2) & f_{2t_2}^{\Delta_2}(t_1,t_2)\end{pmatrix} &= \det\begin{pmatrix}2t_1+1 & t_2^2+2t_2+1\\ 2t_2+1 & t_1(2t_2+1)\end{pmatrix}\\
&=t_1(2t_1+1)(2t_2+1)-(2t_2+1)(t_2^2+2t_2+1)\\
&=(2t_1^2+t_1)(2t_2+1)-2t_2^3-4t_2^2-2t_2\\
&\quad -t_2^2-2t_2-1\\
&=4t_1^2t_2+2t_1t_2+2t_1^2+t_1-2t_2^3\\
&\quad -5t_2^2-4t_2-1,\quad (t_1,t_2)\in U.
\end{aligned}$$

Consequently, the equations of the σ_2-normal are as follows

$$\begin{aligned}
&\frac{X-t_1^2-t_2^2}{-t_1^2-3t_1^2t_2^2-2t_1t_2^2-4t_1^2t_2-3t_1t_2-t_1}\\
&=\frac{Y-t_1t_2^2}{-2t_1^3-t_1^2+4t_1t_2^2+6t_1t_2+2t_1+3t_2+2t_2^2+1}\\
&=\frac{Z-t_1^2t_2}{4t_1^2t_2+2t_1t_2+2t_1^2+t_1-2t_2^3-5t_2^2-4t_2-1},
\end{aligned}$$

$(t_1,t_2)\in U$, provided they exist.

Exercise 2.4 Let $\mathbb{T}_1=2\mathbb{Z}$, $\mathbb{T}_2=3\mathbb{Z}$, $\mathbb{T}_{(1)}=\mathbb{T}_{(2)}=\mathbb{T}_{(3)}=\mathbb{R}$, (U,f) be a local parameterization of a surface S, where $U\subset\mathbb{T}_1\times\mathbb{T}_2$ and

$$f(t_1,t_2)=\left(t_1+\frac{1}{1+t_2^2},t_1t_2^3+t_1^2t_2,t_1t_2-t_1^2\right).$$

Find the equations of the σ_1-normal and the equations of the σ_2-normal to S.

Example 2.11 Let $\mathbb{T}_1=\mathbb{T}_2=2^{\mathbb{N}_0}$, $\mathbb{T}_{(1)}=\mathbb{T}_{(2)}=\mathbb{T}_{(3)}=\mathbb{R}$, $U=\mathbb{T}_1\times\mathbb{T}_2$, S be a surface with local parameterization (U,f), where

$$f(t_1,t_2)=(t_1^2t_2^2,t_1,t_2),\quad (t_1,t_2)\in\mathbb{T}_1\times\mathbb{T}_2.$$

Here

$$\begin{aligned}
\sigma_1(t_1) &= 2t_1, \quad t_1 \in \mathbb{T}_1,\\
\sigma_2(t_2) &= 2t_2, \quad t_2 \in \mathbb{T}_2,\\
f_1(t_1,t_2) &= t_1^2 t_2^2,\\
f_2(t_1,t_2) &= t_1,\\
f_3(t_1,t_2) &= t_2, \quad (t_1,t_2) \in \mathbb{T}_1 \times \mathbb{T}_2.
\end{aligned}$$

Then

$$\begin{aligned}
f_{1t_2}^{\Delta_2}(t_1,t_2) &= (\sigma_1(t_1)+t_1)t_2^2\\
&= (2t_1+t_1)t_2^2\\
&= 3t_1t_2^2,\\
f_{2t_1}^{\Delta_1}(t_1,t_2) &= 1,\\
f_{3t_1}^{\Delta_1}(t_1,t_2) &= 0,\\
f_{1t_1}^{\Delta_1^2}(t_1,t_2) &= 3t_2^2,\\
f_{2t_1}^{\Delta_1^2}(t_1,t_2) &= 0,\\
f_{3t_1}^{\Delta_1^2}(t_1,t_2) &= 0,\\
f_{1t_1t_2}^{\Delta_1\Delta_2}(t_1,t_2) &= 3t_1(\sigma_2(t_2)+t_2)\\
&= 3t_1(2t_2+t_2)\\
&= 9t_1t_2,\\
f_{2t_1t_2}^{\Delta_1\Delta_2}(t_1,t_2) &= 0,\\
f_{3t_1t_2}^{\Delta_1\Delta_2}(t_1,t_2) &= 0,\\
f_{1t_2}^{\Delta_2}(t_1,t_2) &= t_1^2(\sigma_2(t_2)+t_2)\\
&= t_1^2(2t_2+t_2)\\
&= 3t_1^2t_2,\\
f_{2t_2}^{\Delta_2}(t_1,t_2) &= 0,\\
f_{3t_2}^{\Delta_2}(t_1,t_2) &= 1,\\
f_{1t_2}^{\Delta_2^2}(t_1,t_2) &= 3t_1^2,\\
f_{2t_2}^{\Delta_2^2}(t_1,t_2) &= 0,\\
f_{3t_2}^{\Delta_2^2}(t_1,t_2) &= 0, \quad (t_1,t_2) \in \mathbb{T}_1 \times \mathbb{T}_2.
\end{aligned}$$

Therefore

$$\begin{aligned}
f_{t_1}^{\Delta_1}(t_1,t_2) &= \left(f_{1t_1}^{\Delta_1}(t_1,t_2), f_{2t_1}^{\Delta_1}(t_1,t_2), f_{3t_1}^{\Delta_1}(t_1,t_2)\right)\\
&= (3t_1t_2^2,1,0),\\
f_{t_2}^{\Delta_2}(t_1,t_2) &= \left(f_{1t_2}^{\Delta_2}(t_1,t_2), f_{2t_2}^{\Delta_2}(t_1,t_2), f_{3t_2}^{\Delta_2}(t_1,t_2)\right)\\
&= (3t_1^2t_2,0,1),\\
f_{t_1}^{\Delta_1^2}(t_1,t_2) &= \left(f_{1t_1}^{\Delta_1^2}(t_1,t_2), f_{2t_1}^{\Delta_1^2}(t_1,t_2), f_{3t_1}^{\Delta_1^2}(t_1,t_2)\right)\\
&= (3t_2^2,0,0),\\
f_{t_2}^{\Delta_2^2}(t_1,t_2) &= \left(f_{1t_2}^{\Delta_2^2}(t_1,t_2), f_{2t_2}^{\Delta_2^2}(t_1,t_2), f_{3t_2}^{\Delta_2^2}(t_1,t_2)\right)\\
&= (3t_1^2,0,0),\\
f_{t_1t_2}^{\Delta_1\Delta_2}(t_1,t_2) &= \left(f_{1t_1t_2}^{\Delta_1\Delta_2}(t_1,t_2), f_{2t_1t_2}^{\Delta_1\Delta_2}(t_1,t_2), f_{3t_1t_2}^{\Delta_1\Delta_2}(t_1,t_2)\right)\\
&= (9t_1t_2,0,0), \quad (t_1,t_2)\in\mathbb{T}_1\times\mathbb{T}_2.
\end{aligned}$$

From here, we find

$$\begin{aligned}
f_{t_1}^{\Delta_1\sigma_1\sigma_2}(t_1,t_2) &= \left(3\sigma_1(t_1)(\sigma_2(t_2))^2,1,0\right)\\
&= (24t_1t_2^2,1,0),\\
f_{t_2}^{\Delta_2\sigma_1^2\sigma_2}(t_1,t_2) &= \left(3(\sigma_1^2(t_1))^2\sigma_2(t_2),0,1\right)\\
&= (96t_1^2t_2,0,1), \quad (t_1,t_2)\in\mathbb{T}_1\times\mathbb{T}_2,
\end{aligned}$$

and

$$\begin{aligned}
f_{t_1}^{\Delta_1\sigma_1\sigma_2}(t_1,t_2)\times f_{t_2}^{\Delta_2\sigma_1^2\sigma_2}(t_1,t_2) &= (24t_1t_2^2,1,0)\\
&\quad\times(96t_1^2t_2,0,1)\\
&= \left(1,-24t_1t_2^2,-96t_1^2t_2\right),\\
\left\|f_{t_1}^{\Delta_1\sigma_1\sigma_2}(t_1,t_2)\times f_{t_2}^{\Delta_2\sigma_1^2\sigma_2}(t_1,t_2)\right\| &= \left(1+(96t_1^2t_2)^2+(24t_1t_2^2)^2\right)^{\frac{1}{2}}\\
&= \left(1+9216t_1^4t_2^2+576t_1^2t_2^4\right)^{\frac{1}{2}},
\end{aligned}$$

$(t_1,t_2)\in\mathbb{T}_1\times\mathbb{T}_2$, and

$$N_{\sigma_1^2\sigma_2}(t_1,t_2) = \frac{1}{\left(1+9216t_1^4t_2^2+576t_1^2t_2^4\right)^{\frac{1}{2}}}\left(1,-24t_1t_2^2,-96t_1^2t_2\right),$$

$(t_1,t_2)\in\mathbb{T}_1\times\mathbb{T}_2$.

Exercise 2.5 Let $\mathbb{T}_1 = 2^{\mathbb{N}_0}$, $\mathbb{T}_2 = 3^{\mathbb{N}_0}$, $\mathbb{T}_{(1)} = \mathbb{T}_{(2)} = \mathbb{T}_{(3)} = \mathbb{R}$, $U = \mathbb{T}_1\times\mathbb{T}_2$, S be an oriented surface with a local parameterization (U,f), where

$$f(t_1,t_2) = \left((t_1+t_2)^4-2t_1^2t_2^2, t_1t_2^2, t_1^2+t_2^2-3t_1^4t_2^4\right),\ (t_1,t_2)\in\mathbb{T}_1\times\mathbb{T}_2.$$

Find

1. $N_{\sigma_1^2\sigma_2}(t_1,t_2)$,
2. $N_{\sigma_1^2\sigma_2}(t_1,t_2)\cdot\left(2N_{\sigma_1^2\sigma_2}(t_1,t_2)-3N_{\sigma_1^2\sigma_2}(t_1,t_2)\right)$,
3. $N_{\sigma_1^2\sigma_2}(t_1,t_2)\times\left(4N_{\sigma_1^2\sigma_2}(t_1,t_2)-2N_{\sigma_1^2\sigma_2}(t_1,t_2)\right)$.

Example 2.12 Let $\mathbb{T}_1$, $\mathbb{T}_2$, $\mathbb{T}_{(1)}$, $\mathbb{T}_{(2)}$, $\mathbb{T}_{(3)}$, U, S and f be as in Example 2.11. Then

$$\begin{aligned} f_{t_1}^{\Delta_1\sigma_1\sigma_2^2}(t_1,t_2) &= \left(3\sigma_1(t_1)(\sigma_2^2(t_2))^2,1,0\right) \\ &= (96t_1t_2^2,1,0), \\ f_{t_2}^{\Delta_2\sigma_1\sigma_2}(t_1,t_2) &= \left(3(\sigma_1(t_1))^2\sigma_2(t_2),0,1\right) \\ &= (24t_1^2t_2,0,1), \quad (t_1,t_2)\in\mathbb{T}_1\times\mathbb{T}_2, \end{aligned}$$

and

$$\begin{aligned} f_{t_1}^{\Delta_1\sigma_1^2\sigma_2}(t_1,t_2)\times f_{t_2}^{\Delta_2\sigma_1^2\sigma_2}(t_1,t_2) &= (96t_1t_2^2,1,0) \\ &\quad\times(24t_1^2t_2,0,1) \\ &= \left(1,-96t_1t_2^2,-24t_1^2t_2\right), \\ \left\|f_{t_1}^{\Delta_1\sigma_1\sigma_2}(t_1,t_2)\times f_{t_2}^{\Delta_{)}2\sigma_1^2\sigma_2}(t_1,t_2)\right\| &= \left(1+(96t_1t_2^2)^2+(24t_1^2t_2)^2\right)^{\frac{1}{2}} \\ &= \left(1+9216t_1^2t_2^4+576t_1^4t_2^2\right)^{\frac{1}{2}}, \end{aligned}$$

$(t_1,t_2)\in\mathbb{T}_1\times\mathbb{T}_2$, and

$$N_{\sigma_1^2\sigma_2}(t_1,t_2)=\frac{1}{\left(1+9216t_1^2t_2^4+576t_1^4t_2^2\right)^{\frac{1}{2}}}\left(1,-96t_1t_2^2,-24t_1^2t_2\right),$$

$(t_1,t_2)\in\mathbb{T}_1\times\mathbb{T}_2$.

Exercise 2.6 Let $\mathbb{T}_1=2\mathbb{Z}$, $\mathbb{T}_2=3^{\mathbb{N}_0}$, $\mathbb{T}_{(1)}=\mathbb{T}_{(2)}=\mathbb{T}_{(3)}=\mathbb{R}$, $U=\mathbb{T}_1\times\mathbb{T}_2$ and S be an oriented surface with a local parameterization (U,f), where

$$f(t_1,t_2)=\left(\frac{1+t_1}{1+t_1^2+t_2^2},t_1-t_2^2+e_1(t_1,1),\frac{1+2t_1+3t_2+t_1^2}{1+t_1^2+t_2^2+t_1t_2}\right),$$

$(t_1,t_2)\in\mathbb{T}_1\times\mathbb{T}_2$.

Find

1. $N_{\sigma_1\sigma_2^2}(t_1,t_2)$,
2. $N_{\sigma_1^2\sigma_2}(t_1,t_2)$,

3. $N_{\sigma_1^2\sigma_2}(t_1,t_2)\cdot\left(2N_{\sigma_1^2\sigma_2}(t_1,t_2)-3N_{\sigma_1\sigma_2^2}(t_1,t_2)\right)$,
4. $N_{\sigma_1\sigma_2^2}(t_1,t_2)\times\left(4N_{\sigma_1^2\sigma_2}(t_1,t_2)-2N_{\sigma_1\sigma_2^2}(t_1,t_2)\right)$.

2.5 Orientations of a Surface

Definition 2.29 A σ_1-orientation or σ_2-orientation or $\sigma_1^2\sigma_2$-orientation or $\sigma_1\sigma_2^2$-orientation of a σ_1-regular surface or σ_2-regular surface or $\sigma_1^2\sigma_2$-regular surface or $\sigma_1\sigma_2^2$-regular surface is a choice of an orientation in each σ_1-tangent space $T_{a\sigma_1}S$ or σ_2-tangent space $T_{a\sigma_2}$ or $\sigma_1^2\sigma_2$-tangent space $T_{a\sigma_1^2\sigma_2}S$ or $\sigma_1\sigma_2^2$-tangent space $T_{a\sigma_1\sigma_2^2}S$, respectively. The surfaces on which it is possible to define a σ_1-orientation or σ_2-orientation or $\sigma_1^2\sigma_2$-orientation or $\sigma_1\sigma_2^2$-orientation are called σ_1-orientable or σ_2-orientable or $\sigma_1^2\sigma_2$-orientable or $\sigma_1\sigma_2^2$-orientable, respectively. The surface on which a σ_1-orientation or σ_2-orientation or $\sigma_1^2\sigma_2$-orientation or $\sigma_1\sigma_2^2$-orientation are already chosen and are called σ_1-oriented or σ_2-oriented or $\sigma_1^2\sigma_2$-oriented or $\sigma_1\sigma_2^2$-oriented, respectively.

If S is a σ_1-regular surface or σ_2-regular surface or $\sigma_1^2\sigma_2$-regular surface or $\sigma_1\sigma_2^2$-regular surface with the global σ_1-regular parameterization (U,f) or global σ_2-regular parameterization (U,f) or global $\sigma_1^2\sigma_2$-regular parameterization (U,f) or global $\sigma_1\sigma_2^2$-regular parameterization (U,f), respectively, then it can be σ_1-oriented or σ_2-oriented or $\sigma_1^2\sigma_2$-oriented or $\sigma_1\sigma_2^2$-oriented by using the N_{σ_1} or N_{σ_2} or $N_{\sigma_1^2\sigma_2}$ or $N_{\sigma_1\sigma_2^2}$, respectively.

Definition 2.30 If a σ_1-orientation or σ_2-orientation or $\sigma_1^2\sigma_2$-orientation or $\sigma_1\sigma_2^2$-orientation is given by the vector function g, then the vector function $-g$ also defines a σ_1-orientation or σ_2-orientation or $\sigma_1^2\sigma_2$-orientation or $\sigma_1\sigma_2^2$-orientation on S and it is called the opposite σ_1-orientation or opposite σ_2-orientation or opposite $\sigma_1^2\sigma_2$-orientation or opposite $\sigma_1\sigma_2^2$-orientation.

Not any surface is σ_1-orientable or σ_2-orientable or $\sigma_1^2\sigma_2$-orientable or $\sigma_1\sigma_2^2$-orientable. For instance the Möbius band is not orientable (see Figure 2.3)

Definition 2.31 Let S be a σ_1-oriented or σ_2-oriented or $\sigma_1^2\sigma_2$-oriented or $\sigma_1\sigma_2^2$-oriented surface with σ_1-orientation or σ_2-orientation or $\sigma_1^2\sigma_2$-orientation or $\sigma_1\sigma_2^2$-orientation g, respectively. A local parameterization (U,f) of the surface S is called compatible or σ_1-compatible or σ_2-compatible with the orientation or σ_1-orientation or σ_2-orientation g or $\sigma_1^2\sigma_2$-compatible or $\sigma_1\sigma_2^2$-compatible, respectively, if

$$g(a)=N_{\sigma_1}(a)\quad\text{or}\quad g(a)=N_{\sigma_2}\quad\text{or}\quad g(a)=N_{\sigma_1^2\sigma_2}\quad\text{or}\quad g(a)=N_{\sigma_1\sigma_2^2}(a),$$

respectively, at each point $a\in S$.

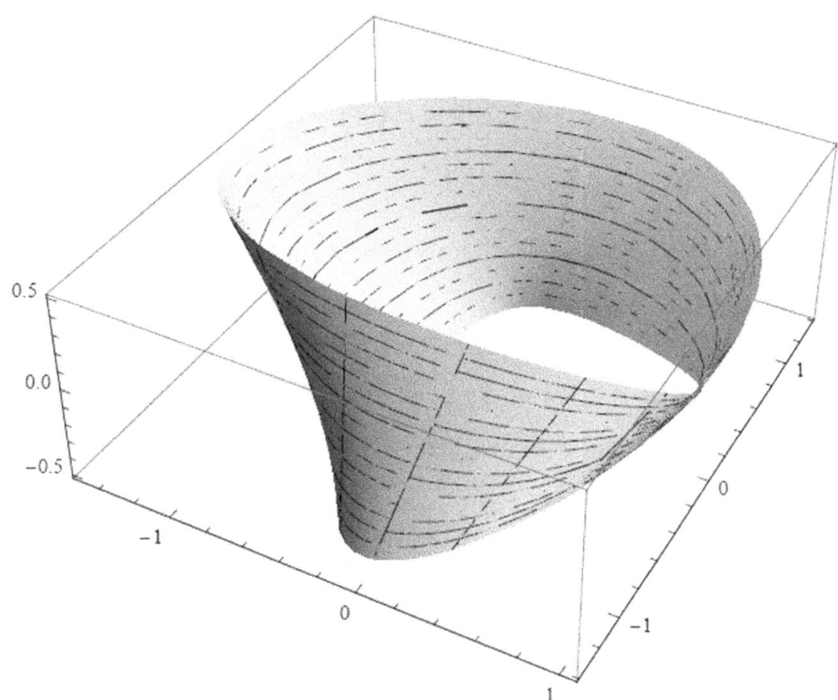

FIGURE 2.3
The Möbius Band.

2.6 Advanced Practical Problems

Problem 2.1 Let $\mathbb{T}_1 = \mathbb{N}_0^2$, $\mathbb{T}_2 = \mathbb{N}_0$, $\mathbb{T}_{(1)} = \mathbb{T}_{(2)} = \mathbb{T}_{(3)} = \mathbb{R}$. Prove that $f : \mathbb{T}_1 \times \mathbb{T}_2 \to \mathbb{R}^3$, given by

$$f(t_1,t_2) = \left(e_1(t_1,1) + t_2^2, t_1^2 + e_4(t_2,1), t_1 + t_2\right), \quad (t_1,t_2) \in \mathbb{T}_1 \times \mathbb{T}_2,$$

is a σ_1-regular parameterized surface.

Problem 2.2 Let $\mathbb{T}_1 = 3^{\mathbb{N}_0}$, $\mathbb{T}_2 = 7^{\mathbb{N}_0}$, S be a surface with a local parameterization (U,f), where $U = [1,81] \times [1,243]$ and

$$f(t_1,t_2) = \left((1-t_1)t_2^2, \frac{1-t_1}{1+t_2+t_2^2+t_2^3t_1^2}, (1-t_2)t_1^2\right), \quad (t_1,t_2) \in U.$$

Find the σ_1-tangent plane and the σ_2-tangent plane of S at the point $(0,0,0)$.

Problem 2.3 Let $\mathbb{T}_1 = \mathbb{Z}$, $\mathbb{T}_2 = 2^{\mathbb{N}_0}$, $\mathbb{T}_{(1)} = \mathbb{T}_{(2)} = \mathbb{T}_{(3)} = \mathbb{R}$, (U,f) be a local parameterization of a surface S, where $U \subset \mathbb{T}_1 \times \mathbb{T}_2$ and

$$f(t_1,t_2) = \left(e_1(t_1,1)(1+t_2), 1 + t_1^2 t_2, t_1^2 t_2^3\right), \quad (t_1,t_2) \in U.$$

Find the equations of the σ_1-normal and the equations of the σ_2-normal to S.

Problem 2.4 Let $\mathbb{T}_1 = 7\mathbb{Z}$, $\mathbb{T}_2 = \mathbb{Z}$, $\mathbb{T}_{(1)} = \mathbb{T}_{(2)} = \mathbb{T}_{(3)} = \mathbb{R}$, (U,f) be a local parameterization of a surface S, where $U \subset \mathbb{T}_1 \times \mathbb{T}_2$ and

$$f(t_1,t_2) = \left(\frac{1+t_1+t_1t_2}{1+t_1^3+t_2^3}, t_1 - t_2, t_1^4 - t_2^2\right), \quad (t_1,t_2) \in U.$$

Find the equations of the σ_1-normal and the equations of the σ_2-normal to S.

Problem 2.5 Let $\mathbb{T}_1 = 4^{\mathbb{N}_0}$, $\mathbb{T}_2 = \mathbb{Z}$, $\mathbb{T}_{(1)} = \mathbb{T}_{(2)} = \mathbb{T}_{(3)} = \mathbb{R}$, (U,f) be a local parameterization of a surface S, where $U \subset \mathbb{T}_1 \times \mathbb{T}_2$ and

$$f(t_1,t_2) = \left(\sin_1(t_1,1) + \cosh_2(t_2,1), 1 = (\cos_3(t_2,1))^2 t_1, 1 + t_1^3 t_2^2\right), \quad (t_1,t_2) \in U.$$

Find the equations of the σ_1-normal and the equations of the σ_2-normal to S.

Problem 2.6 Let $\mathbb{T}_1 = 2^{\mathbb{N}_0}$, $\mathbb{T}_2 = 3^{\mathbb{N}_0}$, $\mathbb{T}_{(1)} = \mathbb{T}_{(2)} = \mathbb{T}_{(3)} = \mathbb{R}$, $U = \mathbb{T}_1 \times \mathbb{T}_2$, S be an oriented surface with a local parameterization (U,f), where

$$f(t_1,t_2) = \left(\frac{t_1+t_2}{1+t_1^2+(t_1-t_2)^6}, t_1^2t_2^2 - t_1t_2, \frac{1+4t_2}{1+t_1^6}\right),$$

$(t_1,t_2) \in \mathbb{T}_1 \times \mathbb{T}_2$. Find

1. $N_{\sigma_1}(t_1,t_2)$,
2. $N_{\sigma_2}(t_1,t_2)$,
3. $N_{\sigma_1\sigma_2^2}(t_1,t_2)$,
4. $N_{\sigma_1^2\sigma_2}(t_1,t_2)$,
5. $4N_{\sigma_1^2\sigma_2}(t_1,t_2) \cdot \left(N_{\sigma_1}(t_1,t_2) + N_{\sigma_1^2\sigma_2}(t_1,t_2) - N_{\sigma_1\sigma_2^2}(t_1,t_2)\right)$,
6. $N_{\sigma_1\sigma_2^2}(t_1,t_2) \times \left(N_{\sigma_1}(t_1,t_2) - N_{\sigma_2}(t_1,t_2) - 3N_{\sigma_1^2\sigma_2}(t_1,t_2) - 4N_{\sigma_1\sigma_2^2}(t_1,t_2)\right)$,
7. $N_{\sigma_1}(t-1,t_2) \times \left(N_{\sigma_2}(t_1,t_2) + 2N_{\sigma_1\sigma_2^2}(t_1,t_2)\right)$.

2.7 Notes and References

In this chapter surfaces have been introduced, along with the local and global parameterization of surfaces. σ_1-tangent planes, σ_2-tangent planes, $\sigma_1^2\sigma_2$-tangent planes, $\sigma_1\sigma_2^2$-tangent planes, σ_1-normals and σ_2-normals, and $\sigma_1^2\sigma_2$-normals, $\sigma_1\sigma_2^2$-normals to a surface have been defined and their equations deduced. Some of the results in this chapter can be found in [2, 10].

3

First Fundamental Forms

Suppose that $k \in \mathbb{N}$, $\mathbb{T}$, $\mathbb{T}_1$, $\mathbb{T}_2$, $\mathbb{T}_{(1)}$, $\mathbb{T}_{(2)}$, $\mathbb{T}_{(3)}$, $\mathbb{T}^{(1)}, \dots, \mathbb{T}^{(k)}$ are time scales with forward jump operators and delta differentiation operators σ, σ_1, σ_2, $\sigma_{(1)}$, $\sigma_{(2)}$, $\sigma_{(3)}$, $\sigma^{(1)}, \dots, \sigma^{(k)}$ and Δ, Δ_1, Δ_2, $\Delta_{(1)}$, $\Delta_{(2)}$, $\Delta_{(3)}$, $\Delta^{(1)}, \dots, \Delta^{(k)}$, respectively. Let $I, J, A \subseteq \mathbb{T}$, $U \subseteq \mathbb{T}_1 \times \mathbb{T}_2$, $W \subseteq \mathbb{T}_{(1)} \times \mathbb{T}_{(2)} \times \mathbb{T}_{(3)}$, $V \subseteq \Lambda^{(k)} = \mathbb{T}^{(1)} \times, \dots, \times \mathbb{T}^{(k)}$.

3.1 Differentiable Maps on a Surface

Definition 3.1 Let S be a surface on $\mathbb{T}_{(1)} \times \mathbb{T}_{(2)} \times \mathbb{T}_{(3)}$. A map: $g : S \to V$ is called smooth if for any parameterization (U, f) of S the map $g \circ f : U \to V$ is smooth. The map $g_f = g \circ f$ is called the expression of g in the curvilinear coordinates (t_1, t_2) or the local representation of g with respect to the parameterization (U, f).

Example 3.1 Let $S \subset V$ be a surface. The inclusion $i : S \to V$, defined as

$$i(a) = a, \quad a \in S,$$

is a smooth map for any local parameterization (U, f) of S. The local representation of i is

$$i_f = i \circ f = f.$$

Example 3.2 Let $S \subset V$ be a surface and $B \in V$ Any map $g : S \to V$, defined as

$$g(a) = B, \quad a \in S,$$

is smooth because its local representation with respect to any local parameterization of S is a constant map and hence, it is a smooth map.

Let S be a surface with local parameterization (U, f) and $g : S \to V$ be a smooth map.

Assume

$$f_1(\mathbb{T}_1, \cdot) = f_1(\cdot, \mathbb{T}_2) = \mathbb{T}_{(1)},$$
$$f_2(\mathbb{T}_1, \cdot) = f_2(\cdot, \mathbb{T}_2) = \mathbb{T}_{(2)},$$
$$f_3(\mathbb{T}_1, \cdot) = f_3(\cdot, \mathbb{T}_2) = \mathbb{T}_{(3)}$$

DOI: 10.1201/9781003205265-3

and

$$\begin{aligned}
f_1(\sigma_1(t_1),t_2) &= f_1(t_1,\sigma_2(t_2)) = \sigma_{(1)}(f_1(t_1,t_2)),\\
f_2(\sigma_1(t_1),t_2) &= f_2(t_1,\sigma_2(t_2)) = \sigma_{(2)}(f_2(t_1,t_2)),\\
f_3(\sigma_1(t_1),t_2) &= f_3(t_1,\sigma_2(t_2)) = \sigma_{(3)}(f_3(t_1,t_2)), \quad (t_1,t_2) \in \mathbb{T}_1 \times \mathbb{T}_2.
\end{aligned}$$

We will deduce the representations of

$$g_{lt_j}^{\Delta_j}(f) = g_{lt_j}^{\Delta_j}(f_1,f_2,f_3), \quad j \in \{1,2,3\}, \quad l \in \{1,\dots,k\}.$$

We have the following cases.

1. Let g be $\sigma_{(1)}$-completely delta differentiable. Then, using the chain rule, we get

$$\begin{aligned}
g_{lt_j}^{\Delta_j}(f) &= g_{lf_1}^{\Delta_{(1)}}(f_1,f_2,f_3)f_{1t_j}^{\Delta_j} + g_{lf_2}^{\Delta_{(2)}\sigma_{(1)}}(f_1,f_2,f_3)f_{2t_j}^{\Delta_j}\\
&\quad + g_{lf_3}^{\Delta_{(3)}\sigma_{(1)}\sigma_{(2)}}(f_1,f_2,f_3)f_{3t_j}^{\Delta_j}, \quad j \in \{1,2\},
\end{aligned}$$

   or

$$\begin{aligned}
g_{lt_j}^{\Delta_j}(f) &= g_{lf_1}^{\Delta_{(1)}}(f_1,f_2,f_3)f_{1t_j}^{\Delta_j} + g_{lf_2}^{\Delta_{(2)}\sigma_{(1)}\sigma_{(3)}}(f_1,f_2,f_3)f_{2t_j}^{\Delta_j}\\
&\quad + g_{lf_3}^{\Delta_{(3)}\sigma_{(1)}}(f_1,f_2,f_3)f_{3t_j}^{\Delta_j}, \quad j \in \{1,2\}.
\end{aligned}$$

2. Let g be $\sigma_{(2)}$-completely delta differentiable. Then, using the chain rule, we get

$$\begin{aligned}
g_{lt_j}^{\Delta_j}(f) &= g_{lf_1}^{\Delta_{(1)}\sigma_{(2)}\sigma_{(3)}}(f_1,f_2,f_3)f_{1t_j}^{\Delta_j} + g_{lf_2}^{\Delta_{(2)}}(f_1,f_2,f_3)f_{2t_j}^{\Delta_j}\\
&\quad + g_{lf_3}^{\Delta_{(3)}\sigma_{(2)}}(f_1,f_2,f_3)f_{3t_j}^{\Delta_j}, \quad j \in \{1,2\},
\end{aligned}$$

   or

$$\begin{aligned}
g_{lt_j}^{\Delta_j}(f) &= g_{lf_1}^{\Delta_{(1)}\sigma_{(2)}}(f_1,f_2,f_3)f_{1t_j}^{\Delta_j} + g_{lf_2}^{\Delta_{(2)}}(f_1,f_2,f_3)f_{2t_j}^{\Delta_j}\\
&\quad + g_{lf_3}^{\Delta_{(3)}\sigma_{(1)}\sigma_{(2)}}(f_1,f_2,f_3)f_{3t_j}^{\Delta_j}, \quad j \in \{1,2\}.
\end{aligned}$$

3. Let g be $\sigma_{(3)}$-completely delta differentiable. Then, using the chain rule, we get

$$\begin{aligned}
g_{lt_j}^{\Delta_j}(f) &= g_{lf_1}^{\Delta_{(1)}\sigma_{(2)}\sigma_{(3)}}(f_1,f_2,f_3)f_{1t_j}^{\Delta_j} + g_{lf_2}^{\Delta_{(2)}\sigma_{(3)}}(f_1,f_2,f_3)f_{2t_j}^{\Delta_j}\\
&\quad + g_{lf_3}^{\Delta_{(3)}}(f_1,f_2,f_3)f_{3t_j}^{\Delta_j}, \quad j \in \{1,2\},
\end{aligned}$$

   or

$$\begin{aligned}
g_{lt_j}^{\Delta_j}(f) &= g_{lf_1}^{\Delta_{(1)}\sigma_{(3)}}(f_1,f_2,f_3)f_{1t_j}^{\Delta_j} + g_{lf_2}^{\Delta_{(2)}\sigma_{(1)}\sigma_{(3)}}(f_1,f_2,f_3)f_{2t_j}^{\Delta_j}\\
&\quad + g_{lf_3}^{\Delta_{(3)}}(f_1,f_2,f_3)f_{3t_j}^{\Delta_j}, \quad j \in \{1,2\}.
\end{aligned}$$

Definition 3.2 Let $S_1, S_2 \subset W$ be two surfaces. A map $F : S_1 \to S_2$ is called smooth if the map

$$F_1 = i \circ F : S_1 \to W$$

is smooth.

Definition 3.3 Let $S_1, S_2 \subset W$ be two surfaces. A map $F : S_1 \to S_2$ is called a diffeomorphism if the map

$$F_1 = i \circ F : S_1 \to W$$

is a diffeomorphism.

3.2 The Differential of a Smooth Map Between Surfaces

Let $g : W \to \mathbb{R}^3$ be a smooth map that is $\sigma_{(1)}$- or $\sigma_{(2)}$- or $\sigma_{(3)}$-completely delta differentiable on W. Let also,

$$g = (g_1, g_2, g_3)$$

and $a \in W$. Define the following matrices.

$$D^1_{12a}(g) = \left(d^1_{12amn}(g)\right) = \begin{pmatrix} g_{1t_{(1)}}^{\Delta_{(1)}}(a) & g_{1t_{(2)}}^{\Delta_{(2)}\sigma_{(1)}}(a) & g_{1t_{(3)}}^{\Delta_{(3)}\sigma_{(1)}\sigma_{(2)}}(a) \\ g_{2t_{(1)}}^{\Delta_{(1)}}(a) & g_{2t_{(2)}}^{\Delta_{(2)}\sigma_{(1)}}(a) & g_{2t_{(3)}}^{\Delta_{(3)}\sigma_{(1)}\sigma_{(2)}}(a) \\ g_{3t_{(1)}}^{\Delta_{(1)}}(a) & g_{3t_{(2)}}^{\Delta_{(2)}\sigma_{(1)}} & g_{3t_{(3)}}^{\Delta_{(3)}\sigma_{(1)}\sigma_{(2)}}(a) \end{pmatrix},$$

$$D^2_{12a}(g) = \left(d^2_{12amn}(g)\right) = \begin{pmatrix} g_{1t_{(1)}}^{\Delta_{(1)}}(a) & g_{1t_{(2)}}^{\Delta_{(2)}\sigma_{(1)}}(a) & g_{1t_{(3)}}^{\Delta_{(3)}\sigma_{(1)}\sigma_{(2)}}(a) \\ g_{2t_{(1)}}^{\Delta_{(1)}\sigma_{(2)}}(a) & g_{2t_{(2)}}^{\Delta_{(2)}}(a) & g_{2t_{(3)}}^{\Delta_{(3)}\sigma_{(1)}\sigma_{(2)}}(a) \\ g_{3t_{(1)}}^{\Delta_{(1)}}(a) & g_{3t_{(2)}}^{\Delta_{(2)}\sigma_{(1)}} & g_{3t_{(3)}}^{\Delta_{(3)}\sigma_{(1)}\sigma_{(2)}}(a) \end{pmatrix},$$

$$D^3_{12a}(g) = \left(d^3_{12amn}(g)\right) = \begin{pmatrix} g_{1t_{(1)}}^{\Delta_{(1)}}(a) & g_{1t_{(2)}}^{\Delta_{(2)}\sigma_{(1)}}(a) & g_{1t_{(3)}}^{\Delta_{(3)}\sigma_{(1)}\sigma_{(2)}}(a) \\ g_{2t_{(1)}}^{\Delta_{(1)}}(a) & g_{2t_{(2)}}^{\Delta_{(2)}\sigma_{(1)}}(a) & g_{2t_{(3)}}^{\Delta_{(3)}\sigma_{(1)}\sigma_{(2)}}(a) \\ g_{3t_{(1)}}^{\Delta_{(1)}\sigma_{(2)}}(a) & g_{3t_{(2)}}^{\Delta_{(2)}} & g_{3t_{(3)}}^{\Delta_{(3)}\sigma_{(1)}\sigma_{(2)}}(a) \end{pmatrix},$$

$$D^4_{12a}(g) = \left(d^4_{12amn}(g)\right) = \begin{pmatrix} g_{1t_{(1)}}^{\Delta_{(1)}}(a) & g_{1t_{(2)}}^{\Delta_{(2)}\sigma_{(1)}}(a) & g_{1t_{(3)}}^{\Delta_{(3)}\sigma_{(1)}\sigma_{(2)}}(a) \\ g_{2t_{(1)}}^{\Delta_{(1)}\sigma_{(2)}}(a) & g_{2t_{(2)}}^{\Delta_{(2)}}(a) & g_{2t_{(3)}}^{\Delta_{(3)}\sigma_{(1)}\sigma_{(2)}}(a) \\ g_{3t_{(1)}}^{\Delta_{(1)}\sigma_{(2)}}(a) & g_{3t_{(2)}}^{\Delta_{(2)}} & g_{3t_{(3)}}^{\Delta_{(3)}\sigma_{(1)}\sigma_{(2)}}(a) \end{pmatrix},$$

$$D^5_{12a}(g) = \left(d^5_{12amn}(g)\right) = \begin{pmatrix} g_{1t_{(1)}}^{\Delta_{(1)}\sigma_{(2)}}(a) & g_{1t_{(2)}}^{\Delta_{(2)}}(a) & g_{1t_{(3)}}^{\Delta_{(3)}\sigma_{(1)}\sigma_{(2)}}(a) \\ g_{2t_{(1)}}^{\Delta_{(1)}}(a) & g_{2t_{(2)}}^{\Delta_{(2)}\sigma_{(1)}}(a) & g_{2t_{(3)}}^{\Delta_{(3)}\sigma_{(1)}\sigma_{(2)}}(a) \\ g_{3t_{(1)}}^{\Delta_{(1)}}(a) & g_{3t_{(2)}}^{\Delta_{(2)}\sigma_{(1)}} & g_{3t_{(3)}}^{\Delta_{(3)}\sigma_{(1)}\sigma_{(2)}}(a) \end{pmatrix},$$

$$D^6_{12a}(g) = \left(d^6_{12amn}(g)\right) = \begin{pmatrix} g_{1t_{(1)}}^{\Delta_{(1)}\sigma_{(2)}}(a) & g_{1t_{(2)}}^{\Delta_{(2)}}(a) & g_{1t_{(3)}}^{\Delta_{(3)}\sigma_{(1)}\sigma_{(2)}}(a) \\ g_{2t_{(1)}}^{\Delta_{(1)}\sigma_{(2)}}(a) & g_{2t_{(2)}}^{\Delta_{(2)}}(a) & g_{2t_{(3)}}^{\Delta_{(3)}\sigma_{(1)}\sigma_{(2)}}(a) \\ g_{3t_{(1)}}^{\Delta_{(1)}}(a) & g_{3t_{(2)}}^{\Delta_{(2)}\sigma_{(1)}} & g_{3t_{(3)}}^{\Delta_{(3)}\sigma_{(1)}\sigma_{(2)}}(a) \end{pmatrix},$$

$$D^7_{12a}(g) = \left(d^7_{12amn}(g)\right) = \begin{pmatrix} g_{1t_{(1)}}^{\Delta_{(1)}\sigma_{(2)}}(a) & g_{1t_{(2)}}^{\Delta_{(2)}}(a) & g_{1t_{(3)}}^{\Delta_{(3)}\sigma_{(1)}\sigma_{(2)}}(a) \\ g_{2t_{(1)}}^{\Delta_{(1)}}(a) & g_{2t_{(2)}}^{\Delta_{(2)}\sigma_{(1)}}(a) & g_{2t_{(3)}}^{\Delta_{(3)}\sigma_{(1)}\sigma_{(2)}}(a) \\ g_{3t_{(1)}}^{\Delta_{(1)}\sigma_{(2)}}(a) & g_{3t_{(2)}}^{\Delta_{(2)}} & g_{3t_{(3)}}^{\Delta_{(3)}\sigma_{(1)}\sigma_{(2)}}(a) \end{pmatrix},$$

$$D^8_{12a}(g) = \left(d^8_{12amn}(g)\right) = \begin{pmatrix} g_{1t_{(1)}}^{\Delta_{(1)}\sigma_{(2)}}(a) & g_{1t_{(2)}}^{\Delta_{(2)}}(a) & g_{1t_{(3)}}^{\Delta_{(3)}\sigma_{(1)}\sigma_{(2)}}(a) \\ g_{2t_{(1)}}^{\Delta_{(1)}\sigma_{(2)}}(a) & g_{2t_{(2)}}^{\Delta_{(2)}}(a) & g_{2t_{(3)}}^{\Delta_{(3)}\sigma_{(1)}\sigma_{(2)}}(a) \\ g_{3t_{(1)}}^{\Delta_{(1)}\sigma_{(2)}}(a) & g_{3t_{(2)}}^{\Delta_{(2)}} & g_{3t_{(3)}}^{\Delta_{(3)}\sigma_{(1)}\sigma_{(2)}}(a) \end{pmatrix},$$

$$D^1_{13a}(g) = \left(d^1_{13amn}(g)\right) = \begin{pmatrix} g_{1t_{(1)}}^{\Delta_{(1)}}(a) & g_{1t_{(2)}}^{\Delta_{(2)}\sigma_{(1)}\sigma_{(3)}}(a) & g_{1t_{(3)}}^{\Delta_{(3)}\sigma_{(1)}}(a) \\ g_{2t_{(1)}}^{\Delta_{(1)}}(a) & g_{2t_{(2)}}^{\Delta_{(2)}\sigma_{(1)}\sigma_{(3)}}(a) & g_{2t_{(3)}}^{\Delta_{(3)}\sigma_{(1)}}(a) \\ g_{3t_{(1)}}^{\Delta_{(1)}}(a) & g_{3t_{(2)}}^{\Delta_{(2)}\sigma_{(1)}\sigma_{(3)}} & g_{3t_{(3)}}^{\Delta_{(3)}\sigma_{(1)}}(a) \end{pmatrix},$$

$$D^2_{13a}(g) = \left(d^2_{13amn}(g)\right) = \begin{pmatrix} g_{1t_{(1)}}^{\Delta_{(1)}}(a) & g_{1t_{(2)}}^{\Delta_{(2)}\sigma_{(1)}\sigma_{(3)}}(a) & g_{1t_{(3)}}^{\Delta_{(3)}\sigma_{(1)}}(a) \\ g_{2t_{(1)}}^{\Delta_{(1)}\sigma_{(3)}}(a) & g_{2t_{(2)}}^{\Delta_{(2)}\sigma_{(1)}\sigma_{(3)}}(a) & g_{2t_{(3)}}^{\Delta_{(3)}}(a) \\ g_{3t_{(1)}}^{\Delta_{(1)}}(a) & g_{3t_{(2)}}^{\Delta_{(2)}\sigma_{(1)}\sigma_{(3)}} & g_{3t_{(3)}}^{\Delta_{(3)}\sigma_{(1)}}(a) \end{pmatrix},$$

$$D^3_{13a}(g) = \left(d^3_{13amn}(g)\right) = \begin{pmatrix} g_{1t_{(1)}}^{\Delta_{(1)}}(a) & g_{1t_{(2)}}^{\Delta_{(2)}\sigma_{(1)}\sigma_{(3)}}(a) & g_{1t_{(3)}}^{\Delta_{(3)}\sigma_{(1)}}(a) \\ g_{2t_{(1)}}^{\Delta_{(1)}}(a) & g_{2t_{(2)}}^{\Delta_{(2)}\sigma_{(1)}\sigma_{(3)}}(a) & g_{2t_{(3)}}^{\Delta_{(3)}\sigma_{(1)}}(a) \\ g_{3t_{(1)}}^{\Delta_{(1)}\sigma_{(3)}}(a) & g_{3t_{(2)}}^{\Delta_{(2)}\sigma_{(1)}\sigma_{(3)}} & g_{3t_{(3)}}^{\Delta_{(3)}}(a) \end{pmatrix},$$

$$D^4_{13a}(g) = \left(d^4_{13amn}(g)\right) = \begin{pmatrix} g_{1t_{(1)}}^{\Delta_{(1)}}(a) & g_{1t_{(2)}}^{\Delta_{(2)}\sigma_{(1)}\sigma_{(3)}}(a) & g_{1t_{(3)}}^{\Delta_{(3)}\sigma_{(1)}}(a) \\ g_{2t_{(1)}}^{\Delta_{(1)}\sigma_{(3)}}(a) & g_{2t_{(2)}}^{\Delta_{(2)}\sigma_{(1)}\sigma_{(3)}}(a) & g_{2t_{(3)}}^{\Delta_{(3)}}(a) \\ g_{3t_{(1)}}^{\Delta_{(1)}\sigma_{(3)}}(a) & g_{3t_{(2)}}^{\Delta_{(2)}\sigma_{(1)}\sigma_{(3)}} & g_{3t_{(3)}}^{\Delta_{(3)}}(a) \end{pmatrix},$$

$$D^5_{13a}(g) = \left(d^5_{13amn}(g)\right) = \begin{pmatrix} g_{1t_{(1)}}^{\Delta_{(1)}\sigma_{(3)}}(a) & g_{1t_{(2)}}^{\Delta_{(2)}\sigma_{(1)}\sigma_{(3)}}(a) & g_{1t_{(3)}}^{\Delta_{(3)}}(a) \\ g_{2t_{(1)}}^{\Delta_{(1)}}(a) & g_{2t_{(2)}}^{\Delta_{(2)}\sigma_{(1)}\sigma_{(3)}}(a) & g_{2t_{(3)}}^{\Delta_{(3)}\sigma_{(1)}}(a) \\ g_{3t_{(1)}}^{\Delta_{(1)}}(a) & g_{3t_{(2)}}^{\Delta_{(2)}\sigma_{(1)}\sigma_{(3)}} & g_{3t_{(3)}}^{\Delta_{(3)}\sigma_{(1)}}(a) \end{pmatrix},$$

$$D^6_{13a}(g) = \left(d^6_{13amn}(g)\right) = \begin{pmatrix} g_{1t_{(1)}}^{\Delta_{(1)}\sigma_{(3)}}(a) & g_{1t_{(2)}}^{\Delta_{(2)}\sigma_{(1)}\sigma_{(3)}}(a) & g_{1t_{(3)}}^{\Delta_{(3)}}(a) \\ g_{2t_{(1)}}^{\Delta_{(1)}\sigma_{(3)}}(a) & g_{2t_{(2)}}^{\Delta_{(2)}\sigma_{(1)}\sigma_{(3)}}(a) & g_{2t_{(3)}}^{\Delta_{(3)}}(a) \\ g_{3t_{(1)}}^{\Delta_{(1)}}(a) & g_{3t_{(2)}}^{\Delta_{(2)}\sigma_{(1)}\sigma_{(3)}} & g_{3t_{(3)}}^{\Delta_{(3)}\sigma_{(1)}}(a) \end{pmatrix},$$

$$D^7_{13a}(g) = \left(d^7_{13amn}(g)\right) = \begin{pmatrix} g_{1t_{(1)}}^{\Delta_{(1)}\sigma_{(3)}}(a) & g_{1t_{(2)}}^{\Delta_{(2)}\sigma_{(1)}\sigma_{(3)}}(a) & g_{1t_{(3)}}^{\Delta_{(3)}}(a) \\ g_{2t_{(1)}}^{\Delta_{(1)}}(a) & g_{2t_{(2)}}^{\Delta_{(2)}\sigma_{(1)}\sigma_{(3)}}(a) & g_{2t_{(3)}}^{\Delta_{(3)}\sigma_{(1)}}(a) \\ g_{3t_{(1)}}^{\Delta_{(1)}\sigma_{(3)}}(a) & g_{3t_{(2)}}^{\Delta_{(2)}\sigma_{(1)}\sigma_{(3)}} & g_{3t_{(3)}}^{\Delta_{(3)}}(a) \end{pmatrix},$$

$$D_{13a}^{8}(g)=\left(d_{13amn}^{8}(g)\right)=\begin{pmatrix} g_{1t_{(1)}}^{\Delta_{(1)}\sigma_{(3)}}(a) & g_{1t_{(2)}}^{\Delta_{(2)}\sigma_{(1)}\sigma_{(3)}}(a) & g_{1t_{(3)}}^{\Delta_{(3)}}(a) \\ g_{2t_{(1)}}^{\Delta_{(1)}\sigma_{(3)}}(a) & g_{2t_{(2)}}^{\Delta_{(2)}\sigma_{(1)}\sigma_{(3)}}(a) & g_{2t_{(3)}}^{\Delta_{(3)}}(a) \\ g_{3t_{(1)}}^{\Delta_{(1)}\sigma_{(3)}}(a) & g_{3t_{(2)}}^{\Delta_{(2)}\sigma_{(1)}\sigma_{(3)}} & g_{3t_{(3)}}^{\Delta_{(3)}}(a) \end{pmatrix},$$

$$D_{23a}^{1}(g)=\left(d_{23amn}^{1}(g)\right)=\begin{pmatrix} g_{1t_{(1)}}^{\Delta_{(1)}\sigma_{(2)}\sigma_{(3)}}(a) & g_{1t_{(2)}}^{\Delta_{(2)}}(a) & g_{1t_{(3)}}^{\Delta_{(3)}\sigma_{(2)}}(a) \\ g_{2t_{(1)}}^{\Delta_{(1)}\sigma_{(2)}\sigma_{(3)}}(a) & g_{2t_{(2)}}^{\Delta_{(2)}}(a) & g_{2t_{(3)}}^{\Delta_{(3)}\sigma_{(2)}}(a) \\ g_{3t_{(1)}}^{\Delta_{(1)}\sigma_{(2)}\sigma_{(3)}}(a) & g_{3t_{(2)}}^{\Delta_{(2)}} & g_{3t_{(3)}}^{\Delta_{(3)}\sigma_{(2)}}(a) \end{pmatrix},$$

$$D_{23a}^{2}(g)=\left(d_{23amn}^{2}(g)\right)=\begin{pmatrix} g_{1t_{(1)}}^{\Delta_{(1)}\sigma_{(2)}\sigma_{(3)}}(a) & g_{1t_{(2)}}^{\Delta_{(2)}}(a) & g_{1t_{(3)}}^{\Delta_{(3)}\sigma_{(2)}}(a) \\ g_{2t_{(1)}}^{\Delta_{(1)}\sigma_{(2)}\sigma_{(3)}}(a) & g_{2t_{(2)}}^{\Delta_{(2)}\sigma_{(3)}}(a) & g_{2t_{(3)}}^{\Delta_{(3)}}(a) \\ g_{3t_{(1)}}^{\Delta_{(1)}\sigma_{(2)}\sigma_{(3)}}(a) & g_{3t_{(2)}}^{\Delta_{(2)}} & g_{3t_{(3)}}^{\Delta_{(3)}\sigma_{(2)}}(a) \end{pmatrix},$$

$$D_{23a}^{3}(g)=\left(d_{23amn}^{3}(g)\right)=\begin{pmatrix} g_{1t_{(1)}}^{\Delta_{(1)}\sigma_{(2)}\sigma_{(3)}}(a) & g_{1t_{(2)}}^{\Delta_{(2)}}(a) & g_{1t_{(3)}}^{\Delta_{(3)}\sigma_{(2)}}(a) \\ g_{2t_{(1)}}^{\Delta_{(1)}\sigma_{(2)}\sigma_{(3)}}(a) & g_{2t_{(2)}}^{\Delta_{(2)}}(a) & g_{2t_{(3)}}^{\Delta_{(3)}\sigma_{(2)}}(a) \\ g_{3t_{(1)}}^{\Delta_{(1)}\sigma_{(2)}\sigma_{(3)}}(a) & g_{3t_{(2)}}^{\Delta_{(2)}\sigma_{(3)}} & g_{3t_{(3)}}^{\Delta_{(3)}}(a) \end{pmatrix},$$

$$D_{23a}^{4}(g)=\left(d_{23amn}^{4}(g)\right)=\begin{pmatrix} g_{1t_{(1)}}^{\Delta_{(1)}\sigma_{(2)}\sigma_{(3)}}(a) & g_{1t_{(2)}}^{\Delta_{(2)}}(a) & g_{1t_{(3)}}^{\Delta_{(3)}\sigma_{(2)}}(a) \\ g_{2t_{(1)}}^{\Delta_{(1)}\sigma_{(2)}\sigma_{(3)}}(a) & g_{2t_{(2)}}^{\Delta_{(2)}\sigma_{(3)}}(a) & g_{2t_{(3)}}^{\Delta_{(3)}}(a) \\ g_{3t_{(1)}}^{\Delta_{(1)}\sigma_{(2)}\sigma_{(3)}}(a) & g_{3t_{(2)}}^{\Delta_{(2)}\sigma_{(3)}} & g_{3t_{(3)}}^{\Delta_{(3)}}(a) \end{pmatrix},$$

$$D_{23a}^{5}(g)=\left(d_{23amn}^{5}(g)\right)=\begin{pmatrix} g_{1t_{(1)}}^{\Delta_{(1)}\sigma_{(2)}\sigma_{(3)}}(a) & g_{1t_{(2)}}^{\Delta_{(2)}\sigma_{(3)}}(a) & g_{1t_{(3)}}^{\Delta_{(3)}}(a) \\ g_{2t_{(1)}}^{\Delta_{(1)}\sigma_{(2)}\sigma_{(3)}}(a) & g_{2t_{(2)}}^{\Delta_{(2)}}(a) & g_{2t_{(3)}}^{\Delta_{(3)}\sigma_{(2)}}(a) \\ g_{3t_{(1)}}^{\Delta_{(1)}\sigma_{(2)}\sigma_{(3)}}(a) & g_{3t_{(2)}}^{\Delta_{(2)}} & g_{3t_{(3)}}^{\Delta_{(3)}\sigma_{(2)}}(a) \end{pmatrix},$$

$$D_{23a}^{6}(g)=\left(d_{23amn}^{6}(g)\right)=\begin{pmatrix} g_{1t_{(1)}}^{\Delta_{(1)}\sigma_{(2)}\sigma_{(3)}}(a) & g_{1t_{(2)}}^{\Delta_{(2)}\sigma_{(3)}}(a) & g_{1t_{(3)}}^{\Delta_{(3)}}(a) \\ g_{2t_{(1)}}^{\Delta_{(1)}\sigma_{(2)}\sigma_{(3)}}(a) & g_{2t_{(2)}}^{\Delta_{(2)}\sigma_{(3)}}(a) & g_{2t_{(3)}}^{\Delta_{(3)}}(a) \\ g_{3t_{(1)}}^{\Delta_{(1)}\sigma_{(2)}\sigma_{(3)}}(a) & g_{3t_{(2)}}^{\Delta_{(2)}} & g_{3t_{(3)}}^{\Delta_{(3)}\sigma_{(2)}}(a) \end{pmatrix},$$

$$D_{23a}^{7}(g)=\left(d_{23amn}^{7}(g)\right)=\begin{pmatrix} g_{1t_{(1)}}^{\Delta_{(1)}\sigma_{(2)}\sigma_{(3)}}(a) & g_{1t_{(2)}}^{\Delta_{(2)}\sigma_{(3)}}(a) & g_{1t_{(3)}}^{\Delta_{(3)}}(a) \\ g_{2t_{(1)}}^{\Delta_{(1)}\sigma_{(2)}\sigma_{(3)}}(a) & g_{2t_{(2)}}^{\Delta_{(2)}}(a) & g_{2t_{(3)}}^{\Delta_{(3)}\sigma_{(2)}}(a) \\ g_{3t_{(1)}}^{\Delta_{(1)}\sigma_{(2)}\sigma_{(3)}}(a) & g_{3t_{(2)}}^{\Delta_{(2)}\sigma_{(3)}} & g_{3t_{(3)}}^{\Delta_{(3)}}(a) \end{pmatrix},$$

$$D_{23a}^{8}(g)=\left(d_{23amn}^{8}(g)\right)=\begin{pmatrix} g_{1t_{(1)}}^{\Delta_{(1)}\sigma_{(2)}\sigma_{(3)}}(a) & g_{1t_{(2)}}^{\Delta_{(2)}\sigma_{(3)}}(a) & g_{1t_{(3)}}^{\Delta_{(3)}}(a) \\ g_{2t_{(1)}}^{\Delta_{(1)}\sigma_{(2)}\sigma_{(3)}}(a) & g_{2t_{(2)}}^{\Delta_{(2)}\sigma_{(3)}}(a) & g_{2t_{(3)}}^{\Delta_{(3)}}(a) \\ g_{3t_{(1)}}^{\Delta_{(1)}\sigma_{(2)}\sigma_{(3)}}(a) & g_{3t_{(2)}}^{\Delta_{(2)}\sigma_{(3)}} & g_{3t_{(3)}}^{\Delta_{(3)}}(a) \end{pmatrix}.$$

Definition 3.4 1. If g is $\sigma_{(1)}$-completely delta differentiable, then the $\sigma_{(1)}$-differential of j-th kind of g at a

$$d_{\sigma_{(1)},j,a}g:\left(\mathbb{T}_{(1)}\times\mathbb{T}_{(2)}\times\mathbb{T}_{(3)}\right)_a\to\left(\mathbb{T}_{(1)}\times\mathbb{T}_{(2)}\times\mathbb{T}_{(3)}\right)_{g(a)}$$

is a linear map whose matrix is the matrix $D_{12a}^{j}(g)$.

2. If g is $\sigma_{(2)}$-completely delta differentiable, then the $\sigma_{(2)}$-differential of j-th kind of g at a

$$d_{\sigma_{(2)},j,a}g : \left(\mathbb{T}_{(1)} \times \mathbb{T}_{(2)} \times \mathbb{T}_{(3)}\right)_a \to \left(\mathbb{T}_{(1)} \times \mathbb{T}_{(2)} \times \mathbb{T}_{(3)}\right)_{g(a)}$$

is a linear map whose matrix is the matrix $D^j_{23a}(g)$.

3. If g is $\sigma_{(3)}$-completely delta differentiable, then the $\sigma_{(3)}$-differential of j-th kind of g at a

$$d_{\sigma_{(3)},j,a}g : \left(\mathbb{T}_{(1)} \times \mathbb{T}_{(2)} \times \mathbb{T}_{(3)}\right)_a \to \left(\mathbb{T}_{(1)} \times \mathbb{T}_{(2)} \times \mathbb{T}_{(3)}\right)_{g(a)}$$

is a linear map whose matrix is the matrix $D^j_{13a}(g)$.

Example 3.3 Let $\mathbb{T}_{(1)} = \mathbb{Z}$, $\mathbb{T}_{(2)} = 2\mathbb{Z}$, $\mathbb{T}_{(3)} = 3\mathbb{Z}$, $g : \mathbb{T}_{(1)} \times \mathbb{T}_{(2)} \times \mathbb{T}_{(3)} \to \mathbb{R}^3$ be given by

$$g(t_{(1)}, t_{(2)}, t_{(3)}) = \left(t_{(1)}t_{(2)}t_{(3)}, t_{(1)}^2 + t_{(2)}^2 + t_{(3)}^2, t_{(1)}^3 - t_{(2)} - 3t_{(3)}^2\right),$$

$(t_{(1)}, t_{(2)}, t_{(3)}) \in \mathbb{T}_{(1)} \times \mathbb{T}_{(2)} \times \mathbb{T}_{(3)}$. Let $a \in \mathbb{T}_{(1)} \times \mathbb{T}_{(2)} \times \mathbb{T}_{(3)}$ be arbitrarily chosen. We will find $D^2_{13a}(g)$ and $D^3_{23a}(g)$.

Here

$$\begin{aligned}
\sigma_{(1)}(t_{(1)}) &= t_{(1)} + 1, \quad t_{(1)} \in \mathbb{T}_{(1)},\\
\sigma_{(2)}(t_{(2)}) &= t_{(2)} + 2, \quad t_{(2)} \in \mathbb{T}_{(2)},\\
\sigma_{(3)}(t_{(3)}) &= t_{(3)} + 3, \quad t_{(3)} \in \mathbb{T}_{(3)},\\
g_1(t_{(1)}, t_{(2)}, t_{(3)}) &= t_{(1)}t_{(2)}t_{(3)},\\
g_2(t_{(1)}, t_{(2)}, t_{(3)}) &= t_{(1)}^2 + t_{(2)}^2 + t_{(3)}^2,\\
g_3(t_{(1)}, t_{(2)}, t_{(3)}) &= t_{(1)}^3 - t_{(2)} - 3t_{(3)}^2, \quad (t_{(1)}, t_{(2)}, t_{(3)}) \in \mathbb{T}_{(1)} \times \mathbb{T}_{(2)} \times \mathbb{T}_{(3)}.
\end{aligned}$$

We have

$$\begin{aligned}
g_{1t_{(1)}}^{\Delta_{(1)}}(t_{(1)}, t_{(2)}, t_{(3)}) &= t_{(2)}t_{(3)},\\
g_{1t_{(2)}}^{\Delta_{(2)}}(t_{(1)}, t_{(2)}, t_{(3)}) &= t_{(1)}t_{(3)},\\
g_{1t_{(3)}}^{\Delta_{(3)}}(t_{(1)}, t_{(2)}, t_{(3)}) &= t_{(1)}t_{(2)},\\
g_{2t_{(1)}}^{\Delta_{(1)}}(t_{(1)}, t_{(2)}, t_{(3)}) &= \sigma_{(1)}(t_{(1)}) + t_{(1)}\\
&= t_{(1)} + 1 + t_{(1)}\\
&= 2t_{(1)} + 1,\\
g_{2t_{(2)}}^{\Delta_{(2)}}(t_{(1)}, t_{(2)}, t_{(3)}) &= \sigma_{(2)}(t_{(2)}) + t_{(2)}\\
&= t_{(2)} + 2 + t_{(2)}\\
&= 2t_{(2)} + 2,
\end{aligned}$$

$$\begin{aligned}
g_{2t_{(3)}}^{\Delta_{(3)}}(t_{(1)},t_{(2)},t_{(3)}) &= \sigma_{(3)}(t_{(3)}) + t_{(3)}\\
&= t_{(3)} + 3 + t_{(3)}\\
&= 2t_{(3)} + 3,\\
g_{3t_{(1)}}^{\Delta_{(1)}}(t_{(1)},t_{(2)},t_{(3)}) &= (\sigma_{(1)}(t_{(1)}))^2 + t_{(1)}\sigma_{(1)}(t_{(1)}) + t_{(1)}^2\\
&= (t_{(1)}+1)^2 + t_{(1)}(t_{(1)}+1) + t_{(1)}^2\\
&= t_{(1)}^2 + 2t_{(1)} + 1 + t_{(1)}^2 + t_{(1)} + t_{(1)}^2\\
&= 3t_{(1)}^2 + 3t_{(1)} + 1,\\
g_{3t_{(2)}}^{\Delta_{(2)}}(t_{(1)},t_{(2)},t_{(3)}) &= -1,\\
g_{3t_{(3)}}^{\Delta_{(3)}}(t_{(1)},t_{(2)},t_{(3)}) &= -3\left((\sigma_{(3)}(t_{(3)}))^2 + t_{(3)}\sigma_{(3)}(t_{(3)}) + t_{(3)}^2\right)\\
&= -3\left((t_{(3)}+3)^2 + t_{(3)}(t_{(3)}+3) + t_{(3)}^2\right)\\
&= -3\left(t_{(3)}^2 + 6t_{(3)} + 9 + t_{(3)}^2 + 3t_{(3)} + t_{(3)}^2\right)\\
&= -3\left(3t_{(3)}^2 + 9t_{(3)} + 9\right)\\
&= -9t_{(3)}^2 - 27t_{(3)} - 27,\\
g_{1t_{(2)}}^{\Delta_{(2)}\sigma_{(1)}\sigma_{(3)}}(t_{(1)},t_{(2)},t_{(3)}) &= \sigma_{(1)}(t_{(1)})\sigma_{(3)}(t_{(3)})\\
&= (t_{(1)}+1)(t_{(3)}+3),\\
g_{1t_{(3)}}^{\Delta_{(3)}\sigma_{(1)}}(t_{(1)},t_{(2)},t_{(3)}) &= \sigma_{(1)}(t_{(1)})t_{(2)}\\
&= (t_{(1)}+1)t_{(2)},\\
g_{2t_{(1)}}^{\Delta_{(1)}\sigma_{(3)}}(t_{(1)},t_{(2)},t_{(3)}) &= 2t_{(1)} + 1,\\
g_{2t_{(2)}}^{\Delta_{(2)}\sigma_{(1)}\sigma_{(3)}} &= 2t_{(2)} + 2,\\
g_{3t_{(2)}}^{\Delta_{(2)}\sigma_{(1)}\sigma_{(3)}}(t_{(1)},t_{(2)},t_{(3)}) &= -1,\\
g_{3t_{(3)}}^{\Delta_{(3)}\sigma_{(1)}}(t_{(1)},t_{(2)},t_{(3)}) &= -9t_{(3)}^2 - 27t_{(3)} - 27,\\
g_{1t_{(1)}}^{\Delta_{(1)}\sigma_{(2)}\sigma_{(3)}}(t_{(1)},t_{(2)},t_{(3)}) &= \sigma_{(2)}(t_{(2)})\sigma_{(1)}(t_{(1)})\\
&= (t_{(2)}+2)(t_{(3)}+3),\\
g_{1t_{(3)}}^{\Delta_{(3)}\sigma_{(2)}}(t_{(1)},t_{(2)},t_{(3)}) &= t_{(1)}\sigma_{(2)}(t_{(2)})\\
&= t_{(1)}(t_{(2)}+2),\\
g_{2t_{(1)}}^{\Delta_{(1)}\sigma_{(2)}\sigma_{(3)}}(t_{(1)},t_{(2)},t_{(3)}) &= 2t_{(1)} + 1,\\
g_{2t_{(3)}}^{\Delta_{(3)}\sigma_{(2)}}(t_{(1)},t_{(2)},t_{(3)}) &= 2t_{(3)} + 3,
\end{aligned}$$

$$g_{3t_{(1)}}^{\Delta_{(1)}\sigma_{(2)}\sigma_{(3)}}(t_{(1)},t_{(2)},t_{(3)}) = 3t_{(1)}^2 + 3t_{(1)} + 1,$$
$$g_{3t_{(2)}}^{\Delta_{(2)}\sigma_{(3)}}(t_{(1)},t_{(2)},t_{(3)}) = -1, \quad (t_{(1)},t_{(2)},t_{(3)}) \in \mathbb{T}_{(1)} \times \mathbb{T}_{(2)} \times \mathbb{T}_{(3)}.$$

Therefore

$$D_{13a}^2(g) = \begin{pmatrix} t_{(2)}t_{(3)} & (t_{(1)}+1)(t_{(3)}+3) & (t_{(1)}+1)t_{(2)} \\ 2t_{(1)}+1 & 2t_{(2)}+2 & 2t_{(3)}+3 \\ 3t_{(1)}^2+3t_{(1)}+1 & -1 & -9t_{(3)}^2-27t_{(3)}-27 \end{pmatrix}$$

and

$$D_{23a}^3(g) = \begin{pmatrix} (t_{(2)}+2)(t_{(3)}+3) & t_{(1)}t_{(3)} & t_{(1)}(t_{(2)}+2) \\ 2t_{(1)}+1 & 2t_{(2)}+2 & 2t_{(3)}+3 \\ 3t_{(1)}^2+3t_{(1)}+1 & -1 & -9t_{(3)}^2-27t_{(3)}-27 \end{pmatrix}.$$

Exercise 3.1 Let $\mathbb{T}_{(1)} = 2\mathbb{Z}$, $\mathbb{T}_{(2)} = 2^{\mathbb{N}_0}$, $\mathbb{T}_{(3)} = \mathbb{N}_0$ and $g : \mathbb{T}_{(1)} \times \mathbb{T}_{(2)} \times \mathbb{T}_{(3)} \to \mathbb{R}^3$ be given by

$$g(t_{(1)},t_{(2)},t_{(3)}) = \left(t_{(1)}^3 - t_{(2)}^3 - t_{(3)}, 1 + \frac{1}{1+t_{(2)}^2} + t_{(3)}^3, t_{(1)}^2 t_{(2)} t_{(3)}^3 \right),$$

$(t_{(1)},t_{(2)},t_{(3)}) \in \mathbb{T}_{(1)} \times \mathbb{T}_{(2)} \times \mathbb{T}_{(3)}$. Find $D_{13a}^2(g)$, $D_{13a}^8(g)$, $D_{23a}^1(g)$.

Let

$$f(t) = (f_1(t), f_2(t), f_3(t)), \quad t \in A,$$

be a parameterized curve such that it is Δ-differentiable on A and

$$\begin{array}{rcl} f_1(\sigma(t)) & = & \sigma_{(1)}(f_1(t)), \\ f_2(\sigma(t)) & = & \sigma_{(2)}(f_2(t)), \\ f_3(\sigma(t)) & = & \sigma_{(3)}(f_3(t)), \quad t \in A, \\ f : A & \to & W. \end{array} \tag{3.1}$$

If g is $\sigma_{(1)}$- or $\sigma_{(2)}$- or $\sigma_{(3)}$-completely delta differentiable, we have

$$g(f(t)) = \Big(g_1(f_1(t), f_2(t), f_3(t)), g_2(f_1(t), f_2(t), f_3(t)),$$
$$g_3(f_1(t), f_2(t), f_3(t)) \Big), \quad t \in A,$$

and

$$(g(f(\cdot)))^{\Delta}(t) = \Bigg(\sum_{j=1}^{3} d_{kla1j}^p(g(f(t))) f_j^{\Delta}(t), \sum_{j=1}^{3} d_{kla2j}^p(g(f(t))) f_j^{\Delta}(t),$$
$$\sum_{j=1}^{3} d_{kla3j}^p(g(f(t))) f_j^{\Delta}(t) \Bigg), \quad t \in A,$$

for some $k \in \{1,2,3\}$, $p \in \{1,\dots,8\}$. Thus, $d_{\sigma_{(k)},j,a}g$, $k \in \{1,2,3\}$, $j \in \{1,\dots,8\}$, assigns to a tangent vector to the path $f(t)$ at a point $t \in A$ the tangent vector to the path $g(f(t))$ at t.

Definition 3.5 Let $S_1, S_2 \subseteq W$ be two surfaces and $a \in S_1$ and $g : S_1 \to S_2$ be a smooth map, that is $\sigma_{(k)}$-completely delta differentiable, $k \in \{1,2,3\}$, and $f : \mathbb{T} \to S_1$ is a parameterized curve that is Δ-differentiable and satisfies (3.1), and $f(t_0) = a$.

The map

$$d_{\sigma(k),j,a} f^{\Delta}(t) = \left(\sum_{j=1}^{3} d^{p}_{kl1j}(g(f(t))) f_j^{\Delta}(t), \sum_{j=1}^{3} d^{p}_{kl2j}(g(f(t))) f_j^{\Delta}(t), \right.$$

$$\left. \sum_{j=1}^{3} d^{p}_{kl3j}(g(f(t))) f_j^{\Delta}(t) \right), \quad t \in A, \quad k \in \{1,2,3\}, \quad j \in \{1,2,\ldots,8\},$$

is called $\sigma_{(k)}$-differential of j-th kind of the smooth map $g : S_1 \to S_2$ at the point a.

3.3 Spherical Maps. Shape Operators

Let $S \subseteq W$ be an oriented surface and S^2 be the unit sphere centered at the origin. Suppose that the σ_1-orientation is given by the σ_1-normal $N_{\sigma_1}(a)$, the σ_2-orientation is given by $N_{\sigma_2}(a)$, the $\sigma_1^2\sigma_2$-orientation is given by the $\sigma_1^2\sigma_2$-normal $N_{\sigma_1^2\sigma_2}(a)$, the $\sigma_1\sigma_2^2$-orientation is given by the $\sigma_1\sigma_2^2$-normal $N_{\sigma_1\sigma_2^2}(a)$. Let (U,f) be a local parameterization of the surface S.

Definition 3.6 Suppose that f is σ_1-completely delta differentiable and σ_1 is Δ_1-differentiable. The map $\Gamma_{\sigma_1} : S \to S^2$ that assigns to each point $a \in S$ the point on S^2 that has as a radius vector $\Gamma_{\sigma_1}(a) = N_{\sigma_1}(a)$, is called the σ_1-spherical map of the surface S.

By the definition of the σ_1-normal, we get

$$(\Gamma_{\sigma_1} \circ f)(t_1,t_2) = N_{\sigma_1}(t_1,t_2) = \frac{f_{t_1}^{\Delta_1}(t_1,t_2) \times f_{t_2}^{\Delta_2\sigma_1}(t_1,t_2)}{\|f_{t_1}^{\Delta_1}(t_1,t_2) \times f_{t_2}^{\Delta_2\sigma_1}(t_1,t_2)\|}.$$

Definition 3.7 Suppose that f is σ_2-completely delta differentiable and σ_2 is Δ_2-differentiable. The map $\Gamma_{\sigma_2} : S \to S^2$ that assigns to each point $a \in S$ the point on S^2 that has as a radius vector $\Gamma_{\sigma_2}(a) = N_{\sigma_2}(a)$, is called the σ_2-spherical map of the surface S.

By the definition of the σ_2-normal, we get

$$(\Gamma_{\sigma_2} \circ f)(t_1,t_2) = N_{\sigma_2}(t_1,t_2) = \frac{f_{t_1}^{\Delta_1\sigma_2}(t_1,t_2) \times f_{t_2}^{\Delta_2}(t_1,t_2)}{\|f_{t_1}^{\Delta_1\sigma_2}(t_1,t_2) \times f_{t_2}^{\Delta_2}(t_1,t_2)\|}.$$

Definition 3.8 Suppose that f is σ_1-completely delta differentiable and σ_1 is Δ_1-differentiable. The map $\Gamma_{\sigma_1^2\sigma_2} : S \to S^2$ that assigns to each point $a \in S$ the point on S^2 that has as a radius vector $\Gamma_{\sigma_1^2\sigma_2}(a) = N_{\sigma_1^2\sigma_2}(a)$, is called the $\sigma_1^2\sigma_2$-spherical map of the surface S.

By the definition of the $\sigma_1^2\sigma_2$-normal, we get

$$(\Gamma_{\sigma_1^2\sigma_2} \circ f)(t_1,t_2) = N_{\sigma_1^2\sigma_2}(t_1,t_2) = \frac{f_{t_1}^{\Delta_1\sigma_1\sigma_2}(t_1,t_2) \times f_{t_2}^{\Delta_2\sigma_1^2\sigma_2}(t_1,t_2)}{\|f_{t_1}^{\Delta_1\sigma_1\sigma_2}(t_1,t_2) \times f_{t_2}^{\Delta_2\sigma_1^2\sigma_2}(t_1,t_2)\|}.$$

Definition 3.9 Suppose that f is σ_2-completely delta differentiable and σ_2 is Δ_2-differentiable. The map $\Gamma_{\sigma_1\sigma_2^2} : S \to S^2$ that assigns to each point $a \in S$ the point on S^2 that has as a radius vector $\Gamma_{\sigma_1\sigma_2^2}(a) = N_{\sigma_1\sigma_2^2}(a)$, is called the $\sigma_1\sigma_2^2$-spherical map of the surface S.

By the definition of the $\sigma_1\sigma_2^2$-normal, we get

$$(\Gamma_{\sigma_1\sigma_2^2} \circ f)(t_1,t_2) = N_{\sigma_1\sigma_2^2}(t_1,t_2) = \frac{f_{t_1}^{\Delta_1\sigma_1\sigma_2^2}(t_1,t_2) \times f_{t_2}^{\Delta_2\sigma_1\sigma_2}(t_1,t_2)}{\|f_{t_1}^{\Delta_1\sigma_1\sigma_2^2}(t_1,t_2) \times f_{t_2}^{\Delta_2\sigma_1\sigma_2}(t_1,t_2)\|}.$$

Definition 3.10 Suppose that σ_1 is Δ_1-differentiable. For each vector $h = (h_1,h_2) \in T_{a\sigma_1}S$, the operator

$$\mathscr{A}_{a\sigma_1}(h) = N_{\sigma_1 t_1}^{\Delta_1} h_1 + N_{\sigma_1 t_2}^{\Delta_2\sigma_1} h_2$$

will be called σ_1-shape operator.

In particular, we have

$$\mathscr{A}_{a\sigma_1}\left(f_{t_1}^{\Delta_1}\right) = N_{t_1}^{\Delta_1},$$
$$\mathscr{A}_{a\sigma_1}\left(f_{t_2}^{\Delta_2\sigma_1}\right) = N_{t_2}^{\Delta_2\sigma_1}.$$

Theorem 3.1 *Suppose that σ_1 is Δ_1-differentiable. Then*

$$f_{t_2}^{\Delta_2\sigma_1} \cdot \mathscr{A}_{a\sigma_1}\left(f_{t_1}^{\Delta_1}\right) = \left(1+\mu_1^{\Delta_1}\right) f_{t_1}^{\Delta_1\sigma_1\sigma_2} \cdot \mathscr{A}_{a\sigma_1}\left(f_{t_2}^{\Delta_2\sigma_1}\right).$$

Proof. We differentiate with respect to t_2 the equality

$$f_{t_1}^{\Delta_1} \cdot N_{\sigma_1} = 0$$

and we get

$$f_{t_1t_2}^{\Delta_1\Delta_2} \cdot N_{\sigma_1} + f_{t_1}^{\Delta_1\sigma_2} \cdot N_{\sigma_1 t_2}^{\Delta_2} = 0. \tag{3.2}$$

Now, we differentiate with respect to t_1 the equation

$$f_{t_2}^{\Delta_2\sigma_1} \cdot N_{\sigma_1} = 0$$

and we find

$$f_{t_2t_1}^{\Delta_2\sigma_1\Delta_1} \cdot N_{\sigma_1}^{\sigma_1} + f_{t_2}^{\Delta_2\sigma_1} \cdot N_{\sigma_1 t_1}^{\Delta_1} = 0$$

or

$$\left(1+\mu_1^{\Delta_1}\right) f_{t_1 t_2}^{\Delta_1 \Delta_2 \sigma_1} \cdot N_{\sigma_1}^{\sigma_1} + f_{t_2}^{\Delta_2 \sigma_1} \cdot N_{\sigma_1 t_1}^{\Delta_1} = 0. \tag{3.3}$$

By (3.2), we get

$$f_{t_1 t_2}^{\Delta_1 \Delta_2 \sigma_1} \cdot N_{\sigma_1}^{\sigma_1} + f_{t_1}^{\Delta_1 \sigma_1 \sigma_2} \cdot N_{\sigma_1 t_2}^{\Delta_2 \sigma_1} = 0$$

and

$$\left(1+\mu_1^{\Delta_1}\right) f_{t_1 t_2}^{\Delta_1 \Delta_2 \sigma_1} \cdot N_{\sigma_1}^{\sigma_1} + \left(1+\mu_1^{\Delta_1}\right) f_{t_1}^{\Delta_1 \sigma_1 \sigma_2} \cdot N_{\sigma_1 t_2}^{\Delta_2 \sigma_1} = 0.$$

Hence, using the above equation and using (3.3), we find

$$f_{t_2}^{\Delta_2 \sigma_1} \cdot N_{\sigma_1 t_1}^{\Delta_1} = \left(1+\mu_1^{\Delta_1}\right) f_{t_1}^{\Delta_1 \sigma_1 \sigma_2} \cdot N_{\sigma_1 t_2}^{\Delta_2 \sigma_1}$$

or

$$f_{t_2}^{\Delta_2 \sigma_1} \cdot \mathscr{A}_{a\sigma_1}\left(f_{t_1}^{\Delta_1}\right) = \left(1+\mu_1^{\Delta_1}\right) f_{t_1}^{\Delta_1 \sigma_1 \sigma_2} \cdot \mathscr{A}_{a\sigma_1}\left(f_{t_2}^{\Delta_2 \sigma_1}\right).$$

This completes the proof.

Now, we will find a 2×2-matrix $\mathscr{A}_{a\sigma_1}$ so that

$$\left(N_{\sigma_1 t_1}^{\Delta_1}, N_{\sigma_1 t_2}^{\Delta_2 \sigma_1}\right) = \left(f_{t_1}^{\Delta_1}, f_{t_2}^{\Delta_2 \sigma_1}\right) \mathscr{A}_{a\sigma_1}.$$

We multiply the above equation, on the left, by the matrix

$$\begin{pmatrix} f_{t_1}^{\Delta_1} \\ f_{t_2}^{\Delta_2 \sigma_1} \end{pmatrix}$$

and we get

$$\begin{pmatrix} f_{t_1}^{\Delta_1} \\ f_{t_2}^{\Delta_2 \sigma_1} \end{pmatrix} \left(N_{\sigma_1 t_1}^{\Delta_1}, N_{\sigma_1 t_2}^{\Delta_2 \sigma_1}\right) = \begin{pmatrix} f_{t_1}^{\Delta_1} \\ f_{t_2}^{\Delta_2 \sigma_1} \end{pmatrix} \left(f_{t_1}^{\Delta_1}, f_{t_2}^{\Delta_2 \sigma_1}\right) \mathscr{A}_{a\sigma_1}$$

or

$$\begin{pmatrix} f_{t_1}^{\Delta_1} \cdot N_{\sigma_1 t_1}^{\Delta_1} & f_{t_1}^{\Delta_1} \cdot N_{\sigma_1 t_2}^{\Delta_2 \sigma_1} \\ f_{t_2}^{\Delta_2 \sigma_1} \cdot N_{\sigma_1 t_1}^{\Delta_1} & f_{t_2}^{\Delta_2 \sigma_1} \cdot N_{\sigma_1 t_2}^{\Delta_2 \sigma_1} \end{pmatrix} = \begin{pmatrix} f_{t_1}^{\Delta_1} \cdot f_{t_1}^{\Delta_1} & f_{t_1}^{\Delta_1} \cdot f_{t_2}^{\Delta_2 \sigma_1} \\ f_{t_2}^{\Delta_2 \sigma_1} \cdot f_{t_1}^{\Delta_1} & f_{t_2}^{\Delta_2 \sigma_1} \cdot f_{t_2}^{\Delta_2 \sigma_1} \end{pmatrix} \mathscr{A}_{a\sigma_1}.$$

Let

$$\mathscr{H}_{a\sigma_1} = \begin{pmatrix} f_{t_1}^{\Delta_1} \cdot N_{\sigma_1 t_1}^{\Delta_1} & f_{t_1}^{\Delta_1} \cdot N_{\sigma_1 t_2}^{\Delta_2 \sigma_1} \\ f_{t_2}^{\Delta_2 \sigma_1} \cdot N_{\sigma_1 t_1}^{\Delta_1} & f_{t_2}^{\Delta_2 \sigma_1} \cdot N_{\sigma_1 t_2}^{\Delta_2 \sigma_1} \end{pmatrix}$$

and

$$\mathscr{G}_{a\sigma_1} = \begin{pmatrix} f_{t_1}^{\Delta_1} \cdot f_{t_1}^{\Delta_1} & f_{t_1}^{\Delta_1} \cdot f_{t_2}^{\Delta_2 \sigma_1} \\ f_{t_2}^{\Delta_2 \sigma_1} \cdot f_{t_1}^{\Delta_1} & f_{t_2}^{\Delta_2 \sigma_1} \cdot f_{t_2}^{\Delta_2 \sigma_1} \end{pmatrix}.$$

Then

$$\mathscr{H}_{a\sigma_1} = \mathscr{G}_{a\sigma_1} \mathscr{A}_{a\sigma_1}.$$

Let

$$\begin{aligned} &E = \left(f_{t_1}^{\Delta_1}\right)^2, \quad F_{\sigma_1} = f_{t_2}^{\Delta_2 \sigma_1} \cdot f_{t_1}^{\Delta_1}, \quad G_{\sigma_1} = \left(f_{t_2}^{\Delta_2 \sigma_1}\right)^2, \\ &L = f_{t_1}^{\Delta_1} \cdot N_{\sigma_1 t_1}^{\Delta_1}, \quad M_{1\sigma_1} = f_{t_1}^{\Delta_1} \cdot N_{\sigma_1 t_2}^{\Delta_2 \sigma_1}, \quad M_{2\sigma_1} = f_{t_2}^{\Delta_2 \sigma_1} \cdot N_{\sigma_1 t_1}^{\Delta_1}, \\ &P_{\sigma_1} = f_{t_2}^{\Delta_2 \sigma_1} \cdot N_{\sigma_1 t_2}^{\Delta_2 \sigma_1}. \end{aligned}$$

Then

$$\mathscr{G}_{a\sigma_1} = \begin{pmatrix} E & F_{\sigma_1} \\ F_{\sigma_1} & G_{\sigma_1} \end{pmatrix}, \quad \mathscr{H}_{a\sigma_1} = \begin{pmatrix} L & M_{1\sigma_1} \\ M_{2\sigma_1} & P_{\sigma_1} \end{pmatrix}$$

and if

$$EG_{\sigma_1} - F_{\sigma_1}^2 \neq 0,$$

then

$$\mathscr{G}_{a\sigma_1}^{-1} = \frac{1}{EG_{\sigma_1} - F_{\sigma_1}^2} \begin{pmatrix} G_{\sigma_1} & -F_{\sigma_1} \\ -F_{\sigma_1} & E \end{pmatrix}$$

and

$$\begin{aligned} \mathscr{A}_{a\sigma_1} &= \mathscr{G}_{a\sigma_1}^{-1} \mathscr{H}_{a\sigma_1} \\ &= \frac{1}{EG_{\sigma_1} - F_{\sigma_1}^2} \begin{pmatrix} G_{\sigma_1} & -F_{\sigma_1} \\ -F_{\sigma_1} & E \end{pmatrix} \begin{pmatrix} L & M_{1\sigma_1} \\ M_{2\sigma_1} & P_{\sigma_1} \end{pmatrix} \\ &= \frac{1}{EG_{\sigma_1} - F_{\sigma_1}^2} \begin{pmatrix} G_{\sigma_1} L - F_{\sigma_1} M_{2\sigma_1} & G_{\sigma_1} M_{1\sigma_1} - F_{\sigma_1} P_{\sigma_1} \\ -F_{\sigma_1} L + EM_{2\sigma_1} & -F_{\sigma_1} M_{1\sigma_1} + EP_{\sigma_1} \end{pmatrix}. \end{aligned}$$

Example 3.4 Let $\mathbb{T}_1 = \mathbb{T}_2 = 2^{\mathbb{N}_0}$, $\mathbb{T}_{(1)} = \mathbb{T}_{(2)} = \mathbb{T}_{(3)} = \mathbb{R}$, S be a surface with a local representation (U,f), where $U \subseteq \mathbb{T}_1 \times \mathbb{T}_2, f : U \to \mathbb{R}^3$ be given by

$$f(t_1,t_2) = (t_1^2 + t_2^2, t_1 + t_2, -t_1 + t_2), \quad (t_1,t_2) \in U.$$

If Figure 3.1 shows f.

Here

$$\begin{aligned} \sigma_1(t_1) &= 2t_1, \quad t_1 \in \mathbb{T}_1, \\ \sigma_2(t_2) &= 2t_2, \quad t_2 \in \mathbb{T}_2, \\ f_1(t_1,t_2) &= t_1^2 + t_2^2, \\ f_2(t_1,t_2) &= t_1 + t_2, \\ f_3(t_1,t_2) &= -t_1 + t_2, \quad (t_1,t_2) \in U. \end{aligned}$$

Then

$$\begin{aligned} f_{1t_1}^{\Delta_1}(t_1,t_2) &= \sigma_1(t_1) + t_1 \\ &= 2t_1 + t_1 \\ &= 3t_1, \\ f_{2t_1}^{\Delta_1}(t_1,t_2) &= 1, \\ f_{3t_1}^{\Delta_1}(t_1,t_2) &= -1, \\ f_{1t_2}^{\Delta_2}(t_1,t_2) &= 3t_2, \\ f_{2t_2}^{\Delta_2}(t_1,t_2) &= 1, \end{aligned}$$

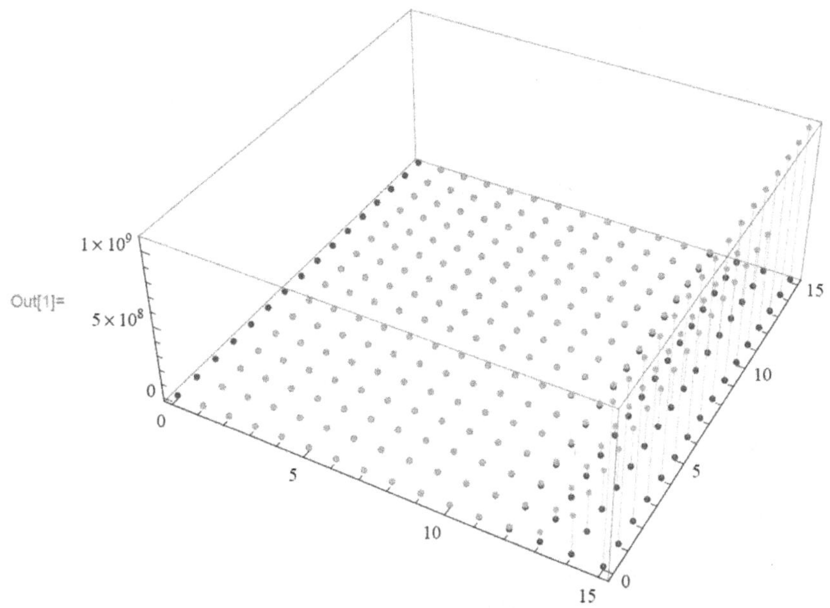

FIGURE 3.1

$$f_{3t_2}^{\Delta_2}(t_1,t_2) = 1, \quad (t_1,t_2) \in U.$$

Therefore

$$\begin{aligned}
f_{t_1}^{\Delta_1}(t_1,t_2) &= \left(f_{1t_1}^{\Delta_1}(t_1,t_2), f_{2t_1}^{\Delta_1}(t_1,t_2), f_{3t_1}^{\Delta_1}(t_1,t_2)\right) \\
&= (3t_1, 1, -1), \\
f_{t_2}^{\Delta_2}(t_1,t_2) &= \left(f_{1t_2}^{\Delta_2}(t_1,t_2), f_{2t_2}^{\Delta_2}(t_1,t_2), f_{3t_2}^{\Delta_2}(t_1,t_2)\right) \\
&= (3t_2, 1, 1), \\
f_{t_2}^{\Delta_2\sigma_1}(t_1,t_2) &= (3t_2, 1, 1), \quad (t_1,t_2) \in U,
\end{aligned}$$

and

$$\begin{aligned}
f_{t_1}^{\Delta_1}(t_1,t_2) \times f_{t_2}^{\Delta_2\sigma_1}(t_1,t_2) &= (3t_1, 1, -1) \times (3t_2, 1, 1) \\
&= (2, -3t_1 - 3t_2, 3t_1 - 3t_2), \\
\|f_{t_1}^{\Delta_1}(t_1,t_2) \times f_{t_2}^{\Delta_2\sigma_1}(t_1,t_2)\| &= \sqrt{4 + (3t_1 + 3t_2)^2 + (3t_1 - 3t_2)^2} \\
&= \sqrt{4 + 9t_1^2 + 18t_1t_2 + 9t_2^2 + 9t_1^2 - 18t_1t_2 + 9t_2^2} \\
&= \sqrt{4 + 18t_1^2 + 18t_2^2}, \quad (t_1,t_2) \in U,
\end{aligned}$$

and

$$N_{\sigma_1}(t_1,t_2)=\left(\frac{2}{\sqrt{4+18t_1^2+18t_2^2}},\frac{-3t_1-3t_2}{\sqrt{4+18t_1^2+18t_2^2}},\frac{3t_1-3t_2}{\sqrt{4+18t_1^2+18t_2^2}}\right),$$

$(t_1,t_2)\in U$. Let

$$h(t_1,t_2)=\sqrt{4+18t_1^2+18t_2^2},\quad (t_1,t_2)\in U.$$

Then

$$\begin{aligned}
h_{t_1}^{\Delta_1}(t_1,t_2)&=\frac{1}{\sigma_1(t_1)-t_1}\left(\sqrt{4+18(\sigma_1(t_1))^2+18t_2^2}-\sqrt{4+18t_1^2+18t_2^2}\right)\\
&=\frac{1}{t_1}\left(\sqrt{4+72t_1^2+18t_2^2}-\sqrt{4+18t_1^2+18t_2^2}\right),\\
h_{t_2}^{\Delta_2}(t_1,t_2)&=\frac{1}{t_2}\left(\sqrt{4+18t_1^2+72t_2^2}-\sqrt{4+18t_1^2+18t_2^2}\right),
\end{aligned}$$

$(t_1,t_2)\in U$. Set

$$\begin{aligned}
N_1(t_1,t_2)&=\frac{2}{\sqrt{4+18t_1^2+18t_2^2}},\\
N_2(t_1,t_2)&=\frac{-3t_1-3t_2}{\sqrt{4+18t_1^2+18t_2^2}},\\
N_3(t_1,t_2)&=\frac{3t_1-3t_2}{\sqrt{4+18t_1^2+18t_2^2}},\quad (t_1,t_2)\in U.
\end{aligned}$$

For their partial derivatives we have the following representations.

$$\begin{aligned}
N_{1t_1}^{\Delta_1}(t_1,t_2)&=-\frac{2(\sqrt{4+72t_1^2+18t_2^2}-\sqrt{4+18t_1^2+18t_2^2})}{t_1\sqrt{4+72t_1^2+18t_2^2}\sqrt{4+18t_1^2+18t_2^2}},\\
N_{2t_1}^{\Delta_1}(t_1,t_2)&=\frac{1}{\sqrt{4+18t_1^2+18t_2^2}\sqrt{4+72t_1^2+18t_2^2}}\left(-3\sqrt{4+18t_1^2+18t_2^2}\right.\\
&\quad\left.+\frac{3t_1+3t_2}{t_1}\sqrt{4+72t_1^2+18t_2^2}-\frac{3t_1+3t_2}{t_1}\sqrt{4+18t_1^2+18t_2^2}\right)\\
&=\frac{(3t_1+3t_2)\sqrt{4+72t_1^2+18t_2^2}-(6t_1+3t_2)\sqrt{4+18t_1^2+18t_2^2}}{t_1\sqrt{4+18t_1^2+18t_2^2}\sqrt{4+72t_1^2+18t_2^2}},\\
N_{3t_1}^{\Delta_1}(t_1,t_2)&=\frac{1}{\sqrt{4+18t_1^2+18t_2^2}\sqrt{4+72t_1^2+18t_2^2}}\left(3\sqrt{4+18t_1^2+18t_2^2}\right.\\
&\quad\left.-\frac{3t_1-3t_2}{t_1}\sqrt{4+72t_1^2+18t_2^2}+\frac{3t_1-3t_2}{t_1}\sqrt{4+18t_1^2+18t_2^2}\right)\\
&=\frac{(-3t_1+3t_2)\sqrt{4+72t_1^2+18t_2^2}+(6t_1-3t_2)\sqrt{4+18t_1^2+18t_2^2}}{t_1\sqrt{4+18t_1^2+18t_2^2}\sqrt{4+72t_1^2+18t_2^2}},
\end{aligned}$$

$$N^{\Delta_2}_{1t_2}(t_1,t_2) = -\frac{2(\sqrt{4+18t_1^2+72t_2^2}-\sqrt{4+18t_1^2+18t_2^2})}{t_2\sqrt{4+18t_1^2+18t_2^2}\sqrt{4+18t_1^2+72t_2^2}},$$

$$N^{\Delta_2}_{2t_2}(t_1,t_2) = \frac{1}{\sqrt{4+18t_1^2+18t_2^2}\sqrt{4+18t_1^2+72t_2^2}}$$

$$\left(-3\sqrt{4+18t_1^2+18t_2^2}+\frac{3t_1+3t_2}{t_2}\left(\sqrt{4+18t_1^2+72t_2^2}\right.\right.$$

$$\left.\left.-\sqrt{4+18t_1^2+18t_2^2}\right)\right)$$

$$=\frac{(-3t_1-6t_2)\sqrt{4+18t_1^2+18t_2^2}+(3t_1+3t_2)\sqrt{4+18t_1^2+72t_2^2}}{t_2\sqrt{4+18t_1^2+18t_2^2}\sqrt{4+18t_1^2+72t_2^2}},$$

$$N^{\Delta_2}_{3t_2}(t_1,t_2) = \frac{1}{\sqrt{4+18t_1^2+18t_2^2}\sqrt{4+18t_1^2+72t_2^2}}\left(-3\sqrt{4+18t_1^2+18t_2^2}\right.$$

$$\left.-\frac{3t_1-3t_2}{t_2}\sqrt{4+18t_1^2+72t_2^2}+\frac{3t_1-3t_2}{t_2}\sqrt{4+18t_1^2+18t_2^2}\right)$$

$$=\frac{(3t_1-6t_2)\sqrt{4+18t_1^2+18t_2^2}+(-3t_1+3t_2)\sqrt{4+18t_1^2+72t_2^2}}{t_2\sqrt{4+18t_1^2+18t_2^2}\sqrt{4+18t_1^2+72t_2^2}},$$

$(t_1,t_2)\in U$. Therefore

$$N^{\Delta_1}_{\sigma_1 t_1}(t_1,t_2) = \left(-\frac{2(\sqrt{4+72t_1^2+18t_2^2}-\sqrt{4+18t_1^2+18t_2^2})}{t_1\sqrt{4+72t_1^2+18t_2^2}\sqrt{4+18t_1^2+18t_2^2}},\right.$$

$$\frac{(3t_1+3t_2)\sqrt{4+72t_1^2+18t_2^2}-(6t_1+3t_2)\sqrt{4+18t_1^2+18t_2^2}}{t_1\sqrt{4+18t_1^2+18t_2^2}\sqrt{4+72t_1^2+18t_2^2}},$$

$$\left.\frac{(-3t_1+3t_2)\sqrt{4+72t_1^2+18t_2^2}+(6t_1-3t_2)\sqrt{4+18t_1^2+18t_2^2}}{t_1\sqrt{4+18t_1^2+18t_2^2}\sqrt{4+72t_1^2+18t_2^2}}\right),$$

$$N^{\Delta_2}_{\sigma_1 t_2}(t_1,t_2) = \left(-\frac{2(\sqrt{4+18t_1^2+72t_2^2}-\sqrt{4+18t_1^2+18t_2^2})}{t_2\sqrt{4+18t_1^2+18t_2^2}\sqrt{4+18t_1^2+72t_2^2}},\right.$$

$$\frac{(-3t_1-6t_2)\sqrt{4+18t_1^2+18t_2^2}+(3t_1+3t_2)\sqrt{4+18t_1^2+72t_2^2}}{t_2\sqrt{4+18t_1^2+18t_2^2}\sqrt{4+18t_1^2+72t_2^2}},$$

$$\left.\frac{(3t_1-6t_2)\sqrt{4+18t_1^2+18t_2^2}+(-3t_1+3t_2)\sqrt{4+18t_1^2+72t_2^2}}{t_2\sqrt{4+18t_1^2+18t_2^2}\sqrt{4+18t_1^2+72t_2^2}}\right),$$

$$N^{\Delta_2\sigma_1}_{\sigma_1 t_2}(t_1,t_2) = \left(-\frac{(2\sqrt{1+18t_1^2+18t_2^2}-\sqrt{4+72t_1^2+18t_2^2})}{t_2\sqrt{4+72t_1^2+18t_2^2}\sqrt{1+18t_1^2+18t_2^2}},\right.$$

$$\frac{(-3t_1-6t_2)\sqrt{4+72t_1^2+18t_2^2}+2(3t_1+3t_2)\sqrt{1+18t_1^2+18t_2^2}}{2t_2\sqrt{4+72t_1^2+18t_2^2}\sqrt{1+18t_1^2+18t_2^2}},$$

$$\left.\frac{(3t_1 - 6t_2)\sqrt{4+72t_1^2+18t_2^2} + 2(-3t_1+3t_2)\sqrt{1+18t_1^2+18t_2^2}}{2t_2\sqrt{4+72t_1^2+18t_2^2}\sqrt{1+18t_1^2+18t_2^2}}\right),$$

$(t_1, t_2) \in U$. Hence,

$$\begin{aligned}
f_{t_1}^{\Delta_1}(t_1,t_2)\cdot N_{\sigma_1 t_1}^{\Delta_1}(t_1,t_2) &= -\frac{6(\sqrt{4+72t_1^2+18t_2^2} - \sqrt{4+18t_1^2+18t_2^2})}{\sqrt{4+72t_1^2+18t_2^2}\sqrt{4+18t_1^2+18t_2^2}} \\
&\quad + \frac{(3t_1+3t_2)\sqrt{4+72t_1^2+18t_2^2} - (6t_1+3t_2)\sqrt{4+18t_1^2+18t_2^2}}{t_1\sqrt{4+18t_1^2+18t_2^2}\sqrt{4+72t_1^2+18t_2^2}} \\
&\quad + \frac{(3t_1-3t_2)\sqrt{4+72t_1^2+18t_2^2} + (-6t_1+3t_2)\sqrt{4+18t_1^2+18t_2^2}}{t_1\sqrt{4+18t_1^2+18t_2^2}\sqrt{4+72t_1^2+18t_2^2}} \\
&= -\frac{6}{\sqrt{4+72t_1^2+18t_2^2}}, \\
f_{t_2}^{\Delta_2\sigma_1}(t_1,t_2)\cdot N_{\sigma_1 t_1}^{\Delta_1}(t_1,t_2) &= -\frac{6t_2(\sqrt{4+72t_1^2+18t_2^2} - \sqrt{4+18t_1^2+18t_2^2})}{t_1\sqrt{4+72t_1^2+18t_2^2}\sqrt{4+18t_1^2+18t_2^2}} \\
&\quad + \frac{(3t_1+3t_2)\sqrt{4+72t_1^2+18t_2^2} - (6t_1+3t_2)\sqrt{4+18t_1^2+18t_2^2}}{t_1\sqrt{4+18t_1^2+18t_2^2}\sqrt{4+72t_1^2+18t_2^2}} \\
&\quad + \frac{(-3t_1+3t_2)\sqrt{4+72t_1^2+18t_2^2} + (6t_1-3t_2)\sqrt{4+18t_1^2+18t_2^2}}{t_1\sqrt{4+18t_1^2+18t_2^2}\sqrt{4+72t_1^2+18t_2^2}} \\
&= 0, \\
&= \frac{4t_2\left(\sqrt{4+72t_1^2+18t_2^2} - \sqrt{4+18t_1^2+18t_2^2}\right)}{t_1\sqrt{4+18t_1^2+18t_2^2}\sqrt{4+72t_1^2+1q8t_2^2}}, \\
f_{t_1}^{\Delta_1}(t_1,t_2)\cdot N_{\sigma_1 t_2}^{\Delta_2\sigma_1}(t_1,t_2) &= -\frac{6t_1\left(2\sqrt{1+18t_1^2+18t_2^2} - \sqrt{4+72t_1^2+18t_2^2}\right)}{2t_2\sqrt{4+72t_1^2+18t_2^2}\sqrt{1+18t_1^2+18t_2^2}} \\
&\quad + \frac{(-3t_1-6t_2)\sqrt{4+72t_1^2+18t_2^2} + 2(3t_1+3t_2)\sqrt{1+18t_1^2+18t_2^2}}{2t_2\sqrt{4+72t_1^2+18t_2^2}\sqrt{1+18t_1^2+18t_2^2}} \\
&\quad + \frac{(-3t_1+6t_2)\sqrt{4+72t_1^2+18t_2^2} + 2(3t_1-3t_2)\sqrt{1+18t_1^2+18t_2^2}}{2t_2\sqrt{4+72t_1^2+18t_2^2}\sqrt{1+18t_1^2+18t_2^2}} \\
&= 0, \\
f_{t_2}^{\Delta_2\sigma_1}(t_1,t_2)\cdot N_{\sigma_1 t_2}^{\Delta_2\sigma_1}(t_1,t_2) &= -\frac{\left(6\sqrt{1+18t_1^2+18t_2^2} - 3\sqrt{4+72t_1^2+18t_2^2}\right)}{\sqrt{4+72t_1^2+18t_2^2}\sqrt{1+18t_1^2+18t_2^2}} \\
&\quad + \frac{(-3t_1-6t_2)\sqrt{4+72t_1^2+18t_2^2} + (6t_1+6t_2)\sqrt{1+18t_1^2+18t_2^2}}{2t_2\sqrt{4+72t_1^2+18t_2^2}\sqrt{1+18t_1^2+18t_2^2}}
\end{aligned}$$

$$+\frac{(3t_1-6t_2)\sqrt{4+72t_1^2+18t_2^2}+(-6t_1+6t_2)\sqrt{1+18t_1^2+18t_2^2}}{2t_2\sqrt{4+72t_1^2+18t_2^2}\sqrt{1+18t_1^2+18t_2^2}}$$

$$=-\frac{\left(6\sqrt{1+18t_1^2+18t_2^2}-3\sqrt{4+72t_1^2+18t_2^2}\right)}{\sqrt{4+72t_1^2+18t_2^2}\sqrt{1+18t_1^2+18t_2^2}}$$

$$+\frac{-6\sqrt{4+72t_1^2+18t_2^2}+6\sqrt{1+18t_1^2+18t_2^2}}{\sqrt{4+72t_1^2+18t_2^2}\sqrt{1+18t_1^2+18t_2^2}}$$

$$=-\frac{3}{\sqrt{1+18t_1^2+18t_2^2}},$$

$$\begin{aligned}
\left(f_{t_1}^{\Delta_1}(t_1,t_2)\right)^2 &= 9t_1^2+1+1\\
&= 9t_1^2+2,\\
\left(f_{t_2}^{\Delta_2\sigma_1}(t_1,t_2)\right)^2 &= 9t_2^2+1+1\\
&= 9t_2^2+2,\\
f_{t_1}^{\Delta_1}(t_1,t_2)\cdot f_{t_2}^{\Delta_2\sigma_1}(t_1,t_2) &= 9t_1t_2+1-1\\
&= 9t_1t_2, \quad (t_1,t_2)\in U.
\end{aligned}$$

From here,

$$\begin{aligned}
&\|f_{t_1}^{\Delta_1}(t_1,t_2)\|^2\|f_{t_2}^{\Delta_2\sigma_1}(t_1,t_2)\|^2-\left(f_{t_1}^{\Delta_1}(t_1,t_2)\cdot f_{t_2}^{\Delta_2\sigma_1}(t_1,t_2)\right)^2\\
&= (2+9t_1^2)(2+9t_2^2)-81t_1^2t_2^2\\
&= 4+18t_1^2+18t_2^2+81t_1^2t_2^2-81t_1^2t_2^2\\
&= 4+18t_1^2+18t_2^2,\\
&\|f_{t_1}^{\Delta_1}(t_1,t_2)\|^2\left(f_{t_1}^{\Delta_1}(t_1,t_2)\cdot N_{\sigma_1t_1}^{\Delta_1}(t_1,t_2)\right)-\left(f_{t_1}^{\Delta_1}(t_1,t_2)\right.\\
&\left.\cdot f_{t_2}^{\Delta_2\sigma_1}(t_1,t_2)\right)\left(f_{t_2}^{\Delta_2\sigma_1}(t_1,t_2)\cdot N_{\sigma_1t_1}^{\Delta_1}(t_1,t_2)\right)\\
&= (2+9t_1^2)\left(-\frac{6}{\sqrt{4+72t_1^2+18t_2^2}}\right)-9t_1t_2\left(-\frac{3}{\sqrt{1+18t_1^2+18t_2^2}}\right)\\
&= -\frac{12+54t_1^2}{\sqrt{4+72t_1^2+18t_2^2}}+\frac{27t_1t_2}{\sqrt{1+18t_1^2+18t_2^2}},\\
&\|f_{t_2}^{\Delta_2\sigma_1}(t_1,t_2)\|^2\left(f_{t_1}^{\Delta_1}(t_1,t_2)\cdot N_{\sigma_1t_2}^{\Delta_2\sigma_1}(t_1,t_2)\right)\\
&\quad-\left(f_{t_1}^{\Delta_1}(t_1,t_2)\cdot f_{t_2}^{\Delta_2\sigma_1}(t_1,t_2)\right)\left(f_{t_2}^{\Delta_2\sigma_1}(t_1,t_2)\cdot N_{\sigma_1t_2}^{\Delta_2\sigma_1}(t_1,t_2)\right)\\
&= 9t_1t_2\left(-\frac{3}{\sqrt{1+18t_1^2+18t_2^2}}\right)\\
&= -\frac{27t_1t_2}{\sqrt{1+18t_1^2+18t_2^2}},
\end{aligned}$$

$$\begin{aligned}
\|f_{t_1}^{\Delta_1} & (t_1,t_2)\|^2 \left(f_{t_2}^{\Delta_2\sigma_1}(t_1,t_2)\cdot N_{\sigma_1 t_1}^{\Delta_1}(t_1,t_2)\right) \\
& - \left(f_{t_1}^{\Delta_1}(t_1,t_2)\cdot f_{t_2}^{\Delta_2\sigma_1}(t_1,t_2)\right)\left(f_{t_1}^{\Delta_1}(t_1,t_2)\cdot N_{\sigma_1 t_1}^{\Delta_1}(t_1,t_2)\right) \\
= & \quad -9t_1t_2\left(-\frac{6}{\sqrt{4+72t_1^2+18t_2^2}}\right) \\
= & \quad \frac{54t_1t_2}{\sqrt{4+72t_1^2+18t_2^2}}, \\
\|f_{t_1}^{\Delta_1} & (t_1,t_2)\|^2 \left(f_{t_2}^{\Delta_2\sigma_1}(t_1,t_2)\cdot N_{\sigma_1 t_2}^{\Delta_2\sigma_1}(t_1,t_2)\right) \\
& - \left(f_{t_1}^{\Delta_1}(t_1,t_2)\cdot f_{t_2}^{\Delta_2\sigma_1}(t_1,t_2)\right)\left(f_{t_1}^{\Delta_1}(t_1,t_2)\cdot N_{\sigma_1 t_2}^{\Delta_2\sigma_1}(t_1,t_2)\right) \\
& = (2+9t_1^2)\left(-\frac{3}{\sqrt{1+18t_1^2+18t_2^2}}\right) \\
& = -\frac{6+27t_1^2}{\sqrt{1+18t_1^2+18t_2^2}}, \quad (t_1,t_2)\in U.
\end{aligned}$$

Consequently

$$\begin{aligned}
\mathscr{A}_{a\sigma_1} &= \frac{1}{4+18t_1^2+18t_2^2} \\
&= \begin{pmatrix} -\frac{12+54t_1^2}{\sqrt{4+72t_1^2+18t_2^2}}+\frac{27t_1t_2}{\sqrt{1+18t_1^2+18t_2^2}} & -\frac{27t_1t_2}{\sqrt{1+18t_1^2+18t_2^2}} \\ \frac{54t_1t_2}{\sqrt{4+72t_1^2+18t_2^2}}+\frac{-8t_2+18t_1^2t_2}{\sqrt{4+72t_1^2+18t_2^2}} & -\frac{6+27t_1^2}{\sqrt{1+18t_1^2+18t_2^2}} \end{pmatrix},
\end{aligned}$$

$(t_1,t_2)\in U$.

Example 3.5 Let $\mathbb{T}_1=\mathbb{T}_2=\mathbb{T}_{(1)}=\mathbb{T}_{(2)}=\mathbb{T}_{(3)}=\mathbb{R}$. Let also,

$$\begin{aligned}
f_1(t_1,t_2) &= t_1\cos t_2, \\
f_2(t_1,t_2) &= t_1\sin t_2, \\
f_3(t_1,t_2) &= t_2, \quad (t_1,t_2)\in\mathbb{R}^2.
\end{aligned}$$

We have

$$\begin{aligned}
f'_{t_1}(t_1,t_2) &= (\cos t_2,\sin t_2,0), \\
f_{t_2}^{\prime\sigma_1}(t_1,t_2) &= (-t_1\sin t_2,t_1\cos t_2,1), \\
f'_{t_1}(t_1,t_2)\times f_{t_2}^{\prime\sigma_2}(t_1,t_2) &= (\sin t_2,\cos t_2,t_1), \\
\|f'_{t_1}(t_1,t_2)\times f_{t_2}^{\prime\sigma_1}(t_1,t_2)\| &= \sqrt{\sin^2 t_2+\cos^2 t_2+t_1^2} \\
&= \sqrt{1+t_1^2}
\end{aligned}$$

and

$$N_{a\sigma_1}(t_1,t_2)=\left(\frac{\sin t_2}{\sqrt{1+t_1^2}},-\frac{\cos t_2}{\sqrt{1+t_1^2}},\frac{t_1}{\sqrt{1+t_1^2}}\right),\quad (t_1,t_2)\in\mathbb{R}^2.$$

Then

$$N'_{\sigma_1 t_1}(t_1,t_2) = \left(-\frac{t_1 \sin t_2}{(1+t_1^2)^{\frac{3}{2}}}, \frac{t_1 \cos t_2}{(1+t_1^2)^{\frac{3}{2}}}, \frac{1}{(1+t_1^2)^{\frac{3}{2}}}\right),$$

$$N'^{\sigma_1}_{\sigma_1 t_2}(t_1,t_2) = \left(\frac{\cos t_2}{(1+t_1^2)^{\frac{1}{2}}}, \frac{\sin t_2}{(1+t_1^2)^{\frac{1}{2}}}, 0\right), \quad (t_1,t_2) \in \mathbb{T}_1 \times \mathbb{T}_2,$$

and

$$\begin{aligned} f'_{t_1}(t_1,t_2) \cdot N'_{\sigma_1 t_1}(t_1,t_2) &= -\frac{t_1 \sin t_2 \cos t_2}{(1+t_1^2)^2} + \frac{t_1 \sin t_2 \cos t_2}{(1+t_1^2)^2} \\ &= 0, \\ f'_{t_1}(t_1,t_2) \cdot N'^{\sigma_1}_{\sigma_1 t_2}(t_1,t_2) &= \frac{\cos^2 t_2}{\sqrt{1+t_1^2}} + \frac{\sin^2 t_2}{\sqrt{1+t_1^2}} \\ &= \frac{1}{\sqrt{1+t_1^2}}, \\ f'^{\sigma_1}_{t_2}(t_1,t_2) \cdot N'_{\sigma_1 t_1}(t_1,t_2) &= \frac{t_1^2 \sin^2 t_2}{(1+t_1^2)^{\frac{3}{2}}} + \frac{t_1^2 \cos^2 t_2}{(1+t_1^2)^{\frac{3}{2}}} + \frac{1}{(1+t_1^2)^{\frac{3}{2}}} \\ &= \frac{1+t_1^2}{(1+t_1^2)^{\frac{3}{2}}} \\ &= \frac{1}{\sqrt{1+t_1^2}}, \\ f'_{t_1}(t_1,t_2) \cdot N'^{\sigma_1}_{\sigma_1 t_2}(t_1,t_2) &= \frac{\cos^2 t_2}{\sqrt{1+t_1^2}} + \frac{\sin^2 t_2}{\sqrt{1+t_1^2}} \\ &= \frac{1}{\sqrt{1+t_1^2}}, \\ f'^{\sigma_1}_{t_2}(t_1,t_2) \cdot N'^{\sigma_1}_{\sigma_1 t_2}(t_1,t_2) &= -\frac{t_1 \sin t_2 \cos t_2}{\sqrt{1+t_1^2}} + \frac{t_1 \sin t_2 \cos t_2}{\sqrt{1+t_1^2}} \\ &= 0. \end{aligned}$$

Therefore

$$\mathscr{H}_{a\sigma_1} = \begin{pmatrix} 0 & \frac{1}{\sqrt{1+t_1^2}} \\ \frac{1}{\sqrt{1+t_1^2}} & 0 \end{pmatrix}.$$

Next,

$$\begin{aligned} f'_{t_1}(t_1,t_2) \cdot f'_{t_1}(t_1,t_2) &= \cos^2 t_2 + \sin^2 t_2 \\ &= 1, \end{aligned}$$

$$\begin{aligned} f'_{t_1}(t_1,t_2)\cdot f'^{\sigma_1}_{t_2}(t_1,t_2) &= -t_1\sin t_2\cos t_2 + t_1\sin t_2\cos t_2 \\ &= 0, \\ f'^{\sigma_1}_{t_2}(t_1,t_2)\cdot f'^{\sigma_1}_{t_2}(t_1,t_2) &= t_1^2\sin^2 t_2 + t_1^2\cos^2 t_2 + 1 \\ &= t_1^2+1, \quad (t_1,t_2)\in\mathbb{R}^2, \end{aligned}$$

and

$$\mathscr{G}_{a\sigma_1} = \begin{pmatrix} 1 & 0 \\ 0 & 1+t_1^2 \end{pmatrix}.$$

Hence,

$$\begin{aligned} \mathscr{G}^{-1}_{a\sigma_1} &= \frac{1}{1+t_1^2}\begin{pmatrix} 1+t_1^2 & 0 \\ 0 & 1 \end{pmatrix} \\ &= \begin{pmatrix} 1 & 0 \\ 0 & \frac{1}{1+t_1^2} \end{pmatrix} \end{aligned}$$

and

$$\begin{aligned} \mathscr{A}_{a\sigma_1} &= \mathscr{G}^{-1}_{a\sigma_1}\mathscr{H}_{a\sigma_1} \\ &= \begin{pmatrix} 1 & 0 \\ 0 & \frac{1}{1+t_1^2} \end{pmatrix}\begin{pmatrix} 0 & \frac{1}{\sqrt{1+t_1^2}} \\ \frac{1}{\sqrt{1+t_1^2}} & 0 \end{pmatrix} \\ &= \begin{pmatrix} 1 & \frac{1}{\sqrt{1+t_1^2}} \\ \frac{1}{(1+t_1^2)^{\frac{3}{2}}} & 0 \end{pmatrix}. \end{aligned}$$

Exercise 3.2 Let $\mathbb{T}_1 = \mathbb{Z}$, $\mathbb{T}_2 = 2\mathbb{Z}$, $\mathbb{T}_{(1)} = \mathbb{T}_{(2)} = \mathbb{T}_{(3)} = \mathbb{R}$ and $f : \mathbb{T}_1 \times \mathbb{T}_2 \to \mathbb{R}^3$ be given by

$$f(t_1,t_2) = \left(1+t_1^2+t_2^3, \frac{1+t_2^2}{1+t_1^2}, t_1^2t_2^2t_3^2\right), \quad (t_1,t_2)\in\mathbb{T}_1\times\mathbb{T}_2.$$

Find the matrices $\mathscr{G}_{a\sigma_1}$, $\mathscr{H}_{a\sigma_1}$ and $\mathscr{A}_{a\sigma_1}$.

Definition 3.11 Suppose that σ_2 is Δ_2-differentiable. For each vector $h = (h_1,h_2) \in \mathbb{T}_{a\sigma_2}S$ the operator

$$\mathscr{A}_{a\sigma_2}(h) = N_{t_1}^{\Delta_1\sigma_2}h_1 + N_{t_2}^{\Delta_2}h_2$$

will be called σ_2-shape operator.

In particular, we have

$$\mathscr{A}_{a\sigma_2}\left(f_{t_1}^{\Delta_1\sigma_2}\right) = N_{t_1}^{\Delta_1\sigma_2} \quad \text{and} \quad \mathscr{A}_{a\sigma_2}\left(f_{t_2}^{\Delta_2}\right) = N_{t_2}^{\Delta_2}.$$

Theorem 3.2 *Suppose that σ_2 is Δ_2-differentiable.*

Then

$$f_{t_1}^{\Delta_1\sigma_2}\cdot\mathscr{A}_{a\sigma_2}\left(f_{t_2}^{\Delta_2}\right)=\left(1+\mu_2^{\Delta_2}\right)f_{t_2}^{\Delta_2\sigma_1\sigma_2}\cdot\mathscr{A}_{a\sigma_2}\left(f_{t_1}^{\Delta_1\sigma_2}\right).$$

Proof. We differentiate with respect to t_2 the equation

$$f_{t_1}^{\Delta_1\sigma_2}\cdot N_{\sigma_2}=0$$

and we get

$$f_{t_1t_2}^{\Delta_1\sigma_2\Delta_2}\cdot N_{\sigma_2}^{\sigma_2}+f_{t_1}^{\Delta_1\sigma_2}\cdot N_{\sigma_2t_2}^{\Delta_2}=0$$

or

$$\left(1+\mu_2^{\Delta_2}\right)f_{t_1t_2}^{\Delta_1\Delta_2\sigma_2}\cdot N_{\sigma_2}^{\sigma_2}+f_{t_1}^{\Delta_1\sigma_2}\cdot N_{\sigma_2t_2}^{\Delta_2}=0. \tag{3.4}$$

Now, we differentiate with respect to t_1 the equation

$$f_{t_2}^{\Delta_2}\cdot N_{\sigma_2}=0$$

and we find

$$f_{t_2t_1}^{\Delta_2\Delta_1}\cdot N_{\sigma_2}+f_{t_2}^{\Delta_2\sigma_1}\cdot N_{\sigma_2t_1}^{\Delta_1}=0,$$

whereupon

$$f_{t_1t_2}^{\Delta_1\Delta_2\sigma_2}\cdot N_{\sigma_2}^{\sigma_2}+f_{t_2}^{\Delta_2\sigma_1\sigma_2}\cdot N_{\sigma_2t_1}^{\Delta_1\sigma_2}=0$$

and

$$\left(1+\mu_2^{\Delta_2}\right)f_{t_1t_2}^{\Delta_1\Delta_2\sigma_2}\cdot N_{\sigma_2}^{\sigma_2}+\left(1+\mu_2^{\Delta_2}\right)f_{t_2}^{\Delta_2\sigma_1\sigma_2}\cdot N_{\sigma_2t_1}^{\Delta_1\sigma_2}=0.$$

By the last equation and (3.4), we obtain

$$f_{t_1}^{\Delta_1\sigma_2}\cdot N_{\sigma_2t_2}^{\Delta_2}=\left(1+\mu_2^{\Delta_2}\right)f_{t_2}^{\Delta_2\sigma_1\sigma_2}\cdot N_{\sigma_2t_1}^{\Delta_1\sigma_2}$$

or

$$f_{t_1}^{\Delta_1\sigma_2}\cdot\mathscr{A}_{a\sigma_2}\left(f_{t_2}^{\Delta_2}\right)=\left(1+\mu_2^{\Delta_2}\right)f_{t_2}^{\Delta_2\sigma_1\sigma_2}\cdot\mathscr{A}_{a\sigma_2}\left(f_{t_1}^{\Delta_1\sigma_2}\right).$$

This completes the proof.

Now, we will search a 2×2-matrix $\mathscr{A}_{a\sigma_2}$ so that

$$\left(N_{\sigma_2t_1}^{\Delta_1\sigma_2},N_{\sigma_2t_2}^{\Delta_2}\right)=\left(f_{t_1}^{\Delta_1\sigma_2},f_{t_2}^{\Delta_2}\right)\mathscr{A}_{a\sigma_2}.$$

We multiply the above equation, on the left, by the vector

$$\begin{pmatrix} f_{t_1}^{\Delta_1\sigma_2}\\ f_{t_2}^{\Delta_2}\end{pmatrix}$$

and we find

$$\begin{pmatrix} f_{t_1}^{\Delta_1\sigma_2}\\ f_{t_2}^{\Delta_2}\end{pmatrix}\left(N_{\sigma_2t_1}^{\Delta_1\sigma_2},N_{\sigma_2t_2}^{\Delta_2}\right)=\begin{pmatrix} f_{t_1}^{\Delta_1\sigma_2}\\ f_{t_2}^{\Delta_2}\end{pmatrix}\left(f_{t_1}^{\Delta_1\sigma_2},f_{t_2}^{\Delta_2}\right)\mathscr{A}_{a\sigma_2}$$

or

$$\begin{pmatrix} f_{t_1}^{\Delta_1\sigma_2}\cdot N_{\sigma_2t_1}^{\Delta_1\sigma_2} & f_{t_1}^{\Delta_1\sigma_2}\cdot N_{\sigma_2t_2}^{\Delta_2}\\ f_{t_2}^{\Delta_2}\cdot N_{\sigma_2t_1}^{\Delta_1\sigma_2} & f_{t_2}^{\Delta_2}\cdot N_{\sigma_2t_2}^{\Delta_2}\end{pmatrix}=\begin{pmatrix} \left(f_{t_1}^{\Delta_1\sigma_2}\right)^2 & f_{t_1}^{\Delta_1\sigma_2}\cdot f_{t_2}^{\Delta_2}\\ f_{t_1}^{\Delta_1\sigma_2}\cdot f_{t_2}^{\Delta_2} & \left(f_{t_2}^{\Delta_2}\right)^2\end{pmatrix}\mathscr{A}_{a\sigma_2}.$$

Let

$$\mathscr{H}_{a\sigma_2} = \begin{pmatrix} f_{t_1}^{\Delta_1\sigma_2} \cdot N_{\sigma_2 t_1}^{\Delta_1\sigma_2} & f_{t_1}^{\Delta_1\sigma_2} \cdot N_{\sigma_2 t_2}^{\Delta_2} \\ f_{t_2}^{\Delta_2} \cdot N_{\sigma_2 t_1}^{\Delta_1\sigma_2} & f_{t_2}^{\Delta_2} \cdot N_{\sigma_2 t_2}^{\Delta_2} \end{pmatrix}$$

and

$$\mathscr{G}_{a\sigma_2} = \begin{pmatrix} \left(f_{t_1}^{\Delta_1\sigma_2}\right)^2 & f_{t_1}^{\Delta_1\sigma_2} \cdot f_{t_2}^{\Delta_2} \\ f_{t_1}^{\Delta_1\sigma_2} \cdot f_{t_2}^{\Delta_2} & \left(f_{t_2}^{\Delta_2}\right)^2 \end{pmatrix}.$$

Let

$$\begin{aligned} &E_{\sigma_2} = \left(f_{t_1}^{\Delta_1\sigma_2}\right)^2, \quad F_{\sigma_2} = f_{t_2}^{\Delta_2} \cdot f_{t_1}^{\Delta_1\sigma_2}, \quad G = \left(f_{t_2}^{\Delta_2}\right)^2, \\ &L_{\sigma_2} = f_{t_1}^{\Delta_1\sigma_2} \cdot N_{\sigma_2 t_1}^{\Delta_1\sigma_2}, \quad M_{1\sigma_2} = f_{t_1}^{\Delta_1\sigma_2} \cdot N_{\sigma_2 t_2}^{\Delta_2}, \quad M_{2\sigma_2} = f_{t_2}^{\Delta_2} \cdot N_{\sigma_2 t_1}^{\Delta_1\sigma_2}, \\ &P = f_{t_2}^{\Delta_2} \cdot N_{\sigma_2 t_2}^{\Delta_2}. \end{aligned}$$

Then

$$\mathscr{G}_{a\sigma_2} = \begin{pmatrix} E_{\sigma_2} & F_{\sigma_2} \\ F_{\sigma_2} & G \end{pmatrix}, \quad \mathscr{H}_{a\sigma_2} = \begin{pmatrix} L_{\sigma_2} & M_{1\sigma_2} \\ M_{2\sigma_2} & P \end{pmatrix}$$

and if

$$E_{\sigma_2} G - F_{\sigma_2}^2 \neq 0,$$

then

$$\mathscr{G}_{a\sigma_2}^{-1} = \frac{1}{E_{\sigma_2} G - F_{\sigma_2}^2} \begin{pmatrix} G & -F_{\sigma_2} \\ -F_{\sigma_2} & E_{\sigma_2} \end{pmatrix}$$

and

$$\begin{aligned} \mathscr{A}_{a\sigma_2} &= \mathscr{G}_{a\sigma_2}^{-1} \mathscr{H}_{a\sigma_2} \\ &= \frac{1}{E_{\sigma_2} G - F_{\sigma_2}^2} \begin{pmatrix} G & -F_{\sigma_2} \\ -F_{\sigma_2} & E_{\sigma_2} \end{pmatrix} \begin{pmatrix} L_{\sigma_2} & M_{1\sigma_2} \\ M_{2\sigma_2} & P \end{pmatrix} \\ &= \frac{1}{E_{\sigma_2} G - F_{\sigma_2}^2} \begin{pmatrix} GL_{\sigma_2} - F_{\sigma_2} M_{2\sigma_2} & GM_{1\sigma_2} - F_{\sigma_2} P \\ -F_{\sigma_2} L_{\sigma_2} + E_{\sigma_2} M_{2\sigma_2} & -F_{\sigma_2} M_{1\sigma_2} + E_{\sigma_2} P \end{pmatrix}. \end{aligned}$$

Definition 3.12 Suppose that σ_1 is Δ_1-differentiable and σ_2 is Δ_2-differentiable. For each vector $h = (h_1, h_2) \in T_{a\sigma_1^2\sigma_2} S$ the operator

$$\mathscr{A}_{a\sigma_1^2\sigma_2}(h) = N_{\sigma_1^2\sigma_2 t_1}^{\Delta_1\sigma_1\sigma_2} h_1 + N_{\sigma_1^2\sigma_2 t_2}^{\Delta_2\sigma_1^2\sigma_2} h_2$$

will be called $\sigma_1^2\sigma_2$-shape operator.

In particular, we have

$$\begin{aligned} \mathscr{A}_{a\sigma_1^2\sigma_2}\left(f_{t_1}^{\Delta_1\sigma_1\sigma_2}\right) &= N_{\sigma_1^2\sigma_2 t_1}^{\Delta_1\sigma_1\sigma_2}, \\ \mathscr{A}_{a\sigma_1^2\sigma_2}\left(f_{t_2}^{\Delta_2\sigma_1^2\sigma_2}\right) &= N_{\sigma_1^2\sigma_2 t_2}^{\Delta_2\sigma_1^2\sigma_2}. \end{aligned}$$

Theorem 3.3 *Let σ_1 be Δ_1-differentiable and σ_2 be Δ_2-differentiable.*

Then

$$\left(1+\mu_1^{\Delta_1}\right)^{2\sigma_1} f_{t_1}^{\Delta_1\sigma_1^3\sigma_2^3} \cdot \mathscr{A}_{a\sigma_1^2\sigma_2}\left(f_{t_2}^{\Delta_2\sigma_1^2\sigma_2}\right) = \left(1+\mu_2^{\Delta_2}\right)^{\sigma_2} f_{t_2}^{\Delta_2\sigma_1^3\sigma_2^2} \mathscr{A}_{a\sigma_1^2\sigma_2}\left(f_{t_1}^{\Delta_1\sigma_1\sigma_2}\right).$$

Proof. We differentiate with respect to t_2 the equation

$$f_{t_1}^{\Delta_1\sigma_1\sigma_2} \cdot N_{\sigma_1^2\sigma_2} = 0$$

and we get

$$f_{t_1t_2}^{\Delta_1\sigma_1\sigma_2\Delta_2} \cdot N_{\sigma_1^2\sigma_2} + f_{t_1}^{\Delta_1\sigma_1\sigma_2^2} \cdot N_{\sigma_1^2\sigma_2t_2}^{\Delta_2} = 0$$

or

$$\left(1+\mu_2^{\Delta_2}\right) f_{t_1t_2}^{\Delta_1\Delta_2\sigma_1\sigma_2} \cdot N_{\sigma_1^2\sigma_2} + f_{t_1}^{\Delta_1\sigma_1\sigma_2^2} \cdot N_{\sigma_1^2\sigma_2t_2}^{\Delta_2} = 0$$

and

$$\left(1+\mu_2^{\Delta_2}\right)^{\sigma_2} f_{t_1t_2}^{\Delta_1\Delta_2\sigma_1^3\sigma_2^2} \cdot N_{\sigma_1^2\sigma_2}^{\sigma_1^2\sigma_2} + f_{t_1}^{\Delta_1\sigma_1^3\sigma_2^3} \cdot N_{\sigma_1^2\sigma_2t_2}^{\Delta_2\sigma_1^2\sigma_2} = 0. \tag{3.5}$$

Now, we differentiate with respect to t_1 the equation

$$f_{t_2}^{\Delta_2\sigma_1^2\sigma_2} \cdot N_{\sigma_1^2\sigma_2} = 0$$

and we find

$$f_{t_2t_1}^{\Delta_2\sigma_1^2\sigma_2\Delta_1} \cdot N_{\sigma_1^2\sigma_2}^{\sigma_1} + f_{t_2}^{\Delta_2\sigma_1^2\sigma_2} \cdot N_{\sigma_1^2\sigma_2t_1}^{\Delta_1} = 0$$

or

$$\left(1+\mu_1^{\Delta_1}\right)^2 f_{t_2t_1}^{\Delta_2\Delta_1\sigma_1^2\sigma_2} \cdot N_{\sigma_1^2\sigma_2}^{\sigma_1} + f_{t_2}^{\Delta_2\sigma_1^2\sigma_2} \cdot N_{\sigma_1^2\sigma_2t_1}^{\Delta_1} = 0$$

and

$$\left(1+\mu_1^{\Delta_1}\right)^{2\sigma_1} f_{t_2t_1}^{\Delta_2\Delta_1\sigma_1^3\sigma_2^2} \cdot N_{\sigma_1^2\sigma_2}^{\sigma_1^2\sigma_2} + f_{t_2}^{\Delta_2\sigma_1^3\sigma_2^2} \cdot N_{\sigma_1^2\sigma_2t_1}^{\Delta_1\sigma_1\sigma_2} = 0.$$

By the last equation and (3.5), we find

$$\left(1+\mu_1^{\Delta_1}\right)^{2\sigma_1}\left(1+\mu_2^{\Delta_2}\right)^{\sigma_2} f_{t_1t_2}^{\Delta_1\Delta_2\sigma_1^3\sigma_2^2} \cdot N_{\sigma_1^2\sigma_2}^{\sigma_1^2\sigma_2} + \left(1+\mu_1^{\Delta_1}\right)^{2\sigma_1} f_{t_1}^{\Delta_1\sigma_1^3\sigma_2^3} \cdot N_{\sigma_1^2\sigma_2t_2}^{\Delta_2\sigma_1^2\sigma_2} = 0$$

and

$$\left(1+\mu_1^{\Delta_1}\right)^{2\sigma_1}\left(1+\mu_2^{\Delta_2}\right)^{\sigma_2} f_{t_2t_1}^{\Delta_2\Delta_1\sigma_1^3\sigma_2^2} \cdot N_{\sigma_1^2\sigma_2}^{\sigma_1^2\sigma_2} + \left(1+\mu_2^{\Delta_2}\right)^{\sigma_2} f_{t_2}^{\Delta_2\sigma_1^3\sigma_2^2} \cdot N_{\sigma_1^2\sigma_2t_1}^{\Delta_1\sigma_1\sigma_2} = 0,$$

whereupon

$$\left(1+\mu_1^{\Delta_1}\right)^{2\sigma_1} f_{t_1}^{\Delta_1\sigma_1^3\sigma_2^3} \cdot N_{\sigma_1^2\sigma_2t_2}^{\Delta_2\sigma_1^2\sigma_2} = \left(1+\mu_2^{\Delta_2}\right)^{\sigma_2} f_{t_2}^{\Delta_2\sigma_1^3\sigma_2^2} \cdot N_{\sigma_1^2\sigma_2t_1}^{\Delta_1\sigma_1\sigma_2}$$

or

$$\left(1+\mu_1^{\Delta_1}\right)^{2\sigma_1} f_{t_1}^{\Delta_1\sigma_1^3\sigma_2^3} \cdot \mathscr{A}_{a\sigma_1^2\sigma_2}\left(f_{t_2}^{\Delta_2\sigma_1^2\sigma_2}\right) = \left(1+\mu_2^{\Delta_2}\right)^{\sigma_2} f_{t_2}^{\Delta_2\sigma_1^3\sigma_2^2} \mathscr{A}_{a\sigma_1^2\sigma_2}\left(f_{t_1}^{\Delta_1\sigma_1\sigma_2}\right).$$

This completes the proof.

Now, we will find a 2×2-matrix $\mathscr{A}_{a\sigma_1^2\sigma_2}$ so that

$$\left(N^{\Delta_1\sigma_1\sigma_2}_{\sigma_1^2\sigma_2t_1}, N^{\Delta_2\sigma_1^2\sigma_2}_{\sigma_1^2\sigma_2t_2}\right) = \left(f^{\Delta_1\sigma_1\sigma_2}_{t_1}, f^{\Delta_2\sigma_1^2\sigma_2}_{t_2}\right)\mathscr{A}_{a\sigma_1^2\sigma_2}.$$

We multiply both sides of the above equation by the matrix

$$\begin{pmatrix} f^{\Delta_1\sigma_1\sigma_2}_{t_1} \\ f^{\Delta_2\sigma_1^2\sigma_2}_{t_2} \end{pmatrix}$$

and we get

$$\begin{pmatrix} f^{\Delta_1\sigma_1\sigma_2}_{t_1} \\ f^{\Delta_2\sigma_1^2\sigma_2}_{t_2} \end{pmatrix}\left(N^{\Delta_1\sigma_1\sigma_2}_{\sigma_1^2\sigma_2t_1}, N^{\Delta_2\sigma_1^2\sigma_2}_{\sigma_1^2\sigma_2t_2}\right) = \begin{pmatrix} f^{\Delta_1\sigma_1\sigma_2}_{t_1} \\ f^{\Delta_2\sigma_1^2\sigma_2}_{t_2} \end{pmatrix}\left(f^{\Delta_1\sigma_1\sigma_2}_{t_1}, f^{\Delta_2\sigma_1^2\sigma_2}_{t_2}\right)\mathscr{A}_{a\sigma_1^2\sigma_2}$$

or

$$\begin{pmatrix} f^{\Delta_1\sigma_1\sigma_2}_{t_1}\cdot N^{\Delta_1\sigma_1\sigma_2}_{\sigma_1^2\sigma_2t_1} & f^{\Delta_1\sigma_1\sigma_2}_{t_1}\cdot N^{\Delta_2\sigma_1^2\sigma_2}_{\sigma_1^2\sigma_2t_2} \\ f^{\Delta_2\sigma_1^2\sigma_2}_{t_2}\cdot N^{\Delta_1\sigma_1\sigma_2}_{\sigma_1^2\sigma_2t_1} & f^{\Delta_2\sigma_1^2\sigma_2}_{t_2}\cdot N^{\Delta_2\sigma_1^2\sigma_2}_{\sigma_1^2\sigma_2t_2} \end{pmatrix} = \begin{pmatrix} f^{\Delta_1\sigma_1\sigma_2}_{t_1}\cdot f^{\Delta_1\sigma_1\sigma_2}_{t_1} & f^{\Delta_1\sigma_1\sigma_2}_{t_1}\cdot f^{\Delta_2\sigma_1^2\sigma_2}_{t_2} \\ f^{\Delta_2\sigma_1^2\sigma_2}_{t_2}\cdot f^{\Delta_1\sigma_1\sigma_2}_{t_1} & f^{\Delta_2\sigma_1^2\sigma_2}_{t_2}\cdot f^{\Delta_2\sigma_1^2\sigma_2}_{t_2} \end{pmatrix}.$$

Let

$$\mathscr{H}_{a\sigma_1^2\sigma_2} = \begin{pmatrix} f^{\Delta_1\sigma_1\sigma_2}_{t_1}\cdot N^{\Delta_1\sigma_1\sigma_2}_{\sigma_1^2\sigma_2t_1} & f^{\Delta_1\sigma_1\sigma_2}_{t_1}\cdot N^{\Delta_2\sigma_1^2\sigma_2}_{\sigma_1^2\sigma_2t_2} \\ f^{\Delta_2\sigma_1^2\sigma_2}_{t_2}\cdot N^{\Delta_1\sigma_1\sigma_2}_{\sigma_1^2\sigma_2t_1} & f^{\Delta_2\sigma_1^2\sigma_2}_{t_2}\cdot N^{\Delta_2\sigma_1^2\sigma_2}_{\sigma_1^2\sigma_2t_2} \end{pmatrix}$$

and

$$\mathscr{G}_{a\sigma_1^2\sigma_2} = \begin{pmatrix} f^{\Delta_1\sigma_1\sigma_2}_{t_1}\cdot f^{\Delta_1\sigma_1\sigma_2}_{t_1} & f^{\Delta_1\sigma_1\sigma_2}_{t_1}\cdot f^{\Delta_2\sigma_1^2\sigma_2}_{t_2} \\ f^{\Delta_2\sigma_1^2\sigma_2}_{t_2}\cdot f^{\Delta_1\sigma_1\sigma_2}_{t_1} & f^{\Delta_2\sigma_1^2\sigma_2}_{t_2}\cdot f^{\Delta_2\sigma_1^2\sigma_2}_{t_2} \end{pmatrix}.$$

Then

$$\mathscr{H}_{a\sigma_1^2\sigma_2} = \mathscr{G}_{a\sigma_1^2\sigma_2}\mathscr{A}_{a\sigma_1^2\sigma_2}.$$

Let

$$E_{\sigma_1^2\sigma_2} = \left(f^{\Delta_1\sigma_1\sigma_2}_{t_1}\right)^2, \quad F_{\sigma_1^2\sigma_2} = f^{\Delta_2\sigma_1^2\sigma_2}_{t_2}\cdot f^{\Delta_1\sigma_1\sigma_2}_{t_1}, \quad G_{\sigma_1^2\sigma_2} = \left(f^{\Delta_2\sigma_1^2\sigma_2}_{t_2}\right)^2,$$

$$L_{\sigma_1^2\sigma_2} = f^{\Delta_1\sigma_1\sigma_2}_{t_1}\cdot N^{\Delta_1\sigma_1\sigma_2}_{\sigma_1^2\sigma_2t_1}, \quad M_{1\sigma_1^2\sigma_2} = f^{\Delta_1\sigma_1\sigma_2}_{t_1}\cdot N^{\Delta_2\sigma_1^2\sigma_2}_{\sigma_1^2\sigma_2t_2}, \quad M_{2\sigma_1^2\sigma_2} = f^{\Delta_2\sigma_1^2\sigma_2}_{t_2}\cdot N^{\Delta_1\sigma_1\sigma_2}_{\sigma_1^2\sigma_2t_1},$$

$$P_{\sigma_1^2\sigma_2} = f^{\Delta_2\sigma_1^2\sigma_2}_{t_2}\cdot N^{\Delta_2\sigma_1^2\sigma_2}_{\sigma_1^2\sigma_2t_2}.$$

Then

$$\mathscr{G}_{a\sigma_1^2\sigma_2} = \begin{pmatrix} E_{\sigma_1^2\sigma_2} & F_{\sigma_1^2\sigma_2} \\ F_{\sigma_1^2\sigma_2} & G_{\sigma_1^2\sigma_2} \end{pmatrix}, \quad \mathscr{H}_{a\sigma_1^2\sigma_2} = \begin{pmatrix} L_{\sigma_1^2\sigma_2} & M_{1\sigma_1^2\sigma_2} \\ M_{2\sigma_1^2\sigma_2} & P_{\sigma_1^2\sigma_2} \end{pmatrix}$$

and if

$$E_{\sigma_1^2\sigma_2}G_{\sigma_1^2\sigma_2} - F^2_{\sigma_1^2\sigma_2} \neq 0,$$

then

$$\mathscr{G}^{-1}_{a\sigma_1^2\sigma_2} = \frac{1}{E_{\sigma_1^2\sigma_2}G_{\sigma_1^2\sigma_2} - F^2_{\sigma_1^2\sigma_2}}\begin{pmatrix} G_{\sigma_1^2\sigma_2} & -F_{\sigma_1^2\sigma_2} \\ -F_{\sigma_1^2\sigma_2} & E_{\sigma_1^2\sigma_2} \end{pmatrix}$$

and

$$\begin{aligned}
\mathscr{A}_{a\sigma_1^2\sigma_2} &= \mathscr{G}^{-1}_{a\sigma_1^2\sigma_2}\mathscr{H}_{a\sigma_1^2\sigma_2}\\
&= \frac{1}{E_{\sigma_1^2\sigma_2}G_{\sigma_1^2\sigma_2}-F^2_{\sigma_1^2\sigma_2}}\begin{pmatrix} G_{\sigma_1^2\sigma_2} & -F_{\sigma_1^2\sigma_2}\\ -F_{\sigma_1^2\sigma_2} & E_{\sigma_1^2\sigma_2}\end{pmatrix}\begin{pmatrix} L_{\sigma_1^2\sigma_2} & M_{1\sigma_1^2\sigma_2}\\ M_{2\sigma_1^2\sigma_2} & P_{\sigma_1^2\sigma_2}\end{pmatrix}\\
&= \frac{1}{E_{\sigma_1^2\sigma_2}G_{\sigma_1^2\sigma_2}-F^2_{\sigma_1^2\sigma_2}}\begin{pmatrix} G_{\sigma_1^2\sigma_2}L_{\sigma_1^2\sigma_2}-F_{\sigma_1^2\sigma_2}M_{2\sigma_1^2\sigma_2} & G_{\sigma_1^2\sigma_2}M_{1\sigma_1^2\sigma_2}-F_{\sigma_1^2\sigma_2}P_{\sigma_1^2\sigma_2}\\ -F_{\sigma_1^2\sigma_2}L_{\sigma_1^2\sigma_2}+E_{\sigma_1^2\sigma_2}M_{2\sigma_1^2\sigma_2} & -F_{\sigma_1^2\sigma_2}M_{1\sigma_1^2\sigma_2}+E_{\sigma_1^2\sigma_2}P_{\sigma_1^2\sigma_2}\end{pmatrix}.
\end{aligned}$$

Definition 3.13 Suppose that σ_1 is Δ_1-differentiable and σ_2 is Δ_2-differentiable. For each vector $h=(h_1,h_2)\in T_{a\sigma_1\sigma_2^2}S$ the operator

$$\mathscr{A}_{a\sigma_1\sigma_2^2}(h) = N^{\Delta_1\sigma_1\sigma_2^2}_{\sigma_1\sigma_2^2t_1}h_1 + N^{\Delta_2\sigma_1\sigma_2}_{\sigma_1\sigma_2^2t_2}h_2$$

will be called the $\sigma_1\sigma_2^2$-shape operator.

In particular, we have

$$\mathscr{A}_{a\sigma_1\sigma_2^2}\left(f^{\Delta_1\sigma_1\sigma_2^2}_{t_1}\right) = N^{\Delta_1\sigma_1\sigma_2^2}_{\sigma_1\sigma_2^2t_1},$$

$$\mathscr{A}_{a\sigma_1\sigma_2^2}\left(f^{\Delta_2\sigma_1\sigma_2}_{t_2}\right) = N^{\Delta_2\sigma_1\sigma_2}_{\sigma_1\sigma_2^2t_2}.$$

Theorem 3.4 *Suppose that σ_1 is Δ_1-differentiable and σ_2 is Δ_2-differentiable. Then*

$$\left(1+\mu_1^{\Delta_1}\right)^{\sigma_1}f^{\Delta_1\sigma_1^2\sigma_2^3}_{t_1}\cdot\mathscr{A}_{a\sigma_1\sigma_2^2}\left(f^{\Delta_2\sigma_1\sigma_2}_{t_2}\right) = \left(1+\mu_2^{\Delta_2}\right)^{2\sigma_2}f^{\Delta_2\sigma_1^3\sigma_2^3}_{t_2}\cdot\mathscr{A}_{a\sigma_1\sigma_2^2}\left(f^{\Delta_1\sigma_1\sigma_2^2}_{t_1}\right).$$

Proof. We differentiate with respect to t_2 the equation

$$f^{\Delta_1\sigma_1\sigma_2^2}_{t_1}\cdot N_{\sigma_1\sigma_2^2}=0$$

and we find

$$f^{\Delta_1\sigma_1\sigma_2^2\Delta_2}_{t_1t_2}\cdot N^{\sigma_2}_{\sigma_1^2\sigma_2} + f^{\Delta_1\sigma_1\sigma_2^2}_{t_1}\cdot N^{\Delta_2}_{\sigma_1\sigma_2^2t_2}=0,$$

whereupon

$$\left(1+\mu_2^{\Delta_2}\right)^2 f^{\Delta_1\Delta_2\sigma_1\sigma_2^2}_{t_1t_2}\cdot N^{\sigma_2}_{\sigma_1\sigma_2^2} + f^{\Delta_1\sigma_1\sigma_2^2}_{t_1}\cdot N^{\Delta_2}_{\sigma_1\sigma_2^2t_2}=0$$

and

$$\left(1+\mu_2^{\Delta_2}\right)^{2\sigma_2} f^{\Delta_1\Delta_2\sigma_1^2\sigma_2^3}_{t_1t_2}\cdot N^{\sigma_1\sigma_2^2}_{\sigma_1\sigma_2^2} + f^{\Delta_1\sigma_1^2\sigma_2^3}_{t_1}\cdot N^{\Delta_2\sigma_1\sigma_2}_{\sigma_1\sigma_2^2t_2}=0. \tag{3.6}$$

We differentiate with respect to t_1 the equation

$$f^{\Delta_2\sigma_1\sigma_2}_{t_2}\cdot N_{\sigma_1\sigma_2^2}=0$$

and we find

$$f^{\Delta_2\sigma_1\sigma_2\Delta_1}_{t_2t_1}\cdot N_{\sigma_1\sigma_2^2} + f^{\Delta_2\sigma_1^2\sigma_2}_{t_2}\cdot N^{\Delta_1}_{\sigma_1\sigma_2^2t_1}=0$$

or

$$\left(1+\mu_1^{\Delta_1}\right)f_{t_1t_2}^{\Delta_1\Delta_2\sigma_1\sigma_2}\cdot N_{\sigma_1\sigma_2}+f_{t_2}^{\Delta_2\sigma_1\sigma_2^2}\cdot N_{\sigma_1\sigma_2^2t_1}^{\Delta_1}=0$$

and

$$(1+\mu_1^{\Delta_1})^{\sigma_1}f_{t_1t_2}^{\Delta_1\Delta_2\sigma_1^2\sigma_2^3}\cdot N_{\sigma_1\sigma_2^2}^{\sigma_1\sigma_2^2}+f_{t_2}^{\Delta_2\sigma_1^3\sigma_2^3}\cdot N_{\sigma_1\sigma_2^2t_1}^{\Delta_1\sigma_1\sigma_2^2}=0.$$

By (3.6) and the above equation, we arrive at

$$\left(1+\mu_2^{\Delta_2}\right)^{2\sigma_2}\left(1+\mu_1^{\Delta_1}\right)^{\sigma_1}f_{t_1t_2}^{\Delta_1\Delta_2\sigma_1^2\sigma_2^3}\cdot N_{\sigma_1\sigma_2^2}^{\sigma_1\sigma_2^2}+\left(1+\mu_1^{\Delta_1}\right)^{\sigma_1}f_{t_1}^{\Delta_1\sigma_1^2\sigma_2^3}\cdot N_{\sigma_1\sigma_2^2t_2}^{\Delta_2\sigma_1\sigma_2}=0$$

and

$$(1+\mu_1^{\Delta_1})^{\sigma_1}\left(1+\mu_2^{\Delta_2}\right)^{2\sigma_2}f_{t_1t_2}^{\Delta_1\Delta_2\sigma_1^2\sigma_2^3}\cdot N_{\sigma_1\sigma_2^2}^{\sigma_1\sigma_2^2}+\left(1+\mu_2^{\Delta_2}\right)^{2\sigma_2}f_{t_2}^{\Delta_2\sigma_1^3\sigma_2^3}\cdot N_{\sigma_1\sigma_2^2t_1}^{\Delta_1\sigma_1\sigma_2^2}=0.$$

Therefore

$$\left(1+\mu_1^{\Delta_1}\right)^{\sigma_1}f_{t_1}^{\Delta_1\sigma_1^2\sigma_2^3}\cdot N_{\sigma_1\sigma_2^2t_2}^{\Delta_2\sigma_1\sigma_2}=\left(1+\mu_2^{\Delta_2}\right)^{2\sigma_2}f_{t_2}^{\Delta_2\sigma_1^3\sigma_2^3}\cdot N_{\sigma_1\sigma_2^2t_1}^{\Delta_1\sigma_1\sigma_2^2}$$

and

$$\left(1+\mu_1^{\Delta_1}\right)^{\sigma_1}f_{t_1}^{\Delta_1\sigma_1^2\sigma_2^3}\cdot\mathscr{A}_{a\sigma_1\sigma_2^2}\left(f_{t_2}^{\Delta_2\sigma_1\sigma_2}\right)=\left(1+\mu_2^{\Delta_2}\right)^{2\sigma_2}f_{t_2}^{\Delta_2\sigma_1^3\sigma_2^3}\cdot\mathscr{A}_{a\sigma_1\sigma_2^2}\left(f_{t_1}^{\Delta_1\sigma_1\sigma_2^2}\right).$$

This completes the proof.

Now, we will search for a 2×2-matrix $\mathscr{A}_{a\sigma_1\sigma_2^2}$ so that

$$\left(N_{\sigma_1\sigma_2^2t_1}^{\Delta_1\sigma_1\sigma_2^2},N_{\sigma_1\sigma_2^2t_2}^{\Delta_2\sigma_1\sigma_2}\right)=\left(f_{t_1}^{\Delta_1\sigma_1\sigma_2^2},f_{t_2}^{\Delta_2\sigma_1\sigma_2}\right)\mathscr{A}_{a\sigma_1\sigma_2^2}.$$

We multiply the last equation, on the left, by the vector

$$\begin{pmatrix} f_{t_1}^{\Delta_1\sigma_1\sigma_2^2}\\ f_{t_2}^{\Delta_2\sigma_1\sigma_2}\end{pmatrix}$$

and we find

$$\begin{pmatrix} f_{t_1}^{\Delta_1\sigma_1\sigma_2^2}\\ f_{t_2}^{\Delta_2\sigma_1\sigma_2}\end{pmatrix}\left(N_{\sigma_1\sigma_2^2t_1}^{\Delta_1\sigma_1\sigma_2^2},N_{\sigma_1\sigma_2^2t_2}^{\Delta_2\sigma_1\sigma_2}\right)=\begin{pmatrix} f_{t_1}^{\Delta_1\sigma_1\sigma_2^2}\\ f_{t_2}^{\Delta_2\sigma_1\sigma_2}\end{pmatrix}\left(f_{t_1}^{\Delta_1\sigma_1\sigma_2^2},f_{t_2}^{\Delta_2\sigma_1\sigma_2}\right)\mathscr{A}_{a\sigma_1\sigma_2^2}$$

or

$$\begin{pmatrix} f_{t_1}^{\Delta_1\sigma_1\sigma_2^2}\cdot N_{\sigma_1\sigma_2^2t_1}^{\Delta_1\sigma_1\sigma_2^2} & f_{t_1}^{\Delta_1\sigma_1\sigma_2^2}\cdot N_{\sigma_1\sigma_2^2t_2}^{\Delta_2\sigma_1\sigma_2}\\ f_{t_2}^{\Delta_2\sigma_1\sigma_2}\cdot N_{\sigma_1\sigma_2^2t_1}^{\Delta_1\sigma_1\sigma_2^2} & f_{t_2}^{\Delta_2\sigma_1\sigma_2}\cdot N_{\sigma_1\sigma_2^2t_2}^{\Delta_2\sigma_1\sigma_2}\end{pmatrix}=\begin{pmatrix}\left(f_{t_1}^{\Delta_1\sigma_1\sigma_2^2}\right)^2 & f_{t_1}^{\Delta_1\sigma_1\sigma_2^2}\cdot f_{t_2}^{\Delta_2\sigma_1\sigma_2}\\ f_{t_1}^{\Delta_1\sigma_1\sigma_2^2}\cdot f_{t_2}^{\Delta_2\sigma_1\sigma_2} & \left(f_{t_2}^{\Delta_2\sigma_1\sigma_2}\right)^2\end{pmatrix}\mathscr{A}_{a\sigma_1\sigma_2^2}.$$

Let

$$\mathscr{H}_{a\sigma_1\sigma_2^2}=\begin{pmatrix} f_{t_1}^{\Delta_1\sigma_1\sigma_2^2}\cdot N_{\sigma_1\sigma_2^2t_1}^{\Delta_1\sigma_1\sigma_2^2} & f_{t_1}^{\Delta_1\sigma_1\sigma_2^2}\cdot N_{\sigma_1\sigma_2^2t_2}^{\Delta_2\sigma_1\sigma_2}\\ f_{t_2}^{\Delta_2\sigma_1\sigma_2}\cdot N_{\sigma_1\sigma_2^2t_1}^{\Delta_1\sigma_1\sigma_2^2} & f_{t_2}^{\Delta_2\sigma_1\sigma_2}\cdot N_{\sigma_1\sigma_2^2t_2}^{\Delta_2\sigma_1\sigma_2}\end{pmatrix}$$

and

$$\mathscr{G}_{a\sigma_1\sigma_2^2} = \begin{pmatrix} \left(f_{t_1}^{\Delta_1\sigma_1\sigma_2^2}\right)^2 & f_{t_1}^{\Delta_1\sigma_1\sigma_2^2}\cdot f_{t_2}^{\Delta_2\sigma_1\sigma_2} \\ f_{t_1}^{\Delta_1\sigma_1\sigma_2^2}\cdot f_{t_2}^{\Delta_2\sigma_1\sigma_2} & \left(f_{t_2}^{\Delta_2\sigma_1\sigma_2}\right)^2 \end{pmatrix}.$$

Let

$$E_{\sigma_1\sigma_2^2} = \left(f_{t_1}^{\Delta_1\sigma_1\sigma_2^2}\right)^2, \quad F_{\sigma_1\sigma_2^2} = f_{t_2}^{\Delta_2\sigma_1\sigma_2}\cdot f_{t_1}^{\Delta_1\sigma_1\sigma_2^2}, \quad G_{\sigma_1\sigma_2^2} = \left(f_{t_2}^{\Delta_2\sigma_1\sigma_2}\right)^2,$$

$$L_{\sigma_1\sigma_2^2} = f_{t_1}^{\Delta_1\sigma_1\sigma_2^2}\cdot N_{\sigma_1\sigma_2^2 t_1}^{\Delta_1\sigma_1\sigma_2^2}, \quad M_{1\sigma_1\sigma_2^2} = f_{t_1}^{\Delta_1\sigma_1\sigma_2^2}\cdot N_{\sigma_1\sigma_2^2 t_2}^{\Delta_2\sigma_1\sigma_2}, \quad M_{2\sigma_1\sigma_2^2} = f_{t_2}^{\Delta_2\sigma_1\sigma_2}\cdot N_{\sigma_1\sigma_2^2 t_1}^{\Delta_1\sigma_1\sigma_2^2},$$

$$P_{\sigma_1\sigma_2^2} = f_{t_2}^{\Delta_2\sigma_1\sigma_2^2}\cdot N_{\sigma_1\sigma_2^2 t_2}^{\Delta_2\sigma_1\sigma_2}.$$

Then

$$\mathscr{G}_{a\sigma_1\sigma_2^2} = \begin{pmatrix} E_{\sigma_1\sigma_2^2} & F_{\sigma_1\sigma_2^2} \\ F_{\sigma_1\sigma_2^2} & G_{\sigma_1\sigma_2^2} \end{pmatrix}, \quad \mathscr{H}_{a\sigma_1\sigma_2^2} = \begin{pmatrix} L_{\sigma_1\sigma_2^2} & M_{1\sigma_1\sigma_2^2} \\ M_{2\sigma_1\sigma_2^2} & P_{\sigma_1\sigma_2^2} \end{pmatrix}$$

and if

$$E_{\sigma_1\sigma_2^2}G_{\sigma_1\sigma_2^2} - F_{\sigma_1\sigma_2^2}^2 \neq 0,$$

then

$$\mathscr{G}_{a\sigma_1\sigma_2^2}^{-1} = \frac{1}{E_{\sigma_1\sigma_2^2}G_{\sigma_1\sigma_2^2} - F_{\sigma_1\sigma_2^2}^2}\begin{pmatrix} G_{\sigma_1\sigma_2^2} & -F_{\sigma_1\sigma_2^2} \\ -F_{\sigma_1\sigma_2^2} & E_{\sigma_1\sigma_2^2} \end{pmatrix}$$

and

$$\begin{aligned}
\mathscr{A}_{a\sigma_1\sigma_2^2} &= \mathscr{G}_{a\sigma_1\sigma_2^2}^{-1}\mathscr{H}_{a\sigma_1\sigma_2^2} \\
&= \frac{1}{E_{\sigma_1\sigma_2^2}G_{\sigma_1\sigma_2^2} - F_{\sigma_1\sigma_2^2}^2}\begin{pmatrix} G_{\sigma_1\sigma_2^2} & -F_{\sigma_1\sigma_2^2} \\ -F_{\sigma_1\sigma_2^2} & E_{\sigma_1\sigma_2^2} \end{pmatrix}\begin{pmatrix} L_{\sigma_1\sigma_2^2} & M_{1\sigma_1\sigma_2^2} \\ M_{2\sigma_1\sigma_2^2} & P_{\sigma_1\sigma_2^2} \end{pmatrix} \\
&= \frac{1}{E_{\sigma_1\sigma_2^2}G_{\sigma_1\sigma_2^2} - F_{\sigma_1\sigma_2^2}^2}\begin{pmatrix} G_{\sigma_1\sigma_2^2}L_{\sigma_1\sigma_2^2} - F_{\sigma_1\sigma_2^2}M_{2\sigma_1\sigma_2^2} & G_{\sigma_1\sigma_2^2}M_{1\sigma_1\sigma_2^2} - F_{\sigma_1\sigma_2^2}P_{\sigma_1\sigma_2^2} \\ -F_{\sigma_1\sigma_2^2}L_{\sigma_1\sigma_2^2} + E_{\sigma_1\sigma_2^2}M_{2\sigma_1\sigma_2^2} & -F_{\sigma_1\sigma_2^2}M_{1\sigma_1\sigma_2^2} + E_{\sigma_1\sigma_2^2}P_{\sigma_1\sigma_2^2} \end{pmatrix}.
\end{aligned}$$

Exercise 3.3 Let $\mathbb{T}_1$, $\mathbb{T}_2$, $\mathbb{T}_{(1)}$, $\mathbb{T}_{(2)}$, $\mathbb{T}_{(3)}$, a, S, U and f be as in Example 3.4. Find $\mathscr{A}_{a\sigma_1\sigma_2^2}$.

3.4 First Fundamental Forms

Definition 3.14 The first σ_1-fundamental form of a σ_1-regular surface S is the function $\phi_{1\sigma_1}$ that associates to each $a \in S$ the restriction of the inner product of $\mathbb{T}_{(1)} \times \mathbb{T}_{(2)} \times \mathbb{T}_{(3)}$ to $T_{a\sigma_1}S$.

Let (U,f) be a local σ_1-parametrization of a σ_1-regular surface $S \subseteq \mathbb{T}_{(1)} \times \mathbb{T}_{(2)} \times \mathbb{T}_{(3)}$. Take $p,q \in T_{a\sigma_1}S$ so that

$$p = p_1 f_{t_1}^{\Delta_1} + p_2 f_{t_2}^{\Delta_2\sigma_1},$$

$$q = q_1 f_{t_1}^{\Delta_1} + q_2 f_{t_2}^{\Delta_2\sigma_1},$$

where $p_1, p_2, q_1, q_2 \in \mathbb{R}$. Then

$$\begin{aligned} p \cdot q &= \left(p_1 f_{t_1}^{\Delta_1} + p_2 f_{t_2}^{\Delta_2\sigma_1}\right) \cdot \left(q_1 f_{t_1}^{\Delta_1} + q_2 f_{t_2}^{\Delta_2\sigma_1}\right) \\ &= p_1 q_1 \left(f_{t_1}^{\Delta_1}\right)^2 + p_1 q_2 \left(f_{t_1}^{\Delta_1} \cdot f_{t_2}^{\Delta_2\sigma_1}\right) + p_2 q_1 \left(f_{t_1}^{\Delta_1} \cdot f_{t_2}^{\Delta_2\sigma_1}\right) \\ &\quad + p_2 q_2 \left(f_{t_2}^{\Delta_2\sigma_1}\right)^2 \\ &= p_1 q_1 \left(f_{t_1}^{\Delta_1}\right)^2 + (p_1 q_2 + p_2 q_1) \left(f_{t_1}^{\Delta_1} \cdot f_{t_2}^{\Delta_2\sigma_1}\right) \\ &\quad + p_2 q_2 \left(f_{t_2}^{\Delta_2\sigma_1}\right)^2 \end{aligned}$$

If

$$p = \begin{pmatrix} p_1 \\ p_2 \end{pmatrix}, \quad q = \begin{pmatrix} q_1 \\ q_2 \end{pmatrix},$$

then

$$p \cdot q = p^T \mathscr{G}_{a\sigma_1} q.$$

Example 3.6 Let $\mathbb{T}_1 = \mathbb{T}_2 = \mathbb{T}_{(1)} = \mathbb{T}_{(2)} = \mathbb{T}_{(3)} = \mathbb{R}$,

$$U = \left(-\frac{\pi}{2}, \frac{\pi}{2}\right) \times \left(-\frac{\pi}{2}, \frac{\pi}{2}\right),$$

$f : U \to \mathbb{R}^3$ be given by

$$f(t_1, t_2) = (\cos t_1 \cos t_2, \cos t_1 \sin t_2, \sin t_1), \quad (t_1, t_2) \in U.$$

We will find the matrix $\mathscr{G}_{a\sigma_1}(t_1, t_2)$, $(t_1, t_2) \in U$. We have

$$\begin{aligned} f_1(t_1, t_2) &= \cos t_1 \cos t_2, \\ f_2(t_1, t_2) &= \cos t_1 \sin t_2, \\ f_3(t_1, t_2) &= \sin t_1, \quad (t_1, t_2) \in U. \end{aligned}$$

Then

$$\begin{aligned} f'_{1t_1}(t_1, t_2) &= -\sin t_1 \cos t_2, \\ f'_{1t_2}(t_1, t_2) &= -\cos t_1 \sin t_2, \\ f'_{2t_1}(t_1, t_2) &= -\sin t_1 \sin t_2, \\ f'_{2t_2}(t_1, t_2) &= \cos t_1 \cos t_2, \\ f'_{3t_1}(t_1, t_2) &= \cos t_1, \\ f'_{3t_2}(t_1, t_2) &= 0, \quad (t_1, t_2) \in U. \end{aligned}$$

Therefore

$$\begin{aligned} f'_{t_1}(t_1, t_2) &= (f'_{1t_1}(t_1, t_2), f'_{2t_1}(t_1, t_2), f'_{3t_1}(t_1, t_2)) \\ &= (-\sin t_1 \cos t_2, -\sin t_1 \sin t_2, \cos t_1), \end{aligned}$$

$$\begin{aligned} f'_{t_2}(t_1,t_2) &= (f'_{1t_2}(t_1,t_2), f'_{2t_2}(t_1,t_2), f'_{3t_2}(t_1,t_2)) \\ &= (-\cos t_1 \sin t_2, \cos t_1 \cos t_2, 0), \quad (t_1,t_2) \in U. \end{aligned}$$

Hence,

$$\begin{aligned} E(t_1,t_2) &= f'_{t_1}(t_1,t_2) \cdot f'_{t_1}(t_1,t_2) \\ &= (-\sin t_1 \cos t_2, -\sin t_1 \sin t_2, \cos t_1) \\ &\quad \cdot (-\sin t_1 \cos t_2, -\sin t_1 \sin t_2, \cos t_1) \\ &= \sin^2 t_1 \cos^2 t_2 + \sin^2 t_1 \sin^2 t_2 + \cos^2 t_1 \\ &= \sin^2 t_1 + \cos^2 t_1 \\ &= 1, \\ F_{\sigma_1}(t_1,t_2) &= f'_{t_1}(t_1,t_2) \cdot f'_{t_2}(t_1,t_2) \\ &= (-\sin t_1 \cos t_2, -\sin t_1 \sin t_2, \cos t_1) \\ &\quad \cdot (-\cos t_1 \sin t_2, \cos t_1 \cos t_2, 0) \\ &= \sin t_1 \cos t_1 \sin t_2 \cos t_2 - \sin t_1 \cos t_1 \sin t_2 \cos t_2 \\ &= 0, \\ G_{\sigma_1}(t_1,t_2) &= f'_{t_2}(t_1,t_2) \cdot f'_{t_2}(t_1,t_2) \\ &= (-\cos t_1 \sin t_2, \cos t_1 \cos t_2, 0) \\ &\quad \cdot (-\cos t_1 \sin t_2, \cos t_1 \cos t_2, 0) \\ &= \cos^2 t_1 \sin^2 t_2 + \cos^2 t_1 \cos^2 t_2 \\ &= \cos^2 t_1, \quad (t_1,t_2) \in U. \end{aligned}$$

Therefore

$$\begin{aligned} \mathcal{G}_{a\sigma_1}(t_1,t_2) &= \begin{pmatrix} E(t_1,t_2) & F_{\sigma_1}(t_1,t_2) \\ F_{\sigma_1}(t_1,t_2) & G_{\sigma_1}(t_1,t_2) \end{pmatrix} \\ &= \begin{pmatrix} 1 & 0 \\ 0 & \cos^2 t_1 \end{pmatrix}, \quad (t_1,t_2) \in U. \end{aligned}$$

Example 3.7 Let $\mathbb{T}_1 = \mathbb{Z}$, $\mathbb{T}_2 = 2^{\mathbb{N}_0}$, $\mathbb{T}_{(1)} = \mathbb{T}_{(2)} = \mathbb{T}_{(3)} = \mathbb{R}$, $U \subseteq \mathbb{T}_1 \times \mathbb{T}_2$, $(0,1) \in U$, $f : U \to \mathbb{T}_{(1)} \times \mathbb{T}_{(2)} \times \mathbb{T}_{(3)}$ be given by

$$f(t_1,t_2) = \left(\frac{1+t_2}{1+t_1^2}, t_1^3 + t_2^2, t_1 - 3t_2^2 \right), \quad (t_1,t_2) \in U.$$

We will find the matrix $\mathcal{G}_{a\sigma_1}(0,1)$. We have

$$\begin{aligned} f_1(t_1,t_2) &= \frac{1+t_2}{1+t_1^2}, \\ f_2(t_1,t_2) &= t_1^3 + t_2^2, \end{aligned}$$

$$f_3(t_1,t_2) = t_1 - 3t_2^2, \quad (t_1,t_2) \in U.$$

Then

$$\begin{aligned}
f_{1t_1}^{\Delta_1}(t_1,t_2) &= -(1+t_2)\frac{\sigma_1(t_1)+t_1}{(1+t_1^2)(1+(\sigma_1(t_1))^2)} \\
&= -(1+t_2)\frac{1+t_1+t_1}{(1+t_1^2)(1+(t_1+1)^2)} \\
&= -\frac{(1+2t_1)(1+t_2)}{(1+t_1^2)(1+t_1^2+2t_1+1)} \\
&= -\frac{(1+2t_1)(1+t_2)}{(1+t_1^2)(2+2t_1+t_1^2)}, \\
f_{1t_2}^{\Delta_2}(t_1,t_2) &= \frac{1}{1+t_1^2}, \\
f_{2t_1}^{\Delta_1}(t_1,t_2) &= (\sigma_1(t_1))^2 + t_1\sigma_1(t_1) + t_1^2 \\
&= (t_1+1)^2 + t_1(t_1+1) + t_1^2 \\
&= t_1^2 + 2t_1 + 1 + t_1^2 + t_1 + t_1^2 \\
&= 3t_1^2 + 3t_1 + 1, \\
f_{2t_2}^{\Delta_2}(t_1,t_2) &= \sigma_2(t_2) + t_2 \\
&= 2t_2 + t_2 \\
&= 3t_2, \\
f_{3t_1}^{\Delta_1}(t_1,t_2) &= 1, \\
f_{3t_2}^{\Delta_2}(t_1,t_2) &= -3(\sigma_2(t_2)+t_2) \\
&= -3(2t_2+t_2) \\
&= -9t_2, \quad (t_1,t_2) \in U.
\end{aligned}$$

Therefore

$$\begin{aligned}
f_{t_1}^{\Delta_1}(t_1,t_2) &= \left(-\frac{(1+2t_1)(1+t_2)}{(1+t_1^2)(2+2t_1+t_1^2)}, 3t_1^2+3t_1+1, 1\right), \\
f_{t_2}^{\Delta_2}(t_1,t_2) &= \left(\frac{1}{1+t_1^2}, 3t_2, -9t_2\right), \quad (t_1,t_2) \in U.
\end{aligned}$$

Hence,

$$f_{t_1}^{\Delta_1}(0,1) = (-1,1,1).$$

Next,

$$f_{t_2}^{\Delta_2\sigma_1}(t_1,t_2) = \left(\frac{1}{1+(\sigma_1(t_1))^2}, 3t_2, -9t_2\right)$$

$$
\begin{aligned}
&= \left(\frac{1}{1+(t_1+1)^2}, 3t_2, -9t_2 \right) \\
&= \left(\frac{1}{1+t_1^2+2t_1+1}, 3t_2, -9t_2 \right) \\
&= \left(\frac{1}{2+2t_1+t_1^2}, 3t_2, -9t_2 \right), \quad (t_1, t_2) \in U.
\end{aligned}
$$

Then

$$
\begin{aligned}
f_{t_2}^{\Delta_2\sigma_1}(0,1) &= \left(\frac{1}{2}, 3, -9 \right), \\
E(0,1) &= 3, \\
F_{\sigma_1}(0,1) &= f_{t_1}^{\Delta_1} 0,1) \cdot f_{t_2}^{\Delta_2\sigma_1}(0,1) \\
&= (-1,1,1) \cdot \left(\frac{1}{2}, 3, -9 \right) \\
&= -\frac{1}{2} + 3 - 9 \\
&= -\frac{1}{2} - 6 \\
&= -\frac{13}{2}, \\
G_{\sigma_1}(0,1) &= f_{t_2}^{\Delta_2\sigma_1}(0,1) \cdot f_{t_2}^{\Delta_2\sigma_1}(0,1) \\
&= \left(\frac{1}{2}, 3, -9 \right) \cdot \left(\frac{1}{2}, 3, -9 \right) \\
&= \frac{1}{4} + 9 + 81 \\
&= \frac{1}{4} + 90 \\
&= \frac{361}{4}
\end{aligned}
$$

and

$$
\begin{aligned}
\mathscr{G}_{a\sigma_1}(0,1) &= \begin{pmatrix} E(0,1) & F_{\sigma_1}(0,1) \\ F_{\sigma_1}(0,1) & G_{\sigma_1}(0,1) \end{pmatrix} \\
&= \begin{pmatrix} 3 & -\frac{13}{2} \\ -\frac{13}{2} & \frac{361}{4} \end{pmatrix}.
\end{aligned}
$$

Exercise 3.4 Let $\mathbb{T}_1 = 7^{\mathbb{N}_0}$, $\mathbb{T}_2 = 4\mathbb{N}_0$, $\mathbb{T}_{(1)} = \mathbb{T}_{(2)} = \mathbb{T}_{(3)} = \mathbb{R}$, $U \subset \mathbb{T}_1 \times \mathbb{T}_2$, $f : U \to \mathbb{T}_{(1)} \times \mathbb{T}_{(2)} \times \mathbb{T}_{(3)}$ be given by

$$f(t_1,t_2) = \left(t_1t_2^3, t_1^4t_2, t_1 + t_2^2 + 3t_1^3t_2^5\right), \quad (t_1,t_2) \in U.$$

Find the matrix $\mathscr{G}_{a\sigma_1}(t_1,t_2)$, $(t_1,t_2) \in U$.

Definition 3.15 The first σ_2-fundamental form of a σ_2-regular surface S is the function $\phi_{1\sigma_2}$ that associates to each $a \in S$ the restriction of the inner product of $\mathbb{T}_{(1)} \times \mathbb{T}_{(2)} \times \mathbb{T}_{(3)}$ to $T_{a\sigma_2}S$.

Let (U,f) be a local σ_2-parametrization of a σ_2-regular surface $S \subseteq \mathbb{T}_{(1)} \times \mathbb{T}_{(2)} \times \mathbb{T}_{(3)}$. Take $p,q \in T_{a\sigma_1}S$ so that

$$p = p_1 f_{t_1}^{\Delta_1\sigma_2} + p_2 f_{t_2}^{\Delta_2},$$
$$q = q_1 f_{t_1}^{\Delta_1\sigma_2} + q_2 f_{t_2}^{\Delta_2},$$

where $p_1, p_2, q_1, q_2 \in \mathbb{R}$.

Then

$$\begin{aligned} p \cdot q &= \left(p_1 f_{t_1}^{\Delta_1\sigma_2} + p_2 f_{t_2}^{\Delta_2}\right) \cdot \left(q_1 f_{t_1}^{\Delta_1\sigma_2} + q_2 f_{t_2}^{\Delta_2}\right) \\ &= p_1q_1 \left(f_{t_1}^{\Delta_1\sigma_2}\right)^2 + p_1q_2 \left(f_{t_1}^{\Delta_1\sigma_2} \cdot f_{t_2}^{\Delta_2}\right) + p_2q_1 \left(f_{t_1}^{\Delta_1\sigma_2} \cdot f_{t_2}^{\Delta_2}\right) \\ &\quad + p_2q_2 \left(f_{t_2}^{\Delta_2}\right)^2 \\ &= p_1q_1 \left(f_{t_1}^{\Delta_1\sigma_2}\right)^2 + (p_1q_2 + p_2q_1)\left(f_{t_1}^{\Delta_1\sigma_2} \cdot f_{t_2}^{\Delta_2}\right) \\ &\quad + p_2q_2 \left(f_{t_2}^{\Delta_2}\right)^2. \end{aligned}$$

If

$$p = \begin{pmatrix} p_1 \\ p_2 \end{pmatrix}, \quad q = \begin{pmatrix} q_1 \\ q_2 \end{pmatrix},$$

then

$$p \cdot q = p^T \mathscr{G}_{a\sigma_2} q.$$

Example 3.8 Consider Example 3.7. We will find $\mathscr{G}_{a\sigma_2}(0,1)$.

We have

$$\begin{aligned} f_{t_1}^{\Delta_1\sigma_2}(t_1,t_2) &= \left(-\frac{(1+2t_1)(1+\sigma_2(t_2))}{(1+t_1^2)(2+2t_1+t_1^2)}, 3t_1^2 + 3t_1 + 1, 1\right) \\ &= \left(-\frac{(1+2t_1)(1+2t_2)}{(1+t_1^2)(2+2t_1+t_1^2)}, 3t_1^2 + 3t_1 + 1, 1\right), \end{aligned}$$

$$f_{t_1}^{\Delta_1\sigma_2}(0,1) = \left(-\frac{3}{2}, 1, 1\right),$$

$$f_{t_2}^{\Delta_2}(0,1) = (1,3,-9).$$

Then

$$\begin{aligned}
E_{\sigma_2}(0,1) &= f_{t_1}^{\Delta_1\sigma_2}(0,1)\cdot f_{t_1}^{\Delta_1\sigma_2}(0,1)\\
&= \left(-\frac{3}{2},1,1\right)\cdot\left(-\frac{3}{2},1,1\right)\\
&= \frac{9}{4}+1+1\\
&= \frac{9}{4}+2\\
&= \frac{17}{4},\\
F_{\sigma_2}(0,1) &= f_{t_1}^{\Delta_1\sigma_2}(0,1)\cdot f_{t_2}^{\Delta_2}(0,1)\\
&= \left(-\frac{3}{2},1,1\right)\cdot(1,3,-9)\\
&= -\frac{3}{2}+3-9\\
&= -\frac{3}{2}-6\\
&= -\frac{15}{2},\\
G(0,1) &= f_{t_2}^{\Delta_2}(0,1)\cdot f_{t_2}^{\Delta_2}(0,1)\\
&= (1,3,-9)\cdot(1,3,-9)\\
&= 1+9+81\\
&= 91
\end{aligned}$$

and

$$\begin{aligned}
\mathcal{G}_{a\sigma_2}(0,1) &= \begin{pmatrix} E_{\sigma_2}(0,1) & F_{\sigma_2}(0,1)\\ F_{\sigma_2}(0,1) & G(0,1)\end{pmatrix}\\
&= \begin{pmatrix} \frac{17}{4} & -\frac{15}{2}\\ -\frac{15}{2} & 91\end{pmatrix}.
\end{aligned}$$

Exercise 3.5 Let $\mathbb{T}_1 = 13\mathbb{N}_0$, $\mathbb{T}_2 = 2^{\mathbb{N}_0}$, $\mathbb{T}_{(1)} = \mathbb{T}_{(2)} = \mathbb{T}_{(3)} = \mathbb{R}$, $U \subset \mathbb{T}_1\times\mathbb{T}_2$, $f: U \to \mathbb{T}_{(1)}\times\mathbb{T}_{(2)}\times\mathbb{T}_{(3)}$ be given by

$$f(t_1,t_2) = \left(\sinh_1(t_2,1)+t_1t_2, t_1+t_2^3, t_2^2-4t_1^7t_2^8\right),\quad (t_1,t_2)\in U.$$

Find the matrix $\mathcal{G}_{a\sigma_2}(t_1,t_2)$, $(t_1,t_2)\in U$.

Definition 3.16 The first $\sigma_1^2\sigma_2$-fundamental form of a $\sigma_1^2\sigma_2$-regular surface S is the function $\phi_{1\sigma_1^2\sigma_2}$ that associates to each $a\in S$ the restriction of the inner product of $\mathbb{T}_{(1)}\times\mathbb{T}_{(2)}\times\mathbb{T}_{(3)}$ to $T_{a\sigma_1^2\sigma_2}S$.

Let (U,f) be a local $\sigma_1^2\sigma_2$-parameterization of a $\sigma_1^2\sigma_2$-regular surface $S \subseteq \mathbb{T}_{(1)} \times \mathbb{T}_{(2)} \times \mathbb{T}_{(3)}$. Take $p,q \in T_{a\sigma_1^2\sigma_2}S$ so that

$$p = p_1 f_{t_1}^{\Delta_1\sigma_1\sigma_2} + p_2 f_{t_2}^{\Delta_2\sigma_1^2\sigma_2},$$
$$q = q_1 f_{t_1}^{\Delta_1\sigma_1\sigma_2} + q_2 f_{t_2}^{\Delta_2\sigma_1^2\sigma_2},$$

where $p_1, p_2, q_1, q_2 \in \mathbb{R}$.

Then

$$\begin{aligned} p \cdot q &= \left(p_1 f_{t_1}^{\Delta_1\sigma_1\sigma_2} + p_2 f_{t_2}^{\Delta_2\sigma_1^2\sigma_2}\right) \cdot \left(q_1 f_{t_1}^{\Delta_1\sigma_1\sigma_2} + q_2 f_{t_2}^{\Delta_2\sigma_1^2\sigma_2}\right) \\ &= p_1 q_1 \left(f_{t_1}^{\Delta_1\sigma_1\sigma_2}\right)^2 + p_1 q_2 \left(f_{t_1}^{\Delta_1\sigma_1\sigma_2} \cdot f_{t_2}^{\Delta_2\sigma_1^2\sigma_2}\right) + p_2 q_1 \left(f_{t_1}^{\Delta_1\sigma_1\sigma_2} \cdot f_{t_2}^{\Delta_2\sigma_1^2\sigma_2}\right) \\ &\quad + p_2 q_2 \left(f_{t_2}^{\Delta_2\sigma_1^2\sigma_2}\right)^2 \\ &= p_1 q_1 \left(f_{t_1}^{\Delta_1\sigma_1\sigma_2}\right)^2 + (p_1 q_2 + p_2 q_1)\left(f_{t_1}^{\Delta_1\sigma_1\sigma_2} \cdot f_{t_2}^{\Delta_2\sigma_1^2\sigma_2}\right) \\ &\quad + p_2 q_2 \left(f_{t_2}^{\Delta_2\sigma_1^2\sigma_2}\right)^2 \end{aligned}$$

If

$$p = \begin{pmatrix} p_1 \\ p_2 \end{pmatrix}, \quad q = \begin{pmatrix} q_1 \\ q_2 \end{pmatrix},$$

then

$$p \cdot q = p^T \mathcal{G} a\sigma_1^2\sigma_2 q.$$

Exercise 3.6 Let $\mathbb{T}_1 = 3^{\mathbb{N}_0}$, $\mathbb{T}_2 = 4^{\mathbb{N}_0}$, $\mathbb{T}_{(1)} = \mathbb{T}_{(2)} = \mathbb{T}_{(3)} = \mathbb{R}$, $U \subset \mathbb{T}_1 \times \mathbb{T}_2$, $f : U \to \mathbb{T}_{(1)} \times \mathbb{T}_{(2)} \times \mathbb{T}_{(3)}$ be given by

$$f(t_1,t_2) = \left(t_1^2 + t_2^2 + t_1 t_2, t_1 + t_2 t_1^3 + t_2, t_2^2 - t_1^2 t_2^3\right), \quad (t_1,t_2) \in U.$$

Find the matrix $\mathcal{G}_{a\sigma_1^2\sigma_2}(t_1,t_2)$, $(t_1,t_2) \in U$.

Definition 3.17 The first $\sigma_1\sigma_2^2$-fundamental form of a $\sigma_1\sigma_2^2$-regular surface S is the function $\phi_{1\sigma_1\sigma_2^2}$ that associates to each $a \in S$ the restriction of the inner product of $\mathbb{T}_{(1)} \times \mathbb{T}_{(2)} \times \mathbb{T}_{(3)}$ to $T_{a\sigma_1\sigma_2^2}S$.

Let (U,f) be a local $\sigma_1\sigma_2^2$-parameterization of a $\sigma_1\sigma_2^2$-regular surface $S \subseteq \mathbb{T}_{(1)} \times \mathbb{T}_{(2)} \times \mathbb{T}_{(3)}$. Take $p,q \in T_{a\sigma_1\sigma_2^2}S$ so that

$$p = p_1 f_{t_1}^{\Delta_1\sigma_1\sigma_2^2} + p_2 f_{t_2}^{\Delta_2\sigma_1\sigma_2},$$
$$q = q_1 f_{t_1}^{\Delta_1\sigma_1\sigma_2^2} + q_2 f_{t_2}^{\Delta_2\sigma_1\sigma_2},$$

where $p_1, p_2, q_1, q_2 \in \mathbb{R}$.

Then

$$\begin{aligned} p\cdot q &= \left(p_1 f_{t_1}^{\Delta_1\sigma_1\sigma_2^2} + p_2 f_{t_2}^{\Delta_2\sigma_1\sigma_2}\right)\cdot\left(q_1 f_{t_1}^{\Delta_1\sigma_1\sigma_2^2} + q_2 f_{t_2}^{\Delta_2\sigma_1\sigma_2}\right)\\ &= p_1q_1\left(f_{t_1}^{\Delta_1\sigma_1\sigma_2^2}\right)^2 + p_1q_2\left(f_{t_1}^{\Delta_1\sigma_1\sigma_2^2}\cdot f_{t_2}^{\Delta_2\sigma_1\sigma_2}\right) + p_2q_1\left(f_{t_1}^{\Delta_1\sigma_1\sigma_2^2}\cdot f_{t_2}^{\Delta_2\sigma_1\sigma_2}\right)\\ &\quad + p_2q_2\left(f_{t_2}^{\Delta_2\sigma_1\sigma_2}\right)^2\\ &= p_1q_1\left(f_{t_1}^{\Delta_1\sigma_1\sigma_2^2}\right)^2 + (p_1q_2+p_2q_1)\left(f_{t_1}^{\Delta_1\sigma_1\sigma_2^2}\cdot f_{t_2}^{\Delta_2\sigma_1\sigma_2}\right)\\ &\quad + p_2q_2\left(f_{t_2}^{\Delta_2\sigma_1\sigma_2}\right)^2. \end{aligned}$$

If

$$p = \begin{pmatrix} p_1 \\ p_2 \end{pmatrix}, \quad q = \begin{pmatrix} q_1 \\ q_2 \end{pmatrix},$$

then

$$p\cdot q = p^T \mathcal{G}_{a\sigma_1\sigma_2^2} q.$$

Exercise 3.7 Let $\mathbb{T}_1 = 3\mathbb{N}_0$, $\mathbb{T}_2 = 2\mathbb{N}_0$, $\mathbb{T}_{(1)} = \mathbb{T}_{(2)} = \mathbb{T}_{(3)} = \mathbb{R}$, $U \subset \mathbb{T}_1 \times \mathbb{T}_2$, $f : U \to \mathbb{T}_{(1)} \times \mathbb{T}_{(2)} \times \mathbb{T}_{(3)}$ be given by

$$f(t_1,t_2) = \left(\frac{1+t_1+t_1^2}{1+t_2+t_2^2}, t_1+t_2, t_2^2 - t_1^3\right), \quad (t_1,t_2)\in U.$$

Find the matrix $\mathcal{G}_{a\sigma_1\sigma_2^2}(t_1,t_2)$, $(t_1,t_2)\in U$.

3.5 Lengths of a Curve on a Surface

Let S is a σ_1- or σ_2- or $\sigma_1^2\sigma_2$- or $\sigma_1\sigma_2^2$-regular surface, (U,f) is a local σ_1- or σ_2- or $\sigma_1^2\sigma_2$- or $\sigma_1\sigma_2^2$- parameterization of S, (I,g) is a parameterized curve with $g(I) \subseteq f(U)$ given by

$$g(t) = (g_1(t), g_2(t)), \quad t\in I,$$

where

$$\begin{aligned} g_1(\sigma(t)) &= \sigma_1(g_1(t)),\\ g_2(\sigma(t)) &= \sigma_2(g_2(t)), \quad t\in I, \end{aligned}$$

$I = [b_1,b_2] \subset \mathbb{T}$, $b_1 < b_2$. We have that

$$g^{\Delta}(t) = (g_1^{\Delta}(t), g_2^{\Delta}(t)), \quad t\in I,$$

is the tangent vector to the curve (I,g).

Let

$$q_1(t) = E(g_1(t), g_2(t))(g_1^\Delta(t))^2 + 2F_{\sigma_1}(g_1(t), g_2(t))g_1^\Delta(t)g_2^\Delta(t)$$
$$+G_{\sigma_1}(g_1(t), g_2(t))\left(g_2^\Delta(t)\right)^2$$

in the case when S is σ_1-regular, and

$$q_2(t) = E_{\sigma_2}(g_1(t), g_2(t))(g_1^\Delta(t))^2 + 2F_{\sigma_2}(g_1(t), g_2(t))g_1^\Delta(t)g_2^\Delta(t)$$
$$+G(g_1(t), g_2(t))\left(g_2^\Delta(t)\right)^2, \quad t \in I,$$

in the case when S is σ_2-regular.

If S is σ_1-regular, then we have

$$(f(g_1, g_2))^\Delta(t) = f_{t_1}^{\Delta_1}(g_1(t), g_2(t))g_1^\Delta(t) + f_{t_2}^{\Delta_2\sigma_1}(g_1(t), g_2(t))g_2^\Delta(t), \tag{3.7}$$

$t \in I$, and

$$\begin{aligned}
\|(f(g_1, g_2))^\Delta(t)\|^2 &= \left(f_{t_1}^{\Delta_1}(g_1(t), g_2(t))g_1^\Delta(t) + f_{t_2}^{\Delta_2\sigma_1}(g_1(t), g_2(t))g_2^\Delta(t)\right)\\
&\quad \cdot \left(f_{t_1}^{\Delta_1}(g_1(t), g_2(t))g_1^\Delta(t) + f_{t_2}^{\Delta_2\sigma_1}(g_1(t), g_2(t))g_2^\Delta(t)\right)\\
&= \left(f_{t_1}^{\Delta_1}(g_1(t), g_2(t))\right)^2\left(g_1^\Delta(t)\right)^2\\
&\quad +2f_{t_1}^{\Delta_1}(g_1(t), g_2(t))f_{t_2}^{\Delta_2\sigma_1}(g_1(t), g_2(t))g_1^\Delta(t)g_2^\Delta(t)\\
&\quad +\left(f_{t_2}^{\Delta_2\sigma_1}(g_1(t), g_2(t))\right)^2\left(g_2^\Delta(t)\right)^2\\
&= E(g_1(t), g_2(t))\left(g_1^\Delta(t)\right)^2 + 2F_{\sigma_1}(g_1(t), g_2(t))g_1^\Delta(t)g_2^\Delta(t)\\
&\quad +G_{\sigma_1}(g_1(t), g_2(t))\left(g_2^\Delta(t)\right)^2\\
&= q_1(t), \quad t \in I.
\end{aligned}$$

If S is σ_2-regular, then we have

$$(f(g_1, g_2))^\Delta(t) = f_{t_1}^{\Delta_1\sigma_2}(g_1(t), g_2(t))g_1^\Delta(t) + f_{t_2}^{\Delta_2}(g_1(t), g_2(t))g_2^\Delta(t), \tag{3.8}$$

$t \in I$, and

$$\begin{aligned}
\|(f(g_1, g_2))^\Delta(t)\|^2 &= \left(f_{t_1}^{\Delta_1\sigma_2}(g_1(t), g_2(t))g_1^\Delta(t) + f_{t_2}^{\Delta_2}(g_1(t), g_2(t))g_2^\Delta(t)\right)\\
&\quad \cdot \left(f_{t_1}^{\Delta_1\sigma_2}(g_1(t), g_2(t))g_1^\Delta(t) + f_{t_2}^{\Delta_2}(g_1(t), g_2(t))g_2^\Delta(t)\right)\\
&= \left(f_{t_1}^{\Delta_1\sigma_2}(g_1(t), g_2(t))\right)^2\left(g_1^\Delta(t)\right)^2\\
&\quad +2f_{t_1}^{\Delta_1\sigma_2}(g_1(t), g_2(t))f_{t_2}^{\Delta_2}(g_1(t), g_2(t))g_1^\Delta(t)g_2^\Delta(t)\\
&\quad +\left(f_{t_2}^{\Delta_2}(g_1(t), g_2(t))\right)^2\left(g_2^\Delta(t)\right)^2\\
&= E_{\sigma_2}(g_1(t), g_2(t))\left(g_1^\Delta(t)\right)^2 + 2F_{\sigma_2}(g_1(t), g_2(t))g_1^\Delta(t)g_2^\Delta(t)\\
&\quad +G(g_1(t), g_2(t))\left(g_2^\Delta(t)\right)^2\\
&= q_2(t), \quad t \in I.
\end{aligned}$$

Definition 3.18 Let S be σ_1-regular. The σ_1-length of the curve (I,g) is defined by

$$l_{b_1,b_2,\sigma_1} = \int_{b_1}^{b_2} \sqrt{q_1(t)}\Delta t.$$

Definition 3.19 Let S be σ_2-regular. The σ_2-length of the curve (I,g) is defined by

$$l_{b_1,b_2,\sigma_2} = \int_{b_1}^{b_2} \sqrt{q_2(t)}\Delta t.$$

Example 3.9 Let $\mathbb{T} = \mathbb{N}_0$, $\mathbb{T}_1 = \mathbb{N}_0$, $\mathbb{T}_2 = \mathbb{N}_0^2$, $\mathbb{T}_{(1)} = \mathbb{T}_{(2)} = \mathbb{T}_{(3)} = \mathbb{R}$, S is a surface with a local parameterization (U,f), where $U \subseteq \mathbb{T}_1 \times \mathbb{T}_2$, $f : U \to \mathbb{R}^3$ be defined by

$$f(t_1,t_2) = \left(t_1 + t_2^2, t_1^2 t_2^2, t_1 + t_2\right), \quad (t_1,t_2) \in \mathbb{T}_1 \times \mathbb{T}_2,$$

$g : \mathbb{T} \to \mathbb{T}_1 \times \mathbb{T}_2$ be a curve given by

$$g(t) = (t,t^2), \quad t \in \mathbb{T}.$$

We will find $l_{0,2,\sigma_1}$ and $l_{0,2,\sigma_2}$.

Here

$$\begin{aligned}
\sigma(t) &= t+1, \quad t \in \mathbb{T},\\
\sigma_1(t_1) &= t_1 + 1, \quad t_1 \in \mathbb{T}_1,\\
\sigma_2(t_2) &= (\sqrt{t_2} + 1)^2, \quad t_2 \in \mathbb{T}_2,\\
g_1(t) &= t,\\
g_2(t) &= t^2, \quad t \in \mathbb{T},\\
f_1(t_1,t_2) &= t_1 + t_2^2,\\
f_2(t_1,t_2) &= t_1^2 t_2^2,\\
f_3(t_1,t_2) &= t_1 + t_2, \quad (t_1,t_2) \in \mathbb{T}_1 \times \mathbb{T}_2,\\
g_1(\sigma(t)) &= \sigma(t)\\
&= t+1\\
&= \sigma_1(g_1(t)),\\
g_2(\sigma(t)) &= (\sigma(t))^2\\
&= (t+1)^2,\\
\sigma_2(g_2(t)) &= \sigma_2(t^2)\\
&= (t+1)^2\\
&= g_2(\sigma(t)), \quad t \in \mathbb{T}.
\end{aligned}$$

We have

$$f_{1t_1}^{\Delta_1}(t_1,t_2) = 1,$$

$$\begin{aligned}
f_{1t_2}^{\Delta_2}(t_1,t_2) &= \sigma_2(t_2)+t_2\\
&= (\sqrt{t_2}+1)^2+t_2\\
&= t_2+2\sqrt{t_2}+1+t_2\\
&= 2t_2+2\sqrt{t_2}+1,\\
f_{2t_1}^{\Delta_1}(t_1,t_2) &= (\sigma_1(t_1)+t_1)t_2^2\\
&= (t_1+1+t_1)t_2^2\\
&= (2t_1+1)t_2^2,\\
f_{2t_2}^{\Delta_2}(t_1,t_2) &= t_1^2(\sigma_2(t_2)+t_2)\\
&= t_1^2((\sqrt{t_2}+1)^2+t_2)\\
&= t_1^2(t_2+2\sqrt{t_2}+1+t_2)\\
&= t_1^2(2t_2+2\sqrt{t_2}+1),\\
f_{3t_1}^{\Delta_1}(t_1,t_2) &= 1,\\
f_{3t_2}^{\Delta_2}(t_1,t_2) &= 1, \quad (t_1,t_2)\in\mathbb{T}_1\times\mathbb{T}_2.
\end{aligned}$$

Then

$$\begin{aligned}
f_{t_1}^{\Delta_1}(t_1,t_2) &= \left(f_{1t_1}^{\Delta_1}(t_1,t_2), f_{2t_1}^{\Delta_1}(t_1,t_2), f_{3t_1}^{\Delta_1}(t_1,t_2)\right)\\
&= \left(1,(2t_1+1)t_2^2,1\right),\\
f_{t_1}^{\Delta_1\sigma_2}(t_1,t_2) &= \left(1,(2t_1+1)(\sigma_2(t_2))^2,1\right)\\
&= \left(1,(2t_1+1)(\sqrt{t_2}+1)^4,1\right),\\
f_{t_2}^{\Delta_2}(t_1,t_2) &= \left(f_{1t_2}^{\Delta_2}(t_1,t_2), f_{2t_2}^{\Delta_2}(t_1,t_2), f_{3t_2}^{\Delta_2}(t_1,t_2)\right)\\
&= \left(2t_2+2\sqrt{t_2}+1, t_1^2(2t_2+2\sqrt{t_2}+1),1\right),\\
f_{t_2}^{\Delta_2\sigma_1}(t_1,t_2) &= \left(2t_2+2\sqrt{t_2}+1,(\sigma_1(t_1))^2(2t_2+2\sqrt{t_2}+1),1\right)\\
&= \left(2t_2+2\sqrt{t_2}+1,(t_1+1)^2(2t_2+2\sqrt{t_2}+1),1\right), \quad (t_1,t_2)\in\mathbb{T}_1\times\mathbb{T}_2,
\end{aligned}$$

and

$$\begin{aligned}
f_{t_1}^{\Delta_1}(g_1(t),g_2(t)) &= \left(1,(2t+1)t^4,1\right),\\
f_{t_1}^{\Delta_1\sigma_2}(g_1(t),g_2(t)) &= \left(1,(2t+1)(t+1)^4,1\right),\\
f_{t_2}^{\Delta_2}(t_1,t_2) &= \left(2t^2+2t+1,t^2(2t^2+2t+1),1\right),\\
f_{t_2}^{\Delta_2\sigma_1}(t_1,t_2) &= (2t^2+2t+1,(t+1)^2(2t^2+2t+1),1), \quad t\in\mathbb{T},
\end{aligned}$$

and

$$g_1^{\Delta}(t)=1,$$

$$\begin{aligned} g_2^{\Delta}(t) &= \sigma(t) + t \\ &= t + 1 + t \\ &= 2t + 1, \quad t \in \mathbb{T}, \end{aligned}$$

and

$$\begin{aligned}
E(g_1(t), g_2(t)) &= f_{t_1}^{\Delta_1}(g_1(t), g_2(t)) \cdot f_{t_1}^{\Delta_1}(g_1(t), g_2(t)) \\
&= \left(1, (2t+1)t^4, 1\right) \cdot \left(1, (2t+1)t^4, 1\right) \\
&= 1 + (2t+1)^2 t^8 + 1 \\
&= 2 + (2t+1)^2 t^8, \\
E_{\sigma_2}(g_1(t), g_2(t)) &= f_{t_1}^{\Delta_1\sigma_2}(g_1(t), g_2(t)) \cdot f_{t_1}^{\Delta_1\sigma_2}(g_1(t), g_2(t)) \\
&= \left(1, (2t+1)(t+1)^4, 1\right) \cdot \left(1, (2t+1)(t+1)^4, 1\right) \\
&= 1 + (2t+1)^2 (t+1)^8 + 1 \\
&= 2 + (2t+1)^2 (t+1)^8, \\
F_{\sigma_1}(g_1(t), g_2(t)) &= f_{t_1}^{\Delta_1}(g_1(t), g_2(t)) \cdot f_{t_2}^{\Delta_2\sigma_1}(g_1(t), g_2(t)) \\
&= \left(1, (2t+1)t^4, 1\right) \cdot (2t^2 + 2t + 1, (t+1)^2(2t^2 + 2t + 1), 1) \\
&= 2t^2 + 2t + 1 + (t+1)^2 t^4 (2t+1)(2t^2 + 2t + 1) + 1 \\
&= 2t^2 + 2t + 2 + (t+1)^2 t^4 (2t+1)(2t^2 + 2t + 1), \\
F_{\sigma_2}(g_1(t), g_2(t)) &= f_{t_1}^{\Delta_1\sigma_2}(g_1(t), g_2(t)) \cdot f_{t_2}^{\Delta_2}(g_1(t), g_2(t)) \\
&= \left(1, (2t+1)(t+1)^4, 1\right) \cdot \left(2t^2 + 2t + 1, t^2(2t^2 + 2t + 1), 1\right) \\
&= 2t^2 + 2t + 1 + t^2(2t+1)(t+1)^4(2t^2 + 2t + 1) + 1 \\
&= 2t^2 + 2t + 2 + t^2(2t+1)(t+1)^4(2t^2 + 2t + 1), \\
G(g_1(t), g_2(t)) &= f_{t_2}^{\Delta_2}(g_1(t), g_2(t)) \cdot f_{t_2}^{\Delta_2}(g_1(t), g_2(t)) \\
&= \left(2t^2 + 2t + 1, t^2(2t^2 + 2t + 1), 1\right) \\
&\quad \cdot \left(2t^2 + 2t + 1, t^2(2t^2 + 2t + 1), 1\right) \\
&= (2t^2 + 2t + 1)^2 + t^4(2t^2 + 2t + 1)^2 + 1 \\
&= (1 + t^4)(2t^2 + 2t + 1)^2 + 1, \\
G_{\sigma_1}(g_1(t), g_2(t)) &= f_{t_2}^{\Delta_2\sigma_1}(g_1(t), g_2(t)) \cdot f_{t_2}^{\Delta_2\sigma_1}(g_1(t), g_2(t)) \\
&= \left(2t^2 + 2t + 1, (t+1)^2(2t^2 + 2t + 1), 1\right) \\
&\quad \cdot \left(2t^2 + 2t + 1, (t+1)^2(2t^2 + 2t + 1), 1\right) \\
&= (2t^2 + 2t + 1)^2 + (t+1)^4(2t^2 + 2t + 1)^2 + 1 \\
&= (1 + (t+1)^4)(2t^2 + 2t + 1)^2 + 1, \quad t \in \mathbb{T}.
\end{aligned}$$

Therefore

$$\begin{aligned}
q_1(t) &= E(g_1(t),g_2(t))(g_1^{\Delta}(t))^2 + 2F_{\sigma_1}(g_1(t),g_2(t))g_1^{\Delta}(t)g_2^{\Delta}(t) \\
&\quad + G_{\sigma_1}(g_1(t),g_2(t))g_2^{\Delta}(t) \\
&= 2 + (2t+1)^2 t^8 \\
&\quad + 2(2t+1)\left(2t^2+2t+1+(t+1)^2 t^4 (2t+1)(2t^2+2t+1)+1\right) \\
&\quad + (2t+1)\left(\left(1+(t+1)^4\right)(2t^2+2t+1)+1\right), \\
q_1(0) &= 2+2(1+1)+(2+1) \\
&= 2+4+3 \\
&= 9, \\
q_1(1) &= 2+9+6(5+4\cdot 3\cdot 5+1) \\
&\quad + 9((1+16)\cdot 5^2+2) \\
&= 11+6(5+60+1)+9(425+1) \\
&= 11+6\cdot 66+9\cdot 426 \\
&= 11+396+3834 \\
&= 4241, \\
q_2(t) &= E_{\sigma_2}(g_1(t),g_2(t))(g_1^{\Delta}(t))^2 + 2F_{\sigma_2}(g_1(t),g_2(t))g_1^{\Delta}(t)g_2^{\Delta}(t) \\
&\quad + G(g_1(t),g_2(t))\left(g_2^{\Delta}(t)\right)^2 \\
&= 2+(2t+1)^2(t+1)^8 \\
&\quad + 2(2t+1)\left(2t^2+2t+2+t^2(2t+1)(t+1)^4(2t^2+2t+1)\right) \\
&\quad + (2t+1)^2\left((1+t^4)(2t^2+2t+1)^2+1\right), \\
q_2(0) &= 2+1+4+2 \\
&= 9, \\
q_2(1) &= 2+9\cdot 2^8+6(5+3\cdot 2^4\cdot 5+1) \\
&\quad + 3^2\cdot(2\cdot 5^2+1) \\
&= 2+9\cdot 256+6(5+15\cdot 16+1)+3^2(50+1) \\
&= 2+2304+6\cdot 246+9\cdot 51 \\
&= 2306+1476+459 \\
&= 4241.
\end{aligned}$$

Consequently

$$l_{0,2,\sigma_1} = \int_0^2 \sqrt{q_1(t)}\Delta t$$

$$\begin{aligned}
&= \mu(0)\sqrt{q_1(0)} + \mu(1)\sqrt{q_1(1)} \\
&= \sqrt{9} + \sqrt{4241} \\
&= 3 + \sqrt{4241}, \\
l_{0,2,\sigma_2} &= \int_0^2 \sqrt{q_2(t)}\Delta t \\
&= \mu(0)\sqrt{q_2(0)} + \mu(1)\sqrt{q_2(1)} \\
&= 3 + \sqrt{4241}.
\end{aligned}$$

Exercise 3.8 Let $\mathbb{T} = \mathbb{N}_0$, $\mathbb{T}_1 = \mathbb{N}_0^2$, $\mathbb{T}_2 = \mathbb{N}_0^3$, $\mathbb{T}_{(1)} = \mathbb{T}_{(2)} = \mathbb{T}_{(3)} = \mathbb{R}$, S be a surface with a local parameterization (U,f), where $U \subset \mathbb{T}_1 \times \mathbb{T}_2$, $f : U \to \mathbb{R}^3$ is defined by

$$f(t_1,t_2) = \left(\frac{1+t_1t_2}{2+t_1+t_2}, t_1t_2, t_1^3t_2^2\right), \quad (t_1,t_2) \in \mathbb{T}_1 \times \mathbb{T}_2,$$

$g : \mathbb{T} \to \mathbb{T}_1 \times \mathbb{T}_2$ is a curve given by

$$g(t) = (t^2, t^3), \quad t \in \mathbb{T}.$$

Find $l_{1,4,\sigma_1}$ and $l_{0,7,\sigma_2}$.

If S is σ_1-regular, we set

$$\begin{aligned}
q_3(t) = {}& E_{\sigma_1^2\sigma_2}(g_1(t),g_2(t))(g_1^{\Delta\sigma}(t))^2 + 2F_{\sigma_1^2\sigma_2}(g_1(t),g_2(t))g_1^{\Delta\sigma}(t)g_2^{\Delta\sigma}(t) \\
&+ G_{\sigma_1^2\sigma_2}(g_1(t),g_2(t))\left(g_2^{\Delta\sigma}(t)\right)^2.
\end{aligned}$$

If S is σ_2-regular, we denote

$$\begin{aligned}
q_4(t) = {}& E_{\sigma_1\sigma_2^2}(g_1(t),g_2(t))(g_1^{\Delta\sigma}(t))^2 + 2F_{\sigma_1\sigma_2^2}(g_1(t),g_2(t))g_1^{\Delta\sigma}(t)g_2^{\Delta\sigma}(t) \\
&+ G_{\sigma_1\sigma_2^2}(g_1(t),g_2(t))\left(g_2^{\Delta\sigma}(t)\right)^2, \quad t \in I.
\end{aligned}$$

In the case when S is σ_1-regular, using (3.7), we have

$$(f(g_1,g_2))^{\Delta\sigma}(t) = f_{t_1}^{\Delta_1\sigma_1\sigma_2}(g_1(t),g_2(t))g_1^{\Delta\sigma}(t) + f_{t_2}^{\Delta_2\sigma_1^2\sigma_2}(g_1(t),g_2(t))g_2^{\Delta\sigma}(t),$$

$t \in I$, and

$$\begin{aligned}
\|(f(g_1,g_2))^{\Delta\sigma}(t)\|^2 &= \left(f_{t_1}^{\Delta_1\sigma_1\sigma_2}(g_1(t),g_2(t))g_1^{\Delta\sigma}(t) + f_{t_2}^{\Delta_2\sigma_1^2\sigma_2}(g_1(t),g_2(t))g_2^{\Delta\sigma}(t)\right) \\
&\quad \cdot \left(f_{t_1}^{\Delta_1\sigma_1\sigma_2}(g_1(t),g_2(t))g_1^{\Delta\sigma}(t) + f_{t_2}^{\Delta_2\sigma_1^2\sigma_2}(g_1(t),g_2(t))g_2^{\Delta\sigma}(t)\right) \\
&= \left(f_{t_1}^{\Delta_1\sigma_1\sigma_2}(g_1(t),g_2(t))\right)^2\left(g_1^{\Delta\sigma}(t)\right)^2 \\
&\quad + 2f_{t_1}^{\Delta_1\sigma_1\sigma_2}(g_1(t),g_2(t))f_{t_2}^{\Delta_2\sigma_1^2\sigma_2}(g_1(t),g_2(t))g_1^{\Delta\sigma}(t)g_2^{\Delta\sigma}(t)
\end{aligned}$$

$$
\begin{aligned}
&+\left(f_{t_2}^{\Delta_2\sigma_1^2\sigma_2}(g_1(t),g_2(t))\right)^2\left(g_2^{\Delta\sigma}(t)\right)^2\\
&=E_{\sigma_1^2\sigma_2}(g_1(t),g_2(t))\left(g_1^{\Delta\sigma}(t)\right)^2+2F_{\sigma_1^2\sigma_2}(g_1(t),g_2(t))g_1^{\Delta\sigma}(t)g_2^{\Delta\sigma}(t)\\
&+G_{\sigma_1^2\sigma_2}(g_1(t),g_2(t))\left(g_2^{\Delta\sigma}(t)\right)^2\\
&=q_3(t),\quad t\in I.
\end{aligned}
$$

Definition 3.20 Let S be σ_1-regular. The $\sigma_1^2\sigma_2$-length of the curve (I,g) is defined by

$$l_{b_1,b_2,\sigma_1^2\sigma_2}=\int_{b_1}^{b_2}\sqrt{q_3(t)}\Delta t.$$

If S is σ_2-regular, then we have

$$(f(g_1,g_2))^{\Delta\sigma}(t)=f_{t_1}^{\Delta_1\sigma_1\sigma_2^2}(g_1(t),g_2(t))g_1^{\Delta\sigma}(t)+f_{t_2}^{\Delta_2\sigma_1\sigma_2}(g_1(t),g_2(t))g_2^{\Delta\sigma}(t),$$

$t\in I$, and

$$
\begin{aligned}
\|(f(g_1,g_2))^{\Delta\sigma}(t)\|^2&=\left(f_{t_1}^{\Delta_1\sigma_1\sigma_2^2}(g_1(t),g_2(t))g_1^{\Delta\sigma}(t)+f_{t_2}^{\Delta_2\sigma_1\sigma_2}(g_1(t),g_2(t))g_2^{\Delta\sigma}(t)\right)\\
&\cdot\left(f_{t_1}^{\Delta_1\sigma_1\sigma_2^2}(g_1(t),g_2(t))g_1^{\Delta\sigma}(t)+f_{t_2}^{\Delta_2\sigma_1\sigma_2}(g_1(t),g_2(t))g_2^{\Delta\sigma}(t)\right)\\
&=\left(f_{t_1}^{\Delta_1\sigma_1\sigma_2^2}(g_1(t),g_2(t))\right)^2\left(g_1^{\Delta\sigma}(t)\right)^2\\
&+2f_{t_1}^{\Delta_1\sigma_1\sigma_2^2}(g_1(t),g_2(t))f_{t_2}^{\Delta_2\sigma_1\sigma_2}(g_1(t),g_2(t))g_1^{\Delta}(t)g_2^{\Delta\sigma}(t)\\
&+\left(f_{t_2}^{\Delta_2\sigma_1\sigma_2}(g_1(t),g_2(t))\right)^2\left(g_2^{\Delta\sigma}(t)\right)^2\\
&=E_{\sigma_1\sigma_2^2}(g_1(t),g_2(t))\left(g_1^{\Delta\sigma}(t)\right)^2+2F_{\sigma_1\sigma_2^2}(g_1(t),g_2(t))g_1^{\Delta\sigma}(t)g_2^{\Delta\sigma}(t)\\
&+G_{\sigma_1\sigma_2^2}(g_1(t),g_2(t))\left(g_2^{\Delta\sigma}(t)\right)^2\\
&=q_4(t),\quad t\in I.
\end{aligned}
$$

Definition 3.21 Let S be σ_2-regular. The $\sigma_1\sigma_2^2$-length of the curve (I,g) is defined by

$$l_{b_1,b_2,\sigma_2}=\int_{b_1}^{b_2}\sqrt{q_4(t)}\Delta t.$$

Exercise 3.9 Let $\mathbb{T}=\mathbb{N}_0^2$, $\mathbb{T}_1=\mathbb{N}_0^6$, $\mathbb{T}_2=\mathbb{N}_0^8$, $\mathbb{T}_{(1)}=\mathbb{T}_{(2)}=\mathbb{T}_{(3)}=\mathbb{R}$, S be a surface with a local parameterization (U,f), where $U\subseteq\mathbb{T}_1\times\mathbb{T}_2$, $f:U\to\mathbb{R}^3$ be defined by

$$f(t_1,t_2)=\left(\frac{t_1+t_2-3}{4+t_1^2+t_2},t_1+t_2^2,t_1^2t_2^3\right),\quad (t_1,t_2)\in\mathbb{T}_1\times\mathbb{T}_2,$$

$g:\mathbb{T}\to\mathbb{T}_1\times\mathbb{T}_2$ be a curve given by

$$g(t)=(t^3,t^4),\quad t\in\mathbb{T}.$$

Find $l_{1,16,\sigma_1^2\sigma_2}$ and $l_{0,49,\sigma_1\sigma_2^2}$.

3.6 Angles between Two Curves on a Surface

Let S be a σ_1- or σ_2- or $\sigma_1^2\sigma_2$-or $\sigma_1\sigma_2^2$-regular surface with a local σ_1- or σ_2- or $\sigma_1^2\sigma_2$ - or $\sigma_1\sigma_2^2$-parameterization (U,f), $f=(f_1,f_2,f_3)$. Let also, (I,g) and (J,h) be two parameterized regular curves with $g(I)\subset f(U)$, $h(J)\subset f(U)$ given by

$$\begin{aligned} g(t)&=(g_1(t),g_2(t)), \quad t\in I,\\ h(t)&=(h_1(t),h_2(t)), \quad t\in J, \end{aligned} \tag{3.9}$$

and

$$\begin{aligned} g_1(\sigma(t))&=\sigma_1(g_1(t)),\\ g_2(\sigma(t))&=\sigma_2(g_2(t)), \quad t\in I,\\ h_1(\sigma(t))&=\sigma_1(h_1(t)),\\ h_2(\sigma(t))&=\sigma_2(h_2(t)), \quad t\in J. \end{aligned} \tag{3.10}$$

Suppose that $t_0\in I$, $s_0\in J$, $(t_1^0,t_2^0)\in\mathbb{T}_1\times\mathbb{T}_2$, are such that

$$g(t_0)=h(s_0)=(t_1^0,t_2^0). \tag{3.11}$$

Definition 3.22 Let S be σ_1-regular surface with a local σ_1-parametrization (U,f). Also, let, (I,g) and (J,h) be two parameterized regular curves that satisfy (3.9), (3.10) and (3.11). The σ_1-angle of the curves (I,g) and (J,h) is defined to be the angle of the tangent vectors with respect to the basis $\{f_{t_1}^{\Delta_1},f_{t_2}^{\Delta_2\sigma_1}\}$.

Suppose that S is σ_1-regular. Then $\{f_{t_1}^{\Delta_1},f_{t_2}^{\Delta_2\sigma_1}\}$ be a basis in $T_{a\sigma_1}S$ and the tangent vectors at t_0 and s_0 of f and h, respectively, with respect to the basis $\{f_{t_1}^{\Delta_1},f_{t_2}^{\Delta_2\sigma_1}\}$ will be

$$\begin{aligned} g^\Delta(t_0)&=g_1^\Delta(t_0)f_{t_1}^{\Delta_1}(t_1^0,t_2^0)+g_2^\Delta(t_0)f_{t_2}^{\Delta_2\sigma_1}(t_1^0,t_2^0),\\ h^\Delta(s_0)&=h_1^\Delta(s_0)f_{t_1}^{\Delta_1}(t_1^0,t_2^0)+h_2^\Delta(s_0)f_{t_2}^{\Delta_2\sigma_1}(t_1^0,t_2^0). \end{aligned}$$

We have that

$$\begin{aligned} g^\Delta(t_0)\cdot h^\Delta(s_0)&=\left(g_1^\Delta(t_0)f_{t_1}^{\Delta_1}(t_1^0,t_2^0)+g_2^\Delta(t_0)f_{t_2}^{\Delta_2\sigma_1}(t_1^0,t_2^0)\right)\\ &\quad\cdot\left(h_1^\Delta(s_0)f_{t_1}^{\Delta_1}(t_1^0,t_2^0)+h_2^\Delta(s_0)f_{t_2}^{\Delta_2\sigma_1}(t_1^0,t_2^0)\right)\\ &=g_1^\Delta(t_0)h_1^\Delta(s_0)\left(f_{t_1}^{\Delta_1}(t_1^0,t_2^0)\right)^2\\ &\quad+g_2^\Delta(t_0)h_1^\Delta(s_0)\left(f_{t_1}^{\Delta_1}(t_1^0,t_2^0)\cdot f_{t_2}^{\Delta_2\sigma_1}(t_1^0,t_2^0)\right)\\ &\quad+g_1^\Delta(t_0)h_2^\Delta(s_0)\left(f_{t_1}^{\Delta_1}(t_1^0,t_2^0)\cdot f_{t_2}^{\Delta_2\sigma_1}(t_1^0,t_2^0)\right)\\ &\quad+g_2^\Delta(t_0)h_2^\Delta(s_0)\left(f_{t_2}^{\Delta_2\sigma_1}(t_1^0,t_2^0)\right)^2\\ &=E(t_1^0,t_2^0)g_1^\Delta(t_0)h_1^\Delta(s_0)\\ &\quad+F_{\sigma_1}(t_1^0,t_2^0)\left(g_1^\Delta(t_0)h_2^\Delta(s_0)+g_2^\Delta(t_0)h_1^\Delta(s_0)\right)\\ &\quad+G_{\sigma_1}(t_1^0,t_2^0)g_2^\Delta(t_0)h_2^\Delta(s_0), \end{aligned}$$

$$
\begin{aligned}
\left(g^{\Delta}(t_0)\right)^2 &= \left(g_1^{\Delta}(t_0)f_{t_1}^{\Delta_1}(t_1^0,t_2^0)+g_2^{\Delta}(t_0)f_{t_2}^{\Delta_2\sigma_1}(t_1^0,t_2^0)\right)\\
&\quad\cdot\left(g_1^{\Delta}(t_0)f_{t_1}^{\Delta_1}(t_1^0,t_2^0)+g_2^{\Delta}(t_0)f_{t_2}^{\Delta_2\sigma_1}(t_1^0,t_2^0)\right)\\
&= \left(g_1^{\Delta}(t_0)\right)^2\left(f_{t_1}^{\Delta_1}(t_1^0,t_2^0)\right)^2+2g_1^{\Delta}(t_0)g_2^{\Delta}(t_0)\left(f_{t_1}^{\Delta_1}(t_1^0,t_2^0)\cdot f_{t_2}^{\Delta_2\sigma_1}(t_1^0,t_2^0)\right)\\
&\quad+\left(g_2^{\Delta}(t_0)\right)^2\left(f_{t_2}^{\Delta_2\sigma_1}(t_1^0,t_2^0)\right)^2\\
&= E(t_1^0,t_2^0)\left(g_1^{\Delta}(t_0)\right)^2+2F_{\sigma_1}(t_1^0,t_2^0)\left(g_1^{\Delta}(t_0)g_2^{\Delta}(t_0)\right)\\
&\quad+G_{\sigma_1}(t_1^0,t_2^0)\left(g_2^{\Delta}(t_0)\right)^2,\\
\left(h^{\Delta}(s_0)\right)^2 &= \left(h_1^{\Delta}(s_0)f_{t_1}^{\Delta_1}(t_1^0,t_2^0)+h_2^{\Delta}(s_0)f_{t_2}^{\Delta_2\sigma_1}(t_1^0,t_2^0)\right)\\
&\quad\cdot\left(h_1^{\Delta}(s_0)f_{t_1}^{\Delta_1}(t_1^0,t_2^0)+h_2^{\Delta}(s_0)f_{t_2}^{\Delta_2\sigma_1}(t_1^0,t_2^0)\right)\\
&= E(t_1^0,t_2^0)\left(h_1^{\Delta}(s_0)\right)^2+2F_{\sigma_1}(t_1^0,t_2^0)\left(h_1^{\Delta}(s_0)h_2^{\Delta}(s_0)\right)\\
&\quad+G_{\sigma_1}(t_1^0,t_2^0)\left(h_2^{\Delta}(s_0)\right)^2.
\end{aligned}
$$

Thus,

$$
\begin{aligned}
\cos\theta_{\sigma_1} &= \frac{g^{\Delta}(t_0)\cdot h^{\Delta}(s_0)}{\|g^{\Delta}(t_0)\|\,\|h^{\Delta}(s_0)\|}\\
&= \Bigg(E(t_1^0,t_2^0)\left(g_1^{\Delta}(t_0)\right)^2+2F_{\sigma_1}(t_1^0,t_2^0)\left(g_1^{\Delta}(t_0)g_2^{\Delta}(t_0)\right)\\
&\quad+G_{\sigma_1}(t_1^0,t_2^0)\left(g_2^{\Delta}(t_0)\right)^2\Bigg)^{-\frac{1}{2}}\\
&\quad\times\Bigg(E(t_1^0,t_2^0)\left(h_1^{\Delta}(s_0)\right)^2+2F_{\sigma_1}(t_1^0,t_2^0)\left(h_1^{\Delta}(s_0)h_2^{\Delta}(s_0)\right)\\
&\quad+G_{\sigma_1}(t_1^0,t_2^0)\left(h_2^{\Delta}(s_0)\right)^2\Bigg)^{-\frac{1}{2}}\\
&\quad\times\Bigg(E(t_1^0,t_2^0)g_1^{\Delta}(t_0)h_1^{\Delta}(s_0)\\
&\quad+F_{\sigma_1}(t_1^0,t_2^0)\left(g_1^{\Delta}(t_0)h_2^{\Delta}(s_0)+g_2^{\Delta}(t_0)h_1^{\Delta}(s_0)\right)\\
&\quad+G_{\sigma_1}(t_1^0,t_2^0)g_2^{\Delta}(t_0)h_2^{\Delta}(s_0)\Bigg).
\end{aligned}
$$

Definition 3.23 Let S be σ_2-regular surface with a local σ_2-parametrization (U,f). Also, let, (I,g) and (J,h) be two parameterized regular curves that satisfy (3.9), (3.10) and (3.11). The σ_2-angle of the curves (I,g) and (J,h) is defined to be the angle of the tangent vectors with respect to the basis $\{f_{t_1}^{\Delta_1\sigma_2}, f_{t_2}^{\Delta_2}\}$.

If S is σ_2-regular, as above, we get

$$
\cos\theta_{\sigma_2} = \frac{g^{\Delta}(t_0)\cdot h^{\Delta}(s_0)}{\|g^{\Delta}(t_0)\|\,\|h^{\Delta}(s_0)\|}
$$

$$= \Big(E_{\sigma_2}(t_1^0, t_2^0) \left(g_1^\Delta(t_0)\right)^2 + 2F_{\sigma_2}(t_1^0, t_2^0) \left(g_1^\Delta(t_0) g_2^\Delta(t_0)\right)$$

$$+ G(t_1^0, t_2^0) \left(g_2^\Delta(t_0)\right)^2 \Big)^{-\frac{1}{2}}$$

$$\times \Big(E_{\sigma_2}(t_1^0, t_2^0) \left(h_1^\Delta(s_0)\right)^2 + 2F_{\sigma_2}(t_1^0, t_2^0) \left(h_1^\Delta(s_0) h_2^\Delta(s_0)\right)$$

$$+ G(t_1^0, t_2^0) \left(h_2^\Delta(s_0)\right)^2 \Big)^{-\frac{1}{2}}$$

$$\times \Big(E_{\sigma_2}(t_1^0, t_2^0) g_1^\Delta(t_0) h_1^\Delta(s_0)$$

$$+ F_{\sigma_2}(t_1^0, t_2^0) \left(g_1^\Delta(t_0) h_2^\Delta(s_0) + g_2^\Delta(t_0) h_1^\Delta(s_0)\right)$$

$$+ G(t_1^0, t_2^0) g_2^\Delta(t_0) h_2^\Delta(s_0) \Big).$$

Definition 3.24 Let S be $\sigma_1^2\sigma_2$-regular surface with a local $\sigma_1^2\sigma_2$-parametrization (U,f). Also, let (I,g) and (J,h) be two parameterized regular curves that satisfy (3.9), (3.10) and (3.11). The $\sigma_1^2\sigma_2$-angle of the curves (I,g) and (J,h) is defined to be the angle of the tangent vectors with respect to the basis $\left\{ f_{t_1}^{\Delta_1\sigma_1\sigma_2}, f_{t_2}^{\Delta_2\sigma_1^2\sigma_2} \right\}$.

If S is $\sigma_1^2\sigma_2$-regular, then

$$\cos\theta_{\sigma_1^2\sigma_2} = \frac{g^\Delta(t_0) \cdot h^\Delta(s_0)}{\|g^\Delta(t_0)\| \, \|h^\Delta(s_0)\|}$$

$$= \Big(E_{\sigma_1^2\sigma_2}(t_1^0, t_2^0) \left(g_1^\Delta(t_0)\right)^2 + 2F_{\sigma_1^2\sigma_2}(t_1^0, t_2^0) \left(g_1^\Delta(t_0) g_2^\Delta(t_0)\right)$$

$$+ G_{\sigma_1^2\sigma_2}(t_1^0, t_2^0) \left(g_2^\Delta(t_0)\right)^2 \Big)^{-\frac{1}{2}}$$

$$\times \Big(E_{\sigma_1^2\sigma_2}(t_1^0, t_2^0) \left(h_1^\Delta(s_0)\right)^2 + 2F_{\sigma_1^2\sigma_2}(t_1^0, t_2^0) \left(h_1^\Delta(s_0) h_2^\Delta(s_0)\right)$$

$$+ G_{\sigma_1^2\sigma_2}(t_1^0, t_2^0) \left(h_2^\Delta(s_0)\right)^2 \Big)^{-\frac{1}{2}}$$

$$\times \Big(E_{\sigma_1^2\sigma_2}(t_1^0, t_2^0) g_1^\Delta(t_0) h_1^\Delta(s_0)$$

$$+ F_{\sigma_1^2\sigma_2}(t_1^0, t_2^0) \left(g_1^\Delta(t_0) h_2^\Delta(s_0) + g_2^\Delta(t_0) h_1^\Delta(s_0)\right)$$

$$+ G_{\sigma_1^2\sigma_2}(t_1^0, t_2^0) g_2^\Delta(t_0) h_2^\Delta(s_0) \Big).$$

Definition 3.25 Let S be a $\sigma_1\sigma_2^2$-regular surface with a local $\sigma_1\sigma_2^2$-parametrization (U,f). Also, let, (I,g) and (J,h) be two parameterized regular curves that satisfy (3.9), (3.10) and (3.11). The $\sigma_1\sigma_2^2$-angle of the curves (I,g) and (J,h) is defined to be the angle of the tangent vectors with respect to the basis $\left\{f_{t_1}^{\Delta_1\sigma_1\sigma_2^2}, f_{t_2}^{\Delta_2\sigma_1\sigma_2}\right\}$.

Let S be $\sigma_1\sigma_2^2$-regular.

Then

$$\begin{aligned}
\cos\theta_{\sigma_1\sigma_2^2} &= \frac{g^{\Delta}(t_0)\cdot h^{\Delta}(s_0)}{\|g^{\Delta}(t_0)\|\,\|h^{\Delta}(s_0)\|}\\
&= \Big(E_{\sigma_1\sigma_2^2}(t_1^0,t_2^0)\left(g_1^{\Delta}(t_0)\right)^2 + 2F_{\sigma_1\sigma_2^2}(t_1^0,t_2^0)\left(g_1^{\Delta}(t_0)g_2^{\Delta}(t_0)\right)\\
&\quad + G_{\sigma_1\sigma_2^2}(t_1^0,t_2^0)\left(g_2^{\Delta}(t_0)\right)^2\Big)^{-\frac{1}{2}}\\
&\quad \times\Big(E_{\sigma_1\sigma_2^2}(t_1^0,t_2^0)\left(h_1^{\Delta}(s_0)\right)^2 + 2F_{\sigma_1\sigma_2^2}(t_1^0,t_2^0)\left(h_1^{\Delta}(s_0)h_2^{\Delta}(s_0)\right)\\
&\quad + G_{\sigma_1\sigma_2^2}(t_1^0,t_2^0)\left(h_2^{\Delta}(s_0)\right)^2\Big)^{-\frac{1}{2}}\\
&\quad \times\Big(E_{\sigma_1\sigma_2^2}(t_1^0,t_2^0)g_1^{\Delta}(t_0)h_1^{\Delta}(s_0)\\
&\quad + F_{\sigma_1\sigma_2^2}(t_1^0,t_2^0)\left(g_1^{\Delta}(t_0)h_2^{\Delta}(s_0) + g_2^{\Delta}(t_0)h_1^{\Delta}(s_0)\right)\\
&\quad + G_{\sigma_1\sigma_2^2}(t_1^0,t_2^0)g_2^{\Delta}(t_0)h_2^{\Delta}(s_0)\Big).
\end{aligned}$$

3.7 Areas of Parameterized Surfaces

Definition 3.26 Suppose that S is a σ_1-regular surface with a local σ_1-parametrization (U,f). The σ_1-area of S is given by

$$A_{\sigma_1}(S) = \int\int_U \|f_{t_1}^{\Delta_1} \times f_{t_2}^{\Delta_2\sigma_1}\|\,\Delta_1 t_1\,\Delta_2 t_2.$$

Suppose that S is a σ_1-regular surface with a local σ_1-parameterization (U,f).

Then

$$\begin{aligned}
f_{t_1}^{\Delta_1} &= \left(f_{1t_1}^{\Delta_1}, f_{2t_1}^{\Delta_1}, f_{3t_1}^{\Delta_1}\right),\\
f_{t_2}^{\Delta_2\sigma_1} &= \left(f_{1t_2}^{\Delta_2\sigma_1}, f_{2t_2}^{\Delta_2\sigma_1}, f_{3t_2}^{\Delta_2\sigma_1}\right)
\end{aligned}$$

and

$$f_{t_1}^{\Delta_1} \times f_{t_2}^{\Delta_2\sigma_1} = \left(f_{2t_1}^{\Delta_1} f_{3t_2}^{\Delta_2\sigma_1} - f_{2t_2}^{\Delta_2\sigma_1} f_{3t_1}^{\Delta_1}, f_{1t_2}^{\Delta_2\sigma_1} f_{3t_1}^{\Delta_1} - f_{1t_1}^{\Delta_1} f_{3t_2}^{\Delta_2\sigma_1}, \right.$$
$$\left. f_{1t_1}^{\Delta_1} f_{2t_2}^{\Delta_2\sigma_1} - f_{1t_2}^{\Delta_2\sigma_1} f_{2t_1}^{\Delta_1} \right).$$

Hence,

$$\begin{aligned}
\left(f_{t_1}^{\Delta_1} \times f_{t_2}^{\Delta_2\sigma_1}\right)^2 &= \left(f_{2t_1}^{\Delta_1} f_{3t_2}^{\Delta_2\sigma_1} - f_{2t_2}^{\Delta_2\sigma_1} f_{3t_1}^{\Delta_1}\right)^2 + \left(f_{1t_2}^{\Delta_2\sigma_1} f_{3t_1}^{\Delta_1} - f_{1t_1}^{\Delta_1} f_{3t_2}^{\Delta_2\sigma_1}\right)^2 \\
&\quad + \left(f_{1t_1}^{\Delta_1} f_{2t_2}^{\Delta_2\sigma_1} - f_{1t_2}^{\Delta_2\sigma_1} f_{2t_1}^{\Delta_1}\right)^2 \\
&= \left(f_{2t_1}^{\Delta_1}\right)^2 \left(f_{3t_2}^{\Delta_2\sigma_1}\right)^2 - 2 f_{2t_1}^{\Delta_1} f_{2t_2}^{\Delta_2\sigma_1} f_{3t_1}^{\Delta_1} f_{3t_2}^{\Delta_2\sigma_1} + \left(f_{2t_2}^{\Delta_2\sigma_1}\right)^2 \left(f_{3t_1}^{\Delta_1}\right)^2 \\
&\quad + \left(f_{1t_2}^{\Delta_2\sigma_1}\right)^2 \left(f_{3t_1}^{\Delta_1}\right)^2 + \left(f_{1t_1}^{\Delta_1}\right)^2 \left(f_{3t_2}^{\Delta_2\sigma_1}\right)^2 - 2 f_{1t_1}^{\Delta_1} f_{1t_2}^{\Delta_2\sigma_1} f_{3t_1}^{\Delta_1} f_{3t_2}^{\Delta_2\sigma_1} \\
&\quad + \left(f_{1t_1}^{\Delta_1}\right)^2 \left(f_{2t_2}^{\Delta_2\sigma_1}\right)^2 + \left(f_{1t_2}^{\Delta_2\sigma_1}\right)^2 \left(f_{2t_1}^{\Delta_1}\right)^2 - 2 f_{1t_2}^{\Delta_2\sigma_1} f_{1t_1}^{\Delta_1} f_{2t_1}^{\Delta_1} f_{2t_2}^{\Delta_2\sigma_1} \\
&= \left(f_{1t_1}^{\Delta_1}\right)^2 \left(f_{1t_2}^{\Delta_2\sigma_1}\right)^2 + \left(f_{1t_1}^{\Delta_1}\right)^2 \left(f_{2t_2}^{\Delta_2\sigma_1}\right)^2 + \left(f_{1t_1}^{\Delta_1}\right)^2 \left(f_{3t_2}^{\Delta_2\sigma_1}\right)^2 \\
&\quad + \left(f_{2t_1}^{\Delta_1}\right)^2 \left(f_{1t_2}^{\Delta_2\sigma_1}\right)^2 + \left(f_{2t_1}^{\Delta_1}\right)^2 \left(f_{2t_2}^{\Delta_2\sigma_1}\right)^2 + \left(f_{2t_1}^{\Delta_1}\right)^2 \left(f_{3t_2}^{\Delta_2\sigma_1}\right)^2 \\
&\quad + \left(f_{3t_1}^{\Delta_1}\right)^2 \left(f_{1t_2}^{\Delta_2\sigma_1}\right)^2 + \left(f_{3t_1}^{\Delta_1}\right)^2 \left(f_{2t_2}^{\Delta_2\sigma_1}\right)^2 + \left(f_{3t_1}^{\Delta_1}\right)^2 \left(f_{3t_2}^{\Delta_2\sigma_1}\right)^2 \\
&\quad - \left(f_{1t_1}^{\Delta_1}\right)^2 \left(f_{1t_2}^{\Delta_2\sigma_1}\right)^2 - \left(f_{2t_1}^{\Delta_1}\right)^2 \left(f_{2t_2}^{\Delta_2\sigma_1}\right)^2 - \left(f_{3t_1}^{\Delta_1}\right)^2 \left(f_{3t_2}^{\Delta_2\sigma_1}\right)^2 \\
&\quad - 2 f_{2t_1}^{\Delta_1} f_{2t_2}^{\Delta_2\sigma_1} f_{3t_1}^{\Delta_1} f_{3t_2}^{\Delta_2\sigma_1} - 2 f_{1t_1}^{\Delta_1} f_{1t_2}^{\Delta_2\sigma_1} f_{3t_1}^{\Delta_1} f_{3t_2}^{\Delta_2\sigma_1} \\
&\quad - 2 f_{1t_2}^{\Delta_2\sigma_1} f_{1t_1}^{\Delta_1} f_{2t_1}^{\Delta_1} f_{2t_2}^{\Delta_2\sigma_1} \\
&= \left(\left(f_{1t_1}^{\Delta_1}\right)^2 + \left(f_{2t_1}^{\Delta_1}\right)^2 + \left(f_{3t_1}^{\Delta_1}\right)^2 \right) \left(\left(f_{1t_2}^{\Delta_2\sigma_1}\right)^2 + \left(f_{2t_2}^{\Delta_2\sigma_1}\right)^2 + \left(f_{3t_2}^{\Delta_2\sigma_1}\right)^2 \right) \\
&\quad - \left(f_{1t_1}^{\Delta_1} f_{1t_2}^{\Delta_2\sigma_1} + f_{2t_1}^{\Delta_1} f_{2t_2}^{\Delta_2\sigma_1} + f_{3t_1}^{\Delta_1} f_{3t_2}^{\Delta_2\sigma_1}\right)^2 \\
&= EG_{\sigma_1} - F_{\sigma_1}^2.
\end{aligned}$$

Therefore

$$A_{\sigma_1}(S) = \int\int_U \sqrt{EG_{\sigma_1} - F_{\sigma_1}^2}\, \Delta_1 t_1 \Delta_2 t_2.$$

Example 3.10 Let $\mathbb{T}_1 = \mathbb{T}_2 = \mathbb{Z}$, $\mathbb{T}_{(1)} = \mathbb{T}_{(2)} = \mathbb{T}_{(3)} = \mathbb{R}$ and

$$U = \{(t_1, t_2) \in \mathbb{T}_1 \times \mathbb{T}_2 : \ 0 \le t_1, t_2 \le 3\}.$$

Consider a surface S with a local parameterization

$$f(t_1, t_2) = (t_1 + 2, t_2 - 2, t_1 + t_2), \quad (t_1, t_2) \in U.$$

Here,

$$f_1(t_1, t_2) = t_1 + 2,$$

$$f_2(t_1,t_2) = t_2 - 2,$$
$$f_3(t_1,t_2) = t_1 + t_2, \quad (t_1,t_2) \in U.$$

Then

$$\begin{aligned}
f_{1t_1}^{\Delta_1}(t_1,t_2) &= 1,\\
f_{1t_2}^{\Delta_2}(t_1,t_2) &= 0,\\
f_{1t_2}^{\Delta_2\sigma_1}(t_1,t_2) &= 1,\\
f_{2t_1}^{\Delta_1}(t_1,t_2) &= 0,\\
f_{2t_2}^{\Delta_2}(t_1,t_2) &= 1,\\
f_{2t_2}^{\Delta_2\sigma_1}(t_1,t_2) &= 1,\\
f_{3t_1}^{\Delta_1}(t_1,t_2) &= 1,\\
f_{3t_2}^{\Delta_2}(t_1,t_2) &= 1,\\
f_{3t_2}^{\Delta_2\sigma_1}(t_1,t_2) &= 1.
\end{aligned}$$

Hence,

$$\begin{aligned}
E &= \left(f_{1t_1}^{\Delta_1}(t_1,t_2)\right)^2 + \left(f_{2t_1}^{\Delta_1}(t_1,t_2)\right)^2 + \left(f_{3t_1}^{\Delta_1}(t_1,t_2)\right)^2\\
&= 2,\\
F_{\sigma_1} &= f_{1t_1}^{\Delta_1}(t_1,t_2) f_{1t_2}^{\Delta_2\sigma_1}(t_1,t_2) + f_{2t_1}^{\Delta_1}(t_1,t_2) f_{2t_2}^{\Delta_2\sigma_1}(t_1,t_2)\\
&\quad + f_{3t_1}^{\Delta_1}(t_1,t_2) f_{3t_2}^{\Delta_2\sigma_1}(t_1,t_2)\\
&= 1,\\
G_{\sigma_1} &= \left(f_{1t_2}^{\Delta_2\sigma_1}(t_1,t_2)\right)^2 + \left(f_{2t_2}^{\Delta_2\sigma_1}(t_1,t_2)\right)^2 + \left(f_{3t_2}^{\Delta_2\sigma_1}(t_1,t_2)\right)^2\\
&= 2,
\end{aligned}$$

so that

$$EG_{\sigma_1} - F_{\sigma_1}^2 = 3$$

and

$$\begin{aligned}
A_{\sigma_1}(S) &= \int\int_U \sqrt{3}\Delta_1 t_1 \Delta_2 t_2\\
&= 9\sqrt{3}.
\end{aligned}$$

Definition 3.27 Suppose that S is a σ_2-regular surface with a local σ_2-parameterization (U,f). The σ_2-area of S is given by

$$\begin{aligned}
A_{\sigma_2}(S) &= \int\int \| f_{t_1}^{\Delta_1\sigma_2} \times f_{t_2}^{\Delta_2} \| \Delta_1 t_1 \Delta_2 t_2\\
&= \int\int_U \sqrt{E_{\sigma_2} G - F_{\sigma_2}^2}\, \Delta_1 t_1 \Delta_2 t_2.
\end{aligned}$$

Example 3.11 Let $\mathbb{T}_1 = \mathbb{Z}$, $\mathbb{T}_2 = 2^{\mathbb{N}_0} \cup \{0\}$, $\mathbb{T}_{(1)} = \mathbb{T}_{(2)} = \mathbb{T}_{(3)} = \mathbb{R}$,

$$U = \{(t_1, t_2) : \; 0 \le t_1 \le 1, \quad 1 \le t_2 \le 2\},$$

and S be a surface with a local parameterization (U, f), where

$$f(t_1, t_2) = (t_1^2 + t_1 t_2^2, t_1^2 + t_2^2, t_1 + t_2), \quad (t_1, t_2) \in U.$$

Here,

$$\begin{aligned}
&\sigma_1(t_1) = t_1 + 1,\\
&\mu_1(t_1) = 1, \quad t_1 \in \mathbb{T}_1,\\
&\sigma_2(t_2) = 2t_2,\\
&\mu_2(t_2) = t_2, \quad t_2 \in \mathbb{T}_2, \quad t_2 \neq 0,
\end{aligned}$$

$$\begin{aligned}
&f_1(t_1, t_2) = t_1^2 + t_1 t_2^2,\\
&f_2(t_1, t_2) = t_1^2 + t_2^2,\\
&f_3(t_1, t_2) = t_1 + t_2, \quad (t_1, t_2) \in U.
\end{aligned}$$

Thus,

$$\begin{aligned}
f_{1t_1}^{\Delta_1}(t_1, t_2) &= \sigma_1(t_1) + t_1 + t_2^2\\
&= t_1 + 1 + t_1 + t_2^2\\
&= 2t_1 + 1 + t_2^2,\\
f_{1t_1}^{\Delta_1 \sigma_2}(t_1, t_2) &= 2t_1 + 1 + (\sigma_2(t_2))^2\\
&= 2t_1 + 1 + 4t_2^2,\\
f_{1t_2}^{\Delta_2}(t_1, t_2) &= t_1(\sigma_2(t_2) + t_2)\\
&= t_1(2t_2 + t_2)\\
&= 3t_1 t_2,
\end{aligned}$$

$$\begin{aligned}
f_{2t_1}^{\Delta_1}(t_1, t_2) &= \sigma_1(t_1) + t_1\\
&= t_1 + 1 + t_1\\
&= 2t_1 + 1,\\
f_{2t_1}^{\Delta_1 \sigma_2}(t_1, t_2) &= 2t_1 + 1,\\
f_{2t_2}^{\Delta_2}(t_1, t_2) &= \sigma_2(t_2) + t_2\\
&= 2t_2 + t_2\\
&= 3t_2,\\
f_{3t_1}^{\Delta_1}(t_1, t_2) &= 1,
\end{aligned}$$

$$f_{3t_1}^{\Delta_1\sigma_2}(t_1,t_2) = 1,$$
$$f_{3t_2}^{\Delta_2}(t_1,t_2) = 1, \quad (t_1,t_2) \in \mathbb{T}_1 \times \mathbb{T}_2.$$

Hence,

$$\begin{aligned}
E_{\sigma_2}(t_1,t_2) &= \left(f_{1t_1}^{\Delta_1\sigma_2}(t_1,t_2)\right)^2 + \left(f_{2t_1}^{\Delta_1\sigma_2}(t_1,t_2)\right)^2 + \left(f_{3t_1}^{\Delta_1\sigma_2}(t_1,t_2)\right)^2 \\
&= (2t_1+1+4t_2^2)^2 + (2t_1+1)^2 + 1 \\
&= 4t_1^2 + 1 + 16t_2^4 + 4t_1 + 16t_1t_2^2 + 8t_2^2 + 4t_1^2 + 4t_1 + 1 + 1 \\
&= 8t_1^2 + 16t_2^4 + 8t_2^2 + 16t_1t_2^2 + 8t_1 + 3, \\
F_{\sigma_2}(t_1,t_2) &= f_{1t_1}^{\Delta_1\sigma_2}(t_1,t_2) f_{1t_2}^{\Delta_2}(t_1,t_2) + f_{2t_1}^{\Delta_1\sigma_2}(t_1,t_2) f_{2t_2}^{\Delta_2}(t_1,t_2) \\
&\quad + f_{3t_1}^{\Delta_1\sigma_2}(t_1,t_2) f_{3t_2}^{\Delta_2}(t_1,t_2) \\
&= 3t_1t_2(1+2t_1+4t_2^2) + (2t_1+1)3t_2 + 1 \\
&= 3t_1t_2 + 6t_1^2t_2 + 12t_1t_2^3 + 6t_1t_2 + 3t_2 + 1 \\
&= 6t_1^2t_2 + 12t_1t_2^3 + 9t_1t_2 + 3t_2 + 1, \\
G(t_1,t_2) &= \left(f_{1t_2}^{\Delta_2}(t_1,t_2)\right)^2 + \left(f_{2t_2}^{\Delta_2}(t_1,t_2)\right)^2 + \left(f_{3t_2}^{\Delta_2}(t_1,t_2)\right)^2 \\
&= 9t_1^2t_2^2 + 9t_2^2 + 1, \quad (t_1,t_2) \in U.
\end{aligned}$$

Let

$$\begin{aligned}
h(t_1,t_2) &= E_{\sigma_2}(t_1,t_2)G(t_1,t_2) - (F_{\sigma_2}(t_1,t_2))^2 \\
&= EG - F^2 \\
&= (8t_1^2 + 16t_2^4 + 8t_2^2 + 16t_1t_2^2 + 8t_1 + 3)(9t_1^2t_2^2 + 9t_2^2 + 1) \\
&\quad - (6t_1^2t_2 + 12t_1t_2^3 + 9t_1t_2 + 3t_2 + 1)^2, \\
l(t_2) &= \sqrt{h(t_1,t_2)\Big|_{t_1=0}} \\
&= \sqrt{(16t_2^4 + 8t_2^2 + 3)(9t_2^2 + 1) - (3t_2 + 1)^2}, \quad (t_1,t_2) \in U.
\end{aligned}$$

Hence,

$$\begin{aligned}
A_{\sigma_2}(S) &= \int_1^2 \int_0^1 \sqrt{h(t_1,t_2)}\,\Delta_1 t_1 \Delta_2 t_2 \\
&= \int_1^2 \sqrt{h(t_1,t_2)}\,\mu_1(t_1)\Big|_{t_1=0} \Delta_2 t_2 \\
&= \int_1^2 l(t_2)\Delta_2 t_2 \\
&= l(t_2)\mu_2(t_2)\Big|_{t_2=1} \\
&= \sqrt{254}.
\end{aligned}$$

Definition 3.28 Suppose that S is a σ_1-regular surface with a local σ_1-parametrization (U,f). The $\sigma_1^2\sigma_2$-area of S is given by

$$\begin{aligned} A_{\sigma_1^2\sigma_2}(S) &= \int\int \| f_{t_1}^{\Delta_1\sigma_1\sigma_2} \times f_{t_2}^{\Delta_2\sigma_1^2\sigma_2} \| \Delta_1 t_1 \Delta_2 t_2 \\ &= \int\int_U \sqrt{E_{\sigma_1^2\sigma_2} G_{\sigma_1^2\sigma_2} - F^2_{\sigma_1^2\sigma_2}} \Delta_1 t_1 \Delta_2 t_2. \end{aligned}$$

Definition 3.29 Suppose that S is a σ_2-regular surface with a local σ_2-parametrization (U,f). The σ_2-area of S is given by

$$\begin{aligned} A_{\sigma_1\sigma_2^2}(S) &= \int\int \| f_{t_1}^{\Delta_1\sigma_1\sigma_2^2} \times f_{t_2}^{\Delta_2\sigma_1\sigma_2} \| \Delta_1 t_1 \Delta_2 t_2 \\ &= \int\int_U \sqrt{E_{\sigma_1\sigma_2^2} G_{\sigma_1\sigma_2^2} - F^2_{\sigma_1\sigma_2^2}} \Delta_1 t_1 \Delta_2 t_2. \end{aligned}$$

Exercise 3.10 Let $\mathbb{T}_1 = 4\mathbb{Z}$, $\mathbb{T}_2 = 2^{\mathbb{N}_0} \cup \{0\}$, $\mathbb{T}_{(1)} = \mathbb{T}_{(2)} = \mathbb{T}_{(3)} = \mathbb{R}$,

$$U = \{(t_1,t_2) : 0 \le t_1 \le 16, \quad 1 \le t_2 \le 2\},$$

and S be a surface with a local parameterization (U,f), where

$$f(t_1,t_2) = (t_1^3 + t_1 + t_1^4 t_2^4, t_1 + t_1^3 + t_2^2, t_1^2 + t_2^2), \quad (t_1,t_2) \in U.$$

Find $A_{\sigma_1}(S)$, $A_{\sigma_2}(S)$, $A_{\sigma_1^2\sigma_2}(S)$, $A_{\sigma_1\sigma_2^2}(S)$.

3.8 Advanced Practical Problems

Problem 3.1 Let $\mathbb{T}_{(1)} = 3^{\mathbb{N}_0}$, $\mathbb{T}_{(2)} = 2^{\mathbb{N}_0}$, $\mathbb{T}_{(3)} = 4^{\mathbb{N}_0}$ and $g : \mathbb{T}_{(1)} \times \mathbb{T}_{(2)} \times \mathbb{T}_{(3)} \to \mathbb{R}^3$ be given by

$$g(t_{(1)},t_{(2)},t_{(3)}) = \left(\frac{t_{(1)} + t_{(2)}^2}{1 + t_{(3)}^4}, \frac{t_{(1)}^2 - t_{(1)}t_{(2)} - t_{(3)}^4}{t_{(1)}^4 + t_{(1)}^2 t_{(2)}^2 + t_{(3)}^8}, t_{(1)}^2 t_{(2)}^4 t_{(3)}^7 \right),$$

$(t_{(1)},t_{(2)},t_{(3)}) \in \mathbb{T}_{(1)} \times \mathbb{T}_{(2)} \times \mathbb{T}_{(3)}$. Find $D_{12a}^4(g)$, $D_{13a}^5(g)$, $D_{23a}^7(g)$.

Problem 3.2 Let $\mathbb{T}_1 = 3^{\mathbb{N}_0}$, $\mathbb{T}_2 = 4^{\mathbb{N}_0}$, $\mathbb{T}_{(1)} = \mathbb{T}_{(2)} = \mathbb{T}_{(3)} = \mathbb{R}$ and $f : \mathbb{T}_1 \times \mathbb{T}_2 \to \mathbb{R}^3$ be given by

$$f(t_1,t_2) = \left(\frac{1 + t_2}{1 + t_2^2 + t_2^4}, e_1(t_1,1)e_2(t_2,1), t_1^2 + t_2^2 + t_3^4 \right), \quad (t_1,t_2) \in \mathbb{T}_1 \times \mathbb{T}_2.$$

Find the matrices $\mathscr{G}_{\sigma_1}$, $\mathscr{H}_{\sigma_1}$, $\mathscr{A}_{\sigma_1}$, $\mathscr{G}_{\sigma_2}$, $\mathscr{H}_{\sigma_2}$ and $\mathscr{A}_{\sigma_2}$.

Problem 3.3 Let $\mathbb{T}_1 = 2^{\mathbb{N}_0}$, $\mathbb{T}_2 = 3^{\mathbb{N}_0}$, $\mathbb{T}_{(1)} = \mathbb{T}_{(2)} = \mathbb{T}_{(3)} = \mathbb{R}$, $U \subseteq \mathbb{T}_1 \times \mathbb{T}_2$, $f: U \to \mathbb{T}_{(1)} \times \mathbb{T}_{(2)} \times \mathbb{T}_{(3)}$ be given by

$$f(t_1,t_2) = \left(\sin_1(t_1,1) + \cos_2(t_2,2), t_1 - t_2^2, (1+t_1)^2 + t_2^4\right), \quad (t_1,t_2) \in U.$$

Find the matrix $B_{\sigma_1\sigma_2^2}(t_1,t_2)$, $(t_1,t_2) \in U$.

Problem 3.4 Let $\mathbb{T}_1 = 2^{\mathbb{N}_0}$, $\mathbb{T}_2 = 11^{\mathbb{N}_0}$, $\mathbb{T}_{(1)} = \mathbb{T}_{(2)} = \mathbb{T}_{(3)} = \mathbb{R}$, $U \subseteq \mathbb{T}_1 \times \mathbb{T}_2$, $f: U \to \mathbb{T}_{(1)} \times \mathbb{T}_{(2)} \times \mathbb{T}_{(3)}$ be given by

$$f(t_1,t_2) = \left(\frac{1+t_1}{1+t_2+t_2^2}, 1 - t_1 - 10t_2^4, t_1^2 + t_2^3\right), \quad (t_1,t_2) \in U.$$

Find the matrix $B_{\sigma_1}(t_1,t_2)$, $(t_1,t_2) \in U$.

Problem 3.5 Let $\mathbb{T}_1 = 3^{\mathbb{N}_0}$, $\mathbb{T}_2 = 4\mathbb{N}_0$, $\mathbb{T}_{(1)} = \mathbb{T}_{(2)} = \mathbb{T}_{(3)} = \mathbb{R}$, $U \subseteq \mathbb{T}_1 \times \mathbb{T}_2$, $f: U \to \mathbb{T}_{(1)} \times \mathbb{T}_{(2)} \times \mathbb{T}_{(3)}$ be given by

$$f(t_1,t_2) = \left(1 + t_1 + t_2, t_1 + t_2^3, (t_1+t_2)^3 - t_1^2 t_2^2\right), \quad (t_1,t_2) \in U.$$

Find the matrix $B_{\sigma_2}(t_1,t_2)$, $(t_1,t_2) \in U$.

Problem 3.6 Let $\mathbb{T} = \mathbb{N}_0$, $\mathbb{T}_1 = \mathbb{N}_0^3$, $\mathbb{T}_2 = \mathbb{N}_0^4$, $\mathbb{T}_{(1)} = \mathbb{T}_{(2)} = \mathbb{T}_{(3)} = \mathbb{R}$, S be a surface with a local parameterization (U,f), where $U \subset \mathbb{T}_1 \times \mathbb{T}_2, f: U \to \mathbb{R}^3$ be defined by

$$f(t_1,t_2) = \left(t_1 + t_1^2 t_2^2 + t_2, t_1 - 3t_2 + 4t_1t_2, t_1^2 + t_2^2 - 7t_1t_2\right), \quad (t_1,t_2) \in \mathbb{T}_1 \times \mathbb{T}_2,$$

$g: \mathbb{T} \to \mathbb{T}_1 \times \mathbb{T}_2$ be a curve given by

$$g(t) = (t^3, t^4), \quad t \in \mathbb{T}.$$

Find $l_{0,3,\sigma_1}$ and $l_{1,4,\sigma_2}$.

Problem 3.7 Let $\mathbb{T}_1 = 2\mathbb{Z}$, $\mathbb{T}_2 = 3^{\mathbb{N}_0} \cup \{0\}$, $\mathbb{T}_{(1)} = \mathbb{T}_{(2)} = \mathbb{T}_{(3)} = \mathbb{R}$,

$$U = \{(t_1,t_2) : 0 \le t_1 \le 8, \quad 1 \le t_2 \le 27\},$$

and S be a surface with a local parameterization (U,f), where

$$f(t_1,t_2) = ((t_1+t_2)^4, t_1^2 + t_2^4, t_1^2 + 3t_2^6), \quad (t_1,t_2) \in U.$$

Find $A_{\sigma_1}(S)$, $A_{\sigma_2}(S)$, $A_{\sigma_1^2\sigma_2}(S)$, $A_{\sigma_1\sigma_2^2}(S)$.

3.9 Notes and References

In this chapter we introduced differentiable maps on surfaces, the differential of a smooth map between surfaces and some of their properties have been deduced. The spherical map, σ_1-, σ_2-, $\sigma_1^2\sigma_2$-, $\sigma_1\sigma_2^2$-spherical maps and shape operator have been defined, σ_1-, σ_2-, $\sigma_1^2\sigma_2$-, $\sigma_1\sigma_2^2$-shape operators have been defined. It has been proved that the shape operators are not symmetric operators when one of the time scales does not coincide with $\mathbb{R}$. In the chapter the first fundamental form, first σ_1-, σ_2-, $\sigma_1^2\sigma_2$-, $\sigma_1\sigma_2^2$-fundamental forms of a surface are defined and σ_1-, σ_2-, $\sigma_1^2\sigma_2$-, $\sigma_1\sigma_2^2$- length of a curve on a surface and area of a parameterized surface are introduced as applications. Some of the results in this chapter can be found in [2], [9] and [10].

4

The Second Fundamental Forms

Suppose that $k \in \mathbb{N}$, $\widetilde{\mathbb{T}}$, $\mathbb{T}$, $\mathbb{T}_1$, $\mathbb{T}_2$, $\mathbb{T}_{(1)}$, $\mathbb{T}_{(2)}$, $\mathbb{T}_{(3)}$, $\mathbb{T}^{(1)}, \ldots, \mathbb{T}^{(k)}$ are time scales with forward jump operators and delta differentiation operators $\widetilde{\sigma}$, σ, σ_1, σ_2, $\sigma_{(1)}$, $\sigma_{(2)}$, $\sigma_{(3)}$, $\sigma^{(1)}, \ldots, \sigma^{(k)}$ and $\widetilde{\Delta}$, Δ, Δ_1, Δ_2, $\Delta_{(1)}$, $\Delta_{(2)}$, $\Delta_{(3)}$, $\Delta^{(1)}, \ldots, \Delta^{(k)}$, respectively. Let

$$I \subseteq \mathbb{T}, \quad J \subseteq \widetilde{\mathbb{T}}, \quad U \subseteq \mathbb{T}_1 \times \mathbb{T}_2, \quad W \subseteq \mathbb{T}_{(1)} \times \mathbb{T}_{(2)} \times \mathbb{T}_{(3)}, \quad V \subseteq \Lambda^{(k)} = \mathbb{T}^{(1)} \times \ldots \times \mathbb{T}^{(k)}.$$

4.1 Definitions

Definition 4.1 Let S be a σ_1-regular surface. The second σ_1-fundamental form of the surface S is given by

$$\phi_{2\sigma_1}(\xi,\eta) = -\phi_{1\sigma_1}(\mathscr{A}_{a\sigma_1}(\xi),\eta) = -\left(\mathscr{A}_{a\sigma_1}(\xi)\cdot\eta\right), \quad \xi,\eta \in T_{a\sigma_1}S.$$

Theorem 4.1 *Let S be a σ_1-regular surface. Then $\phi_{2\sigma_1}$ is a bilinear form.*

Proof. Let $\xi_1,\xi_2,\eta_1,\eta_2 \in T_{a\sigma_1}S$, $\alpha_1,\alpha_2 \in \mathbb{C}$ be arbitrarily chosen. Then

$$\begin{aligned}
\phi_{2\sigma_1}(\alpha_1\xi_1+\alpha_2\xi_2,\eta_1) &= -\left(\mathscr{A}_{a\sigma_1}(\alpha_1\xi_1+\alpha_2\xi_2)\cdot\eta_1\right)\\
&= -\left(\left(\alpha_1\mathscr{A}_{a\sigma_1}(\xi_1)+\alpha_2\mathscr{A}_{a\sigma_1}(\xi_2)\right)\cdot\eta_1\right)\\
&= -\alpha_1\left(\mathscr{A}_{a\sigma_1}(\xi_1)\cdot\eta_1\right)-\alpha_2\left(\mathscr{A}_{a\sigma_1}(\xi_1^2)\cdot\eta_1\right)\\
&= \alpha_1\phi_{2\sigma_1}(\xi_1,\eta_1)+\alpha_2\phi_{2\sigma_1}(\xi_2,\eta_1)
\end{aligned}$$

and

$$\begin{aligned}
\phi_{2\sigma_1}(\xi_1,\alpha_1\eta_1+\alpha_2\eta_2) &= -\left(\mathscr{A}_{a\sigma_1}(\xi_1)\cdot(\alpha_1\eta_1+\alpha_2\eta_2)\right)\\
&= -\alpha_1\left(\mathscr{A}_{a\sigma_1}(\xi_1)\cdot\eta_1\right)-\alpha_2\left(\mathscr{A}_{a\sigma_1}(\xi_1)\cdot\eta_2\right)\\
&= \alpha_1\phi_{2\sigma_1}(\xi_1,\eta_1)+\alpha_2\phi_{2\sigma_1}(\xi_1,\eta_2).
\end{aligned}$$

This completes the proof.

DOI: 10.1201/9781003205265-4

Suppose that S is a σ_1-regular surface with a local σ_1-parameterization (U,f). Then the matrix $[\phi_{2\sigma_1}]$ of the second σ_1-fundamental form with respect to the basis $\{f_{t_1}^{\Delta_1}, f_{t_2}^{\Delta_2\sigma_1}\}$ is given by

$$[\phi_{2\sigma_1}] = \begin{pmatrix} \phi_{2\sigma_1}\left(f_{t_1}^{\Delta_1}, f_{t_1}^{\Delta_1}\right) & \phi_{2\sigma_1}\left(f_{t_2}^{\Delta_2\sigma_1}, f_{t_1}^{\Delta_1}\right) \\ \phi_{2\sigma_1}\left(f_{t_1}^{\Delta_1}, f_{t_2}^{\Delta_2\sigma_1}\right) & \phi_{2\sigma_1}\left(f_{t_2}^{\Delta_2\sigma_1}, f_{t_2}^{\Delta_2\sigma_1}\right) \end{pmatrix}.$$

We have

$$\begin{aligned}
\phi_{2\sigma_1}\left(f_{t_1}^{\Delta_1}, f_{t_1}^{\Delta_1}\right) &= -\phi_{1\sigma_1}\left(\mathscr{A}_{a\sigma_1}\left(f_{t_1}^{\Delta_1}\right), f_{t_1}^{\Delta_1}\right) \\
&= -\left(\mathscr{A}_{a\sigma_1}\left(f_{t_1}^{\Delta_1}\right) \cdot f_{t_1}^{\Delta_1}\right) \\
&= -\left(N_{\sigma_1 t_1}^{\Delta_1} \cdot f_{t_1}^{\Delta_1}\right) \\
&= -L, \\
\phi_{2\sigma_1}\left(f_{t_2}^{\Delta_2\sigma_1}, f_{t_1}^{\Delta_1}\right) &= -\phi_{1\sigma_1}\left(\mathscr{A}_{a\sigma_1}\left(f_{t_2}^{\Delta_2\sigma_1}\right), f_{t_1}^{\Delta_1}\right) \\
&= -\left(\mathscr{A}_{a\sigma_1}\left(f_{t_2}^{\Delta_2\sigma_1}\right) \cdot f_{t_1}^{\Delta_1}\right) \\
&= -\left(N_{\sigma_1 t_2}^{\Delta_2\sigma_1} \cdot f_{t_1}^{\Delta_1}\right) \\
&= -M_{1\sigma_1}, \\
\phi_{2\sigma_1}\left(f_{t_1}^{\Delta_1}, f_{t_2}^{\Delta_2\sigma_1}\right) &= -\phi_{1\sigma_1}\left(\mathscr{A}_{a\sigma_1}\left(f_{t_1}^{\Delta_1}\right), f_{t_2}^{\Delta_2\sigma_1}\right) \\
&= -\left(\mathscr{A}_{a\sigma_1}\left(f_{t_1}^{\Delta_1}\right) \cdot f_{t_2}^{\Delta_2\sigma_1}\right) \\
&= -\left(N_{\sigma_1 t_1}^{\Delta_1} \cdot f_{t_2}^{\Delta_2\sigma_1}\right) \\
&= -M_{2\sigma_1}, \\
\phi_{2\sigma_1}\left(f_{t_1}^{\Delta_1}, f_{t_2}^{\Delta_2\sigma_1}\right) &= -\phi_{1\sigma_1}\left(\mathscr{A}_{a\sigma_1}\left(f_{t_2}^{\Delta_2\sigma_1}\right), f_{t_2}^{\Delta_2\sigma_1}\right) \\
&= -\left(\mathscr{A}_{a\sigma_1}\left(f_{t_2}^{\Delta_2\sigma_1}\right) \cdot f_{t_2}^{\Delta_2\sigma_1}\right) \\
&= -\left(N_{\sigma_1 t_2}^{\Delta_2\sigma_1} \cdot f_{t_2}^{\Delta_2\sigma_1}\right) \\
&= -P_{\sigma_1}.
\end{aligned}$$

Thus,

$$\begin{aligned}
[\phi_{2\sigma_1}] &= -\begin{pmatrix} L & M_{1\sigma_1} \\ M_{2\sigma_1} & P_{\sigma_1} \end{pmatrix} \\
&= -\mathscr{H}_{a\sigma_1}.
\end{aligned}$$

Exercise 4.1 Let $\mathbb{T}_1 = \mathbb{Z}$, $\mathbb{T}_2 = 2\mathbb{Z}$, $\mathbb{T}_{(1)} = \mathbb{T}_{(2)} = \mathbb{T}_{(3)} = \mathbb{R}$, S be a surface with a local parameterization (U,f), $U \subseteq \mathbb{T}_1 \times \mathbb{T}_2$, where $f : U \to \mathbb{R}^3$ is given by

$$f(t_1,t_2) = (t_1^2 + t_1 t_2, t_1 + t_2^2 + 2, t_1 + t_2), \quad (t_1,t_2) \in U.$$

Find $[\phi_{2\sigma_1}]$.

Definition 4.2 Let S be a σ_2-regular surface. The second σ_2-fundamental form of the surface S is given by

$$\phi_{2\sigma_2}(\xi,\eta) = -\phi_{1\sigma_2}(\mathscr{A}_{a\sigma_2}(\xi),\eta) = -\left(\mathscr{A}_{a\sigma_2}(\xi)\cdot\eta\right), \quad \xi,\eta \in T_{a\sigma_2}S.$$

Theorem 4.2 *Let S be a σ_2-regular surface. Then $\phi_{2\sigma_2}$ is a bilinear form.*

Proof. Let $\xi_1, \xi_2, \eta_1, \eta_2 \in T_{a\sigma_2}S$, $\alpha_1, \alpha_2 \in \mathbb{C}$ be arbitrarily chosen. Then

$$\begin{aligned}\phi_{2\sigma_2}(\alpha_1\xi_1 + \alpha_2\xi_2, \eta_1) &= -\left(\mathscr{A}_{a\sigma_2}(\alpha_1\xi_1 + \alpha_2\xi_2) \cdot \eta_1\right)\\ &= -\left((\alpha_1\mathscr{A}_{a\sigma_2}(\xi_1) + \alpha_2\mathscr{A}_{a\sigma_2}(\xi_2)) \cdot \eta_1\right)\\ &= -\alpha_1\left(\mathscr{A}_{a\sigma_2}(\xi_1) \cdot \eta_1\right) - \alpha_2\left(\mathscr{A}_{a\sigma_2}(\xi_2) \cdot \eta_1\right)\\ &= \alpha_1\phi_{2\sigma_2}(\xi_1, \eta_1) + \alpha_2\phi_{2\sigma_2}(\xi_2, \eta_1)\end{aligned}$$

and

$$\begin{aligned}\phi_{2\sigma_2}(\xi_1, \alpha_1\eta_1 + \alpha_2\eta_2) &= -\left(\mathscr{A}_{a\sigma_2}(\xi_1) \cdot (\alpha_1\eta_1 + \alpha_2\eta_2)\right)\\ &= -\alpha_1\left(\mathscr{A}_{a\sigma_2}(\xi_1) \cdot \eta_1\right) - \alpha_2\left(\mathscr{A}_{a\sigma_2}(\xi_1) \cdot \eta_2\right)\\ &= \alpha_1\phi_{2\sigma_2}(\xi_1, \eta_1) + \alpha_2\phi_{2\sigma_2}(\xi_1, \eta_2).\end{aligned}$$

This completes the proof.

Suppose that S is a σ_2-regular surface with a local σ_2-parameterization (U,f). Then the matrix $[\phi_{2\sigma_2}]$ of the second σ_2-fundamental form with respect to the basis $\{f_{t_1}^{\Delta_1\sigma_2}, f_{t_2}^{\Delta_2}\}$ is given by

$$[\phi_{2\sigma_2}] = \begin{pmatrix} \phi_{2\sigma_2}\left(f_{t_1}^{\Delta_1\sigma_2}, f_{t_1}^{\Delta_1\sigma_2}\right) & \phi_{2\sigma_2}\left(f_{t_2}^{\Delta_2}, f_{t_1}^{\Delta_1\sigma_2}\right) \\ \phi_{2\sigma_2}\left(f_{t_1}^{\Delta_1\sigma_2}, f_{t_2}^{\Delta_2}\right) & \phi_{2\sigma_2}\left(f_{t_2}^{\Delta_2}, f_{t_2}^{\Delta_2}\right) \end{pmatrix}.$$

We have

$$\begin{aligned}\phi_{2\sigma_2}\left(f_{t_1}^{\Delta_1\sigma_2}, f_{t_1}^{\Delta_1\sigma_2}\right) &= -\phi_{1\sigma_2}\left(\mathscr{A}_{a\sigma_2}\left(f_{t_1}^{\Delta_1\sigma_2}\right), f_{t_1}^{\Delta_1\sigma_2}\right)\\ &= -\left(\mathscr{A}_{a\sigma_2}\left(f_{t_1}^{\Delta_1\sigma_2}\right) \cdot f_{t_1}^{\Delta_1\sigma_2}\right)\\ &= -\left(N_{\sigma_2 t_1}^{\Delta_1\sigma_2} \cdot f_{t_1}^{\Delta_1\sigma_2}\right)\\ &= -L_{\sigma_2},\\ \phi_{2\sigma_2}\left(f_{t_2}^{\Delta_2}, f_{t_1}^{\Delta_1\sigma_2}\right) &= -\phi_{1\sigma_2}\left(\mathscr{A}_{a\sigma_2}\left(f_{t_2}^{\Delta_2}\right), f_{t_1}^{\Delta_1\sigma_2}\right)\\ &= -\left(\mathscr{A}_{a\sigma_2}\left(f_{t_2}^{\Delta_2}\right) \cdot f_{t_1}^{\Delta_1\sigma_2}\right)\\ &= -\left(N_{\sigma_2 t_2}^{\Delta_2} \cdot f_{t_1}^{\Delta_1\sigma_2}\right)\\ &= -M_{1\sigma_2},\\ \phi_{2\sigma_2}\left(f_{t_1}^{\Delta_1\sigma_2}, f_{t_2}^{\Delta_2}\right) &= -\phi_{1\sigma_2}\left(\mathscr{A}_{a\sigma_2}\left(f_{t_1}^{\Delta_1\sigma_2}\right), f_{t_2}^{\Delta_2}\right)\\ &= -\left(\mathscr{A}_{a\sigma_2}\left(f_{t_1}^{\Delta_1\sigma_2}\right) \cdot f_{t_2}^{\Delta_2}\right)\\ &= -\left(N_{\sigma_2 t_1}^{\Delta_1\sigma_2} \cdot f_{t_2}^{\Delta_2}\right)\\ &= -M_{2\sigma_2},\\ \phi_{2\sigma_2}\left(f_{t_1}^{\Delta_1\sigma_2}, f_{t_2}^{\Delta_2}\right) &= -\phi_{1\sigma_2}\left(\mathscr{A}_{a\sigma_2}\left(f_{t_2}^{\Delta_2}\right), f_{t_2}^{\Delta_2}\right)\\ &= -\left(\mathscr{A}_{a\sigma_2}\left(f_{t_2}^{\Delta_2}\right) \cdot f_{t_2}^{\Delta_2}\right)\end{aligned}$$

$$= -\left(N^{\Delta_2}_{\sigma_2 t_2} \cdot f^{\Delta_2}_{t_2}\right)$$
$$= -P.$$

Thus,

$$[\phi_{2\sigma_2}] = -\begin{pmatrix} L_{\sigma_2} & M_{1\sigma_2} \\ M_{2\sigma_2} & P \end{pmatrix}$$
$$= -\mathcal{H}_{a\sigma_2}.$$

Exercise 4.2 Let $\mathbb{T}_1 = 2^{\mathbb{N}_0}$, $\mathbb{T}_2 = \mathbb{Z}$, $\mathbb{T}_{(1)} = \mathbb{T}_{(2)} = \mathbb{T}_{(3)} = \mathbb{R}$, S be a surface with a local parameterization (U,f), $U \subseteq \mathbb{T}_1 \times \mathbb{T}_2$, where $f : U \to \mathbb{R}^3$ is given by

$$f(t_1,t_2) = (t_1 + t_1 t_2^2, t_1^2 + t_2 + 4, t_1 + t_2 + 3), \quad (t_1,t_2) \in U.$$

Find $[\phi_{2\sigma_2}]$.

Definition 4.3 Let S be a $\sigma_1^2\sigma_2$-regular surface. The second $\sigma_1^2\sigma_2$-fundamental form of the surface S is given by

$$\phi_{2\sigma_1^2\sigma_2}(\xi,\eta) = -\phi_{1\sigma_1^2\sigma_2}(\mathcal{A}_{a\sigma_1^2\sigma_2}(\xi),\eta) = -\left(\mathcal{A}_{a\sigma_1^2\sigma_2}(\xi)\cdot\eta\right), \quad \xi,\eta \in T_{a\sigma_1^2\sigma_2}S.$$

Theorem 4.3 *Let S be a $\sigma_1^2\sigma_2$-regular surface. Then $\phi_{2\sigma_1^2\sigma_2}$ is a bilinear form.*

Proof. Let $\xi_1,\xi_2,\eta_1,\eta_2 \in T_{a\sigma_1^2\sigma_2}S$, $\alpha_1,\alpha_2 \in \mathbb{C}$ be arbitrarily chosen. Then

$$\begin{aligned}
\phi_{2\sigma_1^2\sigma_2}(\alpha_1\xi_1 + \alpha_2\xi_2, \eta_1) &= -\left(\mathcal{A}_{a\sigma_1^2\sigma_2}(\alpha_1\xi_1 + \alpha_2\xi_2)\cdot\eta_1\right) \\
&= -\left(\left(\alpha_1\mathcal{A}_{a\sigma_1^2\sigma_2}(\xi_1) + \alpha_2\mathcal{A}_{a\sigma_1^2\sigma_2}(\xi_2)\right)\cdot\eta_1\right) \\
&= -\alpha_1\left(\mathcal{A}_{a\sigma_1^2\sigma_2}(\xi_1)\cdot\eta_1\right) - \alpha_2\left(\mathcal{A}_{a\sigma_1^2\sigma_2}(\xi_2)\cdot\eta_1\right) \\
&= \alpha_1\phi_{2\sigma_1^2\sigma_2}(\xi_1,\eta_1) + \alpha_2\phi_{2\sigma_1^2\sigma_2}(\xi_2,\eta_1)
\end{aligned}$$

and

$$\begin{aligned}
\phi_{2\sigma_1^2\sigma_2}(\xi_1, \alpha_1\eta_1 + \alpha_2\eta_2) &= -\left(\mathcal{A}_{a\sigma_1^2\sigma_2}(\xi_1)\cdot(\alpha_1\eta_1 + \alpha_2\eta_2)\right) \\
&= -\alpha_1\left(\mathcal{A}_{a\sigma_1^2\sigma_2}(\xi_1)\cdot\eta_1\right) - \alpha_2\left(\mathcal{A}_{a\sigma_1^2\sigma_2}(\xi_1)\cdot\eta_2\right) \\
&= \alpha_1\phi_{2\sigma_1^2\sigma_2}(\xi_1,\eta_1) + \alpha_2\phi_{2\sigma_1^2\sigma_2}(\xi_1,\eta_2).
\end{aligned}$$

This completes the proof.

Suppose that S is a $\sigma_1^2\sigma_2$-regular surface with a local $\sigma_1^2\sigma_2$-parameterization (U,f). Then the matrix $[\phi_{2\sigma_1^2\sigma_2}]$ of the second $\sigma_1^2\sigma_2$-fundamental form with respect to the basis $\{f_{t_1}^{\Delta_1\sigma_1\sigma_2}, f_{t_2}^{\Delta_2\sigma_1^2\sigma_2}\}$ is given by

$$[\phi_{2\sigma_1^2\sigma_2}] = \begin{pmatrix} \phi_{2\sigma_1^2\sigma_2}\left(f_{t_1}^{\Delta_1\sigma_1\sigma_2}, f_{t_1}^{\Delta_1\sigma_1\sigma_2}\right) & \phi_{2\sigma_1^2\sigma_2}\left(f_{t_2}^{\Delta_2\sigma_1^2\sigma_2}, f_{t_1}^{\Delta_1\sigma_1\sigma_2}\right) \\ \phi_{2\sigma_1^2\sigma_2}\left(f_{t_1}^{\Delta_1\sigma_1\sigma_2}, f_{t_2}^{\Delta_2\sigma_1^2\sigma_2}\right) & \phi_{2\sigma_1^2\sigma_2}\left(f_{t_2}^{\Delta_2\sigma_1^2\sigma_2}, f_{t_2}^{\Delta_2\sigma_1^2\sigma_2}\right) \end{pmatrix}.$$

We have

$$\begin{aligned}
\phi_{2\sigma_1^2\sigma_2}\left(f_{t_1}^{\Delta_1\sigma_1\sigma_2}, f_{t_1}^{\Delta_1\sigma_1\sigma_2}\right) &= -\phi_{1\sigma_1^2\sigma_2}\left(\mathscr{A}_{a\sigma_1^2\sigma_2}\left(f_{t_1}^{\Delta_1\sigma_1\sigma_2}\right), f_{t_1}^{\Delta_1\sigma_1\sigma_2}\right)\\
&= -\left(\mathscr{A}_{a\sigma_1^2\sigma_2}\left(f_{t_1}^{\Delta_1\sigma_1\sigma_2}\right)\cdot f_{t_1}^{\Delta_1\sigma_1\sigma_2}\right)\\
&= -\left(N_{\sigma_1^2\sigma_2 t_1}^{\Delta_1\sigma_1\sigma_2}\cdot f_{t_1}^{\Delta_1\sigma_1\sigma_2}\right)\\
&= -L_{\sigma_1^2\sigma_2},\\
\phi_{2\sigma_1^2\sigma_2}\left(f_{t_2}^{\Delta_2\sigma_1^2\sigma_2}, f_{t_1}^{\Delta_1\sigma_1\sigma_2}\right) &= -\phi_{1\sigma_1^2\sigma_2}\left(\mathscr{A}_{a\sigma_1^2\sigma_2}\left(f_{t_2}^{\Delta_2\sigma_1^2\sigma_2}\right), f_{t_1}^{\Delta_1\sigma_1\sigma_2}\right)\\
&= -\left(\mathscr{A}_{a\sigma_1^2\sigma_2}\left(f_{t_2}^{\Delta_2\sigma_1^2\sigma_2}\right)\cdot f_{t_1}^{\Delta_1\sigma_1\sigma_2}\right)\\
&= -\left(N_{\sigma_1^2\sigma_2 t_2}^{\Delta_2\sigma_1^2\sigma_2}\cdot f_{t_1}^{\Delta_1\sigma_1\sigma_2}\right)\\
&= -M_{1\sigma_1^2\sigma_2},\\
\phi_{2\sigma_1^2\sigma_2}\left(f_{t_1}^{\Delta_1\sigma_1\sigma_2}, f_{t_2}^{\Delta_2\sigma_1^2\sigma_2}\right) &= -\phi_{1\sigma_1^2\sigma_2}\left(\mathscr{A}_{a\sigma_1^2\sigma_2}\left(f_{t_1}^{\Delta_1\sigma_1\sigma_2}\right), f_{t_2}^{\Delta_2\sigma_1^2\sigma_2}\right)\\
&= -\left(\mathscr{A}_{a\sigma_1^2\sigma_2}\left(f_{t_1}^{\Delta_1\sigma_1\sigma_2}\right)\cdot f_{t_2}^{\Delta_2\sigma_1^2\sigma_2}\right)\\
&= -\left(N_{\sigma_1^2\sigma_2 t_1}^{\Delta_1\sigma_1\sigma_2}\cdot f_{t_2}^{\Delta_2\sigma_1^2\sigma_2}\right)\\
&= -M_{2\sigma_1^2\sigma_2},\\
\phi_{2\sigma_1^2\sigma_2}\left(f_{t_2}^{\Delta_2\sigma_1^2\sigma_2}, f_{t_2}^{\Delta_2\sigma_1^2\sigma_2}\right) &= -\phi_{1\sigma_1^2\sigma_2}\left(\mathscr{A}_{a\sigma_1^2\sigma_2}\left(f_{t_2}^{\Delta_2\sigma_1^2\sigma_2}\right), f_{t_2}^{\Delta_2\sigma_1^2\sigma_2}\right)\\
&= -\left(\mathscr{A}_{a\sigma_1^2\sigma_2}\left(f_{t_2}^{\Delta_2\sigma_1^2\sigma_2}\right)\cdot f_{t_2}^{\Delta_2\sigma_1^2\sigma_2}\right)\\
&= -\left(N_{\sigma_1^2\sigma_2 t_2}^{\Delta_2\sigma_1^2\sigma_2}\cdot f_{t_2}^{\Delta_2\sigma_1^2\sigma_2}\right)\\
&= -P_{\sigma_1^2\sigma_2}.
\end{aligned}$$

Thus,

$$\begin{aligned}
[\phi_{2\sigma_1^2\sigma_2}] &= -\begin{pmatrix} L_{\sigma_1^2\sigma_2} & M_{1\sigma_1^2\sigma_2} \\ M_{2\sigma_1^2\sigma_2} & P_{\sigma_1^2\sigma_2} \end{pmatrix}\\
&= -\mathscr{H}_{a\sigma_1^2\sigma_2}.
\end{aligned}$$

Exercise 4.3 Let $\mathbb{T}_1 = 3^{\mathbb{N}_0}$, $\mathbb{T}_2 = 2^{\mathbb{N}_0}$, $\mathbb{T}_{(1)} = \mathbb{T}_{(2)} = \mathbb{T}_{(3)} = \mathbb{R}$, S be a surface with a local parameterization (U,f), $U \subseteq \mathbb{T}_1 \times \mathbb{T}_2$, where $f : U \to \mathbb{R}^3$ is given by

$$f(t_1,t_2) = \left(\frac{1+t_1}{1+t_1+t_2}, t_1^2+t_2^2, t_1-2t_2\right), \quad (t_1,t_2) \in U.$$

Find $[\phi_{2\sigma_1^2\sigma_2}]$.

Definition 4.4 Let S be a $\sigma_1\sigma_2^2$-regular surface. The second $\sigma_1\sigma_2^2$-fundamental form of the surface S is given by

$$\phi_{2\sigma_1\sigma_2^2}(\xi,\eta)=-\phi_{1\sigma_1\sigma_2^2}(\mathscr{A}_{a\sigma_1\sigma_2^2}(\xi),\eta)=-\left(\mathscr{A}_{a\sigma_1\sigma_2^2}(\xi)\cdot\eta\right),\quad \xi,\eta\in T_{a\sigma_1\sigma_2^2}S.$$

Theorem 4.4 *Let S be a $\sigma_1\sigma_2^2$-regular surface. Then $\phi_{2\sigma_1\sigma_2^2}$ is a bilinear form.*

Proof. Let $\xi_1,\xi_2,\eta_1,\eta_2\in T_{a\sigma_1\sigma_2^2}S$, $\alpha_1,\alpha_2\in\mathbb{C}$ be arbitrarily chosen. Then

$$\begin{aligned}\phi_{2\sigma_1\sigma_2^2}(\alpha_1\xi_1+\alpha_2\xi_2,\eta_1)&=-\left(\mathscr{A}_{a\sigma_1\sigma_2^2}(\alpha_1\xi_1+\alpha_2\xi_2)\cdot\eta_1\right)\\&=-\left(\left(\alpha_1\mathscr{A}_{a\sigma_1\sigma_2^2}(\xi_1)+\alpha_2\mathscr{A}_{a\sigma_1\sigma_2^2}(\xi_2)\right)\cdot\eta_1\right)\\&=-\alpha_1\left(\mathscr{A}_{a\sigma_1\sigma_2^2}(\xi_1)\cdot\eta_1\right)-\alpha_2\left(\mathscr{A}_{a\sigma_1\sigma_2^2}(\xi_2)\cdot\eta_1\right)\\&=\alpha_1\phi_{2\sigma_1\sigma_2^2}(\xi_1,\eta_1)+\alpha_2\phi_{2\sigma_1\sigma_2^2}(\xi_2,\eta_1)\end{aligned}$$

and

$$\begin{aligned}\phi_{2\sigma_1\sigma_2^2}(\xi_1,\alpha_1\eta_1+\alpha_2\eta_2)&=-\left(\mathscr{A}_{a\sigma_1\sigma_2^2}(\xi_1)\cdot(\alpha_1\eta_1+\alpha_2\eta_2)\right)\\&=-\alpha_1\left(\mathscr{A}_{a\sigma_1\sigma_2^2}(\xi_1)\cdot\eta_1\right)-\alpha_2\left(\mathscr{A}_{a\sigma_1\sigma_2^2}(\xi_1)\cdot\eta_2\right)\\&=\alpha_1\phi_{2\sigma_1\sigma_2^2}(\xi_1,\eta_1)+\alpha_2\phi_{2\sigma_1\sigma_2^2}(\xi_1,\eta_2).\end{aligned}$$

This completes the proof.

Suppose that S is a $\sigma_1\sigma_2^2$-regular surface with a local $\sigma_1\sigma_2^2$-parameterization (U,f). Then the matrix $[\phi_{2\sigma_1\sigma_2^2}]$ of the second $\sigma_1\sigma_2^2$-fundamental form with respect to the basis $\{f_{t_1}^{\Delta_1\sigma_1\sigma_2^2},f_{t_2}^{\Delta_2\sigma_1\sigma_2}\}$ is given by

$$[\phi_{2\sigma_1\sigma_2^2}]=\begin{pmatrix}\phi_{2\sigma_1\sigma_2^2}\left(f_{t_1}^{\Delta_1\sigma_1\sigma_2^2},f_{t_1}^{\Delta_1\sigma_1\sigma_2^2}\right) & \phi_{2\sigma_1\sigma_2^2}\left(f_{t_2}^{\Delta_2\sigma_1\sigma_2},f_{t_1}^{\Delta_1\sigma_1\sigma_2^2}\right)\\ \phi_{2\sigma_1\sigma_2^2}\left(f_{t_1}^{\Delta_1\sigma_1\sigma_2^2},f_{t_2}^{\Delta_2\sigma_1\sigma_2}\right) & \phi_{2\sigma_1\sigma_2^2}\left(f_{t_2}^{\Delta_2\sigma_1\sigma_2},f_{t_2}^{\Delta_2\sigma_1\sigma_2}\right)\end{pmatrix}.$$

We have

$$\begin{aligned}\phi_{2\sigma_1\sigma_2^2}\left(f_{t_1}^{\Delta_1\sigma_1\sigma_2^2},f_{t_1}^{\Delta_1\sigma_1\sigma_2^2}\right)&=-\phi_{1\sigma_1\sigma_2^2}\left(\mathscr{A}_{a\sigma_1\sigma_2^2}\left(f_{t_1}^{\Delta_1\sigma_1\sigma_2^2}\right),f_{t_1}^{\Delta_1\sigma_1\sigma_2^2}\right)\\&=-\left(\mathscr{A}_{a\sigma_1\sigma_2^2}\left(f_{t_1}^{\Delta_1\sigma_1\sigma_2^2}\right)\cdot f_{t_1}^{\Delta_1\sigma_1\sigma_2^2}\right)\\&=-\left(N_{\sigma_1\sigma_2^2t_1}^{\Delta_1\sigma_1\sigma_2^2}\cdot f_{t_1}^{\Delta_1\sigma_1\sigma_2^2}\right)\\&=-L_{\sigma_1\sigma_2^2},\end{aligned}$$

$$\phi_{2\sigma_1\sigma_2^2}\left(f_{t_2}^{\Delta_2\sigma_1\sigma_2},f_{t_1}^{\Delta_1\sigma_1\sigma_2^2}\right)=-\phi_{1\sigma_1\sigma_2^2}\left(\mathscr{A}_{a\sigma_1\sigma_2^2}\left(f_{t_2}^{\Delta_2\sigma_1\sigma_2}\right),f_{t_1}^{\Delta_1\sigma_1\sigma_2^2}\right)$$

$$
\begin{aligned}
&= -\left(\mathscr{A}_{a\sigma_1\sigma_2^2}\left(f_{t_2}^{\Delta_2\sigma_1\sigma_2}\right)\cdot f_{t_1}^{\Delta_1\sigma_1\sigma_2^2}\right)\\
&= -\left(N_{\sigma_1\sigma_2^2 t_2}^{\Delta_2\sigma_1\sigma_2}\cdot f_{t_1}^{\Delta_1\sigma_1\sigma_2^2}\right)\\
&= -M_{1\sigma_1\sigma_2^2},
\end{aligned}
$$

$$
\begin{aligned}
\phi_{2\sigma_1\sigma_2^2}\left(f_{t_1}^{\Delta_1\sigma_1\sigma_2^2}, f_{t_2}^{\Delta_2\sigma_1\sigma_2}\right) &= -\phi_{1\sigma_1\sigma_2^2}\left(\mathscr{A}_{a\sigma_1\sigma_2^2}\left(f_{t_1}^{\Delta_1\sigma_1\sigma_2^2}\right), f_{t_2}^{\Delta_2\sigma_1\sigma_2}\right)\\
&= -\left(\mathscr{A}_{a\sigma_1\sigma_2^2}\left(f_{t_1}^{\Delta_1\sigma_1\sigma_2^2}\right)\cdot f_{t_2}^{\Delta_2\sigma_1\sigma_2}\right)\\
&= -\left(N_{\sigma_1\sigma_2^2 t_1}^{\Delta_1\sigma_1\sigma_2^2}\cdot f_{t_2}^{\Delta_2\sigma_1\sigma_2}\right)\\
&= -M_{2\sigma_1\sigma_2^2},
\end{aligned}
$$

$$
\begin{aligned}
\phi_{2\sigma_1\sigma_2^2}\left(f_{t_2}^{\Delta_2\sigma_1\sigma_2}, f_{t_2}^{\Delta_2\sigma_1\sigma_2}\right) &= -\phi_{1\sigma_1\sigma_2^2}\left(\mathscr{A}_{a\sigma_1\sigma_2^2}\left(f_{t_2}^{\Delta_2\sigma_1\sigma_2}\right), f_{t_2}^{\Delta_2\sigma_1\sigma_2}\right)\\
&= -\left(\mathscr{A}_{a\sigma_1\sigma_2^2}\left(f_{t_2}^{\Delta_2\sigma_1\sigma_2}\right)\cdot f_{t_2}^{\Delta_2\sigma_1\sigma_2}\right)\\
&= -\left(N_{\sigma_1\sigma_2^2 t_2}^{\Delta_2\sigma_1\sigma_2}\cdot f_{t_2}^{\Delta_2\sigma_1\sigma_2}\right)\\
&= -P_{\sigma_1\sigma_2^2}.
\end{aligned}
$$

Thus,

$$
\begin{aligned}
[\phi_{2\sigma_1\sigma_2^2}] &= -\begin{pmatrix} L_{\sigma_1\sigma_2^2} & M_{1\sigma_1\sigma_2^2} \\ M_{2\sigma_1\sigma_2^2} & P_{\sigma_1\sigma_2^2} \end{pmatrix}\\
&= -\mathscr{H}_{a\sigma_1\sigma_2^2}.
\end{aligned}
$$

Exercise 4.4 Let $\mathbb{T}_1 = 2^{\mathbb{N}_0}$, $\mathbb{T}_2 = 3^{\mathbb{N}_0}$, $\mathbb{T}_{(1)} = \mathbb{T}_{(2)} = \mathbb{T}_{(3)} = \mathbb{R}$, S be a surface with a local parameterization (U,f), $U \subseteq \mathbb{T}_1 \times \mathbb{T}_2$, where $f : U \to \mathbb{R}^3$ is given by

$$
f(t_1,t_2) = \left(\frac{1+t_1+t_2}{1+t_2^2}, t_1+t_2, t_1^2-2t_2\right), \quad (t_1,t_2) \in U.
$$

Find $[\phi_{2\sigma_1\sigma_2^2}]$.

4.2 Normal Curvatures. Variant of the Meusnier Theorem

Let S be a σ_1- or σ_2- or $\sigma_1^2\sigma_2$- or $\sigma_1\sigma_2^2$-oriented surface with a local σ_1- or σ_2- or $\sigma_1^2\sigma_2$- or $\sigma_1\sigma_2^2$- parameterization (U,f), $g = g(t)$, $t \in I$, be a regular parameterized curve. Let also, $g_1 = g_1(s)$, $s \in J$, be an equivalent to g naturally parameterized curve, the natural parameter being the arc length. Assume that $g_1(s)$ has this parametrization in the local equations

$$
\begin{aligned}
t_1 &= h_1(s),\\
t_2 &= h_2(s), \quad s \in J,
\end{aligned}
$$

for which

$$h_1(\tilde{\sigma}(s)) = \sigma_1(h_1(s)),$$
$$h_2(\tilde{\sigma}(s)) = \sigma_2(h_2(s)), \quad s \in J,$$

and

$$g_1(s) = f(h_1(s), h_2(s)), \quad s \in J. \tag{4.1}$$

Then the curvature vector of the curve $g_1(s)$ will be $g_1^{\tilde{\Delta}^2}(s)$.

Definition 4.5 Let S be a σ_1-regular surface with a local σ_1-parametrization (U,f) and $g : I \to S$ be a curve lying in S. The projection of the curvature vector $k(t)$ of $g(t)$ on $N_{\sigma_1}^{\sigma_1\sigma_2}(g(t))$ is called σ_1-normal curvature of the curve g at t and it is denoted by $k_{n\sigma_1}(t)$.

Definition 4.6 Let σ_1 be Δ_1-differentiable on $\mathbb{T}_1, f$ be σ_1-completely delta differentiable on $U, f_{t_1}^{\Delta_1}$ and $f_{t_2}^{\Delta_2\sigma_1}$ be σ_1-completely delta differentiable on U. Let also,

$$b_{11n\sigma_1\sigma_1\sigma_1}(t,t_1,t_2) = f_{t_1}^{\Delta_1^2}(t_1,t_2) \cdot N_{\sigma_1}^{\sigma_1\sigma_2}(g(t)),$$
$$b_{12n\sigma_1\sigma_1\sigma_1}(t,t_1,t_2) = f_{t_1t_2}^{\Delta_1\Delta_2\sigma_1}(t_1,t_2) \cdot N_{\sigma_1}^{\sigma_1\sigma_2}(g(t)),$$
$$b_{21n\sigma_1\sigma_1\sigma_1}(t,t_1,t_2) = \left(1 + \mu_1^{\Delta_1}(t_1)\right) b_{12n\sigma_1\sigma_1\sigma_1}(t,t_1,t_2),$$
$$b_{22n\sigma_1\sigma_1\sigma_1}(t,t_1,t_2) = f_{t_2}^{\Delta_2^2\sigma_1^2}(t_1,t_2) \cdot N_{\sigma_1}^{\sigma_1\sigma_2}(g(t)), \quad (t_1,t_2) \in U, \quad t \in I,$$

and

$$\mathscr{B}_{n\sigma_1\sigma_1\sigma_1}(t,t_1,t_2) = \begin{pmatrix} b_{11n\sigma_1\sigma_1\sigma_1}(t,t_1,t_2) & b_{12n\sigma_1\sigma_1\sigma_1}(t,t_1,t_2) \\ b_{21n\sigma_1\sigma_1\sigma_1}(t,t_1,t_2) & b_{22n\sigma_1\sigma_1\sigma_1}(t,t_1,t_2) \end{pmatrix}, \quad (t_1,t_2) \in U, \quad t \in I.$$

For $p, q \in T_{a\sigma_1}S$, define

$$\phi_{\sigma_1\sigma_1\sigma_1}^1(p,q) = p^T \mathscr{B}_{n\sigma_1\sigma_1\sigma_1} q.$$

Theorem 4.5 (Variant of the Meusnier Theorem) *Let σ_1 be Δ_1-differentiable, S be a σ_1-regular surface with a local σ_1-parameterization (U,f) such that f is σ_1-completely delta differentiable, $f_{t_1}^{\Delta_1}$ and $f_{t_2}^{\Delta_2\sigma_1}$ be σ_1-completely delta differentiable on U.*

Then

$$k_{n\sigma_1}(t) = \phi_{\sigma_1\sigma_1\sigma_1}^1\left(\frac{g^\Delta(t)}{\|g^\Delta(t)\|}, \frac{g^\Delta(t)}{\|g^\Delta(t)\|}\right), \quad t \in I.$$

Proof. We have

$$g_1^{\tilde{\Delta}}(s) = f_{t_1}^{\Delta_1}(h_1(s), h_2(s)) h_1^{\tilde{\Delta}}(s) + f_{t_2}^{\Delta_2\sigma_1}(h_1(s), h_2(s)) h_2^{\tilde{\Delta}}(s),$$
$$g_1^{\tilde{\Delta}^2}(s) = f_{t_1}^{\Delta_1}(h_1(\tilde{\sigma}(s)), h_2(\tilde{\sigma}(s))) h_1^{\tilde{\Delta}^2}(s)$$

$$
\begin{aligned}
&+f_{t_2}^{\Delta_2\sigma_1}(h_1(\widetilde{\sigma}(s)),h_2(\widetilde{\sigma}(s)))h_2^{\widetilde{\Delta}^2}(s)\\
&+f_{t_1}^{\Delta_1^2}(h_1(s),h_2(s))\left(h_1^{\widetilde{\Delta}}(s)\right)^2\\
&+f_{t_1t_2}^{\Delta_1\Delta_2\sigma_1}(h_1(s),h_2(s))h_1^{\widetilde{\Delta}}(s)h_2^{\widetilde{\Delta}}(s)\\
&+f_{t_2t_1}^{\Delta_1\sigma_1\Delta_1}(h_1(s),h_2(s))h_1^{\widetilde{\Delta}}(s)h_2^{\widetilde{\Delta}}(s)\\
&+f_{t_2}^{\Delta_2^2\sigma_1^2}(h_1(s),h_2(s))\left(h_2^{\widetilde{\Delta}}(s)\right)^2\\
=&f_{t_1}^{\Delta_1\sigma_1\sigma_2}(h_1(s),h_2(s))h_1^{\widetilde{\Delta}^2}(s)\\
&+f_{t_2}^{\Delta_2\sigma_1^2\sigma_2}(h_1(s),h_2(s))h_2^{\widetilde{\Delta}^2}(s)\\
&+f_{t_1}^{\Delta_1^2}(h_1(s),h_2(s))\left(h_1^{\widetilde{\Delta}}(s)\right)^2\\
&+f_{t_1t_2}^{\Delta_1\Delta_2\sigma_1}(h_1(s),h_2(s))h_1^{\widetilde{\Delta}}(s)h_2^{\widetilde{\Delta}}(s)\\
&+\left(1+\mu_1^{\Delta_1}(t_1)\right)f_{t_1t_2}^{\Delta_1\Delta_2\sigma_1}(h_1(s),h_2(s))h_1^{\widetilde{\Delta}}(s)h_2^{\widetilde{\Delta}}(s)\\
&+f_{t_2}^{\Delta_2^2\sigma_1^2}(h_1(s),h_2(s))\left(h_2^{\widetilde{\Delta}}(s)\right)^2,\quad s\in J,
\end{aligned}
$$

that is

$$
\begin{aligned}
g_1^{\widetilde{\Delta}^2}(s)=&f_{t_1}^{\Delta_1\sigma_1\sigma_2}(h_1(s),h_2(s))h_1^{\widetilde{\Delta}^2}(s)\\
&+f_{t_2}^{\Delta_2\sigma_1^2\sigma_2}(h_1(s),h_2(s))h_2^{\widetilde{\Delta}^2}(s)\\
&+f_{t_1}^{\Delta_1^2}(h_1(s),h_2(s))\left(h_1^{\widetilde{\Delta}}(s)\right)^2\\
&+f_{t_1t_2}^{\Delta_1\Delta_2\sigma_1}(h_1(s),h_2(s))h_1^{\widetilde{\Delta}}(s)h_2^{\widetilde{\Delta}}(s)\\
&+\left(1+\mu_1^{\Delta_1}(t_1)\right)f_{t_1t_2}^{\Delta_1\Delta_2\sigma_1}(h_1(s),h_2(s))h_1^{\widetilde{\Delta}}(s)h_2^{\widetilde{\Delta}}(s)\\
&+f_{t_2}^{\Delta_2^2\sigma_1^2}(h_1(s),h_2(s))\left(h_2^{\widetilde{\Delta}}(s)\right)^2,\quad s\in J.
\end{aligned}
\tag{4.2}
$$

Then

$$
\begin{aligned}
k_{n\sigma_1}(s)&=k(s)\cdot N_{\sigma_1}^{\sigma_1\sigma_2}(g_1(s))\\
&=g^{\widetilde{\Delta}^2}(s)\cdot N_{\sigma_1}^{\sigma_1\sigma_2}(g_1(s))\\
&=\left(f_{t_1}^{\Delta_1\sigma_1\sigma_2}(h_1(s),h_2(s))\cdot N_{\sigma_1}^{\sigma_1\sigma_2}(g_1(s))\right)h_1^{\widetilde{\Delta}^2}(s)\\
&\quad+\left(f_{t_2}^{\Delta_2\sigma_1^2\sigma_2}(h_1(s),h_2(s))\cdot N_{\sigma_1}^{\sigma_1\sigma_2}(g_1(s))\right)h_2^{\widetilde{\Delta}^2}(s)\\
&\quad+\left(f_{t_1}^{\Delta_1^2}(h_1(s),h_2(s))\cdot N_{\sigma_1}^{\sigma_1\sigma_2}(g_1(s))\right)\left(h_1^{\widetilde{\Delta}}(s)\right)^2\\
&\quad+\left(f_{t_1t_2}^{\Delta_1\Delta_2\sigma_1}(h_1(s),h_2(s))\cdot N_{\sigma_1}^{\sigma_1\sigma_2}(g_1(s))\right)h_1^{\widetilde{\Delta}}(s)h_2^{\widetilde{\Delta}}(s)\\
&\quad+\left(1+\mu_1^{\Delta_1}(t_1)\right)\left(f_{t_1t_2}^{\Delta_1\Delta_2\sigma_1}(h_1(s),h_2(s))\cdot N_{\sigma_1}^{\sigma_1\sigma_2}(g_1(s))\right)h_1^{\widetilde{\Delta}}(s)h_2^{\widetilde{\Delta}}(s)
\end{aligned}
$$

$$
\begin{aligned}
&+\left(f_{t_2}^{\Delta_2^2\sigma_1^2}(h_1(s),h_2(s))\cdot N_{\sigma_1}^{\sigma_1\sigma_2}(g_1(s))\right)\left(h_2^{\tilde{\Delta}}(s)\right)^2\\
&=\left(f_{t_1}^{\Delta_1^2}(h_1(s),h_2(s))\cdot N_{\sigma_1}^{\sigma_1\sigma_2}(g_1(s))\right)\left(h_1^{\tilde{\Delta}}(s)\right)^2\\
&+\left(f_{t_1t_2}^{\Delta_1\Delta_2\sigma_1}(h_1(s),h_2(s))\cdot N_{\sigma_1}^{\sigma_1\sigma_2}(g_1(s))\right)h_1^{\tilde{\Delta}}(s)h_2^{\tilde{\Delta}}(s)\\
&+\left(1+\mu_1^{\Delta_1}(t_1)\right)\left(f_{t_1t_2}^{\Delta_1\Delta_2\sigma_1}(h_1(s),h_2(s))\cdot N_{\sigma_1}^{\sigma_1\sigma_2}(g_1(s))\right)h_1^{\tilde{\Delta}}(s)h_2^{\tilde{\Delta}}(s)\\
&+\left(f_{t_2}^{\Delta_2^2\sigma_1^2}(h_1(s),h_2(s))\cdot N_{\sigma_1}^{\sigma_1\sigma_2}(g_1(s))\right)\left(h_2^{\tilde{\Delta}}(s)\right)^2\\
&=\phi_{\sigma_1\sigma_1\sigma_1}^1\left(g_1^{\tilde{\Delta}}(s),g_1^{\tilde{\Delta}}(s)\right),
\end{aligned}
$$

$s\in J$,

that is,

$$k_{n\sigma_1}(s)=\phi_{\sigma_1\sigma_1\sigma_1}^1\left(g_1^{\tilde{\Delta}}(s),g_1^{\tilde{\Delta}}(s)\right),$$

$s\in J$.

Now, we will use the initial parameterized curve. We have

$$g^{\Delta}(t)=g_1^{\tilde{\Delta}}(s(t))s^{\Delta}(t),\quad s\in J,\quad t\in I, \tag{4.3}$$

where

$$s^{\Delta}(t)=\|g^{\Delta}(t)\|,\quad t\in I.$$

Therefore

$$g_1^{\tilde{\Delta}}(s)=\frac{g^{\Delta}(t)}{\|g^{\Delta}(t)\|},\quad s\in J,\quad t\in I. \tag{4.4}$$

From here,

$$k_{n\sigma_1}(t)=\phi_{\sigma_1\sigma_1\sigma_1}^1\left(\frac{g^{\Delta}(t)}{\|g^{\Delta}(t)\|},\frac{g^{\Delta}(t)}{\|g^{\Delta}(t)\|}\right),\quad t\in I.$$

This completes the proof.

Definition 4.7 Let σ_1 be Δ_1-differentiable on $\mathbb{T}_1$, f be σ_1-completely delta differentiable on U, $f_{t_1}^{\Delta_1}$ be σ_1-completely delta differentiable and $f_{t_2}^{\Delta_2\sigma_1}$ be σ_1-completely delta differentiable on U. Let also,

$$
\begin{aligned}
&b_{11n\sigma_1\sigma_1\sigma_2}(t,t_1,t_2)=f_{t_1}^{\Delta_1^2}(t_1,t_2)\cdot N_{\sigma_1}^{\sigma_1\sigma_2}(g(t)),\\
&b_{12n\sigma_1\sigma_1\sigma_2}(t,t_1,t_2)=f_{t_1t_2}^{\Delta_1\Delta_2\sigma_1}(t_1,t_2)\cdot N_{\sigma_1}^{\sigma_1\sigma_2}(g(t)),\\
&b_{21n\sigma_1\sigma_1\sigma_1}(t,t_1,t_2)=\left(1+\mu_1^{\Delta_1}(t_1)\right)f_{t_1t_2}^{\Delta_1\Delta_2\sigma_1\sigma_2}(t_1,t_2)\cdot N_{\sigma_1}^{\sigma_1\sigma_2}(g(t)),\\
&b_{22n\sigma_1\sigma_1\sigma_2}(t,t_1,t_2)=f_{t_2}^{\Delta_2^2\sigma_1}(t_1,t_2)\cdot N_{\sigma_1}^{\sigma_1\sigma_2}(g(t)),\quad (t_1,t_2)\in U,\quad t\in I,
\end{aligned}
$$

and

$$\mathscr{B}_{n\sigma_1\sigma_1\sigma_2}(t,t_1,t_2)=\begin{pmatrix} b_{11n\sigma_1\sigma_1\sigma_2}(t,t_1,t_2) & b_{12n\sigma_1\sigma_1\sigma_2}(t,t_1,t_2)\\ b_{21n\sigma_1\sigma_1\sigma_2}(t,t_1,t_2) & b_{22n\sigma_1\sigma_1\sigma_2}(t,t_1,t_2)\end{pmatrix},\quad (t_1,t_2)\in U,\quad t\in I.$$

For $p,q \in T_{a\sigma_1}S$, define

$$\phi^1_{\sigma_1\sigma_1\sigma_2}(p,q) = p^T \mathscr{B}_{n\sigma_1\sigma_1\sigma_2} q.$$

Theorem 4.6 (Variant of the Meusnier Theorem) *Let σ_1 be Δ_1-differentiable, S be a σ_1-regular surface with a local σ_1-parameterization (U,f) such that f is σ_1-completely delta differentiable, $f^{\Delta_1}_{t_1}$ be σ_1-completely delta differentiable and $f^{\Delta_2\sigma_1}_{t_2}$ be σ_2-completely delta differentiable on U.*

Then

$$k_{n\sigma_1}(t) = \phi^1_{\sigma_1\sigma_1\sigma_2}\left(\frac{g^{\Delta}(t)}{\|g^{\Delta}(t)\|}, \frac{g^{\Delta}(t)}{\|g^{\Delta}(t)\|}\right), \quad t \in I.$$

Proof. By (4.1), we find

$$\begin{aligned}
g_1^{\tilde{\Delta}}(s) &= f^{\Delta_1}_{t_1}(h_1(s),h_2(s))h_1^{\tilde{\Delta}}(s) + f^{\Delta_2\sigma_1}_{t_2}(h_1(s),h_2(s))h_2^{\tilde{\Delta}}(s),\\
g_1^{\tilde{\Delta}^2}(s) &= f^{\Delta_1}_{t_1}(h_1(\tilde{\sigma}(s)),h_2(\tilde{\sigma}(s)))h_1^{\tilde{\Delta}^2}(s)\\
&\quad + f^{\Delta_2\sigma_1}_{t_2}(h_1(\tilde{\sigma}(s)),h_2(\tilde{\sigma}(s)))h_2^{\tilde{\Delta}^2}(s)\\
&\quad + f^{\Delta_1^2}_{t_1}(h_1(s),h_2(s))\left(h_1^{\tilde{\Delta}}(s)\right)^2\\
&\quad + f^{\Delta_1\Delta_2\sigma_1}_{t_1t_2}(h_1(s),h_2(s))h_1^{\tilde{\Delta}}(s)h_2^{\tilde{\Delta}}(s)\\
&\quad + f^{\Delta_2\sigma_1\Delta_1\sigma_2}_{t_2t_1}(h_1(s),h_2(s))h_1^{\tilde{\Delta}}(s)h_2^{\tilde{\Delta}}(s)\\
&\quad + f^{\Delta_2^2\sigma_1}_{t_2}(h_1(s),h_2(s))\left(h_2^{\tilde{\Delta}}(s)\right)^2\\
&= f^{\Delta_1\sigma_1\sigma_2}_{t_1}(h_1(s),h_2(s))h_1^{\tilde{\Delta}^2}(s)\\
&\quad + f^{\Delta_2\sigma_1^2\sigma_2}_{t_2}(h_1(s),h_2(s))h_2^{\tilde{\Delta}^2}(s)\\
&\quad + f^{\Delta_1^2}_{t_1}(h_1(s),h_2(s))\left(h_1^{\tilde{\Delta}}(s)\right)^2\\
&\quad + f^{\Delta_1\Delta_2\sigma_1}_{t_1t_2}(h_1(s),h_2(s))h_1^{\tilde{\Delta}}(s)h_2^{\tilde{\Delta}}(s)\\
&\quad + \left(1+\mu_1^{\Delta_1}(t_1)\right)f^{\Delta_1\Delta_2\sigma_1\sigma_2}_{t_1t_2}(h_1(s),h_2(s))h_1^{\tilde{\Delta}}(s)h_2^{\tilde{\Delta}}(s)\\
&\quad + f^{\Delta_2^2\sigma_1}_{t_2}(h_1(s),h_2(s))\left(h_2^{\tilde{\Delta}}(s)\right)^2, \quad s \in J,
\end{aligned}$$

that is,

$$\begin{aligned}
g_1^{\tilde{\Delta}^2}(s) &= f^{\Delta_1\sigma_1\sigma_2}_{t_1}(h_1(s),h_2(s))h_1^{\tilde{\Delta}^2}(s)\\
&\quad + f^{\Delta_2\sigma_1^2\sigma_2}_{t_2}(h_1(s),h_2(s))h_2^{\tilde{\Delta}^2}(s)\\
&\quad + f^{\Delta_1^2}_{t_1}(h_1(s),h_2(s))\left(h_1^{\tilde{\Delta}}(s)\right)^2
\end{aligned} \tag{4.5}$$

$$+f_{t_1t_2}^{\Delta_1\Delta_2\sigma_1}(h_1(s),h_2(s))h_1^{\tilde{\Delta}}(s)h_2^{\tilde{\Delta}}(s)$$
$$+\left(1+\mu_1^{\Delta_1}(t_1)\right)f_{t_1t_2}^{\Delta_1\Delta_2\sigma_1\sigma_2}(h_1(s),h_2(s))h_1^{\tilde{\Delta}}(s)h_2^{\tilde{\Delta}}(s)$$
$$+f_{t_2}^{\Delta_2^2\sigma_1}(h_1(s),h_2(s))\left(h_2^{\tilde{\Delta}}(s)\right)^2, \quad s\in J.$$

Thus,

$$\begin{aligned}
k_{n\sigma_1}(s) &= k(s)\cdot N_{\sigma_1}^{\sigma_1\sigma_2}(g_1(s))\\
&= g^{\tilde{\Delta}^2}(s)\cdot N_{\sigma_1}^{\sigma_1\sigma_2}(g_1(s))\\
&= \left(f_{t_1}^{\Delta_1\sigma_1\sigma_2}(h_1(s),h_2(s))\cdot N_{\sigma_1}^{\sigma_1\sigma_2}(g_1(s))\right)h_1^{\tilde{\Delta}^2}(s)\\
&\quad+\left(f_{t_2}^{\Delta_2\sigma_1^2\sigma_2}(h_1(s),h_2(s))\cdot N_{\sigma_1}^{\sigma_1\sigma_2}(g_1(s))\right)h_2^{\tilde{\Delta}^2}(s)\\
&\quad+\left(f_{t_1}^{\Delta_1^2}(h_1(s),h_2(s))\cdot N_{\sigma_1}^{\sigma_1\sigma_2}(g_1(s))\right)\left(h_1^{\tilde{\Delta}}(s)\right)^2\\
&\quad+\left(f_{t_1t_2}^{\Delta_1\Delta_2\sigma_1}(h_1(s),h_2(s))\cdot N_{\sigma_1}^{\sigma_1\sigma_2}(g_1(s))\right)h_1^{\tilde{\Delta}}(s)h_2^{\tilde{\Delta}}(s)\\
&\quad+\left(1+\mu_1^{\Delta_1}(t_1)\right)\left(f_{t_1t_2}^{\Delta_1\Delta_2\sigma_1\sigma_2}(h_1(s),h_2(s))\cdot N_{\sigma_1}^{\sigma_1\sigma_2}(g_1(s))\right)h_1^{\tilde{\Delta}}(s)h_2^{\tilde{\Delta}}(s)\\
&\quad+\left(f_{t_2}^{\Delta_2^2\sigma_1}(h_1(s),h_2(s))\cdot N_{\sigma_1}^{\sigma_1\sigma_2}(g_1(s))\right)\left(h_2^{\tilde{\Delta}}(s)\right)^2\\
&= \left(f_{t_1}^{\Delta_1^2}(h_1(s),h_2(s))\cdot N_{\sigma_1}^{\sigma_1\sigma_2}(g_1(s))\right)\left(h_1^{\tilde{\Delta}}(s)\right)^2\\
&\quad+\left(f_{t_1t_2}^{\Delta_1\Delta_2\sigma_1}(h_1(s),h_2(s))\cdot N_{\sigma_1}^{\sigma_1\sigma_2}(g_1(s))\right)h_1^{\tilde{\Delta}}(s)h_2^{\tilde{\Delta}}(s)\\
&\quad+\left(1+\mu_1^{\Delta_1}(t_1)\right)\left(f_{t_1t_2}^{\Delta_1\Delta_2\sigma_1\sigma_2}(h_1(s),h_2(s))\cdot N_{\sigma_1}^{\sigma_1\sigma_2}(g_1(s))\right)h_1^{\tilde{\Delta}}(s)h_2^{\tilde{\Delta}}(s)\\
&\quad+\left(f_{t_2}^{\Delta_2^2\sigma_1}(h_1(s),h_2(s))\cdot N_{\sigma_1}^{\sigma_1\sigma_2}(g_1(s))\right)\left(h_2^{\tilde{\Delta}}(s)\right)^2\\
&= \phi_{\sigma_1\sigma_1\sigma_2}^{1}\left(g_1^{\tilde{\Delta}}(s),g_1^{\tilde{\Delta}}(s)\right), \quad s\in J,
\end{aligned}$$

that is,

$$k_{n\sigma_1}(s)=\phi_{\sigma_1\sigma_1\sigma_2}^{1}\left(g_1^{\tilde{\Delta}}(s),g_1^{\tilde{\Delta}}(s)\right), \quad s\in J.$$

Now, applying (4.4), we obtain

$$k_{n\sigma_1}(t)=\phi_{\sigma_1\sigma_1\sigma_2}^{1}\left(\frac{g^{\Delta}(t)}{\|g^{\Delta}(t)\|},\frac{g^{\Delta}(t)}{\|g^{\Delta}(t)\|}\right), \quad t\in I.$$

This completes the proof.

Definition 4.8 Let σ_1 be Δ_1-differentiable on $\mathbb{T}_1$, f be σ_1-completely delta differentiable on U, $f_{t_1}^{\Delta_1}$ be σ_2-completely delta differentiable and $f_{t_2}^{\Delta_2\sigma_1}$ be σ_1-completely delta differentiable on U. Also let,

$$b_{11n\sigma_1\sigma_2\sigma_1}(t,t_1,t_2)=f_{t_1}^{\Delta_1^2\sigma_2}(t_1,t_2)\cdot N_{\sigma_1}^{\sigma_1\sigma_2}(g(t)),$$

$$b_{12n\sigma_1\sigma_2\sigma_1}(t,t_1,t_2) = f_{t_1t_2}^{\Delta_1\Delta_2}(t_1,t_2)\cdot N_{\sigma_1}^{\sigma_1\sigma_2}(g(t)),$$

$$b_{21n\sigma_1\sigma_2\sigma_1}(t,t_1,t_2) = \left(1+\mu_1^{\Delta_1}(t_1)\right)f_{t_1t_2}^{\Delta_1\Delta_2\sigma_1}\cdot N_{\sigma_1}^{\sigma_1\sigma_2}(g(t)),$$

$$b_{22n\sigma_1\sigma_1\sigma_1}(t,t_1,t_2) = f_{t_2}^{\Delta_2^2\sigma_1^2}(t_1,t_2)\cdot N_{\sigma_1}^{\sigma_1\sigma_2}(g(t)), \quad (t_1,t_2)\in U, \quad t\in I,$$

and

$$\mathscr{B}_{n\sigma_1\sigma_2\sigma_1}(t,t_1,t_2) = \begin{pmatrix} b_{11n\sigma_1\sigma_2\sigma_1}(t,t_1,t_2) & b_{12n\sigma_1\sigma_2\sigma_1}(t,t_1,t_2) \\ b_{21n\sigma_1\sigma_2\sigma_1}(t,t_1,t_2) & b_{22n\sigma_1\sigma_2\sigma_1}(t,t_1,t_2) \end{pmatrix}, \quad (t_1,t_2)\in U, \quad t\in I.$$

For $p,q\in T_{a\sigma_1}S$, define

$$\phi^1_{\sigma_1\sigma_2\sigma_1}(p,q) = p^T\mathscr{B}_{n\sigma_1\sigma_2\sigma_1}q.$$

Theorem 4.7 (Variant of the Meusnier Theorem) *Let σ_1 be Δ_1-differentiable, S be a σ_1-regular surface with a local σ_1-parameterization (U,f) such that f is σ_1-completely delta differentiable, $f_{t_1}^{\Delta_1}$ be σ_2-completely delta differentiable and $f_{t_2}^{\Delta_2\sigma_1}$ be σ_1-completely delta differentiable on U.*

Then

$$k_{n\sigma_1}(t) = \phi^1_{\sigma_1\sigma_2\sigma_1}\left(\frac{g^\Delta(t)}{\|g^\Delta(t)\|}, \frac{g^\Delta(t)}{\|g^\Delta(t)\|}\right), \quad t\in I.$$

Proof. By (4.1), we get

$$\begin{aligned}
g_1^{\tilde{\Delta}}(s) &= f_{t_1}^{\Delta_1}(h_1(s),h_2(s))h_1^{\tilde{\Delta}}(s) + f_{t_2}^{\Delta_2\sigma_1}(h_1(s),h_2(s))h_2^{\tilde{\Delta}}(s),\\
g_1^{\tilde{\Delta}^2}(s) &= f_{t_1}^{\Delta_1}(h_1(\tilde{\sigma}(s)),h_2(\tilde{\sigma}(s)))h_1^{\tilde{\Delta}^2}(s)\\
&\quad + f_{t_2}^{\Delta_2\sigma_1}(h_1(\tilde{\sigma}(s)),h_2(\tilde{\sigma}(s)))h_2^{\tilde{\Delta}^2}(s)\\
&\quad + f_{t_1}^{\Delta_1^2\sigma_2}(h_1(s),h_2(s))\left(h_1^{\tilde{\Delta}}(s)\right)^2\\
&\quad + f_{t_1t_2}^{\Delta_1\Delta_2}(h_1(s),h_2(s))h_1^{\tilde{\Delta}}(s)h_2^{\tilde{\Delta}}(s)\\
&\quad + f_{t_2t_1}^{\Delta_2\sigma_1\Delta_1}(h_1(s),h_2(s))h_1^{\tilde{\Delta}}(s)h_2^{\tilde{\Delta}}(s)\\
&\quad + f_{t_2}^{\Delta_2^2\sigma_1^2}(h_1(s),h_2(s))\left(h_2^{\tilde{\Delta}}(s)\right)^2\\
&= f_{t_1}^{\Delta_1\sigma_1\sigma_2}(h_1(s),h_2(s))h_1^{\tilde{\Delta}^2}(s)\\
&\quad + f_{t_2}^{\Delta_2\sigma_1^2\sigma_2}(h_1(s),h_2(s))h_2^{\tilde{\Delta}^2}(s)\\
&\quad + f_{t_1}^{\Delta_1^2\sigma_2}(h_1(s),h_2(s))\left(h_1^{\tilde{\Delta}}(s)\right)^2\\
&\quad + f_{t_1t_2}^{\Delta_1\Delta_2}(h_1(s),h_2(s))h_1^{\tilde{\Delta}}(s)h_2^{\tilde{\Delta}}(s)\\
&\quad + \left(1+\mu_1^{\Delta_1}(t_1)\right)f_{t_1t_2}^{\Delta_1\Delta_2\sigma_1}(h_1(s),h_2(s))h_1^{\tilde{\Delta}}(s)h_2^{\tilde{\Delta}}(s)\\
&\quad + f_{t_2}^{\Delta_2^2\sigma_1^2}(h_1(s),h_2(s))\left(h_2^{\tilde{\Delta}}(s)\right)^2, \quad s\in J,
\end{aligned}$$

that is,

$$\begin{aligned}g_1^{\tilde{\Delta}^2}(s) = &f_{t_1}^{\Delta_1\sigma_1\sigma_2}(h_1(s),h_2(s))h_1^{\tilde{\Delta}^2}(s)\\ &+f_{t_2}^{\Delta_2\sigma_1^2\sigma_2}(h_1(s),h_2(s))h_2^{\tilde{\Delta}^2}(s)\\ &+f_{t_1}^{\Delta_1^2\sigma_2}(h_1(s),h_2(s))\left(h_1^{\tilde{\Delta}}(s)\right)^2\\ &+f_{t_1t_2}^{\Delta_1\Delta_2}(h_1(s),h_2(s))h_1^{\tilde{\Delta}}(s)h_2^{\tilde{\Delta}}(s)\\ &+\left(1+\mu_1^{\Delta_1}(t_1)\right)f_{t_1t_2}^{\Delta_1\Delta_2\sigma_1}(h_1(s),h_2(s))h_1^{\tilde{\Delta}}(s)h_2^{\tilde{\Delta}}(s)\\ &+f_{t_2}^{\Delta_2^2\sigma_1^2}(h_1(s),h_2(s))\left(h_2^{\tilde{\Delta}}(s)\right)^2, \quad s\in J.\end{aligned} \tag{4.6}$$

Thus,

$$\begin{aligned}k_{n\sigma_1}(s) &= k(s)\cdot N_{\sigma_1}^{\sigma_1\sigma_2}(g_1(s))\\ &= g^{\tilde{\Delta}^2}(s)\cdot N_{\sigma_1}^{\sigma_1\sigma_2}(g_1(s))\\ &= \left(f_{t_1}^{\Delta_1\sigma_1\sigma_2}(h_1(s),h_2(s))\cdot N_{\sigma_1}^{\sigma_1\sigma_2}(g_1(s))\right)h_1^{\tilde{\Delta}^2}(s)\\ &\quad+\left(f_{t_2}^{\Delta_2\sigma_1^2\sigma_2}(h_1(s),h_2(s))\cdot N_{\sigma_1}^{\sigma_1\sigma_2}(g_1(s))\right)h_2^{\tilde{\Delta}^2}(s)\\ &\quad+\left(f_{t_1}^{\Delta_1^2\sigma_2}(h_1(s),h_2(s))\cdot N_{\sigma_1}^{\sigma_1\sigma_2}(g_1(s))\right)\left(h_1^{\tilde{\Delta}}(s)\right)^2\\ &\quad+\left(f_{t_1t_2}^{\Delta_1\Delta_2}(h_1(s),h_2(s))\cdot N_{\sigma_1}^{\sigma_1\sigma_2}(g_1(s))\right)h_1^{\tilde{\Delta}}(s)h_2^{\tilde{\Delta}}(s)\\ &\quad+\left(1+\mu_1^{\Delta_1}(t_1)\right)\left(f_{t_1t_2}^{\Delta_1\Delta_2\sigma_1}(h_1(s),h_2(s))\cdot N_{\sigma_1}^{\sigma_1\sigma_2}(g_1(s))\right)h_1^{\tilde{\Delta}}(s)h_2^{\tilde{\Delta}}(s)\\ &\quad+\left(f_{t_2}^{\Delta_2^2\sigma_1^2}(h_1(s),h_2(s))\cdot N_{\sigma_1}^{\sigma_1\sigma_2}(g_1(s))\right)\left(h_2^{\tilde{\Delta}}(s)\right)^2\\ &= \left(f_{t_1}^{\Delta_1^2\sigma_2}(h_1(s),h_2(s))\cdot N_{\sigma_1}^{\sigma_1\sigma_2}(g_1(s))\right)\left(h_1^{\tilde{\Delta}}(s)\right)^2\\ &\quad+\left(f_{t_1t_2}^{\Delta_1\Delta_2}(h_1(s),h_2(s))\cdot N_{\sigma_1}^{\sigma_1\sigma_2}(g_1(s))\right)h_1^{\tilde{\Delta}}(s)h_2^{\tilde{\Delta}}(s)\\ &\quad+\left(1+\mu_1^{\Delta_1}(t_1)\right)\left(f_{t_1t_2}^{\Delta_1\Delta_2\sigma_1}(h_1(s),h_2(s))\cdot N_{\sigma_1}^{\sigma_1\sigma_2}(g_1(s))\right)h_1^{\tilde{\Delta}}(s)h_2^{\tilde{\Delta}}(s)\\ &\quad+\left(f_{t_2}^{\Delta_2^2\sigma_1^2}(h_1(s),h_2(s))\cdot N_{\sigma_1}^{\sigma_1\sigma_2}(g_1(s))\right)\left(h_2^{\tilde{\Delta}}(s)\right)^2\\ &= \phi_{\sigma_1\sigma_2\sigma_1}^1\left(g_1^{\tilde{\Delta}}(s),g_1^{\tilde{\Delta}}(s)\right), \quad s\in J,\end{aligned}$$

that is,

$$k_{n\sigma_1}(s) = \phi_{\sigma_1\sigma_2\sigma_1}^1\left(g_1^{\tilde{\Delta}}(s),g_1^{\tilde{\Delta}}(s)\right), \quad s\in J.$$

Using (4.4), we find

$$k_{n\sigma_1}(t) = \phi_{\sigma_1\sigma_2\sigma_1}^1\left(\frac{g^\Delta(t)}{\|g^\Delta(t)\|}, \frac{g^\Delta(t)}{\|g^\Delta(t)\|}\right), \quad t\in I.$$

This completes the proof.

Definition 4.9 Let σ_1 be Δ_1-differentiable on $\mathbb{T}_1, f$ be σ_1-completely delta differentiable on U, $f_{t_1}^{\Delta_1}$ and $f_{t_2}^{\Delta_2\sigma_1}$ be σ_2-completely delta differentiable on U. Also, let,

$$\begin{aligned}
b_{11n\sigma_1\sigma_2\sigma_2}(t,t_1,t_2) &= f_{t_1}^{\Delta_1^2\sigma_2}(t_1,t_2)\cdot N_{\sigma_1}^{\sigma_1\sigma_2}(g(t)),\\
b_{12n\sigma_1\sigma_2\sigma_2}(t,t_1,t_2) &= f_{t_1t_2}^{\Delta_1\Delta_2}(t_1,t_2)\cdot N_{\sigma_1}^{\sigma_1\sigma_2}(g(t)),\\
b_{21n\sigma_1\sigma_2\sigma_2}(t,t_1,t_2) &= \left(1+\mu_1^{\Delta_1}(t_1)\right)f_{t_1t_2}^{\Delta_1\Delta_2\sigma_1\sigma_2}\cdot N_{\sigma_1}^{\sigma_1\sigma_2},\\
b_{22n\sigma_1\sigma_2\sigma_2}(t,t_1,t_2) &= f_{t_2}^{\Delta_2^2\sigma_1}(t_1,t_2)\cdot N_{\sigma_1}^{\sigma_1\sigma_2}(g(t)),\quad (t_1,t_2)\in U,\quad t\in I,
\end{aligned}$$

and

$$\mathscr{B}_{n\sigma_1\sigma_2\sigma_2}(t,t_1,t_2)=\begin{pmatrix} b_{11n\sigma_1\sigma_2\sigma_2}(t,t_1,t_2) & b_{12n\sigma_1\sigma_2\sigma_2}(t,t_1,t_2)\\ b_{21n\sigma_1\sigma_2\sigma_2}(t,t_1,t_2) & b_{22n\sigma_1\sigma_2\sigma_2}(t,t_1,t_2)\end{pmatrix},\quad (t_1,t_2)\in U,\quad t\in I.$$

For $p,q\in T_{a\sigma_1}S$, define

$$\phi^1_{\sigma_1\sigma_2\sigma_2}(p,q)=p^T\mathscr{B}_{n\sigma_1\sigma_2\sigma_2}q.$$

Theorem 4.8 (Variant of the Meusnier Theorem) *Let σ_1 be Δ_1-differentiable, S be a σ_1-regular surface with a local σ_1-parameterization (U,f) such that f is σ_1-completely delta differentiable, $f_{t_1}^{\Delta_1}$ and $f_{t_2}^{\Delta_2\sigma_1}$ are σ_2-completely delta differentiable on U.*

Then

$$k_{n\sigma_1}(t)=\phi^1_{\sigma_1\sigma_2\sigma_2}\left(\frac{g^\Delta(t)}{\|g^\Delta(t)\|},\frac{g^\Delta(t)}{\|g^\Delta(t)\|}\right),\quad t\in I.$$

Proof. By (4.1), we get

$$\begin{aligned}
g_1^{\tilde\Delta}(s) &= f_{t_1}^{\Delta_1}(h_1(s),h_2(s))h_1^{\tilde\Delta}(s)+f_{t_2}^{\Delta_2\sigma_1}(h_1(s),h_2(s))h_2^{\tilde\Delta}(s),\\
g_1^{\tilde\Delta^2}(s) &= f_{t_1}^{\Delta_1}(h_1(\tilde\sigma(s)),h_2(\tilde\sigma(s)))h_1^{\tilde\Delta^2}(s)\\
&\quad +f_{t_2}^{\Delta_2\sigma_1}(h_1(\tilde\sigma(s)),h_2(\tilde\sigma(s)))h_2^{\tilde\Delta^2}(s)\\
&\quad +f_{t_1}^{\Delta_1^2\sigma_2^2}(h_1(s),h_2(s))\left(h_1^{\tilde\Delta}(s)\right)^2\\
&\quad +f_{t_1t_2}^{\Delta_1\Delta_2}(h_1(s),h_2(s))h_1^{\tilde\Delta}(s)h_2^{\tilde\Delta}(s)\\
&\quad +f_{t_2t_1}^{\Delta_1\sigma_1\Delta_1\sigma_2}(h_1(s),h_2(s))h_1^{\tilde\Delta}(s)h_2^{\tilde\Delta}(s)\\
&\quad +f_{t_2}^{\Delta_2^2\sigma_1}(h_1(s),h_2(s))\left(h_2^{\tilde\Delta}(s)\right)^2\\
&= f_{t_1}^{\Delta_1\sigma_1\sigma_2}(h_1(s),h_2(s))h_1^{\tilde\Delta^2}(s)\\
&\quad +f_{t_2}^{\Delta_2\sigma_1^2\sigma_2}(h_1(s),h_2(s))h_2^{\tilde\Delta^2}(s)
\end{aligned}$$

$$
\begin{aligned}
&+f_{t_1}^{\Delta_1^2\sigma_2^2}(h_1(s),h_2(s))\left(h_1^{\tilde{\Delta}}(s)\right)^2\\
&+f_{t_1t_2}^{\Delta_1\Delta_2}(h_1(s),h_2(s))h_1^{\tilde{\Delta}}(s)h_2^{\tilde{\Delta}}(s)\\
&+\left(1+\mu_1^{\Delta_1}(t_1)\right)f_{t_1t_2}^{\Delta_1\Delta_2\sigma_1\sigma_2}(h_1(s),h_2(s))h_1^{\tilde{\Delta}}(s)h_2^{\tilde{\Delta}}(s)\\
&+f_{t_2}^{\Delta_2^2\sigma_1}(h_1(s),h_2(s))\left(h_2^{\tilde{\Delta}}(s)\right)^2, \quad s\in J,
\end{aligned}
$$

that is,

$$
\begin{aligned}
g_1^{\tilde{\Delta}^2}(s)=&f_{t_1}^{\Delta_1\sigma_1\sigma_2}(h_1(s),h_2(s))h_1^{\tilde{\Delta}^2}(s)\\
&+f_{t_2}^{\Delta_2\sigma_1^2\sigma_2}(h_1(s),h_2(s))h_2^{\tilde{\Delta}^2}(s)\\
&+f_{t_1}^{\Delta_1^2\sigma_2^2}(h_1(s),h_2(s))\left(h_1^{\tilde{\Delta}}(s)\right)^2\\
&+f_{t_1t_2}^{\Delta_1\Delta_2}(h_1(s),h_2(s))h_1^{\tilde{\Delta}}(s)h_2^{\tilde{\Delta}}(s)\\
&+\left(1+\mu_1^{\Delta_1}(t_1)\right)f_{t_1t_2}^{\Delta_1\Delta_2\sigma_1\sigma_2}(h_1(s),h_2(s))h_1^{\tilde{\Delta}}(s)h_2^{\tilde{\Delta}}(s)\\
&+f_{t_2}^{\Delta_2^2\sigma_1}(h_1(s),h_2(s))\left(h_2^{\tilde{\Delta}}(s)\right)^2, \quad s\in J.
\end{aligned}
\tag{4.7}
$$

Thus,

$$
\begin{aligned}
k_{n\sigma_1}(s)&=k(s)\cdot N_{\sigma_1}^{\sigma_1\sigma_2}(g_1(s))\\
&=g^{\tilde{\Delta}^2}(s)\cdot N_{\sigma_1}^{\sigma_1\sigma_2}(g_1(s))\\
&=\left(f_{t_1}^{\Delta_1\sigma_1\sigma_2}(h_1(s),h_2(s))\cdot N_{\sigma_1}^{\sigma_1\sigma_2}(g_1(s))\right)h_1^{\tilde{\Delta}^2}(s)\\
&\quad+\left(f_{t_2}^{\Delta_2\sigma_1^2\sigma_2}(h_1(s),h_2(s))\cdot N_{\sigma_1}^{\sigma_1\sigma_2}(g_1(s))\right)h_2^{\tilde{\Delta}^2}(s)\\
&\quad+\left(f_{t_1}^{\Delta_1^2\sigma_2}(h_1(s),h_2(s))\cdot N_{\sigma_1}^{\sigma_1\sigma_2}(g_1(s))\right)\left(h_1^{\tilde{\Delta}}(s)\right)^2\\
&\quad+\left(f_{t_1t_2}^{\Delta_1\Delta_2}(h_1(s),h_2(s))\cdot N_{\sigma_1}^{\sigma_1\sigma_2}(g_1(s))\right)h_1^{\tilde{\Delta}}(s)h_2^{\tilde{\Delta}}(s)\\
&\quad+\left(1+\mu_1^{\Delta_1}(t_1)\right)\left(f_{t_1t_2}^{\Delta_1\Delta_2\sigma_1\sigma_2}(h_1(s),h_2(s))\cdot N_{\sigma_1}^{\sigma_1\sigma_2}(g_1(s))\right)h_1^{\tilde{\Delta}}(s)h_2^{\tilde{\Delta}}(s)\\
&\quad+\left(f_{t_2}^{\Delta_2^2\sigma_1}(h_1(s),h_2(s))\cdot N_{\sigma_1}^{\sigma_1\sigma_2}(g_1(s))\right)\left(h_2^{\tilde{\Delta}}(s)\right)^2\\
&=\left(f_{t_1}^{\Delta_1^2\sigma_2}(h_1(s),h_2(s))\cdot N_{\sigma_1}^{\sigma_1\sigma_2}(g_1(s))\right)\left(h_1^{\tilde{\Delta}}(s)\right)^2\\
&\quad+\left(f_{t_1t_2}^{\Delta_1\Delta_2}(h_1(s),h_2(s))\cdot N_{\sigma_1}^{\sigma_1\sigma_2}(g_1(s))\right)h_1^{\tilde{\Delta}}(s)h_2^{\tilde{\Delta}}(s)\\
&\quad+\left(1+\mu_1^{\Delta_1}(t_1)\right)\left(f_{t_1t_2}^{\Delta_1\Delta_2\sigma_1\sigma_2}(h_1(s),h_2(s))\cdot N_{\sigma_1}^{\sigma_1\sigma_2}(g_1(s))\right)h_1^{\tilde{\Delta}}(s)h_2^{\tilde{\Delta}}(s)\\
&\quad+\left(f_{t_2}^{\Delta_2^2\sigma_1}(h_1(s),h_2(s))\cdot N_{\sigma_1}^{\sigma_1\sigma_2}(g_1(s))\right)\left(h_2^{\tilde{\Delta}}(s)\right)^2\\
&=\phi^1_{\sigma_1\sigma_2\sigma_2}\left(g_1^{\tilde{\Delta}}(s),g_1^{\tilde{\Delta}}(s)\right),\ s\in J,
\end{aligned}
$$

that is,

$$k_{n\sigma_1}(s) = \phi^1_{\sigma_1\sigma_2\sigma_2}\left(g_1^{\tilde{\Delta}}(s), g_1^{\tilde{\Delta}}(s)\right), \ s \in J.$$

Now, applying (4.4), we obtain

$$k_{n\sigma_1}(t) = \phi^1_{\sigma_1\sigma_2\sigma_2}\left(\frac{g^\Delta(t)}{\|g^\Delta(t)\|}, \frac{g^\Delta(t)}{\|g^\Delta(t)\|}\right), \quad t \in I.$$

This completes the proof.

Definition 4.10 Let S be a σ_2-regular surface with a local σ_2-parametrization (U,f) and $g: I \to S$ be a curve lying in S. The projection of the curvature vector $k(t)$ of $g(t)$ on $N_{\sigma_2}^{\sigma_1\sigma_2}(g(t))$ is called the σ_2-normal curvature of the curve g at t and it is denoted by $k_{n\sigma_2}(t)$.

Definition 4.11 Let σ_2 be Δ_2-differentiable on $\mathbb{T}_2$, f be σ_1-completely delta differentiable on U, $f_{t_1}^{\Delta_1\sigma_2}$ and $f_{t_2}^{\Delta_2}$ be σ_1-completely delta differentiable on U. Let also,

$$\begin{aligned}
b_{11n\sigma_2\sigma_1\sigma_1}(t,t_1,t_2) &= f_{t_1}^{\Delta_1^2}(t_1,t_2)\cdot N_{\sigma_2}^{\sigma_1\sigma_2}(g(t)),\\
b_{12n\sigma_2\sigma_1\sigma_1}(t,t_1,t_2) &= f_{t_1t_2}^{\Delta_1\Delta_2}(t_1,t_2)\cdot N_{\sigma_2}^{\sigma_1\sigma_2}(g(t)),\\
b_{21n\sigma_2\sigma_1\sigma_1}(t,t_1,t_2) &= \left(1+\mu_2^{\Delta_2}(t_2)\right)f_{t_1t_2}^{\Delta_1\Delta_2\sigma_1\sigma_2}(t_1,t_2)\cdot N_{\sigma_2}^{\sigma_1\sigma_2}(g(t)),\\
b_{22n\sigma_2\sigma_1\sigma_1}(t,t_1,t_2) &= f_{t_2}^{\Delta_2^2\sigma_1}(t_1,t_2)\cdot N_{\sigma_2}^{\sigma_1\sigma_2}(g(t)), \quad (t_1,t_2)\in U, \quad t\in I,
\end{aligned}$$

and

$$\mathscr{B}_{n\sigma_2\sigma_1\sigma_1}(t,t_1,t_2) = \begin{pmatrix} b_{11n\sigma_2\sigma_1\sigma_1}(t,t_1,t_2) & b_{12n\sigma_2\sigma_1\sigma_1}(t,t_1,t_2) \\ b_{21n\sigma_2\sigma_1\sigma_1}(t,t_1,t_2) & b_{22n\sigma_2\sigma_1\sigma_1}(t,t_1,t_2)\end{pmatrix}, \quad (t_1,t_2)\in U, \quad t\in I.$$

For $p,q \in T_{a\sigma_2}S$, define

$$\phi^1_{\sigma_2\sigma_1\sigma_1}(p,q) = p^T \mathscr{B}_{n\sigma_2\sigma_1\sigma_1} q.$$

Theorem 4.9 (Variant of the Meusnier Theorem) *Let σ_2 be Δ_2-differentiable, S be a σ_2-regular surface with a local σ_2-parameterization (U,f) such that f is σ_2-completely delta differentiable, $f_{t_1}^{\Delta_1\sigma_2}$ and $f_{t_2}^{\Delta_2}$ be σ_1-completely delta differentiable on U. Then*

$$k_{n\sigma_2}(t) = \phi^1_{\sigma_2\sigma_1\sigma_1}\left(\frac{g^\Delta(t)}{\|g^\Delta(t)\|}, \frac{g^\Delta(t)}{\|g^\Delta(t)\|}\right), \quad t \in I.$$

Proof. By (4.1), we get

$$\begin{aligned}
g_1^{\tilde{\Delta}}(s) &= f_{t_1}^{\Delta_1\sigma_2}(h_1(s),h_2(s))h_1^{\tilde{\Delta}}(s) + f_{t_2}^{\Delta_2}(h_1(s),h_2(s))h_2^{\tilde{\Delta}}(s),\\
g_1^{\tilde{\Delta}^2}(s) &= f_{t_1}^{\Delta_1\sigma_2}(h_1(\tilde{\sigma}(s)),h_2(\tilde{\sigma}(s)))h_1^{\tilde{\Delta}^2}(s)
\end{aligned}$$

$$
\begin{aligned}
&+f_{t_2}^{\Delta_2}(h_1(\widetilde{\sigma}(s)),h_2(\widetilde{\sigma}(s)))h_2^{\widetilde{\Delta}^2}(s)\\
&+f_{t_1}^{\Delta_1^2\sigma_2}(h_1(s),h_2(s))\left(h_1^{\widetilde{\Delta}}(s)\right)^2\\
&+f_{t_1t_2}^{\Delta_1\sigma_2\Delta_2\sigma_1}(h_1(s),h_2(s))h_1^{\widetilde{\Delta}}(s)h_2^{\widetilde{\Delta}}(s)\\
&+f_{t_1t_2}^{\Delta_1\Delta_2}(h_1(s),h_2(s))h_1^{\widetilde{\Delta}}(s)h_2^{\widetilde{\Delta}}(s)\\
&+f_{t_2}^{\Delta_2^2\sigma_1}(h_1(s),h_2(s))\left(h_2^{\widetilde{\Delta}}(s)\right)^2\\
=&f_{t_1}^{\Delta_1\sigma_1\sigma_2^2}(h_1(s),h_2(s))h_1^{\widetilde{\Delta}^2}(s)\\
&+f_{t_2}^{\Delta_2\sigma_1\sigma_2}(h_1(s),h_2(s))h_2^{\widetilde{\Delta}^2}(s)\\
&+f_{t_1}^{\Delta_1^2\sigma_2}(h_1(s),h_2(s))\left(h_1^{\widetilde{\Delta}}(s)\right)^2\\
&+(1+\mu_2^{\Delta_2}(t_2))f_{t_1t_2}^{\Delta_1\Delta_2\sigma_1\sigma_2}(h_1(s),h_2(s))h_1^{\widetilde{\Delta}}(s)h_2^{\widetilde{\Delta}}(s)\\
&+f_{t_1t_2}^{\Delta_1\Delta_2}(h_1(s),h_2(s))h_1^{\widetilde{\Delta}}(s)h_2^{\widetilde{\Delta}}(s)\\
&+f_{t_2}^{\Delta_2^2\sigma_1}(h_1(s),h_2(s))\left(h_2^{\widetilde{\Delta}}(s)\right)^2, \quad s\in J,
\end{aligned}
$$

that is,

$$
\begin{aligned}
g_1^{\widetilde{\Delta}^2}(s)=&f_{t_1}^{\Delta_1\sigma_1\sigma_2^2}(h_1(s),h_2(s))h_1^{\widetilde{\Delta}^2}(s)\\
&+f_{t_2}^{\Delta_2\sigma_1\sigma_2}(h_1(s),h_2(s))h_2^{\widetilde{\Delta}^2}(s)\\
&+f_{t_1}^{\Delta_1^2\sigma_2}(h_1(s),h_2(s))\left(h_1^{\widetilde{\Delta}}(s)\right)^2\\
&+(1+\mu_2^{\Delta_2}(t_2))f_{t_1t_2}^{\Delta_1\Delta_2\sigma_1\sigma_2}(h_1(s),h_2(s))h_1^{\widetilde{\Delta}}(s)h_2^{\widetilde{\Delta}}(s)\\
&+f_{t_1t_2}^{\Delta_1\Delta_2}(h_1(s),h_2(s))h_1^{\widetilde{\Delta}}(s)h_2^{\widetilde{\Delta}}(s)\\
&+f_{t_2}^{\Delta_2^2\sigma_1}(h_1(s),h_2(s))\left(h_2^{\widetilde{\Delta}}(s)\right)^2, \quad s\in J.
\end{aligned}
\tag{4.8}
$$

Thus,

$$
\begin{aligned}
k_{n\sigma_2}(s)&=k(s)\cdot N_{\sigma_2}^{\sigma_1\sigma_2}(g_1(s))\\
&=g^{\widetilde{\Delta}^2}(s)\cdot N_{\sigma_2}^{\sigma_1\sigma_2}(g_1(s))\\
&=\left(f_{t_1}^{\Delta_1\sigma_1\sigma_2^2}(h_1(s),h_2(s))\cdot N_{\sigma_2}^{\sigma_1\sigma_2}(g_1(s))\right)h_1^{\widetilde{\Delta}^2}(s)\\
&\quad+\left(f_{t_2}^{\Delta_2\sigma_1\sigma_2}(h_1(s),h_2(s))\cdot N_{\sigma_2}^{\sigma_1\sigma_2}(g_1(s))\right)h_2^{\widetilde{\Delta}^2}(s)\\
&\quad+\left(f_{t_1}^{\Delta_1^2\sigma_2}(h_1(s),h_2(s))\cdot N_{\sigma_2}^{\sigma_1\sigma_2}(g_1(s))\right)\left(h_1^{\widetilde{\Delta}}(s)\right)^2\\
&\quad+(1+\mu_2^{\Delta_2}(t_2))\left(f_{t_1t_2}^{\Delta_1\Delta_2\sigma_1\sigma_2}(h_1(s),h_2(s))\cdot N_{\sigma_2}^{\sigma_1\sigma_2}(g_1(s))\right)h_1^{\widetilde{\Delta}}(s)h_2^{\widetilde{\Delta}}(s)\\
&\quad+\left(f_{t_1t_2}^{\Delta_1\Delta_2}(h_1(s),h_2(s))\cdot N_{\sigma_2}^{\sigma_1\sigma_2}(g_1(s))\right)h_1^{\widetilde{\Delta}}(s)h_2^{\widetilde{\Delta}}(s)
\end{aligned}
$$

$$
\begin{aligned}
&+\left(f_{t_2}^{\Delta_2^2\sigma_1}(h_1(s),h_2(s))\cdot N_{\sigma_2}^{\sigma_1\sigma_2}(g_1(s))\right)\left(h_2^{\tilde{\Delta}}(s)\right)^2\\
&=\left(f_{t_1}^{\Delta_1^2\sigma_2}(h_1(s),h_2(s))\cdot N_{\sigma_2}^{\sigma_1\sigma_2}(g_1(s))\right)\left(h_1^{\tilde{\Delta}}(s)\right)^2\\
&\quad+(1+\mu_2^{\Delta_2}(t_2))\left(f_{t_1t_2}^{\Delta_1\Delta_2\sigma_1\sigma_2}(h_1(s),h_2(s))\cdot N_{\sigma_2}^{\sigma_1\sigma_2}(g_1(s))\right)h_1^{\tilde{\Delta}}(s)h_2^{\tilde{\Delta}}(s)\\
&\quad+\left(f_{t_1t_2}^{\Delta_1\Delta_2}(h_1(s),h_2(s))\cdot N_{\sigma_2}^{\sigma_1\sigma_2}(g_1(s))\right)h_1^{\tilde{\Delta}}(s)h_2^{\tilde{\Delta}}(s)\\
&\quad+\left(f_{t_2}^{\Delta_2^2\sigma_1}(h_1(s),h_2(s))\cdot N_{\sigma_2}^{\sigma_1\sigma_2}(g_1(s))\right)\left(h_2^{\tilde{\Delta}}(s)\right)^2\\
&=\phi^1_{\sigma_2\sigma_1\sigma_1}\left(g_1^{\tilde{\Delta}}(s),g_1^{\tilde{\Delta}}(s)\right),\quad s\in J,
\end{aligned}
$$

that is,

$$k_{n\sigma_2}(s)=\phi^1_{\sigma_2\sigma_1\sigma_1}\left(g_1^{\tilde{\Delta}}(s),g_1^{\tilde{\Delta}}(s)\right),$$

$s\in J$. Now, we use (4.4) and we arrive at

$$k_{n\sigma_2}(t)=\phi^1_{\sigma_2\sigma_1\sigma_1}\left(\frac{g^{\Delta}(t)}{\|g^{\Delta}(t)\|},\frac{g^{\Delta}(t)}{\|g^{\Delta}(t)\|}\right),\quad t\in I.$$

This completes the proof.

Definition 4.12 Let σ_2 be Δ_2-differentiable on $\mathbb{T}_2$, f be σ_1-completely delta differentiable on U, $f_{t_1}^{\Delta_1\sigma_2}$ be σ_1-completely delta differentiable and $f_{t_2}^{\Delta_2}$ be σ_2-completely delta differentiable on U. Also let,

$$
\begin{aligned}
b_{11n\sigma_2\sigma_1\sigma_2}(t,t_1,t_2)&=f_{t_1}^{\Delta_1^2}(t_1,t_2)\cdot N_{\sigma_2}^{\sigma_1\sigma_2}(g(t)),\\
b_{12n\sigma_2\sigma_1\sigma_2}(t,t_1,t_2)&=f_{t_1t_2}^{\Delta_1\Delta_2\sigma_2}(t_1,t_2)\cdot N_{\sigma_2}^{\sigma_1\sigma_2}(g(t)),\\
b_{21n\sigma_2\sigma_1\sigma_2}(t,t_1,t_2)&=\left(1+\mu_2^{\Delta_2}(t_2)\right)f_{t_1t_2}^{\Delta_1\Delta_2\sigma_1\sigma_2}(t_1,t_2)\cdot N_{\sigma_2}^{\sigma_1\sigma_2}(g(t)),\\
b_{22n\sigma_2\sigma_1\sigma_2}(t,t_1,t_2)&=f_{t_2}^{\Delta_2^2}(t_1,t_2)\cdot N_{\sigma_2}^{\sigma_1\sigma_2}(g(t)),\quad (t_1,t_2)\in U,\quad t\in I,
\end{aligned}
$$

and

$$\mathscr{B}_{n\sigma_2\sigma_1\sigma_2}(t,t_1,t_2)=\begin{pmatrix} b_{11n\sigma_2\sigma_1\sigma_2}(t,t_1,t_2) & b_{12n\sigma_2\sigma_1\sigma_2}(t,t_1,t_2)\\ b_{21n\sigma_2\sigma_1\sigma_2}(t,t_1,t_2) & b_{22n\sigma_2\sigma_1\sigma_2}(t,t_1,t_2)\end{pmatrix},\quad (t_1,t_2)\in U,\quad t\in I.$$

For $p,q\in T_{a\sigma_2}S$, define

$$\phi^1_{\sigma_2\sigma_1\sigma_2}(p,q)=p^T\mathscr{B}_{n\sigma_2\sigma_1\sigma_2}q.$$

Theorem 4.10 (Variant of the Meusnier Theorem) *Let σ_2 be Δ_2-differentiable, S be a σ_2-regular surface with a local σ_2-parameterization (U,f) such that f is σ_2-completely delta differentiable, $f_{t_1}^{\Delta_1\sigma_2}$ be σ_1-completely delta differentiable and $f_{t_2}^{\Delta_2}$ be σ_2-completely delta differentiable on U.*

Then

$$k_{n\sigma_2}(t)=\phi^1_{\sigma_2\sigma_1\sigma_2}\left(\frac{g^{\Delta}(t)}{\|g^{\Delta}(t)\|},\frac{g^{\Delta}(t)}{\|g^{\Delta}(t)\|}\right),\quad t\in I.$$

Proof. By (4.1), we get

$$
\begin{aligned}
g_1^{\widetilde{\Delta}}(s) &= f_{t_1}^{\Delta_1\sigma_2}(h_1(s),h_2(s))h_1^{\widetilde{\Delta}}(s) + f_{t_2}^{\Delta_2}(h_1(s),h_2(s))h_2^{\widetilde{\Delta}}(s),\\
g_1^{\widetilde{\Delta}^2}(s) &= f_{t_1}^{\Delta_1\sigma_2}(h_1(\widetilde{\sigma}(s)),h_2(\widetilde{\sigma}(s)))h_1^{\widetilde{\Delta}^2}(s)\\
&\quad + f_{t_2}^{\Delta_2}(h_1(\widetilde{\sigma}(s)),h_2(\widetilde{\sigma}(s)))h_2^{\widetilde{\Delta}^2}(s)\\
&\quad + f_{t_1}^{\Delta_1^2\sigma_2}(h_1(s),h_2(s))\left(h_1^{\widetilde{\Delta}}(s)\right)^2\\
&\quad + f_{t_1t_2}^{\Delta_1\sigma_2\Delta_2\sigma_1}(h_1(s),h_2(s))h_1^{\widetilde{\Delta}}(s)h_2^{\widetilde{\Delta}}(s)\\
&\quad + f_{t_1t_2}^{\Delta_1\Delta_2\sigma_2}(h_1(s),h_2(s))h_1^{\widetilde{\Delta}}(s)h_2^{\widetilde{\Delta}}(s)\\
&\quad + f_{t_2}^{\Delta_2^2}(h_1(s),h_2(s))\left(h_2^{\widetilde{\Delta}}(s)\right)^2\\
&= f_{t_1}^{\Delta_1\sigma_1\sigma_2^2}(h_1(s),h_2(s))h_1^{\widetilde{\Delta}^2}(s)\\
&\quad + f_{t_2}^{\Delta_2\sigma_1\sigma_2}(h_1(s),h_2(s))h_2^{\widetilde{\Delta}^2}(s)\\
&\quad + f_{t_1}^{\Delta_1^2\sigma_2}(h_1(s),h_2(s))\left(h_1^{\widetilde{\Delta}}(s)\right)^2\\
&\quad + (1+\mu_2^{\Delta_2}(t_2))f_{t_1t_2}^{\Delta_1\Delta_2\sigma_1\sigma_2}(h_1(s),h_2(s))h_1^{\widetilde{\Delta}}(s)h_2^{\widetilde{\Delta}}(s)\\
&\quad + f_{t_1t_2}^{\Delta_1\Delta_2\sigma_2}(h_1(s),h_2(s))h_1^{\widetilde{\Delta}}(s)h_2^{\widetilde{\Delta}}(s)\\
&\quad + f_{t_2}^{\Delta_2^2}(h_1(s),h_2(s))\left(h_2^{\widetilde{\Delta}}(s)\right)^2, \quad s\in J,
\end{aligned}
$$

that is,

$$
\begin{aligned}
g_1^{\widetilde{\Delta}^2}(s) &= f_{t_1}^{\Delta_1\sigma_1\sigma_2^2}(h_1(s),h_2(s))h_1^{\widetilde{\Delta}^2}(s)\\
&\quad + f_{t_2}^{\Delta_2\sigma_1\sigma_2}(h_1(s),h_2(s))h_2^{\widetilde{\Delta}^2}(s)\\
&\quad + f_{t_1}^{\Delta_1^2\sigma_2}(h_1(s),h_2(s))\left(h_1^{\widetilde{\Delta}}(s)\right)^2\\
&\quad + (1+\mu_2^{\Delta_2}(t_2))f_{t_1t_2}^{\Delta_1\Delta_2\sigma_1\sigma_2}(h_1(s),h_2(s))h_1^{\widetilde{\Delta}}(s)h_2^{\widetilde{\Delta}}(s)\\
&\quad + f_{t_1t_2}^{\Delta_1\Delta_2\sigma_2}(h_1(s),h_2(s))h_1^{\widetilde{\Delta}}(s)h_2^{\widetilde{\Delta}}(s)\\
&\quad + f_{t_2}^{\Delta_2^2}(h_1(s),h_2(s))\left(h_2^{\widetilde{\Delta}}(s)\right)^2, \quad s\in J.
\end{aligned}
\tag{4.9}
$$

Thus,

$$
\begin{aligned}
k_{n\sigma_2}(s) &= k(s)\cdot N_{\sigma_2}^{\sigma_1\sigma_2}(g_1(s))\\
&= g^{\widetilde{\Delta}^2}(s)\cdot N_{\sigma_2}^{\sigma_1\sigma_2}(g_1(s))\\
&= \left(f_{t_1}^{\Delta_1\sigma_1\sigma_2^2}(h_1(s),h_2(s))\cdot N_{\sigma_2}^{\sigma_1\sigma_2}(g_1(s))\right)h_1^{\widetilde{\Delta}^2}(s)\\
&\quad + \left(f_{t_2}^{\Delta_2\sigma_1\sigma_2}(h_1(s),h_2(s))\cdot N_{\sigma_2}^{\sigma_1\sigma_2}(g_1(s))\right)h_2^{\widetilde{\Delta}^2}(s)
\end{aligned}
$$

$$
\begin{aligned}
&+\left(f_{t_1}^{\Delta_1^2\sigma_2}(h_1(s),h_2(s))\cdot N_{\sigma_2}^{\sigma_1\sigma_2}(g_1(s))\right)\left(h_1^{\tilde{\Delta}}(s)\right)^2\\
&+(1+\mu_2^{\Delta_2}(t_2))\left(f_{t_1t_2}^{\Delta_1\Delta_2\sigma_1\sigma_2}(h_1(s),h_2(s))\cdot N_{\sigma_2}^{\sigma_1\sigma_2}(g_1(s))\right)h_1^{\tilde{\Delta}}(s)h_2^{\tilde{\Delta}}(s)\\
&+\left(f_{t_1t_2}^{\Delta_1\Delta_2\sigma_2}(h_1(s),h_2(s))\cdot N_{\sigma_2}^{\sigma_1\sigma_2}(g_1(s))\right)h_1^{\tilde{\Delta}}(s)h_2^{\tilde{\Delta}}(s)\\
&+\left(f_{t_2}^{\Delta_2^2}(h_1(s),h_2(s))\cdot N_{\sigma_2}^{\sigma_1\sigma_2}(g_1(s))\right)\left(h_2^{\tilde{\Delta}}(s)\right)^2\\
&=\left(f_{t_1}^{\Delta_1^2\sigma_2}(h_1(s),h_2(s))\cdot N_{\sigma_2}^{\sigma_1\sigma_2}(g_1(s))\right)\left(h_1^{\tilde{\Delta}}(s)\right)^2\\
&+(1+\mu_2^{\Delta_2}(t_2))\left(f_{t_1t_2}^{\Delta_1\Delta_2\sigma_1\sigma_2}(h_1(s),h_2(s))\cdot N_{\sigma_2}^{\sigma_1\sigma_2}(g_1(s))\right)h_1^{\tilde{\Delta}}(s)h_2^{\tilde{\Delta}}(s)\\
&+\left(f_{t_1t_2}^{\Delta_1\Delta_2\sigma_2}(h_1(s),h_2(s))\cdot N_{\sigma_2}^{\sigma_1\sigma_2}(g_1(s))\right)h_1^{\tilde{\Delta}}(s)h_2^{\tilde{\Delta}}(s)\\
&+\left(f_{t_2}^{\Delta_2^2}(h_1(s),h_2(s))\cdot N_{\sigma_2}^{\sigma_1\sigma_2}(g_1(s))\right)\left(h_2^{\tilde{\Delta}}(s)\right)\\
&=\phi^1_{\sigma_2\sigma_1\sigma_2}\left(g_1^{\tilde{\Delta}}(s),g_1^{\tilde{\Delta}}(s)\right),\quad s\in J,
\end{aligned}
$$

that is,

$$k_{n\sigma_2}(s)=\phi^1_{\sigma_2\sigma_1\sigma_2}\left(g_1^{\tilde{\Delta}}(s),g_1^{\tilde{\Delta}}(s)\right),\quad s\in J.$$

Now, we use (4.4) and we find that

$$k_{n\sigma_2}(t)=\phi^1_{\sigma_2\sigma_1\sigma_2}\left(\frac{g^{\Delta}(t)}{\|g^{\Delta}(t)\|},\frac{g^{\Delta}(t)}{\|g^{\Delta}(t)\|}\right),\quad t\in I.$$

This completes the proof.

Definition 4.13 Let σ_2 be Δ_2-differentiable on $\mathbb{T}_2$, f be σ_2-completely delta differentiable on U, $f_{t_1}^{\Delta_1\sigma_2}$ be σ_2-completely delta differentiable and $f_{t_2}^{\Delta_2}$ be σ_1-completely delta differentiable on U.

Also let,

$$
\begin{aligned}
b_{11n\sigma_2\sigma_2\sigma_1}(t,t_1,t_2)&=f_{t_1}^{\Delta_1^2\sigma_2^2}(t_1,t_2)\cdot N_{\sigma_2}^{\sigma_1\sigma_2}(g(t)),\\
b_{12n\sigma_2\sigma_2\sigma_1}(t,t_1,t_2)&=f_{t_1t_2}^{\Delta_1\Delta_2}(t_1,t_2)\cdot N_{\sigma_2}^{\sigma_1\sigma_2}(g(t)),\\
b_{21n\sigma_2\sigma_2\sigma_1}(t,t_1,t_2)&=\left(1+\mu_2^{\Delta_2}(t_2)\right)f_{t_1t_2}^{\Delta_1\Delta_2\sigma_2}(t_1,t_2)\cdot N_{\sigma_2}^{\sigma_1\sigma_2}(g(t)),\\
b_{22n\sigma_2\sigma_2\sigma_1}(t,t_1,t_2)&=f_{t_2}^{\Delta_2^2\sigma_1}(t_1,t_2)\cdot N_{\sigma_2}^{\sigma_1\sigma_2}(g(t)),\quad (t_1,t_2)\in U,\quad t\in I,
\end{aligned}
$$

and

$$\mathscr{B}_{n\sigma_2\sigma_2\sigma_1}(t,t_1,t_2)=\begin{pmatrix} b_{11n\sigma_2\sigma_2\sigma_1}(t,t_1,t_2) & b_{12n\sigma_2\sigma_2\sigma_1}(t,t_1,t_2)\\ b_{21n\sigma_2\sigma_2\sigma_1}(t,t_1,t_2) & b_{22n\sigma_2\sigma_2\sigma_1}(t,t_1,t_2)\end{pmatrix},\quad (t_1,t_2)\in U,\quad t\in I.$$

For $p,q\in T_{a\sigma_2}S$, define

$$\phi^1_{\sigma_2\sigma_2\sigma_1}(p,q)=p^T\mathscr{B}_{n\sigma_2\sigma_2\sigma_1}q.$$

Theorem 4.11 (Variant of the Meusnier Theorem) *Let σ_2 be Δ_2-differentiable, S be a σ_2-regular surface with a local σ_2-parameterization (U,f) such that f is σ_2-completely delta differentiable, $f_{t_1}^{\Delta_1\sigma_2}$ be σ_2-completely delta differentiable and $f_{t_2}^{\Delta_2}$ be σ_1-completely delta differentiable on U.*
Then

$$k_{n\sigma_2}(t) = \phi^1_{\sigma_2\sigma_2\sigma_1}\left(\frac{g^\Delta(t)}{\|g^\Delta(t)\|}, \frac{g^\Delta(t)}{\|g^\Delta(t)\|}\right), \quad t \in I.$$

Proof. By (4.1), we get

$$\begin{aligned}
g_1^{\tilde{\Delta}}(s) &= f_{t_1}^{\Delta_1\sigma_2}(h_1(s),h_2(s))h_1^{\tilde{\Delta}}(s) + f_{t_2}^{\Delta_2}(h_1(s),h_2(s))h_2^{\tilde{\Delta}}(s),\\
g_1^{\tilde{\Delta}^2}(s) &= f_{t_1}^{\Delta_1\sigma_2}(h_1(\tilde{\sigma}(s)),h_2(\tilde{\sigma}(s)))h_1^{\tilde{\Delta}^2}(s)\\
&\quad + f_{t_2}^{\Delta_2}(h_1(\tilde{\sigma}(s)),h_2(\tilde{\sigma}(s)))h_2^{\tilde{\Delta}^2}(s)\\
&\quad + f_{t_1}^{\Delta_1^2\sigma_2^2}(h_1(s),h_2(s))\left(h_1^{\tilde{\Delta}}(s)\right)^2\\
&\quad + f_{t_1t_2}^{\Delta_1\sigma_2\Delta_2}(h_1(s),h_2(s))h_1^{\tilde{\Delta}}(s)h_2^{\tilde{\Delta}}(s)\\
&\quad + f_{t_1t_2}^{\Delta_1\Delta_2}(h_1(s),h_2(s))h_1^{\tilde{\Delta}}(s)h_2^{\tilde{\Delta}}(s)\\
&\quad + f_{t_2}^{\Delta_2^2\sigma_1}(h_1(s),h_2(s))\left(h_2^{\tilde{\Delta}}(s)\right)^2\\
&= f_{t_1}^{\Delta_1\sigma_1\sigma_2^2}(h_1(s),h_2(s))h_1^{\tilde{\Delta}^2}(s)\\
&\quad + f_{t_2}^{\Delta_2\sigma_1\sigma_2}(h_1(s),h_2(s))h_2^{\tilde{\Delta}^2}(s)\\
&\quad + f_{t_1}^{\Delta_1^2\sigma_2^2}(h_1(s),h_2(s))\left(h_1^{\tilde{\Delta}}(s)\right)^2\\
&\quad + (1+\mu_2^{\Delta_2}(t_2))f_{t_1t_2}^{\Delta_1\Delta_2\sigma_2}(h_1(s),h_2(s))h_1^{\tilde{\Delta}}(s)h_2^{\tilde{\Delta}}(s)\\
&\quad + f_{t_1t_2}^{\Delta_1\Delta_2}(h_1(s),h_2(s))h_1^{\tilde{\Delta}}(s)h_2^{\tilde{\Delta}}(s)\\
&\quad + f_{t_2}^{\Delta_2^2\sigma_1}(h_1(s),h_2(s))\left(h_2^{\tilde{\Delta}}(s)\right)^2, \quad s \in J,
\end{aligned}$$

that is,

$$\begin{aligned}
g_1^{\tilde{\Delta}^2}(s) &= f_{t_1}^{\Delta_1\sigma_1\sigma_2^2}(h_1(s),h_2(s))h_1^{\tilde{\Delta}^2}(s)\\
&\quad + f_{t_2}^{\Delta_2\sigma_1\sigma_2}(h_1(s),h_2(s))h_2^{\tilde{\Delta}^2}(s)\\
&\quad + f_{t_1}^{\Delta_1^2\sigma_2^2}(h_1(s),h_2(s))\left(h_1^{\tilde{\Delta}}(s)\right)^2\\
&\quad + (1+\mu_2^{\Delta_2}(t_2))f_{t_1t_2}^{\Delta_1\Delta_2\sigma_2}(h_1(s),h_2(s))h_1^{\tilde{\Delta}}(s)h_2^{\tilde{\Delta}}(s)\\
&\quad + f_{t_1t_2}^{\Delta_1\Delta_2}(h_1(s),h_2(s))h_1^{\tilde{\Delta}}(s)h_2^{\tilde{\Delta}}(s)\\
&\quad + f_{t_2}^{\Delta_2^2\sigma_1}(h_1(s),h_2(s))\left(h_2^{\tilde{\Delta}}(s)\right)^2, \quad s \in J.
\end{aligned} \tag{4.10}$$

So,

$$\begin{aligned}
k_{n\sigma_2}(s) &= k(s)\cdot N_{\sigma_2}^{\sigma_1\sigma_2}(g_1(s))\\
&= g^{\tilde{\Delta}^2}(s)\cdot N_{\sigma_2}^{\sigma_1\sigma_2}(g_1(s))\\
&= \left(f_{t_1}^{\Delta_1\sigma_1\sigma_2^2}(h_1(s),h_2(s))\cdot N_{\sigma_2}^{\sigma_1\sigma_2}(g_1(s))\right)h_1^{\tilde{\Delta}^2}(s)\\
&\quad+\left(f_{t_2}^{\Delta_2\sigma_1\sigma_2}(h_1(s),h_2(s))\cdot N_{\sigma_2}^{\sigma_1\sigma_2}(g_1(s))\right)h_2^{\tilde{\Delta}^2}(s)\\
&\quad+\left(f_{t_1}^{\Delta_1^2\sigma_2^2}(h_1(s),h_2(s))\cdot N_{\sigma_2}^{\sigma_1\sigma_2}(g_1(s))\right)\left(h_1^{\tilde{\Delta}}(s)\right)^2\\
&\quad+(1+\mu_2^{\Delta_2}(t_2))\left(f_{t_1t_2}^{\Delta_1\Delta_2\sigma_2}(h_1(s),h_2(s))\cdot N_{\sigma_2}^{\sigma_1\sigma_2}(g_1(s))\right)h_1^{\tilde{\Delta}}(s)h_2^{\tilde{\Delta}}(s)\\
&\quad+\left(f_{t_1t_2}^{\Delta_1\Delta_2}(h_1(s),h_2(s))\cdot N_{\sigma_2}^{\sigma_1\sigma_2}(g_1(s))\right)h_1^{\tilde{\Delta}}(s)h_2^{\tilde{\Delta}}(s)\\
&\quad+\left(f_{t_2}^{\Delta_2^2\sigma_1}(h_1(s),h_2(s))\cdot N_{\sigma_2}^{\sigma_1\sigma_2}(g_1(s))\right)\left(h_2^{\tilde{\Delta}}(s)\right)^2\\
&= \left(f_{t_1}^{\Delta_1^2\sigma_2^2}(h_1(s),h_2(s))\cdot N_{\sigma_2}^{\sigma_1\sigma_2}(g_1(s))\right)\left(h_1^{\tilde{\Delta}}(s)\right)^2\\
&\quad+(1+\mu_2^{\Delta_2}(t_2))\left(f_{t_1t_2}^{\Delta_1\Delta_2\sigma_2}(h_1(s),h_2(s))\cdot N_{\sigma_2}^{\sigma_1\sigma_2}(g_1(s))\right)h_1^{\tilde{\Delta}}(s)h_2^{\tilde{\Delta}}(s)\\
&\quad+\left(f_{t_1t_2}^{\Delta_1\Delta_2}(h_1(s),h_2(s))\cdot N_{\sigma_2}^{\sigma_1\sigma_2}(g_1(s))\right)h_1^{\tilde{\Delta}}(s)h_2^{\tilde{\Delta}}(s)\\
&\quad+\left(f_{t_2}^{\Delta_2^2\sigma_1}(h_1(s),h_2(s))\cdot N_{\sigma_2}^{\sigma_1\sigma_2}(g_1(s))\right)\left(h_2^{\tilde{\Delta}}(s)\right)^2\\
&= \phi^1_{\sigma_2\sigma_2\sigma_1}\left(g_1^{\tilde{\Delta}}(s),g_1^{\tilde{\Delta}}(s)\right),\quad s\in J,
\end{aligned}$$

that is,

$$k_{n\sigma_2}(s)=\phi^1_{2\sigma_2\sigma_2\sigma_1}\left(g_1^{\tilde{\Delta}}(s),g_1^{\tilde{\Delta}}(s)\right),\quad s\in J.$$

Now, we use (4.4) and we obtain

$$k_{n\sigma_2}(t)=\phi^1_{\sigma_2\sigma_2\sigma_1}\left(\frac{g^{\Delta}(t)}{\|g^{\Delta}(t)\|},\frac{g^{\Delta}(t)}{\|g^{\Delta}(t)\|}\right),\quad t\in I.$$

This completes the proof.

Definition 4.14 Let σ_2 be Δ_2-differentiable on $\mathbb{T}_2$, f be σ_2-completely delta differentiable on U, $f_{t_1}^{\Delta_1\sigma_2}$ and $f_{t_2}^{\Delta_2}$ be σ_2-completely delta differentiable on U. Also, let

$$\begin{aligned}
b_{11n\sigma_2\sigma_2\sigma_2}(t,t_1,t_2) &= f_{t_1}^{\Delta_1^2\sigma_2^2}(t_1,t_2)\cdot N_{\sigma_2}^{\sigma_1\sigma_2}(g(t)),\\
b_{12n\sigma_2\sigma_2\sigma_2}(t,t_1,t_2) &= f_{t_1t_2}^{\Delta_1\Delta_2\sigma_2}(t_1,t_2)\cdot N_{\sigma_2}^{\sigma_1\sigma_2}(g(t)),\\
b_{21n\sigma_2\sigma_2\sigma_2}(t,t_1,t_2) &= \left(1+\mu_2^{\Delta_2}(t_2)\right)f_{t_1t_2}^{\Delta_1\Delta_2\sigma_2}(t_1,t_2)\cdot N_{\sigma_2}^{\sigma_1\sigma_2}(g(t)),\\
b_{22n\sigma_2\sigma_2\sigma_2}(t,t_1,t_2) &= f_{t_2}^{\Delta_2^2}(t_1,t_2)\cdot N_{\sigma_2}^{\sigma_1\sigma_2}(g(t)),\quad (t_1,t_2)\in U,\quad t\in I,
\end{aligned}$$

and

$$\mathscr{B}_{n\sigma_2\sigma_2\sigma_2}(t,t_1,t_2) = \begin{pmatrix} b_{11n\sigma_2\sigma_2\sigma_2}(t,t_1,t_2) & b_{12n\sigma_2\sigma_2\sigma_2}(t,t_1,t_2) \\ b_{21n\sigma_2\sigma_2\sigma_2}(t,t_1,t_2) & b_{22n\sigma_2\sigma_2\sigma_2}(t,t_1,t_2) \end{pmatrix}, \quad (t_1,t_2) \in U, \quad t \in I.$$

For $p,q \in T_{a\sigma_2}S$, define

$$\phi^1_{\sigma_2\sigma_2\sigma_2}(p,q) = p^T \mathscr{B}_{n\sigma_2\sigma_2\sigma_2} q.$$

Theorem 4.12 (Variant of the Meusnier Theorem) *Let σ_2 be Δ_2-differentiable, S be a σ_2-regular surface with a local σ_2-parameterization (U,f) such that f is σ_2-completely delta differentiable, $f_{t_1}^{\Delta_1\sigma_2}$ and $f_{t_2}^{\Delta_2}$ are σ_2-completely delta differentiable on U. Then*

$$k_{n\sigma_2}(t) = \phi^1_{\sigma_2\sigma_2\sigma_2}\left(\frac{g^\Delta(t)}{\|g^\Delta(t)\|}, \frac{g^\Delta(t)}{\|g^\Delta(t)\|}\right), \quad t \in I.$$

Proof. By (4.1), we get

$$\begin{aligned}
g_1^{\tilde{\Delta}}(s) &= f_{t_1}^{\Delta_1\sigma_2}(h_1(s),h_2(s))h_1^{\tilde{\Delta}}(s) + f_{t_2}^{\Delta_2}(h_1(s),h_2(s))h_2^{\tilde{\Delta}}(s),\\
g_1^{\tilde{\Delta}^2}(s) &= f_{t_1}^{\Delta_1\sigma_2}(h_1(\tilde{\sigma}(s)),h_2(\tilde{\sigma}(s)))h_1^{\tilde{\Delta}^2}(s)\\
&\quad + f_{t_2}^{\Delta_2}(h_1(\tilde{\sigma}(s)),h_2(\tilde{\sigma}(s)))h_2^{\tilde{\Delta}^2}(s)\\
&\quad + f_{t_1}^{\Delta_1^2\sigma_2^2}(h_1(s),h_2(s))\left(h_1^{\tilde{\Delta}}(s)\right)^2\\
&\quad + f_{t_1t_2}^{\Delta_1\sigma_2\Delta_2}(h_1(s),h_2(s))h_1^{\tilde{\Delta}}(s)h_2^{\tilde{\Delta}}(s)\\
&\quad + f_{t_1t_2}^{\Delta_1\Delta_2\sigma_2}(h_1(s),h_2(s))h_1^{\tilde{\Delta}}(s)h_2^{\tilde{\Delta}}(s)\\
&\quad + f_{t_2}^{\Delta_2^2}(h_1(s),h_2(s))\left(h_2^{\tilde{\Delta}}(s)\right)^2\\
&= f_{t_1}^{\Delta_1\sigma_1\sigma_2^2}(h_1(s),h_2(s))h_1^{\tilde{\Delta}^2}(s)\\
&\quad + f_{t_2}^{\Delta_2\sigma_1\sigma_2}(h_1(s),h_2(s))h_2^{\tilde{\Delta}^2}(s)\\
&\quad + f_{t_1}^{\Delta_1^2\sigma_2^2}(h_1(s),h_2(s))\left(h_1^{\tilde{\Delta}}(s)\right)^2\\
&\quad + (1+\mu_2^{\Delta_2}(t_2))f_{t_1t_2}^{\Delta_1\Delta_2\sigma_2}(h_1(s),h_2(s))h_1^{\tilde{\Delta}}(s)h_2^{\tilde{\Delta}}(s)\\
&\quad + f_{t_1t_2}^{\Delta_1\Delta_2\sigma_2}(h_1(s),h_2(s))h_1^{\tilde{\Delta}}(s)h_2^{\tilde{\Delta}}(s)\\
&\quad + f_{t_2}^{\Delta_2^2}(h_1(s),h_2(s))\left(h_2^{\tilde{\Delta}}(s)\right)^2, \quad s \in J,
\end{aligned}$$

$$\begin{aligned}
g_1^{\tilde{\Delta}^2}(s) &= f_{t_1}^{\Delta_1\sigma_1\sigma_2^2}(h_1(s),h_2(s))h_1^{\tilde{\Delta}^2}(s)\\
&\quad + f_{t_2}^{\Delta_2\sigma_1\sigma_2}(h_1(s),h_2(s))h_2^{\tilde{\Delta}^2}(s)
\end{aligned}$$

$$
\begin{aligned}
&+f_{t_1}^{\Delta_1^2\sigma_2^2}(h_1(s),h_2(s))\left(h_1^{\tilde{\Delta}}(s)\right)^2 \\
&+(1+\mu_2^{\Delta_2}(t_2))f_{t_1t_2}^{\Delta_1\Delta_2\sigma_2}(h_1(s),h_2(s))h_1^{\tilde{\Delta}}(s)h_2^{\tilde{\Delta}}(s) \\
&+f_{t_1t_2}^{\Delta_1\Delta_2\sigma_2}(h_1(s),h_2(s))h_1^{\tilde{\Delta}}(s)h_2^{\tilde{\Delta}}(s) \\
&+f_{t_2}^{\Delta_2^2}(h_1(s),h_2(s))\left(h_2^{\tilde{\Delta}}(s)\right)^2, \quad s\in J.
\end{aligned}
\tag{4.11}
$$

From here,

$$
\begin{aligned}
k_{n\sigma_2}(s) &= k(s)\cdot N_{\sigma_2}^{\sigma_1\sigma_2}(g_1(s)) \\
&= g^{\tilde{\Delta}^2}(s)\cdot N_{\sigma_2}^{\sigma_1\sigma_2}(g_1(s)) \\
&= \left(f_{t_1}^{\Delta_1\sigma_1\sigma_2^2}(h_1(s),h_2(s))\cdot N_{\sigma_2}^{\sigma_1\sigma_2}(g_1(s))\right)h_1^{\tilde{\Delta}^2}(s) \\
&\quad+\left(f_{t_2}^{\Delta_2\sigma_1\sigma_2}(h_1(s),h_2(s))\cdot N_{\sigma_2}^{\sigma_1\sigma_2}(g_1(s))\right)h_2^{\tilde{\Delta}^2}(s) \\
&\quad+\left(f_{t_1}^{\Delta_1^2\sigma_2^2}(h_1(s),h_2(s))\cdot N_{\sigma_2}^{\sigma_1\sigma_2}(g_1(s))\right)\left(h_1^{\tilde{\Delta}}(s)\right)^2 \\
&\quad+(1+\mu_2^{\Delta_2}(t_2))\left(f_{t_1t_2}^{\Delta_1\Delta_2\sigma_2}(h_1(s),h_2(s))\cdot N_{\sigma_2}^{\sigma_1\sigma_2}(g_1(s))\right)h_1^{\tilde{\Delta}}(s)h_2^{\tilde{\Delta}}(s) \\
&\quad+\left(f_{t_1t_2}^{\Delta_1\Delta_2\sigma_2}(h_1(s),h_2(s))\cdot N_{\sigma_2}^{\sigma_1\sigma_2}(g_1(s))\right)h_1^{\tilde{\Delta}}(s)h_2^{\tilde{\Delta}}(s) \\
&\quad+\left(f_{t_2}^{\Delta_2^2}(h_1(s),h_2(s))\cdot N_{\sigma_2}^{\sigma_1\sigma_2}(g_1(s))\right)\left(h_2^{\tilde{\Delta}}(s)\right)^2 \\
&= \left(f_{t_1}^{\Delta_1^2\sigma_2^2}(h_1(s),h_2(s))\cdot N_{\sigma_2}^{\sigma_1\sigma_2}(g_1(s))\right)\left(h_1^{\tilde{\Delta}}(s)\right)^2 \\
&\quad+(1+\mu_2^{\Delta_2}(t_2))\left(f_{t_1t_2}^{\Delta_1\Delta_2\sigma_2}(h_1(s),h_2(s))\cdot N_{\sigma_2}^{\sigma_1\sigma_2}(g_1(s))\right)h_1^{\tilde{\Delta}}(s)h_2^{\tilde{\Delta}}(s) \\
&\quad+\left(f_{t_1t_2}^{\Delta_1\Delta_2\sigma_2}(h_1(s),h_2(s))\cdot N_{\sigma_2}^{\sigma_1\sigma_2}(g_1(s))\right)h_1^{\tilde{\Delta}}(s)h_2^{\tilde{\Delta}}(s) \\
&\quad+\left(f_{t_2}^{\Delta_2^2}(h_1(s),h_2(s))\cdot N_{\sigma_2}^{\sigma_1\sigma_2}(g_1(s))\right)\left(h_2^{\tilde{\Delta}}(s)\right)^2 \\
&= \phi_{\sigma_2\sigma_2\sigma_2}^1\left(g_1^{\tilde{\Delta}}(s),g_1^{\tilde{\Delta}}(s)\right), \quad s\in J,
\end{aligned}
$$

that is,

$$
k_{n\sigma_2}(s) = \phi_{\sigma_2\sigma_2\sigma_2}^1\left(g_1^{\tilde{\Delta}}(s),g_1^{\tilde{\Delta}}(s)\right), \quad s\in J.
$$

Now, we use (4.4) and we find

$$
k_{n\sigma_2}(t) = \phi_{\sigma_2\sigma_2\sigma_2}^1\left(\frac{g^{\Delta}(t)}{\|g^{\Delta}(t)\|},\frac{g^{\Delta}(t)}{\|g^{\Delta}(t)\|}\right), \quad t\in I.
$$

This completes the proof.

Corollary 4.1 *Let σ_1 be Δ_1-differentiable, S be a σ_1-regular surface with a local σ_1-parameterization (U,f) such that f is σ_1-completely delta differentiable, $f_{t_1}^{\Delta_1}$ and $f_{t_2}^{\Delta_2\sigma_1}$ be σ_1-completely delta differentiable. If two curves on S have a common point* *and they have the same tangent line at the common point, then the curves have the same σ_1-normal curvature at the common point.*

Proof. Let p and q be the tangent vectors to the two curves at their common point. From the hypothesis, we have that

$$p = \alpha q$$

for some positive constant α.

Then

$$\begin{aligned} k_{n\sigma_1} &= \phi^1_{\sigma_1\sigma_1\sigma_1}\left(\frac{p}{\|p\|}, \frac{p}{\|p\|}\right) \\ &= \phi^1_{\sigma_1\sigma_1\sigma_1}\left(\frac{\alpha p}{\|\alpha p\|}, \frac{\alpha p}{\|\alpha p\|}\right) \\ &= \phi^1_{\sigma_1\sigma_1\sigma_1}\left(\frac{q}{\|q\|}, \frac{q}{\|q\|}\right). \end{aligned}$$

This completes the proof.

As above, one can prove the following corollaries.

Corollary 4.2 *Let σ_1 be Δ_1-differentiable, S be a σ_1-regular surface with a local σ_1-parameterization (U,f) such that f is σ_1-completely delta differentiable, $f_{t_1}^{\Delta_1}$ be σ_1-completely delta differentiable and $f_{t_2}^{\Delta_2\sigma_1}$ be σ_2-completely delta differentiable. If two curves on S have a common point and they have the same tangent line at the common point, then the curves have the same σ_1-normal curvature at the common point.*

Corollary 4.3 *Let σ_1 be Δ_1-differentiable, S be a σ_1-regular surface with a local σ_1-parameterization (U,f) such that f is σ_1-completely delta differentiable, $f_{t_1}^{\Delta_1}$ be σ_2-completely delta differentiable and $f_{t_2}^{\Delta_2\sigma_1}$ be σ_1-completely delta differentiable. If two curves on S have a common point and they have the same tangent line at the common point, then the curves have the same σ_1-normal curvature at the common point.*

Corollary 4.4 *Let σ_1 be Δ_1-differentiable, S be a σ_1-regular surface with a local σ_1-parameterization (U,f) such that f is σ_1-completely delta differentiable, $f_{t_1}^{\Delta_1}$ and $f_{t_2}^{\Delta_2\sigma_1}$ be σ_2-completely delta differentiable. If two curves on S have a common point and they have the same tangent line at the common point, then the curves have the same σ_1-normal curvature at the common point.*

Corollary 4.5 *Let σ_2 be Δ_2-differentiable, S be a σ_2-regular surface with a local σ_2-parameterization (U,f) such that f is σ_2-completely delta differentiable, $f_{t_1}^{\Delta_1\sigma_2}$ and $f_{t_2}^{\Delta_2}$ be σ_1-completely delta differentiable. If two curves on S have a common point and they have the same tangent line at the common point, then the curves have the same σ_2-normal curvature at the common point.*

Corollary 4.6 *Let σ_2 be Δ_2-differentiable, S be a σ_2-regular surface with a local σ_2-parameterization (U,f) such that f is σ_2-completely delta differentiable, $f_{t_1}^{\Delta_1\sigma_2}$ be σ_1-completely delta differentiable and $f_{t_2}^{\Delta_2}$ be σ_2-completely delta differentiable. If two curves on S have a common point and they have the same tangent line at the common point, then the curves have the same σ_2-normal curvature at the common point.*

Corollary 4.7 *Let σ_2 be Δ_2-differentiable, S be a σ_2-regular surface with a local σ_2-parameterization (U,f) such that f is σ_2-completely delta differentiable, $f_{t_1}^{\Delta_1\sigma_2}$ be σ_2-completely delta differentiable and $f_{t_2}^{\Delta_2}$ be σ_1-completely delta differentiable. If two curves on S have a common point and they have the same tangent line at the common point, then the curves have the same σ_2-normal curvature at the common point.*

Corollary 4.8 *Let σ_2 be Δ_2-differentiable, S be a σ_2-regular surface with a local σ_2-parameterization (U,f) such that f is σ_2-completely delta differentiable, $f_{t_1}^{\Delta_1\sigma_2}$ and $f_{t_2}^{\Delta_2}$ be σ_2-completely delta differentiable. If two curves on S have a common point and they have the same tangent line at the common point, then the curves have the same σ_2-normal curvature at the common point.*

Definition 4.15 Let S be a σ_1-regular surface with a local σ_1-parameterization (U,f) and $g: I \to S$ be a curve lying in S. The projection of the curvature vector $k(t)$ of $g(t)$ on $N_{\sigma_1^2\sigma_2}^{\sigma_1\sigma_2}(g(t))$ is called $\sigma_1^2\sigma_2$-normal curvature of the curve g at t and it is denoted by $k_{n\sigma_1^2\sigma_2}(t)$.

Definition 4.16 Let σ_1 be Δ_1-differentiable on $\mathbb{T}_1$, f be σ_1-completely delta differentiable on U, $f_{t_1}^{\Delta_1}$ and $f_{t_2}^{\Delta_2\sigma_1}$ be σ_1-completely delta differentiable on U. Also, let,

$$b_{11n\sigma_1^2\sigma_2\sigma_1\sigma_1}(t,t_1,t_2) = f_{t_1}^{\Delta_1^2}(t_1,t_2)\cdot N_{\sigma_1^2\sigma_2}^{\sigma_1\sigma_2}(g(t)),$$

$$b_{12n\sigma_1^2\sigma_2\sigma_1\sigma_1}(t,t_1,t_2) = f_{t_1t_2}^{\Delta_1\Delta_2\sigma_1}(t_1,t_2)\cdot N_{\sigma_1^2\sigma_2}^{\sigma_1\sigma_2}(g(t)),$$

$$b_{21n\sigma_1^2\sigma_2\sigma_1\sigma_1}(t,t_1,t_2) = \left(1+\mu_1^{\Delta_1}(t_1)\right) b_{12n\sigma_1^2\sigma_2\sigma_1\sigma_1}(t,t_1,t_2),$$

$$b_{22n\sigma_1^2\sigma_2\sigma_1\sigma_1}(t,t_1,t_2) = f_{t_2}^{\Delta_2^2\sigma_1^2}(t_1,t_2)\cdot N_{\sigma_1^2\sigma_2}^{\sigma_1\sigma_2}(g(t)), \quad (t_1,t_2)\in U, \quad t\in I,$$

and

$$\mathscr{B}_{n\sigma_1^2\sigma_2\sigma_1\sigma_1}(t,t_1,t_2) = \begin{pmatrix} b_{11n\sigma_1^2\sigma_2\sigma_1\sigma_1}(t,t_1,t_2) & b_{12n\sigma_1^2\sigma_2\sigma_1\sigma_1}(t,t_1,t_2) \\ b_{21n\sigma_1^2\sigma_2\sigma_1\sigma_1}(t,t_1,t_2) & b_{22n\sigma_1^2\sigma_2\sigma_1\sigma_1}(t,t_1,t_2) \end{pmatrix}, \quad (t_1,t_2)\in U, \quad t\in I.$$

For $p,q \in T_{a\sigma_1^2\sigma_2}S$, define

$$\phi^1_{\sigma_1^2\sigma_2\sigma_1\sigma_1}(p,q) = p^T \mathscr{B}_{n\sigma_1^2\sigma_2\sigma_1\sigma_1} q.$$

Theorem 4.13 (Variant of the Meusnier Theorem) *Let σ_1 be Δ_1-differentiable, S be a σ_1-regular surface with a local σ_1-parameterization (U,f) such that f is*

σ_1-completely delta differentiable, $f_{t_1}^{\Delta_1}$ and $f_{t_2}^{\Delta_2\sigma_1}$ be σ_1-completely delta differentiable on U. Then

$$k_{n\sigma_1^2\sigma_2}(t) = \phi^1_{\sigma_1^2\sigma_2\sigma_1\sigma_1}\left(\frac{g^\Delta(t)}{\|g^\Delta(t)\|}, \frac{g^\Delta(t)}{\|g^\Delta(t)\|}\right), \quad t \in I.$$

Definition 4.17 Let σ_1 be Δ_1-differentiable on $\mathbb{T}_1$, f be σ_1-completely delta differentiable on U, $f_{t_1}^{\Delta_1}$ be σ_1-completely delta differentiable and $f_{t_2}^{\Delta_2\sigma_1}$ be σ_1-completely delta differentiable on U.
Also, let,

$$b_{11n\sigma_1^2\sigma_2\sigma_1\sigma_2}(t,t_1,t_2) = f_{t_1}^{\Delta_1^2}(t_1,t_2)\cdot N^{\sigma_1\sigma_2}_{\sigma_1^2\sigma_2}(g(t)),$$

$$b_{12n\sigma_1^2\sigma_2\sigma_1\sigma_2}(t,t_1,t_2) = f_{t_1t_2}^{\Delta_1\Delta_2\sigma_1}(t_1,t_2)\cdot N^{\sigma_1\sigma_2}_{\sigma_1^2\sigma_2}(g(t)),$$

$$b_{21n\sigma_1^2\sigma_2\sigma_1\sigma_2}(t,t_1,t_2) = \left(1+\mu_1^{\Delta_1}(t_1)\right) f_{t_1t_2}^{\Delta_1\Delta_2\sigma_1\sigma_2} t_1,t_2)\cdot N^{\sigma_1\sigma_2}_{\sigma_1^2\sigma_2}(g(t)),$$

$$b_{22n\sigma_1^2\sigma_2\sigma_1\sigma_2}(t,t_1,t_2) = f_{t_2}^{\Delta_2^2\sigma_1}(t_1,t_2)\cdot N^{\sigma_1\sigma_2}_{\sigma_1^2\sigma_2}(g(t)), \quad (t_1,t_2)\in U, \quad t\in I,$$

and

$$\mathscr{B}_{n\sigma_1^2\sigma_2\sigma_1\sigma_2}(t,t_1,t_2) = \begin{pmatrix} b_{11n\sigma_1^2\sigma_2\sigma_1\sigma_2}(t,t_1,t_2) & b_{12n\sigma_1^2\sigma_2\sigma_1\sigma_2}(t,t_1,t_2) \\ b_{21n\sigma_1^2\sigma_2\sigma_1\sigma_2}(t,t_1,t_2) & b_{22n\sigma_1^2\sigma_2\sigma_1\sigma_2}(t,t_1,t_2) \end{pmatrix}, \quad (t_1,t_2)\in U, \quad t\in I.$$

For $p,q \in T_{a\sigma_1^2\sigma_2}S$, define

$$\phi^1_{\sigma_1^2\sigma_2\sigma_1\sigma_2}(p,q) = p^T \mathscr{B}_{n\sigma_1^2\sigma_2\sigma_1\sigma_2} q.$$

Theorem 4.14 (Variant of the Meusnier Theorem) *Let σ_1 be Δ_1-differentiable, S be a σ_1-regular surface with a local σ_1-parameterization (U,f) such that f is σ_1-completely delta differentiable, $f_{t_1}^{\Delta_1}$ be σ_1-completely delta differentiable and $f_{t_2}^{\Delta_2\sigma_1}$ be σ_2-completely delta differentiable on U. Then*

$$k_{n\sigma_1^2\sigma_2}(t) = \phi^1_{\sigma_1^2\sigma_2\sigma_1\sigma_2}\left(\frac{g^\Delta(t)}{\|g^\Delta(t)\|}, \frac{g^\Delta(t)}{\|g^\Delta(t)\|}\right), \quad t \in I.$$

Definition 4.18 Let σ_1 be Δ_1-differentiable on $\mathbb{T}_1$, f be σ_1-completely delta differentiable on U, $f_{t_1}^{\Delta_1}$ be σ_2-completely delta differentiable and $f_{t_2}^{\Delta_2\sigma_1}$ be σ_1-completely delta differentiable on U.
Also, let

$$b_{11n\sigma_1^2\sigma_2\sigma_2\sigma_1}(t,t_1,t_2) = f_{t_1}^{\Delta_1^2\sigma_2}(t_1,t_2)\cdot N^{\sigma_1\sigma_2}_{\sigma_1^2\sigma_2}(g(t)),$$

$$b_{12n\sigma_1^2\sigma_2\sigma_2\sigma_1}(t,t_1,t_2) = f_{t_1t_2}^{\Delta_1\Delta_2}(t_1,t_2)\cdot N^{\sigma_1\sigma_2}_{\sigma_1^2\sigma_2}(g(t)),$$

$$b_{21n\sigma_1^2\sigma_2\sigma_2\sigma_1}(t,t_1,t_2) = \left(1+\mu_1^{\Delta_1}(t_1)\right) f_{t_1t_2}^{\Delta_1\Delta_2\sigma_1}\cdot N^{\sigma_1\sigma_2}_{\sigma_1^2\sigma_2},$$

$$b_{22n\sigma_1^2\sigma_2\sigma_2\sigma_1}(t,t_1,t_2) = f_{t_2}^{\Delta_2^2\sigma_1^2}(t_1,t_2)\cdot N_{\sigma_1^2\sigma_2}^{\sigma_1\sigma_2}(g(t)), \quad (t_1,t_2)\in U, \quad t\in I,$$

and

$$\mathscr{B}_{n\sigma_1^2\sigma_2\sigma_2\sigma_1}(t,t_1,t_2) = \begin{pmatrix} b_{11n\sigma_1^2\sigma_2\sigma_2\sigma_1}(t,t_1,t_2) & b_{12n\sigma_1^2\sigma_2\sigma_2\sigma_1}(t,t_1,t_2) \\ b_{21n\sigma_1^2\sigma_2\sigma_2\sigma_1}(t,t_1,t_2) & b_{22n\sigma_1^2\sigma_2\sigma_2\sigma_1}(t,t_1,t_2) \end{pmatrix}, \quad (t_1,t_2)\in U, \quad t\in I.$$

For $p,q\in T_{a\sigma_1}S$, define

$$\phi^1_{\sigma_1^2\sigma_2\sigma_2\sigma_1}(p,q) = p^T\mathscr{B}_{n\sigma_1^2\sigma_2\sigma_2\sigma_1}q.$$

Theorem 4.15 (Variant of the Meusnier Theorem) *Let σ_1 be Δ_1-differentiable, S be a σ_1-regular surface with a local σ_1-parameterization (U,f) such that f is σ_1-completely delta differentiable, $f_{t_1}^{\Delta_1}$ be σ_2-completely delta differentiable and $f_{t_2}^{\Delta_2\sigma_1}$ be σ_1-completely delta differentiable on U. Then*

$$k_{n\sigma_1^2\sigma_2}(t) = \phi^1_{\sigma_1^2\sigma_2\sigma_2\sigma_1}\left(\frac{g^\Delta(t)}{\|g^\Delta(t)\|},\frac{g^\Delta(t)}{\|g^\Delta(t)\|}\right), \quad t\in I.$$

Definition 4.19 Let σ_1 be Δ_1-differentiable on $\mathbb{T}_1$, f be σ_1-completely delta differentiable on U, $f_{t_1}^{\Delta_1}$ and $f_{t_2}^{\Delta_2\sigma_1}$ be σ_2-completely delta differentiable on U. Let also,

$$b_{11n\sigma_1^2\sigma_2\sigma_2\sigma_2}(t,t_1,t_2) = f_{t_1}^{\Delta_1^2\sigma_2}(t_1,t_2)\cdot N_{\sigma_1^2\sigma_2}^{\sigma_1\sigma_2}(g(t)),$$

$$b_{12n\sigma_1^2\sigma_2\sigma_2\sigma_2}(t,t_1,t_2) = f_{t_1t_2}^{\Delta_1\Delta_2}(t_1,t_2)\cdot N_{\sigma_1^2\sigma_2}^{\sigma_1\sigma_2}(g(t)),$$

$$b_{21n\sigma_1^2\sigma_2\sigma_2\sigma_2}(t,t_1,t_2) = (1+\mu_1^{\Delta_1}(t_1))f_{t_1t_2}^{\Delta_1\Delta_2\sigma_1\sigma_2}\cdot N_{\sigma_1^2\sigma_2}^{\sigma_1\sigma_2},$$

$$b_{22n\sigma_1^2\sigma_2\sigma_2\sigma_2}(t,t_1,t_2) = f_{t_2}^{\Delta_2^2\sigma_1}(t_1,t_2)\cdot N_{\sigma_1^2\sigma_2}^{\sigma_1\sigma_2}(g(t)), \quad (t_1,t_2)\in U, \quad t\in I,$$

and

$$\mathscr{B}_{n\sigma_1^2\sigma_2\sigma_2\sigma_2}(t,t_1,t_2) = \begin{pmatrix} b_{11n\sigma_1^2\sigma_2\sigma_2\sigma_2}(t,t_1,t_2) & b_{12n\sigma_1^2\sigma_2\sigma_2\sigma_2}(t,t_1,t_2) \\ b_{21n\sigma_1^2\sigma_2\sigma_2\sigma_2}(t,t_1,t_2) & b_{22n\sigma_1^2\sigma_2\sigma_2\sigma_2}(t,t_1,t_2) \end{pmatrix}, \quad (t_1,t_2)\in U, \quad t\in I.$$

For $p,q\in T_{a\sigma_1^2\sigma_2}S$, define

$$\phi^1_{\sigma_1^2\sigma_2\sigma_2\sigma_2}(p,q) = p^T\mathscr{B}_{n\sigma_1^2\sigma_2\sigma_2\sigma_2}q.$$

Theorem 4.16 (Variant of the Meusnier Theorem) *Let σ_1 be Δ_1-differentiable, S be a σ_1-regular surface with a local σ_1-parameterization (U,f) such that f is σ_1-completely delta differentiable, $f_{t_1}^{\Delta_1}$ and $f_{t_2}^{\Delta_2\sigma_1}$ be σ_2-completely delta differentiable on U. Then*

$$k_{n\sigma_1^2\sigma_2}(t) = \phi^1_{\sigma_1^2\sigma_2\sigma_2\sigma_2}\left(\frac{g^\Delta(t)}{\|g^\Delta(t)\|},\frac{g^\Delta(t)}{\|g^\Delta(t)\|}\right), \quad t\in I.$$

Corollary 4.9 *Let σ_1 be Δ_1-differentiable, S be a σ_1-regular surface with a local σ_1-parameterization (U,f) such that f is σ_1-completely delta differentiable, $f_{t_1}^{\Delta_1}$ and $f_{t_2}^{\Delta_2\sigma_1}$ be σ_1-completely delta differentiable. If two curves on S have a common point and they have the same tangent line at the common point, then the curves have the same $\sigma_1^2\sigma_2$-normal curvature at the common point.*

Corollary 4.10 *Let σ_1 be Δ_1-differentiable, S be a σ_1-regular surface with a local σ_1-parameterization (U,f) such that f is σ_1-completely delta differentiable, $f_{t_1}^{\Delta_1}$ be σ_1-completely delta differentiable and $f_{t_2}^{\Delta_2\sigma_1}$ be σ_2-completely delta differentiable. If two curves on S have a common point and they have the same tangent line at the common point, then the curves have the same $\sigma_1^2\sigma_2$-normal curvature at the common point.*

Corollary 4.11 *Let σ_1 be Δ_1-differentiable, S be a σ_1-regular surface with a local σ_1-parameterization (U,f) such that f is σ_1-completely delta differentiable, $f_{t_1}^{\Delta_1}$ be σ_2-completely delta differentiable and $f_{t_2}^{\Delta_2\sigma_1}$ be σ_1-completely delta differentiable. If two curves on S have a common point and they have the same tangent line at the common point, then the curves have the same $\sigma_1^2\sigma_2$-normal curvature at the common point.*

Corollary 4.12 *Let σ_1 be Δ_1-differentiable, S be a σ_1-regular surface with a local σ_1-parameterization (U,f) such that f is σ_1-completely delta differentiable, $f_{t_1}^{\Delta_1}$ and $f_{t_2}^{\Delta_2\sigma_1}$ be σ_2-completely delta differentiable. If two curves on S have a common point and they have the same tangent line at the common point, then the curves have the same $\sigma_1^2\sigma_2$-normal curvature at the common point.*

Definition 4.20 Let S be a σ_2-regular surface with a local σ_2-parametrization (U,f) and $g: I \to S$ be a curve lying in S. The projection of the curvature vector $k(t)$ of $g(t)$ on $N_{\sigma_1\sigma_2^2}^{\sigma_1\sigma_2}(g(t))$ is called the $\sigma_1\sigma_2^2$-normal curvature of the curve g at t and it is denoted by $k_{n\sigma_1\sigma_2^2}(t)$.

Definition 4.21 Let σ_2 be Δ_2-differentiable on $\mathbb{T}_2$, f be σ_1-completely delta differentiable on U, $f_{t_1}^{\Delta_1\sigma_2}$ and $f_{t_2}^{\Delta_2}$ be σ_1-completely delta differentiable on U. Let also,

$$b_{11n\sigma_1\sigma_2^2\sigma_1\sigma_1}(t,t_1,t_2) = f_{t_1}^{\Delta_1^2}(t_1,t_2) \cdot N_{\sigma_1\sigma_2^2}^{\sigma_1\sigma_2}(g(t)),$$

$$b_{12n\sigma_1\sigma_2^2\sigma_1\sigma_1}(t,t_1,t_2) = f_{t_1t_2}^{\Delta_1\Delta_2}(t_1,t_2) \cdot N_{\sigma_1\sigma_2^2}^{\sigma_1\sigma_2}(g(t)),$$

$$b_{21n\sigma_1\sigma_2^2\sigma_1\sigma_1}(t,t_1,t_2) = \left(1+\mu_2^{\Delta_2}(t_2)\right) f_{t_1t_2}^{\Delta_1\Delta_2\sigma_1\sigma_2}(t_1,t_2) \cdot N_{\sigma_1\sigma_2^2}^{\sigma_1\sigma_2}(g(t)),$$

$$b_{22n\sigma_1\sigma_2^2\sigma_1\sigma_1}(t,t_1,t_2) = f_{t_2}^{\Delta_2^2\sigma_1}(t_1,t_2) \cdot N_{\sigma_1\sigma_2^2}^{\sigma_1\sigma_2}(g(t)), \quad (t_1,t_2) \in U, \quad t \in I,$$

and

$$\mathscr{B}_{n\sigma_1\sigma_2^2\sigma_1\sigma_1}(t,t_1,t_2) = \begin{pmatrix} b_{11n\sigma_1\sigma_2^2\sigma_1\sigma_1}(t,t_1,t_2) & b_{12n\sigma_1\sigma_2^2\sigma_1\sigma_1}(t,t_1,t_2) \\ b_{21n\sigma_1\sigma_2^2\sigma_1\sigma_1}(t,t_1,t_2) & b_{22n\sigma_1\sigma_2^2\sigma_1\sigma_1}(t,t_1,t_2) \end{pmatrix}, \quad (t_1,t_2) \in U, \quad t \in I.$$

For $p,q \in T_{a\sigma_1\sigma_2^2}S$, define

$$\phi^1_{\sigma_1\sigma_2^2\sigma_1\sigma_1}(p,q) = p^T \mathscr{B}_{n\sigma_1\sigma_2^2\sigma_1\sigma_1} q.$$

Theorem 4.17 (Variant of the Meusnier Theorem) *Let σ_2 be Δ_2-differentiable, S is a σ_2-regular surface with a local σ_2-parameterization (U,f) such that f is σ_2-completely delta differentiable, $f_{t_1}^{\Delta_1\sigma_2}$ and $f_{t_2}^{\Delta_2}$ are σ_1-completely delta differentiable on U. Then*

$$k_{n\sigma_1\sigma_2^2}(t) = \phi^1_{\sigma_1\sigma_2^2\sigma_1\sigma_1}\left(\frac{g^\Delta(t)}{\|g^\Delta(t)\|}, \frac{g^\Delta(t)}{\|g^\Delta(t)\|}\right), \quad t \in I.$$

Definition 4.22 Let σ_2 be Δ_2-differentiable on $\mathbb{T}_2$, f be σ_1-completely delta differentiable on U, $f_{t_1}^{\Delta_1\sigma_2}$ be σ_1-completely delta differentiable and $f_{t_2}^{\Delta_2}$ be σ_2-completely delta differentiable on U.

Also, let

$$b_{11n\sigma_1\sigma_2^2\sigma_1\sigma_2}(t,t_1,t_2) = f_{t_1}^{\Delta_1^2}(t_1,t_2)\cdot N^{\sigma_1\sigma_2}_{\sigma_1\sigma_2^2}(g(t)),$$

$$b_{12n\sigma_1\sigma_2^2\sigma_1\sigma_2}(t,t_1,t_2) = f_{t_1t_2}^{\Delta_1\Delta_2\sigma_2}(t_1,t_2)\cdot N^{\sigma_1\sigma_2}_{\sigma_1\sigma_2^2}(g(t)),$$

$$b_{21n\sigma_1\sigma_2^2\sigma_1\sigma_1}(t,t_1,t_2) = \left(1+\mu_2^{\Delta_2}(t_2)\right) f_{t_1t_2}^{\Delta_1\Delta_2\sigma_1\sigma_2}(t_1,t_2)\cdot N^{\sigma_1\sigma_2}_{\sigma_1\sigma_2^2}(g(t)),$$

$$b_{22n\sigma_1\sigma_2^2\sigma_1\sigma_1}(t,t_1,t_2) = f_{t_2}^{\Delta_2^2}(t_1,t_2)\cdot N^{\sigma_1\sigma_2}_{\sigma_1\sigma_2^2}(g(t)), \quad (t_1,t_2)\in U, \quad t \in I,$$

and

$$\mathscr{B}_{n\sigma_1\sigma_2^2\sigma_1\sigma_2}(t,t_1,t_2) = \begin{pmatrix} b_{11n\sigma_1\sigma_2^2\sigma_1\sigma_2}(t,t_1,t_2) & b_{12n\sigma_1\sigma_2^2\sigma_1\sigma_2}(t,t_1,t_2) \\ b_{21n\sigma_1\sigma_2^2\sigma_1\sigma_2}(t,t_1,t_2) & b_{22n\sigma_1\sigma_2^2\sigma_1\sigma_2}(t,t_1,t_2) \end{pmatrix}, \quad (t_1,t_2)\in U, \quad t\in I.$$

For $p,q \in T_{a\sigma_2}S$, define

$$\phi^1_{\sigma_1\sigma_2^2\sigma_1\sigma_2}(p,q) = p^T \mathscr{B}_{n\sigma_1\sigma_2^2\sigma_1\sigma_2} q.$$

Theorem 4.18 (Variant of the Meusnier Theorem) *Let σ_2 be Δ_2-differentiable, S be a σ_2-regular surface with a local σ_2-parameterization (U,f) such that f is σ_2-completely delta differentiable, $f_{t_1}^{\Delta_1\sigma_2}$ be σ_1-completely delta differentiable and $f_{t_2}^{\Delta_2}$ be σ_2-completely delta differentiable on U. Then*

$$k_{n\sigma_1\sigma_2^2}(t) = \phi^1_{\sigma_1\sigma_2^2\sigma_1\sigma_2}\left(\frac{g^\Delta(t)}{\|g^\Delta(t)\|}, \frac{g^\Delta(t)}{\|g^\Delta(t)\|}\right), \quad t \in I.$$

Definition 4.23 Let σ_2 be Δ_2-differentiable on $\mathbb{T}_2$, f be σ_2-completely delta differentiable on U, $f_{t_1}^{\Delta_1\sigma_2}$ be σ_2-completely delta differentiable and $f_{t_2}^{\Delta_2}$ be σ_1-completely delta differentiable on U.

Also, let

$$b_{11n\sigma_1\sigma_2^2\sigma_2\sigma_1}(t,t_1,t_2) = f_{t_1}^{\Delta_1^2\sigma_2^2}(t_1,t_2)\cdot N^{\sigma_1\sigma_2}_{\sigma_1\sigma_2^2}(g(t)),$$

$$b_{12n\sigma_1\sigma_2^2\sigma_2\sigma_1}(t,t_1,t_2) = f_{t_1t_2}^{\Delta_1\Delta_2}(t_1,t_2)\cdot N_{\sigma_1\sigma_2^2}^{\sigma_1\sigma_2}(g(t)),$$

$$b_{21n\sigma_1\sigma_2^2\sigma_2\sigma_1}(t,t_1,t_2) = \left(1+\mu_2^{\Delta_2}(t_2)\right)f_{t_1t_2}^{\Delta_1\Delta_2\sigma_2}(t_1,t_2)\cdot N_{\sigma_1\sigma_2^2}^{\sigma_1\sigma_2}(g(t)),$$

$$b_{22n\sigma_1\sigma_2^2\sigma_2\sigma_1}(t,t_1,t_2) = f_{t_2}^{\Delta_2^2\sigma_1}(t_1,t_2)\cdot N_{\sigma_1\sigma_2^2}^{\sigma_1\sigma_2}(g(t)), \quad (t_1,t_2)\in U, \quad t\in I,$$

and

$$\mathscr{B}_{n\sigma_1\sigma_2^2\sigma_2\sigma_1}(t,t_1,t_2) = \begin{pmatrix} b_{11n\sigma_1\sigma_2^2\sigma_2\sigma_1}(t,t_1,t_2) & b_{12n\sigma_1\sigma_2^2\sigma_2\sigma_1}(t,t_1,t_2) \\ b_{21n\sigma_1\sigma_2^2\sigma_2\sigma_1}(t,t_1,t_2) & b_{22n\sigma_1\sigma_2^2\sigma_2\sigma_1}(t,t_1,t_2) \end{pmatrix}, \quad (t_1,t_2)\in U, \quad t\in I.$$

For $p,q\in T_{a\sigma_1\sigma_2^2}S$, define

$$\phi^1_{\sigma_1\sigma_2^2\sigma_2\sigma_1}(p,q) = p^T\mathscr{B}_{n\sigma_1\sigma_2^2\sigma_2\sigma_1}q.$$

Theorem 4.19 (Variant of the Meusnier Theorem) *Let σ_2 be Δ_2-differentiable, S be a σ_2-regular surface with a local σ_2-parameterization (U,f) such that f is σ_2-completely delta differentiable, $f_{t_1}^{\Delta_1\sigma_2}$ be σ_2-completely delta differentiable and $f_{t_2}^{\Delta_2}$ be σ_1-completely delta differentiable on U. Then*

$$k_{n\sigma_1\sigma_2^2}(t) = \phi^1_{\sigma_1\sigma_2^2\sigma_2\sigma_1}\left(\frac{g^\Delta(t)}{\|g^\Delta(t)\|}, \frac{g^\Delta(t)}{\|g^\Delta(t)\|}\right), \quad t\in I.$$

Definition 4.24 Let σ_2 be Δ_2-differentiable on $\mathbb{T}_2$, f be σ_1-completely delta differentiable on U, $f_{t_1}^{\Delta_1\sigma_2}$ and $f_{t_2}^{\Delta_2}$ be σ_2-completely delta differentiable on U. Let also,

$$b_{11n\sigma_1\sigma_2^2\sigma_2\sigma_2}(t,t_1,t_2) = f_{t_1}^{\Delta_1^2\sigma_2^2}(t_1,t_2)\cdot N_{\sigma_1\sigma_2^2}^{\sigma_1\sigma_2}(g(t)),$$

$$b_{12n\sigma_1\sigma_2^2\sigma_2\sigma_2}(t,t_1,t_2) = f_{t_1t_2}^{\Delta_1\Delta_2\sigma_2}(t_1,t_2)\cdot N_{\sigma_1\sigma_2^2}^{\sigma_1\sigma_2}(g(t)),$$

$$b_{21n\sigma_1\sigma_2^2\sigma_2\sigma_2}(t,t_1,t_2) = \left(1+\mu_2^{\Delta_2}(t_2)\right)f_{t_1t_2}^{\Delta_1\Delta_2\sigma_2}(t_1,t_2)\cdot N_{\sigma_1\sigma_2^2}^{\sigma_1\sigma_2}(g(t)),$$

$$b_{22n\sigma_1\sigma_2^2\sigma_2\sigma_2}(t,t_1,t_2) = f_{t_2}^{\Delta_2^2}(t_1,t_2)\cdot N_{\sigma_1\sigma_2^2}^{\sigma_1\sigma_2}(g(t)), \quad (t_1,t_2)\in U, \quad t\in I,$$

and

$$\mathscr{B}_{n\sigma_1\sigma_2^2\sigma_2\sigma_2}(t,t_1,t_2) = \begin{pmatrix} b_{11n\sigma_1\sigma_2^2\sigma_2\sigma_2}(t,t_1,t_2) & b_{12n\sigma_1\sigma_2^2\sigma_2\sigma_2}(t,t_1,t_2) \\ b_{21n\sigma_1\sigma_2^2\sigma_2\sigma_2}(t,t_1,t_2) & b_{22n\sigma_1\sigma_2^2\sigma_2\sigma_2}(t,t_1,t_2) \end{pmatrix}, \quad (t_1,t_2)\in U, \quad t\in I.$$

For $p,q\in T_{a\sigma_1\sigma_2^2}S$, define

$$\phi^1_{\sigma_1\sigma_2^2\sigma_1\sigma_1}(p,q) = p^T\mathscr{B}_{n\sigma_1\sigma_2^2\sigma_2\sigma_2}q.$$

Theorem 4.20 (Variant of the Meusnier Theorem) *Let σ_2 be Δ_2-differentiable, S be a σ_2-regular surface with a local σ_2-parameterization (U,f) such that f is σ_2-completely delta differentiable, $f_{t_1}^{\Delta_1\sigma_2}$ and $f_{t_2}^{\Delta_2}$ be σ_2-completely delta differentiable on U. Then*

$$k_{n\sigma_1\sigma_2^2}(t) = \phi^1_{\sigma_1\sigma_2^2\sigma_2\sigma_2}\left(\frac{g^\Delta(t)}{\|g^\Delta(t)\|}, \frac{g^\Delta(t)}{\|g^\Delta(t)\|}\right), \quad t \in I.$$

Corollary 4.13 *Let σ_2 be Δ_2-differentiable, S be a σ_2-regular surface with a local σ_2-parameterization (U,f) such that f is σ_2-completely delta differentiable, $f_{t_1}^{\Delta_1\sigma_2}$ and $f_{t_2}^{\Delta_2}$ be σ_1-completely delta differentiable. If two curves on S have a common point and they have the same tangent line at the common point, then the curves have the same $\sigma_1\sigma_2^2$-normal curvature at the common point.*

Corollary 4.14 *Let σ_2 be Δ_2-differentiable, S be a σ_2-regular surface with a local σ_2-parameterization (U,f) such that f is σ_2-completely delta differentiable, $f_{t_1}^{\Delta_1\sigma_2}$ be σ_1-completely delta differentiable and $f_{t_2}^{\Delta_2}$ be σ_2-completely delta differentiable. If two curves on S have a common point and they have the same tangent line at the common point, then the curves have the same $\sigma_1\sigma_2^2$-normal curvature at the common point.*

Corollary 4.15 *Let σ_2 be Δ_2-differentiable, S be a σ_2-regular surface with a local σ_2-parameterization (U,f) such that f is σ_2-completely delta differentiable, $f_{t_1}^{\Delta_1\sigma_2}$ be σ_2-completely delta differentiable and $f_{t_2}^{\Delta_2}$ be σ_1-completely delta differentiable. If two curves on S have a common point and they have the same tangent line at the common point, then the curves have the same $\sigma_1\sigma_2^2$-normal curvature at the common point.*

Corollary 4.16 *Let σ_2 be Δ_2-differentiable, S be a σ_2-regular surface with a local σ_2-parameterization (U,f) such that f is σ_2-completely delta differentiable, $f_{t_1}^{\Delta_1\sigma_2}$ and $f_{t_2}^{\Delta_2}$ be σ_2-completely delta differentiable. If two curves on S have a common point and they have the same tangent line at the common point, then the curves have the same $\sigma_1\sigma_2^2$-normal curvature at the common point.*

4.3 Asymptotic Directions. Asymptotic Lines on a Surface

Let S be a σ_1- or σ_2-or $\sigma_1^2\sigma_2$-or $\sigma_1\sigma_2^2$-oriented surface with a local σ_1- or σ_2-or $\sigma_1^2\sigma_2$-or $\sigma_1\sigma_2^2$ parameterization (U,f).

Definition 4.25 Suppose that S is a σ_1-regular surface. A non-vanishing vector $v \in T_{a\sigma_1}S$ is said to have σ_1-asymptotic direction if the σ_1-normal curvature in its direction vanishes. A σ_1- asymptotic line or σ_1-curve on S is a curve on S for which all tangent vectors have σ_1-asymptotic direction.

Theorem 4.21 *Let σ_1 be Δ_1-differentiable, f be σ_1-completely delta differentiable, $f_{t_1}^{\Delta_1}$ and $f_{t_2}^{\Delta_2\sigma_1}$ be σ_1-completely delta differentiable and $p \in S$. Then at this point we have σ_1-asymptotic direction if and only if*

$$(b_{12n\sigma_1\sigma_1\sigma_1} + b_{21n\sigma_1\sigma_1\sigma_1})^2 - 4b_{11n\sigma_1\sigma_1\sigma_1}b_{22n\sigma_1\sigma_1\sigma_1} \geq 0.$$

Proof. Let

$$h = \begin{pmatrix} h_1 \\ h_2 \end{pmatrix} \in T_{a\sigma_1} S$$

be a non-vanishing σ_1-tangent vector to the surface S at the point p. Without loss of generality, suppose that $h_2 \neq 0$. Then h has σ_1-asymptotic direction if and only if

$$\phi^1_{\sigma_1\sigma_1\sigma_1}(h,h) = 0$$

if and only if

$$(h_1, h_2) \begin{pmatrix} b_{11n\sigma_1\sigma_1\sigma_1} & b_{12n\sigma_1\sigma_1\sigma_1} \\ b_{21n\sigma_1\sigma_1\sigma_1} & b_{22n\sigma_1\sigma_1\sigma_1} \end{pmatrix} \begin{pmatrix} h_1 \\ h_2 \end{pmatrix} = 0$$

if and only if

$$(h_1, h_2) \begin{pmatrix} b_{11n\sigma_1\sigma_1\sigma_1} h_1 + b_{12n\sigma_1\sigma_1\sigma_1} h_2 \\ b_{21n\sigma_1\sigma_1\sigma_1} h_1 + b_{22n\sigma_1\sigma_1\sigma_1} h_2 \end{pmatrix} = 0$$

if and only if

$$b_{11n\sigma_1\sigma_1\sigma_1} h_1^2 + (b_{12n\sigma_1\sigma_1\sigma_1} + b_{21n\sigma_1\sigma_1\sigma_1}) h_1 h_2 + b_{22n\sigma_1\sigma_1\sigma_1} h_2^2 = 0$$

if and only if

$$b_{11n\sigma_1\sigma_1\sigma_1} \left(\frac{h_1}{h_2}\right)^2 + (b_{12n\sigma_1\sigma_1\sigma_1} + b_{21n\sigma_1\sigma_1\sigma_1}) \frac{h_1}{h_2} + b_{22n\sigma_1\sigma_1\sigma_1} = 0$$

if and only if

$$(b_{12n\sigma_1\sigma_1\sigma_1} + b_{21n\sigma_1\sigma_1\sigma_1})^2 - 4 b_{11n\sigma_1\sigma_1\sigma_1} b_{22n\sigma_1\sigma_1\sigma_1} \geq 0.$$

This completes the proof.

As above, one can prove the following theorems.

Theorem 4.22 *Let σ_1 be Δ_1-differentiable, f be σ_1-completely delta differentiable, $f_{t_1}^{\Delta_1}$ be σ_1-completely delta differentiable and $f_{t_2}^{\Delta_2\sigma_1}$ be σ_2-completely delta differentiable and $p \in S$. Then at this point we have σ_1-asymptotic direction if and only if*

$$(b_{12n\sigma_1\sigma_1\sigma_2} + b_{21n\sigma_1\sigma_1\sigma_2})^2 - 4 b_{11n\sigma_1\sigma_1\sigma_2} b_{22n\sigma_1\sigma_1\sigma_2} \geq 0.$$

Theorem 4.23 *Let σ_1 be Δ_1-differentiable, f be σ_1-completely delta differentiable, $f_{t_1}^{\Delta_1}$ be σ_2-completely delta differentiable and $f_{t_2}^{\Delta_2\sigma_1}$ be σ_1-completely delta differentiable and $p \in S$. Then at this point we have σ_1-asymptotic direction if and only if*

$$(b_{12n\sigma_1\sigma_2\sigma_1} + b_{21n\sigma_1\sigma_2\sigma_1})^2 - 4 b_{11n\sigma_1\sigma_2\sigma_1} b_{22n\sigma_1\sigma_2\sigma_1} \geq 0.$$

Theorem 4.24 *Let σ_1 be Δ_1-differentiable, f be σ_1-completely delta differentiable, $f_{t_1}^{\Delta_1}$ and $f_{t_2}^{\Delta_2\sigma_1}$ be σ_2-completely delta differentiable and $p \in S$. Then at this point we have σ_1-asymptotic direction if and only if*

$$(b_{12n\sigma_1\sigma_2\sigma_2} + b_{21n\sigma_1\sigma_2\sigma_2})^2 - 4 b_{11n\sigma_1\sigma_2\sigma_2} b_{22n\sigma_1\sigma_2\sigma_2} \geq 0.$$

Definition 4.26 Suppose that S be a σ_2-regular surface. A non-vanishing vector $v \in T_{a\sigma_2}S$ is said to have σ_2-asymptotic direction if the σ_2-normal curvature in its direction vanishes. A σ_2- asymptotic line or σ_2-curve on S is a curve on S for which all tangent vectors have σ_2-asymptotic direction.

Theorem 4.25 *Let σ_2 be Δ_2-differentiable, f be σ_2-completely delta differentiable, $f_{t_1}^{\Delta_1\sigma_2}$ and $f_{t_2}^{\Delta_2}$ be σ_1-completely delta differentiable and $p \in S$. Then at this point we have σ_2-asymptotic direction if and only if*

$$(b_{12n\sigma_2\sigma_1\sigma_1} + b_{21n\sigma_2\sigma_1\sigma_1})^2 - 4b_{11n\sigma_2\sigma_1\sigma_1}b_{22n\sigma_2\sigma_1\sigma_1} \geq 0.$$

Theorem 4.26 *Let σ_2 be Δ_2-differentiable, f be σ_2-completely delta differentiable, $f_{t_1}^{\Delta_1}$ be σ_1-completely delta differentiable and $f_{t_2}^{\Delta_2\sigma_1}$ be σ_2-completely delta differentiable and $p \in S$. Then at this point we have σ_2-asymptotic direction if and only if*

$$(b_{12n\sigma_2\sigma_1\sigma_2} + b_{21n\sigma_2\sigma_1\sigma_2})^2 - 4b_{11n\sigma_2\sigma_1\sigma_2}b_{22n\sigma_2\sigma_1\sigma_2} \geq 0.$$

Theorem 4.27 *Let σ_2 be Δ_2-differentiable, f be σ_2-completely delta differentiable, $f_{t_1}^{\Delta_1}$ be σ_2-completely delta differentiable, $f_{t_2}^{\Delta_2\sigma_1}$ be σ_1-completely delta differentiable and $p \in S$. Then at this point we have σ_2-asymptotic direction if and only if*

$$(b_{12n\sigma_2\sigma_2\sigma_1} + b_{21n\sigma_2\sigma_2\sigma_1})^2 - 4b_{11n\sigma_2\sigma_2\sigma_1}b_{22n\sigma_2\sigma_2\sigma_1} \geq 0.$$

Theorem 4.28 *Let σ_2 be Δ_2-differentiable, f be σ_2-completely delta differentiable, $f_{t_1}^{\Delta_1\sigma_2}$ and $f_{t_2}^{\Delta_2}$ be σ_2-completely delta differentiable and $p \in S$. Then at this point we have σ_2-asymptotic direction if and only if*

$$(b_{12n\sigma_2\sigma_2\sigma_2} + b_{21n\sigma_2\sigma_2\sigma_2})^2 - 4b_{11n\sigma_2\sigma_2\sigma_2}b_{22n\sigma_2\sigma_2\sigma_2} \geq 0.$$

Definition 4.27 Suppose that S is a σ_1-regular surface. A non-vanishing vector $v \in T_{a\sigma_1^2\sigma_2}S$ is said to have $\sigma_1^2\sigma_2$-asymptotic direction if the $\sigma_1^2\sigma_2$-normal curvature in its direction vanishes. A $\sigma_1^2\sigma_2$- asymptotic line or $\sigma_1^2\sigma_2$-curve on S is a curve on S for which all tangent vectors have $\sigma_1^2\sigma_2$-asymptotic direction.

Theorem 4.29 *Let σ_1 be Δ_1-differentiable, f be σ_1-completely delta differentiable, $f_{t_1}^{\Delta_1}$ and $f_{t_2}^{\Delta_2\sigma_1}$ be σ_1-completely delta differentiable and $p \in S$. Then at this point we have $\sigma_1^2\sigma_2$-asymptotic direction if and only if*

$$\left(b_{12n\sigma_1^2\sigma_2\sigma_1\sigma_1} + b_{21n\sigma_1^2\sigma_2\sigma_1\sigma_1}\right)^2 - 4b_{11n\sigma_1^2\sigma_2\sigma_1\sigma_1}b_{22n\sigma_1^2\sigma_2\sigma_1\sigma_1} \geq 0.$$

Theorem 4.30 *Let σ_1 be Δ_1-differentiable, f be σ_1-completely delta differentiable, $f_{t_1}^{\Delta_1}$ be σ_1-completely delta differentiable, $f_{t_2}^{\Delta_2\sigma_1}$ be σ_2-completely delta differentiable and $p \in S$. Then at this point we have $\sigma_1^2\sigma_2$-asymptotic direction if and only if*

$$\left(b_{12n\sigma_1^2\sigma_2\sigma_1\sigma_2} + b_{21n\sigma_1^2\sigma_2\sigma_1\sigma_2}\right)^2 - 4b_{11n\sigma_1^2\sigma_2\sigma_1\sigma_2}b_{22n\sigma_1^2\sigma_2\sigma_1\sigma_2} \geq 0.$$

Theorem 4.31 *Let σ_1 be Δ_1-differentiable, f be σ_1-completely delta differentiable, $f_{t_1}^{\Delta_1}$ be σ_2-completely delta differentiable and $f_{t_2}^{\Delta_2\sigma_1}$ is σ_1-completely delta differentiable and $p \in S$. Then at this point we have $\sigma_1^2\sigma_2$-asymptotic direction if and only if*

$$\left(b_{12n\sigma_1^2\sigma_2\sigma_2\sigma_1} + b_{21n\sigma_1^2\sigma_2\sigma_2\sigma_1}\right)^2 - 4b_{11n\sigma_1^2\sigma_2\sigma_2\sigma_1} b_{22n\sigma_1^2\sigma_2\sigma_2\sigma_1} \geq 0.$$

Theorem 4.32 *Let σ_1 be Δ_1-differentiable, f be σ_1-completely delta differentiable, $f_{t_1}^{\Delta_1}$ and $f_{t_2}^{\Delta_2\sigma_1}$ be σ_2-completely delta differentiable and $p \in S$. Then at this point we have $\sigma_1^2\sigma_2$-asymptotic direction if and only if*

$$\left(b_{12n\sigma_1^2\sigma_2\sigma_2\sigma_2} + b_{21n\sigma_1^2\sigma_2\sigma_2\sigma_2}\right)^2 - 4b_{11n\sigma_1^2\sigma_2\sigma_2\sigma_2} b_{22n\sigma_1^2\sigma_2\sigma_2\sigma_2} \geq 0.$$

Definition 4.28 Suppose that S is a σ_2-regular surface. A non-vanishing vector $v \in T_{a\sigma_2}S$ is said to have σ_2-asymptotic direction if the σ_2-normal curvature in its direction vanishes. A $\sigma_1\sigma_2^2$- asymptotic line or $\sigma_1\sigma_2^2$-curve on S is a curve on S for which all tangent vectors have $\sigma_1\sigma_2^2$-asymptotic direction.

Theorem 4.33 *Let σ_2 be Δ_2-differentiable, f be σ_2-completely delta differentiable, $f_{t_1}^{\Delta_1\sigma_2}$ and $f_{t_2}^{\Delta_2}$ be σ_1-completely delta differentiable and $p \in S$. Then at this point we have $\sigma_1\sigma_2^2$-asymptotic direction if and only if*

$$\left(b_{12n\sigma_1\sigma_2^2\sigma_1\sigma_1} + b_{21n\sigma_1\sigma_2^2\sigma_1\sigma_1}\right)^2 - 4b_{11n\sigma_1\sigma_2^2\sigma_1\sigma_1} b_{22n\sigma_1\sigma_2^2\sigma_1\sigma_1} \geq 0.$$

Theorem 4.34 *Let σ_2 be Δ_2-differentiable, f be σ_2-completely delta differentiable, $f_{t_1}^{\Delta_1}$ be σ_1-completely delta differentiable, $f_{t_2}^{\Delta_2\sigma_1}$ be σ_2-completely delta differentiable and $p \in S$. Then at this point we have $\sigma_1\sigma_2^2$-asymptotic direction if and only if*

$$\left(b_{12n\sigma_1\sigma_2^2\sigma_1\sigma_2} + b_{21n\sigma_1\sigma_2^2\sigma_1\sigma_2}\right)^2 - 4b_{11n\sigma_1\sigma_2^2\sigma_1\sigma_2} b_{22n\sigma_1\sigma_2^2\sigma_1\sigma_2} \geq 0.$$

Theorem 4.35 *Let σ_2 be Δ_2-differentiable, f be σ_2-completely delta differentiable, $f_{t_1}^{\Delta_1}$ be σ_2-completely delta differentiable, $f_{t_2}^{\Delta_2\sigma_1}$ be σ_1-completely delta differentiable and $p \in S$. Then at this point we have $\sigma_1\sigma_2^2$-asymptotic direction if and only if*

$$\left(b_{12n\sigma_1\sigma_2^2\sigma_2\sigma_1} + b_{21n\sigma_1\sigma_2^2\sigma_2\sigma_1}\right)^2 - 4b_{11n\sigma_1\sigma_2^2\sigma_2\sigma_1} b_{22n\sigma_1\sigma_2^2\sigma_2\sigma_1} \geq 0.$$

Theorem 4.36 *Let σ_2 be Δ_2-differentiable, f be σ_2-completely delta differentiable, $f_{t_1}^{\Delta_1\sigma_2}$ and $f_{t_2}^{\Delta_2}$ be σ_2-completely delta differentiable and $p \in S$. Then at this point we have $\sigma_1\sigma_2^2$-asymptotic direction if and only if*

$$\left(b_{12n\sigma_1\sigma_2^2\sigma_2\sigma_2} + b_{21n\sigma_1\sigma_2^2\sigma_2\sigma_2}\right)^2 - 4b_{11n\sigma_1\sigma_2^2\sigma_2\sigma_2} b_{22n\sigma_1\sigma_2^2\sigma_2\sigma_2} \geq 0.$$

Definition 4.29 A point $a \in S$ is called

1. σ_1- or σ_2- or $\sigma_1^2\sigma_2$- or $\sigma_1\sigma_2^2$-elliptic if the second σ_1- or σ_2- or $\sigma_1^2\sigma_2$- or $\sigma_1\sigma_2^2$-fundamental form is positively defined at a.

2. σ_1- or σ_2- or $\sigma_1^2\sigma_2$- or $\sigma_1\sigma_2^2$-hyperbolic if the second σ_1- or σ_2- or $\sigma_1^2\sigma_2$- or $\sigma_1\sigma_2^2$-fundamental form is negatively defined at a.
3. σ_1- or σ_2- or $\sigma_1^2\sigma_2$- or $\sigma_1\sigma_2^2$parabolic if the second σ_1- or σ_2- or $\sigma_1^2\sigma_2$- or $\sigma_1\sigma_2^2$-fundamental form is zero at a but at one of the coefficients is different to zero at a.
4. σ_1- or σ_2- or $\sigma_1^2\sigma_2$- or $\sigma_1\sigma_2^2$-flat if the second σ_1- or σ_2- or $\sigma_1^2\sigma_2$- or $\sigma_1\sigma_2^2$-fundamental form vanishes at a.

4.4 Principal Directions. Gaussian Curvatures. Mean Curvatures

Let S be a σ_1- or σ_2- or $\sigma_1^2\sigma_2$- or $\sigma_1\sigma_2^2$-oriented surface with a local σ_1- or σ_2- or $\sigma_1^2\sigma_2$- or $\sigma_1\sigma_2^2$-parameterization (U,f) and $a \in S$.

Definition 4.30 Let S be a σ_1-regular surface. Also, let σ_1 be Δ_1-differentiable and $f_{t_1}^{\Delta_1}$ and $f_{t_2}^{\Delta_2\sigma_1}$ be σ_1-completely delta differentiable. The σ_1-directions on the σ_1-tangent plane $T_{a\sigma_1}S$ at the point a corresponding to the eigenvectors of the matrix $\mathscr{B}_{n\sigma_1\sigma_1\sigma_1}$ are called $\sigma_1\sigma_1\sigma_1$-principal directions of the surface S at the point a. A curve g on the surface S is called a $\sigma_1\sigma_1\sigma_1$-principal line or $\sigma_1\sigma_1\sigma_1$-curvature line if its tangent at each point has $\sigma_1\sigma_1\sigma_1$-principal direction. A $\sigma_1\sigma_1\sigma_1$-principal curvature of the surface at the point a is the σ_1-normal curvature of S at a in a $\sigma_1\sigma_1\sigma_1$-principal direction.

Theorem 4.37 *Let S be a σ_1-regular surface. Also, let σ_1 be Δ_1-differentiable and $f_{t_1}^{\Delta_1}$ and $f_{t_2}^{\Delta_2\sigma_1}$ be σ_1-completely delta differentiable. Then the $\sigma_1\sigma_1\sigma_1$-principal curvatures of S are the eigenvalues of the matrix $\mathscr{B}_{n\sigma_1\sigma_1\sigma_1}$.*

Proof. Let e be an eigenvector of $\mathscr{B}_{n\sigma_1\sigma_1\sigma_1}$.

Then

$$\mathscr{B}_{n\sigma_1\sigma_1\sigma_1}e = \lambda e,$$

where λ is the eigenvalue corresponding to e.

Hence,

$$\begin{aligned} k_{n\sigma_1}(e) &= \phi^1_{\sigma_1\sigma_1\sigma_1}(e) \\ &= e^T \mathscr{B}_{n\sigma_1\sigma_1\sigma_1} e \\ &= \lambda e^T e \\ &= \lambda. \end{aligned}$$

This completes the proof.

Hereafter, with

$$k_{1\sigma_1\sigma_1\sigma_1} \quad \text{and} \quad k_{2\sigma_1\sigma_1\sigma_1}$$

we will denote the $\sigma_1\sigma_1\sigma_1$-principal curvatures of S and without loss of generality, suppose that

$$k_{1\sigma_1\sigma_1\sigma_1} \geq k_{2\sigma_1\sigma_1\sigma_1}.$$

Definition 4.31 Let S be σ_1-regular and σ_1 be Δ_1-differentiable. Also, let $f_{t_1}^{\Delta_1}$ and $f_{t_2}^{\Delta_2\sigma_1}$ be σ_1-completely delta differentiable. An orthonormal basis $\{e_{1\sigma_1\sigma_1\sigma_1}, e_{2\sigma_1\sigma_1\sigma_1}\}$ of the σ_1-tangent space $T_{a\sigma_1}S$ at the point a is called a basis of the $\sigma_1\sigma_1\sigma_1$-principal directions of the σ_1-tangent space $T_{a\sigma_1}S$ if the vectors of the basis have $\sigma_1\sigma_1\sigma_1$-principal directions. We have

$$\mathscr{B}_{n\sigma_1\sigma_1\sigma_1} e_{1\sigma_1\sigma_1\sigma_1} = k_{1\sigma_1\sigma_1\sigma_1} e_{1\sigma_1\sigma_1\sigma_1}$$

and

$$\mathscr{B}_{n\sigma_1\sigma_1\sigma_1} e_{2\sigma_1\sigma_1\sigma_1} = k_{2\sigma_1\sigma_1\sigma_1} e_{2\sigma_1\sigma_1\sigma_1}.$$

Theorem 4.38 (Euler Formula) *Let S be σ_1-regular, σ_1 be Δ_1-differentiable and $f_{t_1}^{\Delta_1}$ and $f_{t_2}^{\Delta_2\sigma_1}$ be σ_1-completely delta differentiable. Then the σ_1-normal curvature at the point a of the surface S in the direction of a vector e is given by the Euler formula*

$$k_{n\sigma_1}(e) = k_{1\sigma_1\sigma_1\sigma_1} \cos^2\theta + k_{2\sigma_1\sigma_1\sigma_1} \sin^2\theta,$$

where

$$\theta = \angle(e, e_{1\sigma_1\sigma_1\sigma_1}).$$

Proof. Without loss of generality, suppose that $\|e\| = 1$. Then

$$e = e_{1\sigma_1\sigma_1\sigma_1} \cos\theta + e_{2\sigma_1\sigma_1\sigma_1} \sin\theta$$

and

$$\begin{aligned}
k_{n\sigma_1}(e) &= e^T \mathscr{B}_{n\sigma_1\sigma_1\sigma_1} e \\
&= (e_{1\sigma_1\sigma_1\sigma_1}\cos\theta + e_{2\sigma_1\sigma_1\sigma_1}\sin\theta)\mathscr{B}_{n\sigma_1\sigma_1\sigma_1}(e_{1\sigma_1\sigma_1\sigma_1}\cos\theta + e_{2\sigma_1\sigma_1\sigma_1}\sin\theta) \\
&= (e_{1\sigma_1\sigma_1\sigma_1}\cos\theta + e_{2\sigma_1\sigma_1\sigma_1}\sin\theta)(\cos\theta\mathscr{B}_{n\sigma_1\sigma_1\sigma_1}(e_{1\sigma_1\sigma_1\sigma_1}) + \sin\theta\mathscr{B}_{n\sigma_1\sigma_1\sigma_1}(e_{2\sigma_1\sigma_1\sigma_1})) \\
&= (e_{1\sigma_1\sigma_1\sigma_1}\cos\theta + e_{2\sigma_1\sigma_1\sigma_1}\sin\theta)(k_{1\sigma_1\sigma_1\sigma_1}e_{1\sigma_1\sigma_1\sigma_1}\cos\theta + k_{2\sigma_1\sigma_1\sigma_1}e_{2\sigma_1\sigma_1\sigma_1}\sin\theta) \\
&= k_{1\sigma_1\sigma_1\sigma_1}\cos^2\theta + k_{2\sigma_1\sigma_1\sigma_1}\sin^2\theta.
\end{aligned}$$

This completes the proof.

Theorem 4.39 *Let S be σ_1-regular, σ_1 be Δ_1-differentiable and $f_{t_1}^{\Delta_1}$ and $f_{t_2}^{\Delta_2\sigma_1}$ be σ_1-completely delta differentiable. Then the $\sigma_1\sigma_1\sigma_1$-principal curvatures of the surface S at the point a are extremum values of the σ_1-normal curvature of the surface in the direction of a vector e, when the vector e rotates around the origin of the σ_1-tangent space of the surface S at the point a.*

Proof. By the Euler formula, we have

$$\begin{aligned} k_{n\sigma_1}(e) &= k_{1\sigma_1\sigma_1}\cos^2\theta + k_{2\sigma_1\sigma_1}\sin^2\theta \\ &= k_{1\sigma_1\sigma_1}\cos^2\theta + k_{2\sigma_1\sigma_1}(1-\cos^2\theta) \\ &= (k_{1\sigma_1\sigma_1} - k_{2\sigma_1\sigma_1})\cos^2\theta + k_{2\sigma_1\sigma_1}. \end{aligned}$$

Since

$$k_{1\sigma_1\sigma_1} \geq k_{2\sigma_1\sigma_1},$$

the maximum value of the σ_1-curvature is reached for $\theta = 0$ and in this case

$$\begin{aligned} k_{n\sigma_1}(e) &= k_{1\sigma_1\sigma_1} - k_{2\sigma_1\sigma_1} + k_{2\sigma_1\sigma_1} \\ &= k_{1\sigma_1\sigma_1}. \end{aligned}$$

This completes the proof.

Definition 4.32 Let S be σ_1-regular, σ_1 be Δ_1-differentiable and $f_{t_1}^{\Delta_1}$ and $f_{t_2}^{\Delta_2\sigma_1}$ be σ_1-completely delta differentiable.

1. The quantity
$$k_{t\sigma_1\sigma_1} = k_{1\sigma_1\sigma_1}k_{2\sigma_1\sigma_1}$$
is called the $\sigma_1\sigma_1\sigma_1$-total(Gaussian) curvature of the surface S.
2. The quantity
$$k_{m\sigma_1\sigma_1} = \frac{1}{2}(k_{1\sigma_1\sigma_1} + k_{2\sigma_1\sigma_1})$$
is called the $\sigma_1\sigma_1\sigma_1$-mean curvature of the surface S.

Exercise 4.5 Let $\mathbb{T}_1 = \mathbb{T}_2 = 3^{\mathbb{N}_0}$, $\mathbb{T}_{(1)} = \mathbb{T}_{(2)} = \mathbb{T}_{(3)} = \mathbb{R}$, S be a surface with a local parametrization (U,f), where $U \subset \mathbb{T}_1 \times \mathbb{T}_2$ and

$$f(t_1,t_2) = \left(1 + t_1^2 + t_1t_2 + t_2^2, t_1^2t_2^2, t_1^2 - t_2^4\right), \quad (t_1,t_2) \in U.$$

Find

1. $k_{r\sigma_1\sigma_1\sigma_1}$, $r \in \{1,2\}$,
2. $k_{t\sigma_1\sigma_1\sigma_1}$,
3. $k_{m\sigma_1\sigma_1\sigma_1}$.

Definition 4.33 Let S be a σ_1-regular surface. Also, let σ_1 be Δ_1-differentiable, $f_{t_1}^{\Delta_1}$ be σ_1-completely delta differentiable and $f_{t_2}^{\Delta_2\sigma_1}$ be σ_2-completely delta differentiable. The σ_1-directions on the σ_1-tangent plane $T_{a\sigma_1}S$ at the point a corresponding to the eigenvectors of the matrix $\mathscr{B}_{n\sigma_1\sigma_2}$ are called $\sigma_1\sigma_1\sigma_2$-principal directions of the surface S at the point a. A curve g on the surface S is called a $\sigma_1\sigma_1\sigma_2$-principal line or $\sigma_1\sigma_1\sigma_2$-curvature line if its tangent at each point has $\sigma_1\sigma_1\sigma_2$-principal direction. A $\sigma_1\sigma_1\sigma_2$-principal curvature of the surface at the point a is the σ_1-normal curvature of S at a in a $\sigma_1\sigma_1\sigma_2$-principal direction.

Theorem 4.40 *Let S be a σ_1-regular surface. Also, let σ_1 be Δ_1-differentiable, $f_{t_1}^{\Delta_1}$ be σ_1-completely delta differentiable and $f_{t_2}^{\Delta_2\sigma_1}$ be σ_2-completely delta differentiable. Then the $\sigma_1\sigma_1\sigma_2$-principal curvatures of S are the eigenvalues of the matrix $\mathscr{B}_{n\sigma_1\sigma_1\sigma_2}$.*

Hereafter, with

$$k_{1\sigma_1\sigma_1\sigma_2} \quad \text{and} \quad k_{2\sigma_1\sigma_1\sigma_2}$$

we will denote the $\sigma_1\sigma_1\sigma_2$-principal curvatures of S and without loss of generality, suppose that

$$k_{1\sigma_1\sigma_1\sigma_2} \geq k_{2\sigma_1\sigma_1\sigma_2}.$$

Definition 4.34 Let S be σ_1-regular and σ_1 be Δ_1-differentiable. Also, let $f_{t_1}^{\Delta_1}$ be σ_1-completely delta differentiable and $f_{t_2}^{\Delta_2\sigma_1}$ be σ_2-completely delta differentiable. An orthonormal basis $\{e_{1\sigma_1\sigma_1\sigma_2}, e_{2\sigma_1\sigma_1\sigma_2}\}$ of the σ_1-tangent space $T_{a\sigma_1}S$ at the point a is called a basis of the $\sigma_1\sigma_1\sigma_2$-principal directions of the σ_1-tangent space $T_{a\sigma_1}S$ if the vectors of the basis have $\sigma_1\sigma_1\sigma_2$-principal directions. We have

$$\mathscr{B}_{n\sigma_1\sigma_1\sigma_2}e_{1\sigma_1\sigma_1\sigma_2} = k_{1\sigma_1\sigma_1\sigma_2}e_{1\sigma_1\sigma_1\sigma_2}$$

and

$$\mathscr{B}_{n\sigma_1\sigma_1\sigma_2}e_{2\sigma_1\sigma_1\sigma_2} = k_{2\sigma_1\sigma_1\sigma_2}e_{2\sigma_1\sigma_1\sigma_2}.$$

Theorem 4.41 (Euler Formula) *Let S be σ_1-regular, σ_1 be Δ_1-differentiable, $f_{t_1}^{\Delta_1}$ be σ_1-completely delta differentiable and $f_{t_2}^{\Delta_2\sigma_1}$ be σ_2-completely delta differentiable. Then the σ_1-normal curvature at the point a of the surface S in the direction of a vector e is given by the Euler formula*

$$k_{n\sigma_1}(e) = k_{1\sigma_1\sigma_1\sigma_2}\cos^2\theta + k_{2\sigma_1\sigma_1\sigma_2}\sin^2\theta,$$

where

$$\theta = \angle(e, e_{1\sigma_1\sigma_1\sigma_2}).$$

Theorem 4.42 *Let S be σ_1-regular, σ_1 be Δ_1-differentiable, $f_{t_1}^{\Delta_1}$ be σ_1-completely delta differentiable and $f_{t_2}^{\Delta_2\sigma_1}$ be σ_2-completely delta differentiable. Then the $\sigma_1\sigma_1\sigma_2$-principal curvatures of the surface S at the point a are extremum values of the σ_1-normal curvature of the surface in the direction of a vector e when the vector e rotates around the origin of the σ_1-tangent space of the surface S at the point a.*

Definition 4.35 Let S be σ_1-regular, σ_1 be Δ_1-differentiable, $f_{t_1}^{\Delta_1}$ be σ_1-completely delta differentiable and $f_{t_2}^{\Delta_2\sigma_1}$ be σ_2-completely delta differentiable.

1. The quantity

$$k_{t\sigma_1\sigma_1\sigma_2} = k_{1\sigma_1\sigma_1\sigma_2}k_{2\sigma_1\sigma_1\sigma_2}$$

is called the $\sigma_1\sigma_1\sigma_2$-total(Gaussian) curvature of the surface S.

2. The quantity

$$k_{m\sigma_1\sigma_1\sigma_2} = \frac{1}{2}(k_{1\sigma_1\sigma_1\sigma_2} + k_{2\sigma_1\sigma_1\sigma_2})$$

is called the $\sigma_1\sigma_1\sigma_2$-mean curvature of the surface S.

Definition 4.36 Let S be a σ_1-regular surface. Also, let σ_1 be Δ_1-differentiable, $f_{t_1}^{\Delta_1}$ be σ_2-completely delta differentiable and $f_{t_2}^{\Delta_2\sigma_1}$ be σ_1-completely delta differentiable. The σ_1-directions on the σ_1-tangent plane $T_{a\sigma_1}S$ at the point a corresponding to the eigenvectors of the matrix $\mathscr{B}_{n\sigma_1\sigma_2\sigma_1}$ are called $\sigma_1\sigma_2\sigma_1$-principal directions of the surface S at the point a. A curve g on the surface S is called a $\sigma_1\sigma_2\sigma_1$-principal line or $\sigma_1\sigma_2\sigma_1$-curvature line if its tangent at each point has $\sigma_1\sigma_2\sigma_1$-principal direction. A $\sigma_1\sigma_2\sigma_1$-principal curvature of the surface at the point a is the σ_1-normal curvature of S at a in a $\sigma_1\sigma_2\sigma_1$-principal direction.

Theorem 4.43 *Let S be a σ_1-regular surface. Also, let σ_1 be Δ_1-differentiable, $f_{t_1}^{\Delta_1}$ be σ_2-completely delta differentiable and $f_{t_2}^{\Delta_2\sigma_1}$ be σ_1-completely delta differentiable. Then the $\sigma_1\sigma_2\sigma_1$-principal curvatures of S are the eigenvalues of the matrix $\mathscr{B}_{n\sigma_1\sigma_2\sigma_1}$.*

Hereafter, with

$$k_{1\sigma_1\sigma_2\sigma_1} \quad \text{and} \quad k_{2\sigma_1\sigma_2\sigma_1}$$

we will denote the $\sigma_1\sigma_2\sigma_1$-principal curvatures of S and without loss of generality, suppose that

$$k_{1\sigma_1\sigma_2\sigma_1} \geq k_{2\sigma_1\sigma_2\sigma_1}.$$

Definition 4.37 Let S be σ_1-regular and σ_1 be Δ_1-differentiable. Also, let $f_{t_1}^{\Delta_1}$ be σ_2-completely delta differentiable and $f_{t_2}^{\Delta_2\sigma_1}$ be σ_1-completely delta differentiable. An orthonormal basis $\{e_{1\sigma_1\sigma_2\sigma_1}, e_{2\sigma_1\sigma_2\sigma_1}\}$ of the σ_1-tangent space $T_{a\sigma_1}S$ at the point a is called a basis of the $\sigma_1\sigma_2\sigma_1$-principal directions of the σ_1-tangent space $T_{a\sigma_1}S$ if the vectors of the basis have $\sigma_1\sigma_2\sigma_1$-principal directions. We have

$$\mathscr{B}_{n\sigma_1\sigma_2\sigma_1} e_{1\sigma_1\sigma_2\sigma_1} = k_{1\sigma_1\sigma_2\sigma_1} e_{1\sigma_1\sigma_2\sigma_1}$$

and

$$\mathscr{B}_{n\sigma_1\sigma_2\sigma_1} e_{2\sigma_1\sigma_2\sigma_1} = k_{2\sigma_1\sigma_2\sigma_1} e_{2\sigma_1\sigma_2\sigma_1}.$$

Theorem 4.44 (Euler Formula) *Let S be σ_1-regular, σ_1 be Δ_1-differentiable, $f_{t_1}^{\Delta_1}$ be σ_2-completely delta differentiable and $f_{t_2}^{\Delta_2\sigma_1}$ be σ_1-completely delta differentiable. Then the σ_1-normal curvature at the point a of the surface S in the direction of a vector e is given by the Euler formula*

$$k_{n\sigma_1}(e) = k_{1\sigma_1\sigma_2\sigma_1}\cos^2\theta + k_{2\sigma_1\sigma_2\sigma_1}\sin^2\theta,$$

where

$$\theta = \angle(e, e_{1\sigma_1\sigma_2\sigma_1}).$$

Theorem 4.45 *Let S be σ_1-regular, σ_1 be Δ_1-differentiable, $f_{t_1}^{\Delta_1}$ be σ_2-completely delta differentiable and $f_{t_2}^{\Delta_2\sigma_1}$ be σ_1-completely delta differentiable. Then the $\sigma_1\sigma_2\sigma_1$-principal curvatures of the surface S at the point a are extremum values of the σ_1-normal curvature of the surface in the direction of a vector e, when the vector e rotates around the origin of the σ_1-tangent space of the surface S at the point a.*

Definition 4.38 Let S be σ_1-regular, $f_{t_1}^{\Delta_1}$ be σ_2-completely delta differentiable and $f_{t_2}^{\Delta_2\sigma_1}$ be σ_1-completely delta differentiable.

1. The quantity
$$k_{t\sigma_1\sigma_2\sigma_1} = k_{1\sigma_1\sigma_2\sigma_1} k_{2\sigma_1\sigma_2\sigma_1}$$
is called the $\sigma_1\sigma_2\sigma_1$-total(Gaussian) curvature of the surface S.
2. The quantity
$$k_{m\sigma_1\sigma_2\sigma_1} = \frac{1}{2}\left(k_{1\sigma_1\sigma_2\sigma_1} + k_{2\sigma_1\sigma_2\sigma_1}\right)$$
is called the $\sigma_1\sigma_2\sigma_1$-mean curvature of the surface S.

Definition 4.39 Let S be a σ_1-regular surface. Also let, σ_1 be Δ_1-differentiable and $f_{t_1}^{\Delta_1}$ and $f_{t_2}^{\Delta_2\sigma_1}$ be σ_2-completely delta differentiable. The σ_1-directions on the σ_1-tangent plane $T_{a\sigma_1}S$ at the point a corresponding to the eigenvectors of the matrix $\mathscr{B}_{n\sigma_1\sigma_2\sigma_2}$ are called $\sigma_1\sigma_2\sigma_2$-principal directions of the surface S at the point a. A curve g on the surface S is called a $\sigma_1\sigma_2\sigma_2$-principal line or $\sigma_1\sigma_2\sigma_2$-curvature line if its tangent at each point has $\sigma_1\sigma_2\sigma_2$-principal direction. A $\sigma_1\sigma_2\sigma_2$-principal curvature of the surface at the point a is the σ_1-normal curvature of S at a in a $\sigma_1\sigma_2\sigma_2$-principal direction.

Theorem 4.46 *Let S be a σ_1-regular surface. Also, let σ_1 be Δ_1-differentiable and $f_{t_1}^{\Delta_1}$ and $f_{t_2}^{\Delta_2\sigma_1}$ be σ_2-completely delta differentiable. Then the $\sigma_1\sigma_2\sigma_2$-principal curvatures of S are the eigenvalues of the matrix $\mathscr{B}_{n\sigma_1\sigma_2\sigma_2}$.*

Hereafter, with
$$k_{1\sigma_1\sigma_2\sigma_2} \quad \text{and} \quad k_{2\sigma_1\sigma_2\sigma_2}$$
we will denote the $\sigma_1\sigma_2\sigma_2$-principal curvatures of S and without loss of generality, suppose that
$$k_{1\sigma_1\sigma_2\sigma_2} \geq k_{2\sigma_1\sigma_2\sigma_2}.$$

Definition 4.40 Let S be σ_1-regular, σ_1 be Δ_1-differentiable. Also, let σ_1 be Δ_1-differentiable and $f_{t_1}^{\Delta_1}$ and $f_{t_2}^{\Delta_2\sigma_1}$ be σ_2-completely delta differentiable. An orthonormal basis $\{e_{1\sigma_1\sigma_2\sigma_2}, e_{2\sigma_1\sigma_2\sigma_2}\}$ of the σ_1-tangent space $T_{a\sigma_1}S$ at the point a is called a basis of the $\sigma_1\sigma_2\sigma_2$-principal directions of the σ_1-tangent space $T_{a\sigma_1}S$ if the vectors of the basis have $\sigma_1\sigma_2\sigma_2$-principal directions. We have
$$\mathscr{B}_{n\sigma_1\sigma_2\sigma_2} e_{1\sigma_1\sigma_2\sigma_2} = k_{1\sigma_1\sigma_2\sigma_2} e_{1\sigma_1\sigma_2\sigma_2}$$

and

$$\mathscr{B}_{n\sigma_1\sigma_2\sigma_2} e_{2\sigma_1\sigma_2\sigma_2} = k_{2\sigma_1\sigma_2\sigma_2} e_{2\sigma_1\sigma_2\sigma_2}.$$

Theorem 4.47 (Euler Formula) *Let S be σ_1-regular, σ_1 be Δ_1-differentiable and $f_{t_1}^{\Delta_1}$ and $f_{t_2}^{\Delta_2\sigma_1}$ be σ_2-completely delta differentiable. Then the σ_1-normal curvature at the point a of the surface S in the direction of a vector e is given by the Euler formula*

$$k_{n\sigma_1}(e) = k_{1\sigma_1\sigma_2\sigma_2} \cos^2\theta + k_{2\sigma_1\sigma_2\sigma_2} \sin^2\theta,$$

where

$$\theta = \angle(e, e_{1\sigma_1\sigma_2\sigma_2}).$$

Theorem 4.48 *Let S be σ_1-regular, σ_1 be Δ_1-differentiable and $f_{t_1}^{\Delta_1}$ and $f_{t_2}^{\Delta_2\sigma_1}$ be σ_2-completely delta differentiable. Then the $\sigma_1\sigma_2\sigma_2$-principal curvatures of the surface S at the point a are extremum values of the σ_1-normal curvature of the surface in the direction of a vector e, when the vector e rotates around the origin of the σ_1-tangent space of the surface S at the point a.*

Definition 4.41 Let S be σ_1-regular, σ_1 be Δ_1-differentiable and $f_{t_1}^{\Delta_1}$ and $f_{t_2}^{\Delta_2\sigma_1}$ be σ_2-completely delta differentiable.

1. The quantity

$$k_{t\sigma_1\sigma_2\sigma_2} = k_{1\sigma_1\sigma_2\sigma_2} k_{2\sigma_1\sigma_2\sigma_2}$$

   is called the $\sigma_1\sigma_2\sigma_2$-total(Gaussian) curvature of the surface S.
2. The quantity

$$k_{m\sigma_1\sigma_2\sigma_2} = \frac{1}{2}\left(k_{1\sigma_1\sigma_2\sigma_2} + k_{2\sigma_1\sigma_2\sigma_2}\right)$$

   is called the $\sigma_1\sigma_2\sigma_2$-mean curvature of the surface S.

Definition 4.42 Let S be a σ_2-regular surface. Also, let σ_2 be Δ_2-differentiable and $f_{t_1}^{\Delta_1\sigma_2}$ and $f_{t_2}^{\Delta_2}$ be σ_1-completely delta differentiable. The σ_2-directions on the σ_2-tangent plane $T_{a\sigma_2}S$ at the point a corresponding to the eigenvectors of the matrix $\mathscr{B}_{n\sigma_2\sigma_1\sigma_1}$ are called $\sigma_2\sigma_1\sigma_1$-principal directions of the surface S at the point a. A curve g on the surface S is called a $\sigma_2\sigma_1\sigma_1$-principal line or $\sigma_2\sigma_1\sigma_1$-curvature line if its tangent at each point has $\sigma_2\sigma_1\sigma_1$-principal direction. A $\sigma_2\sigma_1\sigma_1$-principal curvature of the surface at the point a is the σ_2-normal curvature of S at a in a $\sigma_2\sigma_1\sigma_1$-principal direction.

Theorem 4.49 *Let S be a σ_2-regular surface. Also, let σ_2 be Δ_2-differentiable and $f_{t_1}^{\Delta_1\sigma_2}$ and $f_{t_2}^{\Delta_2}$ be σ_1-completely delta differentiable. Then the $\sigma_2\sigma_1\sigma_1$-principal curvatures of S are the eigenvalues of the matrix $\mathscr{B}_{n\sigma_2\sigma_1\sigma_1}$.*

Hereafter, with

$$k_{1\sigma_2\sigma_1\sigma_1} \quad \text{and} \quad k_{2\sigma_2\sigma_1\sigma_1}$$

we will denote the $\sigma_2\sigma_1\sigma_1$-principal curvatures of S and without loss of generality, suppose that

$$k_{1\sigma_2\sigma_1\sigma_1} \geq k_{2\sigma_2\sigma_1\sigma_1}.$$

Definition 4.43 Let S be σ_2-regular and σ_2 be Δ_2-differentiable. Also, let $f_{t_1}^{\Delta_1\sigma_2}$ and $f_{t_2}^{\Delta_2}$ be σ_1-completely delta differentiable. An orthonormal basis $\{e_{1\sigma_2\sigma_1\sigma_1}, e_{2\sigma_2\sigma_1\sigma_1}\}$ of the σ_2-tangent space $T_{a\sigma_2}S$ at the point a is called a basis of the $\sigma_2\sigma_1\sigma_1$-principal directions of the σ_2-tangent space $T_{a\sigma_2}S$ if the vectors of the basis have $\sigma_2\sigma_1\sigma_1$-principal directions. We have

$$\mathscr{B}_{n\sigma_2\sigma_1\sigma_1} e_{1\sigma_2\sigma_1\sigma_1} = k_{1\sigma_2\sigma_1\sigma_1} e_{1\sigma_2\sigma_1\sigma_1}$$

and

$$\mathscr{B}_{n\sigma_2\sigma_1\sigma_1} e_{2\sigma_2\sigma_1\sigma_1} = k_{2\sigma_2\sigma_1\sigma_1} e_{2\sigma_2\sigma_1\sigma_1}.$$

Theorem 4.50 (Euler Formula) *Let S be σ_2-regular, σ_2 be Δ_2-differentiable and $f_{t_1}^{\Delta_1\sigma_2}$ and $f_{t_2}^{\Delta_2}$ be σ_1-completely delta differentiable. Then the σ_2-normal curvature at the point a of the surface S in the direction of a vector e is given by the Euler formula*

$$k_{n\sigma_2}(e) = k_{1\sigma_2\sigma_1\sigma_1}\cos^2\theta + k_{2\sigma_2\sigma_1\sigma_1}\sin^2\theta,$$

where

$$\theta = \angle(e, e_{1\sigma_2\sigma_1\sigma_1}).$$

Theorem 4.51 *Let S be σ_2-regular, σ_2 be Δ_2-differentiable and $f_{t_1}^{\Delta_1\sigma_2}$ and $f_{t_2}^{\Delta_2}$ be σ_1-completely delta differentiable. Then the $\sigma_2\sigma_1\sigma_1$-principal curvatures of the surface S at the point a are extremum values of the σ_2-normal curvature of the surface in the direction of a vector e, when the vector e rotates around the origin of the σ_2-tangent space of the surface S at the point a.*

Definition 4.44 Let S be σ_2-regular, σ_2 be Δ_2-differentiable and $f_{t_1}^{\Delta_1\sigma_2}$ and $f_{t_2}^{\Delta_2}$ be σ_1-completely delta differentiable.

1. The quantity

$$k_{t\sigma_2\sigma_1\sigma_1} = k_{1\sigma_2\sigma_1\sigma_1} k_{2\sigma_2\sigma_1\sigma_1}$$

   is called the $\sigma_2\sigma_1\sigma_1$-total(Gaussian) curvature of the surface S.

2. The quantity

$$k_{m\sigma_2\sigma_1\sigma_1} = \frac{1}{2}\left(k_{1\sigma_2\sigma_1\sigma_1} + k_{2\sigma_2\sigma_1\sigma_1}\right)$$

   is called the $\sigma_2\sigma_1\sigma_1$-mean curvature of the surface S.

Definition 4.45 Let S be a σ_2-regular surface. Also let, σ_2 be Δ_2-differentiable, $f_{t_1}^{\Delta_1\sigma_2}$ be σ_1-delta differentiable and $f_{t_2}^{\Delta_2}$ be σ_2-completely delta differentiable. The σ_2-directions on the σ_2-tangent plane $T_{a\sigma_2}S$ at the point a corresponding to the eigenvectors of the matrix $\mathscr{B}_{n\sigma_2\sigma_1\sigma_2}$ are called $\sigma_2\sigma_1\sigma_2$-principal directions of the surface S at the point a. A curve g on the surface S is called a $\sigma_2\sigma_1\sigma_2$-principal line or $\sigma_2\sigma_1\sigma_2$-curvature line if its tangent at each point has $\sigma_2\sigma_1\sigma_2$-principal direction. A $\sigma_2\sigma_1\sigma_2$-principal curvature of the surface at the point a is the σ_2-normal curvature of S at a in a $\sigma_2\sigma_1\sigma_2$-principal direction.

Theorem 4.52 *Let S be a σ_2-regular surface. Also, let σ_2 be Δ_2-differentiable, $f_{t_1}^{\Delta_1\sigma_2}$ be σ_1-delta differentiable and $f_{t_2}^{\Delta_2}$ be σ_2-completely delta differentiable. Then the $\sigma_2\sigma_1\sigma_2$-principal curvatures of S are the eigenvalues of the matrix $\mathscr{B}_{n\sigma_2\sigma_1\sigma_2}$.*

Hereafter, with

$$k_{1\sigma_2\sigma_1\sigma_2} \quad \text{and} \quad k_{2\sigma_2\sigma_1\sigma_2}$$

we will denote the $\sigma_2\sigma_1\sigma_2$-principal curvatures of S and without loss of generality, suppose that

$$k_{1\sigma_2\sigma_1\sigma_2} \geq k_{2\sigma_2\sigma_1\sigma_2}.$$

Definition 4.46 Let S be σ_2-regular and σ_2 be Δ_2-differentiable. Also, let $f_{t_1}^{\Delta_1\sigma_2}$ be σ_1-delta differentiable and $f_{t_2}^{\Delta_2}$ be σ_2-completely delta differentiable. An orthonormal basis $\{e_{1\sigma_2\sigma_1\sigma_2}, e_{2\sigma_2\sigma_1\sigma_2}\}$ of the σ_2-tangent space $T_{a\sigma_2}S$ at the point a is called a basis of the $\sigma_2\sigma_1\sigma_2$-principal directions of the σ_2-tangent space $T_{a\sigma_2}S$ if the vectors of the basis have $\sigma_2\sigma_1\sigma_2$-principal directions. We have

$$\mathscr{B}_{n\sigma_2\sigma_1\sigma_2} e_{1\sigma_2\sigma_1\sigma_2} = k_{1\sigma_2\sigma_1\sigma_2} e_{1\sigma_2\sigma_1\sigma_2}$$

and

$$\mathscr{B}_{n\sigma_2\sigma_1\sigma_2} e_{2\sigma_2\sigma_1\sigma_2} = k_{2\sigma_2\sigma_1\sigma_2} e_{2\sigma_2\sigma_1\sigma_2}.$$

Theorem 4.53 (Euler Formula) *Let S be σ_2-regular, σ_2 be Δ_2-differentiable, $f_{t_1}^{\Delta_1\sigma_2}$ be σ_1-delta differentiable and $f_{t_2}^{\Delta_2}$ be σ_2-completely delta differentiable. Then the σ_2-normal curvature at the point a of the surface S in the direction of a vector e is given by the Euler formula*

$$k_{n\sigma_2}(e) = k_{1\sigma_2\sigma_1\sigma_2} \cos^2\theta + k_{2\sigma_2\sigma_1\sigma_2} \sin^2\theta,$$

where

$$\theta = \angle(e, e_{1\sigma_2\sigma_1\sigma_2}).$$

Theorem 4.54 *Let S be σ_2-regular, σ_2 be Δ_2-differentiable, $f_{t_1}^{\Delta_1\sigma_2}$ be σ_1-delta differentiable and $f_{t_2}^{\Delta_2}$ be σ_2-completely delta differentiable. Then the $\sigma_2\sigma_1\sigma_2$-principal curvatures of the surface S at the point a are extremum values of the σ_2-normal curvature of the surface in the direction of a vector e, when the vector e rotates around the origin of the σ_2-tangent space of the surface S at the point a.*

Definition 4.47 Let S be σ_2-regular, σ_2 be Δ_2-differentiable, $f_{t_1}^{\Delta_1\sigma_2}$ be σ_1-delta differentiable and $f_{t_2}^{\Delta_2}$ be σ_2-completely delta differentiable.

1. The quantity
$$k_{t\sigma_2\sigma_1\sigma_2} = k_{1\sigma_2\sigma_1\sigma_2} k_{2\sigma_2\sigma_1\sigma_2}$$
is called the $\sigma_2\sigma_1\sigma_2$-total(Gaussian) curvature of the surface S.
2. The quantity
$$k_{m\sigma_2\sigma_1\sigma_2} = \frac{1}{2}\left(k_{1\sigma_2\sigma_1\sigma_2} + k_{2\sigma_2\sigma_1\sigma_2}\right)$$
is called the $\sigma_2\sigma_1\sigma_2$-mean curvature of the surface S.

Definition 4.48 Let S be a σ_2-regular surface. Also, let σ_2 be Δ_2-differentiable, $f_{t_1}^{\Delta_1\sigma_2}$ be σ_2-delta differentiable and $f_{t_2}^{\Delta_2}$ be σ_1-completely delta differentiable. The σ_2-directions on the σ_2-tangent plane $T_{a\sigma_2}S$ at the point a corresponding to the eigenvectors of the matrix $\mathscr{B}_{n\sigma_2\sigma_2\sigma_1}$ are called $\sigma_2\sigma_2\sigma_1$-principal directions of the surface S at the point a. A curve g on the surface S is called a $\sigma_2\sigma_2\sigma_1$-principal line or $\sigma_2\sigma_2\sigma_1$-curvature line if its tangent at each point has $\sigma_2\sigma_2\sigma_1$-principal direction. A $\sigma_2\sigma_2\sigma_1$-principal curvature of the surface at the point a is the σ_2-normal curvature of S at a in a $\sigma_2\sigma_2\sigma_1$-principal direction.

Theorem 4.55 *Let S be a σ_2-regular surface. Also, let σ_2 be Δ_2-differentiable, $f_{t_1}^{\Delta_1\sigma_2}$ be σ_2-delta differentiable and $f_{t_2}^{\Delta_2}$ be σ_1-completely delta differentiable. Then the $\sigma_2\sigma_2\sigma_1$-principal curvatures of S are the eigenvalues of the matrix $\mathscr{B}_{n\sigma_2\sigma_2\sigma_1}$.*

Hereafter, with
$$k_{1\sigma_2\sigma_2\sigma_1} \quad \text{and} \quad k_{2\sigma_2\sigma_2\sigma_1}$$
we will denote the $\sigma_2\sigma_2\sigma_1$-principal curvatures of S and without loss of generality, suppose that
$$k_{1\sigma_2\sigma_2\sigma_1} \geq k_{2\sigma_2\sigma_2\sigma_1}.$$

Definition 4.49 Let S be σ_2-regular and σ_2 be Δ_2-differentiable. Also, let $f_{t_1}^{\Delta_1\sigma_2}$ be σ_2-delta differentiable and $f_{t_2}^{\Delta_2}$ be σ_1-completely delta differentiable. An orthonormal basis $\{e_{1\sigma_2\sigma_2\sigma_1}, e_{2\sigma_2\sigma_2\sigma_1}\}$ of the σ_2-tangent space $T_{a\sigma_2}S$ at the point a is called a basis of the $\sigma_2\sigma_1\sigma_2$-principal directions of the σ_2-tangent space $T_{a\sigma_2}S$ if the vectors of the basis have $\sigma_2\sigma_2\sigma_1$-principal directions. We have
$$\mathscr{B}_{n\sigma_2\sigma_2\sigma_1} e_{1\sigma_2\sigma_2\sigma_1} = k_{1\sigma_2\sigma_2\sigma_1} e_{1\sigma_2\sigma_2\sigma_1}$$
and
$$\mathscr{B}_{n\sigma_2\sigma_2\sigma_1} e_{2\sigma_2\sigma_2\sigma_1} = k_{2\sigma_2\sigma_2\sigma_1} e_{2\sigma_2\sigma_2\sigma_1}.$$

Theorem 4.56 (Euler Formula) *Let S be σ_2-regular, σ_2 be Δ_2-differentiable, $f_{t_1}^{\Delta_1\sigma_2}$ be σ_2-delta differentiable and $f_{t_2}^{\Delta_2}$ be σ_1-completely delta differentiable. Then the σ_2-normal curvature at the point a of the surface S in the direction of a vector e is given by the Euler formula*

$$k_{n\sigma_2}(e) = k_{1\sigma_2\sigma_2\sigma_1}\cos^2\theta + k_{2\sigma_2\sigma_2\sigma_1}\sin^2\theta,$$

where

$$\theta = \angle(e, e_{1\sigma_2\sigma_2\sigma_1}).$$

Theorem 4.57 *Let S be σ_2-regular, σ_2 be Δ_2-differentiable, $f_{t_1}^{\Delta_1\sigma_2}$ be σ_2-delta differentiable and $f_{t_2}^{\Delta_2}$ is σ_1-completely delta differentiable. Then the $\sigma_2\sigma_2\sigma_1$-principal curvatures of the surface S at the point a are extremum values of the σ_2-normal curvature of the surface in the direction of a vector e, when the vector e rotates around the origin of the σ_2-tangent space of the surface S at the point a.*

Definition 4.50 Let S be σ_2-regular, σ_2 be Δ_2-differentiable, $f_{t_1}^{\Delta_1\sigma_2}$ be σ_2-delta differentiable and $f_{t_2}^{\Delta_2}$ be σ_1-completely delta differentiable.

1. The quantity
$$k_{t\sigma_2\sigma_2\sigma_1} = k_{1\sigma_2\sigma_2\sigma_1}k_{2\sigma_2\sigma_2\sigma_1}$$
is called the $\sigma_2\sigma_2\sigma_1$-total(Gaussian) curvature of the surface S.
2. The quantity
$$k_{m\sigma_2\sigma_2\sigma_1} = \frac{1}{2}(k_{1\sigma_2\sigma_2\sigma_1} + k_{2\sigma_2\sigma_2\sigma_1})$$
is called the $\sigma_2\sigma_2\sigma_1$-mean curvature of the surface S.

Definition 4.51 Let S be a σ_2-regular surface. Also, let σ_2 be Δ_2-differentiable and $f_{t_1}^{\Delta_1\sigma_2}$ and $f_{t_2}^{\Delta_2}$ be σ_2-completely delta differentiable. The σ_2-directions on the σ_2-tangent plane $T_{a\sigma_2}S$ at the point a corresponding to the eigenvectors of the matrix $\mathscr{B}_{n\sigma_2\sigma_2\sigma_2}$ are called $\sigma_2\sigma_2\sigma_2$-principal directions of the surface S at the point a. A curve g on the surface S is called a $\sigma_2\sigma_2\sigma_2$-principal line or $\sigma_2\sigma_2\sigma_2$-curvature line if its tangent at each point has $\sigma_2\sigma_2\sigma_2$-principal direction. A $\sigma_2\sigma_2\sigma_2$-principal curvature of the surface at the point a is the σ_2-normal curvature of S at a in a $\sigma_2\sigma_2\sigma_2$-principal direction.

Theorem 4.58 *Let S be a σ_2-regular surface. Also, let σ_2 be Δ_2-differentiable and $f_{t_1}^{\Delta_1\sigma_2}$ and $f_{t_2}^{\Delta_2}$ be σ_2-completely delta differentiable. Then the $\sigma_2\sigma_2\sigma_2$-principal curvatures of S are the eigenvalues of the matrix $\mathscr{B}_{n\sigma_2\sigma_2\sigma_2}$.*

Hereafter, with

$$k_{1\sigma_2\sigma_2\sigma_2} \quad \text{and} \quad k_{2\sigma_2\sigma_2\sigma_2}$$

we will denote the $\sigma_2\sigma_2\sigma_2$-principal curvatures of S and without loss of generality, suppose that

$$k_{1\sigma_2\sigma_2\sigma_2} \geq k_{2\sigma_2\sigma_2\sigma_2}.$$

Definition 4.52 Let S be σ_2-regular and σ_2 be Δ_2-differentiable. Also, let $f_{t_1}^{\Delta_1\sigma_2}$ and $f_{t_2}^{\Delta_2}$ be σ_2-completely delta differentiable. An orthonormal basis $\{e_{1\sigma_2\sigma_2\sigma_2}, e_{2\sigma_2\sigma_2\sigma_2}\}$ of the σ_2-tangent space $T_{a\sigma_2}S$ at the point a is called a basis of the $\sigma_2\sigma_2\sigma_2$-principal directions of the σ_2-tangent space $T_{a\sigma_2}S$ if the vectors of the basis have $\sigma_2\sigma_2\sigma_2$-principal directions. We have

$$\mathscr{B}_{n\sigma_2\sigma_2\sigma_2}e_{1\sigma_2\sigma_2\sigma_2} = k_{1\sigma_2\sigma_2\sigma_2}e_{1\sigma_2\sigma_2\sigma_2}$$

and

$$\mathscr{B}_{n\sigma_2\sigma_2\sigma_2}e_{2\sigma_2\sigma_2\sigma_2} = k_{2\sigma_2\sigma_2\sigma_2}e_{2\sigma_2\sigma_2\sigma_2}.$$

Theorem 4.59 (Euler Formula) *Let S be σ_2-regular, σ_2 is Δ_2-differentiable and $f_{t_1}^{\Delta_1\sigma_2}$ and $f_{t_2}^{\Delta_2}$ are σ_2-completely delta differentiable. Then the σ_2-normal curvature at the point a of the surface S in the direction of a vector e is given by the Euler formula*

$$k_{n\sigma_2}(e) = k_{1\sigma_2\sigma_2\sigma_2}\cos^2\theta + k_{2\sigma_2\sigma_2\sigma_2}\sin^2\theta,$$

where

$$\theta = \angle(e, e_{1\sigma_2\sigma_2\sigma_2}).$$

Theorem 4.60 *Let S be σ_2-regular, σ_2 be Δ_2-differentiable and $f_{t_1}^{\Delta_1\sigma_2}$ and $f_{t_2}^{\Delta_2}$ be σ_2-completely delta differentiable. Then the $\sigma_2\sigma_2\sigma_2$-principal curvatures of the surface S at the point a are extremum values of the σ_2-normal curvature of the surface in the direction of a vector e, when the vector e rotates around the origin of the σ_2-tangent space of the surface S at the point a.*

Definition 4.53 Let S be σ_2-regular, σ_2 be Δ_2-differentiable and $f_{t_1}^{\Delta_1\sigma_2}$ and $f_{t_2}^{\Delta_2}$ be σ_2-completely delta differentiable.

1. The quantity

$$k_{t\sigma_2\sigma_2\sigma_2} = k_{1\sigma_2\sigma_2\sigma_2}k_{2\sigma_2\sigma_2\sigma_2}$$

   is called the $\sigma_2\sigma_2\sigma_2$-total(Gaussian) curvature of the surface S.

2. The quantity

$$k_{m\sigma_2\sigma_2\sigma_2} = \frac{1}{2}(k_{1\sigma_2\sigma_2\sigma_2} + k_{2\sigma_2\sigma_2\sigma_2})$$

   is called the $\sigma_2\sigma_2\sigma_2$-mean curvature of the surface S.

Definition 4.54 Let S be a $\sigma_1^2\sigma_2$-regular surface. Also, let σ_1 be Δ_1-differentiable and $f_{t_1}^{\Delta_1}$ and $f_{t_2}^{\Delta_2\sigma_1}$ be σ_1-completely delta differentiable. The σ_1-directions on the $\sigma_1^2\sigma_2$-tangent plane $T_{a\sigma_1^2\sigma_2}S$ at the point a corresponding to the eigenvectors of the matrix $\mathscr{B}_{n\sigma_1^2\sigma_2\sigma_1\sigma_1}$ are called $\sigma_1^2\sigma_2\sigma_1\sigma_1$-principal directions of the surface S at the point a. A curve g on the surface S is called a $\sigma_1^2\sigma_2\sigma_1\sigma_1$-principal line or $\sigma_1^2\sigma_2\sigma_1\sigma_1$-curvature line if its tangent at each point has $\sigma_1^2\sigma_2\sigma_1\sigma_1$-principal direction. A $\sigma_1^2\sigma_2\sigma_1\sigma_1$-principal curvature of the surface at the point a is the σ_1-normal curvature of S at a in a $\sigma_1^2\sigma_2\sigma_1\sigma_1$-principal direction.

Theorem 4.61 *Let S be a $\sigma_1^2\sigma_2$-regular surface. Also, let σ_1 be Δ_1-differentiable and $f_{t_1}^{\Delta_1}$ and $f_{t_2}^{\Delta_2\sigma_1}$ be σ_1-completely delta differentiable. Then the $\sigma_1^2\sigma_2\sigma_1\sigma_1$-principal curvatures of S are the eigenvalues of the matrix $\mathscr{B}_{n\sigma_1^2\sigma_2\sigma_1\sigma_1}$.*

Hereafter, with

$$k_{1\sigma_1^2\sigma_2\sigma_1\sigma_1} \quad \text{and} \quad k_{2\sigma_1^2\sigma_2\sigma_1\sigma_1}$$

we will denote the $\sigma_1^2\sigma_2\sigma_1\sigma_1$-principal curvatures of S and without loss of generality, suppose that

$$k_{1\sigma_1^2\sigma_2\sigma_1\sigma_1} \geq k_{2\sigma_1^2\sigma_2\sigma_1\sigma_1}.$$

Definition 4.55 Let S be $\sigma_1^2\sigma_2$-regular and be Δ_1-differentiable. Let also, σ_1 be Δ_1-differentiable and $f_{t_1}^{\Delta_1}$ and $f_{t_2}^{\Delta_2\sigma_1}$ be σ_1-completely delta differentiable. Also, let σ_1 be Δ_1-differentiable and $f_{t_1}^{\Delta_1}$ and $f_{t_2}^{\Delta_2\sigma_1}$ be σ_1-completely delta differentiable. An orthonormal basis $\{e_{1\sigma_1^2\sigma_2\sigma_1\sigma_1}, e_{2\sigma_1^2\sigma_2\sigma_1\sigma_1}\}$ of the $\sigma_1^2\sigma_2$-tangent space $T_{a\sigma_1^2\sigma_2}S$ at the point a is called a basis of the $\sigma_1^2\sigma_2\sigma_1\sigma_1$-principal directions of the $\sigma_1^2\sigma_2$-tangent space $T_{a\sigma_1^2\sigma_2}S$ if the vectors of the basis have $\sigma_1^2\sigma_2\sigma_1\sigma_1$-principal directions. We have

$$\mathscr{B}_{n\sigma_1^2\sigma_2\sigma_1\sigma_1} e_{1\sigma_1^2\sigma_2\sigma_1\sigma_1} = k_{1\sigma_1^2\sigma_2\sigma_1\sigma_1} e_{1\sigma_1^2\sigma_2\sigma_1\sigma_1}$$

and

$$\mathscr{B}_{n\sigma_1^2\sigma_2\sigma_1\sigma_1} e_{2\sigma_1^2\sigma_2\sigma_1\sigma_1} = k_{2\sigma_1^2\sigma_2\sigma_1\sigma_1} e_{2\sigma_1^2\sigma_2\sigma_1\sigma_1}.$$

Theorem 4.62 **(Euler Formula)** *Let S be $\sigma_1^2\sigma_2$-regular, σ_1 be Δ_1-differentiable and $f_{t_1}^{\Delta_1}$ and $f_{t_2}^{\Delta_2\sigma_1}$ be σ_1-completely delta differentiable. Then the σ_1-normal curvature at the point a of the surface S in the direction of a vector e is given by the Euler formula*

$$k_{n\sigma_1}(e) = k_{1\sigma_1^2\sigma_2\sigma_1\sigma_1} \cos^2\theta + k_{2\sigma_1^2\sigma_2\sigma_1\sigma_1} \sin^2\theta,$$

where

$$\theta = \angle(e, e_{1\sigma_1^2\sigma_2\sigma_1\sigma_1}).$$

Theorem 4.63 *Let S be $\sigma_1^2\sigma_2$-regular, σ_1 be Δ_1-differentiable and $f_{t_1}^{\Delta_1}$ and $f_{t_2}^{\Delta_2\sigma_1}$ be σ_1-completely delta differentiable. Then the $\sigma_1^2\sigma_2\sigma_1\sigma_1$-principal curvatures of the surface S at the point a are extremum values of the σ_1-normal curvature of the surface in the direction of a vector e, when the vector e rotates around the origin of the $\sigma_1^2\sigma_2$-tangent space of the surface S at the point a.*

Definition 4.56 Let S be $\sigma_1^2\sigma_2$-regular and $f_{t_1}^{\Delta_1}$ and $f_{t_2}^{\Delta_2\sigma_1}$ be σ_1-completely delta differentiable.

1. The quantity

$$k_{t\sigma_1^2\sigma_2\sigma_1\sigma_1} = k_{1\sigma_1^2\sigma_2\sigma_1\sigma_1} k_{2\sigma_1^2\sigma_2\sigma_1\sigma_1}$$

is called the $\sigma_1^2\sigma_2\sigma_1\sigma_1$-total(Gaussian) curvature of the surface S.

2. The quantity

$$k_{m\sigma_1^2\sigma_2\sigma_1\sigma_1} = \frac{1}{2}\left(k_{1\sigma_1^2\sigma_2\sigma_1\sigma_1} + k_{2\sigma_1^2\sigma_2\sigma_1\sigma_1}\right)$$

is called the $\sigma_1^2\sigma_2\sigma_1\sigma_1$-mean curvature of the surface S.

Definition 4.57 Let S be a $\sigma_1^2\sigma_2$-regular surface. Also, let σ_1 be Δ_1-differentiable, $f_{t_1}^{\Delta_1}$ be σ_1-completely delta differentiable and $f_{t_2}^{\Delta_2\sigma_1}$ be σ_2-completely delta differentiable. The σ_1-directions on the $\sigma_1^2\sigma_2$-tangent plane $T_{a\sigma_1^2\sigma_2}S$ at the point a corresponding to the eigenvectors of the matrix $\mathscr{B}_{n\sigma_1^2\sigma_2\sigma_1\sigma_2}$ are called $\sigma_1^2\sigma_2\sigma_1\sigma_2$-principal directions of the surface S at the point a. A curve g on the surface S is called a $\sigma_1^2\sigma_2\sigma_1\sigma_2$-principal line or $\sigma_1^2\sigma_2\sigma_1\sigma_2$-curvature line if its tangent at each point has $\sigma_1^2\sigma_2\sigma_1\sigma_2$-principal direction. A $\sigma_1^2\sigma_2\sigma_1\sigma_2$-principal curvature of the surface at the point a is the σ_1-normal curvature of S at a in a $\sigma_1^2\sigma_2\sigma_1\sigma_2$-principal direction.

Theorem 4.64 *Let S be a $\sigma_1^2\sigma_2$-regular surface. Also, let σ_1 be Δ_1-differentiable, $f_{t_1}^{\Delta_1}$ be σ_1-completely delta differentiable and $f_{t_2}^{\Delta_2\sigma_1}$ be σ_2-completely delta differentiable. Then the $\sigma_1^2\sigma_2\sigma_1\sigma_2$-principal curvatures of S are the eigenvalues of the matrix $\mathscr{B}_{n\sigma_1^2\sigma_2\sigma_1\sigma_2}$.*

Hereafter, with

$$k_{1\sigma_1^2\sigma_2\sigma_1\sigma_2} \quad \text{and} \quad k_{2\sigma_1^2\sigma_2\sigma_1\sigma_2}$$

we will denote the $\sigma_1^2\sigma_2\sigma_1\sigma_2$-principal curvatures of S and without loss of generality, suppose that

$$k_{1\sigma_1^2\sigma_2\sigma_1\sigma_2} \geq k_{2\sigma_1^2\sigma_2\sigma_1\sigma_2}.$$

Definition 4.58 Let S be $\sigma_1^2\sigma_2$-regular, σ_1 be Δ_1-differentiable. Also, let σ_1 be Δ_1-differentiable, $f_{t_1}^{\Delta_1}$ be σ_1-completely delta differentiable and $f_{t_2}^{\Delta_2\sigma_1}$ be σ_2-completely delta differentiable. An orthonormal basis $\{e_{1\sigma_1^2\sigma_2\sigma_1\sigma_2}, e_{2\sigma_1^2\sigma_2\sigma_1\sigma_2}\}$ of the $\sigma_1^2\sigma_2$-tangent space $T_{a\sigma_1^2\sigma_2}S$ at the point a is called a basis of the $\sigma_1^2\sigma_2\sigma_1\sigma_2$-principal directions of the $\sigma_1^2\sigma_2$-tangent space $T_{a\sigma_1^2\sigma_2}S$ if the vectors of the basis have $\sigma_1^2\sigma_2\sigma_1\sigma_2$-principal directions. We have

$$\mathscr{B}_{n\sigma_1^2\sigma_2\sigma_1\sigma_2} e_{1\sigma_1^2\sigma_2\sigma_1\sigma_2} = k_{1\sigma_1^2\sigma_2\sigma_1\sigma_2} e_{1\sigma_1^2\sigma_2\sigma_1\sigma_2}$$

and

$$\mathscr{B}_{n\sigma_1^2\sigma_2\sigma_1\sigma_2} e_{2\sigma_1^2\sigma_2\sigma_1\sigma_2} = k_{2\sigma_1^2\sigma_2\sigma_1\sigma_2} e_{2\sigma_1^2\sigma_2\sigma_1\sigma_2}.$$

Theorem 4.65 (Euler Formula) *Let S be $\sigma_1^2\sigma_2$-regular, σ_1 be Δ_1-differentiable, $f_{t_1}^{\Delta_1}$ be σ_1-completely delta differentiable and $f_{t_2}^{\Delta_2\sigma_1}$ be σ_2-completely delta differentiable. Then the σ_1-normal curvature at the point a of the surface S in the direction of a vector e is given by the Euler formula*

$$k_{n\sigma_1}(e) = k_{1\sigma_1^2\sigma_2\sigma_1\sigma_2}\cos^2\theta + k_{2\sigma_1^2\sigma_2\sigma_1\sigma_2}\sin^2\theta,$$

where

$$\theta = \angle(e, e_{1\sigma_1^2\sigma_2\sigma_1\sigma_2}).$$

Theorem 4.66 *Let S be $\sigma_1^2\sigma_2$-regular, σ_1 be Δ_1-differentiable, $f_{t_1}^{\Delta_1}$ be σ_1-completely delta differentiable and $f_{t_2}^{\Delta_2\sigma_1}$ be σ_2-completely delta differentiable. Then the $\sigma_1^2\sigma_2\sigma_1\sigma_2$-principal curvatures of the surface S at the point a are extremum values of the σ_1-normal curvature of the surface in the direction of a vector e, when the vector e rotates around the origin of the $\sigma_1^2\sigma_2$-tangent space of the surface S at the point a.*

Definition 4.59 Let S be $\sigma_1^2\sigma_2$-regular, $f_{t_1}^{\Delta_1}$ be σ_1-completely delta differentiable and $f_{t_2}^{\Delta_2\sigma_1}$ is σ_2-completely delta differentiable.

1. The quantity
$$k_{t\sigma_1^2\sigma_2\sigma_1\sigma_2} = k_{1\sigma_1^2\sigma_2\sigma_1\sigma_2}k_{2\sigma_1^2\sigma_2\sigma_1\sigma_2}$$
is called the $\sigma_1^2\sigma_2\sigma_1\sigma_2$-total(Gaussian) curvature of the surface S.
2. The quantity
$$k_{m\sigma_1^2\sigma_2\sigma_1\sigma_2} = \frac{1}{2}\left(k_{1\sigma_1^2\sigma_2\sigma_1\sigma_2} + k_{2\sigma_1^2\sigma_2\sigma_1\sigma_2}\right)$$
is called the $\sigma_1^2\sigma_2\sigma_1\sigma_2$-mean curvature of the surface S.

Definition 4.60 Let S be a $\sigma_1^2\sigma_2$-regular surface. Also, let σ_1 be Δ_1-differentiable, $f_{t_1}^{\Delta_1}$ be σ_2-completely delta differentiable and $f_{t_2}^{\Delta_2\sigma_1}$ be σ_1-completely delta differentiable. The σ_1-directions on the $\sigma_1^2\sigma_2$-tangent plane $T_{a\sigma_1^2\sigma_2}S$ at the point a corresponding to the eigenvectors of the matrix $\mathscr{B}_{n\sigma_1^2\sigma_2\sigma_2\sigma_1}$ are called $\sigma_1^2\sigma_2\sigma_2\sigma_1$-principal directions of the surface S at the point a. A curve g on the surface S is called a $\sigma_1^2\sigma_2\sigma_2\sigma_1$-principal line or $\sigma_1^2\sigma_2\sigma_2\sigma_1$-curvature line if its tangent at each point has $\sigma_1^2\sigma_2\sigma_2\sigma_1$-principal direction. A $\sigma_1^2\sigma_2\sigma_2\sigma_1$-principal curvature of the surface at the point a is the σ_1-normal curvature of S at a in a $\sigma_1^2\sigma_2\sigma_2\sigma_1$-principal direction.

Theorem 4.67 *Let S be a $\sigma_1^2\sigma_2$-regular surface. Let also, σ_1 be Δ_1-differentiable, $f_{t_1}^{\Delta_1}$ be σ_2-completely delta differentiable and $f_{t_2}^{\Delta_2\sigma_1}$ be σ_1-completely delta differentiable. Then the $\sigma_1^2\sigma_2\sigma_2\sigma_1$-principal curvatures of S are the eigenvalues of the matrix $\mathscr{B}_{n\sigma_1^2\sigma_2\sigma_2\sigma_1}$.*

Hereafter, with

$$k_{1\sigma_1^2\sigma_2\sigma_2\sigma_1} \quad \text{and} \quad k_{2\sigma_1^2\sigma_2\sigma_2\sigma_1}$$

we will denote the $\sigma_1^2\sigma_2\sigma_2\sigma_1$-principal curvatures of S and without loss of generality, suppose that

$$k_{1\sigma_1^2\sigma_2\sigma_2\sigma_1} \geq k_{2\sigma_1^2\sigma_2\sigma_2\sigma_1}.$$

Definition 4.61 Let S be $\sigma_1^2\sigma_2$-regular and σ_1 be Δ_1-differentiable. Also, let σ_1 be Δ_1-differentiable, $f_{t_1}^{\Delta_1}$ be σ_2-completely delta differentiable and $f_{t_2}^{\Delta_2\sigma_1}$ be σ_1-completely delta differentiable. An orthonormal basis $\{e_{1\sigma_1^2\sigma_2\sigma_2\sigma_1}, e_{2\sigma_1^2\sigma_2\sigma_2\sigma_1}\}$ of the $\sigma_1^2\sigma_2$-tangent space $T_{a\sigma_1^2\sigma_2}S$ at the point a is called a basis of the $\sigma_1^2\sigma_2\sigma_2\sigma_1$-principal directions of the $\sigma_1^2\sigma_2$-tangent space $T_{a\sigma_1^2\sigma_2}S$ if the vectors of the basis have $\sigma_1^2\sigma_2\sigma_2\sigma_1$-principal directions. We have

$$\mathscr{B}_{n\sigma_1^2\sigma_2\sigma_2\sigma_1} e_{1\sigma_1^2\sigma_2\sigma_2\sigma_1} = k_{1\sigma_1^2\sigma_2\sigma_2\sigma_1} e_{1\sigma_1^2\sigma_2\sigma_2\sigma_1}$$

and

$$\mathscr{B}_{n\sigma_1^2\sigma_2\sigma_2\sigma_1} e_{2\sigma_1^2\sigma_2\sigma_2\sigma_1} = k_{2\sigma_1^2\sigma_2\sigma_2\sigma_1} e_{2\sigma_1^2\sigma_2\sigma_2\sigma_1}.$$

Theorem 4.68 (Euler Formula) *Let S be $\sigma_1^2\sigma_2$-regular, σ_1 be Δ_1-differentiable, $f_{t_1}^{\Delta_1}$ be σ_2-completely delta differentiable and $f_{t_2}^{\Delta_2\sigma_1}$ be σ_1-completely delta differentiable. Then the σ_1-normal curvature at the point a of the surface S in the direction of a vector e is given by the Euler formula*

$$k_{n\sigma_1}(e) = k_{1\sigma_1^2\sigma_2\sigma_2\sigma_1} \cos^2\theta + k_{2\sigma_1^2\sigma_2\sigma_2\sigma_1} \sin^2\theta,$$

where

$$\theta = \angle(e, e_{1\sigma_1^2\sigma_2\sigma_2\sigma_1}).$$

Theorem 4.69 *Let S be $\sigma_1^2\sigma_2$-regular, σ_1 be Δ_1-differentiable, $f_{t_1}^{\Delta_1}$ be σ_2-completely delta differentiable and $f_{t_2}^{\Delta_2\sigma_1}$ be σ_1-completely delta differentiable. Then the $\sigma_1^2\sigma_2\sigma_2\sigma_1$-principal curvatures of the surface S at the point a are extremum values of the σ_1-normal curvature of the surface in the direction of a vector e, when the vector e rotates around the origin of the $\sigma_1^2\sigma_2$-tangent space of the surface S at the point a.*

Definition 4.62 Let S be $\sigma_1^2\sigma_2$-regular, $f_{t_1}^{\Delta_1}$ be σ_2-completely delta differentiable and $f_{t_2}^{\Delta_2\sigma_1}$ be σ_1-completely delta differentiable.

1. The quantity

$$k_{t\sigma_1^2\sigma_2\sigma_2\sigma_1} = k_{1\sigma_1^2\sigma_2\sigma_2\sigma_1} k_{2\sigma_1^2\sigma_2\sigma_2\sigma_1}$$

is called the $\sigma_1^2\sigma_2\sigma_2\sigma_1$-total(Gaussian) curvature of the surface S.

2. The quantity

$$k_{m\sigma_1^2\sigma_2\sigma_2\sigma_1} = \frac{1}{2}\left(k_{1\sigma_1^2\sigma_2\sigma_2\sigma_1} + k_{2\sigma_1^2\sigma_2\sigma_2\sigma_1}\right)$$

is called the $\sigma_1^2\sigma_2\sigma_2\sigma_1$-mean curvature of the surface S.

Definition 4.63 Let S be a $\sigma_1^2\sigma_2$-regular surface. Also, let σ_1 be Δ_1-differentiable and $f_{t_1}^{\Delta_1}$ and $f_{t_2}^{\Delta_2\sigma_1}$ be σ_2-completely delta differentiable. The σ_1-directions on the $\sigma_1^2\sigma_2$-tangent plane $T_{a\sigma_1^2\sigma_2}S$ at the point a corresponding to the eigenvectors of the matrix $\mathcal{B}_{n\sigma_1^2\sigma_2\sigma_2}$ are called $\sigma_1^2\sigma_2\sigma_2\sigma_2$-principal directions of the surface S at the point a. A curve g on the surface S is called a $\sigma_1^2\sigma_2\sigma_2\sigma_2$-principal line or $\sigma_1^2\sigma_2\sigma_2\sigma_2$-curvature line if its tangent at each point has $\sigma_1^2\sigma_2\sigma_2\sigma_2$-principal direction. A $\sigma_1^2\sigma_2\sigma_2\sigma_2$-principal curvature of the surface at the point a is the σ_1-normal curvature of S at a in a $\sigma_1^2\sigma_2\sigma_2\sigma_2$-principal direction.

Theorem 4.70 *Let S be a $\sigma_1^2\sigma_2$-regular surface. Also, let σ_1 be Δ_1-differentiable and $f_{t_1}^{\Delta_1}$ and $f_{t_2}^{\Delta_2\sigma_1}$ be σ_2-completely delta differentiable. Then the $\sigma_1^2\sigma_2\sigma_2\sigma_2$-principal curvatures of S are the eigenvalues of the matrix $\mathcal{B}_{n\sigma_1^2\sigma_2\sigma_2}$.*

Hereafter, with

$$k_{1\sigma_1^2\sigma_2\sigma_2\sigma_2} \quad \text{and} \quad k_{2\sigma_1^2\sigma_2\sigma_2\sigma_2}$$

we will denote the $\sigma_1^2\sigma_2\sigma_2\sigma_2$-principal curvatures of S and without loss of generality, suppose that

$$k_{1\sigma_1^2\sigma_2\sigma_2\sigma_2} \geq k_{2\sigma_1^2\sigma_2\sigma_2\sigma_2}.$$

Definition 4.64 Let S be $\sigma_1^2\sigma_2$-regular and σ_1 be Δ_1-differentiable. Also, let σ_1 be Δ_1-differentiable and $f_{t_1}^{\Delta_1}$ and $f_{t_2}^{\Delta_2\sigma_1}$ be σ_2-completely delta differentiable. An orthonormal basis $\{e_{1\sigma_1^2\sigma_2\sigma_2\sigma_2}, e_{2\sigma_1^2\sigma_2\sigma_2\sigma_2}\}$ of the $\sigma_1^2\sigma_2$-tangent space $T_{a\sigma_1^2\sigma_2}S$ at the point a is called a basis of the $\sigma_1^2\sigma_2\sigma_2\sigma_2$-principal directions of the $\sigma_1^2\sigma_2$-tangent space $T_{a\sigma_1^2\sigma_2}S$ if the vectors of the basis have $\sigma_1^2\sigma_2\sigma_2\sigma_2$-principal directions. We have

$$\mathcal{B}_{n\sigma_1^2\sigma_2\sigma_2\sigma_2} e_{1\sigma_1^2\sigma_2\sigma_2\sigma_2} = k_{1\sigma_1^2\sigma_2\sigma_2\sigma_2} e_{1\sigma_1^2\sigma_2\sigma_2\sigma_2}$$

and

$$\mathcal{B}_{n\sigma_1^2\sigma_2\sigma_2\sigma_2} e_{2\sigma_1^2\sigma_2\sigma_2\sigma_2} = k_{2\sigma_1^2\sigma_2\sigma_2\sigma_2} e_{2\sigma_1^2\sigma_2\sigma_2\sigma_2}.$$

Theorem 4.71 (Euler Formula) *Let S be $\sigma_1^2\sigma_2$-regular, σ_1 be Δ_1-differentiable and $f_{t_1}^{\Delta_1}$ and $f_{t_2}^{\Delta_2\sigma_1}$ be σ_2-completely delta differentiable. Then the σ_1-normal curvature at the point a of the surface S in the direction of a vector e is given by the Euler formula*

$$k_{n\sigma_1}(e) = k_{1\sigma_1^2\sigma_2\sigma_2\sigma_2}\cos^2\theta + k_{2\sigma_1^2\sigma_2\sigma_2\sigma_2}\sin^2\theta,$$

where

$$\theta = \angle(e, e_{1\sigma_1^2\sigma_2\sigma_2\sigma_2}).$$

Theorem 4.72 *Let S be $\sigma_1^2\sigma_2$-regular, σ_1 be Δ_1-differentiable and $f_{t_1}^{\Delta_1}$ and $f_{t_2}^{\Delta_2\sigma_1}$ be σ_2-completely delta differentiable. Then the $\sigma_1^2\sigma_2\sigma_2\sigma_2$-principal curvatures of the surface S at the point a are extremum values of the σ_1-normal curvature of the surface in the direction of a vector e, when the vector e rotates around the origin of the $\sigma_1^2\sigma_2$-tangent space of the surface S at the point a.*

Definition 4.65 Let S be $\sigma_1^2\sigma_2$-regular, σ_1 be Δ_1-differentiable and $f_{t_1}^{\Delta_1}$ and $f_{t_2}^{\Delta_2\sigma_1}$ be σ_2-completely delta differentiable.

1. The quantity
$$k_{t\sigma_1^2\sigma_2\sigma_2\sigma_2} = k_{1\sigma_1^2\sigma_2\sigma_2\sigma_2}k_{2\sigma_1^2\sigma_2\sigma_2\sigma_2}$$
is called the $\sigma_1^2\sigma_2\sigma_2\sigma_2$-total(Gaussian) curvature of the surface S.
2. The quantity
$$k_{m\sigma_1^2\sigma_2\sigma_2\sigma_2} = \frac{1}{2}\left(k_{1\sigma_1^2\sigma_2\sigma_2\sigma_2} + k_{2\sigma_1^2\sigma_2\sigma_2\sigma_2}\right)$$
is called the $\sigma_1^2\sigma_2\sigma_2\sigma_2$-mean curvature of the surface S.

Definition 4.66 Let S be a $\sigma_1\sigma_2^2$-regular surface. Also, let σ_2 be Δ_2-differentiable and $f_{t_1}^{\Delta_1\sigma_2}$ and $f_{t_2}^{\Delta_2}$ be σ_1-completely delta differentiable. The σ_2-directions on the $\sigma_1\sigma_2^2$-tangent plane $T_{a\sigma_1\sigma_2^2}S$ at the point a corresponding to the eigenvectors of the matrix $\mathscr{B}_{n\sigma_1\sigma_2^2\sigma_1\sigma_1}$ are called $\sigma_1\sigma_2^2\sigma_1\sigma_1$-principal directions of the surface S at the point a. A curve g on the surface S is called a $\sigma_1\sigma_2^2\sigma_1\sigma_1$-principal line or $\sigma_1\sigma_2^2\sigma_1\sigma_1$-curvature line if its tangent at each point has $\sigma_1\sigma_2^2\sigma_1\sigma_1$-principal direction. A $\sigma_1\sigma_2^2\sigma_1\sigma_1$-principal curvature of the surface at the point a is the σ_2-normal curvature of S at a in a $\sigma_1\sigma_2^2\sigma_1\sigma_1$-principal direction.

Theorem 4.73 *Let S be a $\sigma_1\sigma_2^2$-regular surface. Let also, σ_2 be Δ_2-differentiable, $f_{t_1}^{\Delta_1\sigma_2}$ and $f_{t_2}^{\Delta_2}$ be σ_1-completely delta differentiable. Then the $\sigma_1\sigma_2^2\sigma_1\sigma_1$-principal curvatures of S are the eigenvalues of the matrix $\mathscr{B}_{n\sigma_1\sigma_2^2\sigma_1\sigma_1}$.*

Hereafter, with
$$k_{1\sigma_1\sigma_2^2\sigma_1\sigma_1} \quad \text{and} \quad k_{2\sigma_1\sigma_2^2\sigma_1\sigma_1}$$
we will denote the $\sigma_1\sigma_2^2\sigma_1\sigma_1$-principal curvatures of S and without loss of generality, suppose that
$$k_{1\sigma_1\sigma_2^2\sigma_1\sigma_1} \geq k_{2\sigma_1\sigma_2^2\sigma_1\sigma_1}.$$

Definition 4.67 Let S be $\sigma_1\sigma_2^2$-regular, σ_2 be Δ_2-differentiable. Let also, $f_{t_1}^{\Delta_1\sigma_2}$ and $f_{t_2}^{\Delta_2}$ be σ_1-completely delta differentiable. An orthonormal basis $\{e_{1\sigma_1\sigma_2^2\sigma_1\sigma_1}, e_{2\sigma_1\sigma_2^2\sigma_1\sigma_1}\}$ of the $\sigma_1\sigma_2^2$-tangent space $T_{a\sigma_1\sigma_2^2}S$ at the point a is called a basis of the $\sigma_1\sigma_2^2\sigma_1\sigma_1$-principal directions of the $\sigma_1\sigma_2^2$-tangent space $T_{a\sigma_1\sigma_2^2}S$ if the vectors of the basis have $\sigma_1\sigma_2^2\sigma_1\sigma_1$-principal directions. We have
$$\mathscr{B}_{n\sigma_1\sigma_2^2\sigma_1\sigma_1}e_{1\sigma_1\sigma_2^2\sigma_1\sigma_1} = k_{1\sigma_1\sigma_2^2\sigma_1\sigma_1}e_{1\sigma_1\sigma_2^2\sigma_1\sigma_1}$$

and

$$\mathscr{B}_{n\sigma_1\sigma_2^2\sigma_1\sigma_1} e_{2\sigma_1\sigma_2^2\sigma_1\sigma_1} = k_{2\sigma_1\sigma_2^2\sigma_1\sigma_1} e_{2\sigma_1\sigma_2^2\sigma_1\sigma_1}.$$

Theorem 4.74 (Euler Formula) *Let S be σ_2-regular, σ_2 be Δ_2-differentiable, $f_{t_1}^{\Delta_1\sigma_2}$ and $f_{t_2}^{\Delta_2}$ be σ_1-completely delta differentiable. Then the σ_2-normal curvature at the point a of the surface S in the direction of a vector e is given by the Euler formula*

$$k_{n\sigma_2}(e) = k_{1\sigma_1\sigma_2^2\sigma_1\sigma_1} \cos^2\theta + k_{2\sigma_1\sigma_2^2\sigma_1\sigma_1} \sin^2\theta,$$

where

$$\theta = \angle(e, e_{1\sigma_1\sigma_2^2\sigma_1\sigma_1}).$$

Theorem 4.75 *Let S be $\sigma_1\sigma_2^2$-regular, σ_2 be Δ_2-differentiable, $f_{t_1}^{\Delta_1\sigma_2}$ and $f_{t_2}^{\Delta_2}$ be σ_1-completely delta differentiable. Then the $\sigma_1\sigma_2^2\sigma_1\sigma_1$-principal curvatures of the surface S at the point a are extremum values of the σ_2-normal curvature of the surface in the direction of a vector e, when the vector e rotates around the origin of the $\sigma_1\sigma_2^2$-tangent space of the surface S at the point a.*

Definition 4.68 Let S be $\sigma_1\sigma_2^2$-regular, σ_2 be Δ_2-differentiable, $f_{t_1}^{\Delta_1\sigma_2}$ and $f_{t_2}^{\Delta_2}$ be σ_1-completely delta differentiable.

1. The quantity

$$k_{t\sigma_1\sigma_2^2\sigma_1\sigma_1} = k_{1\sigma_1\sigma_2^2\sigma_1\sigma_1} k_{2\sigma_1\sigma_2^2\sigma_1\sigma_1}$$

   is called the $\sigma_1\sigma_2^2\sigma_1\sigma_1$-total(Gaussian) curvature of the surface S.
2. The quantity

$$k_{m\sigma_1\sigma_2^2\sigma_1\sigma_1} = \frac{1}{2}\left(k_{1\sigma_1\sigma_2^2\sigma_1\sigma_1} + k_{2\sigma_1\sigma_2^2\sigma_1\sigma_1}\right)$$

   is called the $\sigma_1\sigma_2^2\sigma_1\sigma_1$-mean curvature of the surface S.

Definition 4.69 Let S be a $\sigma_1\sigma_2^2$-regular surface. Let also, σ_2 be Δ_2-differentiable, $f_{t_1}^{\Delta_1\sigma_2}$ be σ_1-delta differentiable and $f_{t_2}^{\Delta_2}$ be σ_2-completely delta differentiable. The σ_2-directions on the $\sigma_1\sigma_2^2$-tangent plane $T_{a\sigma_1\sigma_2^2}S$ at the point a corresponding to the eigenvectors of the matrix $\mathscr{B}_{n\sigma_1\sigma_2^2\sigma_1\sigma_2}$ are called $\sigma_1\sigma_2^2\sigma_1\sigma_2$-principal directions of the surface S at the point a. A curve g on the surface S is called a $\sigma_1\sigma_2^2\sigma_1\sigma_2$-principal line or $\sigma_1\sigma_2^2\sigma_1\sigma_2$-curvature line if its tangent at each point has $\sigma_1\sigma_2^2\sigma_1\sigma_2$-principal direction. A $\sigma_1\sigma_2^2\sigma_1\sigma_2$-principal curvature of the surface at the point a is the σ_2-normal curvature of S at a in a $\sigma_1\sigma_2^2\sigma_1\sigma_2$-principal direction.

Theorem 4.76 *Let S be a $\sigma_1\sigma_2^2$-regular surface. Let also, σ_2 be Δ_2-differentiable, $f_{t_1}^{\Delta_1\sigma_2}$ be σ_1-delta differentiable and $f_{t_2}^{\Delta_2}$ be σ_2-completely delta differentiable. Then the $\sigma_1\sigma_2^2\sigma_1\sigma_2$-principal curvatures of S are the eigenvalues of the matrix $\mathscr{B}_{n\sigma_1\sigma_2^2\sigma_1\sigma_2}$.*

Hereafter, with

$$k_{1\sigma_1\sigma_2^2\sigma_1\sigma_2} \quad \text{and} \quad k_{2\sigma_1\sigma_2^2\sigma_1\sigma_2}$$

we will denote the $\sigma_1\sigma_2^2\sigma_1\sigma_2$-principal curvatures of S and without loss of generality, suppose that

$$k_{1\sigma_1\sigma_2^2\sigma_1\sigma_2} \geq k_{2\sigma_1\sigma_2^2\sigma_1\sigma_2}.$$

Definition 4.70 Let S be $\sigma_1\sigma_2^2$-regular, σ_2 be Δ_2-differentiable. Let also, $f_{t_1}^{\Delta_1\sigma_2}$ be σ_1-delta differentiable and $f_{t_2}^{\Delta_2}$ be σ_2-completely delta differentiable. An orthonormal basis $\{e_{1\sigma_1\sigma_2^2\sigma_1\sigma_2}, e_{2\sigma_1\sigma_2^2\sigma_1\sigma_2}\}$ of the $\sigma_1\sigma_2^2$-tangent space $T_{a\sigma_1\sigma_2^2}S$ at the point a is called a basis of the $\sigma_1\sigma_2^2\sigma_1\sigma_2$-principal directions of the $\sigma_1\sigma_2^2$-tangent space $T_{a\sigma_1\sigma_2^2}S$ if the vectors of the basis have $\sigma_1\sigma_2^2\sigma_1\sigma_2$-principal directions. We have

$$\mathscr{B}_{n\sigma_1\sigma_2^2\sigma_1\sigma_2}e_{1\sigma_1\sigma_2^2\sigma_1\sigma_2} = k_{1\sigma_1\sigma_2^2\sigma_1\sigma_2}e_{1\sigma_1\sigma_2^2\sigma_1\sigma_2}$$

and

$$\mathscr{B}_{n\sigma_1\sigma_2^2\sigma_1\sigma_2}e_{2\sigma_1\sigma_2^2\sigma_1\sigma_2} = k_{2\sigma_1\sigma_2^2\sigma_1\sigma_2}e_{2\sigma_1\sigma_2^2\sigma_1\sigma_2}.$$

Theorem 4.77 (Euler Formula) *Let S be $\sigma_1\sigma_2^2$-regular, σ_2 be Δ_2-differentiable, $f_{t_1}^{\Delta_1\sigma_2}$ be σ_1-delta differentiable and $f_{t_2}^{\Delta_2}$ be σ_2-completely delta differentiable. Then the σ_2-normal curvature at the point a of the surface S in the direction of a vector e is given by the Euler formula*

$$k_{n\sigma_2}(e) = k_{1\sigma_1\sigma_2^2\sigma_1\sigma_2}\cos^2\theta + k_{2\sigma_1\sigma_2^2\sigma_1\sigma_2}\sin^2\theta,$$

where

$$\theta = \angle(e, e_{1\sigma_1\sigma_2^2\sigma_1\sigma_2}).$$

Theorem 4.78 *Let S be $\sigma_1\sigma_2^2$-regular, σ_2 be Δ_2-differentiable, $f_{t_1}^{\Delta_1\sigma_2}$ be σ_1-delta differentiable and $f_{t_2}^{\Delta_2}$ be σ_2-completely delta differentiable. Then the $\sigma_1\sigma_2^2\sigma_1\sigma_2$-principal curvatures of the surface S at the point a are extremum values of the σ_2-normal curvature of the surface in the direction of a vector e, when the vector e rotates around the origin of the $\sigma_1\sigma_2^2$-tangent space of the surface S at the point a.*

Definition 4.71 Let S be $\sigma_1\sigma_2^2$-regular, σ_2 be Δ_2-differentiable, $f_{t_1}^{\Delta_1\sigma_2}$ be σ_1-delta differentiable and $f_{t_2}^{\Delta_2}$ be σ_2-completely delta differentiable.

1. The quantity

$$k_{t\sigma_1\sigma_2^2\sigma_1\sigma_2} = k_{1\sigma_1\sigma_2^2\sigma_1\sigma_2}k_{2\sigma_1\sigma_2^2\sigma_1\sigma_2}$$

is called the $\sigma_1\sigma_2^2\sigma_1\sigma_2$-total(Gaussian) curvature of the surface S.

2. The quantity

$$k_{m\sigma_1\sigma_2^2\sigma_1\sigma_2} = \frac{1}{2}\left(k_{1\sigma_1\sigma_2^2\sigma_1\sigma_2} + k_{2\sigma_1\sigma_2^2\sigma_1\sigma_2}\right)$$

is called the $\sigma_1\sigma_2^2\sigma_1\sigma_2$-mean curvature of the surface S.

Definition 4.72 Let S be a $\sigma_1\sigma_2^2$-regular surface. Let also, σ_2 be Δ_2-differentiable, $f_{t_1}^{\Delta_1\sigma_2}$ be σ_2-delta differentiable and $f_{t_2}^{\Delta_2}$ be σ_1-completely delta differentiable. The σ_2-directions on the $\sigma_1\sigma_2^2$-tangent plane $T_{a\sigma_1\sigma_2^2}S$ at the point a corresponding to the eigenvectors of the matrix $\mathscr{B}_{n\sigma_1\sigma_2^2\sigma_2\sigma_1}$ are called $\sigma_1\sigma_2^2\sigma_2\sigma_1$-principal directions of the surface S at the point a. A curve g on the surface S is called a $\sigma_1\sigma_2^2\sigma_2\sigma_1$-principal line or $\sigma_1\sigma_2^2\sigma_2\sigma_1$-curvature line if its tangent at each point has $\sigma_1\sigma_2^2\sigma_2\sigma_1$-principal direction. A $\sigma_1\sigma_2^2\sigma_2\sigma_1$-principal curvature of the surface at the point a is the σ_2-normal curvature of S at a in a $\sigma_1\sigma_2^2\sigma_2\sigma_1$-principal direction.

Theorem 4.79 *Let S be a $\sigma_1\sigma_2^2$-regular surface. Let also, σ_2 be Δ_2-differentiable, $f_{t_1}^{\Delta_1\sigma_2}$ be σ_2-delta differentiable and $f_{t_2}^{\Delta_2}$ be σ_1-completely delta differentiable. Then the $\sigma_1\sigma_2^2\sigma_2\sigma_1$-principal curvatures of S are the eigenvalues of the matrix $\mathscr{B}_{n\sigma_1\sigma_2^2\sigma_2\sigma_1}$.*

Hereafter, with

$$k_{1\sigma_1\sigma_2^2\sigma_2\sigma_1} \quad \text{and} \quad k_{2\sigma_1\sigma_2^2\sigma_2\sigma_1}$$

we will denote the $\sigma_1\sigma_2^2\sigma_2\sigma_1$-principal curvatures of S and without loss of generality, suppose that

$$k_{1\sigma_1\sigma_2^2\sigma_2\sigma_1} \geq k_{2\sigma_1\sigma_2^2\sigma_2\sigma_1}.$$

Definition 4.73 Let S be $\sigma_1\sigma_2^2$-regular, σ_2 be Δ_2-differentiable. Let also, $f_{t_1}^{\Delta_1\sigma_2}$ be σ_2-delta differentiable and $f_{t_2}^{\Delta_2}$ be σ_1-completely delta differentiable. An orthonormal basis $\{e_{1\sigma_1\sigma_2^2\sigma_2\sigma_1}, e_{2\sigma_1\sigma_2^2\sigma_2\sigma_1}\}$ of the $\sigma_1\sigma_2^2$-tangent space $T_{a\sigma_1\sigma_2^2}S$ at the point a is called a basis of the $\sigma_1\sigma_2^2\sigma_1\sigma_2$-principal directions of the $\sigma_1\sigma_2^2$-tangent space $T_{a\sigma_1\sigma_2^2}S$ if the vectors of the basis have $\sigma_1\sigma_2^2\sigma_2\sigma_1$-principal directions. We have

$$\mathscr{B}_{n\sigma_1\sigma_2^2\sigma_2\sigma_1} e_{1\sigma_1\sigma_2^2\sigma_2\sigma_1} = k_{1\sigma_1\sigma_2^2\sigma_2\sigma_1} e_{1\sigma_1\sigma_2^2\sigma_2\sigma_1}$$

and

$$\mathscr{B}_{n\sigma_1\sigma_2^2\sigma_2\sigma_1} e_{2\sigma_1\sigma_2^2\sigma_2\sigma_1} = k_{2\sigma_1\sigma_2^2\sigma_2\sigma_1} e_{2\sigma_1\sigma_2^2\sigma_2\sigma_1}.$$

Theorem 4.80 (Euler Formula) *Let S be σ_2-regular, σ_2 be Δ_2-differentiable, $f_{t_1}^{\Delta_1\sigma_2}$ be σ_2-delta differentiable and $f_{t_2}^{\Delta_2}$ be σ_1-completely delta differentiable. Then the σ_2-normal curvature at the point a of the surface S in the direction of a vector e is given by the Euler formula*

$$k_{n\sigma_2}(e) = k_{1\sigma_1\sigma_2^2\sigma_2\sigma_1} \cos^2\theta + k_{2\sigma_1\sigma_2^2\sigma_2\sigma_1} \sin^2\theta,$$

where

$$\theta = \angle(e, e_{1\sigma_1\sigma_2^2\sigma_2\sigma_1}).$$

Theorem 4.81 *Let S be σ_2-regular, σ_2 be Δ_2-differentiable, $f_{t_1}^{\Delta_1\sigma_2}$ be σ_2-delta differentiable and $f_{t_2}^{\Delta_2}$ be σ_1-completely delta differentiable. Then the $\sigma_1\sigma_2^2\sigma_2\sigma_1$-principal* *curvatures of the surface S at the point a are extremum values of the σ_2-normal curvature of the surface in the direction of a vector e, when the vector e rotates around the origin of the $\sigma_1\sigma_2^2$-tangent space of the surface S at the point a.*

Definition 4.74 Let S be σ_2-regular, σ_2 be Δ_2-differentiable, $f_{t_1}^{\Delta_1\sigma_2}$ be σ_2-delta differentiable and $f_{t_2}^{\Delta_2}$ be σ_1-completely delta differentiable.

1. The quantity
$$k_{t\sigma_1\sigma_2^2\sigma_2\sigma_1} = k_{1\sigma_1\sigma_2^2\sigma_2\sigma_1}k_{2\sigma_1\sigma_2^2\sigma_2\sigma_1}$$
is called the $\sigma_1\sigma_2^2\sigma_2\sigma_1$-total(Gaussian) curvature of the surface S.
2. The quantity
$$k_{m\sigma_1\sigma_2^2\sigma_2\sigma_1} = \frac{1}{2}\left(k_{1\sigma_1\sigma_2^2\sigma_2\sigma_1} + k_{2\sigma_1\sigma_2^2\sigma_2\sigma_1}\right)$$
is called the $\sigma_1\sigma_2^2\sigma_2\sigma_1$-mean curvature of the surface S.

Definition 4.75 Let S be a σ_2-regular surface. Let also, σ_2 be Δ_2-differentiable, $f_{t_1}^{\Delta_1\sigma_2}$ and $f_{t_2}^{\Delta_2}$ be σ_2-completely delta differentiable. The σ_2-directions on the $\sigma_1\sigma_2^2$-tangent plane $T_{a\sigma_1\sigma_2^2}S$ at the point a corresponding to the eigenvectors of the matrix $\mathcal{B}_{n\sigma_1\sigma_2^2\sigma_2\sigma_2}$ are called $\sigma_1\sigma_2^2\sigma_2\sigma_2$-principal directions of the surface S at the point a. A curve g on the surface S is called a $\sigma_1\sigma_2^2\sigma_2\sigma_2$-principal line or $\sigma_1\sigma_2^2\sigma_2\sigma_2$-curvature line if its tangent at each point has $\sigma_1\sigma_2^2\sigma_2\sigma_2$-principal direction. A $\sigma_1\sigma_2^2\sigma_2\sigma_2$-principal curvature of the surface at the point a is the σ_2-normal curvature of S at a in a $\sigma_1\sigma_2^2\sigma_2\sigma_2$-principal direction.

Theorem 4.82 *Let S be a σ_2-regular surface. Let also, σ_2 be Δ_2-differentiable, $f_{t_1}^{\Delta_1\sigma_2}$ and $f_{t_2}^{\Delta_2}$ be σ_2-completely delta differentiable. Then the $\sigma_1\sigma_2^2\sigma_2\sigma_1$-principal curvatures of S are the eigenvalues of the matrix $\mathcal{B}_{n\sigma_1\sigma_2^2\sigma_2\sigma_2}$.*

Hereafter, with
$$k_{1\sigma_1\sigma_2^2\sigma_2\sigma_2} \quad \text{and} \quad k_{2\sigma_1\sigma_2^2\sigma_2\sigma_2}$$
we will denote the $\sigma_1\sigma_2^2\sigma_2\sigma_2$-principal curvatures of S and without loss of generality, suppose that
$$k_{1\sigma_1\sigma_2^2\sigma_2\sigma_2} \geq k_{2\sigma_1\sigma_2^2\sigma_2\sigma_2}.$$

Definition 4.76 Let S be σ_2-regular, σ_2 be Δ_2-differentiable. Let also, $f_{t_1}^{\Delta_1\sigma_2}$ and $f_{t_2}^{\Delta_2}$ be σ_2-completely delta differentiable. An orthonormal basis $\{e_{1\sigma_1\sigma_2^2\sigma_2\sigma_2}, e_{2\sigma_1\sigma_2^2\sigma_2\sigma_2}\}$ of the $\sigma_1\sigma_2^2$-tangent space $T_{a\sigma_1\sigma_2^2}S$ at the point a is called a basis of the $\sigma_1\sigma_2^2\sigma_2\sigma_2$-principal directions of the $\sigma_1\sigma_2^2$-tangent space $T_{a\sigma_1\sigma_2^2}S$ if the vectors of the basis have $\sigma_1\sigma_2^2\sigma_2\sigma_2$-principal directions. We have
$$\mathcal{B}_{n\sigma_1\sigma_2^2\sigma_2\sigma_2}e_{1\sigma_1\sigma_2^2\sigma_2\sigma_2} = k_{1\sigma_1\sigma_2^2\sigma_2\sigma_2}e_{1\sigma_1\sigma_2^2\sigma_2\sigma_2}$$
and
$$\mathcal{B}_{n\sigma_1\sigma_2^2\sigma_2\sigma_2}e_{2\sigma_1\sigma_2^2\sigma_2\sigma_2} = k_{2\sigma_1\sigma_2^2\sigma_2\sigma_2}e_{2\sigma_1\sigma_2^2\sigma_2\sigma_2}.$$

Theorem 4.83 (Euler Formula) *Let S be σ_2-regular, σ_2 be Δ_2-differentiable, $f_{t_1}^{\Delta_1\sigma_2}$ and $f_{t_2}^{\Delta_2}$ be σ_2-completely delta differentiable. Then the σ_2-normal curvature at the point a of the surface S in the direction of a vector e is given by the Euler formula*

$$k_{n\sigma_2}(e) = k_{1\sigma_1\sigma_2^2\sigma_2\sigma_2}\cos^2\theta + k_{2\sigma_1\sigma_2^2\sigma_2\sigma_2}\sin^2\theta,$$

where

$$\theta = \angle(e, e_{1\sigma_1\sigma_2^2\sigma_2\sigma_2}).$$

Theorem 4.84 *Let S be σ_2-regular, σ_2 be Δ_2-differentiable, $f_{t_1}^{\Delta_1\sigma_2}$ and $f_{t_2}^{\Delta_2}$ be σ_2-completely delta differentiable. Then the $\sigma_1\sigma_2^2\sigma_2\sigma_2$-principal curvatures of the surface S at the point a are extremum values of the σ_2-normal curvature of the surface in the direction of a vector e, when the vector e rotates around the origin of the $\sigma_1\sigma_2^2$-tangent space of the surface S at the point a.*

Definition 4.77 Let S be σ_2-regular, σ_2 be Δ_2-differentiable, $f_{t_1}^{\Delta_1\sigma_2}$ and $f_{t_2}^{\Delta_2}$ be σ_2-completely delta differentiable.

1. The quantity

$$k_{t\sigma_1\sigma_2^2\sigma_2\sigma_2} = k_{1\sigma_1\sigma_2^2\sigma_2\sigma_2}k_{2\sigma_1\sigma_2^2\sigma_2\sigma_2}$$

is called the $\sigma_1\sigma_2^2\sigma_2\sigma_2$-total(Gaussian) curvature of the surface S.

2. The quantity

$$k_{m\sigma_1\sigma_2^2\sigma_2\sigma_2} = \frac{1}{2}\left(k_{1\sigma_1\sigma_2^2\sigma_2\sigma_2} + k_{2\sigma_1\sigma_2^2\sigma_2\sigma_2}\right)$$

is called the $\sigma_1\sigma_2^2\sigma_2\sigma_2$-mean curvature of the surface S.

4.5 The Joachimstahl Theorem

Theorem 4.85 (The Joachimstahl Theorem) *Let σ_1 be Δ_1-differentiable on $\mathbb{T}_1$, σ_2 be Δ_2-differentiable on $\mathbb{T}_2$, f_r be σ_1-completely delta differentiable on U, $f_{rt_1}^{\Delta_1}$ be σ_j-completely delta differentiable, $f_{rt_2}^{\Delta_2\sigma_1}$ be σ_l-completely delta differentiable on U, $j,l \in \{1,2\}$, $r \in \{1,2\}$. Let also, S_1 and S_2 be two σ_1-oriented surfaces with local σ_1-parameterizations (U,f_1) and (U,f_2), respectively. Assume that $N_{1\sigma_1}$ and $N_{2\sigma_1}$ are two unit σ_1-normals of S_1 and S_2, respectively. Suppose that S_1 and S_2 intersect along the curve $g(t) = (g_1(t), g_2(t))$, $t \in I$, so that*

$$g_1(\sigma(t)) = \sigma_1(g_1(t)),$$
$$g_2(\sigma(t)) = \sigma_2(g_2(t)), \quad t \in I,$$

and

$$N_{1\sigma_1} \cdot N_{2\sigma_1} = const,$$

$$\frac{\Delta}{\Delta t}N_{1\sigma_1}(t)\|g^{\Delta}(t), \quad t \in I,$$

and

$$\frac{\Delta}{\Delta t}N_{2\sigma_1}(t)\|g^{\Delta}(t), \quad t \in I.$$

Then g is a $\sigma_1\sigma_j\sigma_l$-curvature line on S_1 if and only if

$$N_{1\sigma_1}^{\sigma_1\sigma_2}(t) \cdot \frac{\Delta}{\Delta t}N_{2\sigma_1}(t) = 0, \quad t \in I,$$

and g is a $\sigma_1\sigma_j\sigma_l$-curvature line on S_2 if and only if

$$\frac{\Delta}{\Delta t}N_{1\sigma_1}(t) \cdot N_{2\sigma_1}^{\sigma_1\sigma_2}(t) = 0, \quad t \in I.$$

Proof. Let $j, l \in \{1,2\}$. Note that

$$\begin{aligned} N_{1\sigma_1}(t) &= N_{1\sigma_1}(g_1(t), g_2(t)), \\ N_{2\sigma_1}(t) &= N_{2\sigma_1}(g_1(t), g_2(t)), \\ N_{1\sigma_1}^{\sigma}(t) &= N_{1\sigma_1}(g_1(\sigma(t)), g_2(\sigma(t))) \\ &= N_{1\sigma_1}(\sigma_1(g_1(t)), \sigma_2(g_2(t))) \\ &= N_{1\sigma_1}^{\sigma_1\sigma_2}(g_1(t), g_2(t)), \quad t \in I, \end{aligned}$$

and, as above,

$$N_{2\sigma_1}^{\sigma}(t) = N_{2\sigma_1}^{\sigma_1\sigma_2}(g_1(t), g_2(t)), \quad t \in I.$$

Since

$$N_{2\sigma_1}^{\sigma} \cdot N_{2\sigma_1} = \text{const},$$

we get

$$\begin{aligned} 0 &= \frac{\Delta}{\Delta t}N_{1\sigma_1}(t) \cdot N_{2\sigma_1}(t) + N_{1\sigma_1}^{\sigma}(t) \cdot \frac{\Delta}{\Delta t}N_{2\sigma_1}(t) \\ &= \frac{\Delta}{\Delta t}N_{1\sigma_1}(t) \cdot N_{2\sigma_1}(t) + N_{1\sigma_1}^{\sigma_1\sigma_2}(t) \cdot \frac{\Delta}{\Delta t}N_{2\sigma_1}(t), \end{aligned} \tag{4.12}$$

$t \in I$, and

$$0 = \frac{\Delta}{\Delta t}N_{1\sigma_1}(t) \cdot N_{2\sigma_1}^{\sigma_1\sigma_2}(t) + N_{1\sigma_1}(t) \cdot \frac{\Delta}{\Delta t}N_{2\sigma_1}(t), \tag{4.13}$$

$t \in I$. Let g be a $\sigma_1\sigma_j\sigma_l$-principal line on S_1. Because

$$\frac{\Delta}{\Delta t}N_{1\sigma_1}(t)\|g^{\Delta}(t), \quad t \in I,$$

we have

$$\frac{\Delta}{\Delta t}N_{1\sigma_1}(t) = b \cdot g^{\Delta}(t), \quad t \in I, \tag{4.14}$$

where b is a constant. On the other hand, since the curve g lies also on S_2, we have

$$g^{\Delta}(t) \perp N_{2\sigma_1}(t), \quad t \in I.$$

Hence and (4.14), we find

$$\frac{\Delta}{\Delta t} N_{1\sigma_1}(t) \perp N_{2\sigma_1}(t), \quad t \in I,$$

and

$$\frac{\Delta}{\Delta t} N_{1\sigma_1}(t) \cdot N_{2\sigma_1}(t) = 0, \quad t \in I.$$

Now, we apply (4.12) and we find

$$N_{1\sigma_1}^{\sigma_1\sigma_2}(t) \cdot \frac{\Delta}{\Delta t} N_{2\sigma_1}(t) = 0, \quad t \in I.$$

Let g be a $\sigma_1\sigma_j\sigma_l$-principal line on S_2. Since

$$\frac{\Delta}{\Delta t} N_{2\sigma_1}(t) \| g^{\Delta}(t), \quad t \in I,$$

we have

$$\frac{\Delta}{\Delta t} N_{2\sigma_1}(t) = c \cdot g^{\Delta}(t), \quad t \in I, \tag{4.15}$$

where c is a real constant. On the other hand,

$$g^{\Delta}(t) \perp N_{1\sigma_1}(t), \quad t \in I.$$

Now, we apply (4.15) and we find

$$\frac{\Delta}{\Delta t} N_{2\sigma_1}(t) \cdot N_{1\sigma_1}(t) = 0, \quad t \in I.$$

From here and from (4.13), we arrive at

$$\frac{\Delta}{\Delta t} N_{1\sigma_1}(t) \cdot N_{2\sigma_1}^{\sigma_1\sigma_2}(t) = 0, \quad t \in I.$$

This completes the proof.

As above, one can prove the following result.

Theorem 4.86 (The Joachimstahl Theorem) *Let σ_1 be Δ_1-differentiable on $\mathbb{T}_1$, σ_2 be Δ_2-differentiable on $\mathbb{T}_2$, f_r be σ_1-completely delta differentiable on U, $f_{rt_1}^{\Delta_1}$ be σ_j-completely delta differentiable, $f_{rt_2}^{\Delta_2\sigma_1}$ be σ_l-completely delta differentiable on U, $j,l \in \{1,2\}$, $r \in \{1,2\}$. Let also, S_1 and S_2 be two σ_2-oriented surfaces with local σ_2-parameterizations (U,f_1) and (U,f_2), respectively. Assume that $N_{1\sigma_2}$ and $N_{2\sigma_2}$ are two unit σ_2-normals of S_1 and S_2, respectively. Suppose that S_1 and S_2 intersect along the curve $g(t) = (g_1(t), g_2(t))$, $t \in I$, so that*

$$g_1(\sigma(t)) = \sigma_1(g_1(t)),$$

$$g_2(\sigma(t)) = \sigma_2(g_2(t)), \quad t \in I,$$

and

$$N_{1\sigma_2} \cdot N_{2\sigma_2} = const,$$

$$\frac{\Delta}{\Delta t} N_{1\sigma_2}(t) \| g^{\Delta}(t), \quad t \in I,$$

and

$$\frac{\Delta}{\Delta t} N_{2\sigma_2}(t) \| g^{\Delta}(t), \quad t \in I.$$

Then g is a $\sigma_2\sigma_j\sigma_l$-curvature line on S_1 if and only if

$$N_{1\sigma_2}^{\sigma_1\sigma_2}(t) \cdot \frac{\Delta}{\Delta t} N_{2\sigma_2}(t) = 0, \quad t \in I,$$

and g is a $\sigma_2\sigma_j\sigma_l$-curvature line on S_2 if and only if

$$\frac{\Delta}{\Delta t} N_{1\sigma_2}(t) \cdot N_{2\sigma_2}^{\sigma_1\sigma_2}(t) = 0, \quad t \in I.$$

Theorem 4.87 (The Joachimstahl Theorem) *Let σ_1 be Δ_1-differentiable on $\mathbb{T}_1$, σ_2 be Δ_2-differentiable on $\mathbb{T}_2$, f_r be σ_1-completely delta differentiable on U, $f_{rt_1}^{\Delta_1}$ be σ_j-completely delta differentiable, $f_{rt_2}^{\Delta_2\sigma_1}$ be σ_l-completely delta differentiable on U, $j,l \in \{1,2\}$, $r \in \{1,2\}$. Let also, S_1 and S_2 be two $\sigma_1^2\sigma_2$-oriented surfaces with local $\sigma_1^2\sigma_2$-parameterizations (U,f_1) and (U,f_2), respectively. Assume that $N_{1\sigma_1^2\sigma_2}$ and $N_{2\sigma_1^2\sigma_2}$ are two unit $\sigma_1^2\sigma_2$-normals of S_1 and S_2, respectively. Suppose that S_1 and S_2 intersect along the curve $g(t) = (g_1(t), g_2(t))$, $t \in I$, so that*

$$g_1(\sigma(t)) = \sigma_1(g_1(t)),$$
$$g_2(\sigma(t)) = \sigma_2(g_2(t)), \quad t \in I,$$

and

$$N_{1\sigma_1^2\sigma_2} \cdot N_{2\sigma_1^2\sigma_2} = const,$$

$$\frac{\Delta}{\Delta t} N_{1\sigma_1^2\sigma_2}(t) \| g^{\Delta}(t), \quad t \in I,$$

and

$$\frac{\Delta}{\Delta t} N_{2\sigma_1^2\sigma_2}(t) \| g^{\Delta}(t), \quad t \in I.$$

Then g is a $\sigma_1^2\sigma_2\sigma_j\sigma_l$-curvature line on S_1 if and only if

$$N_{1\sigma_1^2\sigma_2}^{\sigma_1\sigma_2}(t) \cdot \frac{\Delta}{\Delta t} N_{2\sigma_1^2\sigma_2}(t) = 0, \quad t \in I,$$

and g is a $\sigma_1^2\sigma_2\sigma_j\sigma_l$-curvature line on S_2 if and only if

$$\frac{\Delta}{\Delta t} N_{1\sigma_1^2\sigma_2}(t) \cdot N_{2\sigma_1^2\sigma_2}^{\sigma_1\sigma_2}(t) = 0, \quad t \in I.$$

Theorem 4.88 (The Joachimstahl Theorem) *Let σ_1 be Δ_1-differentiable on $\mathbb{T}_1$, σ_2 be Δ_2-differentiable on $\mathbb{T}_2$, f_r be σ_1-completely delta differentiable on U, $f_{rt_1}^{\Delta_1}$ be σ_j-completely delta differentiable, $f_{rt_2}^{\Delta_2\sigma_1}$ be $\sigma_l\sigma_2^2$-completely delta differentiable on U, $j,l \in \{1,2\}$, $r \in \{1,2\}$. Let also, S_1 and S_2 be two σ_1-oriented surfaces with local $\sigma_1\sigma_2^2$-parameterizations (U,f_1) and (U,f_2), respectively. Assume that $N_{1\sigma_1\sigma_2^2}$ and $N_{2\sigma_1\sigma_2^2}$ are two unit $\sigma_1\sigma_2^2$-normals of S_1 and S_2, respectively. Suppose that S_1 and S_2 intersect along the curve $g(t) = (g_1(t), g_2(t))$, $t \in I$, so that*

$$g_1(\sigma(t)) = \sigma_1(g_1(t)),$$
$$g_2(\sigma(t)) = \sigma_2(g_2(t)), \quad t \in I,$$

and

$$N_{1\sigma_1\sigma_2^2} \cdot N_{2\sigma_1\sigma_2^2} = const,$$

$$\frac{\Delta}{\Delta t} N_{1\sigma_1\sigma_2^2}(t) \| g^{\Delta}(t), \quad t \in I,$$

and

$$\frac{\Delta}{\Delta t} N_{2\sigma_1\sigma_2^2}(t) \| g^{\Delta}(t), \quad t \in I.$$

Then g is a $\sigma_1\sigma_2^2\sigma_j\sigma_l$-curvature line on S_1 if and only if

$$N_{1\sigma_1\sigma_2^2}^{\sigma_1\sigma_2}(t) \cdot \frac{\Delta}{\Delta t} N_{2\sigma_1\sigma_2^2}(t) = 0, \quad t \in I,$$

and g is a $\sigma_1\sigma_2^2\sigma_j\sigma_l$-curvature line on S_2 if and only if

$$\frac{\Delta}{\Delta t} N_{1\sigma_1\sigma_2^2}(t) \cdot N_{2\sigma_1\sigma_2^2}^{\sigma_1\sigma_2}(t) = 0, \quad t \in I.$$

4.6 The Deteremination of the Lines of Curvatures

Let S be a σ_1- or σ_2- or $\sigma_1^2\sigma_2$-or $\sigma_1\sigma_2^2$-oriented surface with a local σ_1- or σ_2- or $\sigma_1^2\sigma_2$-or $\sigma_1\sigma_2^2$ parameterization (U,f).

Theorem 4.89 *Suppose that S is a σ_1-regular surface, σ_1 is Δ_1-differentiable, $f_{t_1}^{\Delta_1}$ is σ_1-completely delta differentiable and $f_{t_1}^{\Delta_2\sigma_1}$ is σ_1-completely delta differentiable. A σ_1-tangent vector*

$$h = h_1 f_{t_1}^{\Delta_1} + h_2 f_{t_2}^{\Delta_2\sigma_1}$$

has $\sigma_1\sigma_1\sigma_1$-direction if and only if

$$b_{21n\sigma_1\sigma_1\sigma_1} h_1^2 + (b_{22n\sigma_1\sigma_1\sigma_1} - b_{11n\sigma_1\sigma_1\sigma_1}) h_1 h_2 - b_{12n\sigma_1\sigma_1\sigma_1} h_2^2 = 0.$$

Proof. A tangent vector h has a $\sigma_1\sigma_1\sigma_1$-principal direction if and only if it is an eigenvector of the matrix $\mathscr{B}_{n\sigma_1\sigma_1\sigma_1}$, that is, if and only if

$$\mathscr{B}_{n\sigma_1\sigma_1\sigma_1}(h) = \lambda h,$$

where λ is the eigenvalue corresponding to the eigenvector h. Therefore h has $\sigma_1\sigma_1\sigma_1$-principal direction if and only if

$$\mathscr{B}_{n\sigma_1\sigma_1\sigma_1}(h) \times h = 0.$$

By the definition of the matrix $\mathscr{B}_{n\sigma_1\sigma_1\sigma_1}$, we get

$$\begin{aligned}\mathscr{B}_{n\sigma_1\sigma_1\sigma_1}(h) &= \begin{pmatrix} b_{11n\sigma_1\sigma_1\sigma_1} & b_{12n\sigma_1\sigma_1\sigma_1} \\ b_{21n\sigma_1\sigma_1\sigma_1} & b_{22n\sigma_1\sigma_1\sigma_1} \end{pmatrix} \begin{pmatrix} h_1 \\ h_2 \end{pmatrix} \\ &= \begin{pmatrix} b_{11n\sigma_1\sigma_1\sigma_1} h_1 + b_{12n\sigma_1\sigma_1\sigma_1} h_2 \\ b_{21n\sigma_1\sigma_1\sigma_1} h_1 + b_{22n\sigma_1\sigma_1\sigma_1} h_2 \end{pmatrix}\end{aligned}$$

and

$$\begin{aligned}\mathscr{B}_{n\sigma_1\sigma_1\sigma_1}(h) &= (b_{11n\sigma_1\sigma_1\sigma_1} h_1 + b_{12n\sigma_1\sigma_1\sigma_1} h_2) f_{t_1}^{\Delta_1} \\ &\quad + (b_{21n\sigma_1\sigma_1\sigma_1} h_1 + b_{22n\sigma_1\sigma_1\sigma_1} h_2) f_{t_2}^{\Delta_2\sigma_1}.\end{aligned}$$

Hence,

$$\begin{aligned}\mathscr{B}_{n\sigma_1\sigma_1\sigma_1}(h) \times h &= \Big((b_{11n\sigma_1\sigma_1\sigma_1} h_1 + b_{12n\sigma_1\sigma_1\sigma_1} h_2) f_{t_1}^{\Delta_1} \\ &\quad + (b_{21n\sigma_1\sigma_1\sigma_1} h_1 + b_{22n\sigma_1\sigma_1\sigma_1} h_2) f_{t_2}^{\Delta_2\sigma_1} \Big) \\ &\quad \times \left(h_1 f_{t_1}^{\Delta_1} + h_2 f_{t_2}^{\Delta_2\sigma_1} \right) \\ &= \left(b_{21n\sigma_1\sigma_1\sigma_1} h_1^2 + b_{22n\sigma_1\sigma_1\sigma_1} h_1 h_2 \right) f_{t_2}^{\Delta_2\sigma_1} \times f_{t_1}^{\Delta_1} \\ &\quad - \left(b_{11n\sigma_1\sigma_1\sigma_1} h_1 h_2 + b_{12n\sigma_1\sigma_1\sigma_1} h_2^2 \right) f_{t_2}^{\Delta_2\sigma_1} \times f_{t_1}^{\Delta_1} \\ &= \Big(b_{21n\sigma_1\sigma_1\sigma_1} h_1^2 + (b_{22n\sigma_1\sigma_1\sigma_1} - b_{11n\sigma_1\sigma_1\sigma_1}) h_1 h_2 \\ &\quad - b_{12n\sigma_1\sigma_1\sigma_1} h_2^2 \Big) f_{t_2}^{\Delta_2\sigma_1} \times f_{t_1}^{\Delta_1}.\end{aligned}$$

Since

$$f_{t_2}^{\Delta_2\sigma_1} \times f_{t_1}^{\Delta_1} \neq 0,$$

we get

$$b_{21n\sigma_1\sigma_1\sigma_1} h_1^2 + (b_{22n\sigma_1\sigma_1\sigma_1} - b_{11n\sigma_1\sigma_1\sigma_1}) h_1 h_2 - b_{12n\sigma_1\sigma_1\sigma_1} h_2^2 = 0.$$

This completes the proof.

Corollary 4.17 *Suppose that all conditions of Theorem 4.89 hold. Let $g = g(t)$, $t \in I$, be a curve lying in the domain $f(U)$ of the local parameterization (U,f) of the surface S with the local equation*

$$g(t) = f(g_1(t), g_2(t)), \quad t \in I,$$

such that

$$g_1(\sigma(t)) = \sigma_1(g_1(t)),$$
$$g_2(\sigma(t)) = \sigma_2(g_2(t)), \quad t \in I.$$

Then g is a line of $\sigma_1\sigma_1\sigma_1$ curvature on S if and only if

$$b_{21n\sigma_1\sigma_1\sigma_1}\left(g_1^{\Delta}\right)^2 + (b_{22n\sigma_1\sigma_1\sigma_1} - b_{11n\sigma_1\sigma_1\sigma_1})g_1^{\Delta}g_2^{\Delta} - b_{12n\sigma_1\sigma_1\sigma_1}\left(g_2^{\Delta}\right)^2 = 0.$$

Proof. Note that

$$g^{\Delta} = f_{t_1}^{\Delta_1}g_1^{\Delta} + f_{t_2}^{\Delta_2\sigma_1}g_2^{\Delta}.$$

Then we apply Theorem 4.89 for

$$h_1 = g_1^{\Delta} \quad \text{and} \quad h_2 = g_2^{\Delta}.$$

This completes the proof.

Theorem 4.90 *Suppose that S is a σ_1-regular surface, σ_1 is Δ_1-differentiable, $f_{t_1}^{\Delta_1}$ is σ_1-completely delta differentiable and $f_{t_1}^{\Delta_2\sigma_1}$ is σ_2-completely delta differentiable. A σ_1-tangent vector*

$$h = h_1f_{t_1}^{\Delta_1} + h_2f_{t_2}^{\Delta_2\sigma_1}$$

has $\sigma_1\sigma_1\sigma_2$-direction if and only if

$$b_{21n\sigma_1\sigma_1\sigma_2}h_1^2 + (b_{22n\sigma_1\sigma_1\sigma_2} - b_{11n\sigma_1\sigma_1\sigma_2})h_1h_2 - b_{12n\sigma_1\sigma_1\sigma_2}h_2^2 = 0.$$

Corollary 4.18 *Suppose that all conditions of Theorem 4.90 hold. Let $g = g(t)$, $t \in I$, be a curve lying in the domain $f(U)$ of the local parametrization (U,f) of the surface S with the local equation*

$$g(t) = f(g_1(t), g_2(t)), \quad t \in I,$$

such that

$$g_1(\sigma(t)) = \sigma_1(g_1(t)),$$
$$g_2(\sigma(t)) = \sigma_2(g_2(t)), \quad t \in I.$$

Then g is a line of $\sigma_1\sigma_1\sigma_2$ curvature on S if and only if

$$b_{21n\sigma_1\sigma_1\sigma_2}\left(g_1^{\Delta}\right)^2 + (b_{22n\sigma_1\sigma_1\sigma_2} - b_{11n\sigma_1\sigma_1\sigma_2})g_1^{\Delta}g_2^{\Delta} - b_{12n\sigma_1\sigma_1\sigma_2}\left(g_2^{\Delta}\right)^2 = 0.$$

Theorem 4.91 *Suppose that S is a σ_1-regular surface, σ_1 is Δ_1-differentiable, $f_{t_1}^{\Delta_1}$ is σ_2-completely delta differentiable and $f_{t_1}^{\Delta_2\sigma_1}$ is σ_1-completely delta differentiable. A σ_1-tangent vector*

$$h = h_1 f_{t_1}^{\Delta_1} + h_2 f_{t_2}^{\Delta_2\sigma_1}$$

has $\sigma_1\sigma_2\sigma_1$-direction if and only if

$$b_{21n\sigma_1\sigma_2\sigma_1} h_1^2 + (b_{22n\sigma_1\sigma_2\sigma_1} - b_{11n\sigma_1\sigma_2\sigma_1}) h_1 h_2 - b_{12n\sigma_1\sigma_2\sigma_1} h_2^2 = 0.$$

Corollary 4.19 *Suppose that all conditions of Theorem 4.91 hold. Let $g = g(t)$, $t \in I$, be a curve lying in the domain $f(U)$ of the local parametrization (U,f) of the surface S with the local equation*

$$g(t) = f(g_1(t), g_2(t)), \quad t \in I,$$

such that

$$g_1(\sigma(t)) = \sigma_1(g_1(t)),$$
$$g_2(\sigma(t)) = \sigma_2(g_2(t)), \quad t \in I.$$

Then g is a line of $\sigma_1\sigma_2\sigma_1$ curvature on S if and only if

$$b_{21n\sigma_1\sigma_2\sigma_1} \left(g_1^\Delta\right)^2 + (b_{22n\sigma_1\sigma_2\sigma_1} - b_{11n\sigma_1\sigma_2\sigma_1}) g_1^\Delta g_2^\Delta - b_{12n\sigma_1\sigma_2\sigma_1} \left(g_2^\Delta\right)^2 = 0.$$

Theorem 4.92 *Suppose that S is a σ_1-regular surface, σ_1 is Δ_1-differentiable, $f_{t_1}^{\Delta_1}$ is σ_2-completely delta differentiable and $f_{t_1}^{\Delta_2\sigma_1}$ is σ_2-completely delta differentiable. A σ_1-tangent vector*

$$h = h_1 f_{t_1}^{\Delta_1} + h_2 f_{t_2}^{\Delta_2\sigma_1}$$

has $\sigma_1\sigma_2\sigma_2$-direction if and only if

$$b_{21n\sigma_1\sigma_2\sigma_2} h_1^2 + (b_{22n\sigma_1\sigma_2\sigma_2} - b_{11n\sigma_1\sigma_2\sigma_2}) h_1 h_2 - b_{12n\sigma_1\sigma_2\sigma_2} h_2^2 = 0.$$

Corollary 4.20 *Suppose that all conditions of Theorem 4.92 hold. Let $g = g(t)$, $t \in I$, be a curve lying in the domain $f(U)$ of the local parametrization (U,f) of the surface S with the local equation*

$$g(t) = f(g_1(t), g_2(t)), \quad t \in I,$$

such that

$$g_1(\sigma(t)) = \sigma_1(g_1(t)),$$
$$g_2(\sigma(t)) = \sigma_2(g_2(t)), \quad t \in I.$$

Then g is a line of $\sigma_1\sigma_2\sigma_2$ curvature on S if and only if

$$b_{21n\sigma_1\sigma_2\sigma_2} \left(g_1^\Delta\right)^2 + (b_{22n\sigma_1\sigma_2\sigma_2} - b_{11n\sigma_1\sigma_2\sigma_2}) g_1^\Delta g_2^\Delta - b_{12n\sigma_1\sigma_2\sigma_2} \left(g_2^\Delta\right)^2 = 0.$$

Theorem 4.93 *Suppose that S is a σ_2-regular surface, σ_2 is Δ_2-differentiable, $f_{t_1}^{\Delta_1\sigma_2}$ is σ_1-completely delta differentiable and $f_{t_1}^{\Delta_2}$ is σ_1-completely delta differentiable. A σ_2-tangent vector*

$$h = h_1 f_{t_1}^{\Delta_1\sigma_2} + h_2 f_{t_2}^{\Delta_2}$$

has $\sigma_2\sigma_1\sigma_1$-direction if and only if

$$b_{21n\sigma_2\sigma_1\sigma_1} h_1^2 + (b_{22n\sigma_2\sigma_1\sigma_1} - b_{11n\sigma_2\sigma_1\sigma_1}) h_1 h_2 - b_{12n\sigma_2\sigma_1\sigma_1} h_2^2 = 0.$$

Corollary 4.21 *Suppose that all conditions of Theorem 4.93 hold. Let $g = g(t)$, $t \in I$, be a curve lying in the domain $f(U)$ of the local parametrization (U,f) of the surface S with the local equation*

$$g(t) = f(g_1(t), g_2(t)), \quad t \in I,$$

such that

$$g_1(\sigma(t)) = \sigma_1(g_1(t)),$$
$$g_2(\sigma(t)) = \sigma_2(g_2(t)), \quad t \in I.$$

Then g is a line of $\sigma_2\sigma_1\sigma_1$ curvature on S if and only if

$$b_{21n\sigma_2\sigma_1\sigma_1} \left(g_1^{\Delta}\right)^2 + (b_{22n\sigma_2\sigma_1\sigma_1} - b_{11n\sigma_2\sigma_1\sigma_1}) g_1^{\Delta} g_2^{\Delta} - b_{12n\sigma_2\sigma_1\sigma_1} \left(g_2^{\Delta}\right)^2 = 0.$$

Theorem 4.94 *Suppose that S is a σ_2-regular surface, σ_2 is Δ_2-differentiable, $f_{t_1}^{\Delta_1\sigma_2}$ is σ_1-completely delta differentiable and $f_{t_1}^{\Delta_2}$ is σ_2-completely delta differentiable. A σ_2-tangent vector*

$$h = h_1 f_{t_1}^{\Delta_1\sigma_2} + h_2 f_{t_2}^{\Delta_2}$$

has $\sigma_2\sigma_1\sigma_2$-direction if and only if

$$b_{21n\sigma_2\sigma_1\sigma_2} h_1^2 + (b_{22n\sigma_2\sigma_1\sigma_2} - b_{11n\sigma_2\sigma_1\sigma_2}) h_1 h_2 - b_{12n\sigma_2\sigma_1\sigma_2} h_2^2 = 0.$$

Corollary 4.22 *Suppose that all conditions of Theorem 4.94 hold. Let $g = g(t)$, $t \in I$, be a curve lying in the domain $f(U)$ of the local parametrization (U,f) of the surface S with the local equation*

$$g(t) = f(g_1(t), g_2(t)), \quad t \in I,$$

such that

$$g_1(\sigma(t)) = \sigma_1(g_1(t)),$$
$$g_2(\sigma(t)) = \sigma_2(g_2(t)), \quad t \in I.$$

Then g is a line of $\sigma_2\sigma_1\sigma_2$ curvature on S if and only if

$$b_{21n\sigma_2\sigma_1\sigma_2} \left(g_1^{\Delta}\right)^2 + (b_{22n\sigma_2\sigma_1\sigma_2} - b_{11n\sigma_2\sigma_1\sigma_2}) g_1^{\Delta} g_2^{\Delta} - b_{12n\sigma_2\sigma_1\sigma_2} \left(g_2^{\Delta}\right)^2 = 0.$$

Theorem 4.95 *Suppose that S is a σ_2-regular surface, σ_2 is Δ_2-differentiable, $f_{t_1}^{\Delta_1\sigma_2}$ is σ_2-completely delta differentiable and $f_{t_1}^{\Delta_2}$ is σ_1-completely delta differentiable. A σ_2-tangent vector*

$$h = h_1 f_{t_1}^{\Delta_1\sigma_2} + h_2 f_{t_2}^{\Delta_2}$$

has $\sigma_2\sigma_2\sigma_1$-direction if and only if

$$b_{21n\sigma_2\sigma_2\sigma_1} h_1^2 + (b_{22n\sigma_2\sigma_2\sigma_1} - b_{11n\sigma_2\sigma_2\sigma_1}) h_1 h_2 - b_{12n\sigma_2\sigma_2\sigma_1} h_2^2 = 0.$$

Corollary 4.23 *Suppose that all conditions of Theorem 4.95 hold. Let $g = g(t)$, $t \in I$, be a curve lying in the domain $f(U)$ of the local parametrization (U,f) of the surface S with the local equation*

$$g(t) = f(g_1(t), g_2(t)), \quad t \in I,$$

such that

$$g_1(\sigma(t)) = \sigma_1(g_1(t)),$$
$$g_2(\sigma(t)) = \sigma_2(g_2(t)), \quad t \in I.$$

Then g is a line of $\sigma_2\sigma_2\sigma_1$ curvature on S if and only if

$$b_{21n\sigma_2\sigma_2\sigma_1} \left(g_1^\Delta\right)^2 + (b_{22n\sigma_2\sigma_2\sigma_1} - b_{11n\sigma_2\sigma_2\sigma_1}) g_1^\Delta g_2^\Delta - b_{12n\sigma_2\sigma_2\sigma_1} \left(g_2^\Delta\right)^2 = 0.$$

Theorem 4.96 *Suppose that S is a σ_2-regular surface, σ_2 is Δ_2-differentiable, $f_{t_1}^{\Delta_1\sigma_2}$ is σ_2-completely delta differentiable and $f_{t_1}^{\Delta_2}$ is σ_2-completely delta differentiable. A σ_2-tangent vector*

$$h = h_1 f_{t_1}^{\Delta_1\sigma_2} + h_2 f_{t_2}^{\Delta_2}$$

has $\sigma_2\sigma_2\sigma_2$-direction if and only if

$$b_{21n\sigma_2\sigma_2\sigma_2} h_1^2 + (b_{22n\sigma_2\sigma_2\sigma_2} - b_{11n\sigma_2\sigma_2\sigma_2}) h_1 h_2 - b_{12n\sigma_2\sigma_2\sigma_2} h_2^2 = 0.$$

Corollary 4.24 *Suppose that all conditions of Theorem 4.96 hold. Let $g = g(t)$, $t \in I$, be a curve lying in the domain $f(U)$ of the local parametrization (U,f) of the surface S with the local equation*

$$g(t) = f(g_1(t), g_2(t)), \quad t \in I,$$

such that

$$g_1(\sigma(t)) = \sigma_1(g_1(t)),$$
$$g_2(\sigma(t)) = \sigma_2(g_2(t)), \quad t \in I.$$

Then g is a line of $\sigma_2\sigma_2\sigma_2$ curvature on S if and only if

$$b_{21n\sigma_2\sigma_2\sigma_2} \left(g_1^\Delta\right)^2 + (b_{22n\sigma_2\sigma_2\sigma_2} - b_{11n\sigma_2\sigma_2\sigma_2}) g_1^\Delta g_2^\Delta - b_{12n\sigma_2\sigma_2\sigma_2} \left(g_2^\Delta\right)^2 = 0.$$

Theorem 4.97 *Suppose that S is a $\sigma_1^2\sigma_2$-regular surface, σ_1 is Δ_1-differentiable, $f_{t_1}^{\Delta_1\sigma_1\sigma_2}$ is σ_1-completely delta differentiable and $f_{t_1}^{\Delta_2\sigma_1^2\sigma_2}$ is σ_1-completely delta differentiable. A $\sigma_1^2\sigma_2$-tangent vector*

$$h = h_1 f_{t_1}^{\Delta_1\sigma_1\sigma_2} + h_2 f_{t_2}^{\Delta_2\sigma_1^2\sigma_2}$$

has $\sigma_1^2\sigma_2\sigma_1\sigma_1$-direction if and only if

$$b_{21n\sigma_1^2\sigma_2\sigma_1\sigma_1} h_1^2 + \left(b_{22n\sigma_1^2\sigma_2\sigma_1\sigma_1} - b_{11n\sigma_1^2\sigma_2\sigma_1\sigma_1}\right) h_1 h_2 - b_{12n\sigma_1^2\sigma_2\sigma_1\sigma_1} h_2^2 = 0.$$

Corollary 4.25 *Suppose that all conditions of Theorem 4.97 hold. Let $g = g(t)$, $t \in I$, be a curve lying in the domain $f(U)$ of the local parametrization (U,f) of the surface S with the local equation*

$$g(t) = f(g_1(t), g_2(t)), \quad t \in I,$$

such that

$$g_1(\sigma(t)) = \sigma_1(g_1(t)),$$
$$g_2(\sigma(t)) = \sigma_2(g_2(t)), \quad t \in I.$$

Then g is a line of $\sigma_1^2\sigma_2\sigma_1\sigma_1$ curvature on S if and only if

$$b_{21n\sigma_1^2\sigma_2\sigma_1\sigma_1} \left(g_1^\Delta\right)^2 + \left(b_{22n\sigma_1^2\sigma_2\sigma_1\sigma_1} - b_{11n\sigma_1^2\sigma_2\sigma_1\sigma_1}\right) g_1^\Delta g_2^\Delta - b_{12n\sigma_1^2\sigma_2\sigma_1\sigma_1} \left(g_2^\Delta\right)^2 = 0.$$

Theorem 4.98 *Suppose that S is a $\sigma_1^2\sigma_2$-regular surface, σ_1 is Δ_1-differentiable, $f_{t_1}^{\Delta_1\sigma_1\sigma_2}$ is σ_1-completely delta differentiable and $f_{t_1}^{\Delta_2\sigma_1^2\sigma_2}$ is σ_2-completely delta differentiable. A $\sigma_1^2\sigma_2$-tangent vector*

$$h = h_1 f_{t_1}^{\Delta_1\sigma_1\sigma_2} + h_2 f_{t_2}^{\Delta_2\sigma_1^2\sigma_2}$$

has $\sigma_1^2\sigma_2\sigma_1\sigma_2$-direction if and only if

$$b_{21n\sigma_1^2\sigma_2\sigma_1\sigma_2} h_1^2 + \left(b_{22n\sigma_1^2\sigma_2\sigma_1\sigma_2} - b_{11n\sigma_1^2\sigma_2\sigma_1\sigma_2}\right) h_1 h_2 - b_{12n\sigma_1^2\sigma_2\sigma_1\sigma_2} h_2^2 = 0.$$

Corollary 4.26 *Suppose that all conditions of Theorem 4.98 hold. Let $g = g(t)$, $t \in I$, be a curve lying in the domain $f(U)$ of the local parametrization (U,f) of the surface S with the local equation*

$$g(t) = f(g_1(t), g_2(t)), \quad t \in I,$$

such that

$$g_1(\sigma(t)) = \sigma_1(g_1(t)),$$
$$g_2(\sigma(t)) = \sigma_2(g_2(t)), \quad t \in I.$$

Then g is a line of $\sigma_1^2\sigma_2\sigma_1\sigma_2$ curvature on S if and only if

$$b_{21n\sigma_1^2\sigma_2\sigma_1\sigma_2} \left(g_1^\Delta\right)^2 + \left(b_{22n\sigma_1^2\sigma_2\sigma_1\sigma_2} - b_{11n\sigma_1^2\sigma_2\sigma_1\sigma_2}\right) g_1^\Delta g_2^\Delta - b_{12n\sigma_1^2\sigma_2\sigma_1\sigma_2} \left(g_2^\Delta\right)^2 = 0.$$

Theorem 4.99 *Suppose that S is a $\sigma_1^2\sigma_2$-regular surface, σ_1 is Δ_1-differentiable, $f_{t_1}^{\Delta_1\sigma_1\sigma_2}$ is σ_2-completely delta differentiable and $f_{t_1}^{\Delta_2\sigma_1^2\sigma_2}$ is σ_1-completely delta differentiable. A $\sigma_1^2\sigma_2$-tangent vector*

$$h = h_1 f_{t_1}^{\Delta_1\sigma_1\sigma_2} + h_2 f_{t_2}^{\Delta_2\sigma_1^2\sigma_2}$$

has $\sigma_1^2\sigma_2\sigma_2\sigma_1$-direction if and only if

$$b_{21n\sigma_1^2\sigma_2\sigma_2\sigma_1} h_1^2 + \left(b_{22n\sigma_1^2\sigma_2\sigma_2\sigma_1} - b_{11n\sigma_1^2\sigma_2\sigma_2\sigma_1}\right) h_1 h_2 - b_{12n\sigma_1^2\sigma_2\sigma_2\sigma_1} h_2^2 = 0.$$

Corollary 4.27 *Suppose that all conditions of Theorem 4.99 hold. Let $g = g(t)$, $t \in I$, be a curve lying in the domain $f(U)$ of the local parametrization (U,f) of the surface S with the local equation*

$$g(t) = f(g_1(t), g_2(t)), \quad t \in I,$$

such that

$$g_1(\sigma(t)) = \sigma_1(g_1(t)),$$
$$g_2(\sigma(t)) = \sigma_2(g_2(t)), \quad t \in I.$$

Then g is a line of $\sigma_1^2\sigma_2\sigma_2\sigma_1$ curvature on S if and only if

$$b_{21n\sigma_1^2\sigma_2\sigma_2\sigma_1} \left(g_1^\Delta\right)^2 + \left(b_{22n\sigma_1^2\sigma_2\sigma_2\sigma_1} - b_{11n\sigma_1^2\sigma_2\sigma_2\sigma_1}\right) g_1^\Delta g_2^\Delta - b_{12n\sigma_1^2\sigma_2\sigma_2\sigma_1} \left(g_2^\Delta\right)^2 = 0.$$

Theorem 4.100 *Suppose that S is a $\sigma_1^2\sigma_2$-regular surface, σ_1 is Δ_1-differentiable, $f_{t_1}^{\Delta_1\sigma_1\sigma_2}$ is σ_2-completely delta differentiable and $f_{t_1}^{\Delta_2\sigma_1^2\sigma_2}$ is σ_2-completely delta differentiable. A $\sigma_1^2\sigma_2$-tangent vector*

$$h = h_1 f_{t_1}^{\Delta_1\sigma_1\sigma_2} + h_2 f_{t_2}^{\Delta_2\sigma_1^2\sigma_2}$$

has $\sigma_1^2\sigma_2\sigma_2\sigma_2$-direction if and only if

$$b_{21n\sigma_1^2\sigma_2\sigma_2\sigma_2} h_1^2 + \left(b_{22n\sigma_1^2\sigma_2\sigma_2\sigma_2} - b_{11n\sigma_1^2\sigma_2\sigma_2\sigma_2}\right) h_1 h_2 - b_{12n\sigma_1^2\sigma_2\sigma_2\sigma_2} h_2^2 = 0.$$

Corollary 4.28 *Suppose that all conditions of Theorem 4.104 hold. Let $g = g(t)$, $t \in I$, be a curve lying in the domain $f(U)$ of the local parametrization (U,f) of the surface S with the local equation*

$$g(t) = f(g_1(t), g_2(t)), \quad t \in I,$$

such that

$$g_1(\sigma(t)) = \sigma_1(g_1(t)),$$
$$g_2(\sigma(t)) = \sigma_2(g_2(t)), \quad t \in I.$$

Then g is a line of $\sigma_1^2\sigma_2\sigma_2\sigma_2$ curvature on S if and only if

$$b_{21n\sigma_1^2\sigma_2\sigma_2\sigma_2} \left(g_1^\Delta\right)^2 + \left(b_{22n\sigma_1^2\sigma_2\sigma_2\sigma_2} - b_{11n\sigma_1^2\sigma_2\sigma_2\sigma_2}\right) g_1^\Delta g_2^\Delta - b_{12n\sigma_1^2\sigma_2\sigma_2\sigma_2} \left(g_2^\Delta\right)^2 = 0.$$

Theorem 4.101 *Suppose that S is a $\sigma_1\sigma_2^2$-regular surface, σ_2 is Δ_2-differentiable, $f_{t_1}^{\Delta_1\sigma_1\sigma_2^2}$ is σ_1-completely delta differentiable and $f_{t_1}^{\Delta_2\sigma_1\sigma_2}$ is σ_1-completely delta differentiable. A σ_2-tangent vector*

$$h = h_1 f_{t_1}^{\Delta_1\sigma_1\sigma_2^2} + h_2 f_{t_2}^{\Delta_2\sigma_1\sigma_2}$$

has $\sigma_1\sigma_2^2\sigma_1\sigma_1$-direction if and only if

$$b_{21n\sigma_1\sigma_2^2\sigma_1\sigma_1} h_1^2 + \left(b_{22n\sigma_1\sigma_2^2\sigma_1\sigma_1} - b_{11n\sigma_1\sigma_2^2\sigma_1\sigma_1}\right) h_1 h_2 - b_{12n\sigma_1\sigma_2^2\sigma_1\sigma_1} h_2^2 = 0.$$

Corollary 4.29 *Suppose that all conditions of Theorem 4.101 hold. Let $g = g(t)$, $t \in I$, be a curve lying in the domain $f(U)$ of the local parametrization (U,f) of the surface S with the local equation*

$$g(t) = f(g_1(t), g_2(t)), \quad t \in I,$$

such that

$$g_1(\sigma(t)) = \sigma_1(g_1(t)),$$
$$g_2(\sigma(t)) = \sigma_2(g_2(t)), \quad t \in I.$$

Then g is a line of $\sigma_1\sigma_2^2\sigma_1\sigma_1$ curvature on S if and only if

$$b_{21n\sigma_1\sigma_2^2\sigma_1\sigma_1}\left(g_1^{\Delta}\right)^2 + \left(b_{22n\sigma_1\sigma_2^2\sigma_1\sigma_1} - b_{11n\sigma_1\sigma_2^2\sigma_1\sigma_1}\right) g_1^{\Delta} g_2^{\Delta} - b_{12n\sigma_1\sigma_2^2\sigma_1\sigma_1}\left(g_2^{\Delta}\right)^2 = 0.$$

Theorem 4.102 *Suppose that S is a $\sigma_1\sigma_2^2$-regular surface, σ_2 is Δ_2-differentiable, $f_{t_1}^{\Delta_1\sigma_1\sigma_2^2}$ is σ_1-completely delta differentiable and $f_{t_1}^{\Delta_2\sigma_1\sigma_2}$ is σ_2-completely delta differentiable. A σ_2-tangent vector*

$$h = h_1 f_{t_1}^{\Delta_1\sigma_1\sigma_2^2} + h_2 f_{t_2}^{\Delta_2\sigma_1\sigma_2}$$

has $\sigma_1\sigma_2^2\sigma_1\sigma_2$-direction if and only if

$$b_{21n\sigma_1\sigma_2^2\sigma_1\sigma_2} h_1^2 + \left(b_{22n\sigma_1\sigma_2^2\sigma_1\sigma_2} - b_{11n\sigma_1\sigma_2^2\sigma_1\sigma_2}\right) h_1 h_2 - b_{12n\sigma_1\sigma_2^2\sigma_1\sigma_2} h_2^2 = 0.$$

Corollary 4.30 *Suppose that all conditions of Theorem 4.102 hold. Let $g = g(t)$, $t \in I$, be a curve lying in the domain $f(U)$ of the local parametrization (U,f) of the surface S with the local equation*

$$g(t) = f(g_1(t), g_2(t)), \quad t \in I,$$

such that

$$g_1(\sigma(t)) = \sigma_1(g_1(t)),$$
$$g_2(\sigma(t)) = \sigma_2(g_2(t)), \quad t \in I.$$

Then g is a line of $\sigma_1\sigma_2^2\sigma_1\sigma_2$ curvature on S if and only if

$$b_{21n\sigma_1\sigma_2^2\sigma_1\sigma_2}\left(g_1^{\Delta}\right)^2 + \left(b_{22n\sigma_1\sigma_2^2\sigma_1\sigma_2} - b_{11n\sigma_1\sigma_2^2\sigma_1\sigma_2}\right) g_1^{\Delta} g_2^{\Delta} - b_{12n\sigma_1\sigma_2^2\sigma_1\sigma_2}\left(g_2^{\Delta}\right)^2 = 0.$$

Theorem 4.103 *Suppose that S is a $\sigma_1\sigma_2^2$-regular surface, σ_2 is Δ_2-differentiable, $f_{t_1}^{\Delta_1\sigma_1\sigma_2^2}$ is σ_2-completely delta differentiable and $f_{t_1}^{\Delta_2\sigma_1\sigma_2}$ is σ_1-completely delta differentiable. A σ_2-tangent vector*

$$h = h_1 f_{t_1}^{\Delta_1\sigma_1\sigma_2^2} + h_2 f_{t_2}^{\Delta_2\sigma_1\sigma_2}$$

has $\sigma_1\sigma_2^2\sigma_2\sigma_1$-direction if and only if

$$b_{21n\sigma_1\sigma_2^2\sigma_2\sigma_1} h_1^2 + \left(b_{22n\sigma_1\sigma_2^2\sigma_2\sigma_1} - b_{11n\sigma_1\sigma_2^2\sigma_2\sigma_1}\right) h_1 h_2 - b_{12n\sigma_1\sigma_2^2\sigma_2\sigma_1} h_2^2 = 0.$$

Corollary 4.31 *Suppose that all conditions of Theorem 4.103 hold. Let $g = g(t)$, $t \in I$, be a curve lying in the domain $f(U)$ of the local parametrization (U,f) of the surface S with the local equation*

$$g(t) = f(g_1(t), g_2(t)), \quad t \in I,$$

such that

$$g_1(\sigma(t)) = \sigma_1(g_1(t)),$$
$$g_2(\sigma(t)) = \sigma_2(g_2(t)), \quad t \in I.$$

Then g is a line of $\sigma_1\sigma_2^2\sigma_2\sigma_1$ curvature on S if and only if

$$b_{21n\sigma_1\sigma_2^2\sigma_2\sigma_1} \left(g_1^{\Delta}\right)^2 + \left(b_{22n\sigma_1\sigma_2^2\sigma_2\sigma_1} - b_{11n\sigma_1\sigma_2^2\sigma_2\sigma_1}\right) g_1^{\Delta} g_2^{\Delta} - b_{12n\sigma_1\sigma_2^2\sigma_2\sigma_1} \left(g_2^{\Delta}\right)^2 = 0.$$

Theorem 4.104 *Suppose that S is a $\sigma_1\sigma_2^2$-regular surface, σ_2 is Δ_2-differentiable, $f_{t_1}^{\Delta_1\sigma_1\sigma_2^2}$ is σ_2-completely delta differentiable and $f_{t_1}^{\Delta_2\sigma_1\sigma_2}$ is σ_2-completely delta differentiable. A σ_2-tangent vector*

$$h = h_1 f_{t_1}^{\Delta_1\sigma_1\sigma_2^2} + h_2 f_{t_2}^{\Delta_2\sigma_1\sigma_2}$$

has $\sigma_1\sigma_2^2\sigma_2\sigma_2$-direction if and only if

$$b_{21n\sigma_1\sigma_2^2\sigma_2\sigma_2} h_1^2 + \left(b_{22n\sigma_1\sigma_2^2\sigma_2\sigma_2} - b_{11n\sigma_1\sigma_2^2\sigma_2\sigma_2}\right) h_1 h_2 - b_{12n\sigma_1\sigma_2^2\sigma_2\sigma_2} h_2^2 = 0.$$

Corollary 4.32 *Suppose that all conditions of Theorem 4.104 hold. Let $g = g(t)$, $t \in I$, be a curve lying in the domain $f(U)$ of the local parametrization (U,f) of the surface S with the local equation*

$$g(t) = f(g_1(t), g_2(t)), \quad t \in I,$$

such that

$$g_1(\sigma(t)) = \sigma_1(g_1(t)),$$
$$g_2(\sigma(t)) = \sigma_2(g_2(t)), \quad t \in I.$$

Then g is a line of $\sigma_2\sigma_2\sigma_2$ curvature on S if and only if

$$b_{21n\sigma_1\sigma_2^2\sigma_2\sigma_2} \left(g_1^{\Delta}\right)^2 + \left(b_{22n\sigma_1\sigma_2^2\sigma_2\sigma_2} - b_{11n\sigma_1\sigma_2^2\sigma_2\sigma_2}\right) g_1^{\Delta} g_2^{\Delta} - b_{12n\sigma_1\sigma_2^2\sigma_2\sigma_2} \left(g_2^{\Delta}\right)^2 = 0.$$

4.7 The Computation of the Curvatures of a Surface

Let S be a σ_1- or σ_2- or $\sigma_1^2\sigma_2$- or $\sigma_1\sigma_2^2$-oriented surface with a local σ_1- or σ_2- or $\sigma_1^2\sigma_2$- or $\sigma_1\sigma_2^2$- parameterization (U,f).

Theorem 4.105 *Let S be a σ_1-regular surface, σ_1 be Δ_1-differentiable on $\mathbb{T}_1$, $f_{t_1}^{\Delta_1}$ and $f_{t_2}^{\Delta_2\sigma_1}$ be σ_1-completely delta differentiable on U. Then*

$$k_{t\sigma_1\sigma_1\sigma_1} = \left(f_{t_1}^{\Delta_1^2}\cdot N_{\sigma_1}^{\sigma_1\sigma_2}\right)\left(f_{t_2}^{\Delta_2^2\sigma_1^2}\cdot N_{\sigma_1}^{\sigma_1\sigma_2}\right) - \left(1+\mu_1^{\Delta_1}\right)\left(f_{t_1t_2}^{\Delta_1\Delta_2\sigma_1}\cdot N_{\sigma_1}^{\sigma_1\sigma_2}\right)^2,$$

$$k_{m\sigma_1\sigma_1\sigma_1} = \frac{1}{2}\left(f_{t_1}^{\Delta_1^2}+f_{t_2}^{\Delta_2^2\sigma_1^2}\right)\cdot N_{\sigma_1}^{\sigma_1\sigma_2}.$$

Proof. We have

$$k_{t\sigma_1\sigma_1\sigma_1} = k_{1\sigma_1\sigma_1\sigma_1}k_{2\sigma_1\sigma_1\sigma_1}$$

and

$$k_{m\sigma_1\sigma_1\sigma_1} = \frac{1}{2}\left(k_{1\sigma_1\sigma_1\sigma_1}+k_{2\sigma_1\sigma_1\sigma_1}\right),$$

where $k_{1\sigma_1\sigma_1\sigma_1}$ and $k_{2\sigma_1\sigma_1\sigma_1}$ are the eigenvalues of the matrix $\mathscr{B}_{n\sigma_1\sigma_1\sigma_1}$. Therefore

$$\begin{aligned} k_{t\sigma_1\sigma_1\sigma_1} &= \det\begin{pmatrix} f_{t_1}^{\Delta_1^2}\cdot N_{\sigma_1}^{\sigma_1\sigma_2} & f_{t_1t_2}^{\Delta_1\Delta_2\sigma_1}\cdot N_{\sigma_1}^{\sigma_1\sigma_2} \\ (1+\mu_1^{\Delta_1})f_{t_1t_2}^{\Delta_1\Delta_2\sigma_1}\cdot N_{\sigma_1}^{\sigma_1\sigma_2} & f_{t_2}^{\Delta_2^2\sigma_1^2}\cdot N_{\sigma_1}^{\sigma_1\sigma_2}\end{pmatrix} \\ &= \left(f_{t_1}^{\Delta_1^2}\cdot N_{\sigma_1}^{\sigma_1\sigma_2}\right)\left(f_{t_2}^{\Delta_2^2\sigma_1^2}\cdot N_{\sigma_1}^{\sigma_1\sigma_2}\right) - \left(1+\mu_1^{\Delta_1}\right)\left(f_{t_1t_2}^{\Delta_1\Delta_2\sigma_1}\cdot N_{\sigma_1}^{\sigma_1\sigma_2}\right)^2 \end{aligned}$$

and

$$\begin{aligned} k_{m\sigma_1\sigma_1\sigma_1} &= \frac{1}{2}\mathrm{Tr}\begin{pmatrix} f_{t_1}^{\Delta_1^2}\cdot N_{\sigma_1}^{\sigma_1\sigma_2} & f_{t_1t_2}^{\Delta_1\Delta_2\sigma_1}\cdot N_{\sigma_1}^{\sigma_1\sigma_2} \\ (1+\mu_1^{\Delta_1})f_{t_1t_2}^{\Delta_1\Delta_2\sigma_1}\cdot N_{\sigma_1}^{\sigma_1\sigma_2} & f_{t_2}^{\Delta_2^2\sigma_1^2}\cdot N_{\sigma_1}^{\sigma_1\sigma_2}\end{pmatrix} \\ &= \frac{1}{2}\left(f_{t_1}^{\Delta_1^2}+f_{t_2}^{\Delta_2^2\sigma_1^2}\right)\cdot N_{\sigma_1}^{\sigma_1\sigma_2}. \end{aligned}$$

This completes the proof.

Example 4.1 Let $\mathbb{T}_1=\mathbb{T}_2=\mathbb{N}_0^2$, $\mathbb{T}_{(1)}=\mathbb{T}_{(2)}=\mathbb{T}_{(3)}=\mathbb{R}$, S be an oriented surface with a local parameterization (U,f), where $U\subseteq\mathbb{T}_1\times\mathbb{T}_2$, $f:U\to\mathbb{R}^3$ is given by

$$f(t_1,t_2)=(t_1t_2,t_1,t_2),\quad (t_1,t_2)\in\mathbb{T}_1\times\mathbb{T}_2.$$

Here

$$\begin{aligned} \sigma_1(t_1) &= (\sqrt{t_1}+1)^2, \\ \mu_1(t_1) &= \sigma_1(t_1)-t_1 \\ &= t_1+2\sqrt{t_1}+1-t_1 \\ &= 2\sqrt{t_1}+1,\quad t_1\in\mathbb{T}_1, \end{aligned}$$

$$\begin{aligned}
\sigma_2(t_2) &= (\sqrt{t_2}+1)^2,\\
\mu_2(t_2) &= 2\sqrt{t_2}+1, \quad t_2 \in \mathbb{T}_2,\\
f_1(t_1,t_2) &= t_1 t_2,\\
f_2(t_1,t_2) &= t_1,\\
f_3(t_1,t_2) &= t_2, \quad (t_1,t_2) \in U.
\end{aligned}$$

Then

$$\begin{aligned}
\mu_1^{\Delta_1}(t_1) &= \frac{2\sqrt{\sigma_1(t_1)}+1-2\sqrt{t_1}-1}{\sigma_1(t_1)-t_1}\\
&= \frac{2\sqrt{(\sqrt{t_1}+1)^2}-2\sqrt{t_1}}{2\sqrt{t_1}+1}\\
&= \frac{2(\sqrt{t_1}+1)-2\sqrt{t_1}}{2\sqrt{t_1}+1}\\
&= \frac{2}{2\sqrt{t_1}+1}, \quad t_1 \in \mathbb{T}_1,\\
\mu_2^{\Delta_2}(t_2) &= \frac{2}{2\sqrt{t_2}+1}, \quad t_2 \in \mathbb{T}_2,\\
f_{1t_1}^{\Delta_1}(t_1,t_2) &= t_2,\\
f_{2t_1}^{\Delta_1}(t_1,t_2) &= 1,\\
f_{3t_1}^{\Delta_1}(t_1,t_2) &= 0,\\
f_{1t_2}^{\Delta_2}(t_1,t_2) &= t_1,\\
f_{2t_2}^{\Delta_2}(t_1,t_2) &= 0,\\
f_{3t_2}^{\Delta_2}(t_1,t_2) &= 1,\\
f_{1t_1}^{\Delta_1^2}(t_1,t_2) &= 0,\\
f_{2t_1}^{\Delta_1^2}(t_1,t_2) &= 0,\\
f_{3t_1}^{\Delta_1^2}(t_1,t_2) &= 0,\\
f_{1t_2}^{\Delta_2^2}(t_1,t_2) &= 0,\\
f_{2t_2}^{\Delta_2^2}(t_1,t_2) &= 0,\\
f_{3t_2}^{\Delta_2^2}(t_1,t_2) &= 0,\\
f_{1t_1t_2}^{\Delta_1\Delta_2}(t_1,t_2) &= 1,\\
f_{2t_1t_2}^{\Delta_1\Delta_2}(t_1,t_2) &= 0,
\end{aligned}$$

$$f_{3t_1t_2}^{\Delta_1\Delta_2}(t_1,t_2) = 0, \quad (t_1,t_2) \in U.$$

Consequently

$$\begin{aligned}
f_{t_1}^{\Delta_1}(t_1,t_2) &= \left(f_{1t_1}^{\Delta_1}(t_1,t_2), f_{2t_1}^{\Delta_1}(t_1,t_2), f_{3t_1}^{\Delta_1}(t_1,t_2)\right) \\
&= (t_2,1,0), \\
f_{t_2}^{\Delta_2}(t_1,t_2) &= \left(f_{1t_2}^{\Delta_2}(t_1,t_2), f_{2t_2}^{\Delta_2}(t_1,t_2), f_{3t_2}^{\Delta_2}(t_1,t_2)\right) \\
&= (t_1,0,1), \\
f_{t_1}^{\Delta_1\sigma_2}(t_1,t_2) &= (\sigma_2(t_2),0,1) \\
&= ((\sqrt{t_2}+1)^2,0,1), \\
f_{t_2}^{\Delta_2\sigma_1}(t_1,t_2) &= (\sigma_1(t_1),0,1) \\
&= ((\sqrt{t_1}+1)^2,0,1), \\
f_{t_1}^{\Delta_1^2}(t_1,t_2) &= \left(f_{1t_1}^{\Delta_1^2}(t_1,t_2), f_{2t_1}^{\Delta_1^2}(t_1,t_2), f_{3t_1}^{\Delta_1^2}(t_1,t_2)\right) \\
&= (0,0,0), \\
f_{t_2}^{\Delta_2^2}(t_1,t_2) &= \left(f_{1t_2}^{\Delta_2^2}(t_1,t_2), f_{2t_2}^{\Delta_2^2}(t_1,t_2), f_{3t_2}^{\Delta_2^2}(t_1,t_2)\right) \\
&= (0,0,0), \\
f_{t_1t_2}^{\Delta_1\Delta_2}(t_1,t_2) &= \left(f_{1t_1t_2}^{\Delta_1\Delta_2}(t_1,t_2), f_{2t_1t_2}^{\Delta_1\Delta_2}(t_1,t_2), f_{3t_1t_2}^{\Delta_1\Delta_2}(t_1,t_2)\right) \\
&= (1,0,0), \quad (t_1,t_2) \in U,
\end{aligned}$$

and

$$\begin{aligned}
f_{t_1}^{\Delta_1}(t_1,t_2) \times f_{t_2}^{\Delta_2\sigma_1}(t_1,t_2) &= (t_2,1,0) \times \left((\sqrt{t_1}+1)^2,0,1\right) \\
&= (1,-t_2,-(\sqrt{t_1}+1)^2), \\
\|f_{t_1}^{\Delta_1}(t_1,t_2) \times f_{t_2}^{\Delta_2\sigma_1}(t_1,t_2)\| &= \sqrt{1+t_2^2+(\sqrt{t_1}+1)^4}, \\
N_{\sigma_1}(t_1,t_2) &= \frac{f_{t_1}^{\Delta_1}(t_1,t_2) \times f_{t_2}^{\Delta_2\sigma_1}(t_1,t_2)}{\|f_{t_1}^{\Delta_1}(t_1,t_2) \times f_{t_2}^{\Delta_2\sigma_1}(t_1,t_2)\|} \\
&= \frac{1}{\sqrt{1+t_2^2+(\sqrt{t_1}+1)^4}}\left(1,-t_2,-(\sqrt{t_1}+1)^2\right), \\
N_{\sigma_1}^{\sigma_1\sigma_2}(t_1,t_2) &= \frac{1}{\sqrt{1+(\sigma_2(t_2))^2+(\sqrt{\sigma_1(t_1)}+1)^4}} \\
&\quad \left(1,-\sigma_2(t_2),-(\sqrt{\sigma_1(t_1)}+1)^4\right) \\
&= \frac{1}{\sqrt{1+(\sqrt{t_1}+2)^4+(\sqrt{t_2}+1)^4}}\left(1,-(\sqrt{t_2}+1)^2,-(\sqrt{t_1}+1+1)^4\right)
\end{aligned}$$

$$= \frac{1}{\sqrt{1+(\sqrt{t_1}+2)^4+(\sqrt{t_2}+1)^4}}\left(1,-(1+\sqrt{t_2})^2,-(2+\sqrt{t_1})^4\right),$$

$(t_1,t_2) \in U.$

From here,

$$\begin{aligned} f_{t_1}^{\Delta_1^2}(t_1,t_2)\cdot N_{\sigma_1}^{\sigma_1\sigma_2}(t_1,t_2) &= 0, \\ f_{t_2}^{\Delta_2^2\sigma_1^2}(t_1,t_2)\cdot N_{\sigma_1}^{\sigma_1\sigma_2}(t_1,t_2) &= 0, \\ f_{t_1t_2}^{\Delta_1\Delta_2\sigma_1}(t_1,t_2)\cdot N_{\sigma_1}^{\sigma_1\sigma_2}(t_1,t_2) &= \frac{1}{\sqrt{1+(\sqrt{t_1}+2)^4+(\sqrt{t_2}+1)^4}}, \\ 1+\mu_1^{\Delta_1}(t_1) &= 1+\frac{2}{2\sqrt{t_1}+1} \\ &= \frac{2\sqrt{t_1}+3}{2\sqrt{t_1}+1}, \quad (t_1,t_2)\in U. \end{aligned}$$

Consequently

$$\begin{aligned} k_{t\sigma_1\sigma_1\sigma_1} &= \left(f_{t_1}^{\Delta_1^2}\cdot N_{\sigma_1}^{\sigma_1\sigma_2}\right)\left(f_{t_2}^{\Delta_2^2\sigma_1^2}\cdot N_{\sigma_1}^{\sigma_1\sigma_2}\right) - \left(1+\mu_1^{\Delta_1}\right)\left(f_{t_1t_2}^{\Delta_1\Delta_2\sigma_1}\cdot N_{\sigma_1}^{\sigma_1\sigma_2}\right)^2, \\ &= -\frac{2\sqrt{t_1}+3}{(2\sqrt{t_1}+1)\left(1+(\sqrt{t_1}+2)^4+(\sqrt{t_1}+1)^2\right)}, \\ k_{m\sigma_1\sigma_1\sigma_1} &= \frac{1}{2}\left(f_{t_1}^{\Delta_1^2}+f_{t_2}^{\Delta_2^2\sigma_1^2}\right)\cdot N_{\sigma_1}^{\sigma_1\sigma_2} \\ &= 0, \quad (t_1,t_2)\in U. \end{aligned}$$

Example 4.2 Let $\mathbb{T}_1 = \mathbb{T}_2 = 2^{\mathbb{N}_0}$, $\mathbb{T}_{(1)} = \mathbb{T}_{(2)} = \mathbb{T}_{(3)} = \mathbb{R}$, S be a surface with a local parameterization (U,f), where $U \subseteq \mathbb{T}_1 \times \mathbb{T}_2, f: U \to \mathbb{R}^3$ is given by

$$f(t_1,t_2) = (t_1^2+t_1t_2+t_2^2, t_1, t_2), \quad (t_1,t_2)\in U.$$

Here

$$\begin{aligned} \sigma_1(t_1) &= 2t_1, \\ \mu_1(t_1) &= \sigma_1(t_1)-t_1 \\ &= 2t_1-t_1 \\ &= t_1, \quad t_1\in\mathbb{T}_1, \\ \sigma_2(t_2) &= 2t_2, \\ \mu_2(t_2) &= t_2, \quad t_2\in\mathbb{T}_2, \\ f_1(t_1,t_2) &= t_1^2+t_1t_2+t_2^2, \\ f_2(t_1,t_2) &= t_1, \\ f_3(t_1,t_2) &= t_2, \quad (t_1,t_2)\in U. \end{aligned}$$

Then

$$\begin{aligned}
\mu_1^{\Delta_1}(t_1) &= 1, \quad t_1 \in \mathbb{T}_1, \\
\mu_2^{\Delta_2}(t_2) &= 1, \quad t_2 \in \mathbb{T}_2, \\
f_{1t_1}^{\Delta_1}(t_1,t_2) &= \sigma_1(t_1) + t_1 + t_2 \\
&= 2t_1 + t_1 + t_2 \\
&= 3t_1 + t_2, \\
f_{2t_1}^{\Delta_1}(t_1,t_2) &= 1, \\
f_{3t_1}^{\Delta_1}(t_1,t_2) &= 0, \\
f_{1t_2}^{\Delta_2}(t_1,t_2) &= t_1 + \sigma_2(t_2) + t_2 \\
&= t_1 + 2t_2 + t_2 \\
&= t_1 + 3t_2, \\
f_{2t_2}^{\Delta_2}(t_1,t_2) &= 0, \\
f_{3t_2}^{\Delta_2}(t_1,t_2) &= 1, \\
f_{1t_1}^{\Delta_1^2}(t_1,t_2) &= 3, \\
f_{2t_1}^{\Delta_1^2}(t_1,t_2) &= 0, \\
f_{3t_1}^{\Delta_1^2}(t_1,t_2) &= 0, \\
f_{1t_2}^{\Delta_2^2}(t_1,t_2) &= 3, \\
f_{2t_2}^{\Delta_2^2}(t_1,t_2) &= 0, \\
f_{3t_2}^{\Delta_2^2}(t_1,t_2) &= 0, \\
f_{1t_1t_2}^{\Delta_1\Delta_2}(t_1,t_2) &= 1, \\
f_{2t_1t_2}^{\Delta_1\Delta_2}(t_1,t_2) &= 0, \\
f_{3t_1t_2}^{\Delta_1\Delta_2}(t_1,t_2) &= 0, \quad (t_1,t_2) \in U.
\end{aligned}$$

Therefore

$$\begin{aligned}
f_{t_1}^{\Delta_1}(t_1,t_2) &= \left(f_{1t_1}^{\Delta_1}(t_1,t_2), f_{2t_1}^{\Delta_1}(t_1,t_2), f_{3t_1}^{\Delta_1}(t_1,t_2)\right) \\
&= (3t_1 + t_2, 1, 0), \\
f_{t_1}^{\Delta_1\sigma_2}(t_1,t_2) &= (3t_1 + \sigma_2(t_2), 1, 0) \\
&= (3t_1 + 2t_2, 1, 0), \\
f_{t_2}^{\Delta_2}(t_1,t_2) &= \left(f_{1t_2}^{\Delta_2}(t_1,t_2), f_{2t_2}^{\Delta_2}(t_1,t_2), f_{3t_2}^{\Delta_2}(t_1,t_2)\right)
\end{aligned}$$

$$
\begin{aligned}
&= (t_1+3t_2,0,1),\\
f_{t_2}^{\Delta_2\sigma_1}(t_1,t_2) &= (\sigma_1(t_1)+3t_2,0,1)\\
&= (2t_1+3t_2,0,1),\\
f_{t_1}^{\Delta_1^2}(t_1,t_2) &= \left(f_{1t_1}^{\Delta_1^2}(t_1,t_2), f_{2t_1}^{\Delta_1^2}(t_1,t_2), f_{3t_1}^{\Delta_1^2}(t_1,t_2)\right)\\
&= (3,0,0),\\
f_{t_2}^{\Delta_2^2}(t_1,t_2) &= \left(f_{1t_2}^{\Delta_2^2}(t_1,t_2), f_{2t_2}^{\Delta_2^2}(t_1,t_2), f_{3t_2}^{\Delta_2^2}(t_1,t_2)\right)\\
&= (3,0,0),\\
f_{t_1t_2}^{\Delta_1\Delta_2}(t_1,t_2) &= (f_{1t_1t_2}^{\Delta_1\Delta_2}(t_1,t_2), f_{2t_1t_2}^{\Delta_1\Delta_2}(t_1,t_2), f_{3t_1t_2}^{\Delta_1\Delta_2}(t_1,t_2))\\
&= (1,0,0), \quad (t_1,t_2)\in U,
\end{aligned}
$$

and

$$
\begin{aligned}
f_{t_1}^{\Delta_1}(t_1,t_2)\times f_{t_2}^{\Delta_2\sigma_1}(t_1,t_2) &= (3t_1+t_2,1,0)\times(2t_1+3t_2,0,1)\\
&= (1,-3t_1-t_2,-2t_1-3t_2),\\
\|f_{t_1}^{\Delta_1}(t_1,t_2)\times f_{t_2}^{\Delta_2\sigma_1}(t_1,t_2)\| &= \sqrt{1+(3t_1+t_2)^2+(2t_1+3t_2)^2}\\
&= \sqrt{1+9t_1^2+6t_1t_2+t_2^2+4t_1^2+12t_1t_2+9t_2^2}\\
&= \sqrt{1+13t_1^2+18t_1t_2+10t_2^2},\\
N_{\sigma_1}(t_1,t_2) &= \frac{f_{t_1}^{\Delta_1}(t_1,t_2)\times f_{t_2}^{\Delta_2\sigma_1}(t_1,t_2)}{\|f_{t_1}^{\Delta_1}(t_1,t_2)\times f_{t_2}^{\Delta_2\sigma_1}(t_1,t_2)\|}\\
&= \frac{1}{\sqrt{1+13t_1^2+18t_1t_2+10t_2^2}}(1,-3t_1-t_2,-2t_1-3t_2),\\
N_{\sigma_1}^{\sigma_1\sigma_2}(t_1,t_2) &= \frac{1}{\sqrt{1+13(\sigma_1(t_1))^2+18\sigma_1(t_1)\sigma_2(t_2)+10(\sigma_2(t_2))^2}}\\
&\quad (1,-3\sigma_1(t_1)-\sigma_2(t_2),-2\sigma_1(t_1)-3\sigma_2(t_2))\\
&= \frac{1}{\sqrt{1+52t_1^2+72t_1t_2+40t_2^2}}(1,-6t_1-2t_2,-4t_1-6t_2),
\end{aligned}
$$

$(t_1,t_2)\in U$, and

$$
\begin{aligned}
f_{t_1}^{\Delta_1^2}(t_1,t_2)\cdot N_{\sigma_1}^{\sigma_1\sigma_2}(t_1,t_2) &= \frac{3}{\sqrt{1+52t_1^2+72t_1t_2+40t_2^2}},\\
f_{t_2}^{\Delta_2^2\sigma_1^2}(t_1,t_2)\cdot N_{\sigma_1}^{\sigma_1\sigma_2}(t_1,t_2) &= \frac{3}{\sqrt{1+52t_1^2+72t_1t_2+40t_2^2}},\\
f_{t_1t_2}^{\Delta_1\Delta_2\sigma_1}(t_1,t_2)\cdot N_{\sigma_1}^{\sigma_1\sigma_2}(t_1,t_2) &= \frac{1}{\sqrt{1+52t_1^2+72t_1t_2+40t_2^2}},
\end{aligned}
$$

$(t_1,t_2)\in U$.

Consequently

$$k_{t\sigma_1\sigma_1\sigma_1} = \left(f_{t_1}^{\Delta_1^2}\cdot N_{\sigma_1}^{\sigma_1\sigma_2}\right)\left(f_{t_2}^{\Delta_2^2\sigma_1^2}\cdot N_{\sigma_1}^{\sigma_1\sigma_2}\right) - \left(1+\mu_1^{\Delta_1}\right)\left(f_{t_1t_2}^{\Delta_1\Delta_2\sigma_1}\cdot N_{\sigma_1}^{\sigma_1\sigma_2}\right)^2$$
$$= \frac{9}{1+52t_1^2+72t_1t_2+40t_2^2}$$
$$-\frac{2}{1+52t_1^2+72t_1t_2+40t_2^2}$$
$$=\frac{7}{1+52t_1^2+72t_1t_2+40t_2^2},$$
$$k_{m\sigma_1\sigma_1\sigma_1} = \frac{1}{2}\left(f_{t_1}^{\Delta_1^2}+f_{t_2}^{\Delta_2^2\sigma_1^2}\right)\cdot N_{\sigma_1}^{\sigma_1\sigma_2}$$
$$=\frac{1}{2}\left(\frac{3}{\sqrt{1+52t_1^2+72t_1t_2+40t_2^2}}+\frac{3}{\sqrt{1+52t_1^2+72t_1t_2+40t_2^2}}\right)$$
$$=\frac{3}{\sqrt{1+52t_1^2+72t_1t_2+40t_2^2}},\quad (t_1,t_2)\in U.$$

Example 4.3 Let $\mathbb{T}_1=\mathbb{Z}$, $\mathbb{T}_2=2^{\mathbb{N}_0}$, $\mathbb{T}_{(1)}=\mathbb{T}_{(2)}=\mathbb{T}_{(3)}=\mathbb{R}$, S be an oriented surface with a local parameterization (U,f), where $U\subseteq\mathbb{T}_1\times\mathbb{T}_2, f:U\to\mathbb{R}^3$ is given by

$$f(t_1,t_2)=(e_1(t_1,1)t_2+t_1e_1(t_2,1),t_1,t_2),\quad (t_1,t_2)\in U.$$

Here

$$\sigma_1(t_1)=t_1+1,$$
$$\mu_1(t_1)=\sigma_1(t_1)-t_1$$
$$=t_1+1-t_1$$
$$=1,\quad t_1\in\mathbb{T}_1,$$
$$\sigma_2(t_2)=2t_2,$$
$$\mu_2(t_2)=\sigma_2(t_2)-t_2$$
$$=2t_2-t_2$$
$$=t_2,\quad t_2\in\mathbb{T}_2,$$
$$f_1(t_1,t_2)=t_2e_1(t_1,1)+t_1e_1(t_2,1),$$
$$f_2(t_1,t_2)=t_1,$$
$$f_3(t_1,t_2)=t_2,\quad (t_1,t_2)\in U.$$

Then

$$\mu_1^{\Delta_1}(t_1)=0,\quad t_1\in\mathbb{T}_1,$$

$$\begin{aligned}
\mu_2^{\Delta_2}(t_2) &= 1, \quad t_2 \in \mathbb{T}_2,\\
f_{1t_1}^{\Delta_1}(t_1,t_2) &= t_2 e_1(t_1,1) + e_1(t_2,1),\\
f_{2t_1}^{\Delta_1}(t_1,t_2) &= 1,\\
f_{3t_1}^{\Delta_1}(t_1,t_2) &= 0,\\
f_{1t_1}^{\Delta_1^2}(t_1,t_2) &= t_2 e_1(t_1,1),\\
f_{2t_1}^{\Delta_1^2}(t_1,t_2) &= 0,\\
f_{3t_1}^{\Delta_1^2}(t_1,t_2) &= 0,\\
f_{1t_1t_2}^{\Delta_1\Delta_2}(t_1,t_2) &= e_1(t_1,1) + e_1(t_2,1),\\
f_{2t_1t_2}^{\Delta_1\Delta_2}(t_1,t_2) &= 0,\\
f_{3t_1t_2}^{\Delta_1\Delta_2}(t_1,t_2) &= 0,\\
f_{1t_2}^{\Delta_2}(t_1,t_2) &= e_1(t_1,1) + t_1 e_1(t_2,1),\\
f_{2t_2}^{\Delta_2}(t_1,t_2) &= 0,\\
f_{3t_2}^{\Delta_2}(t_1,t_2) &= 1,\\
f_{1t_2}^{\Delta_2^2}(t_1,t_2) &= t_1 e_1(t_2,1),\\
f_{2t_2}^{\Delta_2^2}(t_1,t_2) &= 0,\\
f_{3t_2}^{\Delta_2^2}(t_1,t_2) &= 0, \quad (t_1,t_2) \in U.
\end{aligned}$$

Therefore

$$\begin{aligned}
f_{t_1}^{\Delta_1}(t_1,t_2) &= \left(f_{1t_1}^{\Delta_1}(t_1,t_2), f_{2t_1}^{\Delta_1}(t_1,t_2), f_{3t_1}^{\Delta_1}(t_1,t_2)\right)\\
&= (t_2 e_1(t_1,1) + e_1(t_2,1), 1, 0),\\
f_{t_1}^{\Delta_1\sigma_2}(t_1,t_2) &= (\sigma_2(t_2) e_1(t_1,1) + e_1(\sigma_2(t_2),1), 1, 0)\\
&= (2t_2 e_1(t_1,1) + e_1(2t_2,1), 1, 0),\\
f_{t_1}^{\Delta_1^2}(t_1,t_2) &= \left(f_{1t_1}^{\Delta_1^2}(t_1,t_2), f_{2t_1}^{\Delta_1^2}(t_1,t_2), f_{3t_1}^{\Delta_1^2}(t_1,t_2)\right)\\
&= (t_2 e_1(t_1,1), 0, 0),\\
f_{t_1t_2}^{\Delta_1\Delta_2}(t_1,t_2) &= \left(f_{1t_1t_2}^{\Delta_1\Delta_2}(t_1,t_2), f_{2t_1t_2}^{\Delta_1\Delta_2}(t_1,t_2), f_{3t_1t_2}^{\Delta_1\Delta_2}(t_1,t_2)\right)\\
&= (e_1(t_1,1) + e_1(t_2,1), 0, 0),\\
f_{t_1t_2}^{\Delta_1\Delta_2\sigma_1}(t_1,t_2) &= (e_1(\sigma_1(t_1),1) + e_1(t_2,1), 0, 0)\\
&= (e_1(t_1+1,1) + e_1(t_2,1), 0, 0),\\
f_{t_2}^{\Delta_2}(t_1,t_2) &= \left(f_{1t_2}^{\Delta_2}(t_1,t_2), f_{2t_2}^{\Delta_2}(t_1,t_2), f_{3t_2}^{\Delta_2}(t_1,t_2)\right)\\
&= (e_1(t_1,1) + t_1 e_1(t_2,1), 0, 1),
\end{aligned}$$

$$\begin{aligned}
f_{t_2}^{\Delta_2\sigma_1}(t_1,t_2) &= (e_1(\sigma_1(t_1),1)+\sigma_1(t_1)e_1(t_2,1),0,1)\\
&= (e_1(t_1+1,1)+(t_1+1)e_1(t_2,1),0,1),\\
f_{t_2}^{\Delta_2^2}(t_1,t_2) &= \left(f_{1t_2}^{\Delta_2^2}(t_1,t_2), f_{2t_2}^{\Delta_2^2}(t_1,t_2), f_{3t_2}^{\Delta_2^2}(t_1,t_2)\right)\\
&= (t_1e_1(t_2,1),0,0),\\
f_{t_2}^{\Delta_2^2\sigma_1^2}(t_1,t_2) &= (\sigma_1(\sigma_1(t_1))e_1(t_2,1),0,0)\\
&= ((t_1+2)e_1(t_2,1),0,0), \quad (t_1,t_2)\in U,
\end{aligned}$$

and

$$\begin{aligned}
f_{t_1}^{\Delta_1}(t_1,t_2)\times f_{t_2}^{\Delta_2\sigma_1}(t_1,t_2) &= (t_2e_1(t_1,1)+e_1(t_2,1),1,0)\\
&\quad \times e_1(t_1+1,1)+(t_1+1)e_1(t_2,1),0,1)\\
&= (1,-t_2e_1(t_1,1)-e_1(t_2,1),\\
&\quad -e_1(t_1+1,1)-(t_1+1)e_1(t_2,1)),\\
\|f_{t_1}^{\Delta_1}(t_1,t_2)\times f_{t_2}^{\Delta_2\sigma_1}(t_1,t_2)\| &= \Big(1+(t_2e_1(t_1,1)+e_1(t_2,1))^2\\
&\quad +(e_1(t_1+1,1)+(t_1+1)e_1(t_2,1))^2\Big)^{\frac{1}{2}},\\
N_{\sigma_1}(t_1,t_2) &= \frac{f_{t_1}^{\Delta_1}(t_1,t_2)\times f_{t_2}^{\Delta_2\sigma_1}(t_1,t_2)}{\|f_{t_1}^{\Delta_1}(t_1,t_2)\times f_{t_2}^{\Delta_2\sigma_1}(t_1,t_2)\|}\\
&= \Big(1+(t_2e_1(t_1,1)+e_1(t_2,1))^2\\
&\quad +(e_1(t_1+1,1)+(t_1+1)e_1(t_2,1))^2\Big)^{-\frac{1}{2}}\\
&\quad (1,-t_2e_1(t_1,1)-e_1(t_2,1),\\
&\quad -e_1(t_1+1,1)-(t_1+1)e_1(t_2,1)),\\
N_{\sigma_1}^{\sigma_1\sigma_2}(t_1,t_2) &= \Big(1+(\sigma_2(t_2)e_1(\sigma_1(t_1),1)+e_1(\sigma_2(t_2),1))^2\\
&\quad +(e_1(\sigma_1(t_1)+1,1)+(\sigma_1(t_1)+1)e_1(\sigma_2(t_2),1))^2\Big)^{-\frac{1}{2}}\\
&\quad (1,-\sigma_2(t_2)e_1(\sigma_1(t_1),1)-e_1(\sigma_2(t_2),1),\\
&\quad -e_1(\sigma_1(t_1)+1,1)-(\sigma_1(t_1)+1)e_1(\sigma_2(t_2),1))\\
&= \Big(1+(2t_2e_1(t_1+1,1)+e_1(2t_2,1))^2
\end{aligned}$$

$$+(e_1(t_1+2,1)+(t_1+2)e_1(2t_2,1))^2\Big)^{-\frac{1}{2}}$$
$$(1,-2t_2e_1(t_1+1,1)-e_1(2t_2,1),$$
$$-e_1(t_1+2,1)-(t_1+2)e_1(2t_2,1)),\quad (t_1,t_2)\in U,$$

and

$$f_{t_1}^{\Delta_1^2}(t_1,t_2)\cdot N_{\sigma_1}^{\sigma_1\sigma_2}(t_1,t_2)=\Big(1+(2t_2e_1(t_1+1,1)+e_1(2t_2,1))^2$$
$$+(e_1(t_1+2,1)+(t_1+2)e_1(2t_2,1))^2\Big)^{-\frac{1}{2}}t_2e_1(t_1,1),$$

$$f_{t_2}^{\Delta_2^2\sigma_1^2}(t_1,t_2)\cdot N_{\sigma_1}^{\sigma_1\sigma_2}(t_1,t_2)=\Big(1+(2t_2e_1(t_1+1,1)+e_1(2t_2,1))^2$$
$$+(e_1(t_1+2,1)+(t_1+2)e_1(2t_2,1))^2\Big)^{-\frac{1}{2}}(t_1+2)e_1(t_2,1),$$

$$f_{t_1t_2}^{\Delta_1\Delta_2\sigma_1}(t_1,t_2)\cdot N_{\sigma_1}^{\sigma_1\sigma_2}(t_1,t_2)=\Big(1+(2t_2e_1(t_1+1,1)+e_1(2t_2,1))^2$$
$$+(e_1(t_1+2,1)+(t_1+2)e_1(2t_2,1))^2\Big)^{-\frac{1}{2}}(e_1(t_1+1,1)+e_1(t_2,1)),$$

$(t_1,t_2)\in U$.

Consequently

$$k_{t\sigma_1\sigma_1\sigma_1}=\left(f_{t_1}^{\Delta_1^2}\cdot N_{\sigma_1}^{\sigma_1\sigma_2}\right)\left(f_{t_2}^{\Delta_2^2\sigma_1^2}\cdot N_{\sigma_1}^{\sigma_1\sigma_2}\right)-\left(1+\mu_1^{\Delta_1}\right)\left(f_{t_1t_2}^{\Delta_1\Delta_2\sigma_1}\cdot N_{\sigma_1}^{\sigma_1\sigma_2}\right)^2,$$
$$=\Big(1+(2t_2e_1(t_1+1,1)+e_1(2t_2,1))^2$$
$$+(e_1(t_1+2,1)+(t_1+2)e_1(2t_2,1))^2\Big)^{-1}t_2(t_1+2)e_1(t_1,1)e_1(t_2,1)$$
$$-\Big(1+(2t_2e_1(t_1+1,1)+e_1(2t_2,1))^2$$
$$+(e_1(t_1+2,1)+(t_1+2)e_1(2t_2,1))^2\Big)^{-1}(e_1(t_1+1,1)+e_1(t_2,1))^2$$
$$=\Big(1+(2t_2e_1(t_1+1,1)+e_1(2t_2,1))^2$$
$$+(e_1(t_1+2,1)+(t_1+2)e_1(2t_2,1))^2\Big)^{-1}\Big(t_2(t_1+2)e_1(t_1,1)e_1(t_2,1)$$
$$+(e_1(t_1+1,1)+e_1(t_2,1))^2\Big),$$

$$
\begin{aligned}
k_{m\sigma_1\sigma_1\sigma_1} &= \frac{1}{2}\left(f_{t_1}^{\Delta_1^2} + f_{t_2}^{\Delta_2^2\sigma_1^2}\right)\cdot N_{\sigma_1}^{\sigma_1\sigma_2} \\
&= \frac{1}{2}\Bigg(1 + (2t_2e_1(t_1+1,1) + e_1(2t_2,1))^2 \\
&\quad + (e_1(t_1+2,1) + (t_1+2)e_1(2t_2,1))^2\Bigg)^{-\frac{1}{2}} \\
&\quad \left(t_2e_1(t_1,1) + (t_1+2)e_1(t_2,1)\right),
\end{aligned}
$$

$(t_1,t_2) \in U$.

Exercise 4.6 Let $\mathbb{T}_1 = 2\mathbb{Z}$, $\mathbb{T}_2 = 3^{\mathbb{N}_0}$. Find

$$
k_{t\sigma_1\sigma_1\sigma_1} \quad \text{and} \quad k_{m\sigma_1\sigma_1\sigma_1},
$$

if

1. f is as in Example 4.1.
2. f is as in Example 4.2.
3. f is as in Example 4.3.

Theorem 4.106 *Let S be a σ_1-regular surface, σ_1 be Δ_1-differentiable on $\mathbb{T}_1$, $f_{t_1}^{\Delta_1}$ be σ_1-completely delta differentiable and $f_{t_2}^{\Delta_2\sigma_1}$ be σ_2-completely delta differentiable on U. Then*

$$
\begin{aligned}
k_{t\sigma_1\sigma_1\sigma_2} &= \left(f_{t_1}^{\Delta_1^2}\cdot N_{\sigma_1}^{\sigma_1\sigma_2}\right)\left(f_{t_2}^{\Delta_2^2\sigma_1}\cdot N_{\sigma_1}^{\sigma_1\sigma_2}\right) \\
&\quad - \left(1+\mu_1^{\Delta_1}\right)\left(f_{t_1t_2}^{\Delta_1\Delta_2\sigma_1\sigma_2}\cdot N_{\sigma_1}^{\sigma_1\sigma_2}\right)\left(f_{t_1t_2}^{\Delta_1\Delta_2\sigma_1}\cdot N_{\sigma_1}^{\sigma_1\sigma_2}\right), \\
k_{m\sigma_1\sigma_1\sigma_2} &= \frac{1}{2}\left(f_{t_1}^{\Delta_1^2} + f_{t_2}^{\Delta_2^2\sigma_1}\right)\cdot N_{\sigma_1}^{\sigma_1\sigma_2}.
\end{aligned}
$$

Proof. We have

$$
k_{t\sigma_1\sigma_1\sigma_2} = k_{1\sigma_1\sigma_1\sigma_2}k_{2\sigma_1\sigma_1\sigma_2}
$$

and

$$
k_{m\sigma_1\sigma_1\sigma_2} = \frac{1}{2}\left(k_{1\sigma_1\sigma_1\sigma_2} + k_{2\sigma_1\sigma_1\sigma_2}\right),
$$

where $k_{1\sigma_1\sigma_1\sigma_2}$ and $k_{2\sigma_1\sigma_1\sigma_2}$ are the eigenvalues of the matrix $\mathscr{B}_{n\sigma_1\sigma_1\sigma_2}$. Therefore

$$
\begin{aligned}
k_{t\sigma_1\sigma_1\sigma_2} &= \det\begin{pmatrix} f_{t_1}^{\Delta_1^2}\cdot N_{\sigma_1}^{\sigma_1\sigma_2} & f_{t_1t_2}^{\Delta_1\Delta_2\sigma_1}\cdot N_{\sigma_1}^{\sigma_1\sigma_2} \\ (1+\mu_1^{\Delta_1})f_{t_1t_2}^{\Delta_1\Delta_2\sigma_1\sigma_2}\cdot N_{\sigma_1}^{\sigma_1\sigma_2} & f_{t_2}^{\Delta_2^2\sigma_1}\cdot N_{\sigma_1}^{\sigma_1\sigma_2} \end{pmatrix} \\
&= \left(f_{t_1}^{\Delta_1^2}\cdot N_{\sigma_1}^{\sigma_1\sigma_2}\right)\left(f_{t_2}^{\Delta_2^2\sigma_1}\cdot N_{\sigma_1}^{\sigma_1\sigma_2}\right) \\
&\quad - \left(1+\mu_1^{\Delta_1}\right)\left(f_{t_1t_2}^{\Delta_1\Delta_2\sigma_1\sigma_2}\cdot N_{\sigma_1}^{\sigma_1\sigma_2}\right)\left(f_{t_1t_2}^{\Delta_1\Delta_2\sigma_1}\cdot N_{\sigma_1}^{\sigma_1\sigma_2}\right)
\end{aligned}
$$

and

$$k_{m\sigma_1\sigma_1\sigma_2} = \frac{1}{2}\mathrm{Tr}\begin{pmatrix} f_{t_1}^{\Delta_1^2} \cdot N_{\sigma_1}^{\sigma_1\sigma_2} & f_{t_1t_2}^{\Delta_1\Delta_2\sigma_1} \cdot N_{\sigma_1}^{\sigma_1\sigma_2} \\ (1+\mu_1^{\Delta_1})f_{t_1t_2}^{\Delta_1\Delta_2\sigma_1\sigma_2} \cdot N_{\sigma_1}^{\sigma_1\sigma_2} & f_{t_2}^{\Delta_2^2\sigma_1} \cdot N_{\sigma_1}^{\sigma_1\sigma_2} \end{pmatrix}$$
$$= \frac{1}{2}\left(f_{t_1}^{\Delta_1^2} + f_{t_2}^{\Delta_2^2\sigma_1}\right) \cdot N_{\sigma_1}^{\sigma_1\sigma_2}.$$

This completes the proof.

Example 4.4 Let $\mathbb{T}_1$, $\mathbb{T}_2$, $\mathbb{T}_{(1)}$, $\mathbb{T}_{(2)}$, $\mathbb{T}_{(3)}$, S, U and f be as in Example 4.1. Then

$$f_{t_1t_2}^{\Delta_1\Delta_2\sigma_1\sigma_2}(t_1,t_2) = f_{t_1t_2}^{\Delta_1\Delta_2\sigma_1}(t_1,t_2),$$
$$f_{t_2}^{\Delta_2^2\sigma_1^2}(t_1,t_2) = f_{t_2}^{\Delta_2^2\sigma_1}(t_1,t_2), \quad (t_1,t_2) \in U.$$

Then

$$k_{t\sigma_1\sigma_1\sigma_2} = k_{t\sigma_1\sigma_1\sigma_1},$$
$$k_{m\sigma_1\sigma_1\sigma_2} = k_{m\sigma_1\sigma_1\sigma_1}.$$

Example 4.5 Let $\mathbb{T}_1$, $\mathbb{T}_2$, $\mathbb{T}_{(1)}$, $\mathbb{T}_{(2)}$, $\mathbb{T}_{(3)}$, S, U and f be as in Example 4.2. Then

$$f_{t_1t_2}^{\Delta_1\Delta_2\sigma_1\sigma_2}(t_1,t_2) = f_{t_1t_2}^{\Delta_1\Delta_2\sigma_1}(t_1,t_2),$$
$$f_{t_2}^{\Delta_2^2\sigma_1^2}(t_1,t_2) = f_{t_2}^{\Delta_2^2\sigma_1}(t_1,t_2), \quad (t_1,t_2) \in U.$$

Then

$$k_{t\sigma_1\sigma_1\sigma_2} = k_{t\sigma_1\sigma_1\sigma_1},$$
$$k_{m\sigma_1\sigma_1\sigma_2} = k_{m\sigma_1\sigma_1\sigma_1}.$$

Example 4.6 Let $\mathbb{T}_1$, $\mathbb{T}_2$, $\mathbb{T}_{(1)}$, $\mathbb{T}_{(2)}$, $\mathbb{T}_{(3)}$, S, U and f be as in Example 4.3. Then

$$f_{t_1t_2}^{\Delta_1\Delta_2\sigma_1\sigma_2}(t_1,t_2) = (e_1(t_1+1,1) + e_1(\sigma_2(t_2),1),0,0)$$
$$= (e_1(t_1+1,1) + e_2(2t_2,1),0,0),$$
$$f_{t_2}^{\Delta_2^2\sigma_1}(t_1,t_2) = (\sigma_1(t_1)e_1(t_2,1),0,0)$$
$$= ((t_1+1)e_1(t_2,1),0,0), \quad (t_1,t_2) \in U.$$

Therefore

$$(1+\mu_1^{\Delta_1}(t_1))f_{t_1t_2}^{\Delta_1\Delta_2\sigma_1\sigma_2}(t_1,t_2) \cdot N_{\sigma_1}^{\sigma_1\sigma_2}(t_1,t_2) = \Big(1 + (2t_2e_1(t_1+1,1)e_1(2t_2,1))^2$$
$$+e_1(t_1+2,1) + (t_1+2)e_1(2t_2,1))^2\Big)^{-\frac{1}{2}}$$
$$e_1(t_1+1,1) + e_1(2t_2,1)),$$

$$f_{t_2}^{\Delta_2^2\sigma_1}(t_1,t_2)\cdot N_{\sigma_1}^{\sigma_1\sigma_2}(t_1,t_2)=\left(1+(2t_2e_1(t_1+1,1)e_1(2t_2,1))^2\right.$$
$$\left.+(e_1(t_1+2,1)+(t_1+2)e_1(2t_2,1))^2\right)^{-\frac{1}{2}}$$
$$(t_1+1)e_1(t_2,1),\quad (t_1,t_2)\in U.$$

From here, we find

$$\begin{aligned}
k_{t\sigma_1\sigma_1\sigma_2} &= \left(f_{t_1}^{\Delta_1^2}\cdot N_{\sigma_1}^{\sigma_1\sigma_2}\right)\left(f_{t_2}^{\Delta_2^2\sigma_1}\cdot N_{\sigma_1}^{\sigma_1\sigma_2}\right)\\
&\quad -\left(1+\mu_1^{\Delta_1}\right)\left(f_{t_1t_2}^{\Delta_1\Delta_2\sigma_1\sigma_2}\cdot N_{\sigma_1}^{\sigma_1\sigma_2}\right)\left(f_{t_1t_2}^{\Delta_1\Delta_2\sigma_1}\cdot N_{\sigma_1}^{\sigma_1\sigma_2}\right)\\
&= \left(1+(2t_2e_1(t_1+1,1)+e_1(2t_2,1))^2\right.\\
&\quad \left.+(e_1(t_1+2,1)+(t_1+2)e_1(2t_2,1))^2\right)^{-1}t_2(t_1+1)e_1(t_1,1)e_1(t_2,1)\\
&\quad -\left(1+(2t_2e_1(t_1+1,1)+e_1(2t_2,1))^2\right.\\
&\quad \left.+(e_1(t_1+1,1)+(t_1+2)e_1(2t_2,1))^2\right)^{-1}(e_1(t_1+1,1)+e_2(2t_2,1))\\
&\quad (e_1(t_1+1,1)+e_1(t_2,1)),\\
k_{m\sigma_1\sigma_1\sigma_2} &= \frac{1}{2}\left(f_{t_1}^{\Delta_1^2}+f_{t_2}^{\Delta_2^2\sigma_1}\right)\cdot N_{\sigma_1}^{\sigma_1\sigma_2}\\
&= \frac{1}{2}\left(1+(2t_2e_1(t_1+1,1)+e_1(2t_2,1))^2\right.\\
&\quad \left.+(e_1(t_1+2,1)+(t_1+2)e_1(2t_2,1))^2\right)^{-\frac{1}{2}}(t_2e_1(t_1,1)+(t_1+1)e_1(t_2,1)),
\end{aligned}$$

$(t_1,t_2)\in U$.

Exercise 4.7 Let $\mathbb{T}_1=4^{\mathbb{N}_0}$, $\mathbb{T}_2=7\mathbb{N}_0$. Find

$$k_{t\sigma_1\sigma_1\sigma_2}\quad\text{and}\quad k_{m\sigma_1\sigma_1\sigma_2},$$

if

1. f is as in Example 4.1.
2. f is as in Example 4.2.
3. f is as in Example 4.3.

Theorem 4.107 *Let S be a σ_1-regular surface, σ_1 be Δ_1-differentiable on $\mathbb{T}_1$, $f_{t_1}^{\Delta_1}$ be σ_2-completely delta differentiable and let $f_{t_2}^{\Delta_2\sigma_1}$ be σ_1-completely delta differentiable on*

U. Then

$$k_{t\sigma_1\sigma_2\sigma_1} = \left(f_{t_1}^{\Delta_1^2\sigma_2} \cdot N_{\sigma_1}^{\sigma_1\sigma_2}\right)\left(f_{t_2}^{\Delta_2^2\sigma_1^2} \cdot N_{\sigma_1}^{\sigma_1\sigma_2}\right) - \left(1+\mu_1^{\Delta_1}\right)\left(f_{t_1t_2}^{\Delta_1\Delta_2\sigma_1} \cdot N_{\sigma_1}^{\sigma_1\sigma_2}\right)$$
$$\left(f_{t_1t_2}^{\Delta_1\Delta_2} \cdot N_{\sigma_1}^{\sigma_1\sigma_2}\right),$$
$$k_{m\sigma_1\sigma_2\sigma_1} = \frac{1}{2}\left(f_{t_1}^{\Delta_1^2\sigma_2} + f_{t_2}^{\Delta_2^2\sigma_1^2}\right) \cdot N_{\sigma_1}^{\sigma_1\sigma_2}.$$

Proof. We have

$$k_{t\sigma_1\sigma_2\sigma_1} = k_{1\sigma_1\sigma_2\sigma_1} k_{2\sigma_1\sigma_2\sigma_1}$$

and

$$k_{m\sigma_1\sigma_2\sigma_1} = \frac{1}{2}\left(k_{1\sigma_1\sigma_2\sigma_1} + k_{2\sigma_1\sigma_2\sigma_1}\right),$$

where $k_{1\sigma_1\sigma_2\sigma_1}$ and $k_{2\sigma_1\sigma_2\sigma_1}$ are the eigenvalues of the matrix $\mathscr{B}_{n\sigma_1\sigma_2\sigma_1}$.

Therefore

$$k_{t\sigma_1\sigma_2\sigma_1} = \det\begin{pmatrix} f_{t_1}^{\Delta_1^2\sigma_2} \cdot N_{\sigma_1}^{\sigma_1\sigma_2} & f_{t_1t_2}^{\Delta_1\Delta_2} \cdot N_{\sigma_1}^{\sigma_1\sigma_2} \\ (1+\mu_1^{\Delta_1}) f_{t_1t_2}^{\Delta_1\Delta_2\sigma_1} \cdot N_{\sigma_1}^{\sigma_1\sigma_2} & f_{t_2}^{\Delta_2^2\sigma_1^2} \cdot N_{\sigma_1}^{\sigma_1\sigma_2} \end{pmatrix}$$
$$= \left(f_{t_1}^{\Delta_1^2\sigma_2} \cdot N_{\sigma_1}^{\sigma_1\sigma_2}\right)\left(f_{t_2}^{\Delta_2^2\sigma_1^2} \cdot N_{\sigma_1}^{\sigma_1\sigma_2}\right) - \left(1+\mu_1^{\Delta_1}\right)\left(f_{t_1t_2}^{\Delta_1\Delta_2\sigma_1} \cdot N_{\sigma_1}^{\sigma_1\sigma_2}\right)$$
$$\left(f_{t_1t_2}^{\Delta_1\Delta_2} \cdot N_{\sigma_1}^{\sigma_1\sigma_2}\right)$$

and

$$k_{m\sigma_1\sigma_2\sigma_1} = \frac{1}{2}\operatorname{Tr}\begin{pmatrix} f_{t_1}^{\Delta_1^2\sigma_2} \cdot N_{\sigma_1}^{\sigma_1\sigma_2} & f_{t_1t_2}^{\Delta_1\Delta_2} \cdot N_{\sigma_1}^{\sigma_1\sigma_2} \\ (1+\mu_1^{\Delta_1}) f_{t_1t_2}^{\Delta_1\Delta_2\sigma_1} \cdot N_{\sigma_1}^{\sigma_1\sigma_2} & f_{t_2}^{\Delta_2^2\sigma_1^2} \cdot N_{\sigma_1}^{\sigma_1\sigma_2} \end{pmatrix}$$
$$= \frac{1}{2}\left(f_{t_1}^{\Delta_1^2\sigma_2} + f_{t_2}^{\Delta_2^2\sigma_1^2}\right) \cdot N_{\sigma_1}^{\sigma_1\sigma_2}.$$

This completes the proof.

Example 4.7 Let $\mathbb{T}_1$, $\mathbb{T}_2$, $\mathbb{T}_{(1)}$, $\mathbb{T}_{(2)}$, $\mathbb{T}_{(3)}$, S, U and f be as in Example 4.1. Then

$$f_{t_1t_2}^{\Delta_1\Delta_2\sigma_1}(t_1,t_2) = f_{t_1t_2}^{\Delta_1\Delta_2}(t_1,t_2),$$
$$f_{t_1}^{\Delta_1^2\sigma_2}(t_1,t_2) = f_{t_1}^{\Delta_1^2}(t_1,t_2), \quad (t_1,t_2) \in U.$$

Then

$$k_{t\sigma_1\sigma_2\sigma_1} = k_{t\sigma_1\sigma_1\sigma_1},$$
$$k_{m\sigma_1\sigma_2\sigma_1} = k_{m\sigma_1\sigma_1\sigma_1}.$$

Example 4.8 Let $\mathbb{T}_1$, $\mathbb{T}_2$, $\mathbb{T}_{(1)}$, $\mathbb{T}_{(2)}$, $\mathbb{T}_{(3)}$, S, U and f be as in Example 4.2. Then

$$f_{t_1t_2}^{\Delta_1\Delta_2\sigma_1}(t_1,t_2) = f_{t_1t_2}^{\Delta_1\Delta_2}(t_1,t_2),$$
$$f_{t_1}^{\Delta_1^2\sigma_2}(t_1,t_2) = f_{t_1}^{\Delta_1^2}(t_1,t_2), \quad (t_1,t_2) \in U.$$

Then

$$k_{t\sigma_1\sigma_2\sigma_1} = k_{t\sigma_1\sigma_1\sigma_1},$$
$$k_{m\sigma_1\sigma_2\sigma_1} = k_{m\sigma_1\sigma_1\sigma_1}.$$

Example 4.9 Let $\mathbb{T}_1$, $\mathbb{T}_2$, $\mathbb{T}_{(1)}$, $\mathbb{T}_{(2)}$, $\mathbb{T}_{(3)}$, S, U and f be as in Example 4.3. Then

$$\begin{aligned} f_{t_1}^{\Delta_1^2\sigma_2}(t_1,t_2) &= (\sigma_2(t_2)e_1(t_1,1),0,0) \\ &= (2t_2e_1(t_1,1),0,0), \quad (t_1,t_2)\in U, \end{aligned}$$

and

$$\begin{aligned} f_{t_1}^{\Delta_1^2\sigma_2}(t_1,t_2)\cdot N_{\sigma_1}^{\sigma_1\sigma_2}(t_1,t_2) &= 2\Big(1+(2t_2e_1(t_1+1,1)+e_1(2t_2,1))^2 \\ &\quad +(e_1(t_1+2,1)+(t_1+2)e_1(2t_2,1))^2\Big)^{-\frac{1}{2}} t_2e_1(t_1,1), \\ f_{t_1t_2}^{\Delta_1\Delta_2}(t_1,t_2)\cdot N_{\sigma_1}^{\sigma_1\sigma_2}(t_1,t_2) &= \Big(1+(2t_2e_1(t_1+1,1)+e_1(2t_2,1))^2 \\ &\quad +(e_1(t_1+2,1)+(t_1+2)e_1(2t_2,1))^2\Big)^{-\frac{1}{2}} (e_1(t_1,1)+e_1(t_2,1)). \end{aligned}$$

$(t_1,t_2)\in U$, and

$$\begin{aligned} k_{t\sigma_1\sigma_2\sigma_1} &= 2\Big(1+(2t_2e_1(t_1+1,1)+e_1(2t_2,1))^2 \\ &\quad +(e_1(t_1+2,1)+(t_1+2)e_1(2t_2,1))^2\Big)^{-1} t_2(t_1+2)e_1(t_1,1)e_1(t_2,1) \\ &\quad -\Big(1+(2t_2e_1(t_1+1,1)+e_1(2t_2,1))^2 \\ &\quad +(e_1(t_1+2,1)+(t_1+2)e_1(2t_2,1))^2\Big)^{-1} (e_1(t_1,1)+e_1(t_2,1)) \\ &\quad (e_1(t_1+1,1)+e_1(t_2,1)), \\ k_{m\sigma_1\sigma_2\sigma_1} &= \frac{1}{2}\Big(1+(2t_2e_1(t_1+1,1)+e_1(2t_2,1))^2 \\ &\quad +(e_1(t_1+2,1)+(t_1+2)e_1(2t_2,1))^2\Big)^{-\frac{1}{2}} \Big(2t_2e_1(t_1,1) \\ &\quad +(t_1+2)e_1(t_2,1)\Big), \end{aligned}$$

$(t_1,t_2)\in U$.

Exercise 4.8 Let $\mathbb{T}_1 = \mathbb{N}_0^{\,3}$, $\mathbb{T}_2 = 2\mathbb{N}_0$. Find

$$k_{t\sigma_1\sigma_2\sigma_1} \quad \text{and} \quad k_{m\sigma_1\sigma_2\sigma_1},$$

if

1. f is as in Example 4.1.
2. f is as in Example 4.2.
3. f is as in Example 4.3.

Theorem 4.108 *Let S be a σ_1-regular surface, σ_1 be Δ_1-differentiable on $\mathbb{T}_1$, $f_{t_1}^{\Delta_1}$ and $f_{t_2}^{\Delta_2\sigma_1}$ be σ_2-completely delta differentiable on U. Then*

$$\begin{aligned} k_{t\sigma_1\sigma_2\sigma_2} &= \left(f_{t_1}^{\Delta_1^2\sigma_2}\cdot N_{\sigma_1}^{\sigma_1\sigma_2}\right)\left(f_{t_2}^{\Delta_2^2\sigma_1}\cdot N_{\sigma_1}^{\sigma_1\sigma_2}\right) - \left(1+\mu_1^{\Delta_1}\right)\left(f_{t_1t_2}^{\Delta_1\Delta_2\sigma_1\sigma_2}\cdot N_{\sigma_1}^{\sigma_1\sigma_2}\right)\\ &\quad \left(f_{t_1t_2}^{\Delta_1\Delta_2}\cdot N_{\sigma_1}^{\sigma_1\sigma_2}\right),\\ k_{m\sigma_1\sigma_2\sigma_2} &= \frac{1}{2}\left(f_{t_1}^{\Delta_1^2\sigma_2}+f_{t_2}^{\Delta_2^2\sigma_1}\right)\cdot N_{\sigma_1}^{\sigma_1\sigma_2}. \end{aligned}$$

Proof. We have

$$k_{t\sigma_1\sigma_2\sigma_2} = k_{1\sigma_1\sigma_2\sigma_2}k_{2\sigma_1\sigma_2\sigma_2}$$

and

$$k_{m\sigma_1\sigma_2\sigma_2} = \frac{1}{2}\left(k_{1\sigma_1\sigma_2\sigma_2}+k_{2\sigma_1\sigma_2\sigma_2}\right),$$

where $k_{1\sigma_1\sigma_2\sigma_2}$ and $k_{2\sigma_1\sigma_2\sigma_2}$ are the eigenvalues of the matrix $\mathscr{B}_{n\sigma_1\sigma_2\sigma_2}$. Therefore

$$\begin{aligned} k_{t\sigma_1\sigma_2\sigma_2} &= \det\begin{pmatrix} f_{t_1}^{\Delta_1^2\sigma_2}\cdot N_{\sigma_1}^{\sigma_1\sigma_2} & f_{t_1t_2}^{\Delta_1\Delta_2}\cdot N_{\sigma_1}^{\sigma_1\sigma_2}\\ \left(1+\mu_1^{\Delta_1}\right)f_{t_1t_2}^{\Delta_1\Delta_2\sigma_1\sigma_2}\cdot N_{\sigma_1}^{\sigma_1\sigma_2} & f_{t_2}^{\Delta_2^2\sigma_1}\cdot N_{\sigma_1}^{\sigma_1\sigma_2}\end{pmatrix}\\ &= \left(f_{t_1}^{\Delta_1^2\sigma_2}\cdot N_{\sigma_1}^{\sigma_1\sigma_2}\right)\left(f_{t_2}^{\Delta_2^2\sigma_1}\cdot N_{\sigma_1}^{\sigma_1\sigma_2}\right) - \left(1+\mu_1^{\Delta_1}\right)\left(f_{t_1t_2}^{\Delta_1\Delta_2\sigma_1\sigma_2}\cdot N_{\sigma_1}^{\sigma_1\sigma_2}\right)\\ &\quad \left(f_{t_1t_2}^{\Delta_1\Delta_2}\cdot N_{\sigma_1}^{\sigma_1\sigma_2}\right) \end{aligned}$$

and

$$\begin{aligned} k_{m\sigma_1\sigma_2\sigma_2} &= \frac{1}{2}\mathrm{Tr}\begin{pmatrix} f_{t_1}^{\Delta_1^2\sigma_2}\cdot N_{\sigma_1}^{\sigma_1\sigma_2} & f_{t_1t_2}^{\Delta_1\Delta_2}\cdot N_{\sigma_1}^{\sigma_1\sigma_2}\\ \left(1+\mu_1^{\Delta_1}\right)f_{t_1t_2}^{\Delta_1\Delta_2\sigma_1\sigma_2}\cdot N_{\sigma_1}^{\sigma_1\sigma_2} & f_{t_2}^{\Delta_2^2\sigma_1}\cdot N_{\sigma_1}^{\sigma_1\sigma_2}\end{pmatrix}\\ &= \frac{1}{2}\left(f_{t_1}^{\Delta_1^2\sigma_2}+f_{t_2}^{\Delta_2^2\sigma_1}\right)\cdot N_{\sigma_1}^{\sigma_1\sigma_2}. \end{aligned}$$

This completes the proof.

Example 4.10 Let $\mathbb{T}_1$, $\mathbb{T}_2$, $\mathbb{T}_{(1)}$, $\mathbb{T}_{(2)}$, $\mathbb{T}_{(3)}$, S, U and f be as in Example 4.1. Then

$$\begin{aligned} k_{t\sigma_1\sigma_2\sigma_2} &= k_{t\sigma_1\sigma_1\sigma_1},\\ k_{m\sigma_1\sigma_2\sigma_2} &= k_{m\sigma_1\sigma_1\sigma_1}. \end{aligned}$$

Example 4.11 Let $\mathbb{T}_1$, $\mathbb{T}_2$, $\mathbb{T}_{(1)}$, $\mathbb{T}_{(2)}$, $\mathbb{T}_{(3)}$, S, U and f be as in Example 4.2. Then

$$k_{t\sigma_1\sigma_2\sigma_2} = k_{t\sigma_1\sigma_1\sigma_1},$$
$$k_{m\sigma_1\sigma_2\sigma_2} = k_{m\sigma_1\sigma_1\sigma_1}.$$

Example 4.12 Let $\mathbb{T}_1$, $\mathbb{T}_2$, $\mathbb{T}_{(1)}$, $\mathbb{T}_{(2)}$, $\mathbb{T}_{(3)}$, S, U and f be as in Example 4.3. Then

$$\begin{aligned} k_{t\sigma_1\sigma_2\sigma_2} &= 2\Big(1 + (2t_2e_1(t_1+1,1) + e_1(2t_2,1))^2 \\ &\quad + (e_1(t_1+2,1) + (t_1+2)e_1(2t_2,1))^2\Big)^{-1} t_2(t_1+1)e_1(t_1,1)e_1(t_2,1) \\ &\quad - \Big(1 + (2t_2e_1(t_1+1,1) + e_1(2t_2,1))^2 \\ &\quad + (e_1(t_1+2,1) + (t_1+2)e_1(2t_2,1))^2\Big)^{-1} (e_1(t_1,1) + e_1(t_2,1)) \\ &\quad e_1(t_1+1,1) + e_1(2t_2,1)), \\ k_{m\sigma_1\sigma_2\sigma_2} &= \frac{1}{2}\Big(1 + (2t_2e_1(t_1+1,1) + e_1(2t_2,1))^2 \\ &\quad + e_1(t_1+2,1) + (t_1+2)e_1(2t_2,1))^2\Big)\Big(2t_2e_1(t_1,1) \\ &\quad + (t_1+1)e_1(t_2,1)\Big), \end{aligned}$$

$(t_1,t_2) \in U$.

Exercise 4.9 Let $\mathbb{T}_1 = \mathbb{N}_0$, $\mathbb{T}_2 = 4\mathbb{N}_0$. Find

$$k_{t\sigma_1\sigma_2\sigma_2} \quad \text{and} \quad k_{m\sigma_1\sigma_2\sigma_2},$$

if

1. f is as in Example 4.1.
2. f is as in Example 4.2.
3. f is as in Example 4.3.

Theorem 4.109 *Let S be a σ_2-regular surface, σ_2 be Δ_2-differentiable on $\mathbb{T}_2$, $f_{t_1}^{\Delta_1\sigma_2}$ and $f_{t_2}^{\Delta_2}$ be σ_1-completely delta differentiable on U. Then*

$$\begin{aligned} k_{t\sigma_2\sigma_1\sigma_1} &= \left(f_{t_1}^{\Delta_1^2} \cdot N_{\sigma_2}^{\sigma_1\sigma_2}\right)\left(f_{t_2}^{\Delta_2^2\sigma_1} \cdot N_{\sigma_2}^{\sigma_1\sigma_2}\right) \\ &\quad - \left((1+\mu_2^{\Delta_2})f_{t_1t_2}^{\Delta_1\Delta_2\sigma_1\sigma_2} \cdot N_{\sigma_2}^{\sigma_1\sigma_2}\right)\left(f_{t_1t_2}^{\Delta_1\Delta_2} \cdot N_{\sigma_2}^{\sigma_1\sigma_2}\right), \\ k_{m\sigma_2\sigma_1\sigma_1} &= \frac{1}{2}\left(f_{t_1}^{\Delta_1^2} + f_{t_2}^{\Delta_2^2\sigma_1}\right) \cdot N_{\sigma_2}^{\sigma_1\sigma_2}. \end{aligned}$$

Proof. We have

$$k_{t\sigma_2\sigma_1\sigma_1} = k_{1\sigma_2\sigma_1\sigma_1} k_{2\sigma_2\sigma_1\sigma_1}$$

and

$$k_{m\sigma_2\sigma_1\sigma_1} = \frac{1}{2}\left(k_{1\sigma_2\sigma_1\sigma_1} + k_{2\sigma_2\sigma_1\sigma_1}\right),$$

where $k_{1\sigma_2\sigma_1\sigma_1}$ and $k_{2\sigma_2\sigma_1\sigma_1}$ are the eigenvalues of the matrix $\mathscr{B}_{n\sigma_2\sigma_1\sigma_1}$. Therefore

$$\begin{aligned} k_{t\sigma_2\sigma_1\sigma_1} &= \det\begin{pmatrix} f_{t_1}^{\Delta_1^2} \cdot N_{\sigma_2}^{\sigma_1\sigma_2} & f_{t_1 t_2}^{\Delta_1\Delta_2} \cdot N_{\sigma_2}^{\sigma_1\sigma_2} \\ (1+\mu_2^{\Delta_2}) f_{t_1 t_2}^{\Delta_1\Delta_2\sigma_1\sigma_2} \cdot N_{\sigma_2}^{\sigma_1\sigma_2} & f_{t_2}^{\Delta_2^2\sigma_1} \cdot N_{\sigma_2}^{\sigma_1\sigma_2} \end{pmatrix} \\ &= \left(f_{t_1}^{\Delta_1^2} \cdot N_{\sigma_2}^{\sigma_1\sigma_2}\right)\left(f_{t_2}^{\Delta_2^2\sigma_1} \cdot N_{\sigma_2}^{\sigma_1\sigma_2}\right) \\ &\quad - \left((1+\mu_2^{\Delta_2}) f_{t_1 t_2}^{\Delta_1\Delta_2\sigma_1\sigma_2} \cdot N_{\sigma_2}^{\sigma_1\sigma_2}\right)\left(f_{t_1 t_2}^{\Delta_1\Delta_2} \cdot N_{\sigma_2}^{\sigma_1\sigma_2}\right), \end{aligned}$$

and

$$\begin{aligned} k_{m\sigma_2\sigma_1\sigma_1} &= \frac{1}{2}\mathrm{Tr}\begin{pmatrix} f_{t_1}^{\Delta_1^2} \cdot N_{\sigma_2}^{\sigma_1\sigma_2} & f_{t_1 t_2}^{\Delta_1\Delta_2} \cdot N_{\sigma_2}^{\sigma_1\sigma_2} \\ (1+\mu_2^{\Delta_2}) f_{t_1 t_2}^{\Delta_1\Delta_2\sigma_1\sigma_2} \cdot N_{\sigma_2}^{\sigma_1\sigma_2} & f_{t_2}^{\Delta_2^2\sigma_1} \cdot N_{\sigma_2}^{\sigma_1\sigma_2} \end{pmatrix} \\ &= \frac{1}{2}\left(f_{t_1}^{\Delta_1^2} + f_{t_2}^{\Delta_2^2\sigma_1}\right) \cdot N_{\sigma_2}^{\sigma_1\sigma_2}. \end{aligned}$$

This completes the proof.

Example 4.13 Let $\mathbb{T}_1$, $\mathbb{T}_2$, $\mathbb{T}_{(1)}$, $\mathbb{T}_{(2)}$, $\mathbb{T}_{(3)}$, S, U and f be as in Example 4.1. Then

$$\begin{aligned} f_{t_1}^{\Delta_1\sigma_2}(t_1,t_2) &= (\sigma_2(t_2),1,0) \\ &= ((\sqrt{t_2}+1)^2,1,0), \quad (t_1,t_2) \in U, \end{aligned}$$

and

$$f_{t_1}^{\Delta_1\sigma_2}(t_1,t_2) \times f_{t_2}^{\Delta_2}(t_1,t_2) = \left(\left(\sqrt{t_2}+1\right)^2,1,0\right) \times (t_1,0,1)$$

$$= \left(1,-(\sqrt{t_2}+1)^2,-t_1\right),$$

$$\|f_{t_1}^{\Delta_1\sigma_2}(t_1,t_2) \times f_{t_2}^{\Delta_2}(t_1,t_2)\| = \sqrt{1+(\sqrt{t_2}+1)^4+t_1^2},$$

$(t_1,t_2) \in U$, and

$$\begin{aligned} N_{\sigma_2}(t_1,t_2) &= \frac{f_{t_1}^{\Delta_1\sigma_2}(t_1,t_2) \times f_{t_2}^{\Delta_2}(t_1,t_2)}{\|f_{t_1}^{\Delta_1\sigma_2}(t_1,t_2) \times f_{t_2}^{\Delta_2}(t_1,t_2)\|} \\ &= \frac{1}{\sqrt{1+(\sqrt{t_2}+1)^4+t_1^2}}\left(1,-(\sqrt{t_2}+1)^2,-t_1\right), \end{aligned}$$

$$N_{\sigma_2}^{\sigma_1\sigma_2}(t_1,t_2) = \frac{1}{\sqrt{1+(\sqrt{\sigma_2(t_2)}+1)^4+(\sigma_1(t_1))^2}}\left(1,-(\sqrt{\sigma_2(t_2)}+1)^2,-\sigma_1(t_1)\right)$$

$$= \frac{1}{\sqrt{1+(\sqrt{t_2}+2)^4+(\sqrt{t_1}+1)^2}}\left(1,-(\sqrt{t_2}+2)^2,-(\sqrt{t_1}+1)^2\right),$$

$(t_1,t_2) \in U$, and

$$\begin{aligned}
f_{t_1}^{\Delta_1^2}(t_1,t_2) &= (0,0,0),\\
f_{t_1}^{\Delta_1^2\sigma_2}(t_1,t_2) &= (0,0,0),\\
f_{t_2}^{\Delta_2^2}(t_1,t_2) &= (0,0,0),\\
f_{t_2}^{\Delta_2^2\sigma_1}(t_1,t_2) &= (0,0,0),\\
f_{t_1t_2}^{\Delta_1\Delta_2}(t_1,t_2) &= (1,0,0),\\
f_{t_1t_2}^{\Delta_1\Delta_2\sigma_1\sigma_2}(t_1,t_2) &= (1,0,0),\\
\mu_2^{\Delta_2}(t_2) &= \frac{2}{2\sqrt{t_2}+1},\\
1+\mu_2^{\Delta_2}(t_2) &= 1+\frac{2}{2\sqrt{t_2}+1}\\
&= \frac{2\sqrt{t_2}+3}{2\sqrt{t_2}+1}, \quad (t_1,t_2) \in U.
\end{aligned}$$

Hence,

$$\begin{aligned}
f_{t_1}^{\Delta_1^2}(t_1,t_2)\cdot N_{\sigma_2}^{\sigma_1\sigma_2}(t_1,t_2) &= 0,\\
\left(1+\mu_2^{\Delta_2}(t_2)\right)f_{t_1t_2}^{\Delta_1\Delta_2\sigma_1\sigma_2}(t_1,t_2)\cdot N_{\sigma_2}^{\sigma_1\sigma_2}(t_1,t_2) &= \frac{2\sqrt{t_2}+3}{(2\sqrt{t_2}+1)\sqrt{1+(\sqrt{t_2}+2)^4+(\sqrt{t_1}+1)^2}},\\
f_{t_1t_2}^{\Delta_1\Delta_2}(t_1,t_2)\cdot N_{\sigma_2}^{\sigma_1\sigma_2}(t_1,t_2) &= \frac{1}{\sqrt{1+(\sqrt{t_2}+2)^4+(\sqrt{t_1}+1)^2}},\\
f_{t_2}^{\Delta_2^2\sigma_1}(t_1,t_2)\cdot N_{\sigma_2}^{\sigma_1\sigma_2}(t_1,t_2) &= 0, \quad (t_1,t_2) \in U.
\end{aligned}$$

Consequently

$$\begin{aligned}
k_{t\sigma_2\sigma_1\sigma_1} &= \left(f_{t_1}^{\Delta_1^2}\cdot N_{\sigma_2}^{\sigma_1\sigma_2}\right)\left(f_{t_2}^{\Delta_2^2\sigma_1}\cdot N_{\sigma_2}^{\sigma_1\sigma_2}\right)\\
&\quad -\left((1+\mu_2^{\Delta_2})f_{t_1t_2}^{\Delta_1\Delta_2\sigma_1\sigma_2}\cdot N_{\sigma_2}^{\sigma_1\sigma_2}\right)\left(f_{t_1t_2}^{\Delta_1\Delta_2}\cdot N_{\sigma_2}^{\sigma_1\sigma_2}\right)\\
&= -\frac{2\sqrt{t_2}+3}{(2\sqrt{t_2}+1)(1+(\sqrt{t_2}+2)^4+(\sqrt{t_1}+1)^2)},\\
k_{m\sigma_2\sigma_1\sigma_1} &= \frac{1}{2}\left(f_{t_1}^{\Delta_1^2}+f_{t_2}^{\Delta_2^2\sigma_1}\right)\cdot N_{\sigma_2}^{\sigma_1\sigma_2}\\
&= 0, \quad (t_1,t_2) \in U.
\end{aligned}$$

Example 4.14 Let $\mathbb{T}_1$, $\mathbb{T}_2$, $\mathbb{T}_{(1)}$, $\mathbb{T}_{(2)}$, $\mathbb{T}_{(3)}$, S, U and f be as in Example 4.2. Then

$$\begin{aligned}
\mu_2^{\Delta_2}(t_2) &= 1,\\
f_{t_1}^{\Delta_1\sigma_2}(t_1,t_2) &= (3t_1+2t_2,1,0),\\
f_{t_2}^{\Delta_2}(t_1,t_2) &= (t_1+3t_2,0,1),\\
f_{t_1}^{\Delta_1^2}(t_1,t_2) &= (3,0,0),\\
f_{t_1t_2}^{\Delta_1\Delta_2}(t_1,t_2) &= (1,0,0),\\
f_{t_2}^{\Delta_2^2}(t_1,t_2) &= (3,0,0), \quad (t_1,t_2)\in U.
\end{aligned}$$

Hence,

$$\begin{aligned}
f_{t_1}^{\Delta_1\sigma_2}(t_1,t_2)\times f_{t_2}^{\Delta_2}(t_1,t_2) &= (1,-3t_1-2t_2,-t_1-3t_2),\\
\|f_{t_1}^{\Delta_1\sigma_2}(t_1,t_2)\times f_{t_2}^{\Delta_2}(t_1,t_2)\| &= \sqrt{1+(3t_1+2t_2)^2+(t_1+3t_2)^2}\\
&= \sqrt{1+9t_1^2+12t_1t_2+4t_2^2+t_1^2+6t_1t_2+9t_2^2}\\
&= \sqrt{1+10t_1^2+18t_1t_2+13t_2^2},
\end{aligned}$$

$(t_1,t_2)\in U$, and

$$\begin{aligned}
N_{\sigma_2}(t_1,t_2) &= \frac{f_{t_1}^{\Delta_1\sigma_2}(t_1,t_2)\times f_{t_2}^{\Delta_2}(t_1,t_2)}{\|f_{t_1}^{\Delta_1\sigma_2}(t_1,t_2)\times f_{t_2}^{\Delta_2}(t_1,t_2)\|}\\
&= \frac{1}{\sqrt{1+10t_1^2+18t_1t_2+13t_2^2}}(1,-3t_1-2t_2,-t_1-3t_2),\\
N_{\sigma_2}^{\sigma_1\sigma_2}(t_1,t_2) &= \frac{1}{\sqrt{1+10(\sigma_1(t_1))^2+18\sigma_1(t_1)\sigma_2(t_2)+13(\sigma_2(t_2))^2}}\cdot\\
&\quad (1,-3\sigma_1(t_1)-2\sigma_2(t_2),-\sigma_1(t_1)-3\sigma_2(t_2))\\
&= \frac{1}{\sqrt{1+40t_1^2+72t_1t_2+52t_2^2}}(1,-6t_1-4t_2,-2t_1-6t_2),
\end{aligned}$$

$(t_1,t_2)\in U$, and

$$\begin{aligned}
f_{t_1}^{\Delta_1^2}(t_1,t_2)\cdot N_{\sigma_2}^{\sigma_1\sigma_2}(t_1,t_2) &= \frac{3}{\sqrt{1+40t_1^2+72t_1t_2+52t_2^2}},\\
(1+\mu_2^{\Delta_2}(t_2))f_{t_1t_2}^{\Delta_1\Delta_2\sigma_1\sigma_2}(t_1,t_2)\cdot N_{\sigma_2}^{\sigma_1\sigma_2}(t_1,t_2) &= \frac{2}{\sqrt{1+40t_1^2+72t_1t_2+52t_2^2}},\\
f_{t_1t_2}^{\Delta_1\Delta_2}(t_1,t_2)\cdot N_{\sigma_2}^{\sigma_1\sigma_2}(t_1,t_2) &= \frac{1}{\sqrt{1+40t_1^2+72t_1t_2+52t_2^2}},\\
f_{t_2}^{\Delta_2^2\sigma_1}(t_1,t_2)\cdot N_{\sigma_2}^{\sigma_1\sigma_2}(t_1,t_2) &= \frac{3}{\sqrt{1+40t_1^2+72t_1t_2+52t_2^2}},
\end{aligned}$$

$(t_1, t_2) \in U$.

Consequently

$$\begin{aligned} k_{t\sigma_2\sigma_1\sigma_1} &= \left(f_{t_1}^{\Delta_1^2} \cdot N_{\sigma_2}^{\sigma_1\sigma_2}\right)\left(f_{t_2}^{\Delta_2^2\sigma_1} \cdot N_{\sigma_2}^{\sigma_1\sigma_2}\right) \\ &\quad - \left((1+\mu_2^{\Delta_2}) f_{t_1 t_2}^{\Delta_1\Delta_2\sigma_1\sigma_2} \cdot N_{\sigma_2}^{\sigma_1\sigma_2}\right)\left(f_{t_1 t_2}^{\Delta_1\Delta_2} \cdot N_{\sigma_2}^{\sigma_1\sigma_2}\right) \\ &= \frac{9}{1+40t_1^2+72t_1t_2+52t_2^2} - \frac{2}{1+40t_1^2+72t_1t_2+52t_2^2} \\ &= \frac{7}{1+40t_1^2+72t_1t_2+52t_2^2}, \\ k_{m\sigma_2\sigma_1\sigma_1} &= \frac{1}{2}\left(f_{t_1}^{\Delta_1^2} + f_{t_2}^{\Delta_2^2\sigma_1}\right) \cdot N_{\sigma_2}^{\sigma_1\sigma_2} \\ &= \frac{1}{2}\left(\frac{3}{\sqrt{1+40t_1^2+72t_1t_2+52t_2^2}} + \frac{3}{\sqrt{1+40t_1^2+72t_1t_2+52t_2^2}}\right) \\ &= \frac{3}{\sqrt{1+40t_1^2+72t_1t_2+52t_2^2}}, \end{aligned}$$

$(t_1, t_2) \in U$.

Example 4.15 Let $\mathbb{T}_1$, $\mathbb{T}_2$, $\mathbb{T}_{(1)}$, $\mathbb{T}_{(2)}$, $\mathbb{T}_{(3)}$, S, U and f be as in Example 4.3. Then

$$\begin{aligned} \mu_2^{\Delta_2}(t_2) &= 1, \\ f_{t_1}^{\Delta_1}(t_1, t_2) &= (t_2 e_1(t_1, 1) + e_1(t_2, 1), 1, 0), \\ f_{t_2}^{\Delta_2}(t_1, t_2) &= (e_1(t_1, 1) + t_1 e_1(t_2, 1), 0, 1), \\ f_{t_1}^{\Delta_1^2}(t_1, t_2) &= (t_2 e_1(t_1, 1), 0, 0), \\ f_{t_1 t_2}^{\Delta_1\Delta_2}(t_1, t_2) &= (e_1(t_1, 1) + e_1(t_2, 1), 0, 0), \\ f_{t_2}^{\Delta_2^2}(t_1, t_2) &= (t_1 e_1(t_2, 1), 0, 0), \quad (t_1, t_2) \in U. \end{aligned}$$

From here,

$$\begin{aligned} f_{t_1}^{\Delta_1\sigma_2}(t_1, t_2) &= (\sigma_2(t_2) e_1(t_1, 1) + e_1(\sigma_2(t_2), 1), 1, 0) \\ &= (2t_2 e_1(t_1, 1) + e_1(2t_2, 1), 1, 0), \end{aligned}$$

$(t_1, t_2) \in U$, and

$$\begin{aligned} f_{t_1}^{\Delta_1\sigma_2}(t_1, t_2) \times f_{t_2}^{\Delta_2}(t_1, t_2) &= (1, -2t_2 e_1(t_1, 1) - e_1(2t_2, 1), \\ &\qquad -e_1(t_1, 1) - t_1 e_1(t_2, 1)), \\ \|f_{t_1}^{\Delta_1\sigma_2}(t_1, t_2) \times f_{t_2}^{\Delta_2}(t_1, t_2)\| &= \Big(1 + (2t_2 e_1(t_1, 1) + e_1(2t_2, 1))^2 \\ &\qquad + (e_1(t_1, 1) + t_1 e_1(t_2, 1))^2\Big)^{\frac{1}{2}}, \end{aligned}$$

$$\begin{aligned}
N_{\sigma_2}(t_1,t_2) &= \frac{f_{t_1}^{\Delta_1\sigma_2}(t_1,t_2)\times f_{t_2}^{\Delta_2}(t_1,t_2)}{\|f_{t_1}^{\Delta_1\sigma_2}(t_1,t_2)\times f_{t_2}^{\Delta_2}(t_1,t_2)\|} \\
&= \Big(1+(2t_2e_1(t_1,1)+e_1(2t_2,1))^2 \\
&\quad +(e_1(t_1,1)+t_1e_1(t_2,1))^2\Big)^{-\frac{1}{2}} \\
&\quad (1,-2t_2e_1(t_1,1)-e_1(2t_2,1), \\
&\quad -e_1(t_1,1)-t_1e_1(t_2,1)), \\
N_{\sigma_2}^{\sigma_1\sigma_2}(t_1,t_2) &= \Big(1+(2\sigma_2(t_2)e_1(\sigma_1(t_1),1)+e_1(2\sigma_2(t_2),1))^2 \\
&\quad +(e_1(\sigma_1(t_1),1)+\sigma_1(t_1)e_1(\sigma_2(t_2),1))^2\Big)^{-\frac{1}{2}} \\
&\quad (1,-2\sigma_2(t_2)e_1(\sigma_1(t_1),1)-e_1(2\sigma_2(t_2),1), \\
&\quad -e_1(\sigma_1(t_1),1)-\sigma_1(t_1)e_1(\sigma_2(t_2),1)) \\
&= \Big(1+(4t_2e_1(t_1+1,1)+e_1(4t_2,1))^2 \\
&\quad +(e_1(t_1+1,1)+(t_1+1)e_1(2t_2,1))^2\Big)^{-\frac{1}{2}} \\
&\quad (1,-4t_2e_1(t_1+1,1)-e_1(4t_2,1), \\
&\quad -e_1(t_1+1,1)-(t_1+1)e_1(2t_2,1)),
\end{aligned}$$

$(t_1,t_2)\in U$, and

$$\begin{aligned}
f_{t_2}^{\Delta_2^2\sigma_1}(t_1,t_2) &= (\sigma_1(t_1)e_1(t_2,1),0,0) \\
&= ((t_1+1)e_1(t_2,1),0,0), \\
f_{t_1t_2}^{\Delta_1\Delta_2\sigma_1\sigma_2}(t_1,t_2) &= (e_1(\sigma_1(t_1),1)+e_1(\sigma_2(t_2),1),0,0) \\
&= (e_1(t_1+1,1)+e_1(2t_2,1),0,0),
\end{aligned}$$

$(t_1,t_2)\in U$, and

$$\begin{aligned}
f_{t_1}^{\Delta_1^2}(t_1,t_2)\cdot N_{\sigma_2}^{\sigma_1\sigma_2}(t_1,t_2) &= \Big(1+(4t_2e_1(t_1+1,1)+e_1(4t_2,1))^2 \\
&\quad +(e_1(t_1+1,1)+(t_1+1)e_1(2t_2,1))^2\Big)^{-\frac{1}{2}} \\
&\quad t_2e_1(t_1,1), \\
f_{t_2}^{\Delta_2^2\sigma_1}(t_1,t_2)\cdot N_{\sigma_2}^{\sigma_1\sigma_2}(t_1,t_2) &= \Big(1+(4t_2e_1(t_1+1,1)+e_1(4t_2,1))^2
\end{aligned}$$

$$+(e_1(t_1+1,1)+(t_1+1)e_1(2t_2,1))^2\Big)^{-\frac{1}{2}}$$

$$(t_1+1)e_1(t_2,1),$$

$$(1+\mu_2^{\Delta_2}(t_2))f_{t_1t_2}^{\Delta_1\Delta_2\sigma_1\sigma_2}(t_1,t_2)\cdot N_{\sigma_2}^{\sigma_1\sigma_2}(t_1,t_2)=\Big(1+(4t_2e_1(t_1+1,1)+e_1(4t_2,1))^2$$

$$+(e_1(t_1+1,1)+(t_1+1)e_1(2t_2,1))^2\Big)^{-\frac{1}{2}}$$

$$(e_1(t_1+1,1)+e_1(2t_2,1)),$$

$$f_{t_1t_2}^{\Delta_1\Delta_2}(t_1,t_2)\cdot N_{\sigma_2}^{\sigma_1\sigma_2}(t_1,t_2)=\Big(1+(4t_2e_1(t_1+1,1)+e_1(4t_2,1))^2$$

$$+(e_1(t_1+1,1)+(t_1+1)e_1(2t_2,1))^2\Big)^{-\frac{1}{2}}$$

$$(e_1(t_1,1)+e_1(t_2,1)),$$

$(t_1,t_2)\in U.$

Consequently

$$\begin{aligned}
k_{t\sigma_2\sigma_1\sigma_1} &= \left(f_{t_1}^{\Delta_1^2}\cdot N_{\sigma_2}^{\sigma_1\sigma_2}\right)\left(f_{t_2}^{\Delta_2^2\sigma_1}\cdot N_{\sigma_2}^{\sigma_1\sigma_2}\right)\\
&\quad -\left((1+\mu_2^{\Delta_2})f_{t_1t_2}^{\Delta_1\Delta_2\sigma_1\sigma_2}\cdot N_{\sigma_2}^{\sigma_1\sigma_2}\right)\left(f_{t_1t_2}^{\Delta_1\Delta_2}\cdot N_{\sigma_2}^{\sigma_1\sigma_2}\right)\\
&= \Big(1+(4t_2e_1(t_1+1,1)+e_1(4t_2,1))^2\\
&\quad +(e_1(t_1+1,1)+(t_1+1)e_1(2t_2,1))^2\Big)^{-1}\Big(t_2(t_1+1)e_1(t_1,1)e_1(t_2,1)\\
&\quad -2(e_1(t_1+1,1)+e_1(2t_2,1))(e_1(t_1,1)+e_1(t_2,1))\Big),\\
k_{m\sigma_2\sigma_1\sigma_1} &= \frac{1}{2}\left(f_{t_1}^{\Delta_1^2}+f_{t_2}^{\Delta_2^2\sigma_1}\right)\cdot N_{\sigma_2}^{\sigma_1\sigma_2}\\
&= \frac{1}{2}\Big(1+(4t_2e_1(t_1+1,1)+e_1(4t_2,1))^2\\
&\quad +(e_1(t_1+1,1)+(t_1+1)e_1(2t_2,1))^2\Big)^{-\frac{1}{2}}\Big((t_1+1)e_1(t_2,1)\\
&\quad +t_2e_1(t_1,1)\Big),\quad (t_1,t_2)\in U.
\end{aligned}$$

Exercise 4.10 Let $\mathbb{T}_1=\mathbb{T}_2=6\mathbb{Z}$. Find

$$k_{t\sigma_2\sigma_1\sigma_1} \quad \text{and} \quad k_{m\sigma_2\sigma_1\sigma_1}$$

1. if f is as in Example 4.1.
2. if f is as in Example 4.2.
3. if f is as in Example 4.3.

Theorem 4.110 *Let S_2 be a σ_2-regular surface, σ_2 be Δ_2-differentiable on $\mathbb{T}_2$, $f_{t_1}^{\Delta_1\sigma_2}$ be σ_1-completely delta differentiable and let $f_{t_2}^{\Delta_2}$ be σ_2-completely delta differentiable on U. Then*

$$k_{t\sigma_2\sigma_1\sigma_2} = \left(f_{t_1}^{\Delta_1^2}\cdot N_{\sigma_2}^{\sigma_1\sigma_2}\right)\left(f_{t_2}^{\Delta_2^2}\cdot N_{\sigma_2}^{\sigma_1\sigma_2}\right) - \left(\left(1+\mu_2^{\Delta_2}\right)f_{t_1t_2}^{\Delta_1\Delta_2\sigma_1\sigma_2}\cdot N_{\sigma_2}^{\sigma_1\sigma_2}\right)\left(f_{t_1t_2}^{\Delta_1\Delta_2\sigma_2}\cdot N_{\sigma_2}^{\sigma_1\sigma_2}\right),$$

$$k_{m\sigma_2\sigma_1\sigma_2} = \frac{1}{2}\left(f_{t_1}^{\Delta_1^2}+f_{t_2}^{\Delta_2^2}\right)\cdot N_{\sigma_2}^{\sigma_1\sigma_2}.$$

Proof. We have

$$k_{t\sigma_2\sigma_1\sigma_2} = k_{1\sigma_2\sigma_1\sigma_2}k_{2\sigma_2\sigma_1\sigma_2}$$

and

$$k_{m\sigma_2\sigma_1\sigma_2} = \frac{1}{2}\left(k_{1\sigma_2\sigma_1\sigma_2}+k_{2\sigma_2\sigma_1\sigma_2}\right),$$

where $k_{1\sigma_2\sigma_1\sigma_2}$ and $k_{2\sigma_2\sigma_1\sigma_2}$ are the eigenvalues of the matrix $\mathscr{B}_{n\sigma_2\sigma_1\sigma_2}$.

Therefore

$$\begin{aligned} k_{t\sigma_2\sigma_1\sigma_2} &= \det\begin{pmatrix} f_{t_1}^{\Delta_1^2}\cdot N_{\sigma_2}^{\sigma_1\sigma_2} & f_{t_1t_2}^{\Delta_1\Delta_2\sigma_2}\cdot N_{\sigma_2}^{\sigma_1\sigma_2} \\ \left(1+\mu_2^{\Delta_2}\right)f_{t_1t_2}^{\Delta_1\Delta_2\sigma_1\sigma_2}\cdot N_{\sigma_2}^{\sigma_1\sigma_2} & f_{t_2}^{\Delta_2^2}\cdot N_{\sigma_2}^{\sigma_1\sigma_2} \end{pmatrix} \\ &= \left(f_{t_1}^{\Delta_1^2}\cdot N_{\sigma_2}^{\sigma_1\sigma_2}\right)\left(f_{t_2}^{\Delta_2^2}\cdot N_{\sigma_2}^{\sigma_1\sigma_2}\right) \\ &\quad - \left(\left(1+\mu_2^{\Delta_2}\right)f_{t_1t_2}^{\Delta_1\Delta_2\sigma_1\sigma_2}\cdot N_{\sigma_2}^{\sigma_1\sigma_2}\right)\left(f_{t_1t_2}^{\Delta_1\Delta_2\sigma_2}\cdot N_{\sigma_2}^{\sigma_1\sigma_2}\right) \end{aligned}$$

and

$$\begin{aligned} k_{m\sigma_2\sigma_1\sigma_2} &= \frac{1}{2}\mathrm{Tr}\begin{pmatrix} f_{t_1}^{\Delta_1^2}\cdot N_{\sigma_2}^{\sigma_1\sigma_2} & f_{t_1t_2}^{\Delta_1\Delta_2\sigma_2}\cdot N_{\sigma_2}^{\sigma_1\sigma_2} \\ \left(1+\mu_2^{\Delta_2}\right)f_{t_1t_2}^{\Delta_1\Delta_2\sigma_1\sigma_2}\cdot N_{\sigma_2}^{\sigma_1\sigma_2} & f_{t_2}^{\Delta_2^2}\cdot N_{\sigma_2}^{\sigma_1\sigma_2} \end{pmatrix} \\ &= \frac{1}{2}\left(f_{t_1}^{\Delta_1^2}+f_{t_2}^{\Delta_2^2}\right)\cdot N_{\sigma_2}^{\sigma_1\sigma_2}. \end{aligned}$$

This completes the proof.

Example 4.16 Let $\mathbb{T}_1$, $\mathbb{T}_2$, $\mathbb{T}_{(1)}$, $\mathbb{T}_{(2)}$, $\mathbb{T}_{(3)}$, U, S and f be as in Example 4.1. Then

$$k_{t\sigma_2\sigma_1\sigma_2} = k_{t\sigma_2\sigma_1\sigma_1},$$

$$k_{m\sigma_2\sigma_1\sigma_2} = k_{m\sigma_2\sigma_1\sigma_1}.$$

Example 4.17 Let $\mathbb{T}_1$, $\mathbb{T}_2$, $\mathbb{T}_{(1)}$, $\mathbb{T}_{(2)}$, $\mathbb{T}_{(3)}$, U, S and f be as in Example 4.2. Then

$$k_{t\sigma_2\sigma_1\sigma_2} = k_{t\sigma_2\sigma_1\sigma_1},$$

$$k_{m\sigma_2\sigma_1\sigma_2} = k_{m\sigma_2\sigma_1\sigma_1}.$$

Example 4.18 Let $\mathbb{T}_1$, $\mathbb{T}_2$, $\mathbb{T}_{(1)}$, $\mathbb{T}_{(2)}$, $\mathbb{T}_{(3)}$, U, S and f be as in Example 4.3. Then

$$\begin{aligned} f_{t_1t_2}^{\Delta_1\Delta_2\sigma_2}(t_1,t_2) &= (e_1(t_1,1)+e_1(\sigma_2(t_2),1),1,0) \\ &= (e_1(t_1,1)+e_1(2t_2,1),1,0), \end{aligned}$$

$(t_1,t_2) \in U$, and

$$\begin{aligned} f_{t_2}^{\Delta_2^2}(t_1,t_2)\cdot N_{\sigma_2}^{\sigma_1\sigma_2}(t_1,t_2) &= \Big(1+(4t_2e_1(t_1+1,1)+e_1(4t_2,1))^2 \\ &\quad +(e_1(t_1+1,1)+(t_1+1)e_1(2t_2,1))^2\Big)^{-\frac{1}{2}} \\ &\quad t_1e_1(t_2,1), \\ f_{t_1t_2}^{\Delta_1\Delta_2\sigma_2}(t_1,t_2)\cdot N_{\sigma_2}^{\sigma_1\sigma_2}(t_1,t_2) &= \Big(1+(4t_2e_1(t_1+1,1)+e_1(4t_2,1))^2 \\ &\quad +(e_1(t_1+1,1)+(t_1+1)e_1(2t_2,1))^2\Big)^{-\frac{1}{2}} \\ &\quad (e_1(t_1,1)+t_1e_1(2t_2,1)), \end{aligned}$$

$(t_1,t_2) \in U$. Consequently

$$\begin{aligned} k_{t\sigma_2\sigma_1\sigma_2} &= \Big(1+(4t_2e_1(t_1+1,1)+e_1(4t_2,1))^2 \\ &\quad +(e_1(t_1+1,1)+(t_1+1)e_1(2t_2,1))^2\Big)^{-1} \\ &\quad \Big(t_1t_2e_1(t_1,1)e_1(t_2,1) \\ &\quad -(e_1(t_1,1)+e_1(2t_2,1))(e_1(t_1+1,1)+e_1(2t_2,1))\Big), \\ k_{m\sigma_2\sigma_1\sigma_2} &= \frac{1}{2}\Big(1+(4t_2e_1(t_1+1,1)+e_1(4t_2,1))^2 \\ &\quad +(e_1(t_1+1,1)+(t_1+1)e_1(2t_2,1))^2\Big)^{-\frac{1}{2}} \\ &\quad \Big(t_2e_1(t_1,1)+t_1e_1(t_2,1)\Big), \end{aligned}$$

$(t_1,t_2) \in U$.

Exercise 4.11 Let $\mathbb{T}_1 = \left(\frac{1}{2}\right)^{\mathbb{N}_0}$, $\mathbb{T}_2 = 2\mathbb{Z}$. Find

$$k_{t\sigma_2\sigma_1\sigma_2} \quad \text{and} \quad k_{m\sigma_2\sigma_1\sigma_2}$$

1. if f is as in Example 4.1.
2. if f is as in Example 4.2.
3. if f is as in Example 4.3.

Theorem 4.111 *Let S be a σ_2-regular surface, σ_2 be Δ_2-differentiable on $\mathbb{T}_2$, $f_{t_1}^{\Delta_1\sigma_2}$ be σ_2-completely delta differentiable and let $f_{t_2}^{\Delta_2}$ be σ_1-completely delta differentiable on U. Then*

$$\begin{aligned} k_{t\sigma_2\sigma_2\sigma_1} &= \left(f_{t_1}^{\Delta_1^2\sigma_2^2}\cdot N_{\sigma_2}^{\sigma_1\sigma_2}\right)\left(f_{t_2}^{\Delta_2^2\sigma_1}\cdot N_{\sigma_2}^{\sigma_1\sigma_2}\right)\\ &\quad -\left((1+\mu_2^{\Delta_2})f_{t_1t_2}^{\Delta_1\Delta_2\sigma_2}\cdot N_{\sigma_2}^{\sigma_1\sigma_2}\right)\left(f_{t_1t_2}^{\Delta_1\Delta_2}\cdot N_{\sigma_2}^{\sigma_1\sigma_2}\right),\\ k_{m\sigma_2\sigma_2\sigma_1} &= \frac{1}{2}\left(f_{t_1}^{\Delta_1^2\sigma_2^2}+f_{t_2}^{\Delta_2^2\sigma_1}\right)\cdot N_{\sigma_2}^{\sigma_1\sigma_2}. \end{aligned}$$

Proof. We have

$$k_{t\sigma_2\sigma_2\sigma_1} = k_{1\sigma_2\sigma_2\sigma_1}k_{2\sigma_2\sigma_2\sigma_1}$$

and

$$k_{m\sigma_2\sigma_2\sigma_1} = \frac{1}{2}\left(k_{1\sigma_2\sigma_2\sigma_1}+k_{2\sigma_2\sigma_2\sigma_1}\right),$$

where $k_{1\sigma_2\sigma_2\sigma_1}$ and $k_{2\sigma_2\sigma_2\sigma_1}$ are the eigenvalues of the matrix $\mathscr{B}_{n\sigma_2\sigma_2\sigma_1}$. Therefore

$$\begin{aligned} k_{t\sigma_2\sigma_2\sigma_1} &= \det\begin{pmatrix} f_{t_1}^{\Delta_1^2\sigma_2^2}\cdot N_{\sigma_2}^{\sigma_1\sigma_2} & f_{t_1t_2}^{\Delta_1\Delta_2}\cdot N_{\sigma_2}^{\sigma_1\sigma_2}\\ (1+\mu_2^{\Delta_2})f_{t_1t_2}^{\Delta_1\Delta_2\sigma_2}\cdot N_{\sigma_2}^{\sigma_1\sigma_2} & f_{t_2}^{\Delta_2^2\sigma_1}\cdot N_{\sigma_2}^{\sigma_1\sigma_2}\end{pmatrix}\\ &= \left(f_{t_1}^{\Delta_1^2\sigma_2^2}\cdot N_{\sigma_2}^{\sigma_1\sigma_2}\right)\left(f_{t_2}^{\Delta_2^2\sigma_1}\cdot N_{\sigma_2}^{\sigma_1\sigma_2}\right)\\ &\quad -\left((1+\mu_2^{\Delta_2})f_{t_1t_2}^{\Delta_1\Delta_2\sigma_2}\cdot N_{\sigma_2}^{\sigma_1\sigma_2}\right)\left(f_{t_1t_2}^{\Delta_1\Delta_2}\cdot N_{\sigma_2}^{\sigma_1\sigma_2}\right), \end{aligned}$$

and

$$\begin{aligned} k_{m\sigma_2\sigma_2\sigma_1} &= \frac{1}{2}\mathrm{Tr}\begin{pmatrix} f_{t_1}^{\Delta_1^2\sigma_2^2}\cdot N_{\sigma_2}^{\sigma_1\sigma_2} & f_{t_1t_2}^{\Delta_1\Delta_2}\cdot N_{\sigma_2}^{\sigma_1\sigma_2}\\ (1+\mu_2^{\Delta_2})f_{t_1t_2}^{\Delta_1\Delta_2\sigma_2}\cdot N_{\sigma_2}^{\sigma_1\sigma_2} & f_{t_2}^{\Delta_2^2\sigma_1}\cdot N_{\sigma_2}^{\sigma_1\sigma_2}\end{pmatrix}\\ &= \frac{1}{2}\left(f_{t_1}^{\Delta_1^2\sigma_2^2}+f_{t_2}^{\Delta_2^2\sigma_1}\right)\cdot N_{\sigma_2}^{\sigma_1\sigma_2}. \end{aligned}$$

This completes the proof.

Example 4.19 Let $\mathbb{T}_1$, $\mathbb{T}_2$, $\mathbb{T}_{(1)}$, $\mathbb{T}_{(2)}$, $\mathbb{T}_{(3)}$, U, S and f be as in Example 4.1. Then

$$\begin{aligned} k_{t\sigma_2\sigma_2\sigma_1} &= k_{t\sigma_2\sigma_1\sigma_1},\\ k_{m\sigma_2\sigma_2\sigma_1} &= k_{m\sigma_2\sigma_1\sigma_1}. \end{aligned}$$

Example 4.20 Let $\mathbb{T}_1$, $\mathbb{T}_2$, $\mathbb{T}_{(1)}$, $\mathbb{T}_{(2)}$, $\mathbb{T}_{(3)}$, U, S and f be as in Example 4.2. Then

$$\begin{aligned} k_{t\sigma_2\sigma_2\sigma_1} &= k_{t\sigma_2\sigma_1\sigma_1},\\ k_{m\sigma_2\sigma_2\sigma_1} &= k_{m\sigma_2\sigma_1\sigma_1}. \end{aligned}$$

Example 4.21 Let $\mathbb{T}_1$, $\mathbb{T}_2$, $\mathbb{T}_{(1)}$, $\mathbb{T}_{(2)}$, $\mathbb{T}_{(3)}$, U, S and f be as in Example 4.3. Then

$$\begin{aligned} f_{t_1}^{\Delta_1^2\sigma_2^2}(t_1,t_2) &= \left(\sigma_2^2(t_2)e_1(t_1,1),0,0\right) \\ &= (4t_2e_1(t_1,1),0,0), \\ f_{t_1t_2}^{\Delta_1\Delta_2\sigma_2}(t_1,t_2) &= (e_1(t_1,1)+e_1(\sigma_2(t_2),1),0,0) \\ &= (e_1(t_1,1)+e_1(2t_2,1),0,0), \end{aligned}$$

$(t_1,t_2)\in U$, and

$$\begin{aligned} f_{t_1}^{\Delta_1^2\sigma_2^2}(t_1,t_2)\cdot N_{\sigma_2}^{\sigma_1\sigma_2}(t_1,t_2) &= 4\Big(1+(4t_2e_1(t_1+1,1)+e_1(4t_2,1))^2 \\ &\quad +(e_1(t_1+1,1)+(t_1+1)e_1(2t_2,1))^2\Big)^{-\frac{1}{2}} \\ &\quad t_2e_1(t_1,1), \\ f_{t_1t_2}^{\Delta_1\Delta_2\sigma_2}(t_1,t_2)\cdot N_{\sigma_2}^{\sigma_1\sigma_2}(t_1,t_2) &= \Big(1+(4t_2e_1(t_1+1,1)+e_1(4t_2,1))^2 \\ &\quad +(e_1(t_1+1,1)+(t_1+1)e_1(2t_2,1))^2\Big)^{-\frac{1}{2}} \\ &\quad (e_1(t_1,1)+e_1(2t_2,1)), \end{aligned}$$

$(t_1,t_2)\in U$. Consequently

$$\begin{aligned} k_{t\sigma_2\sigma_2\sigma_1} &= 4\Big(1+(4t_2e_1(t_1+1,1)+e_1(4t_2,1))^2 \\ &\quad +(e_1(t_1+1,1)+(t_1+1)e_1(2t_2,1))^2\Big)^{-1}(t_1+1)t_2e_1(t_1,1)e_1(t_2,1) \\ &\quad -2\Big(1+(4t_2e_1(t_1+1,1)+e_1(4t_2,1))^2 \\ &\quad +(e_1(t_1+1,1)+(t_1+1)e_1(2t_2,1))^2\Big)^{-1}(e_1(t_1,1)+e_1(2t_2,1)) \\ &\quad (e_1(t_1,1)+e_1(t_2,1)), \\ k_{m\sigma_2\sigma_2\sigma_1} &= \frac{1}{2}\Big(1+(4t_2e_1(t_1+1,1)+e_1(4t_2,1))^2 \\ &\quad +(e_1(t_1+1,1)+(t_1+1)e_1(2t_2,1))^2\Big)^{-\frac{1}{2}}\Big(4t_2e_1(t_1,1) \\ &\quad +(t_1+1)e_1(t_2,1)\Big), \end{aligned}$$

$(t_1,t_2)\in U$.

Exercise 4.12 Let $\mathbb{T}_1 4^{\mathbb{N}_0}$, $\mathbb{T}_2 = \left(\frac{1}{3}\right)^{\mathbb{N}_0}$. Find

$$k_{t\sigma_2\sigma_2\sigma_1} \quad \text{and} \quad k_{m\sigma_2\sigma_2\sigma_1}$$

1. if f is as in Example 4.1.
2. if f is as in Example 4.2.
3. if f is as in Example 4.3.

Theorem 4.112 *Let S be a σ_2-regular surface, σ_2 be Δ_2-differentiable on $\mathbb{T}_2$ and let $f_{t_1}^{\Delta_1\sigma_2}$ and $f_{t_2}^{\Delta_2}$ be σ_2-completely delta differentiable on U. Then*

$$\begin{aligned} k_{t\sigma_2\sigma_2\sigma_2} &= \left(f_{t_1}^{\Delta_1^2\sigma_2^2}\cdot N_{\sigma_2}^{\sigma_1\sigma_2}\right)\left(f_{t_2}^{\Delta_2^2}\cdot N_{\sigma_2}^{\sigma_1\sigma_2}\right) \\ &\quad -\left(1+\mu_2^{\Delta_2}\right)\left(f_{t_1t_2}^{\Delta_1\Delta_2\sigma_2}\cdot N_{\sigma_2}^{\sigma_1\sigma_2}\right)^2, \\ k_{m\sigma_2\sigma_2\sigma_2} &= \frac{1}{2}\left(f_{t_1}^{\Delta_1^2\sigma_2^2}+f_{t_2}^{\Delta_2^2}\right)\cdot N_{\sigma_2}^{\sigma_1\sigma_2}. \end{aligned}$$

Proof. We have

$$k_{t\sigma_2\sigma_2\sigma_2} = k_{1\sigma_2\sigma_2\sigma_2}k_{2\sigma_2\sigma_2\sigma_2}$$

and

$$k_{m\sigma_2\sigma_2\sigma_2} = \frac{1}{2}\left(k_{1\sigma_2\sigma_2\sigma_2}+k_{2\sigma_2\sigma_2\sigma_2}\right),$$

where $k_{2\sigma_2\sigma_2\sigma_2}$ and $k_{2\sigma_2\sigma_2\sigma_2}$ are the eigenvalues of the matrix $\mathscr{B}_{n\sigma_2\sigma_2\sigma_2}$. Therefore

$$\begin{aligned} k_{t\sigma_2\sigma_2\sigma_2} &= \det\begin{pmatrix} f_{t_1}^{\Delta_1^2\sigma_2^2}\cdot N_{\sigma_2}^{\sigma_1\sigma_2} & f_{t_1t_2}^{\Delta_1\Delta_2\sigma_2}\cdot N_{\sigma_2}^{\sigma_1\sigma_2} \\ (1+\mu_2^{\Delta_2})f_{t_1t_2}^{\Delta_1\Delta_2\sigma_2}\cdot N_{\sigma_2}^{\sigma_1\sigma_2} & f_{t_2}^{\Delta_2^2}\cdot N_{\sigma_2}^{\sigma_1\sigma_2} \end{pmatrix} \\ &= \left(f_{t_1}^{\Delta_1^2\sigma_2^2}\cdot N_{\sigma_2}^{\sigma_1\sigma_2}\right)\left(f_{t_2}^{\Delta_2^2}\cdot N_{\sigma_2}^{\sigma_1\sigma_2}\right) \\ &\quad -\left(1+\mu_2^{\Delta_2}\right)\left(f_{t_1t_2}^{\Delta_1\Delta_2\sigma_2}\cdot N_{\sigma_2}^{\sigma_1\sigma_2}\right)^2, \end{aligned}$$

and

$$\begin{aligned} k_{m\sigma_2\sigma_2\sigma_2} &= \frac{1}{2}\mathrm{Tr}\begin{pmatrix} f_{t_1}^{\Delta_1^2\sigma_2^2}\cdot N_{\sigma_2}^{\sigma_1\sigma_2} & f_{t_1t_2}^{\Delta_1\Delta_2\sigma_2}\cdot N_{\sigma_2}^{\sigma_1\sigma_2} \\ (1+\mu_2^{\Delta_2})f_{t_1t_2}^{\Delta_1\Delta_2\sigma_2}\cdot N_{\sigma_2}^{\sigma_1\sigma_2} & f_{t_2}^{\Delta_2^2}\cdot N_{\sigma_2}^{\sigma_1\sigma_2} \end{pmatrix} \\ &= \frac{1}{2}\left(f_{t_1}^{\Delta_1^2\sigma_2^2}+f_{t_2}^{\Delta_2^2}\right)\cdot N_{\sigma_2}^{\sigma_1\sigma_2}. \end{aligned}$$

This completes the proof.

Example 4.22 Let $\mathbb{T}_1$, $\mathbb{T}_2$, $\mathbb{T}_{(1)}$, $\mathbb{T}_{(2)}$, $\mathbb{T}_{(3)}$, U, S and f be as in Example 4.1. Then

$$\begin{aligned} k_{t\sigma_2\sigma_2\sigma_2} &= k_{t\sigma_2\sigma_1\sigma_1}, \\ k_{m\sigma_2\sigma_2\sigma_2} &= k_{m\sigma_2\sigma_1\sigma_1}. \end{aligned}$$

Example 4.23 Let $\mathbb{T}_1$, $\mathbb{T}_2$, $\mathbb{T}_{(1)}$, $\mathbb{T}_{(2)}$, $\mathbb{T}_{(3)}$, U, S and f be as in Example 4.2. Then

$$k_{t\sigma_2\sigma_2\sigma_2} = k_{t\sigma_2\sigma_1\sigma_1},$$
$$k_{m\sigma_2\sigma_2\sigma_2} = k_{m\sigma_2\sigma_1\sigma_1}.$$

Example 4.24 Let $\mathbb{T}_1$, $\mathbb{T}_2$, $\mathbb{T}_{(1)}$, $\mathbb{T}_{(2)}$, $\mathbb{T}_{(3)}$, U, S and f be as in Example 4.3. Then

$$\begin{aligned}
k_{t\sigma_2\sigma_2\sigma_2} &= \left(f_{t_1}^{\Delta_1^2\sigma_2^2} \cdot N_{\sigma_2}^{\sigma_1\sigma_2}\right)\left(f_{t_2}^{\Delta_2^2} \cdot N_{\sigma_2}^{\sigma_1\sigma_2}\right) \\
&\quad - \left(1+\mu_2^{\Delta_2}\right)\left(f_{t_1t_2}^{\Delta_1\Delta_2\sigma_2} \cdot N_{\sigma_2}^{\sigma_1\sigma_2}\right)^2 \\
&= \Bigg(1+(4t_2e_1(t_1+1,1)+e_1(4t_2,1))^2 \\
&\quad +(e_1(t_1+1,1)+(t_1+1)e_1(2t_2,1))^2\Bigg)^{-1}\Bigg(4t_1t_2e_1(t_1,1)e_1(t_2,1) \\
&\quad -2(e_1(t_1,1)+e_1(2t_2,1))^2\Bigg), \\
k_{m\sigma_2\sigma_2\sigma_2} &= \frac{1}{2}\left(f_{t_1}^{\Delta_1^2\sigma_2^2} + f_{t_2}^{\Delta_2^2}\right) \cdot N_{\sigma_2}^{\sigma_1\sigma_2} \\
&= \frac{1}{2}\Bigg(1+(4t_2e_1(t_1+1,1)+e_1(4t_2,1))^2 \\
&\quad +(e_1(t_1+1,1)+(t_1+1)e_1(2t_2,1))^2\Bigg)^{-\frac{1}{2}}\Bigg(4t_2e_1(t_1,1) \\
&\quad +t_1e_1(t_2,1)\Bigg),
\end{aligned}$$

$(t_1,t_2) \in U$.

Exercise 4.13 Let $\mathbb{T}_1 = \left(\frac{1}{7}\right)^{\mathbb{N}_0}$, $\mathbb{T}_2 = \mathbb{Z}$. Find

$$k_{t\sigma_2\sigma_2\sigma_2} \quad \text{and} \quad k_{m\sigma_2\sigma_2\sigma_2}$$

1. if f is as in Example 4.1.
2. if f is as in Example 4.2.
3. if f is as in Example 4.3.

As above, one can prove the following results.

Theorem 4.113 *Let S be a $\sigma_1^2\sigma_2$-regular surface, σ_1 be Δ_1-differentiable on $\mathbb{T}_1$ and let $f_{t_1}^{\Delta_1}$ and $f_{t_2}^{\Delta_2\sigma_1}$ be σ_1-completely delta differentiable on U. Then*

$$k_{t\sigma_1^2\sigma_2\sigma_1\sigma_1} = \left(f_{t_1}^{\Delta_1^2} \cdot N_{\sigma_1\sigma_2^2}^{\sigma_1\sigma_2}\right)\left(f_{t_2}^{\Delta_2^2\sigma_1^2} \cdot N_{\sigma_1\sigma_2^2}^{\sigma_1\sigma_2}\right) - \left(1+\mu_1^{\Delta_1}\right)\left(f_{t_1t_2}^{\Delta_1\Delta_2\sigma_1} \cdot N_{\sigma_1\sigma_2^2}^{\sigma_1\sigma_2}\right)^2,$$

$$k_{m\sigma_1^2\sigma_2\sigma_1\sigma_1} = \frac{1}{2}\left(f_{t_1}^{\Delta_1^2} + f_{t_2}^{\Delta_2^2\sigma_1^2}\right)\cdot N^{\sigma_1\sigma_2}_{\sigma_1\sigma_2^2}.$$

Theorem 4.114 *Let S be a $\sigma_1^2\sigma_2$-regular surface, σ_1 be Δ_1-differentiable on $\mathbb{T}_1$ and let $f_{t_1}^{\Delta_1}$ be σ_1-completely delta differentiable, $f_{t_2}^{\Delta_2\sigma_1}$ is σ_2-completely delta differentiable on U. Then*

$$\begin{aligned} k_{t\sigma_1^2\sigma_2\sigma_1\sigma_2} &= \left(f_{t_1}^{\Delta_1^2}\cdot N^{\sigma_1\sigma_2}_{\sigma_1\sigma_2^2}\right)\left(f_{t_2}^{\Delta_2^2\sigma_1}\cdot N^{\sigma_1\sigma_2}_{\sigma_1\sigma_2^2}\right)\\ &\quad -\left(1+\mu_1^{\Delta_1}\right)\left(f_{t_1t_2}^{\Delta_1\Delta_2\sigma_1\sigma_2}\cdot N_{\sigma_1\sigma_2^2}\right)\left(f_{t_1t_2}^{\Delta_1\Delta_2\sigma_1}\cdot N^{\sigma_1\sigma_2}_{\sigma_1^2\sigma_2}\right),\\ k_{m\sigma_1^2\sigma_2\sigma_1\sigma_2} &= \frac{1}{2}\left(f_{t_1}^{\Delta_1^2} + f_{t_2}^{\Delta_2^2\sigma_1}\right)\cdot N^{\sigma_1\sigma_2}_{\sigma_1\sigma_2^2}. \end{aligned}$$

Theorem 4.115 *Let S be a $\sigma_1^2\sigma_2$-regular surface, σ_1 be Δ_1-differentiable on $\mathbb{T}_1$, $f_{t_1}^{\Delta_1}$ be σ_2-completely delta differentiable and let $f_{t_2}^{\Delta_2\sigma_1}$ be σ_1-completely delta differentiable on U. Then*

$$\begin{aligned} k_{t\sigma_1^2\sigma_2\sigma_2\sigma_1} &= \left(f_{t_1}^{\Delta_1^2\sigma_2}\cdot N^{\sigma_1\sigma_2}_{\sigma_1\sigma_2^2}\right)\left(f_{t_2}^{\Delta_2^2\sigma_1^2}\cdot N^{\sigma_1\sigma_2}_{\sigma_1\sigma_2^2}\right) - \left(1+\mu_1^{\Delta_1}\right)\left(f_{t_1t_2}^{\Delta_1\Delta_2\sigma_1}\cdot N^{\sigma_1\sigma_2}_{\sigma_1\sigma_2^2}\right)\\ &\quad \left(f_{t_1t_2}^{\Delta_1\Delta_2}\cdot N^{\sigma_1\sigma_2}_{\sigma_1\sigma_2^2}\right),\\ k_{m\sigma_1^2\sigma_1\sigma_2\sigma_1} &= \frac{1}{2}\left(f_{t_1}^{\Delta_1^2\sigma_2} + f_{t_2}^{\Delta_2^2\sigma_1^2}\right)\cdot N^{\sigma_1\sigma_2}_{\sigma_1\sigma_2^2}. \end{aligned}$$

Theorem 4.116 *Let S be a $\sigma_1^2\sigma_2$-regular surface, σ_1 be Δ_1-differentiable on $\mathbb{T}_1$ and let $f_{t_1}^{\Delta_1}$ and $f_{t_2}^{\Delta_2\sigma_1}$ be σ_2-completely delta differentiable on U. Then*

$$\begin{aligned} k_{t\sigma_1^2\sigma_2\sigma_2\sigma_2} &= \left(f_{t_1}^{\Delta_1^2\sigma_2}\cdot N^{\sigma_1\sigma_2}_{\sigma_1^2\sigma_2}\right)\left(f_{t_2}^{\Delta_2^2\sigma_1}\cdot N^{\sigma_1\sigma_2}_{\sigma_1^2\sigma_2}\right) - \left(1+\mu_1^{\Delta_1}\right)\left(f_{t_1t_2}^{\Delta_1\Delta_2\sigma_1\sigma_2}\cdot N^{\sigma_1\sigma_2}_{\sigma_1^2\sigma_2}\right)\\ &\quad \left(f_{t_1t_2}^{\Delta_1\Delta_2}\cdot N^{\sigma_1\sigma_2}_{\sigma_1^2\sigma_2}\right),\\ k_{m\sigma_1^2\sigma_2\sigma_2\sigma_2} &= \frac{1}{2}\left(f_{t_1}^{\Delta_1^2\sigma_2} + f_{t_2}^{\Delta_2^2\sigma_1}\right)\cdot N^{\sigma_1\sigma_2}_{\sigma_1^2\sigma_2}. \end{aligned}$$

Theorem 4.117 *Let S be a $\sigma_1\sigma_2^2$-regular surface, σ_2 be Δ_2-differentiable on $\mathbb{T}_2$ and let $f_{t_1}^{\Delta_1\sigma_2}$ and $f_{t_2}^{\Delta_2}$ be σ_1-completely delta differentiable on U. Then*

$$\begin{aligned} k_{ta\sigma_1\sigma_2^2\sigma_1\sigma_1} &= \left(f_{t_1}^{\Delta_1^2}\cdot N^{\sigma_1\sigma_2}_{\sigma_1\sigma_2^2}\right)\left(f_{t_2}^{\Delta_2^2\sigma_1}\cdot N^{\sigma_1\sigma_2}_{\sigma_1\sigma_2^2}\right)\\ &\quad -\left(\left(1+\mu_2^{\Delta_2}\right)f_{t_1t_2}^{\Delta_1\Delta_2\sigma_1\sigma_2}\cdot N^{\sigma_1\sigma_2}_{\sigma_1\sigma_2^2}\right)\left(f_{t_1t_2}^{\Delta_1\Delta_2}\cdot N^{\sigma_1\sigma_2}_{\sigma_1\sigma_2^2}\right),\\ k_{m\sigma_1\sigma_2^2\sigma_1\sigma_1} &= \frac{1}{2}\left(f_{t_1}^{\Delta_1^2} + f_{t_2}^{\Delta_2^2\sigma_1}\right)\cdot N^{\sigma_1\sigma_2}_{\sigma_1\sigma_2^2}. \end{aligned}$$

Theorem 4.118 *Let S be a $\sigma_1\sigma_2^2$-regular surface, σ_2 be Δ_2-differentiable on $\mathbb{T}_2$, $f_{t_1}^{\Delta_1\sigma_2}$ be σ_1-completely delta differentiable and let $f_{t_2}^{\Delta_2}$ be σ_2-completely delta differentiable on U. Then*

$$k_{t\sigma_1\sigma_2^2\sigma_1\sigma_2} = \left(f_{t_1}^{\Delta_1^2}\cdot N^{\sigma_1\sigma_2}_{\sigma_1\sigma_2^2}\right)\left(f_{t_2}^{\Delta_2^2}\cdot N^{\sigma_1\sigma_2}_{\sigma_1\sigma_2^2}\right)$$

$$-\left(\left(1+\mu_2^{\Delta_2}\right)f_{t_1t_2}^{\Delta_1\Delta_2\sigma_1\sigma_2}\cdot N_{\sigma_1\sigma_2^2}^{\sigma_1\sigma_2}\right)\left(f_{t_1t_2}^{\Delta_1\Delta_2\sigma_2}\cdot N_{\sigma_1\sigma_2^2}^{\sigma_1\sigma_2}\right),$$

$$k_{m\sigma_1\sigma_2^2\sigma_1\sigma_2}=\frac{1}{2}\left(f_{t_1}^{\Delta_1^2}+f_{t_2}^{\Delta_2^2}\right)\cdot N_{\sigma_1\sigma_2^2}^{\sigma_1\sigma_2}.$$

Theorem 4.119 *Let S be a $\sigma_1\sigma_2^2$-regular surface, σ_2 be Δ_2-differentiable on $\mathbb{T}_2$, $f_{t_1}^{\Delta_1\sigma_2}$ be σ_2-completely delta differentiable and let it $f_{t_2}^{\Delta_2}$ be σ_1-completely delta differentiable on U. Then*

$$k_{t\sigma_1\sigma_2^2\sigma_2\sigma_1}=\left(f_{t_1}^{\Delta_1^2\sigma_2^2}\cdot N_{\sigma_1\sigma_2^2}^{\sigma_1\sigma_2}\right)\left(f_{t_2}^{\Delta_2^2\sigma_1}\cdot N_{\sigma_1\sigma_2^2}^{\sigma_1\sigma_2}\right)$$

$$-\left(\left(1+\mu_2^{\Delta_2}\right)f_{t_1t_2}^{\Delta_1\Delta_2\sigma_2}\cdot N_{\sigma_1\sigma_2^2}^{\sigma_1\sigma_2}\right)\left(f_{t_1t_2}^{\Delta_1\Delta_2}\cdot N_{\sigma_1\sigma_2^2}^{\sigma_1\sigma_2}\right),$$

$$k_{m\sigma_1\sigma_2^2\sigma_2\sigma_1}=\frac{1}{2}\left(f_{t_1}^{\Delta_1^2\sigma_2^2}+f_{t_2}^{\Delta_2^2\sigma_1}\right)\cdot N_{\sigma_1\sigma_2^2}^{\sigma_1\sigma_2}.$$

Theorem 4.120 *Let S be a $\sigma_1\sigma_2^2$-regular surface, σ_2 be Δ_2-differentiable on $\mathbb{T}_2$ and let $f_{t_1}^{\Delta_1\sigma_2}$ and $f_{t_2}^{\Delta_2}$ be σ_2-completely delta differentiable on U. Then*

$$k_{t\sigma_1\sigma_2^2\sigma_2\sigma_2}=\left(f_{t_1}^{\Delta_1^2\sigma_2^2}\cdot N_{\sigma_1\sigma_2^2}^{\sigma_1\sigma_2}\right)\left(f_{t_2}^{\Delta_2^2}\cdot N_{\sigma_1\sigma_2^2}^{\sigma_1\sigma_2}\right)$$

$$-\left(\left(1+\mu_2^{\Delta_2}\right)f_{t_1t_2}^{\Delta_1\Delta_2\sigma_2}\cdot N_{\sigma_1\sigma_2^2}^{\sigma_1\sigma_2}\right)\left(f_{t_1t_2}^{\Delta_1\Delta_2\sigma_2}\cdot N_{\sigma_1\sigma_2^2}^{\sigma_1\sigma_2}\right),$$

$$k_{m\sigma_1\sigma_2^2\sigma_2\sigma_2}=\frac{1}{2}\left(f_{t_1}^{\Delta_1^2\sigma_2^2}+f_{t_2}^{\Delta_2^2}\right)\cdot N_{\sigma_1\sigma_2^2}^{\sigma_1\sigma_2}.$$

Exercise 4.14 Let $\mathbb{T}_1=\mathbb{Z}$, $\mathbb{T}_2=3^{\mathbb{N}_0}$, $\mathbb{T}_{(1)}=\mathbb{T}_{(2)}=\mathbb{T}_{(3)}=\mathbb{R}$ and let S be an oriented surface with a local parametrization (U,f), where $U\subset\mathbb{T}_1\times\mathbb{T}_2$, $f:U\to\mathbb{R}^3$ is given by

$$f(t)=(t_1t_2^2,t_1^2+t_2^2,t_1),\quad (t_1,t_2)\in U.$$

Find

1. $k_{t\sigma_1^2\sigma_2\sigma_1\sigma_2}$.
2. $k_{t\sigma_1^2\sigma_2\sigma_1\sigma_2}$.
3. $k_{m\sigma_1^2\sigma_2\sigma_2\sigma_2}$.

4.8 Advanced Practical Problems

Problem 4.1 Let $\mathbb{T}_1=2\mathbb{Z}$, $\mathbb{T}_2=3\mathbb{Z}$, $\mathbb{T}_{(1)}=\mathbb{T}_{(2)}=\mathbb{T}_{(3)}=\mathbb{R}$ and let S be a surface with a local parameterization (U,f), $U\subseteq\mathbb{T}_1\times\mathbb{T}_2$, where $f:U\to\mathbb{R}^3$ is given by

$$f(t_1,t_2)=(t_1t_2+4t_2^2,t_1t_2^2-2t_1,t_1+3t_2),\quad (t_1,t_2)\in U.$$

Find $\left[\phi_{2\sigma_1}\right]$, $\left[\phi_{2\sigma_2}\right]$.

Problem 4.2 Let $\mathbb{T}_1 = 6^{\mathbb{N}}$, $\mathbb{T}_2 = \mathbb{N}_0^2$, $\mathbb{T}_{(1)} = \mathbb{T}_{(2)} = \mathbb{T}_{(3)} = \mathbb{R}$ and let S be a surface with a local parameterization (U,f), $U \subseteq \mathbb{T}_1 \times \mathbb{T}_2$, where $f : U \to \mathbb{R}^3$ is given by

$$f(t_1,t_2) = \left(\frac{1+t_1t_2}{t_1+t_2}, t_1^2 t_2^3, \frac{1+t_1}{1+t_2} \right), \quad (t_1,t_2) \in U.$$

Find $\left[\phi_{2\sigma}\right]$, $\left[\phi_{2\sigma_1}\right]$.

Problem 4.3 Let $\mathbb{T}_1 = 4\mathbb{Z}$, $\mathbb{T}_2 = 2^{\mathbb{N}_0}$, $\mathbb{T}_{(1)} = \mathbb{T}_{(2)} = \mathbb{T}_{(3)} = \mathbb{R}$, $U = \mathbb{T}_1 \times \mathbb{T}_2$ and let S be an oriented surface with a local parameterization (U,f), where

$$f(t_1,t_2) = \left(1+t_1+t_2+(t_1-t_2)^3, t_1t_2^2 - t_1^4 - t_2^3, \frac{1}{1+t_1^2+t_2^2} \right),$$

$(t_1,t_2) \in \mathbb{T}_1 \times \mathbb{T}_2$. Find

1. $N_{\sigma_1\sigma_2^2}(t_1,t_2)$,
2. $N_{\sigma_1^2\sigma_2}(t_1,t_2)$,
3. $4N_{\sigma_1^2\sigma_2}(t_1,t_2) \cdot \left(3N_{\sigma_1^2\sigma_2}(t_1,t_2) - 2N_{\sigma_1\sigma_2^2}(t_1,t_2)\right)$,
4. $N_{\sigma_1\sigma_2^2}(t_1,t_2) \times \left(7N_{\sigma_1^2\sigma_2}(t_1,t_2) - 4N_{\sigma_1\sigma_2^2}(t_1,t_2)\right)$.

Problem 4.4 Let $\mathbb{T}_1 = \mathbb{T}_2 = 3^{\mathbb{N}_0}$, $\mathbb{T}_{(1)} = \mathbb{T}_{(2)} = \mathbb{T}_{(3)} = \mathbb{R}$ and let S be a surface with a local parametrization (U,f), where $U \subset \mathbb{T}_1 \times \mathbb{T}_2$ and

$$f(t_1,t_2) = \left(1+t_1+t_2, (t_1+t_2)^2, t_1^2+t_2^4+t_1^2t_2^2\right), \quad (t_1,t_2) \in U.$$

Find

1. $k_{r\sigma_1^2\sigma_2\sigma_2\sigma_1}$, $r \in \{1,2\}$,
2. $k_{t\sigma_1^2\sigma_2\sigma_2\sigma_1}$,
3. $k_{m\sigma_1^2\sigma_2\sigma_2\sigma_1}$,
4. $k_{1\sigma_1^2\sigma_2\sigma_2\sigma_1}$,
5. $k_{2\sigma_1^2\sigma_2\sigma_2\sigma_2}$,
6. $k_{1\sigma_1\sigma_2^2\sigma_1\sigma_1}$,
7. $k_{t\sigma_1\sigma_2^2\sigma_1\sigma_1}$,
8. $k_{m\sigma_1\sigma_2^2\sigma_2\sigma_1}$,
9. $k_{1\sigma_1\sigma_2^2\sigma_2\sigma_1}$,
10. $k_{2\sigma_1\sigma_2^2\sigma_1\sigma_2}$.

Problem 4.5 Let $\mathbb{T}_1 = 2^{\mathbb{N}_0}$, $\mathbb{T}_2 = \mathbb{N}_0$, $\mathbb{T}_{(1)} = \mathbb{T}_{(2)} = \mathbb{T}_{(3)} = \mathbb{R}$ and let S be an oriented surface with a local parameterization (U,f), where $U \subset \mathbb{T}_1' \times \mathbb{T}_2$, $f : U \to \mathbb{R}^3$ is given by

$$f(t) = (t_1 - t_2^2, t_1^2t_2^2, t_1+t_2), \quad (t_1,t_2) \in U.$$

Find

1. $k_{t\sigma_1^2\sigma_2\sigma_2\sigma_1}$.
2. $k_{t\sigma_1^2\sigma_2\sigma_1\sigma_1}$.
3. $k_{m\sigma_1^2\sigma_2\sigma_2\sigma_1}$.
4. $k_{t\sigma_1^2\sigma_2\sigma_2\sigma_1} + 2k_{m\sigma_1\sigma_2^2\sigma_2\sigma_1}$.

4.9 Notes and References

In this chapter we introduce second fundamental forms for oriented surfaces. They are deduced from some cases of the Meusnier theorem. Normal curvatures, asymptotic directions for oriented surfaces, asymptotic curves on surfaces, principal directions, principal curvatures, Gauss curvatures and mean curvatures of oriented surfaces are introduced. A classification of the points on oriented surfaces is made. Also, some variants of the Joachimstahl Theorem are formulated and proved.

5

The Fundamental Equations of a Surface

Suppose that $k \in \mathbb{N}$, $\widetilde{\mathbb{T}}$, $\mathbb{T}$, $\mathbb{T}_1$, $\mathbb{T}_2$, $\mathbb{T}_{(1)}$, $\mathbb{T}_{(2)}$, $\mathbb{T}_{(3)}$, $\mathbb{T}^{(1)},\ldots,\mathbb{T}^{(k)}$ are time scales with forward jump operators and delta differentiation operators $\widetilde{\sigma}$, σ, σ_1, σ_2, $\sigma_{(1)}$, $\sigma_{(2)}$, $\sigma_{(3)}$, $\sigma^{(1)},\ldots,\sigma^{(k)}$ and $\widetilde{\Delta}$, Δ, Δ_1, Δ_2, $\Delta_{(1)}$, $\Delta_{(2)}$, $\Delta_{(3)}$, $\Delta^{(1)},\ldots,\Delta^{(k)}$, respectively. Let

$$U \subseteq \mathbb{T}_1 \times \mathbb{T}_2, \quad W \subseteq \mathbb{T}_{(1)} \times \mathbb{T}_{(2)} \times \mathbb{T}_{(3)}, \quad V \subseteq \Lambda^{(k)} = \mathbb{T}^{(1)} \times \ldots \times \mathbb{T}^{(k)}$$

and $I \subseteq \mathbb{T}$, $J \subseteq \widetilde{\mathbb{T}}$.

5.1 The Differentiation Rules. The Christoffel Coefficients

Suppose that f is a continuous function on U and that, $f_{t_2}^{\Delta_2}$, $f_{t_1t_2}^{\Delta_1\Delta_2}$, $f_{t_1}^{\Delta_1^2}$, $f_{t_2}^{\Delta_2^2}$ exist and are continuous on U. Assume that S is a σ_1-or σ_2- or $\sigma_1^2\sigma_2$- or $\sigma_1\sigma_2^2$-oriented surface with a local σ_1-or σ_2- or $\sigma_1^2\sigma_2$- or $\sigma_1\sigma_2^2$-parameterization (U,f).

1. Let S be a σ_1-regular surface. We will represent $f_{t_1}^{\Delta_1^2}$, $f_{t_1t_2}^{\Delta_1\Delta_2}$ and $f_{t_2}^{\Delta_2^2}$ in terms of the basis $\{f_{t_1}^{\Delta_1}, f_{t_2}^{\Delta_2\sigma_1}, N_{\sigma_1}\}$. These expressions will be of the form

$$\begin{aligned} f_{t_1}^{\Delta_1^2} &= \Gamma_{11\sigma_1}^1 f_{t_1}^{\Delta_1} + \Gamma_{11\sigma_1}^2 f_{t_2}^{\Delta_2\sigma_1} + A_{\sigma_1} N_{\sigma_1} \\ f_{t_1t_2}^{\Delta_1\Delta_2} &= \Gamma_{12\sigma_1}^1 f_{t_1}^{\Delta_1} + \Gamma_{12\sigma_1}^2 f_{t_2}^{\Delta_2\sigma_1} + B_{\sigma_1} N_{\sigma_1} \\ f_{t_2}^{\Delta_2^2} &= \Gamma_{22\sigma_1}^1 f_{t_1}^{\Delta_1} + \Gamma_{22\sigma_1}^2 f_{t_2}^{\Delta_2\sigma_1} + C_{\sigma_1} N_{\sigma_1}, \end{aligned} \tag{5.1}$$

where

$$A_{\sigma_1} = f_{t_1}^{\Delta_1^2} \cdot N_{\sigma_1}$$

$$B_{\sigma_1} = f_{t_1t_2}^{\Delta_1\Delta_2} \cdot N_{\sigma_1}$$

$$C_{\sigma_1} = f_{t_2}^{\Delta_2^2} \cdot N_{\sigma_1}.$$

Definition 5.1 The coefficients $\Gamma_{ij\sigma_1}^k$, $i,j,k \in \{1,2\}$, will be called the σ_1-Christoffel coefficients.

We multiply the first equation of (5.1) with $f_{t_1}^{\Delta_1}$ and $f_{t_2}^{\Delta_2\sigma_1}$ and we get

$$f_{t_1}^{\Delta_1} \cdot f_{t_1}^{\Delta_1^2} = E\Gamma_{11\sigma_1}^1 + F_{\sigma_1}\Gamma_{11\sigma_1}^2$$

DOI: 10.1201/9781003205265-5

$$f_{t_2}^{\Delta_2\sigma_1} \cdot f_{t_1}^{\Delta_1^2} = F_{\sigma_1}\Gamma^1_{11\sigma_1} + G_{\sigma_1}\Gamma^2_{11\sigma_1},$$

whereupon

$$\Gamma^1_{11\sigma_1} = \frac{\left(f_{t_1}^{\Delta_1} \cdot f_{t_1}^{\Delta_1^2}\right) G_{\sigma_1} - \left(f_{t_2}^{\Delta_2\sigma_1} \cdot f_{t_1}^{\Delta_1^2}\right) F_{\sigma_1}}{EG_{\sigma_1} - (F_{\sigma_1})^2}$$

$$\Gamma^2_{11\sigma_1} = \frac{\left(f_{t_1}^{\Delta_1} \cdot f_{t_1}^{\Delta_1^2}\right) F_{\sigma_1} - \left(f_{t_2}^{\Delta_2\sigma_1} \cdot f_{t_1}^{\Delta_1^2}\right) E}{(F_{\sigma_1})^2 - EG_{\sigma_1}}.$$

As above,

$$\Gamma^1_{12\sigma_1} = \frac{\left(f_{t_1}^{\Delta_1} \cdot f_{t_1t_2}^{\Delta_1\Delta_2}\right) G_{\sigma_1} - \left(f_{t_2}^{\Delta_2\sigma_1} \cdot f_{t_1t_2}^{\Delta_1\Delta_2}\right) F_{\sigma_1}}{EG_{\sigma_1} - (F_{\sigma_1})^2}$$

$$\Gamma^2_{12\sigma_1} = \frac{\left(f_{t_1}^{\Delta_1} \cdot f_{t_1t_2}^{\Delta_1\Delta_2}\right) F_{\sigma_1} - \left(f_{t_2}^{\Delta_2\sigma_1} \cdot f_{t_1t_2}^{\Delta_1\Delta_2}\right) E}{(F_{\sigma_1})^2 - EG_{\sigma_1}}$$

$$\Gamma^1_{22\sigma_1} = \frac{\left(f_{t_1}^{\Delta_1} \cdot f_{t_2}^{\Delta_2^2}\right) G_{\sigma_1} - \left(f_{t_2}^{\Delta_2\sigma_1} \cdot f_{t_2}^{\Delta_2^2}\right) F_{\sigma_1}}{EG_{\sigma_1} - (F_{\sigma_1})^2}$$

$$\Gamma^2_{22\sigma_1} = \frac{\left(f_{t_1}^{\Delta_1} \cdot f_{t_2}^{\Delta_2^2}\right) F_{\sigma_1} - \left(f_{t_2}^{\Delta_2\sigma_1} \cdot f_{t_2}^{\Delta_2^2}\right) E}{(F_{\sigma_1})^2 - EG_{\sigma_1}}.$$

Example 5.1 We will express

$$f_{t_1}^{\Delta_1\sigma_1}, \quad f_{t_1}^{\Delta_1\sigma_2}, \quad f_{t_2}^{\Delta_2}, \quad f_{t_2}^{\Delta_2\sigma_2}$$

in the terms of the σ_1-Christoffel coefficients. We have

$$\begin{aligned}
f_{t_1}^{\Delta_1\sigma_1} &= f_{t_1}^{\Delta_1} + \mu_1 f_{t_1}^{\Delta_1^2} \\
&= f_{t_1}^{\Delta_1} + \mu_1\left(\Gamma^1_{11\sigma_1} f_{t_1}^{\Delta_1} + \Gamma^2_{11\sigma_1} f_{t_2}^{\Delta_2\sigma_1} + A_{\sigma_1}N_{\sigma_1}\right) \\
&= \left(1 + \mu_1\Gamma^1_{11\sigma_1}\right) f_{t_1}^{\Delta_1} + \mu_1\Gamma^2_{11\sigma_1} f_{t_2}^{\Delta_2\sigma_1} + \mu_1 A_{\sigma_1}N_{\sigma_1}, \\
f_{t_1}^{\Delta_1\sigma_2} &= f_{t_1}^{\Delta_1} + \mu_2 f_{t_1t_2}^{\Delta_1\Delta_2} \\
&= f_{t_1}^{\Delta_1} + \mu_2\left(\Gamma^1_{12\sigma_1} f_{t_1}^{\Delta_1} + \Gamma^2_{12\sigma_1} f_{t_2}^{\Delta_2\sigma_1} + B_{\sigma_1}N_{\sigma_1}\right) \\
&= \left(1 + \mu_2\Gamma^1_{12\sigma_1}\right) f_{t_1}^{\Delta_1} + \mu_2\Gamma^2_{12\sigma_1} f_{t_2}^{\Delta_2\sigma_1} + \mu_2 B_{\sigma_1}N_{\sigma_1}, \\
f_{t_2}^{\Delta_2\sigma_1} &= f_{t_2}^{\Delta_2} + \mu_1 f_{t_1t_2}^{\Delta_1\Delta_2}, \\
f_{t_2}^{\Delta_2} &= f_{t_2}^{\Delta_2\sigma_1} - \mu_1 f_{t_1t_2}^{\Delta_1\Delta_2} \\
&= -\mu_1\Gamma^1_{12\sigma_1} f_{t_1}^{\Delta_1} + \left(1 - \mu_1\Gamma^2_{12\sigma_1}\right) f_{t_2}^{\Delta_2\sigma_1} - \mu_1 B_{\sigma_1}N_{\sigma_1}, \\
f_{t_2}^{\Delta_2\sigma_2} &= f_{t_2}^{\Delta_2} + \mu_2 f_{t_2}^{\Delta_2^2}
\end{aligned}$$

$$
\begin{aligned}
&= -\mu_1\mu_2\Gamma^1_{12\sigma_1}f_{t_1}^{\Delta_1} + \mu_2\left(1-\mu_1\Gamma^2_{12\sigma_1}\right)f_{t_2}^{\Delta_2\sigma_1} - \mu_1\mu_2 B_{\sigma_1}N_{\sigma_1}\\
&\quad +\mu_2\Gamma^1_{22\sigma_1}f_{t_1}^{\Delta_1} + \mu_2\Gamma^2_{22\sigma_1}f_{t_2}^{\Delta_2\sigma_1} + \mu_2 C_{\sigma_1}N_{\sigma_1}\\
&= \mu_2\left(\Gamma^1_{22\sigma_1} - \mu_1\Gamma^1_{12\sigma_1}\right)f_{t_1}^{\Delta_1}\\
&\quad +\mu_2\left(\Gamma^2_{22\sigma_1} + 1 - \mu_1\Gamma^2_{12\sigma_1}\right)f_{t_2}^{\Delta_2\sigma_1}\\
&\quad +\mu_2\left(C_{\sigma_1} - \mu_1 B_{\sigma_1}\right)N_{\sigma_1}.
\end{aligned}
$$

Example 5.2 Let $f \in \mathscr{C}^3$. We will express

$$
f_{t_1}^{\Delta_1^2\sigma_1},\quad f_{t_1}^{\Delta_1^2\sigma_2},\quad f_{t_2}^{\Delta_2^2\sigma_1},\quad f_{t_2}^{\Delta_2^2\sigma_2},\quad f_{t_1t_2}^{\Delta_1\Delta_2\sigma_1},\quad f_{t_1t_2}^{\Delta_1\Delta_2\sigma_2}
$$

in the terms of the σ_1-Christoffel coefficients and the third derivatives of f.

We have

$$
\begin{aligned}
f_{t_1}^{\Delta_1^2\sigma_1} &= f_{t_1}^{\Delta_1^2} + \mu_1 f_{t_1}^{\Delta_1^3}\\
&= \mu_1 f_{t_1}^{\Delta_1^3} + \Gamma^1_{11\sigma_1}f_{t_1}^{\Delta_1} + \Gamma^2_{11\sigma_1}f_{t_2}^{\Delta_2\sigma_1} + A_{\sigma_1}N_{\sigma_1},\\
f_{t_1}^{\Delta_1^2\sigma_2} &= f_{t_1}^{\Delta_1^2} + \mu_2 f_{t_1t_2}^{\Delta_1^2\Delta_2}\\
&= \mu_2 f_{t_1t_2}^{\Delta_1^2\Delta_2} + \Gamma^1_{11\sigma_1}f_{t_1}^{\Delta_1} + \Gamma^2_{11\sigma_1}f_{t_2}^{\Delta_2\sigma_1} + A_{\sigma_1}N_{\sigma_1},\\
f_{t_2}^{\Delta_2^2\sigma_1} &= f_{t_2}^{\Delta_2^2} + \mu_1 f_{t_2t_1}^{\Delta_2^2\Delta_1}\\
&= \mu_1 f_{t_2t_1}^{\Delta_2^2\Delta_1} + \Gamma^1_{22\sigma_1}f_{t_1}^{\Delta_1} + \Gamma^2_{22\sigma_1}f_{t_2}^{\Delta_2\sigma_1} + C_{\sigma_1}N_{\sigma_1},\\
f_{t_2}^{\Delta_2^2\sigma_2} &= f_{t_2}^{\Delta_2^2} + \mu_2 f_{t_2}^{\Delta_2^3}\\
&= \mu_2 f_{t_2}^{\Delta_2^3} + \Gamma^1_{22\sigma_1}f_{t_1}^{\Delta_1} + \Gamma^2_{22\sigma_1}f_{t_2}^{\Delta_2\sigma_1} + C_{\sigma_1}N_{\sigma_1},\\
f_{t_1t_2}^{\Delta_1\Delta_2\sigma_1} &= f_{t_1t_2}^{\Delta_1\Delta_2} + \mu_1 f_{t_1t_2}^{\Delta_1^2\Delta_2}\\
&= \mu_1 f_{t_1t_2}^{\Delta_1^2\Delta_2} + \Gamma^1_{12\sigma_1}f_{t_1}^{\Delta_1} + \Gamma^2_{12\sigma_1}f_{t_2}^{\Delta_2\sigma_1} + B_{\sigma_1}N_{\sigma_1},\\
f_{t_1t_2}^{\Delta_1\Delta_2\sigma_2} &= f_{t_1t_2}^{\Delta_1\Delta_2} + \mu_2 f_{t_1t_2}^{\Delta_1\Delta_2^2}\\
&= \mu_2 f_{t_1t_2}^{\Delta_1\Delta_2^2} + \Gamma^1_{12\sigma_1}f_{t_1}^{\Delta_1} + \Gamma^2_{12\sigma_1}f_{t_2}^{\Delta_2\sigma_1} + B_{\sigma_1}N_{\sigma_1}.
\end{aligned}
$$

Exercise 5.1 Find in the terms of the σ_1-Christoffel coefficients

1. $f_{t_1}^{\Delta_1} + 3f_{t_1}^{\Delta_1\sigma_2} - 5f_{t_2}^{\Delta_2\sigma_1}$.
2. $\Gamma^1_{11\sigma_1}f_{t_1}^{\Delta_1} + \Gamma^2_{12\sigma_1}f_{t_2}^{\Delta_2\sigma_1} + f_{t_1}^{\Delta_1\sigma_1}$.
3. $\Gamma^2_{22\sigma_1}f_{t_2}^{\Delta_2\sigma_2} + \Gamma^2_{22\sigma_1}f_{t_1}^{\Delta_1} + f_{t_2}^{\Delta_2\sigma_1}$.

2. Let S be a σ_2-regular surface. Then $f_{t_1}^{\Delta_1^2}$, $f_{t_1t_2}^{\Delta_1\Delta_2}$ and $f_{t_2}^{\Delta_2^2}$ can be represented in the terms of the basis $\{f_{t_1}^{\Delta_1\sigma_2}, f_{t_2}^{\Delta_2}, N_{\sigma_2}\}$ in the following way.

$$f_{t_1}^{\Delta_1^2} = \Gamma_{11\sigma_2}^1 f_{t_1}^{\Delta_1\sigma_2} + \Gamma_{11\sigma_2}^2 f_{t_2}^{\Delta_2} + A_{\sigma_2} N_{\sigma_2}$$
$$f_{t_1t_2}^{\Delta_1\Delta_2} = \Gamma_{12\sigma_2}^1 f_{t_1}^{\Delta_1\sigma_2} + \Gamma_{12\sigma_2}^2 f_{t_2}^{\Delta_2} + B_{\sigma_2} N_{\sigma_2}$$
$$f_{t_2}^{\Delta_2^2} = \Gamma_{22\sigma_2}^1 f_{t_1}^{\Delta_1\sigma_2} + \Gamma_{22\sigma_2}^2 f_{t_2}^{\Delta_2} + C_{\sigma_2} N_{\sigma_2},$$

where

$$A_{\sigma_2} = f_{t_1}^{\Delta_1^2} \cdot N_{\sigma_2}$$
$$B_{\sigma_2} = f_{t_1t_2}^{\Delta_1\Delta_2} \cdot N_{\sigma_2}$$
$$C_{\sigma_2} = f_{t_2}^{\Delta_2^2} \cdot N_{\sigma_2}$$

and

$$\Gamma_{11\sigma_2}^1 = \frac{\left(f_{t_1}^{\Delta_1\sigma_2} \cdot f_{t_1}^{\Delta_1^2}\right) G - \left(f_{t_2}^{\Delta_2} \cdot f_{t_1}^{\Delta_1^2}\right) F_{\sigma_2}}{E_{\sigma_2} G - (F_{\sigma_2})^2}$$

$$\Gamma_{11\sigma_2}^2 = \frac{\left(f_{t_1}^{\Delta_1\sigma_2} \cdot f_{t_1}^{\Delta_1^2}\right) F_{\sigma_2} - \left(f_{t_2}^{\Delta_2} \cdot f_{t_1}^{\Delta_1^2}\right) E_{\sigma_2}}{(F_{\sigma_2})^2 - E_{\sigma_2} G}$$

$$\Gamma_{12\sigma_2}^1 = \frac{\left(f_{t_1}^{\Delta_1\sigma_2} \cdot f_{t_1t_2}^{\Delta_1\Delta_2}\right) G - \left(f_{t_2}^{\Delta_2} \cdot f_{t_1t_2}^{\Delta_1\Delta_2}\right) F_{\sigma_2}}{E_{\sigma_2} G - (F_{\sigma_2})^2}$$

$$\Gamma_{12\sigma_2}^2 = \frac{\left(f_{t_1}^{\Delta_1\sigma_2} \cdot f_{t_1t_2}^{\Delta_1\Delta_2}\right) F_{\sigma_2} - \left(f_{t_2}^{\Delta_2} \cdot f_{t_1t_2}^{\Delta_1\Delta_2}\right) E_{\sigma_2}}{(F_{\sigma_2})^2 - E_{\sigma_2} G}$$

$$\Gamma_{22\sigma_2}^1 = \frac{\left(f_{t_1}^{\Delta_1\sigma_2} \cdot f_{t_2}^{\Delta_2^2}\right) G - \left(f_{t_2}^{\Delta_2} \cdot f_{t_2}^{\Delta_2^2}\right) F_{\sigma_2}}{E_{\sigma_2} G - (F_{\sigma_2})^2}$$

$$\Gamma_{22\sigma_2}^2 = \frac{\left(f_{t_1}^{\Delta_1\sigma_2} \cdot f_{t_2}^{\Delta_2^2}\right) F_{\sigma_2} - \left(f_{t_2}^{\Delta_2} \cdot f_{t_2}^{\Delta_2^2}\right) E_{\sigma_2}}{(F_{\sigma_2})^2 - E_{\sigma_2} G}.$$

Definition 5.2 The coefficients $\Gamma_{ij\sigma_2}^k$, $i,j,k \in \{1,2\}$, will be called the σ_2-Christoffel coefficients.

Example 5.3 We will find expressions for

$$f_{t_1}^{\Delta_1}, \quad f_{t_1}^{\Delta_1\sigma_1}, \quad f_{t_2}^{\Delta_2\sigma_2}, \quad f_{t_2}^{\Delta_2\sigma_1}$$

in the terms of the σ_2-Christoffel coefficients.

We have

$$f_{t_1}^{\Delta_1} = f_{t_1}^{\Delta_1\sigma_2} - \mu_2 f_{t_1t_2}^{\Delta_1\Delta_2}$$

$$
\begin{aligned}
&= f_{t_1}^{\Delta_1\sigma_2} - \mu_2\left(\Gamma^1_{12\sigma_2} f_{t_1}^{\Delta_1\sigma_2} + \Gamma^2_{12\sigma_2} f_{t_2}^{\Delta_2} + B_{\sigma_2} N_{\sigma_2}\right)\\
&= \left(1 - \mu_2\Gamma^1_{12\sigma_2}\right) f_{t_1}^{\Delta_1\sigma_2} - \mu_2\Gamma^2_{12\sigma_2} f_{t_2}^{\Delta_2} - \mu_2 B_{\sigma_2} N_{\sigma_2},\\
f_{t_1}^{\Delta_1\sigma_1} &= f_{t_1}^{\Delta_1} + \mu_1 f_{t_1}^{\Delta_1^2}\\
&= \left(1 - \mu_2\Gamma^1_{12\sigma_2}\right) f_{t_1}^{\Delta_1\sigma_2} - \mu_2\Gamma^2_{12\sigma_2} f_{t_2}^{\Delta_2} - \mu_2 B_{\sigma_2} N_{\sigma_2}\\
&\quad + \mu_1\Gamma^1_{11\sigma_2} f_{t_1}^{\Delta_1\sigma_2} + \mu_1\Gamma^2_{11\sigma_2} f_{t_2}^{\Delta_2} + \mu_1 A_{\sigma_2} N_{\sigma_2}\\
&= \left(1 - \mu_2\Gamma^1_{12\sigma_2} + \mu_1\Gamma^1_{11\sigma_2}\right) f_{t_1}^{\Delta_1\sigma_2}\\
&\quad + \left(\mu_1\Gamma^2_{11\sigma_2} - \mu_2\Gamma^2_{12\sigma_2}\right) f_{t_2}^{\Delta_2} + \left(\mu_1 A_{\sigma_2} - \mu_2 B_{\sigma_2}\right) N_{\sigma_2},\\
f_{t_2}^{\Delta_2\sigma_2} &= f_{t_2}^{\Delta_2} + \mu_2 f_{t_2}^{\Delta_2^2}\\
&= f_{t_2}^{\Delta_2} + \mu_2\left(\Gamma^1_{22\sigma_2} f_{t_1}^{\Delta_1\sigma_2} + \Gamma^2_{22\sigma_2} f_{t_2}^{\Delta_2} + C_{\sigma_2} N_{\sigma_2}\right)\\
&= \mu_2\Gamma^1_{22\sigma_2} f_{t_1}^{\Delta_1\sigma_2} + \left(1 + \mu_2\Gamma^2_{22\sigma_2}\right) f_{t_2}^{\Delta_2} + \mu_2 C_{\sigma_2} N_{\sigma_2},\\
f_{t_2}^{\Delta_2\sigma_1} &= f_{t_2}^{\Delta_2} + \mu_1 f_{t_1 t_2}^{\Delta_1\Delta_2}\\
&= f_{t_2}^{\Delta_2} + \mu_1\left(\Gamma^1_{12\sigma_2} f_{t_1}^{\Delta_1\sigma_2} + \Gamma^2_{12\sigma_2} f_{t_2}^{\Delta_2} + B_{\sigma_2} N_{\sigma_2}\right)\\
&= \mu_1\Gamma^1_{12\sigma_2} f_{t_1}^{\Delta_1\sigma_2} + \left(1 + \mu_1\Gamma^2_{12\sigma_2}\right) f_{t_2}^{\Delta_2} + \mu_1 B_{\sigma_2} N_{\sigma_2}.
\end{aligned}
$$

Exercise 5.2 Find expressions for

1. $2f_{t_1}^{\Delta_1} - 7\Gamma^1_{12\sigma_2} f_{t_2}^{\Delta_2\sigma_1} + f_{t_1}^{\Delta_1\sigma_2}$.
2. $4f_{t_1}^{\Delta_1\sigma_2} - 2\Gamma_{11\sigma_2} f_{t_1}^{\Delta_1} - \Gamma^1_{11\sigma_2} f_{t_2}^{\Delta_2\sigma_1}$.
3. $f_{t_1}^{\Delta_1} + 2\Gamma^1_{22\sigma_2} f_{t_1}^{\Delta_1\sigma_2} - f_{t_2}^{\Delta_2\sigma_2}$.

3. Let S be a $\sigma_1^2\sigma_2$-regular surface. Then $f_{t_1}^{\Delta_1^2\sigma_1\sigma_2}$, $f_{t_1t_2}^{\Delta_1\Delta_2\sigma_1\sigma_2}$ and $f_{t_2}^{\Delta_2^2\sigma_1\sigma_2}$ can be represented in the terms of the basis $\{f_{t_1}^{\Delta_1\sigma_1\sigma_2}, f_{t_2}^{\Delta_2\sigma_1^2\sigma_2}, N_{\sigma_1^2\sigma_2}\}$ in the following way.

$$
\begin{aligned}
f_{t_1}^{\Delta_1^2\sigma_1\sigma_2} &= \Gamma^1_{11\sigma_1^2\sigma_2} f_{t_1}^{\Delta_1\sigma_1\sigma_2} + \Gamma^2_{11\sigma_1^2\sigma_2} f_{t_2}^{\Delta_2\sigma_1^2\sigma_2} + A_{\sigma_1^2\sigma_2} N_{\sigma_1^2\sigma_2}\\
f_{t_1t_2}^{\Delta_1\Delta_2\sigma_1\sigma_2} &= \Gamma^1_{12\sigma_1^2\sigma_2} f_{t_1}^{\Delta_1\sigma_1\sigma_2} + \Gamma^2_{12\sigma_1^2\sigma_2} f_{t_2}^{\Delta_2\sigma_1^2\sigma_2} + B_{\sigma_1^2\sigma_2} N_{\sigma_1^2\sigma_2}\\
f_{t_2}^{\Delta_2^2\sigma_1\sigma_2} &= \Gamma^1_{22\sigma_1^2\sigma_2} f_{t_1}^{\Delta_1\sigma_1\sigma_2} + \Gamma^2_{22\sigma_1^2\sigma_2} f_{t_2}^{\Delta_2\sigma_1^2\sigma_2} + C_{\sigma_1^2\sigma_2} N_{\sigma_1^2\sigma_2},
\end{aligned}
$$

where

$$
\begin{aligned}
A_{\sigma_1^2\sigma_2} &= f_{t_1}^{\Delta_1^2\sigma_1\sigma_2} \cdot N_{\sigma_1^2\sigma_2}\\
B_{\sigma_1^2\sigma_2} &= f_{t_1t_2}^{\Delta_1\Delta_2\sigma_1\sigma_2} \cdot N_{\sigma_1^2\sigma_2}
\end{aligned}
$$

$$C_{\sigma_1^2\sigma_2} = f_{t_2}^{\Delta_2^2\sigma_1\sigma_2} \cdot N_{\sigma_1^2\sigma_2}$$

and

$$\Gamma^1_{11\sigma_1^2\sigma_2} = \frac{\left(f_{t_1}^{\Delta_1\sigma_1\sigma_2} \cdot f_{t_1}^{\Delta_1^2\sigma_1\sigma_2}\right) G_{\sigma_1^2\sigma_2} - \left(f_{t_2}^{\Delta_2\sigma_1^2\sigma_2} \cdot f_{t_1}^{\Delta_1^2\sigma_1\sigma_2}\right) F_{\sigma_1^2\sigma_2}}{E_{\sigma_1^2\sigma_2} G_{\sigma_1^2\sigma_2} - \left(F_{\sigma_1^2\sigma_2}\right)^2}$$

$$\Gamma^2_{11\sigma_1^2\sigma_2} = \frac{\left(f_{t_1}^{\Delta_1\sigma_1\sigma_2} \cdot f_{t_1}^{\Delta_1^2\sigma_1\sigma_2}\right) F_{\sigma_1^2\sigma_2} - \left(f_{t_2}^{\Delta_2\sigma_1^2\sigma_2} \cdot f_{t_1}^{\Delta_1^2\sigma_1\sigma_2}\right) E_{\sigma_1^2\sigma_2}}{\left(F_{\sigma_1^2\sigma_2}\right)^2 - E_{\sigma_1^2\sigma_2} G_{\sigma_1^2\sigma_2}}$$

$$\Gamma^1_{12\sigma_1^2\sigma_2} = \frac{\left(f_{t_1}^{\Delta_1\sigma_1\sigma_2} \cdot f_{t_1t_2}^{\Delta_1\Delta_2\sigma_1\sigma_2}\right) G_{\sigma_1^2\sigma_2} - \left(f_{t_2}^{\Delta_2\sigma_1^2\sigma_2} \cdot f_{t_1t_2}^{\Delta_1\Delta_2\sigma_1\sigma_2}\right) F_{\sigma_1^2\sigma_2}}{E_{\sigma_1^2\sigma_2} G_{\sigma_1^2\sigma_2} - \left(F_{\sigma_1^2\sigma_2}\right)^2}$$

$$\Gamma^2_{12\sigma_1^2\sigma_2} = \frac{\left(f_{t_1}^{\Delta_1\sigma_1\sigma_2} \cdot f_{t_1t_2}^{\Delta_1\Delta_2\sigma_1\sigma_2}\right) F_{\sigma_1^2\sigma_2} - \left(f_{t_2}^{\Delta_2\sigma_1^2\sigma_2} \cdot f_{t_1t_2}^{\Delta_1\Delta_2\sigma_1\sigma_2}\right) E_{\sigma_1^2\sigma_2}}{\left(F_{\sigma_1^2\sigma_2}\right)^2 - E_{\sigma_1^2\sigma_2} G_{\sigma_1^2\sigma_2}}$$

$$\Gamma^1_{22\sigma_1^2\sigma_2} = \frac{\left(f_{t_1}^{\Delta_1\sigma_1\sigma_2} \cdot f_{t_2}^{\Delta_2^2\sigma_1\sigma_2}\right) G_{\sigma_1^2\sigma_2} - \left(f_{t_2}^{\Delta_2\sigma_1^2\sigma_2} \cdot f_{t_2}^{\Delta_2^2\sigma_1\sigma_2}\right) F_{\sigma_1^2\sigma_2}}{E_{\sigma_1^2\sigma_2} G_{\sigma_1^2\sigma_2} - \left(F_{\sigma_1^2\sigma_2}\right)^2}$$

$$\Gamma^2_{22\sigma_1^2\sigma_2} = \frac{\left(f_{t_1}^{\Delta_1\sigma_1\sigma_2} \cdot f_{t_2}^{\Delta_2^2\sigma_1\sigma_2}\right) F_{\sigma_1^2\sigma_2} - \left(f_{t_2}^{\Delta_2\sigma_1^2\sigma_2} \cdot f_{t_2}^{\Delta_2^2\sigma_1\sigma_2}\right) E_{\sigma_1^2\sigma_2}}{\left(F_{\sigma_1^2\sigma_2}\right)^2 - E_{\sigma_1^2\sigma_2} G_{\sigma_1^2\sigma_2}}.$$

Definition 5.3 The coefficients $\Gamma^k_{ij\sigma_1^2\sigma_2}$, $i,j,k \in \{1,2\}$, will be called the $\sigma_1^2\sigma_2$-Christoffel coefficients.

Example 5.4 Suppose that σ_1 is Δ_1-differentiable and σ_2 is Δ_2-differentiable. We will express

$$f_{t_1}^{\Delta_1\sigma_1^2\sigma_2} \quad \text{and} \quad f_{t_1}^{\Delta_1\sigma_1\sigma_2^2}$$

with the basis $\left\{f_{t_1}^{\Delta_1\sigma_1\sigma_2}, f_{t_2}^{\Delta_2\sigma_1^2\sigma_2}, N_{\sigma_1^2\sigma_2}\right\}$.

We have

$$\begin{aligned} f_{t_1}^{\Delta_1\sigma_1^2\sigma_2} &= f_{t_1}^{\Delta_1\sigma_1\sigma_2} + \mu_1 f_{t_1}^{\Delta_1\sigma_1\sigma_2\Delta_1} \\ &= f_{t_1}^{\Delta_1\sigma_1\sigma_2} + \mu_1\left(1+\mu_1^{\Delta_1}\right) f_{t_1}^{\Delta_1^2\sigma_1\sigma_2} \\ &= f_{t_1}^{\Delta_1\sigma_1\sigma_2} + \mu_1\left(1+\mu_1^{\Delta_1}\right)\left(\Gamma^1_{11\sigma_1^2\sigma_2} f_{t_1}^{\Delta_1\sigma_1\sigma_2} + \Gamma^2_{11\sigma_1^2\sigma_2} f_{t_2}^{\Delta_2\sigma_1^2\sigma_2} + A_{\sigma_1^2\sigma_2} N_{\sigma_1^2\sigma_2}\right) \end{aligned}$$

$$= \left(1+\mu_1\left(1+\mu_1^{\Delta_1}\right)\Gamma^1_{11\sigma_1^2\sigma_2}\right)f_{t_1}^{\Delta_1\sigma_1\sigma_2}+\mu_1\left(1+\mu_1^{\Delta_1}\right)\Gamma^2_{11\sigma_1^2\sigma_2}f_{t_2}^{\Delta_2\sigma_1^2\sigma_2}$$
$$+\mu_1\left(1+\mu_1^{\Delta_1}\right)A_{\sigma_1^2\sigma_2}N_{\sigma_1^2\sigma_2}$$

and

$$f_{t_1}^{\Delta_1\sigma_1\sigma_2^2}=f_{t_1}^{\Delta_1\sigma_1\sigma_2}+f_{t_1t_2}^{\Delta_1\sigma_1\sigma_2\Delta_2}$$
$$=f_{t_1}^{\Delta_1\sigma_1\sigma_2}+\mu_2\left(1+\mu_2^{\Delta_2}\right)f_{t_1t_2}^{\Delta_1\Delta_2\sigma_1\sigma_2}$$
$$=f_{t_1}^{\Delta_1\sigma_1\sigma_2}+\mu_2\left(1+\mu_2^{\Delta_2}\right)\left(\Gamma^1_{12\sigma_1^2\sigma_2}f_{t_1}^{\Delta_1\sigma_1\sigma_2}+\Gamma^2_{12\sigma_1^2\sigma_2}f_{t_2}^{\Delta_2\sigma_1^2\sigma_2}+B_{\sigma_1^2\sigma_2}N_{\sigma_1^2\sigma_2}\right)$$
$$=\left(1+\mu_2\left(1+\mu_2^{\Delta_2}\right)\Gamma^1_{12\sigma_1^2\sigma_2}\right)f_{t_1}^{\Delta_1\sigma_1\sigma_2}+\mu_2\left(1+\mu_2^{\Delta_2}\right)\Gamma^2_{12\sigma_1^2\sigma_2}f_{t_2}^{\Delta_2\sigma_1^2\sigma_2}$$
$$+\mu_2\left(1+\mu_2^{\Delta_2}\right)B_{\sigma_1^2\sigma_2}N_{\sigma_1^2\sigma_2}.$$

Exercise 5.3 Suppose that σ_1 is Δ_1-differentiable and σ_2 is Δ_2-differentiable. Find

1. $f_{t_1}^{\Delta_1\sigma_1\sigma_2}+7f_{t_1}^{\Delta_1\sigma_1^2\sigma_2}-3f_{t_1}^{\Delta_1\sigma_1\sigma_2^2}$.
2. $f_{t_2}^{\Delta_1\sigma_1\sigma_2^2}-f_{t_1}^{\Delta_1\sigma_1^2\sigma_2}+f_{t_1}^{\Delta_1\sigma_1\sigma_2^2}$.
3. $f_{t_2}^{\Delta_2\sigma_1^2\sigma_2}-f_{t_1}^{\Delta_1\sigma_1\sigma_2}-5f_{t_1}^{\Delta_1\sigma_1^2\sigma_2}+2f_{t_1}^{\Delta_1\sigma_1\sigma_2^2}$.

4. Let S be a $\sigma_1\sigma_2^2$-regular surface. Then $f_{t_1}^{\Delta_1^2\sigma_1\sigma_2}$, $f_{t_1t_2}^{\Delta_1\Delta_2\sigma_1\sigma_2}$ and $f_{t_2}^{\Delta_2^2\sigma_1\sigma_2}$ can be represented in the terms of the basis $\{f_{t_1}^{\Delta_1\sigma_1\sigma_2^2}, f_{t_2}^{\Delta_2\sigma_1\sigma_2}, N_{\sigma_1\sigma_2^2}\}$ in the following way:

$$f_{t_1}^{\Delta_1^2\sigma_1\sigma_2}=\Gamma^1_{11\sigma_1\sigma_2^2}f_{t_1}^{\Delta_1\sigma_1\sigma_2^2}+\Gamma^2_{11\sigma_1\sigma_2^2}f_{t_2}^{\Delta_2\sigma_1\sigma_2}+A_{\sigma_1\sigma_2^2}N_{\sigma_1\sigma_2^2}$$
$$f_{t_1t_2}^{\Delta_1\Delta_2\sigma_1\sigma_2}=\Gamma^1_{12\sigma_1\sigma_2^2}f_{t_1}^{\Delta_1}+\Gamma^2_{12\sigma_1\sigma_2^2}f_{t_2}^{\Delta_2\sigma_1\sigma_2}+B_{\sigma_1\sigma_2^2}N_{\sigma_1\sigma_2^2}$$
$$f_{t_2}^{\Delta_2^2\sigma_1\sigma_2}=\Gamma^1_{22\sigma_1\sigma_2^2}f_{t_1}^{\Delta_1\sigma_1\sigma_2^2}+\Gamma^2_{22\sigma_1\sigma_2^2}f_{t_2}^{\Delta_2\sigma_1\sigma_2}+C_{\sigma_1\sigma_2^2}N_{\sigma_1\sigma_2^2},$$

where

$$A_{\sigma_1\sigma_2^2}=f_{t_1}^{\Delta_1^2\sigma_1\sigma_2}\cdot N_{\sigma_1\sigma_2^2},$$
$$B_{\sigma_1\sigma_2^2}=f_{t_1t_2}^{\Delta_1\Delta_2\sigma_1\sigma_2}\cdot N_{\sigma_1\sigma_2^2},$$
$$C_{\sigma_1\sigma_2^2}=f_{t_2}^{\Delta_2^2\sigma_1\sigma_2}\cdot N_{\sigma_1\sigma_2^2}$$

and

$$\Gamma^1_{11\sigma_1\sigma_2^2}=\frac{\left(f_{t_1}^{\Delta_1\sigma_1\sigma_2^2}\cdot f_{t_1}^{\Delta_1^2\sigma_1\sigma_2}\right)G_{\sigma_1\sigma_2^2}-\left(f_{t_2}^{\Delta_2\sigma_1\sigma_2}\cdot f_{t_1}^{\Delta_1^2\sigma_1\sigma_2}\right)F_{\sigma_1\sigma_2^2}}{E_{\sigma_1\sigma_2^2}G_{\sigma_1\sigma_2^2}-\left(F_{\sigma_1\sigma_2^2}\right)^2}$$

$$\Gamma^2_{11\sigma_1\sigma_2^2} = \frac{\left(f_{t_1}^{\Delta_1\sigma_1\sigma_2^2} \cdot f_{t_1}^{\Delta_1^2\sigma_1\sigma_2}\right) F_{\sigma_1\sigma_2^2} - \left(f_{t_2}^{\Delta_2\sigma_1\sigma_2} \cdot f_{t_1}^{\Delta_1^2\sigma_1\sigma_2}\right) E_{\sigma_1\sigma_2^2}}{\left(F_{\sigma_1\sigma_2^2}\right)^2 - E_{\sigma_1\sigma_2^2} G_{\sigma_1\sigma_2^2}}$$

$$\Gamma^1_{12\sigma_1\sigma_2^2} = \frac{\left(f_{t_1}^{\Delta_1\sigma_1\sigma_2^2} \cdot f_{t_1t_2}^{\Delta_1\Delta_2\sigma_1\sigma_2}\right) G_{\sigma_1\sigma_2^2} - \left(f_{t_2}^{\Delta_2\sigma_1\sigma_2} \cdot f_{t_1t_2}^{\Delta_1\Delta_2\sigma_1\sigma_2}\right) F_{\sigma_1\sigma_2^2}}{E_{\sigma_1\sigma_2^2} G_{\sigma_1\sigma_2^2} - \left(F_{\sigma_1\sigma_2^2}\right)^2}$$

$$\Gamma^2_{12\sigma_1\sigma_2^2} = \frac{\left(f_{t_1}^{\Delta_1\sigma_1\sigma_2^2} \cdot f_{t_1t_2}^{\Delta_1\Delta_2\sigma_1\sigma_2}\right) F_{\sigma_1\sigma_2^2} - \left(f_{t_2}^{\Delta_2\sigma_1\sigma_2} \cdot f_{t_1t_2}^{\Delta_1\Delta_2\sigma_1\sigma_2}\right) E_{\sigma_1\sigma_2^2}}{\left(F_{\sigma_1\sigma_2^2}\right)^2 - E_{\sigma_1\sigma_2^2} G_{\sigma_1\sigma_2^2}}$$

$$\Gamma^1_{22\sigma_1\sigma_2^2} = \frac{\left(f_{t_1}^{\Delta_1\sigma_1\sigma_2^2} \cdot f_{t_2}^{\Delta_2^2\sigma_1\sigma_2}\right) G_{\sigma_1\sigma_2^2} - \left(f_{t_2}^{\Delta_2\sigma_1\sigma_2} \cdot f_{t_2}^{\Delta_2^2\sigma_1\sigma_2}\right) F_{\sigma_1\sigma_2^2}}{E_{\sigma_1\sigma_2^2} G_{\sigma_1\sigma_2^2} - \left(F_{\sigma_1\sigma_2^2}\right)^2}$$

$$\Gamma^2_{22\sigma_1\sigma_2^2} = \frac{\left(f_{t_1}^{\Delta_1\sigma_1\sigma_2^2} \cdot f_{t_2}^{\Delta_2^2\sigma_1\sigma_2}\right) F_{\sigma_1\sigma_2^2} - \left(f_{t_2}^{\Delta_2\sigma_1\sigma_2} \cdot f_{t_2}^{\Delta_2^2\sigma_1\sigma_2}\right) E_{\sigma_1\sigma_2^2}}{\left(F_{\sigma_1\sigma_2^2}\right)^2 - E_{\sigma_1\sigma_2^2} G_{\sigma_1\sigma_2^2}}.$$

Definition 5.4 The coefficients $\Gamma^k_{ij\sigma_1\sigma_2^2}$, $i,j,k \in \{1,2\}$, will be called the $\sigma_1\sigma_2^2$-Christoffel coefficients.

Example 5.5 Suppose that σ_1 is Δ_1-differentiable and σ_2 is Δ_2-differentiable. We will express $f_{t_2}^{\Delta_2\sigma_1^2\sigma_2}$ and $f_{t_2}^{\Delta_2\sigma_1\sigma_2^2}$ using the basis $\left\{f_{t_1}^{\Delta_1\sigma_1\sigma_2^2}, f_{t_2}^{\Delta_2\sigma_1\sigma_2}, N_{\sigma_1\sigma_2^2}\right\}$. We have

$$\begin{aligned}
f_{t_2}^{\Delta_2\sigma_1^2\sigma_2} &= f_{t_2}^{\Delta_2\sigma_1\sigma_2} + \mu_1 f_{t_2t_1}^{\Delta_2\sigma_1\sigma_2\Delta_1} \\
&= f_{t_2}^{\Delta_2\sigma_1\sigma_2} + \mu_1\left(1+\mu_1^{\Delta_1}\right) f_{t_1t_2}^{\Delta_1\Delta_2\sigma_1\sigma_2} \\
&= f_{t_2}^{\Delta_2\sigma_1\sigma_2} + \mu_1\left(1+\mu_1^{\Delta_1}\right)\left(\Gamma^1_{12\sigma_1\sigma_2^2} f_{t_1}^{\Delta_1\sigma_1\sigma_2^2} + \Gamma^2_{12\sigma_1\sigma_2^2} f_{t_2}^{\Delta_2\sigma_1\sigma_2} + B_{\sigma_1\sigma_2^2} N_{\sigma_1\sigma_2^2}\right) \\
&= \mu_1\left(1+\mu_1^{\Delta_1}\right)\Gamma^1_{12\sigma_1\sigma_2^2} f_{t_1}^{\Delta_1\sigma_1\sigma_2^2} + \left(1+\mu_1\left(1+\mu_1^{\Delta_1}\right)\Gamma^2_{12\sigma_1\sigma_2^2}\right) f_{t_2}^{\Delta_2\sigma_1\sigma_2} \\
&\quad + \mu_1\left(1+\mu_1^{\Delta_1}\right) B_{\sigma_1\sigma_2^2} N_{\sigma_1\sigma_2^2}
\end{aligned}$$

and

$$\begin{aligned}
f_{t_2}^{\Delta_2\sigma_1\sigma_2^2} &= f_{t_2}^{\Delta_2\sigma_1\sigma_2} + \mu_2 f_{t_2}^{\Delta_2\sigma_1\sigma_2\Delta_2} \\
&= f_{t_2}^{\Delta_2\sigma_1\sigma_2} + \mu_2\left(1+\mu_2^{\Delta_2}\right) f_{t_2}^{\Delta_2^2\sigma_1\sigma_2} \\
&= f_{t_2}^{\Delta_2\sigma_1\sigma_2} + \mu_2\left(1+\mu_2^{\Delta_2}\right)\left(\Gamma^1_{22\sigma_1\sigma_2^2} f_{t_1}^{\Delta_1\sigma_1\sigma_2^2} + \Gamma^2_{22\sigma_1\sigma_2^2} f_{t_2}^{\Delta_2\sigma_1\sigma_2} + C_{\sigma_1\sigma_2^2} N_{\sigma_1\sigma_2^2}\right)
\end{aligned}$$

$$= \mu_2\left(1+\mu_2^{\Delta_2}\right)\Gamma^1_{22\sigma_1\sigma_2^2} f_{t_1}^{\Delta_1\sigma_1\sigma_2^2} + \left(1+\mu_2\left(1+\mu_2^{\Delta_2}\right)\Gamma^2_{22\sigma_1\sigma_2}\right) f_{t_2}^{\Delta_2\sigma_1\sigma_2}$$
$$+\mu_2\left(1+\mu_2^{\Delta_2}\right) C_{\sigma_1\sigma_2^2} N_{\sigma_1\sigma_2^2}.$$

Exercise 5.4 Suppose that σ_1 is Δ_1-differentiable and σ_2 is Δ_2-differentiable.
Find

1. $f_{t_2}^{\Delta_2\sigma_1\sigma_2} + f_{t_2}^{\Delta_1\sigma_1^2\sigma_2} + f_{t_2}^{\Delta_2\sigma_1^2\sigma_2} + f_{t_2}^{\Delta_2\sigma_1\sigma_2^2}$.
2. $-2f_{t_2}^{\Delta_2\sigma_1\sigma_2} + 4f_{t_2}^{\Delta_1\sigma_1^2\sigma_2} + 3f_{t_2}^{\Delta_2\sigma_1^2\sigma_2}$.
3. $3f_{t_2}^{\Delta_2\sigma_1\sigma_2} - 4f_{t_2}^{\Delta_2\sigma_1^2\sigma_2} + 2f_{t_2}^{\Delta_2\sigma_1\sigma_2^2}$.

5.2 The Weingarten Coefficients

Suppose that f is a continuous function on U and that $f_{t_1}^{\Delta_1}$, $f_{t_2}^{\Delta_2}$, $f_{t_1t_2}^{\Delta_1\Delta_2}$, $f_{t_1}^{\Delta_1^2}$, $f_{t_2}^{\Delta_2^2}$ exist and are continuous on U. Assume that S is a σ_1- or σ_2- or $\sigma_1^2\sigma_2$- or $\sigma_1\sigma_2^2$-oriented surface with a local parameterization (U,f).

1. Let S be a σ_1-regular surface and σ_1 be Δ_1-differentiable. We will represent $N_{\sigma_1t_1}^{\Delta_1}$ and $N_{\sigma_1t_2}^{\Delta_2}$ in the following form.

$$\begin{aligned} N_{\sigma_1t_1}^{\Delta_1} &= W^1_{1\sigma_1} f_{t_1}^{\Delta_1} + W^2_{1\sigma_1} f_{t_2}^{\Delta_2\sigma_1} \\ N_{\sigma_1t_2}^{\Delta_2} &= W^1_{2\sigma_1} f_{t_1}^{\Delta_1} + W^2_{2\sigma_1} f_{t_2}^{\Delta_2\sigma_1}. \end{aligned} \tag{5.2}$$

Definition 5.5 The coefficients $W^k_{j\sigma_1}$, $j,k \in \{1,2\}$, will be called σ_1-Weingarten coefficients.

Let

$$A_{\sigma_1\sigma_1} = N_{\sigma_1}^{\sigma_1} \cdot f_{t_1}^{\Delta_1^2},$$
$$A_{\sigma_2\sigma_1} = N_{\sigma_1}^{\sigma_2} \cdot f_{t_1}^{\Delta_1^2},$$
$$B_{\sigma_1\sigma_1} = N_{\sigma_1}^{\sigma_1} \cdot f_{t_1t_2}^{\Delta_1\Delta_2},$$
$$B_{\sigma_2\sigma_1} = N_{\sigma_1}^{\sigma_2} \cdot f_{t_1t_2}^{\Delta_1\Delta_2},$$
$$B_{\sigma_1\sigma_1\sigma_1} = N_{\sigma_1} \cdot f_{t_1t_2}^{\Delta_1\Delta_2\sigma_1},$$
$$B_{\sigma_1\sigma_1\sigma_1\sigma_1} = N_{\sigma_1}^{\sigma_1} \cdot f_{t_1t_2}^{\Delta_1\Delta_2\sigma_1},$$
$$C_{\sigma_1\sigma_1} = N_{\sigma_1} \cdot f_{t_2}^{\Delta_2^2\sigma_1},$$
$$C_{\sigma_2\sigma_1} = N_{\sigma_1}^{\sigma_2} \cdot f_{t_2}^{\Delta_2^2\sigma_1}.$$

Since $N_{\sigma_1} \perp f_{t_1}^{\Delta_1}$, we have

$$N_{\sigma_1} \cdot f_{t_1}^{\Delta_1} = 0.$$

We differentiate the last equation with respect to t_1 and with respect to t_2 and we find

$$\begin{aligned}0 &= N_{\sigma_1 t_1}^{\Delta_1} \cdot f_{t_1}^{\Delta_1} + N_{\sigma_1}^{\sigma_1} \cdot f_{t_1}^{\Delta_1^2}\\ &= N_{\sigma_1 t_1}^{\Delta_1} \cdot f_{t_1}^{\Delta_1\sigma_1} + N_{\sigma_1} \cdot f_{t_1}^{\Delta_1^2}\end{aligned}$$

and

$$\begin{aligned}0 &= N_{\sigma_1 t_2}^{\Delta_2} \cdot f_{t_1}^{\Delta_1} + N_{\sigma_1}^{\sigma_2} \cdot f_{t_1 t_2}^{\Delta_1\Delta_2}\\ &= N_{\sigma_1 t_2}^{\Delta_2} \cdot f_{t_1}^{\Delta_1\sigma_2} + N_{\sigma_1} \cdot f_{t_1 t_2}^{\Delta_1\Delta_2}.\end{aligned}$$

Therefore

$$\begin{aligned}N_{\sigma_1 t_1}^{\Delta_1} \cdot f_{t_1}^{\Delta_1\sigma_1} &= -A_{\sigma_1}\\ N_{\sigma_1 t_1}^{\Delta_1} \cdot f_{t_1}^{\Delta_1} &= -A_{\sigma_1\sigma_1}\\ N_{\sigma_1 t_2}^{\Delta_2} \cdot f_{t_1}^{\Delta_1\sigma_2} &= -B_{\sigma_1}\\ N_{\sigma_1 t_2}^{\Delta_2} \cdot f_{t_2}^{\Delta_2} &= -B_{\sigma_2\sigma_1}.\end{aligned}$$

Because $N_{\sigma_1} \perp f_{t_2}^{\Delta_2\sigma_1}$, we have that

$$N_{\sigma_1} \cdot f_{t_2}^{\Delta_2\sigma_1} = 0.$$

We differentiate the last equation with respect to t_1 and with respect to t_2 and we find

$$\begin{aligned}0 &= N_{\sigma_1 t_1}^{\Delta_1} \cdot f_{t_2}^{\Delta_2\sigma_1^2} + N_{\sigma_1} \cdot f_{t_2 t_1}^{\Delta_2\sigma_1\Delta_1}\\ &= N_{\sigma_1 t_1}^{\Delta_1} \cdot f_{t_2}^{\Delta_2\sigma_1} + N_{\sigma_1}^{\sigma_1} \cdot f_{t_2 t_1}^{\Delta_2\sigma_1\Delta_1}\end{aligned}$$

and

$$\begin{aligned}0 &= N_{\sigma_1 t_2}^{\Delta_2} \cdot f_{t_2}^{\Delta_2\sigma_1\sigma_2} + N_{\sigma_1} \cdot f_{t_2}^{\Delta_2^2\sigma_1}\\ &= N_{\sigma_1 t_2}^{\Delta_2} \cdot f_{t_2}^{\Delta_2\sigma_1} + N_{\sigma_1}^{\sigma_2} \cdot f_{t_2}^{\Delta_2^2\sigma_1}.\end{aligned}$$

Therefore

$$\begin{aligned}N_{\sigma_1 t_1}^{\Delta_1} \cdot f_{t_2}^{\Delta_2\sigma_1^2} &= -N_{\sigma_1} \cdot f_{t_2 t_1}^{\Delta_2\sigma_1\Delta_1}\\ &= -\left(1+\mu_1^{\Delta_1}\right) N_{\sigma_1} \cdot f_{t_1 t_2}^{\Delta_1\Delta_2\sigma_1}\\ &= -\left(1+\mu_1^{\Delta_1}\right) B_{\sigma_1\sigma_1\sigma_1},\\ N_{\sigma_1 t_1}^{\Delta_1} \cdot f_{t_2}^{\Delta_2\sigma_1} &= -N_{\sigma_1}^{\sigma_1} \cdot f_{t_2 t_1}^{\Delta_2\sigma_1\Delta_1}\\ &= -\left(1+\mu_1^{\Delta_1}\right) N_{\sigma_1}^{\sigma_1} \cdot f_{t_1 t_2}^{\Delta_1\Delta_2\sigma_1}\end{aligned}$$

$$
\begin{aligned}
&= -\left(1+\mu_1^{\Delta_1}\right) B_{\sigma_1\sigma_1\sigma_1\sigma_1},\\
N_{\sigma_1 t_2}^{\Delta_2} \cdot f_{t_2}^{\Delta_2\sigma_1\sigma_2} &= -N_{\sigma_1} \cdot f_{t_2}^{\Delta_2^2\sigma_1}\\
&= -C_{\sigma_1\sigma_1},\\
N_{\sigma_1 t_2}^{\Delta_2} \cdot f_{t_2}^{\Delta_2\sigma_1} &= -N_{\sigma_1}^{\sigma_2} \cdot f_{t_2}^{\Delta_2^2\sigma_1}\\
&= -C_{\sigma_2\sigma_1}.
\end{aligned}
$$

a. Consider the first equation of the system (5.2). We multiply it by $f_{t_1}^{\Delta_1\sigma_1}$ and $f_{t_1}^{\Delta_1}$ and we get

$$
\begin{aligned}
N_{\sigma_1 t_1}^{\Delta_1} \cdot f_{t_1}^{\Delta_1\sigma_1} &= W_{1\sigma_1}^{1}\left(f_{t_1}^{\Delta_1} \cdot f_{t_1}^{\Delta_1\sigma_1}\right) + W_{1\sigma_1}^{2}\left(f_{t_2}^{\Delta_2\sigma_1} \cdot f_{t_1}^{\Delta_1\sigma_1}\right)\\
N_{\sigma_1 t_1}^{\Delta_1} \cdot f_{t_1}^{\Delta_1} &= W_{1\sigma_1}^{1}\left(f_{t_1}^{\Delta_1} \cdot f_{t_1}^{\Delta_1}\right) + W_{1\sigma_1}^{2}\left(f_{t_2}^{\Delta_2\sigma_1} \cdot f_{t_1}^{\Delta_1}\right)
\end{aligned}
$$

or

$$
\begin{aligned}
-A_{\sigma_1} &= W_{1\sigma_1}^{1}\left(f_{t_1}^{\Delta_1} \cdot f_{t_1}^{\Delta_1\sigma_1}\right) + W_{1\sigma_1}^{2}\left(f_{t_2}^{\Delta_2\sigma_1} \cdot f_{t_1}^{\Delta_1\sigma_1}\right)\\
-A_{\sigma_1\sigma_1} &= W_{1\sigma_1}^{1}\left(f_{t_1}^{\Delta_1} \cdot f_{t_1}^{\Delta_1}\right) + W_{1\sigma_1}^{2}\left(f_{t_2}^{\Delta_2\sigma_1} \cdot f_{t_1}^{\Delta_1}\right).
\end{aligned}
$$

Then

$$
\begin{aligned}
W_{1\sigma_1}^{1} &= -\frac{A_{\sigma_1}\left(f_{t_2}^{\Delta_2\sigma_1} \cdot f_{t_1}^{\Delta_1}\right) - A_{\sigma_1\sigma_1}\left(f_{t_2}^{\Delta_2\sigma_1} \cdot f_{t_1}^{\Delta_1\sigma_1}\right)}{\left(f_{t_1}^{\Delta_1} \cdot f_{t_1}^{\Delta_1\sigma_1}\right)\left(f_{t_2}^{\Delta_2\sigma_1} \cdot f_{t_1}^{\Delta_1}\right) - \left(f_{t_1}^{\Delta_1}\right)^2\left(f_{t_2}^{\Delta_2\sigma_1} \cdot f_{t_1}^{\Delta_1\sigma_1}\right)}\\
W_{1\sigma_1}^{2} &= -\frac{A_{\sigma_1\sigma_1}\left(f_{t_2}^{\Delta_1} \cdot f_{t_1}^{\Delta_1\sigma_1}\right) - A_{\sigma_1}\left(f_{t_1}^{\Delta_1}\right)^2}{\left(f_{t_1}^{\Delta_1} \cdot f_{t_1}^{\Delta_1\sigma_1}\right)\left(f_{t_2}^{\Delta_2\sigma_1} \cdot f_{t_1}^{\Delta_1}\right) - \left(f_{t_1}^{\Delta_1}\right)^2\left(f_{t_2}^{\Delta_2\sigma_1} \cdot f_{t_1}^{\Delta_1\sigma_1}\right)}.
\end{aligned}
$$

Now, we multiply the first equation of the system (5.2) by $f_{t_2}^{\Delta_2\sigma_1^2}$ and we get the following system.

$$
\begin{aligned}
-A_{\sigma_1} &= W_{1\sigma_1}^{1}\left(f_{t_1}^{\Delta_1} \cdot f_{t_1}^{\Delta_1\sigma_1}\right) + W_{1\sigma_1}^{2}\left(f_{t_2}^{\Delta_2\sigma_1} \cdot f_{t_1}^{\Delta_1\sigma_1}\right)\\
-\left(1+\mu_1^{\Delta_1}\right) B_{\sigma_1\sigma_1\sigma_1} &= W_{1\sigma_1}^{1}\left(f_{t_1}^{\Delta_1} \cdot f_{t_2}^{\Delta_2\sigma_1^2}\right) + W_{1\sigma_1}^{2}\left(f_{t_2}^{\Delta_2\sigma_1} \cdot f_{t_2}^{\Delta_2\sigma_1^2}\right).
\end{aligned}
$$

Thus,

$$
\begin{aligned}
W_{1\sigma_1}^{1} &= -\frac{A_{\sigma_1}\left(f_{t_2}^{\Delta_2\sigma_1} \cdot f_{t_2}^{\Delta_2\sigma_1^2}\right) - \left(1+\mu_1^{\Delta_1}\right) B_{\sigma_1\sigma_1\sigma_1}\left(f_{t_2}^{\Delta_2\sigma_1} \cdot f_{t_1}^{\Delta_1\sigma_1}\right)}{\left(f_{t_1}^{\Delta_1} \cdot f_{t_1}^{\Delta_1\sigma_1}\right)\left(f_{t_2}^{\Delta_2\sigma_1} \cdot f_{t_2}^{\Delta_2\sigma_1^2}\right) - \left(f_{t_1}^{\Delta_1} \cdot f_{t_2}^{\Delta_2\sigma_1^2}\right)\left(f_{t_2}^{\Delta_2\sigma_1} \cdot f_{t_1}^{\Delta_1\sigma_1}\right)}\\
W_{1\sigma_1}^{2} &= -\frac{\left(1+\mu_1^{\Delta_1}\right) B_{\sigma_1\sigma_1\sigma_1}\left(f_{t_1}^{\Delta_1} \cdot f_{t_1}^{\Delta_1\sigma_1}\right) - A_{\sigma_1}\left(f_{t_1}^{\Delta_1} \cdot f_{t_2}^{\Delta_2\sigma_1^2}\right)}{\left(f_{t_1}^{\Delta_1} \cdot f_{t_1}^{\Delta_1\sigma_1}\right)\left(f_{t_2}^{\Delta_2\sigma_1} \cdot f_{t_2}^{\Delta_2\sigma_1^2}\right) - \left(f_{t_1}^{\Delta_1} \cdot f_{t_2}^{\Delta_2\sigma_1^2}\right)\left(f_{t_2}^{\Delta_2\sigma_1} \cdot f_{t_1}^{\Delta_1\sigma_1}\right)}.
\end{aligned}
$$

Now, we multiply the first equation of the system (5.2) by $f_{t_2}^{\Delta_2\sigma_1}$ and we obtain the system

$$
-A_{\sigma_1} = W_{1\sigma_1}^{1}\left(f_{t_1}^{\Delta_1} \cdot f_{t_1}^{\Delta_1\sigma_1}\right) + W_{1\sigma_1}^{2}\left(f_{t_2}^{\Delta_2\sigma_1} \cdot f_{t_1}^{\Delta_1\sigma_1}\right)
$$

$$-\left(1+\mu_1^{\Delta_1}\right)B_{\sigma_1\sigma_1\sigma_1\sigma_1} = W^1_{1\sigma_1}\left(f_{t_1}^{\Delta_1}\cdot f_{t_2}^{\Delta_2\sigma_1}\right) + W^2_{1\sigma_1}\left(f_{t_2}^{\Delta_2\sigma_1}\right)^2,$$

whereupon

$$W^1_{1\sigma_1} = -\frac{A_{\sigma_1}\left(f_{t_2}^{\Delta_2\sigma_1}\right)^2 - \left(1+\mu_1^{\Delta_1}\right)B_{\sigma_1\sigma_1\sigma_1\sigma_1}\left(f_{t_2}^{\Delta_2\sigma_1}\cdot f_{t_1}^{\Delta_1\sigma_1}\right)}{\left(f_{t_1}^{\Delta_1}\cdot f_{t_1}^{\Delta_1\sigma_1}\right)\left(f_{t_2}^{\Delta_2\sigma_1}\right)^2 - \left(f_{t_1}^{\Delta_1}\cdot f_{t_2}^{\Delta_2\sigma_1}\right)\left(f_{t_2}^{\Delta_2\sigma_1}\cdot f_{t_1}^{\Delta_1\sigma_1}\right)}$$

$$W^2_{1\sigma_1} = -\frac{\left(1+\mu_1^{\Delta_1}\right)B_{\sigma_1\sigma_1\sigma_1\sigma_1}\left(f_{t_1}^{\Delta_1\sigma_1}\cdot f_{t_1}^{\Delta_1}\right) - A_{\sigma_1}\left(f_{t_1}^{\Delta_1}\cdot f_{t_2}^{\Delta_2\sigma_1}\right)}{\left(f_{t_1}^{\Delta_1}\cdot f_{t_1}^{\Delta_1\sigma_1}\right)\left(f_{t_2}^{\Delta_2\sigma_1}\right)^2 - \left(f_{t_1}^{\Delta_1}\cdot f_{t_2}^{\Delta_2\sigma_1}\right)\left(f_{t_2}^{\Delta_2\sigma_1}\cdot f_{t_1}^{\Delta_1\sigma_1}\right)}.$$

Next, we get the system

$$-A_{\sigma_1\sigma_1} = W^1_{1\sigma_1}\left(f_{t_1}^{\Delta_1}\cdot f_{t_1}^{\Delta_1}\right) + W^2_{1\sigma_1}\left(f_{t_2}^{\Delta_2\sigma_1}\cdot f_{t_1}^{\Delta_1}\right)$$

$$-\left(1+\mu_1^{\Delta_1}\right)B_{\sigma_1\sigma_1\sigma_1} = W^1_{1\sigma_1}\left(f_{t_1}^{\Delta_1}\cdot f_{t_2}^{\Delta_2\sigma_1^2}\right) + W^2_{1\sigma_1}\left(f_{t_2}^{\Delta_2\sigma_1}\cdot f_{t_2}^{\Delta_2\sigma_1^2}\right),$$

whereupon

$$W^1_{1\sigma_1} = -\frac{A_{\sigma_1\sigma_1}\left(f_{t_2}^{\Delta_2\sigma_1}\cdot f_{t_2}^{\Delta_2\sigma_1^2}\right) - \left(1+\mu_1^{\Delta_1}\right)B_{\sigma_1\sigma_1\sigma_1}\left(f_{t_2}^{\Delta_2\sigma_1}\cdot f_{t_1}^{\Delta_1}\right)}{\left(f_{t_1}^{\Delta_1}\right)^2\left(f_{t_2}^{\Delta_2\sigma_1}\cdot f_{t_2}^{\Delta_2\sigma_1^2}\right) - \left(f_{t_1}^{\Delta_1}\cdot f_{t_2}^{\Delta_2\sigma_1}\right)\left(f_{t_1}^{\Delta_1}\cdot f_{t_2}^{\Delta_2\sigma_1^2}\right)}$$

$$W^2_{1\sigma_1} = -\frac{\left(1+\mu_1^{\Delta_1}\right)B_{\sigma_1\sigma_1\sigma_1}\left(f_{t_1}^{\Delta_1}\right)^2 - A_{\sigma_1\sigma_1}\left(f_{t_1}^{\Delta_1}\cdot f_{t_2}^{\Delta_2\sigma_1^2}\right)}{\left(f_{t_1}^{\Delta_1}\right)^2\left(f_{t_2}^{\Delta_2\sigma_1}\cdot f_{t_2}^{\Delta_2\sigma_1^2}\right) - \left(f_{t_1}^{\Delta_1}\cdot f_{t_2}^{\Delta_2\sigma_1}\right)\left(f_{t_1}^{\Delta_1}\cdot f_{t_2}^{\Delta_2\sigma_1^2}\right)}.$$

Next, we obtain the system

$$-A_{\sigma_1\sigma_1} = W^1_{1\sigma_1}\left(f_{t_1}^{\Delta_1}\cdot f_{t_1}^{\Delta_1}\right) + W^2_{1\sigma_1}\left(f_{t_2}^{\Delta_2\sigma_1}\cdot f_{t_1}^{\Delta_1}\right)$$

$$-\left(1+\mu_1^{\Delta_1}\right)B_{\sigma_1\sigma_1\sigma_1\sigma_1} = W^1_{1\sigma_1}\left(f_{t_1}^{\Delta_1}\cdot f_{t_2}^{\Delta_2\sigma_1}\right) + W^2_{1\sigma_1}\left(f_{t_2}^{\Delta_2\sigma_1}\right)^2.$$

Therefore

$$W^1_{1\sigma_1} = -\frac{A_{\sigma_1\sigma_1}\left(f_{t_2}^{\Delta_2\sigma_1}\right)^2 - \left(1+\mu_1^{\Delta_1}\right)B_{\sigma_1\sigma_1\sigma_1\sigma_1}\left(f_{t_2}^{\Delta_2\sigma_1}\cdot f_{t_1}^{\Delta_1}\right)}{\left(f_{t_1}^{\Delta_1}\right)^2\left(f_{t_2}^{\Delta_2\sigma_1}\right)^2 - \left(f_{t_1}^{\Delta_1}\cdot f_{t_2}^{\Delta_2\sigma_1}\right)^2}$$

$$W^2_{1\sigma_1} = -\frac{\left(1+\mu_1^{\Delta_1}\right)B_{\sigma_1\sigma_1\sigma_1\sigma_1}\left(f_{t_1}^{\Delta_1}\right)^2 - A_{\sigma_1\sigma_1}\left(f_{t_1}^{\Delta_1}\cdot f_{t_2}^{\Delta_2\sigma_1}\right)}{\left(f_{t_1}^{\Delta_1}\right)^2\left(f_{t_2}^{\Delta_2\sigma_1}\right)^2 - \left(f_{t_1}^{\Delta_1}\cdot f_{t_2}^{\Delta_2\sigma_1}\right)^2}.$$

Also, we have the system

$$-\left(1+\mu_1^{\Delta_1}\right)B_{\sigma_1\sigma_1\sigma_1} = W^1_{1\sigma_1}\left(f_{t_1}^{\Delta_1}\cdot f_{t_2}^{\Delta_2\sigma_1^2}\right) + W^2_{1\sigma_1}\left(f_{t_2}^{\Delta_2\sigma_1}\cdot f_{t_2}^{\Delta_2\sigma_1^2}\right)$$

$$-\left(1+\mu_1^{\Delta_1}\right)B_{\sigma_1\sigma_1\sigma_1\sigma_1} = W^1_{1\sigma_1}\left(f_{t_1}^{\Delta_1}\cdot f_{t_2}^{\Delta_2\sigma_1}\right) + W^2_{1\sigma_1}\left(f_{t_2}^{\Delta_2\sigma_1}\right)^2.$$

Consequently

$$W^1_{1\sigma_1} = -\left(1+\mu_1^{\Delta_1}\right)\frac{B_{\sigma_1\sigma_1\sigma_1}\left(f_{t_2}^{\Delta_2\sigma_1}\right)^2 - B_{\sigma_1\sigma_1\sigma_1\sigma_1}\left(f_{t_2}^{\Delta_2\sigma_1}\cdot f_{t_2}^{\Delta_2\sigma_1^2}\right)}{\left(f_{t_1}^{\Delta_1}\cdot f_{t_2}^{\Delta_2\sigma_1^2}\right)\left(f_{t_2}^{\Delta_2\sigma_1}\right)^2 - \left(f_{t_1}^{\Delta_1}\cdot f_{t_2}^{\Delta_2\sigma_1}\right)\left(f_{t_2}^{\Delta_2\sigma_1}\cdot f_{t_2}^{\Delta_2\sigma_1^2}\right)}$$

$$W_{1\sigma_1}^2 = -\left(1+\mu_1^{\Delta_1}\right)\frac{B_{\sigma_1\sigma_1\sigma_1\sigma_1}\left(f_{t_1}^{\Delta_1}\cdot f_{t_2}^{\Delta_2\sigma_1^2}\right) - B_{\sigma_1\sigma_1\sigma_1}\left(f_{t_1}^{\Delta_1}\cdot f_{t_2}^{\Delta_2\sigma_1}\right)}{\left(f_{t_1}^{\Delta_1}\cdot f_{t_2}^{\Delta_2\sigma_1^2}\right)\left(f_{t_2}^{\Delta_2\sigma_1}\right)^2 - \left(f_{t_1}^{\Delta_1}\cdot f_{t_2}^{\Delta_2\sigma_1}\right)\left(f_{t_2}^{\Delta_2\sigma_1}\cdot f_{t_2}^{\Delta_2\sigma_1^2}\right)}.$$

b. Now, we consider the second equation of the system (5.2). We multiply it by $f_{t_1}^{\Delta_1\sigma_2}$ and $f_{t_1}^{\Delta_1}$ and we get

$$N_{\sigma_1 t_2}^{\Delta_2}\cdot f_{t_1}^{\Delta_1\sigma_2} = W_{2\sigma_1}^1\left(f_{t_1}^{\Delta_1}\cdot f_{t_1}^{\Delta_1\sigma_2}\right) + W_{2\sigma_1}^2\left(f_{t_2}^{\Delta_2\sigma_1}\cdot f_{t_1}^{\Delta_1\sigma_1}\right)$$

$$N_{\sigma_1 t_2}^{\Delta_2}\cdot f_{t_1}^{\Delta_1} = W_{2\sigma_1}^1\left(f_{t_1}^{\Delta_1}\right)^2 + W_{2\sigma_1}^2\left(f_{t_2}^{\Delta_2\sigma_1}\cdot f_{t_1}^{\Delta_1}\right)$$

or

$$-B_{\sigma_1} = W_{2\sigma_1}^1\left(f_{t_1}^{\Delta_1}\cdot f_{t_1}^{\Delta_1\sigma_2}\right) + W_{2\sigma_1}^2\left(f_{t_2}^{\Delta_2\sigma_1}\cdot f_{t_1}^{\Delta_1\sigma_1}\right)$$

$$-B_{\sigma_2\sigma_1} = W_{2\sigma_1}^1\left(f_{t_1}^{\Delta_1}\right)^2 + W_{2\sigma_1}^2\left(f_{t_2}^{\Delta_2\sigma_1}\cdot f_{t_1}^{\Delta_1}\right).$$

By the last system, we find

$$W_{2\sigma_1}^1 = -\frac{B_{\sigma_1}\left(f_{t_2}^{\Delta_2\sigma_1}\cdot f_{t_1}^{\Delta_1}\right) - B_{\sigma_2\sigma_1}\left(f_{t_2}^{\Delta_2\sigma_1}\cdot f_{t_1}^{\Delta_1\sigma_1}\right)}{\left(f_{t_1}^{\Delta_1}\cdot f_{t_1}^{\Delta_1\sigma_2}\right)\left(f_{t_2}^{\Delta_2\sigma_1}\cdot f_{t_1}^{\Delta_1}\right) - \left(f_{t_2}^{\Delta_2\sigma_1}\cdot f_{t_1}^{\Delta_1\sigma_1}\right)\left(f_{t_1}^{\Delta_1}\right)^2}$$

$$W_{2\sigma_1}^1 = -\frac{B_{\sigma_2\sigma_1}\left(f_{t_1}^{\Delta_1}\cdot f_{t_1}^{\Delta_1\sigma_2}\right) - B_{\sigma_1}\left(f_{t_1}^{\Delta_1}\right)^2}{\left(f_{t_1}^{\Delta_1}\cdot f_{t_1}^{\Delta_1\sigma_2}\right)\left(f_{t_2}^{\Delta_2\sigma_1}\cdot f_{t_1}^{\Delta_1}\right) - \left(f_{t_2}^{\Delta_2\sigma_1}\cdot f_{t_1}^{\Delta_1\sigma_1}\right)\left(f_{t_1}^{\Delta_1}\right)^2}.$$

Now, we multiply the second equation of the system (5.2) by $f_{t_2}^{\Delta_2\sigma_1\sigma_2}$ and we find

$$-B_{\sigma_1} = W_{2\sigma_1}^1\left(f_{t_1}^{\Delta_1}\cdot f_{t_1}^{\Delta_1\sigma_2}\right) + W_{2\sigma_1}^2\left(f_{t_2}^{\Delta_2\sigma_1}\cdot f_{t_1}^{\Delta_1\sigma_1}\right)$$

$$-C_{\sigma_1\sigma_1} = W_{2\sigma_1}^1\left(f_{t_1}^{\Delta_1}\cdot f_{t_2}^{\Delta_2\sigma_1\sigma_2}\right) + W_{2\sigma_1}^2\left(f_{t_2}^{\Delta_2\sigma_1}\cdot f_{t_2}^{\Delta_2\sigma_1\sigma_2}\right),$$

whereupon

$$W_{2\sigma_1}^1 = -\frac{B_{\sigma_1}\left(f_{t_2}^{\Delta_2\sigma_1}\cdot f_{t_2}^{\Delta_2\sigma_1\sigma_2}\right) - C_{\sigma_1\sigma_1}\left(f_{t_2}^{\Delta_2\sigma_1}\cdot f_{t_1}^{\Delta_1\sigma_1}\right)}{\left(f_{t_1}^{\Delta_1}\cdot f_{t_1}^{\Delta_1\sigma_2}\right)\left(f_{t_2}^{\Delta_2\sigma_1}\cdot f_{t_2}^{\Delta_2\sigma_1\sigma_2}\right) - \left(f_{t_1}^{\Delta_1}\cdot f_{t_2}^{\Delta_2\sigma_1\sigma_2}\right)\left(f_{t_2}^{\Delta_2\sigma_1}\cdot f_{t_1}^{\Delta_1\sigma_1}\right)}$$

$$W_{2\sigma_1}^2 = -\frac{C_{\sigma_1\sigma_1}\left(f_{t_1}^{\Delta_1}\cdot f_{t_1}^{\Delta_1\sigma_2}\right) - B_{\sigma_1}\left(f_{t_1}^{\Delta_1}\cdot f_{t_2}^{\Delta_2\sigma_1\sigma_2}\right)}{\left(f_{t_1}^{\Delta_1}\cdot f_{t_1}^{\Delta_1\sigma_2}\right)\left(f_{t_2}^{\Delta_2\sigma_1}\cdot f_{t_2}^{\Delta_2\sigma_1\sigma_2}\right) - \left(f_{t_1}^{\Delta_1}\cdot f_{t_2}^{\Delta_2\sigma_1\sigma_2}\right)\left(f_{t_2}^{\Delta_2\sigma_1}\cdot f_{t_1}^{\Delta_1\sigma_1}\right)}.$$

Now, we multiply the second equation of the system (5.2) by $f_{t_2}^{\Delta_2\sigma_1}$ and we arrive at the system

$$-B_{\sigma_1} = W_{2\sigma_1}^1\left(f_{t_1}^{\Delta_1}\cdot f_{t_1}^{\Delta_1\sigma_2}\right) + W_{2\sigma_1}^2\left(f_{t_2}^{\Delta_2\sigma_1}\cdot f_{t_1}^{\Delta_1\sigma_1}\right)$$

$$-C_{\sigma_2\sigma_1} = W_{2\sigma_1}^1\left(f_{t_1}^{\Delta_1}\cdot f_{t_2}^{\Delta_2\sigma_1}\right) + W_{2\sigma_1}^2\left(f_{t_2}^{\Delta_2\sigma_1}\right)^2.$$

By the last system, we find

$$W_{2\sigma_1}^1 = -\frac{B_{\sigma_1}\left(f_{t_2}^{\Delta_2\sigma_1}\right)^2 - C_{\sigma_2\sigma_1}\left(f_{t_2}^{\Delta_2\sigma_1}\cdot f_{t_1}^{\Delta_1\sigma_1}\right)}{\left(f_{t_1}^{\Delta_1}\cdot f_{t_1}^{\Delta_1\sigma_2}\right)\left(f_{t_2}^{\Delta_2\sigma_1}\right)^2 - \left(f_{t_1}^{\Delta_1}\cdot f_{t_2}^{\Delta_2\sigma_1}\right)\left(f_{t_2}^{\Delta_2\sigma_1}\cdot f_{t_1}^{\Delta_1\sigma_1}\right)}$$

$$W^2_{2\sigma_1} = -\frac{C_{\sigma_2\sigma_1}\left(f^{\Delta_1}_{t_1}\cdot f^{\Delta_1\sigma_2}_{t_1}\right) - B_{\sigma_1}\left(f^{\Delta_1}_{t_1}\cdot f^{\Delta_2\sigma_1}_{t_2}\right)}{\left(f^{\Delta_1}_{t_1}\cdot f^{\Delta_1\sigma_2}_{t_1}\right)\left(f^{\Delta_2\sigma_1}_{t_2}\right)^2 - \left(f^{\Delta_1}_{t_1}\cdot f^{\Delta_2\sigma_1}_{t_2}\right)\left(f^{\Delta_2\sigma_1}_{t_2}\cdot f^{\Delta_1\sigma_1}_{t_1}\right)}.$$

Next, we obtain the system

$$\begin{aligned}
-B_{\sigma_2\sigma_1} &= W^1_{2\sigma_1}\left(f^{\Delta_1}_{t_1}\right)^2 + W^2_{2\sigma_1}\left(f^{\Delta_2\sigma_1}_{t_2}\cdot f^{\Delta_1}_{t_1}\right)\\
-C_{\sigma_1\sigma_1} &= W^1_{2\sigma_1}\left(f^{\Delta_1}_{t_1}\cdot f^{\Delta_2\sigma_1\sigma_2}_{t_2}\right) + W^2_{2\sigma_1}\left(f^{\Delta_2\sigma_1}_{t_2}\cdot f^{\Delta_2\sigma_1\sigma_2}_{t_2}\right),
\end{aligned}$$

whereupon

$$\begin{aligned}
W^1_{2\sigma_1} &= -\frac{B_{\sigma_2\sigma_1}\left(f^{\Delta_2\sigma_1}_{t_2}\cdot f^{\Delta_2\sigma_1\sigma_2}_{t_2}\right) - C_{\sigma_1\sigma_1}\left(f^{\Delta_2\sigma_1}_{t_2}\cdot f^{\Delta_1}_{t_1}\right)}{\left(f^{\Delta_1}_{t_1}\right)^2\left(f^{\Delta_2\sigma_1}_{t_2}\cdot f^{\Delta_2\sigma_1\sigma_2}_{t_2}\right) - \left(f^{\Delta_1}_{t_1}\cdot f^{\Delta_2\sigma_1\sigma_2}_{t_2}\right)\left(f^{\Delta_2\sigma_1}_{t_2}\cdot f^{\Delta_1}_{t_1}\right)}\\
W^2_{2\sigma_1} &= -\frac{C_{\sigma_1\sigma_1}\left(f^{\Delta_1}_{t_1}\right)^2 - B_{\sigma_2\sigma_1}\left(f^{\Delta_1}_{t_1}\cdot f^{\Delta_2\sigma_1\sigma_2}_{t_2}\right)}{\left(f^{\Delta_1}_{t_1}\right)^2\left(f^{\Delta_2\sigma_1}_{t_2}\cdot f^{\Delta_2\sigma_1\sigma_2}_{t_2}\right) - \left(f^{\Delta_1}_{t_1}\cdot f^{\Delta_2\sigma_1\sigma_2}_{t_2}\right)\left(f^{\Delta_2\sigma_1}_{t_2}\cdot f^{\Delta_1}_{t_1}\right)}
\end{aligned}$$

Next, we have the system

$$\begin{aligned}
-B_{\sigma_2\sigma_1} &= W^1_{2\sigma_1}\left(f^{\Delta_1}_{t_1}\right)^2 + W^2_{2\sigma_1}\left(f^{\Delta_2\sigma_1}_{t_2}\cdot f^{\Delta_1}_{t_1}\right)\\
-C_{\sigma_2\sigma_1} &= W^1_{2\sigma_1}\left(f^{\Delta_1}_{t_1}\cdot f^{\Delta_2\sigma_1}_{t_2}\right) + W^2_{2\sigma_1}\left(f^{\Delta_2\sigma_1}_{t_2}\right)^2,
\end{aligned}$$

whereupon

$$\begin{aligned}
W^1_{2\sigma_1} &= -\frac{B_{\sigma_2\sigma_1}\left(f^{\Delta_2\sigma_1}_{t_2}\right)^2 - C_{\sigma_2\sigma_1}\left(f^{\Delta_2\sigma_1}_{t_2}\cdot f^{\Delta_1}_{t_1}\right)}{\left(f^{\Delta_1}_{t_1}\right)^2\left(f^{\Delta_2\sigma_1}_{t_2}\right)^2 - \left(f^{\Delta_1}_{t_1}\cdot f^{\Delta_2\sigma_1}_{t_2}\right)^2}\\
W^2_{2\sigma_1} &= -\frac{C_{\sigma_2\sigma_1}\left(f^{\Delta_1}_{t_1}\right)^2 - B_{\sigma_2\sigma_1}\left(f^{\Delta_1}_{t_1}\cdot f^{\Delta_2\sigma_2}_{t_2}\right)}{\left(f^{\Delta_1}_{t_1}\right)^2\left(f^{\Delta_2\sigma_1}_{t_2}\right)^2 - \left(f^{\Delta_1}_{t_1}\cdot f^{\Delta_2\sigma_1}_{t_2}\right)^2}.
\end{aligned}$$

Also, we have the system

$$\begin{aligned}
-C_{\sigma_1\sigma_1} &= W^1_{2\sigma_1}\left(f^{\Delta_1}_{t_1}\cdot f^{\Delta_2\sigma_1\sigma_2}_{t_2}\right) + W^2_{2\sigma_1}\left(f^{\Delta_2\sigma_1}_{t_2}\cdot f^{\Delta_2\sigma_1\sigma_2}_{t_2}\right)\\
-C_{\sigma_2\sigma_1} &= W^1_{2\sigma_1}\left(f^{\Delta_1}_{t_1}\cdot f^{\Delta_2\sigma_1}_{t_2}\right) + W^2_{2\sigma_1}\left(f^{\Delta_2\sigma_1}_{t_2}\right)^2,
\end{aligned}$$

from where

$$\begin{aligned}
W^1_{2\sigma_1} &= -\frac{C_{\sigma_1\sigma_1}\left(f^{\Delta_2\sigma_1}_{t_2}\right)^2 - C_{\sigma_2\sigma_1}\left(f^{\Delta_2\sigma_1}_{t_2}\cdot f^{\Delta_2\sigma_1\sigma_2}_{t_2}\right)}{\left(f^{\Delta_1}_{t_1}\cdot f^{\Delta_2\sigma_1\sigma_2}_{t_2}\right)\left(f^{\Delta_2\sigma_1}_{t_2}\right)^2 - \left(f^{\Delta_1}_{t_1}\cdot f^{\Delta_2\sigma_1}_{t_2}\right)\left(f^{\Delta_2\sigma_1}_{t_2}\cdot f^{\Delta_2\sigma_1\sigma_2}_{t_2}\right)}\\
W^2_{2\sigma_1} &= -\frac{C_{\sigma_2\sigma_1}\left(f^{\Delta_1}_{t_1}\cdot f^{\Delta_2\sigma_1\sigma_2}_{t_2}\right) - C_{\sigma_1\sigma_1}\left(f^{\Delta_1}_{t_1}\cdot f^{\Delta_2\sigma_2}_{t_2}\right)}{\left(f^{\Delta_1}_{t_1}\cdot f^{\Delta_2\sigma_1\sigma_2}_{t_2}\right)\left(f^{\Delta_2\sigma_1}_{t_2}\right)^2 - \left(f^{\Delta_1}_{t_1}\cdot f^{\Delta_2\sigma_1}_{t_2}\right)\left(f^{\Delta_2\sigma_1}_{t_2}\cdot f^{\Delta_2\sigma_1\sigma_2}_{t_2}\right)}.
\end{aligned}$$

Now, we will express $N^{\sigma_1}_{\sigma_1}$ and $N^{\sigma_2}_{\sigma_1}$. We have

$$N^{\sigma_1}_{\sigma_1} = N_{\sigma_1} + \mu_1 N^{\Delta_1}_{\sigma_1 t_1}$$

$$
\begin{aligned}
&= N_{\sigma_1} + \mu_1 \left(W^1_{1\sigma_1} f^{\Delta_1}_{t_1} + W^2_{1\sigma_1} f^{\Delta_2 \sigma_1}_{t_2} \right) \\
&= \mu_1 W^1_{1\sigma_1} f^{\Delta_1}_{t_1} + \mu_1 W^2_{1\sigma_1} f^{\Delta_2 \sigma_1}_{t_2} + N_{\sigma_1}, \\
N^{\sigma_2}_{\sigma_1} &= N_{\sigma_1} + \mu_2 N^{\Delta_2}_{\sigma_1 t_2} \\
&= N_{\sigma_1} + \mu_2 \left(W^1_{2\sigma_1} f^{\Delta_1}_{t_1} + W^2_{2\sigma_1} f^{\Delta_2 \sigma_1}_{t_2} \right) \\
&= \mu_2 W^1_{2\sigma_1} f^{\Delta_1}_{t_1} + \mu_2 W^2_{2\sigma_1} f^{\Delta_2 \sigma_1}_{t_2} + N_{\sigma_1}.
\end{aligned}
$$

Below, suppose that

$$
1 - \mu_1 \left(1 + \mu_1^{\Delta_1}\right) \Gamma^{2\sigma_1}_{12\sigma_1} \neq 0. \tag{5.3}
$$

Set

$$
\begin{aligned}
V^1_{1\sigma_1} &= \frac{\mu_1 \left(1 + \mu_1^{\Delta_1}\right) \left(\Gamma^{1\sigma_1}_{12\sigma_1} \left(1 + \mu_1 \Gamma^1_{11\sigma_1}\right) + \mu_1 B^{\sigma_1}_{\sigma_1} W^1_{1\sigma_1} \right)}{1 - \mu_1 \left(1 + \mu_1^{\Delta_1}\right) \Gamma^{2\sigma_1}_{12\sigma_1}}, \\
V^2_{1\sigma_1} &= \frac{1 + \mu_1^2 \left(1 + \mu_1^{\Delta_1}\right) \Gamma^{1\sigma_1}_{12\sigma_1} \Gamma^2_{11\sigma_1} + \mu_1^2 \left(1 + \mu_1^{\Delta_1}\right) B^{\sigma_1}_{\sigma_1} W^2_{1\sigma_1}}{1 - \mu_1 \left(1 + \mu_1^{\Delta_1}\right) \Gamma^{2\sigma_1}_{12\sigma_1}}, \\
V_{1\sigma_1} &= \frac{\mu_1 \left(1 + \mu_1^{\Delta_1}\right) \left(\mu_1 A_{\sigma_1} \Gamma^{1\sigma_1}_{12\sigma_1} + B^{\sigma_1}_{\sigma_1} \right)}{1 - \mu_1 \left(1 + \mu_1^{\Delta_1}\right) \Gamma^{2\sigma_1}_{12\sigma_1}}, \\
V^1_{2\sigma_1} &= \mu_2 \Gamma^{1\sigma_1}_{22\sigma_1} \left(1 + \mu_1 \Gamma^1_{11\sigma_1}\right) + \mu_1 \mu_2 C^{\sigma_1}_{\sigma_1} W^1_{1\sigma_1} \\
&\quad + \mu_2 \Gamma^{2\sigma_1}_{22\sigma_1} \frac{\mu_1 \left(1 + \mu_1^{\Delta_1}\right) \left(\left(1 + \mu_1 \Gamma^1_{11\sigma_1}\right) + \mu_1 B^{\sigma_1}_{\sigma_1} W^1_{1\sigma_1} \right)}{1 - \mu_1 \left(1 + \mu_1^{\Delta_1}\right) \Gamma^{2\sigma_1}_{12\sigma_1}}, \\
V^2_{2\sigma_1} &= 1 + \mu_1 \mu_2 \Gamma^{1\sigma_1}_{22\sigma_1} \Gamma^2_{11\sigma_1} + \mu_1 \mu_2 C^{\sigma_1}_{\sigma_1} W^2_{1\sigma_1} \\
&\quad + \mu_2 \Gamma^{2\sigma_1}_{22\sigma_1} \frac{1 + \mu_1^2 \left(1 + \mu_1^{\Delta_1}\right) \Gamma^{1\sigma_1}_{12\sigma_1} \Gamma^2_{11\sigma_1} + \mu_1^2 \left(1 + \mu_1^{\Delta_1}\right) B^{\sigma_1}_{\sigma_1} W^2_{1\sigma_1}}{1 - \mu_1 \left(1 + \mu_1^{\Delta_1}\right) \Gamma^{2\sigma_1}_{12\sigma_1}}, \\
V_{2\sigma_1} &= \mu_1 \mu_2 \Gamma^{1\sigma_1}_{22\sigma_1} A_{\sigma_1} + \mu_2 C^{\sigma_1}_{\sigma_1} \\
&\quad + \mu_2 \Gamma^{2\sigma_1}_{22\sigma_1} \frac{\mu_1 \left(1 + \mu_1^{\Delta_1}\right) \left(\mu_1 A_{\sigma_1} \Gamma^{1\sigma_1}_{12\sigma_1} + B^{\sigma_1}_{\sigma_1} \right)}{1 - \mu_1 \left(1 + \mu_1^{\Delta_1}\right) \Gamma^{2\sigma_1}_{12\sigma_1}}.
\end{aligned}
$$

Consider $f^{\Delta_2 \sigma_1^2}_{t_2}$ and $f^{\Delta_2 \sigma_1 \sigma_2}_{t_2}$. For them we obtain the following expressions.

$$
\begin{aligned}
f^{\Delta_2 \sigma_1^2}_{t_2} &= f^{\Delta_2 \sigma_1}_{t_2} + \mu_1 f^{\Delta_2 \sigma_1 \Delta_1}_{t_2} \\
&= f^{\Delta_2 \sigma_1}_{t_2} + \mu_1 \left(1 + \mu_1^{\Delta_1}\right) f^{\Delta_1 \Delta_2 \sigma_1}_{t_1 t_2} \\
&= f^{\Delta_2 \sigma_1}_{t_2} + \mu_1 \left(1 + \mu_1^{\Delta_1}\right) \left(\Gamma^1_{12\sigma_1} f^{\Delta_1}_{t_1} + \Gamma^2_{12\sigma_1} f^{\Delta_2 \sigma_1}_{t_2} + B_{\sigma_1} N_{\sigma_1} \right)^{\sigma_1} \\
&= f^{\Delta_2 \sigma_1}_{t_2} + \mu_1 \left(1 + \mu_1^{\Delta_1}\right) \left(\Gamma^{1\sigma_1}_{12\sigma_1} f^{\Delta_1 \sigma_1}_{t_1} + \Gamma^{2\sigma_1}_{12\sigma_1} f^{\Delta_2 \sigma_1^2}_{t_2} + B^{\sigma_1}_{\sigma_1} N^{\sigma_1}_{\sigma_1} \right)
\end{aligned}
$$

$$
\begin{aligned}
&= f_{t_2}^{\Delta_2\sigma_1} + \mu_1\left(1+\mu_1^{\Delta_1}\right)\Gamma_{12\sigma_1}^{1\sigma_1} f_{t_1}^{\Delta_1\sigma_1} + \mu_1\left(1+\mu_1^{\Delta_1}\right)\Gamma_{12\sigma_1}^{2\sigma_1} f_{t_2}^{\Delta_2\sigma_1^2} \\
&\quad + \mu_1\left(1+\mu_1^{\Delta_1}\right) B_{\sigma_1}^{\sigma_1} N_{\sigma_1}^{\sigma_1} \\
&= f_{t_2}^{\Delta_2\sigma_1} + \mu_1\left(1+\mu_1^{\Delta_1}\right)\Gamma_{12\sigma_1}^{1\sigma_1}\left(\left(1+\mu_1\Gamma_{11\sigma_1}^{1}\right) f_{t_1}^{\Delta_1} + \mu_1\Gamma_{11\sigma_1}^{2} f_{t_2}^{\Delta_2\sigma_1} + \mu_1 A_{\sigma_1} N_{\sigma_1}\right) \\
&\quad + \mu_1\left(1+\mu_1^{\Delta_1}\right)\Gamma_{12\sigma_1}^{2\sigma_1} f_{t_2}^{\Delta_2\sigma_1^2} \\
&\quad + \mu_1\left(1+\mu_1^{\Delta_1}\right) B_{\sigma_1}^{\sigma_1}\left(\mu_1 W_{1\sigma_1}^{1} f_{t_1}^{\Delta_1} + \mu_1 W_{1\sigma_1}^{2} f_{t_2}^{\Delta_2\sigma_1} + N_{\sigma_1}\right) \\
&= \mu_1\left(1+\mu_1^{\Delta_1}\right)\left(\Gamma_{12\sigma_1}^{1\sigma_1}\left(1+\mu_1\Gamma_{11\sigma_1}^{1}\right) + \mu_1 B_{\sigma_1}^{\sigma_1} W_{1\sigma_1}^{1}\right) f_{t_1}^{\Delta_1} \\
&\quad + \left(1+\mu_1^2\left(1+\mu_1^{\Delta_1}\right)\Gamma_{12\sigma_1}^{1\sigma_1}\Gamma_{11\sigma_1}^{2} + \mu_1^2\left(1+\mu_1^{\Delta_1}\right) B_{\sigma_1}^{\sigma_1} W_{1\sigma_1}^{2}\right) f_{t_2}^{\Delta_2\sigma_1} \\
&\quad + \mu_1\left(1+\mu_1^{\Delta_1}\right)\left(\mu_1 A_{\sigma_1}\Gamma_{12\sigma_1}^{1\sigma_1} + B_{\sigma_1}^{\sigma_1}\right) N_{\sigma_1} \\
&\quad + \mu_1\left(1+\mu_1^{\Delta_1}\right)\Gamma_{12\sigma_1}^{2\sigma_1} f_{t_2}^{\Delta_2\sigma_1^2},
\end{aligned}
$$

whereupon

$$
\begin{aligned}
&\left(1-\mu_1\left(1+\mu_1^{\Delta_1}\right)\Gamma_{12\sigma_1}^{2\sigma_1}\right) f_{t_2}^{\Delta_2\sigma_1^2} \\
&= \mu_1\left(1+\mu_1^{\Delta_1}\right)\left(\Gamma_{12\sigma_1}^{1\sigma_1}\left(1+\mu_1\Gamma_{11\sigma_1}^{1}\right) + \mu_1 B_{\sigma_1}^{\sigma_1} W_{1\sigma_1}^{1}\right) f_{t_1}^{\Delta_1} \\
&\quad + \left(1+\mu_1^2\left(1+\mu_1^{\Delta_1}\right)\Gamma_{12\sigma_1}^{1\sigma_1}\Gamma_{11\sigma_1}^{2} + \mu_1^2\left(1+\mu_1^{\Delta_1}\right) B_{\sigma_1}^{\sigma_1} W_{1\sigma_1}^{2}\right) f_{t_2}^{\Delta_2\sigma_1} \\
&\quad + \mu_1\left(1+\mu_1^{\Delta_1}\right)\left(\mu_1 A_{\sigma_1}\Gamma_{12\sigma_1}^{1\sigma_1} + B_{\sigma_1}^{\sigma_1}\right) N_{\sigma_1}
\end{aligned}
$$

and

$$
\begin{aligned}
f_{t_2}^{\Delta_2\sigma_1^2} &= \frac{\mu_1\left(1+\mu_1^{\Delta_1}\right)\left(\Gamma_{12\sigma_1}^{1\sigma_1}\left(1+\mu_1\Gamma_{11\sigma_1}^{1}\right) + \mu_1 B_{\sigma_1}^{\sigma_1} W_{1\sigma_1}^{1}\right)}{1-\mu_1\left(1+\mu_1^{\Delta_1}\right)\Gamma_{12\sigma_1}^{2\sigma_1}} f_{t_1}^{\Delta_1} \\
&\quad + \frac{1+\mu_1^2\left(1+\mu_1^{\Delta_1}\right)\Gamma_{12\sigma_1}^{1\sigma_1}\Gamma_{11\sigma_1}^{2} + \mu_1^2\left(1+\mu_1^{\Delta_1}\right) B_{\sigma_1}^{\sigma_1} W_{1\sigma_1}^{2}}{1-\mu_1\left(1+\mu_1^{\Delta_1}\right)\Gamma_{12\sigma_1}^{2\sigma_1}} f_{t_2}^{\Delta_2\sigma_1} \\
&\quad + \frac{\mu_1\left(1+\mu_1^{\Delta_1}\right)\left(\mu_1 A_{\sigma_1}\Gamma_{12\sigma_1}^{1\sigma_1} + B_{\sigma_1}^{\sigma_1}\right)}{1-\mu_1\left(1+\mu_1^{\Delta_1}\right)\Gamma_{12\sigma_1}^{2\sigma_1}} N_{\sigma_1} \\
&= V_{1\sigma_1}^{1} f_{t_1}^{\Delta_1} + V_{1\sigma_1}^{2} f_{t_2}^{\Delta_2\sigma_1} + V_{1\sigma_1} N_{\sigma_1}.
\end{aligned}
$$

Next,

$$
\begin{aligned}
f_{t_2}^{\Delta_2\sigma_1\sigma_2} &= f_{t_2}^{\Delta_2\sigma_1} + \mu_2 f_{t_2}^{\Delta_2\sigma_1\Delta_2} \\
&= f_{t_2}^{\Delta_2\sigma_1} + \mu_2 f_{t_2}^{\Delta_2^2\sigma_1} \\
&= f_{t_2}^{\Delta_2\sigma_1} + \mu_2\left(\Gamma_{22\sigma_1}^{1} f_{t_1}^{\Delta_1} + \Gamma_{22\sigma_1}^{2} f_{t_2}^{\Delta_2\sigma_1} + C_{\sigma_1} N_{\sigma_1}\right)^{\sigma_1}
\end{aligned}
$$

$$
\begin{aligned}
&= f_{t_2}^{\Delta_2\sigma_1} + \mu_2 \Gamma_{22\sigma_1}^{1\sigma_1} f_{t_1}^{\Delta_1\sigma_1} + \mu_2 \Gamma_{22\sigma_1}^{2\sigma_1} f_{t_2}^{\Delta_2\sigma_1^2} + \mu_2 C_{\sigma_1}^{\sigma_1} N_{\sigma_1}^{\sigma_1} \\
&= f_{t_2}^{\Delta_2\sigma_1} + \mu_2 \Gamma_{22\sigma_1}^{1\sigma_1} \left(\left(1 + \mu_1 \Gamma_{11\sigma_1}^{1}\right) f_{t_1}^{\Delta_1} + \mu_1 \Gamma_{11\sigma_1}^{2} f_{t_2}^{\Delta_2\sigma_1} + \mu_1 A_{\sigma_1} N_{\sigma_1} \right) \\
&\quad + \mu_2 \Gamma_{22\sigma_1}^{2\sigma_1} \frac{\mu_1 \left(1 + \mu_1^{\Delta_1}\right) \left(\Gamma_{12\sigma_1}^{1\sigma_1} \left(1 + \mu_1 \Gamma_{11\sigma_1}^{1}\right) + \mu_1 B_{\sigma_1}^{\sigma_1} W_{1\sigma_1}^{1} \right)}{1 - \mu_1 \left(1 + \mu_1^{\Delta_1}\right) \Gamma_{12\sigma_1}^{2\sigma_1}} f_{t_1}^{\Delta_1} \\
&\quad + \mu_2 \Gamma_{22\sigma_1}^{2\sigma_1} \frac{1 + \mu_1^2 \left(1 + \mu_1^{\Delta_1}\right) \Gamma_{12\sigma_1}^{1\sigma_1} \Gamma_{11\sigma_1}^{2} + \mu_1^2 \left(1 + \mu_1^{\Delta_1}\right) B_{\sigma_1}^{\sigma_1} W_{1\sigma_1}^{2}}{1 - \mu_1 \left(1 + \mu_1^{\Delta_1}\right) \Gamma_{12\sigma_1}^{2\sigma_1}} f_{t_2}^{\Delta_2\sigma_1} \\
&\quad + \mu_2 \Gamma_{22\sigma_1}^{2\sigma_1} \frac{\mu_1 \left(1 + \mu_1^{\Delta_1}\right) \left(\mu_1 A_{\sigma_1} \Gamma_{12\sigma_1}^{1\sigma_1} + B_{\sigma_1}^{\sigma_1} \right)}{1 - \mu_1 \left(1 + \mu_1^{\Delta_1}\right) \Gamma_{12\sigma_1}^{2\sigma_1}} N_{\sigma_1} \\
&\quad + C_{\sigma_1}^{\sigma_1} \mu_1 \mu_2 W_{1\sigma_1}^{1} f_{t_1}^{\Delta_1} + C_{\sigma_1}^{\sigma_1} \mu_1 \mu_2 W_{1\sigma_1}^{2} f_{t_2}^{\Delta_2\sigma_1} + \mu_2 C_{\sigma_1}^{\sigma_1} N_{\sigma_1}^{\sigma_1} \\
&= \Bigg(\mu_2 \Gamma_{22\sigma_1}^{1\sigma_1} \left(1 + \mu_1 \Gamma_{11\sigma_1}^{1}\right) + \mu_1 \mu_2 C_{\sigma_1}^{\sigma_1} W_{1\sigma_1}^{1} \\
&\quad + \mu_2 \Gamma_{22\sigma_1}^{2\sigma_1} \frac{\mu_1 \left(1 + \mu_1^{\Delta_1}\right) \left(\Gamma_{12\sigma_1}^{1\sigma_1} \left(1 + \mu_1 \Gamma_{11\sigma_1}^{1}\right) + \mu_1 B_{\sigma_1}^{\sigma_1} W_{1\sigma_1}^{1} \right)}{1 - \mu_1 \left(1 + \mu_1^{\Delta_1}\right) \Gamma_{12\sigma_1}^{2\sigma_1}} \Bigg) f_{t_1}^{\Delta_1} \\
&\quad + \Bigg(1 + \mu_1 \mu_2 \Gamma_{22\sigma_1}^{1\sigma_1} \Gamma_{11\sigma_1}^{2} + \mu_1 \mu_2 C_{\sigma_1}^{\sigma_1} W_{1\sigma_1}^{2} \\
&\quad + \mu_2 \Gamma_{22\sigma_1}^{2\sigma_1} \frac{1 + \mu_1^2 \left(1 + \mu_1^{\Delta_1}\right) \Gamma_{12\sigma_1}^{1\sigma_1} \Gamma_{11\sigma_1}^{2} + \mu_1^2 \left(1 + \mu_1^{\Delta_1}\right) B_{\sigma_1}^{\sigma_1} W_{1\sigma_1}^{2}}{1 - \mu_1 \left(1 + \mu_1^{\Delta_1}\right) \Gamma_{12\sigma_1}^{2\sigma_1}} \Bigg) f_{t_2}^{\Delta_2\sigma_1} \\
&\quad + \Bigg(\mu_1 \mu_2 \Gamma_{22\sigma_1}^{1\sigma_1} A_{\sigma_1} + \mu_2 C_{\sigma_1}^{\sigma_1} \\
&\quad + \mu_2 \Gamma_{22\sigma_1}^{2\sigma_1} \frac{\mu_1 \left(1 + \mu_1^{\Delta_1}\right) \left(\mu_1 A_{\sigma_1} \Gamma_{12\sigma_1}^{1\sigma_1} + B_{\sigma_1}^{\sigma_1} \right)}{1 - \mu_1 \left(1 + \mu_1^{\Delta_1}\right) \Gamma_{12\sigma_1}^{2\sigma_1}} \Bigg) N_{\sigma_1} \\
&= V_{2\sigma_1}^{1} f_{t_1}^{\Delta_1} + V_{2\sigma_1}^{2} f_{t_2}^{\Delta_2\sigma_1} + V_{2\sigma_1} N_{\sigma_1}.
\end{aligned}
$$

Denote

$$
\begin{aligned}
U_{1\sigma_1}^{1} &= \left(1 + \mu_1 \Gamma_{11\sigma_1}^{1}\right) \Gamma_{11\sigma_1}^{1} + \Gamma_{11\sigma_1}^{2\sigma_1} V_{1\sigma_1}^{1} + \mu_1 A_{\sigma_1}^{\sigma_1} W_{1\sigma_1}^{1}, \\
U_{1\sigma_1}^{2} &= \mu_1 \Gamma_{11\sigma_1}^{2} \Gamma_{11\sigma_1}^{1\sigma_1} + \Gamma_{11\sigma_1}^{2\sigma_1} V_{1\sigma_1}^{2} + \mu_1 A_{\sigma_1}^{\sigma_1} W_{1\sigma_1}^{2}, \\
U_{1\sigma_1}^{3} &= \mu_1 A_{\sigma_1} \Gamma_{11\sigma_1}^{1\sigma_1} + V_{1\sigma_1} \Gamma_{11\sigma_1}^{2\sigma_1} + A_{\sigma_1}^{\sigma_1}, \\
U_{2\sigma_1}^{1} &= \left(1 + \mu_2 \Gamma_{12\sigma_1}^{1}\right) \Gamma_{11\sigma_1}^{1\sigma_2} + \Gamma_{11\sigma_1}^{2\sigma_2} V_{2\sigma_1}^{1} + \mu_2 A_{\sigma_1}^{\sigma_2} W_{2\sigma_1}^{1}, \\
U_{2\sigma_1}^{2} &= \mu_2 \Gamma_{12\sigma_1}^{2} \Gamma_{11\sigma_1}^{2\sigma_2} + \Gamma_{11\sigma_1}^{1\sigma_2} V_{2\sigma_1}^{2} + \mu_2 A_{\sigma_1}^{\sigma_2} W_{2\sigma_1}^{2}, \\
U_{2\sigma_1}^{3} &= \mu_2 B_{\sigma_1} \Gamma_{11\sigma_1}^{1\sigma_2} + \Gamma_{11\sigma_1}^{2\sigma_2} V_{2\sigma_1} + A_{\sigma_1}^{\sigma_2},
\end{aligned}
$$

$$\begin{aligned}
U^1_{3\sigma_1} &= \left(1+\mu_1\Gamma^1_{11\sigma_1}\right)\Gamma^{1\sigma_1}_{12\sigma_1} + \Gamma^{2\sigma_1}_{12\sigma_1}V^1_{1\sigma_1} + \mu_1 B^{\sigma_1}_{\sigma_1}W^1_{1\sigma_1},\\
U^2_{3\sigma_1} &= \mu_1\Gamma^{1\sigma_1}_{12\sigma_1}\Gamma^2_{11\sigma_1} + \Gamma^{2\sigma_1}_{12\sigma_1}V^2_{1\sigma_1} + \mu_1 B^{\sigma_1}_{\sigma_1}W^2_{1\sigma_1},\\
U^3_{3\sigma_1} &= \mu_1 A_{\sigma_1}\Gamma^{1\sigma_1}_{12\sigma_1} + \Gamma^{2\sigma_1}_{12\sigma_1}V_{1\sigma_1} + B^{\sigma_1}_{\sigma_1},\\
U^1_{4\sigma_1} &= \left(1+\mu_2\Gamma^1_{12\sigma_1}\right)\Gamma^{1\sigma_2}_{12\sigma_1} + V^1_{2\sigma_1}\Gamma^{2\sigma_2}_{12\sigma_1} + \mu_2 B^{\sigma_2}_{\sigma_1}W^1_{2\sigma_1},\\
U^2_{4\sigma_1} &= \mu_2\Gamma^2_{12\sigma_1}\Gamma^{1\sigma_2}_{12\sigma_1} + \Gamma^{2\sigma_2}_{12\sigma_1}V^2_{2\sigma_1} + \mu_2 B^{\sigma_2}_{\sigma_1}W^2_{2\sigma_1},\\
U^3_{4\sigma_1} &= \mu_2 B_{\sigma_1}\Gamma^{1\sigma_2}_{12\sigma_1} + V_{2\sigma_1}\Gamma^{2\sigma_2}_{12\sigma_1} + B^{\sigma_2}_{\sigma_1},\\
U^1_{5\sigma_1} &= \left(1+\mu_1\Gamma^1_{11\sigma_1}\right)\Gamma^{1\sigma_1}_{22\sigma_1} + \Gamma^{2\sigma_1}_{22\sigma_1}V^1_{1\sigma_1} + \mu_1 C^{\sigma_1}_{\sigma_1}W^1_{1\sigma_1},\\
U^2_{5\sigma_1} &= \mu_1\Gamma^2_{11\sigma_1}\Gamma^{1\sigma_1}_{22\sigma_1} + \Gamma^{2\sigma_1}_{22\sigma_1}V^2_{1\sigma_1} + \mu_1 C^{\sigma_1}_{\sigma_1}W^2_{1\sigma_1},\\
U^3_{5\sigma_1} &= \mu_1 A_{\sigma_1}\Gamma^{1\sigma_1}_{22\sigma_1} + V_{1\sigma_1}\Gamma^{2\sigma_1}_{22\sigma_1} + C^{\sigma_1}_{\sigma_1},\\
U^1_{6\sigma_1} &= \left(1+\mu_2\Gamma^1_{12\sigma_1}\right)\Gamma^{1\sigma_2}_{22\sigma_1} + V^1_{2\sigma_1}\Gamma^{2\sigma_2}_{22\sigma_1} + \mu_2 C^{\sigma_2}_{\sigma_1}W^1_{2\sigma_1},\\
U^2_{6\sigma_1} &= \mu_2\Gamma^1_{12\sigma_1}\Gamma^{1\sigma_2}_{22\sigma_1} + \Gamma^{2\sigma_2}_{22\sigma_1}V^2_{2\sigma_1} + \mu_2 C^{\sigma_2}_{\sigma_1}W^2_{2\sigma_1},\\
U^3_{6\sigma_1} &= \mu_2 B_{\sigma_1}\Gamma^{1\sigma_2}_{22\sigma_1} + \Gamma^{2\sigma_2}_{22\sigma_1}V_{2\sigma_1} + C^{\sigma_2}_{\sigma_1}.
\end{aligned}$$

Now, we will express

$$f_{t_1}^{\Delta_1^2\sigma_1},\quad f_{t_1}^{\Delta_1^2\sigma_2},\quad f_{t_1t_2}^{\Delta_1\Delta_2\sigma_1},\quad f_{t_1t_2}^{\Delta_1\Delta_2\sigma_2},\quad f_{t_2}^{\Delta_2^2\sigma_1},\quad f_{t_2}^{\Delta_2^2\sigma_2}$$

in the terms of the basis $\left\{f_{t_1}^{\Delta_1}, f_{t_2}^{\Delta_2\sigma_1}, N_{\sigma_1}\right\}$.

We have

$$\begin{aligned}
f_{t_1}^{\Delta_1^2\sigma_1} &= \left(\Gamma^1_{11\sigma_1}f_{t_1}^{\Delta_1} + \Gamma^2_{11\sigma_1}f_{t_2}^{\Delta_2\sigma_1} + A_{\sigma_1}N_{\sigma_1}\right)^{\sigma_1}\\
&= \Gamma^{1\sigma_1}_{11\sigma_1}f_{t_1}^{\Delta_1\sigma_1} + \Gamma^{2\sigma_1}_{11\sigma_1}f_{t_2}^{\Delta_2\sigma_1^2} + A^{\sigma_1}_{\sigma_1}N^{\sigma_1}_{\sigma_1}\\
&= \Gamma^{1\sigma_1}_{11\sigma_1}\left(\left(1+\mu_1\Gamma^1_{11\sigma_1}\right)f_{t_1}^{\Delta_1} + \mu_1\Gamma^2_{11\sigma_1}f_{t_2}^{\Delta_2\sigma_1} + \mu_1 A_{\sigma_1}N_{\sigma_1}\right)\\
&\quad+\Gamma^{2\sigma_1}_{11\sigma_1}\left(V^1_{1\sigma_1}f_{t_1}^{\Delta_1} + V^2_{1\sigma_1}f_{t_2}^{\Delta_2\sigma_1} + V_{1\sigma_1}N_{\sigma_1}\right)\\
&\quad+A^{\sigma_1}_{\sigma_1}\left(\mu_1 W^1_{1\sigma_1}f_{t_1}^{\Delta_1} + \mu_1 W^2_{1\sigma_1}f_{t_2}^{\Delta_2\sigma_1} + N_{\sigma_1}\right)\\
&= \left(\left(1+\mu_1\Gamma^1_{11\sigma_1}\right)\Gamma^1_{11\sigma_1} + \Gamma^{2\sigma_1}_{11\sigma_1}V^1_{1\sigma_1} + \mu_1 A^{\sigma_1}_{\sigma_1}W^1_{1\sigma_1}\right)f_{t_1}^{\Delta_1}\\
&\quad+\left(\mu_1\Gamma^2_{11\sigma_1}\Gamma^{1\sigma_1}_{11\sigma_1} + \Gamma^{2\sigma_1}_{11\sigma_1}V^2_{1\sigma_1} + \mu_1 A^{\sigma_1}_{\sigma_1}W^2_{1\sigma_1}\right)f_{t_2}^{\Delta_2\sigma_1}\\
&\quad+\left(\mu_1 A_{\sigma_1}\Gamma^{1\sigma_1}_{11\sigma_1} + V_{1\sigma_1}\Gamma^{2\sigma_1}_{11\sigma_1} + A^{\sigma_1}_{\sigma_1}\right)N_{\sigma_1}\\
&= U^1_{1\sigma_1}f_{t_1}^{\Delta_1} + U^2_{1\sigma_1}f_{t_2}^{\Delta_2\sigma_1} + U^3_{1\sigma_1}N_{\sigma_1},\\
f_{t_1}^{\Delta_1^2\sigma_2} &= \left(\Gamma^1_{11\sigma_1}f_{t_1}^{\Delta_1} + \Gamma^2_{11\sigma_1}f_{t_2}^{\Delta_2\sigma_1} + A_{\sigma_1}N_{\sigma_1}\right)^{\sigma_2}\\
&= \Gamma^{1\sigma_2}_{11\sigma_1}f_{t_1}^{\Delta_1\sigma_2} + \Gamma^{2\sigma_2}_{11\sigma_1}f_{t_2}^{\Delta_2\sigma_1\sigma_2} + A^{\sigma_2}_{\sigma_1}N^{\sigma_2}_{\sigma_1}
\end{aligned}$$

$$
\begin{aligned}
&= \Gamma^{1\sigma_2}_{11\sigma_1}\left(\left(1+\mu_2\Gamma^1_{12\sigma_1}\right)f^{\Delta_1}_{t_1} + \mu_2\Gamma^2_{12\sigma_1}f^{\Delta_2\sigma_1}_{t_2} + \mu_2 B_{\sigma_1}N_{\sigma_1}\right)\\
&\quad + \Gamma^{2\sigma_2}_{11\sigma_1}\left(V^1_{2\sigma_1}f^{\Delta_1}_{t_1} + V^2_{2\sigma_1}f^{\Delta_2\sigma_1}_{t_2} + V_{2\sigma_1}N_{\sigma_1}\right)\\
&\quad + A^{\sigma_2}_{\sigma_1}\left(\mu_2 W^1_{2\sigma_1}f^{\Delta_1}_{t_1} + \mu_2 W^2_{2\sigma_1}f^{\Delta_2\sigma_1}_{t_2} + N_{\sigma_1}\right)\\
&= \left(\left(1+\mu_2\Gamma^1_{12\sigma_1}\right)\Gamma^{1\sigma_2}_{11\sigma_1} + \Gamma^{2\sigma_2}_{11\sigma_1}V^1_{2\sigma_1} + \mu_2 A^{\sigma_2}_{\sigma_1}W^1_{2\sigma_1}\right)f^{\Delta_1}_{t_1}\\
&\quad + \left(\mu_2\Gamma^2_{12\sigma_1}\Gamma^{1\sigma_2}_{11\sigma_1} + \Gamma^{2\sigma_2}_{11\sigma_1}V^2_{2\sigma_1} + \mu_2 A^{\sigma_2}_{\sigma_1}W^2_{2\sigma_1}\right)f^{\Delta_2\sigma_1}_{t_2}\\
&\quad + \left(\mu_2 B_{\sigma_1}\Gamma^{1\sigma_2}_{11\sigma_1} + \Gamma^{2\sigma_2}_{11\sigma_1}V_{2\sigma_1} + A^{\sigma_2}_{\sigma_1}\right)N_{\sigma_1}\\
&= U^1_{2\sigma_1}f^{\Delta_1}_{t_1} + U^2_{2\sigma_1}f^{\Delta_2\sigma_1}_{t_2} + U^3_{2\sigma_1}N_{\sigma_1},
\end{aligned}
$$

$$
\begin{aligned}
f^{\Delta_1\Delta_2\sigma_1}_{t_1t_2} &= \left(\Gamma^1_{12\sigma_1}f^{\Delta_1}_{t_1} + \Gamma^2_{12\sigma_1}f^{\Delta_2\sigma_1}_{t_2} + B_{\sigma_1}N_{\sigma_1}\right)^{\sigma_1}\\
&= \Gamma^{1\sigma_1}_{12\sigma_1}f^{\Delta_1\sigma_1}_{t_1} + \Gamma^{2\sigma_1}_{12\sigma_1}f^{\Delta_2\sigma_1^2}_{t_2} + B^{\sigma_1}_{\sigma_1}N^{\sigma_1}_{\sigma_1}\\
&= \Gamma^{1\sigma_1}_{12\sigma_1}\left(\left(1+\mu_1\Gamma^1_{11\sigma_1}\right)f^{\Delta_1}_{t_1} + \mu_1\Gamma^2_{11\sigma_1}f^{\Delta_2\sigma_1}_{t_2} + \mu_1 A_{\sigma_1}N_{\sigma_1}\right)\\
&\quad + \Gamma^{2\sigma_1}_{12\sigma_1}\left(V^1_{1\sigma_1}f^{\Delta_1}_{t_1} + V^2_{1\sigma_1}f^{\Delta_2\sigma_1}_{t_2} + V_{1\sigma_1}N_{\sigma_1}\right)\\
&\quad + B^{\sigma_1}_{\sigma_1}\left(\mu_1 W^1_{1\sigma_1}f^{\Delta_1}_{t_1} + \mu_1 W^2_{2\sigma_1}f^{\Delta_2\sigma_1}_{t_2} + N_{\sigma_1}\right)\\
&= \left(\left(1+\mu_1\Gamma^1_{11\sigma_1}\right)\Gamma^{1\sigma_1}_{12\sigma_1} + \Gamma^{2\sigma_1}_{12\sigma_1}V^1_{1\sigma_1} + \mu_1 B^{\sigma_1}_{\sigma_1}W^1_{1\sigma_1}\right)f^{\Delta_1}_{t_1}\\
&\quad + \left(\mu_1 V^1_{2\sigma_1}\Gamma^2_{11\sigma_1} + \Gamma^{2\sigma_1}_{12\sigma_1}V^2_{1\sigma_1} + \mu_1 B^{\sigma_1}_{\sigma_1}W^2_{1\sigma_1}\right)f^{\Delta_2\sigma_1}_{t_2}\\
&\quad + \left(\mu_1 A_{\sigma_1}\Gamma^{1\sigma_1}_{12\sigma_1} + \Gamma^{2\sigma_1}_{12\sigma_1}V_{1\sigma_1} + B^{\sigma_1}_{\sigma_1}\right)N_{\sigma_1}\\
&= U^1_{3\sigma_1}f^{\Delta_1}_{t_1} + U^2_{3\sigma_1}f^{\Delta_2\sigma_1}_{t_2} + U^3_{3\sigma_1}N_{\sigma_1},
\end{aligned}
$$

$$
\begin{aligned}
f^{\Delta_1\Delta_2\sigma_2}_{t_1t_2} &= \left(\Gamma^1_{12\sigma_1}f^{\Delta_1}_{t_1} + \Gamma^2_{12\sigma_1}f^{\Delta_2\sigma_1}_{t_2} + B_{\sigma_1}N_{\sigma_1}\right)^{\sigma_2}\\
&= \Gamma^{1\sigma_2}_{12\sigma_1}f^{\Delta_1\sigma_2}_{t_1} + \Gamma^{2\sigma_2}_{12\sigma_1}f^{\Delta_2\sigma_1\sigma_2}_{t_2} + B^{\sigma_2}_{\sigma_1}N^{\sigma_2}_{\sigma_1}\\
&= \Gamma^{1\sigma_2}_{12\sigma_1}\left(\left(1+\mu_2\Gamma^1_{12\sigma_1}\right)f^{\Delta_1}_{t_1} + \mu_2\Gamma^2_{12\sigma_1}f^{\Delta_2\sigma_1}_{t_2} + \mu_2 B_{\sigma_1}N_{\sigma_1}\right)\\
&\quad + \Gamma^{2\sigma_2}_{12\sigma_1}\left(V^1_{2\sigma_1}f^{\Delta_1}_{t_1} + V^2_{2\sigma_1}f^{\Delta_2\sigma_1}_{t_2} + V_{2\sigma_1}N_{\sigma_1}\right)\\
&\quad + B^{\sigma_2}_{\sigma_1}\left(\mu_2 W^1_{2\sigma_1}f^{\Delta_1}_{t_1} + \mu_2 W^2_{2\sigma_1}f^{\Delta_2\sigma_1}_{t_2} + N_{\sigma_1}\right)\\
&= \left(\left(1+\mu_2\Gamma^1_{12\sigma_1}\right)\Gamma^{1\sigma_2}_{12\sigma_1} + V^1_{2\sigma_1}\Gamma^{2\sigma_2}_{12\sigma_1} + \mu_2 B^{\sigma_2}_{\sigma_1}W^1_{2\sigma_1}\right)f^{\Delta_1}_{t_1}\\
&\quad + \left(\mu_2\Gamma^2_{12\sigma_1}\Gamma^{1\sigma_2}_{12\sigma_1} + \Gamma^{2\sigma_2}_{12\sigma_1}V^2_{2\sigma_1} + \mu_2 B^{\sigma_2}_{\sigma_1}W^2_{2\sigma_1}\right)f^{\Delta_2\sigma_1}_{t_2}\\
&\quad + \left(\mu_2 B_{\sigma_1}\Gamma^{1\sigma_2}_{12\sigma_1} + V_{2\sigma_1}\Gamma^{2\sigma_2}_{12\sigma_1} + B^{\sigma_2}_{\sigma_1}\right)N_{\sigma_1}\\
&= U^1_{4\sigma_1}f^{\Delta_1}_{t_1} + U^2_{4\sigma_1}f^{\Delta_2\sigma_1}_{t_2} + U^3_{4\sigma_1}N_{\sigma_1},
\end{aligned}
$$

$$
f^{\Delta_2^2\sigma_1}_{t_2} = \left(\Gamma^1_{22\sigma_1}f^{\Delta_1}_{t_1} + \Gamma^2_{22\sigma_1}f^{\Delta_2\sigma_1}_{t_2} + C_{\sigma_1}N_{\sigma_1}\right)^{\sigma_1}
$$

$$
\begin{aligned}
&= \Gamma^{1\sigma_1}_{22\sigma_1} f^{\Delta_1\sigma_1}_{t_1} + \Gamma^{2\sigma_1}_{22\sigma_1} f^{\Delta_2\sigma_1^2}_{t_2} + C^{\sigma_1}_{\sigma_1} N^{\sigma_1}_{\sigma_1} \\
&= \Gamma^{1\sigma_1}_{22\sigma_1} \left((1+\mu_1\Gamma^1_{11\sigma_1}) f^{\Delta_1}_{t_1} + \mu_1 \Gamma^2_{11\sigma_1} f^{\Delta_2\sigma_1}_{t_2} + \mu_1 A_{\sigma_1} N_{\sigma_1} \right) \\
&\quad + \Gamma^{2\sigma_1}_{22\sigma_1} \left(V^1_{1\sigma_1} f^{\Delta_1}_{t_1} + V^2_{1\sigma_1} f^{\Delta_2\sigma_1}_{t_2} + V_{1\sigma_1} N_{\sigma_1} \right) \\
&\quad + C^{\sigma_1}_{\sigma_1} \left(\mu_1 W^1_{1\sigma_1} f^{\Delta_1}_{t_1} + \mu_1 W^2_{1\sigma_1} f^{\Delta_2\sigma_1}_{t_2} + N_{\sigma_1} \right) \\
&= \left((1+\mu_1\Gamma^1_{11\sigma_1}) \Gamma^{1\sigma_1}_{22\sigma_1} + \Gamma^{2\sigma_1}_{22\sigma_1} V^1_{1\sigma_1} + \mu_1 C^{\sigma_1}_{\sigma_1} W^1_{1\sigma_1} \right) f^{\Delta_1}_{t_1} \\
&\quad + \left(\mu_1 \Gamma^2_{11\sigma_1} \Gamma^{1\sigma_1}_{22\sigma_1} + \Gamma^{2\sigma_1}_{22\sigma_1} V^1_{1\sigma_1} + \mu_1 C^{\sigma_1}_{\sigma_1} W^1_{1\sigma_1} \right) f^{\Delta_2\sigma_1}_{t_2} \\
&\quad + \left(\mu_1 A_{\sigma_1} \Gamma^{1\sigma_1}_{22\sigma_1} + V_{1\sigma_1} \Gamma^{2\sigma_1}_{22\sigma_1} + C^{\sigma_1}_{\sigma_1} \right) N_{\sigma_1} \\
&= U^1_{5\sigma_1} f^{\Delta_1}_{t_1} + U^2_{5\sigma_1} f^{\Delta_2\sigma_1}_{t_2} + U^3_{5\sigma_1} N_{\sigma_1},
\end{aligned}
$$

$$
\begin{aligned}
f^{\Delta_2^2\sigma_2}_{t_2} &= \left(\Gamma^1_{22\sigma_1} f^{\Delta_1}_{t_1} + \Gamma^2_{22\sigma_1} f^{\Delta_2\sigma_1}_{t_2} + C_{\sigma_1} N_{\sigma_1} \right)^{\sigma_2} \\
&= \Gamma^{1\sigma_2}_{22\sigma_1} f^{\Delta_1\sigma_2}_{t_1} + \Gamma^{2\sigma_2}_{22\sigma_1} f^{\Delta_2\sigma_1\sigma_2}_{t_2} + C^{\sigma_2}_{\sigma_1} N^{\sigma_2}_{\sigma_1} \\
&= \Gamma^{1\sigma_2}_{22\sigma_1} \left((1+\mu_2\Gamma^1_{12\sigma_1}) f^{\Delta_1}_{t_1} + \mu_2 \Gamma^2_{12\sigma_1} f^{\Delta_2\sigma_1}_{t_2} + \mu_2 B_{\sigma_1} N_{\sigma_1} \right) \\
&\quad + \Gamma^{2\sigma_2}_{22\sigma_1} \left(V^1_{2\sigma_1} f^{\Delta_1}_{t_1} + V^2_{2\sigma_1} f^{\Delta_2\sigma_1}_{t_2} + V_{2\sigma_1} N_{\sigma_1} \right) \\
&\quad + C^{\sigma_2}_{\sigma_1} \left(\mu_2 W^1_{2\sigma_1} f^{\Delta_1}_{t_1} + \mu_2 W^2_{2\sigma_1} f^{\Delta_2\sigma_1}_{t_2} + N_{\sigma_1} \right) \\
&= \left((1+\mu_2\Gamma^1_{12\sigma_1}) \Gamma^{1\sigma_2}_{22\sigma_1} + V^1_{2\sigma_1} \Gamma^{2\sigma_2}_{22\sigma_1} + \mu_2 C^{\sigma_2}_{\sigma_1} W^1_{2\sigma_1} \right) f^{\Delta_1}_{t_1} \\
&\quad + \left(\mu_2 \Gamma^1_{12\sigma_1} \Gamma^{1\sigma_2}_{22\sigma_1} + \Gamma^{2\sigma_2}_{22\sigma_1} V^2_{2\sigma_1} + \mu_2 C^{\sigma_2}_{\sigma_1} W^2_{2\sigma_1} \right) f^{\Delta_2\sigma_1}_{t_2} \\
&\quad + \left(\mu_2 B_{\sigma_1} \Gamma^{1\sigma_2}_{22\sigma_1} + \Gamma^{2\sigma_2}_{22\sigma_1} V_{2\sigma_1} + C^{\sigma_2}_{\sigma_1} \right) N_{\sigma_1} \\
&= U^1_{6\sigma_1} f^{\Delta_1}_{t_1} + U^2_{6\sigma_1} f^{\Delta_2\sigma_1}_{t_2} + U^3_{6\sigma_1} N_{\sigma_1}.
\end{aligned}
$$

2. Suppose S is a σ_2-regular surface and σ_2 is Δ_2-differentiable. We will search for a representation of $N^{\Delta_1}_{\sigma_2 t_1}$ and $N^{\Delta_2}_{\sigma_2 t_2}$ in the following form

$$
\begin{aligned}
N^{\Delta_1}_{\sigma_2 t_1} &= W^1_{1\sigma_2} f^{\Delta_1\sigma_2}_{t_1} + W^2_{1\sigma_2} f^{\Delta_2}_{t_2} \\
N^{\Delta_2}_{\sigma_2 t_2} &= W^1_{2\sigma_2} f^{\Delta_1\sigma_2}_{t_1} + W^2_{2\sigma_2} f^{\Delta_2}_{t_2}.
\end{aligned} \tag{5.4}
$$

Definition 5.6 The coefficients $W^k_{j\sigma_2}$, $j,k \in \{1,2\}$, will be called σ_2-Weingarten coefficients.

Denote

$$
\begin{aligned}
A_{\sigma_2\sigma_2} &= N_{\sigma_2} \cdot f^{\Delta_1^2\sigma_2}_{t_1}, \\
A_{\sigma_1\sigma_2\sigma_2} &= N^{\sigma_1}_{\sigma_2} \cdot f^{\Delta_1^2\sigma_2}_{t_1}, \\
B_{\sigma_2\sigma_2} &= N_{\sigma_2} \cdot f^{\Delta_1\Delta_2\sigma_2}_{t_1t_2}, \\
B_{\sigma_2\sigma_2\sigma_2} &= N^{\sigma_2}_{\sigma_2} \cdot f^{\Delta_1\Delta_2\sigma_2}_{t_1t_2},
\end{aligned}
$$

$$B_{\sigma_1\sigma_2} = N_{\sigma_2}^{\sigma_1} \cdot f_{t_1t_2}^{\Delta_1\Delta_2},$$

$$C_{\sigma_2\sigma_2} = N_{\sigma_2}^{\sigma_2} \cdot f_{t_2}^{\Delta_2^2}.$$

Since $N_{\sigma_2} \perp f_{t_1}^{\Delta_1\sigma_2}$, we have

$$N_{\sigma_2} \cdot f_{t_1}^{\Delta_1\sigma_2} = 0.$$

We differentiate the last equation with respect to t_1 and with respect to t_2 and we find

$$\begin{aligned} 0 &= N_{\sigma_2t_1}^{\Delta_1} \cdot f_{t_1}^{\Delta_1\sigma_1\sigma_2} + N_{\sigma_2} \cdot f_{t_1}^{\Delta_1^2\sigma_2} \\ &= N_{\sigma_2t_1}^{\Delta_1} \cdot f_{t_1}^{\Delta_1\sigma_2} + N_{\sigma_2}^{\sigma_1} \cdot f_{t_1}^{\Delta_1^2\sigma_2} \end{aligned}$$

and

$$\begin{aligned} 0 &= N_{\sigma_2t_2}^{\Delta_2} \cdot f_{t_1}^{\Delta_1\sigma_1\sigma_2} + N_{\sigma_2} \cdot f_{t_1t_2}^{\Delta_1\sigma_2\Delta_2} \\ &= N_{\sigma_2t_2}^{\Delta_2} \cdot f_{t_1}^{\Delta_1\sigma_1} + N_{\sigma_2}^{\sigma_2} \cdot f_{t_1t_2}^{\Delta_1\sigma_2\Delta_2}, \end{aligned}$$

from where

$$\begin{aligned} N_{\sigma_2t_1}^{\Delta_1} \cdot f_{t_1}^{\Delta_1\sigma_1\sigma_2} &= -N_{\sigma_2} \cdot f_{t_1}^{\Delta_1^2\sigma_2} \\ &= -A_{\sigma_2\sigma_2}, \\ N_{\sigma_2t_1}^{\Delta_1} \cdot f_{t_1}^{\Delta_1\sigma_2} &= -N_{\sigma_2}^{\sigma_1} \cdot f_{t_1}^{\Delta_1^2\sigma_2} \\ &= -A_{\sigma_1\sigma_2\sigma_2}, \\ N_{\sigma_2t_2}^{\Delta_2} \cdot f_{t_1}^{\Delta_1\sigma_1\sigma_2} &= -N_{\sigma_2} \cdot f_{t_1}^{\Delta_1\sigma_2\Delta_2} \\ &= -\left(1 + \mu_2^{\Delta_2}\right) N_{\sigma_2} \cdot f_{t_1t_2}^{\Delta_1\Delta_2\sigma_2} \\ &= -\left(1 + \mu_2^{\Delta_2}\right) B_{\sigma_2\sigma_2}, \\ N_{\sigma_2t_2}^{\Delta_2} \cdot f_{t_1}^{\Delta_1\sigma_2} &= -N_{\sigma_2}^{\sigma_2} \cdot f_{t_1t_2}^{\Delta_1\sigma_2\Delta_2} \\ &= -\left(1 + \mu_2^{\Delta_2}\right) N_{\sigma_2}^{\sigma_2} \cdot f_{t_1t_2}^{\Delta_1\Delta_2\sigma_2} \\ &= -\left(1 + \mu_2^{\Delta_2}\right) B_{\sigma_2\sigma_2\sigma_2}. \end{aligned}$$

Because $N_{\sigma_2} \perp f_{t_2}^{\Delta_2}$, we have

$$N_{\sigma_2} \cdot f_{t_2}^{\Delta_2} = 0,$$

which we differentiate with respect to t_1 and with respect to t_2 and we find

$$\begin{aligned} 0 &= N_{\sigma_2t_1}^{\Delta_1} \cdot f_{t_2}^{\Delta_2\sigma_1} + N_{\sigma_2} \cdot f_{t_1t_2}^{\Delta_1\Delta_2} \\ &= N_{\sigma_2t_1}^{\Delta_1} \cdot f_{t_2}^{\Delta_2} + N_{\sigma_2}^{\sigma_1} \cdot f_{t_1t_2}^{\Delta_1\Delta_2} \end{aligned}$$

and

$$\begin{aligned} 0 &= N^{\Delta_2}_{\sigma_2 t_2} \cdot f^{\Delta_2\sigma_2}_{t_2} + N_{\sigma_2} \cdot f^{\Delta_2^2}_{t_2} \\ &= N^{\Delta_2}_{\sigma_2 t_2} \cdot f^{\Delta_2}_{t_2} + N^{\sigma_2}_{\sigma_2} \cdot f^{\Delta_2^2}_{t_2}. \end{aligned}$$

Therefore

$$\begin{aligned} N^{\Delta_1}_{\sigma_2 t_1} \cdot f^{\Delta_2\sigma_1}_{t_2} &= -B_{\sigma_2} \\ N^{\Delta_1}_{\sigma_2 t_1} \cdot f^{\Delta_2}_{t_2} &= -B_{\sigma_1\sigma_2} \\ N^{\Delta_2}_{\sigma_2 t_2} \cdot f^{\Delta_2\sigma_2}_{t_2} &= -C_{\sigma_2} \\ N^{\Delta_2}_{\sigma_2 t_2} \cdot f^{\Delta_2}_{t_2} &= -C_{\sigma_2\sigma_2}. \end{aligned}$$

a. Consider the first equation of the system (5.4). We multiply it by $f^{\Delta_1\sigma_1\sigma_2}_{t_1}$ and $f^{\Delta_1\sigma_2}_{t_1}$ and we get

$$\begin{aligned} N^{\Delta_1}_{\sigma_2 t_1} \cdot f^{\Delta_1\sigma_1\sigma_2}_{t_1} &= W^1_{1\sigma_2}\left(f^{\Delta_1\sigma_2}_{t_1} \cdot f^{\Delta_1\sigma_1\sigma_2}_{t_1}\right) + W^2_{1\sigma_2}\left(f^{\Delta_2}_{t_2} \cdot f^{\Delta_1\sigma_1\sigma_2}_{t_1}\right) \\ N^{\Delta_1}_{\sigma_2 t_1} \cdot f^{\Delta_1\sigma_2}_{t_1} &= W^1_{1\sigma_2}\left(f^{\Delta_1\sigma_2}_{t_1} \cdot f^{\Delta_1\sigma_2}_{t_1}\right) + W^2_{1\sigma_2}\left(f^{\Delta_2}_{t_2} \cdot f^{\Delta_1\sigma_2}_{t_1}\right) \end{aligned}$$

or

$$\begin{aligned} -A_{\sigma_2\sigma_2} &= W^1_{1\sigma_2}\left(f^{\Delta_1\sigma_2}_{t_1} \cdot f^{\Delta_1\sigma_1\sigma_2}_{t_1}\right) + W^2_{1\sigma_2}\left(f^{\Delta_2}_{t_2} \cdot f^{\Delta_1\sigma_1\sigma_2}_{t_1}\right) \\ -A_{\sigma_1\sigma_2\sigma_2} &= W^1_{1\sigma_2}\left(f^{\Delta_1\sigma_2}_{t_1} \cdot f^{\Delta_1\sigma_2}_{t_1}\right) + W^2_{1\sigma_2}\left(f^{\Delta_2}_{t_2} \cdot f^{\Delta_1\sigma_2}_{t_1}\right). \end{aligned}$$

By the last system, we obtain

$$\begin{aligned} W^1_{1\sigma_2} &= -\frac{A_{\sigma_2\sigma_2}\left(f^{\Delta_2}_{t_2} \cdot f^{\Delta_1\sigma_2}_{t_1}\right) - A_{\sigma_1\sigma_2\sigma_2}\left(f^{\Delta_2}_{t_2} \cdot f^{\Delta_1\sigma_1\sigma_2}_{t_1}\right)}{\left(f^{\Delta_1\sigma_2}_{t_1} \cdot f^{\Delta_1\sigma_1\sigma_2}_{t_1}\right)\left(f^{\Delta_2}_{t_2} \cdot f^{\Delta_1\sigma_2}_{t_1}\right) - \left(f^{\Delta_1\sigma_2}_{t_1}\right)^2\left(f^{\Delta_2}_{t_2} \cdot f^{\Delta_1\sigma_1\sigma_2}_{t_1}\right)} \\ W^2_{1\sigma_2} &= -\frac{A_{\sigma_1\sigma_2\sigma_2}\left(f^{\Delta_1\sigma_2}_{t_1} \cdot f^{\Delta_1\sigma_1\sigma_2}_{t_1}\right) - A_{\sigma_2\sigma_2}\left(f^{\Delta_1\sigma_2}_{t_1}\right)^2}{\left(f^{\Delta_1\sigma_2}_{t_1} \cdot f^{\Delta_1\sigma_1\sigma_2}_{t_1}\right)\left(f^{\Delta_2}_{t_2} \cdot f^{\Delta_1\sigma_2}_{t_1}\right) - \left(f^{\Delta_1\sigma_2}_{t_1}\right)^2\left(f^{\Delta_2}_{t_2} \cdot f^{\Delta_1\sigma_1\sigma_2}_{t_1}\right)}. \end{aligned}$$

Now, we multiply the first equation of the system (5.4) by $f^{\Delta_2\sigma_1}_{t_2}$ and we find

$$\begin{aligned} -A_{\sigma_2\sigma_2} &= W^1_{1\sigma_2}\left(f^{\Delta_1\sigma_2}_{t_1} \cdot f^{\Delta_1\sigma_1\sigma_2}_{t_1}\right) + W^2_{1\sigma_2}\left(f^{\Delta_2}_{t_2} \cdot f^{\Delta_1\sigma_1\sigma_2}_{t_1}\right) \\ -B_{\sigma_2} &= W^1_{1\sigma_2}\left(f^{\Delta_1\sigma_2}_{t_1} \cdot f^{\Delta_2\sigma_1}_{t_2}\right) + W^2_{1\sigma_2}\left(f^{\Delta_2}_{t_2} \cdot f^{\Delta_2\sigma_1}_{t_2}\right). \end{aligned}$$

By the last system, we get

$$\begin{aligned} W^1_{1\sigma_2} &= -\frac{A_{\sigma_2\sigma_2}\left(f^{\Delta_2}_{t_2} \cdot f^{\Delta_2\sigma_1}_{t_2}\right) - B_{\sigma_2}\left(f^{\Delta_2}_{t_2} \cdot f^{\Delta_1\sigma_1\sigma_2}_{t_1}\right)}{\left(f^{\Delta_1\sigma_2}_{t_1} \cdot f^{\Delta_1\sigma_1\sigma_2}_{t_1}\right)\left(f^{\Delta_2}_{t_2} \cdot f^{\Delta_2\sigma_1}_{t_2}\right) - \left(f^{\Delta_2}_{t_2} \cdot f^{\Delta_1\sigma_1\sigma_2}_{t_1}\right)\left(f^{\Delta_1\sigma_2}_{t_1} \cdot f^{\Delta_2\sigma_1}_{t_2}\right)} \\ W^2_{1\sigma_2} &= -\frac{B_{\sigma_2}\left(f^{\Delta_1\sigma_2}_{t_1} \cdot f^{\Delta_1\sigma_1\sigma_2}_{t_1}\right) - A_{\sigma_2\sigma_2}\left(f^{\Delta_1\sigma_2}_{t_1} \cdot f^{\Delta_2\sigma_1}_{t_2}\right)}{\left(f^{\Delta_1\sigma_2}_{t_1} \cdot f^{\Delta_1\sigma_1\sigma_2}_{t_1}\right)\left(f^{\Delta_2}_{t_2} \cdot f^{\Delta_2\sigma_1}_{t_2}\right) - \left(f^{\Delta_2}_{t_2} \cdot f^{\Delta_1\sigma_1\sigma_2}_{t_1}\right)\left(f^{\Delta_1\sigma_2}_{t_1} \cdot f^{\Delta_2\sigma_1}_{t_2}\right)}. \end{aligned}$$

Now, we multiply the first equation of the system (5.4) with $f_{t_2}^{\Delta_2}$ and we find

$$-A_{\sigma_2\sigma_2} = W_{1\sigma_2}^1 \left(f_{t_1}^{\Delta_1\sigma_2} \cdot f_{t_1}^{\Delta_1\sigma_1\sigma_2}\right) + W_{1\sigma_2}^2 \left(f_{t_2}^{\Delta_2} \cdot f_{t_1}^{\Delta_1\sigma_1\sigma_2}\right)$$
$$-B_{\sigma_1\sigma_2} = W_{1\sigma_2}^1 \left(f_{t_1}^{\Delta_1\sigma_2} \cdot f_{t_2}^{\Delta_2}\right) + W_{1\sigma_2}^2 \left(f_{t_2}^{\Delta_2}\right)^2.$$

By the last system, we get

$$W_{1\sigma_2}^1 = -\frac{A_{\sigma_2\sigma_2}\left(f_{t_2}^{\Delta_2}\right)^2 - B_{\sigma_1\sigma_2}\left(f_{t_2}^{\Delta_2} \cdot f_{t_1}^{\Delta_1\sigma_1\sigma_2}\right)}{\left(f_{t_2}^{\Delta_2}\right)^2\left(f_{t_1}^{\Delta_1\sigma_2} \cdot f_{t_1}^{\Delta_1\sigma_1\sigma_2}\right) - \left(f_{t_2}^{\Delta_2} \cdot f_{t_1}^{\Delta_1\sigma_1\sigma_2}\right)\left(f_{t_1}^{\Delta_1\sigma_2} \cdot f_{t_1}^{\Delta_1}\right)}$$
$$W_{1\sigma_2}^2 = -\frac{B_{\sigma_1\sigma_2}\left(f_{t_1}^{\Delta_1\sigma_2} \cdot f_{t_1}^{\Delta_1\sigma_1\sigma_2}\right) - A_{\sigma_2\sigma_2}\left(f_{t_1}^{\Delta_1\sigma_2} \cdot f_{t_2}^{\Delta_2}\right)}{\left(f_{t_2}^{\Delta_2}\right)^2\left(f_{t_1}^{\Delta_1\sigma_2} \cdot f_{t_1}^{\Delta_1\sigma_1\sigma_2}\right) - \left(f_{t_2}^{\Delta_2} \cdot f_{t_1}^{\Delta_1\sigma_1\sigma_2}\right)\left(f_{t_1}^{\Delta_1\sigma_2} \cdot f_{t_1}^{\Delta_1}\right)}.$$

Next, we get the system

$$-A_{\sigma_1\sigma_2\sigma_2} = W_{1\sigma_2}^1 \left(f_{t_1}^{\Delta_1\sigma_2} \cdot f_{t_1}^{\Delta_1\sigma_2}\right) + W_{1\sigma_2}^2 \left(f_{t_2}^{\Delta_2} \cdot f_{t_1}^{\Delta_1\sigma_2}\right)$$
$$-B_{\sigma_2} = W_{1\sigma_2}^1 \left(f_{t_1}^{\Delta_1\sigma_2} \cdot f_{t_2}^{\Delta_2\sigma_1}\right) + W_{1\sigma_2}^2 \left(f_{t_2}^{\Delta_2} \cdot f_{t_2}^{\Delta_2\sigma_1}\right).$$

Then

$$W_{1\sigma_2}^1 = -\frac{A_{\sigma_1\sigma_2\sigma_2}\left(f_{t_2}^{\Delta_2} \cdot f_{t_2}^{\Delta_2\sigma_1}\right) - B_{\sigma_2}\left(f_{t_2}^{\Delta_2} \cdot f_{t_1}^{\Delta_1\sigma_2}\right)}{\left(f_{t_1}^{\Delta_1\sigma_2}\right)^2\left(f_{t_2}^{\Delta_2} \cdot f_{t_2}^{\Delta_2\sigma_1}\right) - \left(f_{t_1}^{\Delta_1\sigma_2} \cdot f_{t_2}^{\Delta_2\sigma_1}\right)\left(f_{t_2}^{\Delta_2} \cdot f_{t_1}^{\Delta_1\sigma_2}\right)}$$
$$W_{1\sigma_2}^2 = -\frac{B_{\sigma_2}\left(f_{t_1}^{\Delta_1\sigma_2}\right)^2 - A_{\sigma_1\sigma_2\sigma_2}\left(f_{t_1}^{\Delta_1\sigma_2} \cdot f_{t_2}^{\Delta_2\sigma_1}\right)}{\left(f_{t_1}^{\Delta_1\sigma_2}\right)^2\left(f_{t_2}^{\Delta_2} \cdot f_{t_2}^{\Delta_2\sigma_1}\right) - \left(f_{t_1}^{\Delta_1\sigma_2} \cdot f_{t_2}^{\Delta_2\sigma_1}\right)\left(f_{t_2}^{\Delta_2} \cdot f_{t_1}^{\Delta_1\sigma_2}\right)}.$$

Next, we have the system

$$-A_{\sigma_1\sigma_2\sigma_2} = W_{1\sigma_2}^1 \left(f_{t_1}^{\Delta_1\sigma_2} \cdot f_{t_1}^{\Delta_1\sigma_2}\right) + W_{1\sigma_2}^2 \left(f_{t_2}^{\Delta_2} \cdot f_{t_1}^{\Delta_1\sigma_2}\right)$$
$$-B_{\sigma_1\sigma_2} = W_{1\sigma_2}^1 \left(f_{t_1}^{\Delta_1\sigma_2} \cdot f_{t_2}^{\Delta_2}\right) + W_{1\sigma_2}^2 \left(f_{t_2}^{\Delta_2}\right)^2,$$

from where

$$W_{1\sigma_2}^1 = -\frac{A_{\sigma_1\sigma_2\sigma_2}\left(f_{t_2}^{\Delta_2}\right)^2 - B_{\sigma_1\sigma_2}\left(f_{t_2}^{\Delta_2} \cdot f_{t_1}^{\Delta_1\sigma_2}\right)}{\left(f_{t_1}^{\Delta_1\sigma_2}\right)^2\left(f_{t_2}^{\Delta_2}\right)^2 - \left(f_{t_1}^{\Delta_1\sigma_2} \cdot f_{t_2}^{\Delta_2}\right)^2}$$
$$W_{1\sigma_2}^2 = -\frac{B_{\sigma_1\sigma_2}\left(f_{t_1}^{\Delta_1\sigma_2}\right)^2 - A_{\sigma_1\sigma_2\sigma_2}\left(f_{t_1}^{\Delta_1\sigma_2} \cdot f_{t_2}^{\Delta_2}\right)}{\left(f_{t_1}^{\Delta_1\sigma_2}\right)^2\left(f_{t_2}^{\Delta_2}\right)^2 - \left(f_{t_1}^{\Delta_1\sigma_2} \cdot f_{t_2}^{\Delta_2}\right)^2}.$$

Next, we have

$$-B_{\sigma_2} = W_{1\sigma_2}^1 \left(f_{t_1}^{\Delta_1\sigma_2} \cdot f_{t_2}^{\Delta_2\sigma_1}\right) + W_{1\sigma_2}^2 \left(f_{t_2}^{\Delta_2} \cdot f_{t_2}^{\Delta_2\sigma_1}\right)$$
$$-B_{\sigma_1\sigma_2} = W_{1\sigma_2}^1 \left(f_{t_1}^{\Delta_1\sigma_2} \cdot f_{t_2}^{\Delta_2}\right) + W_{1\sigma_2}^2 \left(f_{t_2}^{\Delta_2}\right)^2$$

and

$$W^1_{1\sigma_2} = -\frac{B_{\sigma_2}\left(f^{\Delta_2}_{t_2}\right)^2 - B_{\sigma_1\sigma_2}\left(f^{\Delta_2}_{t_2}\cdot f^{\Delta_2\sigma_1}_{t_2}\right)}{\left(f^{\Delta_1\sigma_2}_{t_1}\cdot f^{\Delta_2\sigma_1}_{t_2}\right)\left(f^{\Delta_2}_{t_2}\right)^2 - \left(f^{\Delta_1\sigma_2}_{t_1}\cdot f^{\Delta_2}_{t_2}\right)\left(f^{\Delta_2}_{t_2}\cdot f^{\Delta_2\sigma_1}_{t_2}\right)}$$

$$W^2_{1\sigma_2} = -\frac{B_{\sigma_1\sigma_2}\left(f^{\Delta_1\sigma_2}_{t_1}\cdot f^{\Delta_2\sigma_1}_{t_2}\right) - B_{\sigma_2}\left(f^{\Delta_1\sigma_2}_{t_1}\cdot f^{\Delta_2}_{t_2}\right)}{\left(f^{\Delta_1\sigma_2}_{t_1}\cdot f^{\Delta_2\sigma_1}_{t_2}\right)\left(f^{\Delta_2}_{t_2}\right)^2 - \left(f^{\Delta_1\sigma_2}_{t_1}\cdot f^{\Delta_2}_{t_2}\right)\left(f^{\Delta_2}_{t_2}\cdot f^{\Delta_2\sigma_1}_{t_2}\right)}.$$

b. Consider the second equation of the system (5.4). We multiply it by $f^{\Delta_1\sigma_1\sigma_2}_{t_1}$ and $f^{\Delta_1\sigma_1}_{t_1}$ and we get

$$N^{\Delta_2}_{\sigma_2 t_2}\cdot f^{\Delta_1\sigma_1\sigma_2}_{t_1} = W^1_{2\sigma_2}\left(f^{\Delta_1\sigma_2}_{t_1}\cdot f^{\Delta_1\sigma_1\sigma_2}_{t_1}\right) + W^2_{2\sigma_2}\left(f^{\Delta_2}_{t_2}\cdot f^{\Delta_1\sigma_1\sigma_2}_{t_1}\right)$$

$$N^{\Delta_2}_{\sigma_2 t_2}\cdot f^{\Delta_1\sigma_1}_{t_1} = W^1_{2\sigma_2}\left(f^{\Delta_1\sigma_2}_{t_1}\cdot f^{\Delta_1\sigma_1}_{t_1}\right) + W^2_{2\sigma_2}\left(f^{\Delta_2}_{t_2}\cdot f^{\Delta_1\sigma_1}_{t_1}\right)$$

or

$$-\left(1+\mu^{\Delta_2}_2\right)B_{\sigma_2\sigma_2} = W^1_{2\sigma_2}\left(f^{\Delta_1\sigma_2}_{t_1}\cdot f^{\Delta_1\sigma_1\sigma_2}_{t_1}\right) + W^2_{2\sigma_2}\left(f^{\Delta_2}_{t_2}\cdot f^{\Delta_1\sigma_1\sigma_2}_{t_1}\right)$$

$$-\left(1+\mu^{\Delta_2}_2\right)B_{\sigma_2\sigma_2\sigma_2} = W^1_{2\sigma_2}\left(f^{\Delta_1\sigma_2}_{t_1}\cdot f^{\Delta_1\sigma_1}_{t_1}\right) + W^2_{2\sigma_2}\left(f^{\Delta_2}_{t_2}\cdot f^{\Delta_1\sigma_1}_{t_1}\right).$$

By the last system, we find

$$W^1_{2\sigma_2} = -\left(1+\mu^{\Delta_2}_2\right)\frac{B_{\sigma_2\sigma_2}\left(f^{\Delta_2}_{t_2}\cdot f^{\Delta_1\sigma_1}_{t_1}\right) - B_{\sigma_2\sigma_2\sigma_2}\left(f^{\Delta_2}_{t_2}\cdot f^{\Delta_1\sigma_1\sigma_2}_{t_1}\right)}{\left(f^{\Delta_1\sigma_2}_{t_1}\cdot f^{\Delta_1\sigma_1\sigma_2}_{t_1}\right)\left(f^{\Delta_2}_{t_2}\cdot f^{\Delta_1\sigma_1}_{t_1}\right) - \left(f^{\Delta_1\sigma_2}_{t_1}\cdot f^{\Delta_1\sigma_1}_{t_1}\right)\left(f^{\Delta_2}_{t_2}\cdot f^{\Delta_1\sigma_1\sigma_2}_{t_1}\right)}$$

$$W^2_{2\sigma_2} = -\left(1+\mu^{\Delta_2}_2\right)\frac{B_{\sigma_2\sigma_2\sigma_2}\left(f^{\Delta_1\sigma_2}_{t_1}\cdot f^{\Delta_1\sigma_1\sigma_2}_{t_1}\right) - B_{\sigma_2\sigma_2}\left(f^{\Delta_1\sigma_2}_{t_1}\cdot f^{\Delta_1\sigma_1}_{t_1}\right)}{\left(f^{\Delta_1\sigma_2}_{t_1}\cdot f^{\Delta_1\sigma_1\sigma_2}_{t_1}\right)\left(f^{\Delta_2}_{t-2}\cdot f^{\Delta_1\sigma_1}_{t_1}\right) - \left(f^{\Delta_1\sigma_2}_{t_1}\cdot f^{\Delta_1\sigma_1}_{t_1}\right)\left(f^{\Delta_2}_{t_2}\cdot f^{\Delta_1\sigma_1\sigma_2}_{t_1}\right)}.$$

Now, we multiply the second equation of the system (5.4) by $f^{\Delta_2\sigma_2}_{t_2}$ and we obtain

$$-\left(1+\mu^{\Delta_2}_2\right)B_{\sigma_2\sigma_2} = W^1_{2\sigma_2}\left(f^{\Delta_1\sigma_2}_{t_1}\cdot f^{\Delta_1\sigma_1\sigma_2}_{t_1}\right) + W^2_{2\sigma_2}\left(f^{\Delta_2}_{t_2}\cdot f^{\Delta_1\sigma_1\sigma_2}_{t_1}\right)$$

$$-C_{\sigma_2} = W^1_{2\sigma_2}\left(f^{\Delta_1\sigma_2}_{t_1}\cdot f^{\Delta_2\sigma_2}_{t_2}\right) + W^2_{2\sigma_2}\left(f^{\Delta_2}_{t_2}\cdot f^{\Delta_2\sigma_2}_{t_2}\right),$$

from where

$$W^1_{2\sigma_2} = -\frac{\left(1+\mu^{\Delta_2}_2\right)B_{\sigma_2\sigma_2}\left(f^{\Delta_2}_{t_2}\cdot f^{\Delta_2\sigma_2}_{t_2}\right) - C_{\sigma_2}\left(f^{\Delta_2}_{t_2}\cdot f^{\Delta_1\sigma_1\sigma_2}_{t_1}\right)}{\left(f^{\Delta_1\sigma_2}_{t_1}\cdot f^{\Delta_1\sigma_1\sigma_2}_{t_1}\right)\left(f^{\Delta_2}_{t_2}\cdot f^{\Delta_2\sigma_2}_{t_2}\right) - \left(f^{\Delta_1\sigma_2}_{t_1}\cdot f^{\Delta_2\sigma_2}_{t_2}\right)\left(f^{\Delta_2}_{t_2}\cdot f^{\Delta_1\sigma_1\sigma_2}_{t_1}\right)}$$

$$W^2_{2\sigma_2} = -\frac{C_{\sigma_2}\left(f^{\Delta_1\sigma_2}_{t_1}\cdot f^{\Delta_1\sigma_1\sigma_2}_{t_1}\right) - \left(1+\mu^{\Delta_2}_2\right)B_{\sigma_2\sigma_2}\left(f^{\Delta_1\sigma_2}_{t_1}\cdot f^{\Delta_2\sigma_2}_{t_2}\right)}{\left(f^{\Delta_1\sigma_2}_{t_1}\cdot f^{\Delta_1\sigma_1\sigma_2}_{t_1}\right)\left(f^{\Delta_2}_{t_2}\cdot f^{\Delta_2\sigma_2}_{t_2}\right) - \left(f^{\Delta_1\sigma_2}_{t_1}\cdot f^{\Delta_2\sigma_2}_{t_2}\right)\left(f^{\Delta_2}_{t_2}\cdot f^{\Delta_1\sigma_1\sigma_2}_{t_1}\right)}.$$

We multiply the second equation of the system (5.4) by $f^{\Delta_2}_{t_2}$ and we get

$$-\left(1+\mu^{\Delta_2}_2\right)B_{\sigma_2\sigma_2} = W^1_{2\sigma_2}\left(f^{\Delta_1\sigma_2}_{t_1}\cdot f^{\Delta_1\sigma_1\sigma_2}_{t_1}\right) + W^2_{2\sigma_2}\left(f^{\Delta_2\sigma_1}_{t_2}\cdot f^{\Delta_1\sigma_1\sigma_2}_{t_1}\right)$$

$$-C_{\sigma_2\sigma_2} = W^1_{2\sigma_2}\left(f^{\Delta_1\sigma_2}_{t_1}\cdot f^{\Delta_2}_{t_2}\right) + W^2_{2\sigma_2}\left(f^{\Delta_2}_{t_2}\right)^2.$$

By the last system, we find

$$W^1_{2\sigma_2} = -\frac{\left(1+\mu_2^{\Delta_2}\right) B_{\sigma_2\sigma_2}\left(f_{t_2}^{\Delta_2}\right)^2 - C_{\sigma_2\sigma_2}\left(f_{t_2}^{\Delta_2\sigma_1}\cdot f_{t_1}^{\Delta_1\sigma_1\sigma_2}\right)}{\left(f_{t_1}^{\Delta_1\sigma_2}\cdot f_{t_1}^{\Delta_1\sigma_1\sigma_2}\right)\left(f_{t_2}^{\Delta_2}\right)^2 - \left(f_{t_1}^{\Delta_1\sigma_2}\cdot f_{t_2}^{\Delta_2}\right)\left(f_{t_2}^{\Delta_2\sigma_1}\cdot f_{t_1}^{\Delta_1\sigma_1\sigma_2}\right)}$$

$$W^2_{2\sigma_2} = -\frac{C_{\sigma_2\sigma_2}\left(f_{t_1}^{\Delta_1\sigma_2}\cdot f_{t_1}^{\Delta_1\sigma_1\sigma_2}\right) - \left(1+\mu_2^{\Delta_2}\right) B_{\sigma_2\sigma_2}\left(f_{t_1}^{\Delta_1\sigma_2}\cdot f_{t_2}^{\Delta_2}\right)}{\left(f_{t_1}^{\Delta_1\sigma_2}\cdot f_{t_1}^{\Delta_1\sigma_1\sigma_2}\right)\left(f_{t_2}^{\Delta_2}\right)^2 - \left(f_{t_1}^{\Delta_1\sigma_2}\cdot f_{t_2}^{\Delta_2}\right)\left(f_{t_2}^{\Delta_2}\cdot f_{t_1}^{\Delta_1\sigma_1\sigma_2}\right)}.$$

Next, we have the system

$$-\left(1+\mu_2^{\Delta_2}\right) B_{\sigma_2\sigma_2\sigma_2} = W^1_{2\sigma_2}\left(f_{t_1}^{\Delta_1\sigma_2}\cdot f_{t_1}^{\Delta_1\sigma_1}\right) + W^2_{2\sigma_2}\left(f_{t_2}^{\Delta_2}\cdot f_{t_1}^{\Delta_1\sigma_1}\right)$$

$$-C_{\sigma_2} = W^1_{2\sigma_2}\left(f_{t_1}^{\Delta_1\sigma_2}\cdot f_{t_2}^{\Delta_2\sigma_2}\right) + W^2_{2\sigma_2}\left(f_{t_2}^{\Delta_2}\cdot f_{t_2}^{\Delta_2\sigma_2}\right).$$

Hence, we find

$$W^1_{2\sigma_2} = -\frac{\left(1+\mu_2^{\Delta_2}\right) B_{\sigma_2\sigma_2\sigma_2}\left(f_{t_2}^{\Delta_2}\cdot f_{t_2}^{\Delta_2\sigma_2}\right) - C_{\sigma_2}\left(f_{t_2}^{\Delta_2}\cdot f_{t_1}^{\Delta_1\sigma_1}\right)}{\left(f_{t_1}^{\Delta_1\sigma_2}\cdot f_{t_1}^{\Delta_1\sigma_1}\right)\left(f_{t_2}^{\Delta_2}\cdot f_{t_2}^{\Delta_2\sigma_1}\right) - \left(f_{t_1}^{\Delta_1\sigma_2}\cdot f_{t_2}^{\Delta_2\sigma_2}\right)\left(f_{t_2}^{\Delta_2}\cdot f_{t_1}^{\Delta_1\sigma_1}\right)}$$

$$W^2_{2\sigma_2} = -\frac{C_{\sigma_2}\left(f_{t_1}^{\Delta_1\sigma_2}\cdot f_{t_1}^{\Delta_1\sigma_1}\right) - \left(1+\mu_2^{\Delta_2}\right) B_{\sigma_2\sigma_2\sigma_2}\left(f_{t_1}^{\Delta_1\sigma_2}\cdot f_{t_2}^{\Delta_2\sigma_2}\right)}{\left(f_{t_1}^{\Delta_1\sigma_2}\cdot f_{t_1}^{\Delta_1\sigma_1}\right)\left(f_{t_2}^{\Delta_2}\cdot f_{t_2}^{\Delta_2\sigma_1}\right) - \left(f_{t_1}^{\Delta_1\sigma_2}\cdot f_{t_2}^{\Delta_2\sigma_2}\right)\left(f_{t_2}^{\Delta_2}\cdot f_{t_1}^{\Delta_1\sigma_1}\right)}.$$

Next, we get the system

$$-\left(1+\mu_2^{\Delta_2}\right) B_{\sigma_2\sigma_2\sigma_2} = W^1_{2\sigma_2}\left(f_{t_1}^{\Delta_1\sigma_2}\cdot f_{t_1}^{\Delta_1\sigma_1}\right) + W^2_{2\sigma_2}\left(f_{t_2}^{\Delta_2}\cdot f_{t_1}^{\Delta_1\sigma_1}\right)$$

$$-C_{\sigma_2\sigma_2} = W^1_{2\sigma_2}\left(f_{t_1}^{\Delta_1\sigma_2}\cdot f_{t_2}^{\Delta_2}\right) + W^2_{2\sigma_2}\left(f_{t_2}^{\Delta_2}\right)^2,$$

whereupon

$$W^1_{2\sigma_2} = -\frac{\left(1+\mu_2^{\Delta_2}\right) B_{\sigma_2\sigma_2\sigma_2}\left(f_{t_2}^{\Delta_2}\right)^2 - C_{\sigma_2\sigma_2}\left(f_{t_2}^{\Delta_2}\cdot f_{t_1}^{\Delta_1\sigma_1}\right)}{\left(f_{t_1}^{\Delta_1\sigma_2}\cdot f_{t_1}^{\Delta_1\sigma_1}\right)\left(f_{t_2}^{\Delta_2}\right)^2 - \left(f_{t_1}^{\Delta_1\sigma_2}\cdot f_{t_2}^{\Delta_2}\right)\left(f_{t_2}^{\Delta_2}\cdot f_{t_1}^{\Delta_1\sigma_1}\right)}$$

$$W^2_{2\sigma_2} = -\frac{C_{\sigma_2\sigma_2}\left(f_{t_1}^{\Delta_1\sigma_2}\cdot f_{t_1}^{\Delta_1\sigma_1}\right) - \left(1+\mu_2^{\Delta_2}\right) B_{\sigma_2\sigma_2\sigma_2}\left(f_{t_1}^{\Delta_1\sigma_2}\cdot f_{t_2}^{\Delta_2}\right)}{\left(f_{t_1}^{\Delta_1\sigma_2}\cdot f_{t_1}^{\Delta_1\sigma_1}\right)\left(f_{t_2}^{\Delta_2}\right)^2 - \left(f_{t_1}^{\Delta_1\sigma_2}\cdot f_{t_2}^{\Delta_2}\right)\left(f_{t_2}^{\Delta_2}\cdot f_{t_1}^{\Delta_1\sigma_1}\right)}.$$

Next, we have the system

$$-C_{\sigma_2} = W^1_{2\sigma_2}\left(f_{t_1}^{\Delta_1\sigma_2}\cdot f_{t_2}^{\Delta_2\sigma_2}\right) + W^2_{2\sigma_2}\left(f_{t_2}^{\Delta_2}\cdot f_{t_2}^{\Delta_2\sigma_2}\right)$$

$$-C_{\sigma_2\sigma_2} = W^1_{2\sigma_2}\left(f_{t_1}^{\Delta_1\sigma_2}\cdot f_{t_2}^{\Delta_2}\right) + W^2_{2\sigma_2}\left(f_{t_2}^{\Delta_2}\right)^2$$

and

$$W^1_{2\sigma_2} = -\frac{C_{\sigma_2}\left(f_{t_2}^{\Delta_2}\right)^2 - C_{\sigma_2\sigma_2}\left(f_{t_2}^{\Delta_2}\cdot f_{t_2}^{\Delta_2\sigma_2}\right)}{\left(f_{t_1}^{\Delta_1\sigma_2}\cdot f_{t_2}^{\Delta_2\sigma_2}\right)\left(f_{t_2}^{\Delta_2}\right)^2 - \left(f_{t_2}^{\Delta_2}\cdot f_{t_2}^{\Delta_2\sigma_2}\right)\left(f_{t_1}^{\Delta_1\sigma_2}\cdot f_{t_2}^{\Delta_2}\right)}$$

$$W^2_{2\sigma_2} = -\frac{C_{\sigma_2\sigma_2}\left(f_{t_1}^{\Delta_1\sigma_2}\cdot f_{t_2}^{\Delta_2\sigma_2}\right) - C_{\sigma_2}\left(f_{t_1}^{\Delta_1\sigma_2}\cdot f_{t_2}^{\Delta_2}\right)}{\left(f_{t_1}^{\Delta_1\sigma_2}\cdot f_{t_2}^{\Delta_2\sigma_2}\right)\left(f_{t_2}^{\Delta_2}\right)^2 - \left(f_{t_2}^{\Delta_2}\cdot f_{t_2}^{\Delta_2\sigma_2}\right)\left(f_{t_1}^{\Delta_1\sigma_2}\cdot f_{t_2}^{\Delta_2}\right)}.$$

Note that

$$\begin{aligned} N_{\sigma_2}^{\sigma_1} &= N_{\sigma_2} + \mu_1 N_{\sigma_2 t_1}^{\Delta_1} \\ &= N_{\sigma_2} + \mu_1 \left(W_{1\sigma_2}^1 f_{t_1}^{\Delta_1\sigma_2} + W_{1\sigma_2}^2 f_{t_2}^{\Delta_2} \right) \\ &= \mu_1 W_{1\sigma_2}^1 f_{t_1}^{\Delta_1\sigma_2} + \mu_1 W_{1\sigma_2}^2 f_{t_2}^{\Delta_2} + N_{\sigma_2} \end{aligned}$$

and

$$\begin{aligned} N_{\sigma_2}^{\sigma_2} &= N_{\sigma_2} + \mu_2 N_{\sigma_2 t_2}^{\Delta_2} \\ &= N_{\sigma_2} + \mu_2 \left(W_{2\sigma_2}^1 f_{t_1}^{\Delta_1\sigma_2} + W_{2\sigma_2}^2 f_{t_2}^{\Delta_2} \right) \\ &= \mu_2 W_{2\sigma_2}^1 f_{t_1}^{\Delta_1\sigma_2} + \mu_2 W_{2\sigma_2}^2 f_{t_2}^{\Delta_2} + N_{\sigma_2}. \end{aligned}$$

Suppose

$$1 - \mu_2 \Gamma_{12\sigma_2}^{1\sigma_1} \neq 0, \quad 1 - \left(1 + \mu_2^{Delta_2}\right) \Gamma_{12\sigma_2}^{\sigma_2} \neq 0. \tag{5.5}$$

Set

$$\begin{aligned} V_{1\sigma_2}^1 &= \frac{1 - \mu_2 \Gamma_{12\sigma_2}^1 + \mu_1 \Gamma_{11\sigma_2}^1 + \mu_1\mu_2 \Gamma_{12\sigma_2}^1 \Gamma_{12\sigma_2}^{2\sigma_1} + \mu_1\mu_2 W_{1\sigma_2}^1}{1 - \mu_2 \Gamma_{12\sigma_2}^{1\sigma_1}}, \\ V_{1\sigma_2}^2 &= \frac{\mu_1\mu_2 W_{1\sigma_2}^2 + \mu_1 \Gamma_{11\sigma_2}^2 - \mu_2 \Gamma_{12\sigma_2}^2 + \mu_2 \Gamma_{12\sigma_2}^{2\sigma_1} \left(1 + \mu_1 \Gamma_{12\sigma_2}^2\right)}{1 - \mu_2 \Gamma_{12\sigma_2}^{1\sigma_1}}, \\ V_{1\sigma_2} &= \frac{\mu_2 + \mu_1\mu_2 \Gamma_{12\sigma_2}^{2\sigma_1} B_{\sigma_2} + \mu_1 A_{\sigma_2} - \mu_2 B_{\sigma_2}}{1 - \mu_2 \Gamma_{12\sigma_2}^{1\sigma_1}}, \\ V_{2\sigma_2}^1 &= \frac{1 + \mu_2 \left(1 + \mu_2^{\Delta_2}\right) \Gamma_{12\sigma_2}^{2\sigma_2} \Gamma_{22\sigma_2}^1 + \mu_2 \left(1 + \mu_2^{\Delta_2}\right) B_{\sigma_2}^{\sigma_2} W_{2\sigma_2}^1}{1 - \left(1 + \mu_2^{\Delta_2}\right) \Gamma_{12\sigma_2}^{1\sigma_2}}, \\ V_{2\sigma_2}^2 &= \frac{\left(1 + \mu_2^{\Delta_2}\right) \Gamma_{12\sigma_2}^{2\sigma_2} \left(1 + \mu_2 \Gamma_{22\sigma_2}^2\right) + \mu_2 \left(1 + \mu_2^{\Delta_2}\right) B_{\sigma_2}^{\sigma_2} W_{2\sigma_2}^2}{1 - \left(1 + \mu_2^{\Delta_2}\right) \Gamma_{12\sigma_2}^{1\sigma_2}}, \\ V_{2\sigma_2} &= \frac{\left(1 + \mu_2^{\Delta_2}\right) \Gamma_{12\sigma_2}^{2\sigma_2} \mu_2 C_{\sigma_2} + \left(1 + \mu_2^{\Delta_2}\right) B_{\sigma_2}^{\sigma_2}}{1 - \left(1 + \mu_2^{\Delta_2}\right) \Gamma_{12\sigma_2}^{1\sigma_2}}. \end{aligned}$$

Now, we will express $f_{t_1}^{\Delta_1\sigma_1\sigma_2}$ and $f_{t_1}^{\Delta_1\sigma_2^2}$.

We have

$$\begin{aligned} f_{t_1}^{\Delta_1\sigma_1\sigma_2} &= f_{t_1}^{\Delta_1\sigma_1} + \mu_2 f_{t_1t_2}^{\Delta_1\sigma_1\Delta_2} \\ &= f_{t_1}^{\Delta_1\sigma_1} + \mu_2 f_{t_1t_2}^{\Delta_1\Delta_2\sigma_1} \\ &= \left(1 - \mu_2 \Gamma_{12\sigma_2}^1 + \mu_1 \Gamma_{11\sigma_2}^1\right) f_{t_1}^{\Delta_1\sigma_2} \\ &\quad + \left(\mu_1 \Gamma_{11\sigma_2}^2 - \mu_2 \Gamma_{12\sigma_2}^2\right) f_{t_2}^{\Delta_2} + \left(\mu_1 A_{\sigma_2} - \mu_2 B_{\sigma_2}\right) N_{\sigma_2} \end{aligned}$$

$$
\begin{aligned}
&+\mu_2\left(\Gamma^1_{12\sigma_2}f_{t_1}^{\Delta_1\sigma_2}+\Gamma^2_{12\sigma_2}f_{t_2}^{\Delta_2}+N_{\sigma_2}\right)^{\sigma_1}\\
&=\left(1-\mu_2\Gamma^1_{12\sigma_2}+\mu_1\Gamma^1_{11\sigma_2}\right)f_{t_1}^{\Delta_1\sigma_2}\\
&\quad+\left(\mu_1\Gamma^2_{11\sigma_2}-\mu_2\Gamma^2_{12\sigma_2}\right)f_{t_2}^{\Delta_2}+\left(\mu_1A_{\sigma_2}-\mu_2B_{\sigma_2}\right)N_{\sigma_2}\\
&\quad+\mu_2\left(\Gamma^{1\sigma_1}_{12\sigma_2}f_{t_1}^{\Delta_1\sigma_1\sigma_2}+\Gamma^{2\sigma_1}_{12\sigma_2}f_{t_2}^{\Delta_2\sigma_1}+N^{\sigma_1}_{\sigma_2}\right)\\
&=\left(1-\mu_2\Gamma^1_{12\sigma_2}+\mu_1\Gamma^1_{11\sigma_2}\right)f_{t_1}^{\Delta_1\sigma_2}+\left(\mu_1\Gamma^2_{11\sigma_2}-\mu_2\Gamma^2_{12\sigma_2}\right)f_{t_2}^{\Delta_2}\\
&\quad+\left(\mu_1A_{\sigma_2}-\mu_2B_{\sigma_2}\right)N_{\sigma_2}+\mu_2\Gamma^{1\sigma_1}_{12\sigma_2}f_{t_1}^{\Delta_1\sigma_1\sigma_2}\\
&\quad+\mu_2\Gamma^{2\sigma_1}_{12\sigma_2}\left(\mu_1\Gamma^1_{12\sigma_2}f_{t_1}^{\Delta_1\sigma_2}+\left(1+\mu_1\Gamma^2_{12\sigma_2}\right)f_{t_2}^{\Delta_2}+\mu_1B_{\sigma_2}N_{\sigma_2}\right)\\
&\quad+\mu_2\left(\mu_1W^1_{1\sigma_2}f_{t_1}^{\Delta_1\sigma_2}+\mu_1W^2_{1\sigma_2}f_{t_2}^{\Delta_2}+N_{\sigma_2}\right)\\
&=\left(1-\mu_2\Gamma^1_{12\sigma_2}+\mu_1\Gamma^1_{11\sigma_2}+\mu_1\mu_2\Gamma^1_{12\sigma_2}\Gamma^{2\sigma_1}_{12\sigma_2}+\mu_1\mu_2W^1_{1\sigma_2}\right)f_{t_1}^{\Delta_1\sigma_2}\\
&\quad+\left(\mu_1\mu_2W^2_{1\sigma_2}+\mu_1\Gamma^2_{11\sigma_2}-\mu_2\Gamma^2_{12\sigma_2}+\mu_2\Gamma^{2\sigma_1}_{12\sigma_2}\left(1+\mu_1\Gamma^2_{12\sigma_2}\right)\right)f_{t_2}^{\Delta_2}\\
&\quad+\left(\mu_2+\mu_1\mu_2\Gamma^{2\sigma_1}_{12\sigma_2}B_{\sigma_2}+\mu_1A_{\sigma_2}-\mu_2B_{\sigma_2}\right)N_{\sigma_2}+\mu_2\Gamma^{1\sigma_1}_{12\sigma_2}f_{t_1}^{\Delta_1\sigma_1\sigma_2},
\end{aligned}
$$

whereupon

$$
\begin{aligned}
f_{t_1}^{\Delta_1\sigma_1\sigma_2}&=\frac{1-\mu_2\Gamma^1_{12\sigma_2}+\mu_1\Gamma^1_{11\sigma_2}+\mu_1\mu_2\Gamma^1_{12\sigma_2}\Gamma^{2\sigma_1}_{12\sigma_2}+\mu_1\mu_2W^1_{1\sigma_2}}{1-\mu_2\Gamma^{1\sigma_1}_{12\sigma_2}}f_{t_1}^{\Delta_1\sigma_2}\\
&\quad+\frac{\mu_1\mu_2W^2_{1\sigma_2}+\mu_1\Gamma^2_{11\sigma_2}-\mu_2\Gamma^2_{12\sigma_2}+\mu_2\Gamma^{2\sigma_1}_{12\sigma_2}\left(1+\mu_1\Gamma^2_{12\sigma_2}\right)}{1-\mu_2\Gamma^{1\sigma_1}_{12\sigma_2}}f_{t_2}^{\Delta_2}\\
&\quad+\frac{\mu_2+\mu_1\mu_2\Gamma^{2\sigma_1}_{12\sigma_2}B_{\sigma_2}+\mu_1A_{\sigma_2}-\mu_2B_{\sigma_2}}{1-\mu_2\Gamma^{1\sigma_1}_{12\sigma_2}}N_{\sigma_2}\\
&=V^1_{1\sigma_2}f_{t_1}^{\Delta_1\sigma_2}+V^2_{1\sigma_2}f_{t_2}^{\Delta_2}+V_{1\sigma_2}N_{\sigma_2}.
\end{aligned}
$$

Next,

$$
\begin{aligned}
f_{t_1}^{\Delta_1\sigma_2^2}&=f_{t_1}^{\Delta_1\sigma_2}+f_{t_1t_2}^{\Delta_1\sigma_2\Delta_2}\\
&=f_{t_1}^{\Delta_1\sigma_2}+\left(1+\mu_2^{\Delta_2}\right)f_{t_1t_2}^{\Delta_1\Delta_2\sigma_2}\\
&=f_{t_1}^{\Delta_1\sigma_2}+\left(1+\mu_2^{\Delta_2}\right)\left(\Gamma^1_{12\sigma_2}f_{t_1}^{\Delta_1\sigma_2}+\Gamma^2_{12\sigma_2}f_{t_2}^{\Delta_2}+B_{\sigma_2}N_{\sigma_2}\right)^{\sigma_2}\\
&=f_{t_1}^{\Delta_1\sigma_2}+\left(1+\mu_2^{\Delta_2}\right)\Gamma^{1\sigma_2}_{12\sigma_2}f_{t_1}^{\Delta_1\sigma_2^2}+\left(1+\mu_2^{\Delta_2}\right)\Gamma^{2\sigma_2}_{12\sigma_2}f_{t_2}^{\Delta_2\sigma_2}+\left(1+\mu_2^{\Delta_2}\right)B^{\sigma_2}_{\sigma_2}N^{\sigma_2}_{\sigma_2}\\
&=f_{t_1}^{\Delta_1\sigma_2}+\left(1+\mu_2^{\Delta_2}\right)\Gamma^{2\sigma_2}_{12\sigma_2}\left(\mu_2\Gamma^1_{22\sigma_2}f_{t_1}^{\Delta_1\sigma_2}+\left(1+\mu_2\Gamma^2_{22\sigma_2}\right)f_{t_2}^{\Delta_2}+\mu_2C_{\sigma_2}N_{\sigma_2}\right)\\
&\quad+\left(1+\mu_2^{\Delta_2}\right)B^{\sigma_2}_{\sigma_2}\left(\mu_2W^1_{2\sigma_2}f_{t_1}^{\Delta_1\sigma_2}+\mu_2W^2_{2\sigma_2}f_{t_2}^{\Delta_2}+N_{\sigma_2}\right)+\left(1+\mu_2^{\Delta_2}\right)\Gamma^{1\sigma_2}_{12\sigma_2}f_{t_1}^{\Delta_1\sigma_2^2}\\
&=\left(1+\mu_2\left(1+\mu_2^{\Delta_2}\right)\Gamma^{2\sigma_2}_{12\sigma_2}\Gamma^1_{22\sigma_2}+\mu_2\left(1+\mu_2^{\Delta_2}\right)B^{\sigma_2}_{\sigma_2}W^1_{2\sigma_2}\right)f_{t_1}^{\Delta_1\sigma_2}
\end{aligned}
$$

$$+\left(\left(1+\mu_2^{\Delta_2}\right)\Gamma_{12\sigma_2}^{2\sigma_2}\left(1+\mu_2\Gamma_{22\sigma_2}^{2}\right)+\mu_2\left(1+\mu_2^{\Delta_2}\right)B_{\sigma_2}^{\sigma_2}W_{2\sigma_2}^{2}\right)f_{t_2}^{\Delta_2}$$
$$+\left(\left(1+\mu_2^{\Delta_2}\right)\Gamma_{12\sigma_2}^{2\sigma_2}\mu_2 C_{\sigma_2}+\left(1+\mu_2^{\Delta_2}\right)B_{\sigma_2}^{\sigma_2}\right)N_{\sigma_2}$$
$$+\left(1+\mu_2^{\Delta_2}\right)\Gamma_{12\sigma_2}^{1\sigma_2}f_{t_1}^{\Delta_1\sigma_2^2},$$

whereupon

$$\begin{aligned}
f_{t_1}^{\Delta_1\sigma_2^2} &= \frac{1+\mu_2\left(1+\mu_2^{\Delta_2}\right)\Gamma_{12\sigma_2}^{2\sigma_2}\Gamma_{22\sigma_2}^{1}+\mu_2\left(1+\mu_2^{\Delta_2}\right)B_{\sigma_2}^{\sigma_2}W_{2\sigma_2}^{1}}{1-\left(1+\mu_2^{\Delta_2}\right)\Gamma_{12\sigma_2}^{1\sigma_2}}f_{t_1}^{\Delta_1\sigma_2} \\
&\quad+\frac{\left(1+\mu_2^{\Delta_2}\right)\Gamma_{12\sigma_2}^{2\sigma_2}\left(1+\mu_2\Gamma_{22\sigma_2}^{2}\right)+\mu_2\left(1+\mu_2^{\Delta_2}\right)B_{\sigma_2}^{\sigma_2}W_{2\sigma_2}^{2}}{1-\left(1+\mu_2^{\Delta_2}\right)\Gamma_{12\sigma_2}^{1\sigma_2}}f_{t_2}^{\Delta_2} \\
&\quad+\frac{\left(1+\mu_2^{\Delta_2}\right)\Gamma_{12\sigma_2}^{2\sigma_2}\mu_2 C_{\sigma_2}+\left(1+\mu_2^{\Delta_2}\right)B_{\sigma_2}^{\sigma_2}}{1-\left(1+\mu_2^{\Delta_2}\right)\Gamma_{12\sigma_2}^{1\sigma_2}}N_{\sigma_2} \\
&= V_{2\sigma_2}^{1}f_{t_1}^{\Delta_1\sigma_2}+V_{2\sigma_2}^{2}f_{t_2}^{\Delta_2}+V_{2\sigma_2}N_{\sigma_2}.
\end{aligned}$$

Now, we set

$$\begin{aligned}
U_{1\sigma_2}^{1} &= \Gamma_{11\sigma_2}^{1\sigma_1}V_{1\sigma_2}^{1}+\mu_1\Gamma_{11\sigma_2}^{2\sigma_1}\Gamma_{12\sigma_2}^{1}+\mu_1 A_{\sigma_2}^{\sigma_1}W_{1\sigma_2}^{1}, \\
U_{1\sigma_2}^{2} &= \Gamma_{11\sigma_2}^{1\sigma_1}V_{1\sigma_2}^{2}+\Gamma_{11\sigma_2}^{2\sigma_1}\left(1+\mu_1\Gamma_{12\sigma_2}^{2}\right)+\mu_1 A_{\sigma_2}^{\sigma_1}W_{2\sigma_2}^{2}, \\
U_{1\sigma_2}^{3} &= \Gamma_{11\sigma_2}^{1\sigma_1}V_{1\sigma_2}+\mu_1\Gamma_{11\sigma_2}^{2\sigma_1}B_{\sigma_2}+A_{\sigma_2}^{\sigma_1}, \\
U_{2\sigma_2}^{1} &= \Gamma_{11\sigma_2}^{1\sigma_2}V_{2\sigma_2}^{1}+\mu_2\Gamma_{11\sigma_2}^{2\sigma_2}\Gamma_{22\sigma_2}^{1}+\mu_2 A_{\sigma_2}^{\sigma_2}W_{2\sigma_2}^{1}, \\
U_{2\sigma_2}^{2} &= \Gamma_{11\sigma_2}^{1\sigma_2}V_{2\sigma_2}^{2}+\left(1+\mu_2\Gamma_{22\sigma_2}^{2}\right)\Gamma_{11\sigma_2}^{2\sigma_2}+\mu_2 A_{\sigma_2}^{\sigma_2}W_{2\sigma_2}^{2}, \\
U_{2\sigma_2}^{3} &= \Gamma_{11\sigma_2}^{1\sigma_2}V_{2\sigma_2}+\mu_2 C_{\sigma_2}\Gamma_{11\sigma_2}^{2\sigma_2}+A_{\sigma_2}^{\sigma_2}, \\
U_{3\sigma_2}^{1} &= \Gamma_{12\sigma_2}^{1\sigma_1}V_{1\sigma_2}^{1}+\mu_1\Gamma_{12\sigma_2}^{2\sigma_1}\Gamma_{12\sigma_2}^{1}+\mu_1 B_{\sigma_2}^{\sigma_1}W_{1\sigma_2}^{1}, \\
U_{3\sigma_2}^{2} &= \Gamma_{12\sigma_2}^{1\sigma_1}V_{1\sigma_2}^{2}+\Gamma_{12\sigma_2}^{2\sigma_1}\left(1+\mu_1\Gamma_{12\sigma_2}^{2}\right)+\mu_2 B_{\sigma_2}^{\sigma_1}W_{1\sigma_2}^{2}, \\
U_{3\sigma_2}^{3} &= \Gamma_{12\sigma_2}^{1\sigma_1}V_{1\sigma_2}+\mu_1\Gamma_{12\sigma_2}^{2\sigma_1}B_{\sigma_2}+B_{\sigma_2}^{\sigma_1}, \\
U_{4\sigma_2}^{1} &= \Gamma_{12\sigma_2}^{1\sigma_2}V_{2\sigma_2}^{1}+\mu_2\Gamma_{12\sigma_2}^{2}\Gamma_{22\sigma_2}^{1}+\mu_2 B_{\sigma_2}^{\sigma_2}W_{2\sigma_2}^{1}, \\
U_{4\sigma_2}^{2} &= \Gamma_{12\sigma_2}^{1\sigma_2}V_{2\sigma_2}^{2}+\left(1+\mu_2\Gamma_{22\sigma_2}^{2}\right)\Gamma_{12\sigma_2}^{2\sigma_2}+\mu_2 B_{\sigma_2}^{\sigma_2}W_{2\sigma_2}^{2}, \\
U_{4\sigma_2}^{3} &= \Gamma_{12\sigma_2}^{1\sigma_2}V_{2\sigma_2}+\mu_2 C_{\sigma_2}\Gamma_{12\sigma_2}^{2\sigma_2}+B_{\sigma_2}^{\sigma_2}, \\
U_{5\sigma_2}^{1} &= \Gamma_{22\sigma_2}^{1\sigma_1}V_{1\sigma_2}^{1}+\mu_1\Gamma_{22\sigma_2}^{2\sigma_1}\Gamma_{12\sigma_2}^{1}+\mu_1 C_{\sigma_2}^{\sigma_1}W_{1\sigma_2}^{1}, \\
U_{5\sigma_2}^{2} &= \Gamma_{22\sigma_2}^{1\sigma_1}V_{1\sigma_2}^{2}+\Gamma_{22\sigma_2}^{2\sigma_1}\left(1+\mu_1\Gamma_{12\sigma_2}^{2}\right)+\mu_1 C_{\sigma_2}^{\sigma_1}W_{1\sigma_2}^{2}, \\
U_{5\sigma_2}^{3} &= \Gamma_{22\sigma_2}^{1\sigma_1}V_{1\sigma_2}+\mu_1\Gamma_{22\sigma_2}^{2\sigma_1}B_{\sigma_2}+C_{\sigma_2}^{\sigma_1}, \\
U_{6\sigma_2}^{1} &= \Gamma_{22\sigma_2}^{1\sigma_2}V_{2\sigma_2}^{1}+\mu_2\Gamma_{22\sigma_2}^{2\sigma_2}\Gamma_{22\sigma_2}^{1}+\mu_2 C_{\sigma_2}^{\sigma_2}W_{2\sigma_2}^{1},
\end{aligned}$$

$$U^2_{6\sigma_2} = \Gamma^{1\sigma_2}_{22\sigma_2} V^2_{2\sigma_2} + \left(1 + \mu_2 \Gamma^2_{22\sigma_2}\right) \Gamma^{2\sigma_2}_{22\sigma_2} + \mu_2 C^{\sigma_2}_{\sigma_2} W^2_{2\sigma_2},$$
$$U^3_{6\sigma_2} = \Gamma^{1\sigma_2}_{22\sigma_2} V_{2\sigma_2} + \mu_2 C_{\sigma_2} \Gamma^{2\sigma_2}_{22\sigma_2} + C^{\sigma_2}_{\sigma_2}.$$

Then

$$\begin{aligned}
f_{t_1}^{\Delta_1^2\sigma_1} &= \left(\Gamma^1_{11\sigma_2} f_{t_1}^{\Delta_1\sigma_2} + \Gamma^2_{11\sigma_2} f_{t_2}^{\Delta_2} + A_{\sigma_2} N_{\sigma_2}\right)^{\sigma_1} \\
&= \left(\Gamma^1_{11\sigma_2} f_{t_1}^{\Delta_1\sigma_2}\right)^{\sigma_1} + \left(\Gamma^2_{11\sigma_2} f_{t_2}^{\Delta_2}\right)^{\sigma_1} + \left(A_{\sigma_2} N_{\sigma_2}\right)^{\sigma_1} \\
&= \Gamma^{1\sigma_1}_{11\sigma_2} f_{t_1}^{\Delta_1\sigma_1\sigma_2} + \Gamma^{2\sigma_1}_{11\sigma_2} f_{t_2}^{\Delta_2\sigma_1} + A^{\sigma_1}_{\sigma_2} N^{\sigma_1}_{\sigma_2} \\
&= \Gamma^{1\sigma_1}_{11\sigma_2} \left(V^1_{1\sigma_2} f_{t_1}^{\Delta_1\sigma_2} + V^2_{1\sigma_2} f_{t_2}^{\Delta_2} + V_{1\sigma_2} N_{\sigma_2}\right) \\
&\quad + \Gamma^{2\sigma_1}_{11\sigma_2} \left(\mu_1 \Gamma^1_{12\sigma_2} f_{t_1}^{\Delta_1\sigma_2} + \left(1 + \mu_1 \Gamma^2_{12\sigma_2}\right) f_{t_2}^{\Delta_2} + \mu_1 B_{\sigma_2} N_{\sigma_2}\right) \\
&\quad + A^{\sigma_1}_{\sigma_2} \left(\mu_1 W^1_{1\sigma_2} f_{t_1}^{\Delta_1\sigma_2} + \mu_1 W^2_{2\sigma_2} f_{t_2}^{\Delta_2} + N_{\sigma_2}\right) \\
&= \left(\Gamma^{1\sigma_1}_{11\sigma_2} V^1_{1\sigma_2} + \mu_1 \Gamma^{2\sigma_1}_{11\sigma_2} \Gamma^1_{12\sigma_2} + \mu_1 A^{\sigma_1}_{\sigma_2} W^1_{1\sigma_2}\right) f_{t_1}^{\Delta_1\sigma_2} \\
&\quad + \left(\Gamma^{1\sigma_1}_{11\sigma_2} V^2_{1\sigma_2} + \Gamma^{2\sigma_1}_{11\sigma_2} \left(1 + \mu_1 \Gamma^2_{12\sigma_2}\right) + \mu_1 A^{\sigma_1}_{\sigma_2} W^2_{2\sigma_2}\right) f_{t_2}^{\Delta_2} \\
&\quad + \left(\Gamma^{1\sigma_1}_{11\sigma_2} V_{1\sigma_2} + \mu_1 \Gamma^{2\sigma_1}_{11\sigma_2} B_{\sigma_2} + A^{\sigma_1}_{\sigma_2}\right) N_{\sigma_2} \\
&= U^1_{1\sigma_2} f_{t_1}^{\Delta_1\sigma_2} + U^2_{1\sigma_2} f_{t_2}^{\Delta_2} + U^3_{1\sigma_2} N_{\sigma_2},
\end{aligned}$$

that is,

$$f_{t_1}^{\Delta_1^2\sigma_1} = U^1_{1\sigma_1} f_{t_1}^{\Delta_1\sigma_2} + U^2_{1\sigma_2} f_{t_2}^{\Delta_2} + U^3_{1\sigma_2} N_{\sigma_2},$$

and

$$\begin{aligned}
f_{t_1}^{\Delta_1^2\sigma_2} &= \left(\Gamma^1_{11\sigma_2} f_{t_1}^{\Delta_1\sigma_2} + \Gamma^2_{11\sigma_2} f_{t_2}^{\Delta_2} + A_{\sigma_2} N_{\sigma_2}\right)^{\sigma_2} \\
&= \left(\Gamma^1_{11\sigma_2} f_{t_1}^{\Delta_1\sigma_2}\right)^{\sigma_2} + \left(\Gamma^2_{11\sigma_2} f_{t_2}^{\Delta_2}\right)^{\sigma_2} + \left(A_{\sigma_2} N_{\sigma_2}\right)^{\sigma_2} \\
&= \Gamma^{1\sigma_2}_{11\sigma_2} f_{t_1}^{\Delta_1\sigma_2^2} + \Gamma^{2\sigma_2}_{11\sigma_2} f_{t_2}^{\Delta_2\sigma_2} + A^{\sigma_2}_{\sigma_2} N^{\sigma_2}_{\sigma_2} \\
&= \Gamma^{1\sigma_2}_{11\sigma_2} \left(V^1_{2\sigma_2} f_{t_1}^{\Delta_1\sigma_2} + V^2_{2\sigma_2} f_{t_2}^{\Delta_2} + V_{2\sigma_2} N_{\sigma_2}\right) \\
&\quad + \Gamma^{2\sigma_2}_{11\sigma_2} \left(\mu_2 \Gamma^1_{22\sigma_2} f_{t_1}^{\Delta_1\sigma_2} + \left(1 + \mu_2 \Gamma^2_{22\sigma_2}\right) f_{t_2}^{\Delta_2} + \mu_2 C_{\sigma_2} N_{\sigma_2}\right) \\
&\quad + A^{\sigma_2}_{\sigma_2} \left(\mu_2 W^1_{2\sigma_2} f_{t_1}^{\Delta_1\sigma_2} + \mu_2 W^2_{2\sigma_2} f_{t_2}^{\Delta_2} + N_{\sigma_2}\right) \\
&= \left(\Gamma^{1\sigma_2}_{11\sigma_2} V^1_{2\sigma_2} + \mu_2 \Gamma^{2\sigma_2}_{11\sigma_2} \Gamma^1_{22\sigma_2} + \mu_2 A^{\sigma_2}_{\sigma_2} W^1_{2\sigma_2}\right) f_{t_1}^{\Delta_1\sigma_2} \\
&\quad + \left(\Gamma^{1\sigma_2}_{11\sigma_2} V^2_{2\sigma_2} + \left(1 + \mu_2 \Gamma^2_{22\sigma_2}\right) \Gamma^{2\sigma_2}_{11\sigma_2} + \mu_2 A^{\sigma_2}_{\sigma_2} W^2_{2\sigma_2}\right) f_{t_2}^{\Delta_2} \\
&\quad + \left(\Gamma^{1\sigma_2}_{11\sigma_2} V_{2\sigma_2} + \mu_2 C_{\sigma_2} \Gamma^{2\sigma_2}_{11\sigma_2} + A^{\sigma_2}_{\sigma_2}\right) N_{\sigma_2},
\end{aligned}$$

that is,

$$f_{t_1}^{\Delta_1^2\sigma_2} = U^1_{2\sigma_2} f_{t_1}^{\Delta_1\sigma_2} + U^2_{2\sigma_2} f_{t_2}^{\Delta_2} + U^3_{2\sigma_2} N_{\sigma_2},$$

and

$$\begin{aligned}
f_{t_1t_2}^{\Delta_1\Delta_2\sigma_1} &= \left(\Gamma^1_{12\sigma_2}f_{t_1}^{\Delta_1\sigma_2} + \Gamma^2_{12\sigma_2}f_{t_2}^{\Delta_2} + B_{\sigma_2}N_{\sigma_2}\right)^{\sigma_1}\\
&= \left(\Gamma^1_{12\sigma_2}f_{t_1}^{\Delta_1\sigma_2}\right)^{\sigma_1} + \left(\Gamma^2_{12\sigma_2}f_{t_2}^{\Delta_2}\right)^{\sigma_1} + \left(B_{\sigma_2}N_{\sigma_2}\right)^{\sigma_1}\\
&= \Gamma^{1\sigma_1}_{12\sigma_2}f_{t_1}^{\Delta_1\sigma_1\sigma_2} + \Gamma^{2\sigma_1}_{12\sigma_2}f_{t_2}^{\Delta_2\sigma_1} + B^{\sigma_1}_{\sigma_2}N^{\sigma_1}_{\sigma_2}\\
&= \Gamma^{1\sigma_1}_{12\sigma_2}\left(V^1_{1\sigma_2}f_{t_1}^{\Delta_1\sigma_2} + V^2_{1\sigma_2}f_{t_2}^{\Delta_2} + V_{1\sigma_2}N_{\sigma_2}\right)\\
&\quad + \Gamma^{2\sigma_1}_{12\sigma_2}\left(\mu_1\Gamma^1_{12\sigma_2}f_{t_1}^{\Delta_1\sigma_2} + \left(1+\mu_1\Gamma^2_{12\sigma_2}\right)f_{t_2}^{\Delta_2} + \mu_1 B_{\sigma_2}N_{\sigma_2}\right)\\
&\quad + B^{\sigma_1}_{\sigma_2}\left(\mu_1 W^1_{1\sigma_2}f_{t_1}^{\Delta_1\sigma_2} + \mu_1 W^2_{1\sigma_2}f_{t_2}^{\Delta_2} + N_{\sigma_2}\right)\\
&= \left(\Gamma^{1\sigma_1}_{12\sigma_2}V^1_{1\sigma_2} + \mu_1\Gamma^{2\sigma_1}_{12\sigma_2}\Gamma^1_{12\sigma_2} + \mu_1 B^{\sigma_1}_{\sigma_2}W^1_{1\sigma_2}\right)f_{t_1}^{\Delta_1\sigma_2}\\
&\quad + \left(\Gamma^{1\sigma_1}_{12\sigma_2}V^2_{1\sigma_2} + \Gamma^{2\sigma_1}_{12\sigma_2}\left(1+\mu_1\Gamma^2_{12\sigma_2}\right) + \mu_1 B^{\sigma_1}_{\sigma_2}W^2_{1\sigma_2}\right)f_{t_2}^{\Delta_2}\\
&\quad + \left(\Gamma^{1\sigma_1}_{12\sigma_2}V_{1\sigma_2} + \mu_1\Gamma^{2\sigma_1}_{12\sigma_2}B_{\sigma_2} + B^{\sigma_1}_{\sigma_2}\right)N_{\sigma_2}\\
&= U^1_{3\sigma_2}f_{t_1}^{\Delta_1\sigma_2} + U^2_{3\sigma_2}f_{t_2}^{\Delta_2} + U^3_{3\sigma_2}N_{\sigma_2},
\end{aligned}$$

that is,

$$f_{t_1t_2}^{\Delta_1\Delta_2\sigma_1} = U^1_{3\sigma_1}f_{t_1}^{\Delta_1\sigma_2} + U^2_{3\sigma_2}f_{t_2}^{\Delta_2} + U^3_{3\sigma_2}N_{\sigma_2},$$

and

$$\begin{aligned}
f_{t_1t_2}^{\Delta_1\Delta_2\sigma_2} &= \left(\Gamma^1_{12\sigma_2}f_{t_1}^{\Delta_1\sigma_2} + \Gamma^2_{12\sigma_2}f_{t_2}^{\Delta_2} + B_{\sigma_2}N_{\sigma_2}\right)^{\sigma_2}\\
&= \left(\Gamma^1_{12\sigma_2}f_{t_1}^{\Delta_1\sigma_2}\right)^{\sigma_2} + \left(\Gamma^2_{12\sigma_2}f_{t_2}^{\Delta_2}\right)^{\sigma_2} + \left(B_{\sigma_2}N_{\sigma_2}\right)^{\sigma_2}\\
&= \Gamma^{1\sigma_2}_{12\sigma_2}f_{t_1}^{\Delta_1\sigma_2^2} + \Gamma^{2\sigma_2}_{12\sigma_2}f_{t_2}^{\Delta_2\sigma_2} + B^{\sigma_2}_{\sigma_2}N^{\sigma_2}_{\sigma_2}\\
&= \Gamma^{1\sigma_2}_{12\sigma_2}\left(V^1_{2\sigma_2}f_{t_1}^{\Delta_1\sigma_2} + V^2_{2\sigma_2}f_{t_2}^{\Delta_2} + V_{2\sigma_2}N_{\sigma_2}\right)\\
&\quad + \Gamma^2_{12\sigma_2}\left(\mu_2\Gamma^1_{22\sigma_2}f_{t_1}^{\Delta_1\sigma_2} + \left(1+\mu_2\Gamma^2_{22\sigma_2}\right)f_{t_2}^{\Delta_2} + \mu_2 C_{\sigma_2}N_{\sigma_2}\right)\\
&\quad + B^{\sigma_2}_{\sigma_2}\left(\mu_2 W^1_{2\sigma_2}f_{t_1}^{\Delta_1\sigma_2} + \mu_2 W^2_{2\sigma_2}f_{t_2}^{\Delta_2} + N_{\sigma_2}\right)\\
&= \left(\Gamma^{1\sigma_2}_{12\sigma_2}V^1_{2\sigma_2} + \mu_2\Gamma^2_{12\sigma_2}\Gamma^1_{22\sigma_2} + \mu_2 B^{\sigma_2}_{\sigma_2}W^1_{2\sigma_2}\right)f_{t_1}^{\Delta_1\sigma_2}\\
&\quad + \left(\Gamma^{1\sigma_2}_{12\sigma_2}V^2_{2\sigma_2} + \left(1+\mu_2\Gamma^2_{22\sigma_2}\right)\Gamma^{2\sigma_2}_{12\sigma_2} + \mu_2 B^{\sigma_2}_{\sigma_2}W^2_{2\sigma_2}\right)f_{t_2}^{\Delta_2}\\
&\quad + \left(\Gamma^{1\sigma_2}_{12\sigma_2}V_{2\sigma_2} + \mu_2 C_{\sigma_2}\Gamma^{2\sigma_2}_{12\sigma_2} + B^{\sigma_2}_{\sigma_2}\right)N_{\sigma_2},
\end{aligned}$$

that is,

$$f_{t_1t_2}^{\Delta_1\Delta_2\sigma_2} = U^1_{4\sigma_2}f_{t_1}^{\Delta_1\sigma_2} + U^2_{4\sigma_2}f_{t_2}^{\Delta_2} + U^3_{4\sigma_2}N_{\sigma_2},$$

and

$$\begin{aligned}
f_{t_2}^{\Delta_2^2\sigma_1} &= \left(\Gamma^1_{22\sigma_2}f_{t_1}^{\Delta_1\sigma_2} + \Gamma^2_{22\sigma_2}f_{t_2}^{\Delta_2} + C_{\sigma_2}N_{\sigma_2}\right)^{\sigma_1}\\
&= \left(\Gamma^1_{22\sigma_2}f_{t_1}^{\Delta_1\sigma_2}\right)^{\sigma_1} + \left(\Gamma^2_{22\sigma_2}f_{t_2}^{\Delta_2}\right)^{\sigma_1} + \left(C_{\sigma_2}N_{\sigma_2}\right)^{\sigma_1}
\end{aligned}$$

$$\begin{aligned}
&= \Gamma^{1\sigma_1}_{22\sigma_2} f_{t_1}^{\Delta_1\sigma_1\sigma_2} + \Gamma^{2\sigma_1}_{22\sigma_2} f_{t_2}^{\Delta_2\sigma_1} + C^{\sigma_1}_{\sigma_2} N^{\sigma_1}_{\sigma_2}\\
&= \Gamma^{1\sigma_1}_{22\sigma_2}\left(V^1_{1\sigma_2} f_{t_1}^{\Delta_1\sigma_2} + V^2_{1\sigma_2} f_{t_2}^{\Delta_2} + V_{1\sigma_2} N_{\sigma_2}\right)\\
&\quad + \Gamma^{2\sigma_1}_{22\sigma_2}\left(\mu_1 \Gamma^1_{12\sigma_2} f_{t_1}^{\Delta_1\sigma_2} + \left(1+\mu_1\Gamma^2_{12\sigma_2}\right) f_{t_2}^{\Delta_2} + \mu_1 B_{\sigma_2} N_{\sigma_2}\right)\\
&\quad + C^{\sigma_1}_{\sigma_2}\left(\mu_1 W^1_{1\sigma_2} f_{t_1}^{\Delta_1\sigma_2} + \mu_1 W^2_{1\sigma_2} f_{t_2}^{\Delta_2} + N_{\sigma_2}\right)\\
&= \left(\Gamma^{1\sigma_1}_{22\sigma_2} V^1_{1\sigma_2} + \mu_1 \Gamma^{2\sigma_1}_{22\sigma_2}\Gamma^1_{12\sigma_2} + \mu_1 C^{\sigma_1}_{\sigma_2} W^1_{1\sigma_2}\right) f_{t_1}^{\Delta_1\sigma_2}\\
&\quad + \left(\Gamma^{1\sigma_1}_{22\sigma_2} V^2_{1\sigma_2} + \Gamma^{2\sigma_1}_{22\sigma_2}\left(1+\mu_1\Gamma^2_{12\sigma_2}\right) + \mu_1 C^{\sigma_1}_{\sigma_2} W^2_{1\sigma_2}\right) f_{t_2}^{\Delta_2}\\
&\quad + \left(\Gamma^{1\sigma_1}_{22\sigma_2} V_{1\sigma_2} + \mu_1 \Gamma^{2\sigma_1}_{22\sigma_2} B_{\sigma_2} + C^{\sigma_1}_{\sigma_2}\right) N_{\sigma_2}\\
&= U^1_{5\sigma_2} f_{t_1}^{\Delta_1\sigma_2} + U^2_{5\sigma_2} f_{t_2}^{\Delta_2} + U^3_{5\sigma_2} N_{\sigma_2},
\end{aligned}$$

that is,

$$f_{t_2}^{\Delta_2^2\sigma_1} = U^1_{5\sigma_1} f_{t_1}^{\Delta_1\sigma_2} + U^2_{5\sigma_2} f_{t_2}^{\Delta_2} + U^3_{5\sigma_2} N_{\sigma_2},$$

and

$$\begin{aligned}
f_{t_2}^{\Delta_2^2\sigma_2} &= \left(\Gamma^1_{22\sigma_2} f_{t_1}^{\Delta_1\sigma_2} + \Gamma^2_{22\sigma_2} f_{t_2}^{\Delta_2} + C_{\sigma_2} N_{\sigma_2}\right)^{\sigma_2}\\
&= \left(\Gamma^1_{22\sigma_2} f_{t_1}^{\Delta_1\sigma_2}\right)^{\sigma_2} + \left(\Gamma^2_{22\sigma_2} f_{t_2}^{\Delta_2}\right)^{\sigma_2} + \left(C_{\sigma_2} N_{\sigma_2}\right)^{\sigma_2}\\
&= \Gamma^{1\sigma_2}_{22\sigma_2} f_{t_1}^{\Delta_1\sigma_2^2} + \Gamma^{2\sigma_2}_{22\sigma_2} f_{t_2}^{\Delta_2\sigma_2} + C^{\sigma_2}_{\sigma_2} N^{\sigma_2}_{\sigma_2}\\
&= \Gamma^{1\sigma_2}_{22\sigma_2}\left(V^1_{2\sigma_2} f_{t_1}^{\Delta_1\sigma_2} + V^2_{2\sigma_2} f_{t_2}^{\Delta_2} + V_{2\sigma_2} N_{\sigma_2}\right)\\
&\quad + \Gamma^{2\sigma_2}_{22\sigma_2}\left(\mu_2 \Gamma^1_{22\sigma_2} f_{t_1}^{\Delta_1\sigma_2} + \left(1+\mu_2\Gamma^2_{22\sigma_2}\right) f_{t_2}^{\Delta_2} + \mu_2 C_{\sigma_2} N_{\sigma_2}\right)\\
&\quad + C^{\sigma_2}_{\sigma_2}\left(\mu_2 W^1_{2\sigma_2} f_{t_1}^{\Delta_1\sigma_2} + \mu_2 W^2_{2\sigma_2} f_{t_2}^{\Delta_2} + N_{\sigma_2}\right)\\
&= \left(\Gamma^{1\sigma_2}_{22\sigma_2} V^1_{2\sigma_2} + \mu_2 \Gamma^{2\sigma_2}_{22\sigma_2}\Gamma^1_{22\sigma_2} + \mu_2 C^{\sigma_2}_{\sigma_2} W^1_{2\sigma_2}\right) f_{t_1}^{\Delta_1\sigma_2}\\
&\quad + \left(\Gamma^{1\sigma_2}_{22\sigma_2} V^2_{2\sigma_2} + \left(1+\mu_2\Gamma^2_{22\sigma_2}\right)\Gamma^{2\sigma_2}_{22\sigma_2} + \mu_2 C^{\sigma_2}_{\sigma_2} W^2_{2\sigma_2}\right) f_{t_2}^{\Delta_2}\\
&\quad + \left(\Gamma^{1\sigma_2}_{22\sigma_2} V_{2\sigma_2} + \mu_2 C_{\sigma_2}\Gamma^{2\sigma_2}_{22\sigma_2} + C^{\sigma_2}_{\sigma_2}\right) N_{\sigma_2},
\end{aligned}$$

that is,

$$f_{t_2}^{\Delta_2^2\sigma_2} = U^1_{6\sigma_2} f_{t_1}^{\Delta_1\sigma_2} + U^2_{6\sigma_2} f_{t_2}^{\Delta_2} + U^3_{6\sigma_2} N_{\sigma_2}.$$

Exercise 5.5 Suppose S is a $\sigma_1^2\sigma_2$-regular surface, σ_1 is Δ_1-differentiable and σ_2 is Δ_2-differentiable. Find a representation of $N^{\Delta_1}_{\sigma_1^2\sigma_2 t_1}$ and $N^{\Delta_2}_{\sigma_1^2\sigma_2 t_2}$ in the following form

$$\begin{aligned}
N^{\Delta_1}_{\sigma_1^2\sigma_2 t_1} &= W^1_{1\sigma_1^2\sigma_2} f_{t_1}^{\Delta_1\sigma_1\sigma_2} + W^2_{1\sigma_1^2\sigma_2} f_{t_2}^{\Delta_2\sigma_1^2\sigma_2}\\
N^{\Delta_2}_{\sigma_1^2\sigma_2 t_2} &= W^1_{2\sigma_1^2\sigma_2} f_{t_1}^{\Delta_1\sigma_1\sigma_2} + W^2_{2\sigma_1^2\sigma_2} f_{t_2}^{\Delta_2\sigma_1^2\sigma_2}.
\end{aligned}$$

Definition 5.7 The coefficients $W^k_{j\sigma_1^2\sigma_2}$, $j,k \in \{1,2\}$, will be called $\sigma_1^2\sigma_2$-Weingarten coefficients.

Example 5.6 We will express $N^{\sigma_1}_{\sigma_1^2\sigma_2}$ and $N^{\sigma_2}_{\sigma_1^2\sigma_2}$ using the basis $\left\{f_{t_1}^{\Delta_1\sigma_1\sigma_2}, f_{t_2}^{\Delta_2\sigma_1\sigma_2^2}, N_{\sigma_1^2\sigma_2}\right\}$.

We have

$$\begin{aligned} N^{\sigma_1}_{\sigma_1^2\sigma_2} &= N_{\sigma_1^2\sigma_2} + \mu_1 N^{\Delta_1}_{\sigma_1^2\sigma_2 t_1} \\ &= N_{\sigma_1^2\sigma_2} + \mu_1\left(W^1_{1\sigma_1^2\sigma_2} f_{t_1}^{\Delta_1\sigma_1\sigma_2} + W^2_{1\sigma_1^2\sigma_2} f_{t_2}^{\Delta_2\sigma_1^2\sigma_2}\right) \\ &= \left(\mu_1 W^1_{1\sigma_1^2\sigma_2}\right) f_{t_1}^{\Delta_1\sigma_1\sigma_2} + \left(\mu_1 W^2_{1\sigma_1^2\sigma_2}\right) f_{t_2}^{\Delta_2\sigma_1^2\sigma_2} + N_{\sigma_1^2\sigma_2} \end{aligned}$$

and

$$\begin{aligned} N^{\sigma_2}_{\sigma_1^2\sigma_2} &= N_{\sigma_1^2\sigma_2} + \mu_2 N^{\Delta_2}_{\sigma_1^2\sigma_2 t_2} \\ &= N_{\sigma_1^2\sigma_2} + \mu_2\left(W^1_{2\sigma_1^2\sigma_2} f_{t_1}^{\Delta_1\sigma_1\sigma_2} + W^2_{2\sigma_1^2\sigma_2} f_{t_2}^{\Delta_2\sigma_1^2\sigma_2}\right) \\ &= \left(\mu_2 W^1_{2\sigma_1^2\sigma_2}\right) f_{t_1}^{\Delta_1\sigma_1\sigma_2} + \left(\mu_2 W^2_{2\sigma_1^2\sigma_2}\right) f_{t_2}^{\Delta_2\sigma_1^2\sigma_2} + N_{\sigma_1^2\sigma_2}. \end{aligned}$$

Exercise 5.6 Suppose S is a $\sigma_1^2\sigma_2$-regular surface, σ_1 is Δ_1-differentiable and σ_2 is Δ_2-differentiable. Find

a. $N_{\sigma_1^2\sigma_2} + N^{\sigma_1}_{\sigma_1^2\sigma_2} + N^{\sigma_2}_{\sigma_1^2\sigma_2}$.

b. $4N_{\sigma_1^2\sigma_2} - f_{t_1}^{\Delta_1\sigma_1\sigma_2} - 5f_{t_2}^{\Delta_2\sigma_1^2\sigma_2} + N^{\sigma_2}_{\sigma_1^2\sigma_2}$.

c. $3N^{\sigma_2}_{\sigma_1^2\sigma_2} + 2W^1_{1\sigma_1^2\sigma_2} f_{t_1}^{\Delta_1\sigma_1\sigma_2} + 4f_{t_2}^{\Delta_2\sigma_1^2\sigma_2}$.

Exercise 5.7 Suppose S is a $\sigma_1\sigma_2^2$-regular surface, σ_1 is Δ_1-differentiable and σ_2 is Δ_2-differentiable. Find a representation of $N^{\Delta_1}_{\sigma_1\sigma_2^2 t_1}$ and $N^{\Delta_2}_{\sigma_1\sigma_2^2 t_2}$ in the following form

$$\begin{aligned} N^{\Delta_1}_{\sigma_1\sigma_2^2 t_1} &= W^1_{1\sigma_1\sigma_2^2} f_{t_1}^{\Delta_1\sigma_1\sigma_2^2} + W^2_{1\sigma_1\sigma_2^2} f_{t_2}^{\Delta_2\sigma_1\sigma_2} \\ N^{\Delta_2}_{\sigma_1\sigma_2^2 t_2} &= W^1_{2\sigma_1\sigma_2^2} f_{t_1}^{\Delta_1\sigma_1\sigma_2^2} + W^2_{2\sigma_1\sigma_2^2} f_{t_2}^{\Delta_2\sigma_1\sigma_2}. \end{aligned}$$

Definition 5.8 The coefficients $W^k_{j\sigma_1\sigma_2^2}$, $j,k \in \{1,2\}$, will be called $\sigma_1\sigma_2^2$-Weingarten coefficients.

Example 5.7 Suppose S is a $\sigma_1\sigma_2^2$-regular surface, σ_1 is Δ_1-differentiable and σ_2 is Δ_2-differentiable. We will express $N^{\sigma_1}_{\sigma_1\sigma_2^2}$ and $N^{\sigma_2}_{\sigma_1\sigma_2^2}$ using the basis

$\left\{f_{t_1}^{\Delta_1\sigma_1\sigma_2^2}, f_{t_2}^{\Delta_2\sigma_1\sigma_2}, N_{\sigma_1\sigma_2^2}\right\}$. We have

$$\begin{aligned} N^{\sigma_1}_{\sigma_1\sigma_2^2} &= N_{\sigma_1\sigma_2^2} + \mu_1 N^{\Delta_1}_{\sigma_1\sigma_2^2 t_1} \\ &= N_{\sigma_1\sigma_2^2} + \mu_1\left(W^1_{1\sigma_1\sigma_2^2} f_{t_1}^{\Delta_1\sigma_1\sigma_2^2} + W^2_{1\sigma_1\sigma_2^2} f_{t_2}^{\Delta_2\sigma_1\sigma_2}\right) \\ &= \left(\mu_1 W^1_{1\sigma_1\sigma_2^2}\right) f_{t_1}^{\Delta_1\sigma_1\sigma_2^2} + \left(\mu_1 W^2_{1\sigma_1\sigma_2^2}\right) f_{t_2}^{\Delta_2\sigma_1\sigma_2} + N_{\sigma_1\sigma_2^2} \end{aligned}$$

and

$$\begin{aligned} N^{\sigma_2}_{\sigma_1\sigma_2^2} &= N_{\sigma_1\sigma_2^2} + \mu_2 N^{\Delta_2}_{\sigma_1\sigma_2^2 t_2} \\ &= N_{\sigma_1\sigma_2^2} + \mu_2\left(W^1_{2\sigma_1\sigma_2^2} f_{t_1}^{\Delta_1\sigma_1\sigma_2^2} + W^2_{2\sigma_1\sigma_2^2} f_{t_2}^{\Delta_2\sigma_1\sigma_2}\right) \\ &= \left(\mu_2 W^1_{2\sigma_1\sigma_2^2}\right) f_{t_1}^{\Delta_1\sigma_1\sigma_2^2} + \left(\mu_2 W^2_{2\sigma_1\sigma_2^2}\right) f_{t_2}^{\Delta_2\sigma_1\sigma_2} + N_{\sigma_1\sigma_2^2}. \end{aligned}$$

Exercise 5.8 Suppose S is a $\sigma_1\sigma_2^2$-regular surface, σ_1 is Δ_1-differentiable and σ_2 is Δ_2-differentiable. Find

1. $N_{\sigma_1\sigma_2^2} + N^{\sigma_1}_{\sigma_1\sigma_2^2} + N^{\sigma_2}_{\sigma_1\sigma_2^2}$.
2. $2N_{\sigma_1\sigma_2^2} + 2f_{t_2}^{\Delta_2\sigma_1\sigma_2} + 3f_{t_1}^{\Delta_1\sigma_1\sigma_2^2} - 4N^{\sigma_2}_{\sigma_1\sigma_2^2}$.
3. $N^{\sigma_1}_{\sigma_1\sigma_2^2} + 4W^1_{2\sigma_1\sigma_2^2} f_{t_1}^{\Delta_1\sigma_1\sigma_2^2} + 3f_{t_2}^{\Delta_2\sigma_1\sigma_2}$.

5.3 The Gauss and Godazzi-Mainardi Equations for a Surface

In this section, we will deduce the Gauss and Godazzi-Mainardi equations for a surface. Suppose that S is a σ_1- or σ_2- or $\sigma_1^2\sigma_2$- or $\sigma_1\sigma_2^2$-oriented surface with a local σ_1- or σ_2- or $\sigma_1^2\sigma_2$- or $\sigma_1\sigma_2^2$-presentation (U,f).

Theorem 5.1 *Suppose $f \in \mathscr{C}^3$ and S is a σ_1-regular surface. Then*

$$\begin{aligned} \left(\Gamma^1_{12\sigma_1}\right)^{\Delta_1}_{t_1} - \left(\Gamma^1_{11\sigma_1}\right)^{\Delta_2}_{t_2} &= -\Gamma^{1\sigma_1}_{12\sigma_1}\Gamma^1_{11\sigma_1} - \left(1+\mu_1^{\Delta_1}\right)\Gamma^{2\sigma_1}_{12\sigma_1}U^1_{3\sigma_1} - B^{\sigma_1}_{\sigma_1}W^1_{1\sigma_1} \\ &\quad + \Gamma^{1\sigma_2}_{11\sigma_1}\Gamma^1_{12\sigma_1} + \Gamma^{2\sigma_2}_{11\sigma_1}U^1_{5\sigma_1} + A^{\sigma_2}_{\sigma_1}W^1_{2\sigma_1} \\ \left(\Gamma^2_{12\sigma_1}\right)^{\Delta_1}_{t_1} - \left(\Gamma^2_{11\sigma_1}\right)^{\Delta_2}_{t_2} &= -\Gamma^{1\sigma_1}_{12\sigma_1}\Gamma^2_{11\sigma_1} - \left(1+\mu_1^{\Delta_1}\right)\Gamma^{2\sigma_1}_{12\sigma_1}U^2_{3\sigma_1} - B^{\sigma_1}_{\sigma_1}W^2_{1\sigma_1} \\ &\quad + \Gamma^{1\sigma_2}_{11\sigma_1}\Gamma^2_{12\sigma_1} + \Gamma^{2\sigma_1}_{11\sigma_1}U^2_{5\sigma_1} + A^{\sigma_1}_{\sigma_1}W^2_{2\sigma_1} \\ \left(\Gamma^1_{12\sigma_1}\right)^{\Delta_2}_{t_2} - \left(\Gamma^1_{22\sigma_1}\right)^{\Delta_1}_{t_1} &= -\Gamma^{1\sigma_2}_{12\sigma_1}\Gamma^1_{12\sigma_1} - \Gamma^{2\sigma_2}_{12\sigma_1}U^1_{5\sigma_1} - B^{\sigma_2}_{\sigma_1}W^1_{2\sigma_1} \\ &\quad + \Gamma^{1\sigma_1}_{22\sigma_1}\Gamma^1_{11\sigma_1} + \left(1+\mu_1^{\Delta_1}\right)U^1_{3\sigma_1} + C^{\sigma_1}_{\sigma_1}W^1_{1\sigma_1} \\ \left(\Gamma^2_{12\sigma_1}\right)^{\Delta_2}_{t_2} - \left(\Gamma^2_{22\sigma_1}\right)^{\Delta_1}_{t_1} &= -\Gamma^{1\sigma_2}_{12\sigma_1}\Gamma^2_{12\sigma_1} - \Gamma^{2\sigma_2}_{12\sigma_1}U^2_{5\sigma_1} - B^{\sigma_2}_{\sigma_1}W^2_{2\sigma_1} \\ &\quad + \Gamma^{1\sigma_1}_{22\sigma_1}\Gamma^2_{11\sigma_1} + \left(1+\mu_1^{\Delta_1}\right)\Gamma^{2\sigma_1}_{22\sigma_1}U^2_{3\sigma_1} + C^{\sigma_1}_{\sigma_1}W^2_{1\sigma_1} \end{aligned} \tag{5.6}$$

and

$$\begin{aligned}(B_{\sigma_1})_{t_1}^{\Delta_1}-(A_{\sigma_1})_{t_2}^{\Delta_2} &= -\Gamma_{12\sigma_1}^{1\sigma_1}A_{\sigma_1}-\left(1+\mu_1^{\Delta_1}\right)\Gamma_{12\sigma_1}^{2\sigma_1}U_{3\sigma_1}^3+\Gamma_{11\sigma_1}^{1\sigma_2}B_{\sigma_1}\\ &\quad +\Gamma_{11\sigma_1}^{2\sigma_2}U_{5\sigma_1}^3\\ (B_{\sigma_1})_{t_2}^{\Delta_2}-(C_{\sigma_1})_{t_1}^{\Delta_1} &= -\Gamma_{12\sigma_1}^{1\sigma_2}B_{\sigma_1}-\Gamma_{12\sigma_1}^{2\sigma_2}U_{5\sigma_1}^3+\Gamma_{22\sigma_1}^{1\sigma_1}A_{\sigma_1}\\ &\quad +\left(1+\mu_1^{\Delta_1}\right)\Gamma_{22\sigma_1}^{2\sigma_1}U_{3\sigma_1}^3.\end{aligned} \tag{5.7}$$

Definition 5.9 The equations (5.6) will be called the σ_1-Gauss equations.

Definition 5.10 The equations (5.7) will be called the σ_1-Godazzi-Mainardi equations.

Proof. To deduce the σ_1-Gauss and σ_1-Godazzi-Mainardi equations, we will use the relations

$$f_{t_1t_2}^{\Delta_1^2\Delta_2}=f_{t_1t_2t_1}^{\Delta_1\Delta_2\Delta_1} \tag{5.8}$$

and

$$f_{t_2t_1}^{\Delta_2^2\Delta_1}=f_{t_1t_2}^{\Delta_1\Delta_2^2}. \tag{5.9}$$

We have

$$\begin{aligned}f_{t_1t_2}^{\Delta_1^2\Delta_2} &= \left(\Gamma_{11\sigma_1}^1f_{t_1}^{\Delta_1}+\Gamma_{11\sigma_1}^2f_{t_2}^{\Delta_2\sigma_1}+A_{\sigma_1}N_{\sigma_1}\right)_{t_2}^{\Delta_2}\\ &= \left(\Gamma_{11\sigma_1}^1f_{t_1}^{\Delta_1}\right)_{t_2}^{\Delta_2}+\left(\Gamma_{11\sigma_1}^2f_{t_2}^{\Delta_2\sigma_1}\right)_{t_2}^{\Delta_2}+(A_{\sigma_1}N_{\sigma_1})_{t_2}^{\Delta_2}.\end{aligned}$$

Observe that

$$\begin{aligned}\left(\Gamma_{11\sigma_1}f_{t_1}^{\Delta_1}\right)_{t_2}^{\Delta_2} &= \left(\Gamma_{11\sigma_1}^1\right)_{t_2}^{\Delta_2}f_{t_1}^{\Delta_1}+\Gamma_{11\sigma_1}^{1\sigma_2}f_{t_1t_2}^{\Delta_1\Delta_2}\\ &= \left(\Gamma_{11\sigma_1}^1\right)_{t_2}^{\Delta_2}f_{t_1}^{\Delta_1}+\Gamma_{11\sigma_1}^{1\sigma_2}\left(\Gamma_{12\sigma_1}^1f_{t_1}^{\Delta_1}+\Gamma_{12\sigma_1}^2f_{t_2}^{\Delta_2\sigma_1}+B_{\sigma_1}N_{\sigma_1}\right)\\ &= \left(\left(\Gamma_{11\sigma_1}^1\right)_{t_2}^{\Delta_2}+\Gamma_{11\sigma_1}^{1\sigma_2}\Gamma_{12\sigma_1}^1\right)f_{t_1}^{\Delta_1}+\Gamma_{11\sigma_1}^{1\sigma_2}\Gamma_{12\sigma_1}^2f_{t_2}^{\Delta_2\sigma_1}+\Gamma_{11\sigma_1}^{1\sigma_2}B_{\sigma_1}N_{\sigma_1},\\ \left(\Gamma_{11\sigma_1}^2f_{t_2}^{\Delta_2\sigma_1}\right)_{t_2}^{\Delta_2} &= \left(\Gamma_{11\sigma_1}^2\right)_{t_2}^{\Delta_2}f_{t_2}^{\Delta_2\sigma_1}+\Gamma_{11\sigma_1}^{2\sigma_2}f_{t_2}^{\Delta_2^2\sigma_1}\\ &= \left(\Gamma_{11\sigma_1}^2\right)_{t_2}^{\Delta_2}f_{t_2}^{\Delta_2\sigma_1}+\Gamma_{11\sigma_1}^{2\sigma_2}\left(U_{5\sigma_1}^1f_{t_1}^{\Delta_1}+U_{5\sigma_1}^2f_{t_2}^{\Delta_2\sigma_1}+U_{5\sigma_1}^3N_{\sigma_1}\right)\\ &= \Gamma_{11\sigma_1}^{2\sigma_2}U_{5\sigma_1}^1f_{t_1}^{\Delta_1}+\left(\left(\Gamma_{11\sigma_1}^2\right)_{t_2}^{\Delta_2}+\Gamma_{11\sigma_1}^{2\sigma_2}U_{5\sigma_1}^2\right)f_{t_2}^{\Delta_2\sigma_1}+\Gamma_{11\sigma_1}^{2\sigma_2}U_{5\sigma_1}^3N_{\sigma_1},\\ (A_{\sigma_1}N_{\sigma_1})_{t_2}^{\Delta_2} &= (A_{\sigma_1})_{t_2}^{\Delta_2}N_{\sigma_1}+A_{\sigma_1}^{\sigma_2}(N_{\sigma_1})_{t_2}^{\Delta_2}\\ &= (A_{\sigma_1})_{t_2}^{\Delta_2}N_{\sigma_1}+A_{\sigma_1}^{\sigma_2}\left(W_{2\sigma_1}^1f_{t_1}^{\Delta_1}+W_{2\sigma_1}^2f_{t_2}^{\Delta_2\sigma_1}\right)\\ &= \left(A_{\sigma_1}^{\sigma_2}W_{2\sigma_1}^1\right)f_{t_1}^{\Delta_1}+\left(A_{\sigma_1}^{\sigma_2}W_{2\sigma_1}^2\right)f_{t_2}^{\Delta_2\sigma_1}+(A_{\sigma_1})_{t_2}^{\Delta_2}N_{\sigma_1}.\end{aligned}$$

Therefore

$$\begin{aligned}f_{t_1t_2}^{\Delta_1^2\Delta_2} &= \left(\left(\Gamma_{11\sigma_1}^1\right)_{t_2}^{\Delta_2}+\Gamma_{11\sigma_1}^{1\sigma_2}\Gamma_{12\sigma_1}^1+\Gamma_{11\sigma_1}^{2\sigma_2}U_{5\sigma_1}^1+A_{\sigma_1}^{\sigma_2}W_{2\sigma_1}^1\right)f_{t_1}^{\Delta_1}\\ &\quad +\left(\Gamma_{11\sigma_1}^{1\sigma_2}\Gamma_{12\sigma_1}^2+\left(\Gamma_{11\sigma_1}^2\right)_{t_2}^{\Delta_2}+\Gamma_{11\sigma_1}^{2\sigma_2}U_{5\sigma_1}^2+A_{\sigma_1}^{\sigma_2}W_{2\sigma_1}^2\right)f_{t_2}^{\Delta_2\sigma_1}\\ &\quad +\left(\Gamma_{11\sigma_1}^{1\sigma_2}B_{\sigma_1}+\Gamma_{11\sigma_1}^{2\sigma_2}U_{5\sigma_1}^3+(A_{\sigma_1})_{t_2}^{\Delta_2}\right)N_{\sigma_1}.\end{aligned} \tag{5.10}$$

Next,

$$f_{t_1t_2t_1}^{\Delta_1\Delta_2\Delta_1} = \left(\Gamma^1_{12\sigma_1}f_{t_1}^{\Delta_1} + \Gamma^2_{12\sigma_1}f_{t_2}^{\Delta_2\sigma_1} + B_{\sigma_1}N_{\sigma_1}\right)_{t_1}^{\Delta_1}$$
$$= \left(\Gamma^1_{12\sigma_1}f_{t_1}^{\Delta_1}\right)_{t_1}^{\Delta_1} + \left(\Gamma^2_{12\sigma_1}f_{t_2}^{\Delta_2\sigma_1}\right)_{t_1}^{\Delta_1} + (B_{\sigma_1}N_{\sigma_1})_{t_1}^{\Delta_1}$$

and

$$\left(\Gamma^1_{12\sigma_1}f_{t_1}^{\Delta_1}\right)_{t_1}^{\Delta_1} = \left(\Gamma^1_{12\sigma_1}\right)_{t_1}^{\Delta_1}f_{t_1}^{\Delta_1} + \Gamma^{1\sigma_1}_{12\sigma_1}f_{t_1}^{\Delta_1^2}$$
$$= \left(\Gamma^1_{12\sigma_1}\right)_{t_1}^{\Delta_1}f_{t_1}^{\Delta_1} + \Gamma^{1\sigma_1}_{12\sigma_1}\left(\Gamma^1_{11\sigma_1}f_{t_1}^{\Delta_1} + \Gamma^2_{11\sigma_1}f_{t_2}^{\Delta_2\sigma_1} + A_{\sigma_1}N_{\sigma_1}\right)$$
$$= \left(\left(\Gamma^1_{12\sigma_1}\right)_{t_1}^{\Delta_1} + \Gamma^{1\sigma_1}_{12\sigma_1}\Gamma^1_{11\sigma_1}\right)f_{t_1}^{\Delta_1} + \left(\Gamma^{1\sigma_1}_{12\sigma_1}\Gamma^2_{11\sigma_1}\right)f_{t_2}^{\Delta_2\sigma_1} + \left(\Gamma^{1\sigma_1}_{12\sigma_1}A_{\sigma_1}\right)N_{\sigma_1},$$

$$\left(\Gamma^2_{12\sigma_1}f_{t_2}^{\Delta_2\sigma_1}\right)_{t_1}^{\Delta_1} = \left(\Gamma^2_{12\sigma_1}\right)_{t_1}^{\Delta_1}f_{t_2}^{\Delta_2\sigma_1} + \Gamma^{2\sigma_1}_{12\sigma_1}f_{t_2t_1}^{\Delta_2\sigma_1\Delta_1}$$
$$= \left(\Gamma^2_{12\sigma_1}\right)_{t_1}^{\Delta_1}f_{t_2}^{\Delta_2\sigma_1} + \Gamma^{2\sigma_1}_{12\sigma_1}\left(1+\mu_1^{\Delta_1}\right)f_{t_1t_2}^{\Delta_1\Delta_2\sigma_1}$$
$$= \left(\Gamma^2_{12\sigma_1}\right)_{t_1}^{\Delta_1}f_{t_2}^{\Delta_2\sigma_1} + \Gamma^{2\sigma_1}_{12\sigma_1}\left(1+\mu_1^{\Delta_1}\right)\left(U^1_{3\sigma_1}f_{t_1}^{\Delta_1} + U^2_{3\sigma_1}f_{t_2}^{\Delta_2\sigma_1} + U^3_{3\sigma_1}N_{\sigma_1}\right)$$
$$= \left(1+\mu_1^{\Delta_1}\right)\Gamma^{2\sigma_1}_{12\sigma_1}U^1_{3\sigma_1}f_{t_1}^{\Delta_1} + \left(\left(\Gamma^2_{12\sigma_1}\right)_{t_1}^{\Delta_1} + \left(1+\mu_1^{\Delta_1}\right)\Gamma^{2\sigma_1}_{12\sigma_1}U^2_{3\sigma_1}\right)f_{t_2}^{\Delta_2\sigma_1}$$
$$+ \left(1+\mu_1^{\Delta_1}\right)\Gamma^{2\sigma_1}_{12\sigma_1}U^3_{3\sigma_1}N_{\sigma_1},$$

$$(B_{\sigma_1}N_{\sigma_1})_{t_1}^{\Delta_1} = (B_{\sigma_1})_{t_1}^{\Delta_1}N_{\sigma_1} + B_{\sigma_1}^{\sigma_1}(N_{\sigma_1})_{t_1}^{\Delta_1}$$
$$= (B_{\sigma_1})_{t_1}^{\Delta_1}N_{\sigma_1} + B_{\sigma_1}^{\sigma_1}\left(W^1_{1\sigma_1}f_{t_1}^{\Delta_1} + W^2_{1\sigma_1}f_{t_2}^{\Delta_2\sigma_1}\right)$$
$$= \left(B_{\sigma_1}^{\sigma_1}W^1_{1\sigma_1}\right)f_{t_1}^{\Delta_1} + \left(B_{\sigma_1}^{\sigma_1}W^2_{1\sigma_1}\right)f_{t_2}^{\Delta_2\sigma_1} + (B_{\sigma_1})_{t_1}^{\Delta_1}N_{\sigma_1}.$$

Consequently

$$f_{t_1t_2t_1}^{\Delta_1\Delta_2\Delta_1} = \left(\left(\Gamma^1_{12\sigma_1}\right)_{t_1}^{\Delta_1} + \Gamma^{1\sigma_1}_{12\sigma_1}\Gamma^1_{11\sigma_1} + \left(1+\mu_1^{\Delta_1}\right)\Gamma^{2\sigma_1}_{12\sigma_1}U^1_{3\sigma_1} + B_{\sigma_1}^{\sigma_1}W^1_{1\sigma_1}\right)f_{t_1}^{\Delta_1}$$
$$+ \left(\Gamma^{1\sigma_1}_{12\sigma_1}\Gamma^2_{11\sigma_1} + \left(\Gamma^2_{12\sigma_1}\right)_{t_1}^{\Delta_1} + \left(1+\mu_1^{\Delta_1}\right)\Gamma^{2\sigma_1}_{12\sigma_1}U^2_{3\sigma_1} + B_{\sigma_1}^{\sigma_1}W^2_{1\sigma_1}\right)f_{t_2}^{\Delta_2\sigma_1}$$
$$+ \left(\Gamma^{1\sigma_1}_{12\sigma_1}A_{\sigma_1} + \left(1+\mu_1^{\Delta_1}\right)\Gamma^{2\sigma_1}_{12\sigma_1}U^3_{3\sigma_1} + (B_{\sigma_1})_{t_1}^{\Delta_1}\right)N_{\sigma_1}.$$

Now, we equate the last equation and (5.10) and using that $f_{t_1}^{\Delta_1}$, $f_{t_2}^{\Delta_2\sigma_1}$ and N_{σ_1} are linearly independent, we find the first two equations of the σ_1-Gauss equations and the first equation of the σ_1-Godazzi-Mainardi equations.

Next,

$$f_{t_2t_1}^{\Delta_2^2\Delta_1} = \left(\Gamma^1_{22\sigma_1}f_{t_1}^{\Delta_1} + \Gamma^2_{22\sigma_1}f_{t_2}^{\Delta_2\sigma_1} + C_{\sigma_1}N_{\sigma_1}\right)_{t_1}^{\Delta_1}$$
$$= \left(\Gamma^1_{22\sigma_1}f_{t_1}^{\Delta_1}\right)_{t_1}^{\Delta_1} + \left(\Gamma^2_{22\sigma_1}f_{t_2}^{\Delta_2\sigma_1}\right)_{t_1}^{\Delta_1} + (C_{\sigma_1}N_{\sigma_1})_{t_1}^{\Delta_1},$$

$$\left(\Gamma^1_{22\sigma_1}f_{t_1}^{\Delta_1}\right)_{t_1}^{\Delta_1} = \left(\Gamma^1_{22\sigma_1}\right)_{t_1}^{\Delta_1}f_{t_1}^{\Delta_1} + \Gamma^{1\sigma_1}_{22\sigma_1}f_{t_1}^{\Delta_1^2}$$

$$
\begin{aligned}
&= \left(\Gamma^1_{22\sigma_1}\right)^{\Delta_1}_{t_1} f^{\Delta_1}_{t_1} + \Gamma^{1\sigma_1}_{22\sigma_1}\left(\Gamma^1_{11\sigma_1} f^{\Delta_1}_{t_1} + \Gamma^2_{11\sigma_1} f^{\Delta_2\sigma_1}_{t_2} + A_{\sigma_1} N_{\sigma_1}\right)\\
&= \left(\left(\Gamma^1_{22\sigma_1}\right)^{\Delta_1}_{t_1} f^{\Delta_1}_{t_1} + \Gamma^{1\sigma_1}_{22\sigma_1}\Gamma^1_{11\sigma_1}\right) f^{\Delta_1}_{t_1} + \left(\Gamma^{1\sigma_1}_{22\sigma_1}\Gamma^2_{11\sigma_1}\right) f^{\Delta_2\sigma_1}_{t_2} + \Gamma^{1\sigma_1}_{22\sigma_1} A_{\sigma_1} N_{\sigma_1},\\
\left(\Gamma^2_{22\sigma_1} f^{\Delta_2\sigma_1}_{t_2}\right)^{\Delta_1}_{t_1} &= \left(\Gamma^2_{22\sigma_1}\right)^{\Delta_1}_{t_1} f^{\Delta_2\sigma_1}_{t_2} + \Gamma^{2\sigma_1}_{22\sigma_1} f^{\Delta_2\sigma_1\Delta_1}_{t_2 t_1}\\
&= \left(\Gamma^2_{22\sigma_1}\right)^{\Delta_1}_{t_1} f^{\Delta_2\sigma_1}_{t_2} + \left(1+\mu^{\Delta_1}_1\right)\Gamma^{2\sigma_1}_{22\sigma_1} f^{\Delta_2\sigma_1\Delta_1}_{t_2 t_1}\\
&= \left(\Gamma^2_{22\sigma_1}\right)^{\Delta_1}_{t_1} f^{\Delta_2\sigma_1}_{t_2} + \left(1+\mu^{\Delta_1}_1\right)\Gamma^{2\sigma_1}_{22\sigma_1}\left(U^1_{3\sigma_1} f^{\Delta_1}_{t_1} + U^2_{3\sigma_1} f^{\Delta_2\sigma_1}_{t_2} + U^3_{3\sigma_1} N_{\sigma_1}\right)\\
&= \left(\left(1+\mu^{\Delta_1}_1\right)\Gamma^{2\sigma_1}_{22\sigma_1} U^1_{3\sigma_1}\right) f^{\Delta_1}_{t_1} + \left(\left(\Gamma^2_{22\sigma_1}\right)^{\Delta_1}_{t_1} + \left(1+\mu^{\Delta_1}_1\right)\Gamma^{2\sigma_1}_{22\sigma_1} U^2_{3\sigma_1}\right) f^{\Delta_2\sigma_1}_{t_2}\\
&\quad + \left(1+\mu^{\Delta_1}_1\right)\Gamma^{2\sigma_1}_{22\sigma_1} U^3_{3\sigma_1} N_{\sigma_1},\\
\left(C_{\sigma_1} N_{\sigma_1}\right)^{\Delta_1}_{t_1} &= \left(C_{\sigma_1}\right)^{\Delta_1}_{t_1} N_{\sigma_1} + C^{\sigma_1}_{\sigma_1} N^{\Delta_1}_{\sigma_1 t_1}\\
&= \left(C_{\sigma_1}\right)^{\Delta_1}_{t_1} N_{\sigma_1} + C^{\sigma_1}_{\sigma_1}\left(W^1_{1\sigma_1} f^{\Delta_1}_{t_1} + W^2_{1\sigma_1} f^{\Delta_2\sigma_1}_{t_2}\right)\\
&= \left(C^{\sigma_1}_{\sigma_1} W^1_{1\sigma_1}\right) f^{\Delta_1}_{t_1} + \left(C^{\sigma_1}_{\sigma_1} W^2_{1\sigma_1}\right) f^{\Delta_2\sigma_1}_{t_2} + \left(C_{\sigma_1}\right)^{\Delta_1}_{t_1} N_{\sigma_1}.
\end{aligned}
$$

Therefore

$$
\begin{aligned}
f^{\Delta_2^2\Delta_1}_{t_2 t_1} &= \left(\left(\Gamma^1_{22\sigma_1}\right)^{\Delta_1}_{t_1} + \Gamma^{1\sigma_1}_{22\sigma_1}\Gamma^1_{11\sigma_1} + \left(1+\mu^{\Delta_1}_1\right) U^1_{3\sigma_1} + C^{\sigma_1}_{\sigma_1} W^1_{1\sigma_1}\right) f^{\Delta_1}_{t_1}\\
&\quad + \left(\Gamma^{1\sigma_1}_{22\sigma_1}\Gamma^2_{11\sigma_1} + \left(\Gamma^2_{22\sigma_1}\right)^{\Delta_1}_{t_1} + \left(1+\mu^{\Delta_1}_1\right)\Gamma^{2\sigma_1}_{22\sigma_1} U^2_{3\sigma_1} + C^{\sigma_1}_{\sigma_1} W^2_{1\sigma_1}\right) f^{\Delta_2\sigma_1}_{t_2} \qquad (5.11)\\
&\quad + \left(\Gamma^{1\sigma_1}_{22\sigma_1} A_{\sigma_1} + \left(1+\mu^{\Delta_1}_1\right)\Gamma^{2\sigma_1}_{22\sigma_1} U^3_{3\sigma_1} + \left(C_{\sigma_1}\right)^{\Delta_1}_{t_1}\right) N_{\sigma_1}.
\end{aligned}
$$

Also,

$$
\begin{aligned}
f^{\Delta_1\Delta_2^2}_{t_1 t_2} &= \left(\Gamma^1_{12\sigma_1} f^{\Delta_1}_{t_1} + \Gamma^2_{12\sigma_1} f^{\Delta_2\sigma_1}_{t_2} + B_{\sigma_1} N_{\sigma_1}\right)^{\Delta_2}_{t_2}\\
&= \left(\Gamma^1_{12\sigma_1} f^{\Delta_1}_{t_1}\right)^{\Delta_2}_{t_2} + \left(\Gamma^2_{12\sigma_1} f^{\Delta_2\sigma_1}_{t_2}\right)^{\Delta_2}_{t_2} + \left(B_{\sigma_1} N_{\sigma_1}\right)^{\Delta_2}_{t_2}
\end{aligned}
$$

and

$$
\begin{aligned}
\left(\Gamma^1_{12\sigma_1} f^{\Delta_1}_{t_1}\right)^{\Delta_2}_{t_2} &= \left(\Gamma^1_{12\sigma_1}\right)^{\Delta_2}_{t_2} f^{\Delta_1}_{t_1} + \Gamma^{1\sigma_2}_{12\sigma_1} f^{\Delta_1\Delta_2}_{t_1 t_2}\\
&= \left(\Gamma^1_{12\sigma_1}\right)^{\Delta_2}_{t_2} f^{\Delta_1}_{t_1} + \Gamma^{1\sigma_2}_{12\sigma_1}\left(\Gamma^1_{12\sigma_1} f^{\Delta_1}_{t_1} + \Gamma^2_{12\sigma_1} f^{\Delta_2\sigma_1}_{t_2} + B_{\sigma_1} N_{\sigma_1}\right)\\
&= \left(\left(\Gamma^1_{12\sigma_1}\right)^{\Delta_2}_{t_2} + \Gamma^{1\sigma_2}_{12\sigma_1}\Gamma^1_{12\sigma_1}\right) f^{\Delta_1}_{t_1} + \left(\Gamma^{1\sigma_2}_{12\sigma_1}\Gamma^2_{12\sigma_1}\right) f^{\Delta_2\sigma_1}_{t_2} + \left(\Gamma^{1\sigma_2}_{12\sigma_1} B_{\sigma_1}\right) N_{\sigma_1},\\
\left(\Gamma^2_{12\sigma_1} f^{\Delta_2\sigma_1}_{t_2}\right)^{\Delta_2}_{t_2} &= \left(\Gamma^2_{12\sigma_1}\right)^{\Delta_2}_{t_2} f^{\Delta_2\sigma_1}_{t_2} + \Gamma^{2\sigma_2}_{12\sigma_1} f^{\Delta_2^2\sigma_1}_{t_2}\\
&= \left(\Gamma^2_{12\sigma_1}\right)^{\Delta_2}_{t_2} f^{\Delta_2\sigma_1}_{t_2} + \Gamma^{2\sigma_2}_{12\sigma_1}\left(U^1_{5\sigma_1} f^{\Delta_1}_{t_1} + U^2_{5\sigma_1} f^{\Delta_2\sigma_1}_{t_2} + U^3_{5\sigma_1} N_{\sigma_1}\right)\\
&= \left(\Gamma^{2\sigma_2}_{12\sigma_1} U^1_{5\sigma_1}\right) f^{\Delta_1}_{t_1} + \left(\left(\Gamma^2_{12\sigma_1}\right)^{\Delta_2}_{t_2} + \Gamma^{2\sigma_2}_{12\sigma_1} U^2_{5\sigma_1}\right) f^{\Delta_2\sigma_1}_{t_2} + \left(\Gamma^{2\sigma_2}_{12\sigma_1} U^3_{5\sigma_1}\right) N_{\sigma_1},\\
\left(B_{\sigma_1} N_{\sigma_1}\right)^{\Delta_2}_{t_2} &= \left(B_{\sigma_1}\right)^{\Delta_2}_{t_2} N_{\sigma_1} + B^{\sigma_2}_{\sigma_1}\left(N_{\sigma_1}\right)^{\Delta_2}_{t_2}
\end{aligned}
$$

$$= (B_{\sigma_1})_{t_2}^{\Delta_2} N_{\sigma_1} + B_{\sigma_1}^{\sigma_2} \left(W_{2\sigma_1}^1 f_{t_1}^{\Delta_1} + W_{2\sigma_1}^2 f_{t_2}^{\Delta_2\sigma_1}\right)$$
$$= \left(B_{\sigma_1}^{\sigma_2} W_{2\sigma_1}^1\right) f_{t_1}^{\Delta_1} + \left(B_{\sigma_1}^{\sigma_2} W_{2\sigma_1}^2\right) f_{t_2}^{\Delta_2\sigma_1} + (B_{\sigma_1})_{t_2}^{\Delta_2} N_{\sigma_1}.$$

Consequently

$$\begin{aligned} f_{t_1t_2}^{\Delta_1\Delta_2^2} &= \left(\left(\Gamma_{12\sigma_1}^1\right)_{t_2}^{\Delta_2} + \Gamma_{12\sigma_1}^{1\sigma_2}\Gamma_{12\sigma_1}^1 + \Gamma_{12\sigma_1}^{2\sigma_1} U_{5\sigma_1}^1 + B_{\sigma_1}^{\sigma_2} W_{2\sigma_1}^1\right) f_{t_1}^{\Delta_1} \\ &+ \left(\Gamma_{12\sigma_1}^{1\sigma_2}\Gamma_{12\sigma_1}^2 + \left(\Gamma_{12\sigma_1}^2\right)_{t_2}^{\Delta_2} + \Gamma_{12\sigma_1}^{2\sigma_2} U_{5\sigma_1}^2 + B_{\sigma_1}^{\sigma_2} W_{2\sigma_1}^2\right) f_{t_2}^{\Delta_2\sigma_1} \\ &+ \left(\Gamma_{12\sigma_1}^{1\sigma_2} B_{\sigma_1} + \Gamma_{12\sigma_1}^{2\sigma_2} U_{5\sigma_1}^3 + (B_{\sigma_1})_{t_2}^{\Delta_2}\right) N_{\sigma_1}. \end{aligned} \tag{5.12}$$

By the last equation and (5.11), (5.9), using that $f_{t_1}^{\Delta_1}$, $f_{t_2}^{\Delta_2\sigma_1}$ and N_{σ_1} are linearly independent, we get the third and fourth equations of the σ_1-Gauss equations and the second equation of the σ_1-Godazzi-Mainardi equations. This completes the proof.

Remark 5.1 Suppose $f \in \mathscr{C}^3$ and S is a σ_1-regular surface. If we take

$$\begin{aligned} (B_{\sigma_1} N_{\sigma_1})_{t_2}^{\Delta_2} &= (B_{\sigma_1})_{t_2}^{\Delta_2} N_{\sigma_1}^{\sigma_2} + B_{\sigma_1} (N_{\sigma_1})_{t_2}^{\Delta_2} \\ &= (B_{\sigma_1})_{t_2}^{\Delta_2} \left(\mu_2 W_{2\sigma_1}^1 f_{t_1}^{\Delta_1} + \mu_2 W_{2\sigma_1}^2 f_{t_2}^{\Delta_2\sigma_1} + N_{\sigma_1}\right) \\ &+ B_{\sigma_1} \left(W_{2\sigma_1}^1 f_{t_1}^{\Delta_1} + W_{2\sigma_1}^2 f_{t_2}^{\Delta_2\sigma_1}\right) \\ &= \left(\mu_2 (B_{\sigma_1})_{t_2}^{\Delta_2} W_{2\sigma_1}^1 + B_{\sigma_1} W_{2\sigma_1}^1\right) f_{t_1}^{\Delta_1} + \left(\mu_2 (B_{\sigma_1})_{t_2}^{\Delta_2} W_{2\sigma_1}^2 + B_{\sigma_1} W_{2\sigma_1}^2\right) f_{t_2}^{\Delta_2\sigma_1} \\ &+ (B_{\sigma_1})_{t_2}^{\Delta_2} N_{\sigma_1}, \end{aligned}$$

then (5.12) takes the form

$$\begin{aligned} f_{t_1t_2}^{\Delta_1\Delta_2^2} &= \left(\left(\Gamma_{12\sigma_1}^1\right)_{t_2}^{\Delta_2} + \Gamma_{12\sigma_1}^{1\sigma_2}\Gamma_{12\sigma_1}^1 + \Gamma_{12\sigma_1}^{2\sigma_1} U_{5\sigma_1}^1 + \mu_2 (B_{\sigma_1})_{t_2}^{\Delta_2} W_{2\sigma_1}^1 + B_{\sigma_1} W_{2\sigma_1}^1\right) f_{t_1}^{\Delta_1} \\ &+ \left(\Gamma_{12\sigma_1}^{1\sigma_2}\Gamma_{12\sigma_1}^2 + \left(\Gamma_{12\sigma_1}^2\right)_{t_2}^{\Delta_2} + \Gamma_{12\sigma_1}^{2\sigma_2} U_{5\sigma_1}^2 + \mu_2 (B_{\sigma_1})_{t_2}^{\Delta_2} W_{2\sigma_1}^2 + B_{\sigma_1} W_{2\sigma_1}^2\right) f_{t_2}^{\Delta_2\sigma_1} \\ &+ \left(\Gamma_{12\sigma_1}^{1\sigma_2} B_{\sigma_1} + \Gamma_{12\sigma_1}^{2\sigma_2} U_{5\sigma_1}^3 + (B_{\sigma_1})_{t_2}^{\Delta_2}\right) N_{\sigma_2}. \end{aligned}$$

and the third and fourth equations of the σ_1-Gauss equations take the form

$$\begin{aligned} \left(\Gamma_{12\sigma_1}^1\right)_{t_2}^{\Delta_2} - \left(\Gamma_{22\sigma_1}^1\right)_{t_1}^{\Delta_1} &= -\Gamma_{12\sigma_1}^{1\sigma_2}\Gamma_{12\sigma_1}^1 - \Gamma_{12\sigma_1}^{2\sigma_2} U_{5\sigma_1}^1 - \mu_2 (B_{\sigma_1})_{t_2}^{\Delta_2} W_{2\sigma_1}^1 - B_{\sigma_1} W_{2\sigma_1}^1 \\ &+ \Gamma_{22\sigma_1}^{1\sigma_1}\Gamma_{11\sigma_1}^1 + \left(1 + \mu_1^{\Delta_1}\right) U_{3\sigma_1}^1 + C_{\sigma_1}^{\sigma_1} W_{1\sigma_1}^1 \\ \left(\Gamma_{12\sigma_1}^2\right)_{t_2}^{\Delta_2} - \left(\Gamma_{22\sigma_1}^2\right)_{t_1}^{\Delta_1} &= -\Gamma_{12\sigma_1}^{1\sigma_2}\Gamma_{12\sigma_1}^2 - \Gamma_{12\sigma_1}^{2\sigma_2} U_{5\sigma_1}^2 - \mu_2 (B_{\sigma_1})_{t_2}^{\Delta_2} W_{2\sigma_1}^2 - B_{\sigma_1} W_{2\sigma_1}^2 \\ &+ \Gamma_{22\sigma_1}^{1\sigma_1}\Gamma_{11\sigma_1}^2 + \left(1 + \mu_1^{\Delta_1}\right) \Gamma_{22\sigma_1}^{2\sigma_1} U_{3\sigma_1}^2 + C_{\sigma_1}^{\sigma_1} W_{1\sigma_1}^2. \end{aligned}$$

Exercise 5.9 Suppose $f \in \mathscr{C}^3$ and S is a σ_1-regular surface. Use

$$\left(\Gamma_{11\sigma_1}^1 f_{t_1}^{\Delta_1}\right)_{t_2}^{\Delta_2} = \left(\Gamma_{11\sigma_1}^1\right)_{t_2}^{\Delta_2} f_{t_1}^{\Delta_1\sigma_2} + \Gamma_{11\sigma_1}^1 f_{t_1t_2}^{\Delta_1\Delta_2}$$

or

$$\left(\Gamma^2_{11\sigma_1}f^{\Delta_2\sigma_1}_{t_2}\right)^{\Delta_2}_{t_2} = \left(\Gamma^2_{11\sigma_1}\right)^{\Delta_2}_{t_2} f^{\Delta_2\sigma_1\sigma_2}_{t_2} + \Gamma^2_{11\sigma_1}f^{\Delta_2^2\sigma_1}_{t_2},$$

or

$$(A_{\sigma_1}N_{\sigma_1})^{\Delta_2}_{t_2} = (A_{\sigma_1})^{\Delta_2}_{t_2} N^{\sigma_2}_{\sigma_1} + A_{\sigma_1}(N_{\sigma_1})^{\Delta_2}_{t_2},$$

or

$$\left(\Gamma^1_{12\sigma_1}f^{\Delta_1}_{t_1}\right)^{\Delta_1}_{t_1} = \left(\Gamma^1_{12\sigma_1}\right)^{\Delta_1}_{t_1} f^{\Delta_1\sigma_1}_{t_1} + \Gamma^1_{12\sigma_1}f^{\Delta_1^2}_{t_1},$$

or

$$\left(\Gamma^2_{12\sigma_1}f^{\Delta_2\sigma_1}_{t_2}\right)^{\Delta_1}_{t_1} = \left(\Gamma^2_{12\sigma_1}\right)^{\Delta_1}_{t_1} f^{\Delta_2\sigma_1\sigma_1}_{t_2} + \Gamma^2_{12\sigma_1}f^{\Delta_2\sigma_1\Delta_1}_{t_2t_1},$$

or

$$(B_{\sigma_1}N_{\sigma_1})^{\Delta_1}_{t_1} = (B_{\sigma_1})^{\Delta_1}_{t_1} N^{\sigma_1}_{\sigma_1} + B_{\sigma_1}(N_{\sigma_1})^{\Delta_1}_{t_1},$$

or

$$\left(\Gamma^1_{22\sigma_1}f^{\Delta_1}_{t_1}\right)^{\Delta_1}_{t_1} = \left(\Gamma^1_{22\sigma_1}\right)^{\Delta_1}_{t_1} f^{\Delta_1\sigma_1}_{t_1} + \Gamma^1_{22\sigma_1}f^{\Delta_1^2}_{t_1},$$

or

$$\left(\Gamma^2_{22\sigma_1}f^{\Delta_2\sigma_1}_{t_2}\right)^{\Delta_1}_{t_1} = \left(\Gamma^2_{22\sigma_1}\right)^{\Delta_1}_{t_1} f^{\Delta_2\sigma_1\sigma_1}_{t_2} + \Gamma^2_{22\sigma_1}f^{\Delta_2\sigma_1\Delta_1}_{t_2},$$

or

$$(C_{\sigma_1}N_{\sigma_1})^{\Delta_1}_{t_1} = (C_{\sigma_1})^{\Delta_1}_{t_1} N^{\sigma_1}_{\sigma_1} + C_{\sigma_1}N^{\Delta_1}_{\sigma_1},$$

or

$$\left(\Gamma^1_{12\sigma_1}f^{\Delta_1}_{t_1}\right)^{\Delta_2}_{t_2} = \left(\Gamma^1_{12\sigma_1}\right)^{\Delta_2}_{t_2} f^{\Delta_1\sigma_2}_{t_1} + \Gamma^1_{12\sigma_1}f^{\Delta_1\Delta_2}_{t_1t_2},$$

or

$$\left(\Gamma^2_{12\sigma_1}f^{\Delta_2\sigma_1}_{t_2}\right)^{\Delta_2}_{t_2} = \left(\Gamma^2_{12\sigma_1}\right)^{\Delta_2}_{t_2} f^{\Delta_2\sigma_1\sigma_2}_{t_2} + \Gamma^2_{12\sigma_1}f^{\Delta_2^2\sigma_1}_{t_2},$$

or

$$(B_{\sigma_1}N_{\sigma_1})^{\Delta_2}_{t_2} = (B_{\sigma_1})^{\Delta_2}_{t_2} N^{\sigma_2}_{\sigma_1} + B_{\sigma_1}(N_{\sigma_1})^{\Delta_2}_{t_2}$$

to deduce the other σ_1-Gauss equations and the σ_1-Godazzi-Mainardi equations.

Remark 5.2 Suppose $f \in \mathscr{C}^3$, S is a σ_1-regular surface and σ_1 is Δ_1-differentiable and σ_2 is Δ_2-differentiable. Then we can use the relations

$$\left(1+\mu_1^{\Delta_1}\right)f^{\Delta_1^2\sigma_1\Delta_2}_{t_1t_2} = f^{\Delta_1\Delta_2\sigma_1\Delta_1}_{t_1},$$

$$\left(1+\mu_2^{\Delta_2}\right)f^{\Delta_1^2\Delta_2\sigma_2}_{t_1t_2} = f^{\Delta_1^2\sigma_2\Delta_2}_{t_1t_2},$$

and

$$\left(1+\mu_1^{\Delta_1}\right)f^{\Delta_1\Delta_2^2\sigma_1}_{t_1t_2} = f^{\Delta_1\Delta_2\sigma_1\Delta_1}_{t_1t_2},$$

$$\left(1+\mu_2^{\Delta_2}\right)f^{\Delta_1\Delta_2^2\sigma_2}_{t_1t_2} = f^{\Delta_1\Delta_2\sigma_2\Delta_2}_{t_1t_2}$$

to deduce other kinds of the σ_1-Gauss equations and the σ_1-Godazzi-Mainardi edquations.

Theorem 5.2 *Suppose $f \in \mathscr{C}^3$ and S is a σ_1-regular surface. Then*

$$\begin{aligned} \mu_1\left(U^j_{2\sigma_1} - \Gamma^j_{11\sigma_1}\right) &= \mu_2\left(U^j_{3\sigma_1} - \Gamma^j_{12\sigma_1}\right) \\ \mu_1\left(U^j_{4\sigma_1} - \Gamma^j_{12\sigma_1}\right) &= \mu_2\left(U^j_{5\sigma_1} - \Gamma^j_{22\sigma_1}\right), \end{aligned} \tag{5.13}$$

$j \in \{1,2\}$, and

$$\begin{aligned} \mu_1\left(U^3_{2\sigma_1} - A_{\sigma_1}\right) &= \mu_2\left(U^3_{3\sigma_1} - B_{\sigma_1}\right) \\ \mu_1\left(U^3_{4\sigma_1} - B_{\sigma_1}\right) &= \mu_2\left(U^3_{5\sigma_1} - C_{\sigma_1}\right). \end{aligned} \tag{5.14}$$

Definition 5.11 The equations (5.13) will be called the $\sigma_1\mu_1\mu_2$-Gauss equations.

Definition 5.12 The equations (5.14) will be called the $\sigma_1\mu_1\mu_2$-Godazzi-Mainardi equations.

Proof. We will use the relations (5.8) and (5.9). We have

$$\begin{aligned} f^{\Delta_1^2\sigma_2}_{t_1} &= U^1_{2\sigma_1} f^{\Delta_1}_{t_1} + U^2_{2\sigma_1} f^{\Delta_2\sigma_1}_{t_2} + U^3_{2\sigma_1} N_{\sigma_1} \\ &= f^{\Delta_1^2}_{t_1} + \mu_2 f^{\Delta_1^2\Delta_2}_{t_1t_2} \\ &= \Gamma^1_{11\sigma_1} f^{\Delta_1}_{t_1} + \Gamma^2_{11\sigma_1} f^{\Delta_2\sigma_1}_{t_2} + A_{\sigma_1} N_{\sigma_1} + \mu_2 f^{\Delta_1^2\Delta_2}_{t_1t_2}. \end{aligned}$$

Therefore

$$\left(U^1_{2\sigma_1} - \Gamma^1_{11\sigma_1}\right) f^{\Delta_1}_{t_1} + \left(U^2_{2\sigma_1} - \Gamma^2_{11\sigma_1}\right) f^{\Delta_2\sigma_1}_{t_2} + \left(U^3_{2\sigma_1} - A_{\sigma_1}\right) N_{\sigma_1} = \mu_2 f^{\Delta_1^2\Delta_2}_{t_1t_2}$$

and

$$\mu_1\left(U^1_{2\sigma_1} - \Gamma^1_{11\sigma_1}\right) f^{\Delta_1}_{t_1} + \mu_1\left(U^2_{2\sigma_1} - \Gamma^2_{11\sigma_1}\right) f^{\Delta_2\sigma_1}_{t_2} + \mu_1\left(U^3_{2\sigma_1} - A_{\sigma_1}\right) N_{\sigma_1} = \mu_1\mu_2 f^{\Delta_1^2\Delta_2}_{t_1t_2}. \tag{5.15}$$

Next,

$$f^{\Delta_1\Delta_2\sigma_1}_{t_1t_2} = U^1_{3\sigma_1} f^{\Delta_1}_{t_1} + U^2_{3\sigma_1} f^{\Delta_2\sigma_1}_{t_2} + U^3_{3\sigma_1} N_{\sigma_1}$$

and

$$\begin{aligned} f^{\Delta_1\Delta_2\sigma_1}_{t_1t_2} &= f^{\Delta_1\Delta_2}_{t_1t_2} + \mu_1 f^{\Delta_1^2\Delta_2}_{t_1t_2} \\ &= \Gamma^1_{12\sigma_1} f^{\Delta_1}_{t_1} + \Gamma^2_{12\sigma_1} f^{\Delta_2\sigma_1}_{t_2} + B_{\sigma_1} N_{\sigma_1} + \mu_1 f^{\Delta_1^2\Delta_2}_{t_1t_2}. \end{aligned}$$

Thus,

$$\left(U^1_{3\sigma_1} - \Gamma^1_{12\sigma_1}\right) f^{\Delta_1}_{t_1} + \left(U^2_{3\sigma_1} - \Gamma^2_{12\sigma_1}\right) f^{\Delta_2\sigma_1}_{t_2} + \left(U^3_{3\sigma_1} - B_{\sigma_1}\right) N_{\sigma_1} = \mu_1 f^{\Delta_1^2\Delta_2}_{t_1t_2}$$

and

$$\mu_2\left(U^1_{3\sigma_1} - \Gamma^1_{12\sigma_1}\right) f^{\Delta_1}_{t_1} + \mu_2\left(U^2_{3\sigma_1} - \Gamma^2_{12\sigma_1}\right) f^{\Delta_2\sigma_1}_{t_2} + \mu_2\left(U^3_{3\sigma_1} - B_{\sigma_1}\right) N_{\sigma_1} = \mu_1\mu_2 f^{\Delta_1^2\Delta_2}_{t_1t_2}.$$

By the last equation and by (5.15), using that $f^{\Delta_1}_{t_1}$, $f^{\Delta_2\sigma_1}_{t_2}$ and N_{σ_1} are linearly independent, we find the first and second equations of the $\sigma_1\mu_1\mu_2$-Gauss equations and the first equation of the $\sigma_1\mu_1\mu_2$-Godazzi-Mainardi equations. Moreover,

$$\begin{aligned} f_{t_2}^{\Delta_2^2\sigma_1} &= f_{t_2}^{\Delta_2^2} + \mu_1 f_{t_2 t_1}^{\Delta_2^2\Delta_1} \\ &= \Gamma^1_{22\sigma_1} f_{t_1}^{\Delta_1} + \Gamma^2_{22\sigma_1} f_{t_2}^{\Delta_2\sigma_1} + C_{\sigma_1} N_{\sigma_1} + \mu_1 f_{t_2 t_1}^{\Delta_2^2\Delta_1} \\ &= U^1_{5\sigma_1} f_{t_1}^{\Delta_1} + U^2_{5\sigma_1} f_{t_2}^{\Delta_2\sigma_1} + U^3_{5\sigma_1} N_{\sigma_1}, \end{aligned}$$

from where

$$\mu_2\left(U^1_{5\sigma_1} - \Gamma^1_{22\sigma_1}\right) f_{t_1}^{\Delta_1} + \mu_2\left(U^2_{5\sigma_1} - \Gamma^2_{22\sigma_1}\right) f_{t_2}^{\Delta_2\sigma_1} + \mu_2\left(U^3_{5\sigma_1} - C_{\sigma_1}\right) N_{\sigma_1} = \mu_1\mu_2 f_{t_2 t_1}^{\Delta_2^2\Delta_1}. \tag{5.16}$$

Also,

$$\begin{aligned} f_{t_1 t_2}^{\Delta_1\Delta_2\sigma_2} &= f_{t_1 t_2}^{\Delta_1\Delta_2} + \mu_2 f_{t_1 t_2}^{\Delta_1\Delta_2^2} \\ &= \Gamma^1_{12\sigma_1} f_{t_1}^{\Delta_1} + \Gamma^2_{12\sigma_1} f_{t_2}^{\Delta_2\sigma_1} + B_{\sigma_1} N_{\sigma_1} + \mu_2 f_{t_1 t_2}^{\Delta_1\Delta_2^2} \\ &= U^1_{4\sigma_1} f_{t_1}^{\Delta_1} + U^2_{4\sigma_1} f_{t_2}^{\Delta_2\sigma_1} + U^3_{4\sigma_1} N_{\sigma_1}, \end{aligned}$$

whereupon

$$\mu_1\left(U^1_{4\sigma_1} - \Gamma^1_{12\sigma_1}\right) f_{t_1}^{\Delta_1} + \mu_1\left(U^2_{4\sigma_1} - \Gamma^2_{12\sigma_1}\right) f_{t_2}^{\Delta_2\sigma_1} + \mu_1\left(U^3_{4\sigma_1} - B_{\sigma_1}\right) = \mu_1\mu_2 f_{t_1 t_2}^{\Delta_1\Delta_2^2}.$$

By the last equation, (5.16) and using that $f_{t_1}^{\Delta_1}, f_{t_2}^{\Delta_2\sigma_1}$ and N_{σ_1} are linearly independent, we get the third and fourth equations of the $\sigma_1\mu_1\mu_2$-Gauss equations and the second equation of the $\sigma_1\mu_1\mu_2$-Godazzi-Mainardi equations. This completes the proof.

Theorem 5.3 *Suppose $f \in \mathscr{C}^3$, S is a σ_2-regular surface, σ_1 is Δ_1-differentiable and σ_2 is Δ_2-differentiable. Then we have the following equations*

$$\begin{aligned} \left(\Gamma^1_{11\sigma_2}\right)_{t_2}^{\Delta_2} - \left(\Gamma^1_{12\sigma_2}\right)_{t_1}^{\Delta_1} &= -\left(1+\mu_2^{\Delta_2}\right)\Gamma^{1\sigma_2}_{11\sigma_2} U^1_{4\sigma_2} - \Gamma^{2\sigma_2}_{11\sigma_2}\Gamma^1_{22\sigma_2} - A^{\sigma_2}_{\sigma_2} W^1_{2\sigma_2} \\ &\quad + \Gamma^{1\sigma_1}_{12\sigma_2} U^1_{2\sigma_2} + \Gamma^{2\sigma_1}_{12\sigma_2}\Gamma^1_{12\sigma_2} + B^{\sigma_1}_{\sigma_2} W^1_{1\sigma_2} \\ \left(\Gamma^2_{12\sigma_2}\right)_{t_2}^{\Delta_2} - \left(\Gamma^2_{12\sigma_2}\right)_{t_1}^{\Delta_1} &= -\left(1+\mu_2^{\Delta_2}\right)\Gamma^{1\sigma_2}_{11\sigma_2} U^2_{4\sigma_2} - \Gamma^{2\sigma_2}_{11\sigma_2}\Gamma^2_{22\sigma_2} - A^{\sigma_2}_{\sigma_2} W^2_{2\sigma_2} \\ &\quad + \Gamma^{1\sigma_1}_{12\sigma_2} U^2_{2\sigma_2} + \Gamma^{2\sigma_1}_{12\sigma_2}\Gamma^2_{12\sigma_2} + B^{\sigma_1}_{\sigma_2} W^2_{1\sigma_2} \\ \left(\Gamma^1_{22\sigma_2}\right)_{t_1}^{\Delta_1} - \left(\Gamma^1_{12\sigma_2}\right)_{t_2}^{\Delta_2} &= -\Gamma^{1\sigma_1}_{22\sigma_2} U^1_{2\sigma_2} - \Gamma^{2\sigma_1}_{22\sigma_2}\Gamma^2_{12\sigma_2} - C^{\sigma_1}_{\sigma_2} W^2_{1\sigma_2} \\ &\quad + \left(1+\mu_2^{\Delta_2}\right)\Gamma^{1\sigma_2}_{12\sigma_2} U^1_{4\sigma_2} + \Gamma^{2\sigma_2}_{12\sigma_2}\Gamma^1_{22\sigma_2} + B^{\sigma_2}_{\sigma_2} W^1_{2\sigma_2} \\ \left(\Gamma^1_{22\sigma_2}\right)_{t_1}^{\Delta_1} - \left(\Gamma^2_{12\sigma_2}\right)_{t_2}^{\Delta_2} &= -\Gamma^{1\sigma_1}_{22\sigma_2} U^2_{2\sigma_2} - \Gamma^{2\sigma_1}_{22\sigma_2}\Gamma^2_{12\sigma_2} - C^{\sigma_1}_{\sigma_2} W^2_{1\sigma_2} \\ &\quad + \left(1+\mu_2^{\Delta_2}\right)\Gamma^{1\sigma_1}_{12\sigma_2} U^2_{4\sigma_2} + \Gamma^{2\sigma_2}_{12\sigma_2}\Gamma^2_{22\sigma_2} + B^{\sigma_2}_{\sigma_2} W^2_{2\sigma_2} \end{aligned} \tag{5.17}$$

and

$$\left(A_{\sigma_2}\right)_{t_2}^{\Delta_2} - \left(B_{\sigma_2}\right)_{t_1}^{\Delta_1} = -\left(1+\mu_2^{\Delta_2}\right)\Gamma^{1\sigma_2}_{11\sigma_2} U^3_{4\sigma_2} - C_{\sigma_2}\Gamma^{2\sigma_2}_{11\sigma_2}$$

$$
\begin{aligned}
&+\Gamma^{1\sigma_1}_{12\sigma_2}U^3_{2\sigma_2}+\Gamma^{2\sigma_1}_{12\sigma_2}B_{\sigma_2}\\
(C_{\sigma_2})^{\Delta_1}_{t_1}-(B_{\sigma_2})^{\Delta_2}_{t_2} &= -\Gamma^{1\sigma_1}_{22\sigma_2}U^3_{2\sigma_2}-\Gamma^{2\sigma_1}_{22\sigma_2}B_{\sigma_2}\\
&+\left(1+\mu_2^{\Delta_2}\right)\Gamma^{1\sigma_2}_{12\sigma_2}U^3_{4\sigma_2}+\Gamma^{2\sigma_2}_{12\sigma_2}C_{\sigma_2}.
\end{aligned}
\tag{5.18}
$$

Definition 5.13 The equations (5.17) will be called the σ_2-Gauss equations.

Definition 5.14 The equations (5.18) will be called the σ_2-Godazzi-Mainardi equations.

Proof. We have

$$
\begin{aligned}
f^{\Delta_1^2\Delta_2}_{t_1t_2} &= \left(\Gamma^1_{11\sigma_2}f^{\Delta_1\sigma_2}_{t_1}+\Gamma^2_{11\sigma_2}f^{\Delta_2}_{t_2}+A_{\sigma_2}N_{\sigma_2}\right)^{\Delta_2}_{t_2}\\
&= \left(\Gamma^1_{11\sigma_2}f^{\Delta_1\sigma_2}_{t_1}\right)^{\Delta_2}_{t_2}+\left(\Gamma^2_{11\sigma_2}f^{\Delta_2}_{t_2}\right)^{\Delta_2}_{t_2}+\left(A_{\sigma_2}N_{\sigma_2}\right)^{\Delta_2}_{t_2}\\
&= \left(\Gamma^1_{11\sigma_2}\right)^{\Delta_2}_{t_2}f^{\Delta_1\sigma_2}_{t_1}+\Gamma^{1\sigma_2}_{11\sigma_2}f^{\Delta_1\sigma_2\Delta_2}_{t_1t_2}+\left(\Gamma^2_{11\sigma_2}\right)^{\Delta_2}_{t_2}f^{\Delta_2}_{t_2}\\
&\quad+\Gamma^{2\sigma_2}_{11\sigma_2}f^{\Delta_2^2}_{t_2}+\left(A_{\sigma_2}\right)^{\Delta_2}_{t_2}N_{\sigma_2}+A^{\sigma_2}_{\sigma_2}N^{\Delta_2}_{\sigma_2t_2}\\
&= \left(\Gamma^1_{11\sigma_2}\right)^{\Delta_2}_{t_2}f^{\Delta_1\sigma_2}_{t_1}+\left(1+\mu_2^{\Delta_2}\right)\Gamma^{1\sigma_2}_{11\sigma_2}f^{\Delta_1\Delta_2\sigma_2}_{t_1t_2}+\left(\Gamma^2_{11\sigma_2}\right)^{\Delta_2}_{t_2}f^{\Delta_2}_{t_2}\\
&\quad+\Gamma^{2\sigma_2}_{11\sigma_2}\left(\Gamma^1_{22\sigma_2}f^{\Delta_1\sigma_2}_{t_1}+\Gamma^2_{22\sigma_2}f^{\Delta_2}_{t_2}+C_{\sigma_2}N_{\sigma_2}\right)\\
&\quad+\left(A_{\sigma_2}\right)^{\Delta_2}_{t_2}N_{\sigma_2}+A^{\sigma_2}_{\sigma_2}\left(W^1_{2\sigma_2}f^{\Delta_1\sigma_2}_{t_1}+W^2_{2\sigma_2}f^{\Delta_2}_{t_2}\right)\\
&= \left(\Gamma^1_{11\sigma_2}\right)^{\Delta_2}_{t_2}f^{\Delta_1\sigma_2}_{t_1}+\left(1+\mu_2^{\Delta_2}\right)\Gamma^{1\sigma_2}_{11\sigma_2}\left(U^1_{4\sigma_2}f^{\Delta_1\sigma_2}_{t_1}+U^2_{4\sigma_2}f^{\Delta_2}_{t_2}+U^3_{4\sigma_2}N_{\sigma_2}\right)\\
&\quad+\left(\Gamma^2_{11\sigma_2}\right)^{\Delta_2}_{t_2}f^{\Delta_2}_{t_2}+\Gamma^{2\sigma_2}_{11\sigma_2}\left(\Gamma^1_{22\sigma_2}f^{\Delta_1\sigma_2}_{t_1}+\Gamma^2_{22\sigma_2}f^{\Delta_2}_{t_2}+C_{\sigma_2}N_{\sigma_2}\right)\\
&\quad+\left(A_{\sigma_2}\right)^{\Delta_2}_{t_2}N_{\sigma_2}+A^{\sigma_2}_{\sigma_2}\left(W^1_{2\sigma_2}f^{\Delta_1\sigma_2}_{t_1}+W^2_{2\sigma_2}f^{\Delta_2}_{t_2}\right)\\
&= \left(\left(\Gamma^1_{11\sigma_2}\right)^{\Delta_2}_{t_2}+\left(1+\mu_2^{\Delta_2}\right)\Gamma^{1\sigma_2}_{11\sigma_2}U^1_{4\sigma_2}+\Gamma^{2\sigma_2}_{11\sigma_2}\Gamma^1_{22\sigma_2}+A^{\sigma_2}_{\sigma_2}W^1_{2\sigma_2}\right)f^{\Delta_1\sigma_2}_{t_1}\\
&\quad+\left(\left(\Gamma^2_{11\sigma_2}\right)^{\Delta_2}_{t_2}+\left(1+\mu_2^{\Delta_2}\right)\Gamma^{1\sigma_2}_{11\sigma_2}U^2_{4\sigma_2}+\Gamma^{2\sigma_2}_{11\sigma_2}\Gamma^2_{22\sigma_2}+A^{\sigma_2}_{\sigma_2}W^2_{2\sigma_2}\right)f^{\Delta_2}_{t_2}\\
&\quad+\left(\left(A_{\sigma_2}\right)^{\Delta_2}_{t_2}+\left(1+\mu_2^{\Delta_2}\right)\Gamma^{1\sigma_2}_{11\sigma_2}U^3_{4\sigma_2}+C_{\sigma_2}\Gamma^{2\sigma_2}_{11\sigma_2}\right)N_{\sigma_2}
\end{aligned}
$$

and

$$
\begin{aligned}
f^{\Delta_1\Delta_2\Delta_1}_{t_1t_2t_1} &= \left(\Gamma^1_{12\sigma_2}f^{\Delta_1\sigma_2}_{t_1}+\Gamma^2_{12\sigma_2}f^{\Delta_2}_{t_2}+B_{\sigma_2}N_{\sigma_2}\right)^{\Delta_1}_{t_1}\\
&= \left(\Gamma^1_{12\sigma_2}f^{\Delta_1\sigma_2}_{t_1}\right)^{\Delta_1}_{t_1}+\left(\Gamma^2_{12\sigma_2}f^{\Delta_2}_{t_2}\right)^{\Delta_1}_{t_1}+\left(B_{\sigma_2}N_{\sigma_2}\right)^{\Delta_1}_{t_1}\\
&= \left(\Gamma^1_{12\sigma_2}\right)^{\Delta_1}_{t_1}f^{\Delta_1\sigma_2}_{t_1}+\Gamma^{1\sigma_1}_{12\sigma_2}f^{\Delta_1^2\sigma_2}_{t_1}+\left(\Gamma^2_{12\sigma_2}\right)^{\Delta_1}_{t_1}f^{\Delta_2}_{t_2}+\Gamma^{2\sigma_1}_{12\sigma_2}f^{\Delta_1\Delta_2}_{t_1t_2}\\
&\quad+\left(B_{\sigma_2}\right)^{\Delta_1}_{t_1}N_{\sigma_2}+B^{\sigma_1}_{\sigma_2}\left(N_{\sigma_2}\right)^{\Delta_1}_{t_1}\\
&= \left(\Gamma^1_{12\sigma_2}\right)^{\Delta_1}_{t_1}f^{\Delta_1\sigma_2}_{t_1}+\Gamma^{1\sigma_1}_{12\sigma_2}\left(U^1_{2\sigma_2}f^{\Delta_1\sigma_2}_{t_1}+U^1_{2\sigma_2}f^{\Delta_2}_{t_2}+U^3_{2\sigma_2}N_{\sigma_2}\right)\\
&\quad+\left(\Gamma^2_{12\sigma_2}\right)^{\Delta_1}_{t_1}f^{\Delta_2}_{t_2}+\Gamma^{2\sigma_1}_{12\sigma_2}\left(\Gamma^1_{12\sigma_2}f^{\Delta_1\sigma_2}_{t_1}+\Gamma^2_{12\sigma_2}f^{\Delta_2}_{t_2}+B_{\sigma_2}N_{\sigma_2}\right)
\end{aligned}
$$

$$
\begin{aligned}
&+(B_{\sigma_2})_{t_1}^{\Delta_1} N_{\sigma_2} + B_{\sigma_2}^{\sigma_1}\left(W_{1\sigma_2}^1 f_{t_1}^{\Delta_1\sigma_2} + W_{1\sigma_2}^2 f_{t_2}^{\Delta_2}\right)\\
&= \left(\left(\Gamma_{12\sigma_2}^1\right)_{t_1}^{\Delta_1} + \Gamma_{12\sigma_2}^{1\sigma_1} U_{2\sigma_2}^1 + \Gamma_{12\sigma_2}^{2\sigma_1}\Gamma_{12\sigma_2}^1 + B_{\sigma_2}^{\sigma_1} W_{1\sigma_2}^1\right) f_{t_1}^{\Delta_1\sigma_2}\\
&+ \left(\left(\Gamma_{12\sigma_2}^2\right)_{t_1}^{\Delta_1} + \Gamma_{12\sigma_2}^{1\sigma_1} U_{2\sigma_2}^2 + \Gamma_{12\sigma_2}^{2\sigma_1}\Gamma_{12\sigma_2}^2 + B_{\sigma_2}^{\sigma_1} W_{1\sigma_2}^2\right) f_{t_2}^{\Delta_2}\\
&+ \left((B_{\sigma_2})_{t_1}^{\Delta_1} + \Gamma_{12\sigma_2}^{1\sigma_1} U_{2\sigma_2}^3 + \Gamma_{12\sigma_2}^{2\sigma_1} B_{\sigma_2}\right) N_{\sigma_2}.
\end{aligned}
$$

Now, we equate the last two equations and using that $f_{t_1}^{\Delta_1\sigma_2}$, $f_{t_2}^{\Delta_2}$ and N_{σ_2} are linearly independent, we get the first and the second equations of the σ_2-Gauss equations and the first equation of the σ_2-Godazzi-Mainardi equations. Next,

$$
\begin{aligned}
f_{t_2t_1}^{\Delta_2^2\Delta_1} &= \left(\Gamma_{22\sigma_2}^1 f_{t_1}^{\Delta_1\sigma_2} + \Gamma_{22\sigma_2}^2 f_{t_2}^{\Delta_2} + C_{\sigma_2} N_{\sigma_2}\right)_{t_1}^{\Delta_1}\\
&= \left(\Gamma_{22\sigma_2}^1 f_{t_1}^{\Delta_1\sigma_2}\right)_{t_1}^{\Delta_1} + \left(\Gamma_{22\sigma_2}^2 f_{t_2}^{\Delta_2}\right)_{t_1}^{\Delta_1} + \left(C_{\sigma_2} N_{\sigma_2}\right)_{t_1}^{\Delta_1}\\
&= \left(\Gamma_{22\sigma_2}^1\right)_{t_1}^{\Delta_1} f_{t_1}^{\Delta_1\sigma_2} + \Gamma_{22\sigma_2}^{1\sigma_1} f_{t_1}^{\Delta_1^2\sigma_2} + \left(\Gamma_{22\sigma_2}^2\right)_{t_1}^{\Delta_1} f_{t_2}^{\Delta_2} + \Gamma_{22\sigma_2}^{2\sigma_1} f_{t_1t_2}^{\Delta_1\Delta_2}\\
&+ (C_{\sigma_2})_{t_1}^{\Delta_1} N_{\sigma_2} + C_{\sigma_2}^{\sigma_1} N_{\sigma_2 t_1}^{\Delta_1}\\
&= \left(\Gamma_{22\sigma_2}^1\right)_{t_1}^{\Delta_1} f_{t_1}^{\Delta_1\sigma_2} + \Gamma_{22\sigma_2}^{1\sigma_1}\left(U_{2\sigma_2}^1 f_{t_1}^{\Delta_1\sigma_2} + U_{2\sigma_2}^2 f_{t_2}^{\Delta_2} + U_{2\sigma_2}^3 N_{\sigma_2}\right)\\
&+ \left(\Gamma_{22\sigma_2}^2\right)_{t_1}^{\Delta_1} f_{t_2}^{\Delta_2} + \Gamma_{22\sigma_2}^{2\sigma_1}\left(\Gamma_{12\sigma_2}^1 f_{t_1}^{\Delta_1\sigma_2} + \Gamma_{12\sigma_2}^2 f_{t_2}^{\Delta_2} + B_{\sigma_2} N_{\sigma_2}\right)\\
&+ (C_{\sigma_2})_{t_1}^{\Delta_1} N_{\sigma_2} + C_{\sigma_2}^{\sigma_1}\left(W_{1\sigma_2}^1 f_{t_1}^{\Delta_1\sigma_2} + W_{1\sigma_2}^2 f_{t_2}^{\Delta_2}\right)\\
&= \left(\left(\Gamma_{22\sigma_2}^1\right)_{t_1}^{\Delta_1} + \Gamma_{22\sigma_2}^{1\sigma_1} U_{2\sigma_2}^1 + \Gamma_{22\sigma_2}^{2\sigma_1}\Gamma_{12\sigma_2}^1 + C_{\sigma_2}^{\sigma_1} W_{1\sigma_2}^1\right) f_{t_1}^{\Delta_1\sigma_2}\\
&+ \left(\left(\Gamma_{22\sigma_2}^2\right)_{t_1}^{\Delta_1} + \Gamma_{22\sigma_2}^{1\sigma_1} U_{2\sigma_2}^2 + \Gamma_{22\sigma_2}^{2\sigma_1}\Gamma_{12\sigma_2}^2 + C_{\sigma_2}^{\sigma_1} W_{1\sigma_2}^2\right) f_{t_2}^{\Delta_2}\\
&+ \left((C_{\sigma_2})_{t_1}^{\Delta_1} + \Gamma_{22\sigma_2}^{1\sigma_1} U_{2\sigma_2}^3 + \Gamma_{22\sigma_2}^{2\sigma_1} B_{\sigma_2}\right) N_{\sigma_2}
\end{aligned}
$$

and

$$
\begin{aligned}
f_{t_1t_2}^{\Delta_1\Delta_2^2} &= \left(\Gamma_{12\sigma_2}^1 f_{t_1}^{\Delta_1\sigma_2} + \Gamma_{12\sigma_2}^2 f_{t_2}^{\Delta_2} + B_{\sigma_2} N_{\sigma_2}\right)_{t_2}^{\Delta_2}\\
&= \left(\Gamma_{12\sigma_2}^1 f_{t_1}^{\Delta_1\sigma_2}\right)_{t_2}^{\Delta_2} + \left(\Gamma_{12\sigma_2}^2 f_{t_2}^{\Delta_2}\right)_{t_2}^{\Delta_2} + \left(B_{\sigma_2} N_{\sigma_2}\right)_{t_2}^{\Delta_2}\\
&= \left(\Gamma_{12\sigma_2}^1\right)_{t_2}^{\Delta_2} f_{t_1}^{\Delta_1\sigma_2} + \Gamma_{12\sigma_2}^{1\sigma_2} f_{t_1t_2}^{\Delta_1\sigma_2\Delta_2} + \left(\Gamma_{12\sigma_2}^2\right)_{t_2}^{\Delta_2} f_{t_2}^{\Delta_2} + \Gamma_{12\sigma_2}^{2\sigma_2} f_{t_2}^{\Delta_2^2}\\
&+ (B_{\sigma_2})_{t_2}^{\Delta_2} N_{\sigma_2} + B_{\sigma_2}^{\sigma_2} N_{\sigma_2 t_2}^{\Delta_2}\\
&= \left(\Gamma_{12\sigma_2}^1\right)_{t_2}^{\Delta_2} f_{t_1}^{\Delta_1\sigma_2} + \left(1 + \mu_2^{\Delta_2}\right)\Gamma_{12\sigma_2}^{1\sigma_2}\left(U_{4\sigma_2}^1 f_{t_1}^{\Delta_1\sigma_2} + U_{4\sigma_2}^2 f_{t_2}^{\Delta_2} + U_{4\sigma_2}^3 N_{\sigma_2}\right)\\
&+ \left(\Gamma_{12\sigma_2}^2\right)_{t_2}^{\Delta_2} f_{t_2}^{\Delta_2} + \Gamma_{12\sigma_2}^{2\sigma_2}\left(\Gamma_{22\sigma_2}^1 f_{t_1}^{\Delta_1\sigma_2} + \Gamma_{22\sigma_2}^2 f_{t_2}^{\Delta_2} + C_{\sigma_2} N_{\sigma_2}\right)\\
&+ (B_{\sigma_2})_{t_2}^{\Delta_2} N_{\sigma_2} + B_{\sigma_2}^{\sigma_2}\left(W_{2\sigma_2}^1 f_{t_1}^{\Delta_1\sigma_2} + W_{2\sigma_2}^2 f_{t_2}^{\Delta_2}\right)
\end{aligned}
$$

$$= \left(\left(\Gamma^1_{12\sigma_2}\right)^{\Delta_2}_{t_2} + \left(1 + \mu_2^{\Delta_2}\right) \Gamma^{1\sigma_2}_{12\sigma_2} U^1_{4\sigma_2} + \Gamma^{2\sigma_2}_{12\sigma_2} \Gamma^1_{22\sigma_2} + B^{\sigma_2}_{\sigma_2} W^1_{2\sigma_2} \right) f_{t_1}^{\Delta_1 \sigma_2}$$

$$+ \left(\left(\Gamma^2_{12\sigma_2}\right)^{\Delta_2}_{t_2} + \left(1 + \mu_2^{\Delta_2}\right) \Gamma^{1\sigma_2}_{12\sigma_2} U^2_{4\sigma_2} + \Gamma^{2\sigma_2}_{12\sigma_2} \Gamma^2_{22\sigma_2} + B^{\sigma_2}_{\sigma_2} W^2_{2\sigma_2} \right) f_{t_2}^{\Delta_2}$$

$$+ \left((B_{\sigma_2})^{\Delta_2}_{t_2} + \left(1 + \mu_2^{\Delta_2}\right) \Gamma^{1\sigma_2}_{12\sigma_2} U^3_{4\sigma_2} + \Gamma^{2\sigma_2}_{12\sigma_2} C_{\sigma_2} \right) N_{\sigma_2}.$$

We equate the last two equations and we obtain the third and fourth equations of the σ_2-Gauss equations and the second equation of the σ_2-Godazzi-Mainardi equations. This completes the proof.

Exercise 5.10 Suppose $f \in \mathscr{C}^3$, S is a σ_2-regular surface, σ_1 is Δ_1-differentiable and σ_2 is Δ_2-differentiable. Use the following relations

$$\left(\Gamma^1_{11\sigma_2} f_{t_2}^{\Delta_1\sigma_2}\right)^{\Delta_2}_{t_2} = \left(\Gamma^1_{11\sigma_2}\right)^{\Delta_2}_{t_2} f_{t_1}^{\Delta_1\sigma_2^2} + \Gamma^1_{11\sigma_2} \left(f_{t_1}^{\Delta_1\sigma_2}\right)^{\Delta_2}_{t_2},$$

or

$$\left(\Gamma^2_{11\sigma_2} f_{t_2}^{\Delta_2}\right)^{\Delta_2}_{t_2} = \left(\Gamma^2_{11\sigma_2}\right)^{\Delta_2}_{t_2} f_{t_2}^{\Delta_2\sigma_2} + \Gamma^2_{11\sigma_2} f_{t_2}^{\Delta_2^2},$$

or

$$(A_{\sigma_2} N_{\sigma_2})^{\Delta_2}_{t_2} = (A_{\sigma_2})^{\Delta_2}_{t_2} N^{\sigma_2}_{\sigma_2} + A_{\sigma_2} N^{\Delta_2}_{\sigma_2 t_2},$$

or

$$\left(\Gamma^1_{12\sigma_2} f_{t_1}^{\Delta_1\sigma_2}\right)^{\Delta_1}_{t_1} = \left(\Gamma^1_{12\sigma_2}\right)^{\Delta_1}_{t_1} f_{t_1}^{\Delta_1\sigma_1\sigma_2} + \Gamma^1_{12\sigma_2} f_{t_1}^{\Delta_1^2\sigma_2},$$

or

$$\left(\Gamma^2_{12\sigma_2} f_{t_2}^{\Delta_2}\right)^{\Delta_1}_{t_1} = \left(\Gamma^2_{12\sigma_2}\right)^{\Delta_1}_{t_1} f_{t_2}^{\Delta_2\sigma_1} + \Gamma^2_{12\sigma_2} f_{t_1 t_2}^{\Delta_1\Delta_2},$$

or

$$(B_{\sigma_2} N_{\sigma_2})^{\Delta_1}_{t_1} = (B_{\sigma_2})^{\Delta_1}_{t_1} N^{\sigma_1}_{\sigma_2} + B_{\sigma_2} N^{\Delta_1}_{\sigma_2 t_1},$$

$$\left(\Gamma^1_{22\sigma_2} f_{t_1}^{\Delta_1\sigma_2}\right)^{\Delta_1}_{t_1} = \left(\Gamma^1_{22\sigma_2}\right)^{\Delta_1}_{t_1} f_{t_1}^{\Delta_1\sigma_1\sigma_2} + \Gamma^1_{22\sigma_2} f_{t_1}^{\Delta_1^2\sigma_2},$$

or

$$\left(\Gamma^2_{22\sigma_2} f_{t_2}^{\Delta_2}\right)^{\Delta_1}_{t_1} = \left(\Gamma^2_{22\sigma_2}\right)^{\Delta_1}_{t_1} f_{t_2}^{\Delta_2\sigma_1} + \Gamma^2_{22\sigma_2} f_{t_1 t_2}^{\Delta_1\Delta_2},$$

or

$$(C_{\sigma_2} N_{\sigma_2})^{\Delta_1}_{t_1} = (C_{\sigma_2})^{\Delta_1}_{t_1} N^{\sigma_1}_{\sigma_2} + C_{\sigma_2} N^{\Delta_1}_{\sigma_2 t_1},$$

or

$$\left(\Gamma^1_{12\sigma_2} f_{t_1}^{\Delta_1\sigma_2}\right)^{\Delta_2}_{t_2} = \left(\Gamma^1_{12\sigma_2}\right)^{\Delta_2}_{t_2} f_{t_1}^{\Delta_1\sigma_2^2} + \Gamma^1_{12\sigma_2} f_{t_1}^{\Delta_1\sigma_2\Delta_2},$$

or

$$\left(\Gamma^2_{12\sigma_2} f_{t_2}^{\Delta_2}\right)^{\Delta_2}_{t_2} = \left(\Gamma^2_{12\sigma_2}\right)^{\Delta_2}_{t_2} f_{t_2}^{\Delta_2\sigma_2} + \Gamma^2_{12\sigma_2} f_{t_2}^{\Delta_2^2},$$

or

$$(B_{\sigma_2} N_{\sigma_2})^{\Delta_2}_{t_2} = (B_{\sigma_2})^{\Delta_2}_{t_2} N^{\sigma_2}_{\sigma_2} + B_{\sigma_2} N^{\Delta_2}_{\sigma_2 t_2},$$

to deduce the other kinds of the σ_2-Gauss equations and σ_2-Godazzi-Mainardi equations.

Theorem 5.4 *Suppose $f \in \mathscr{C}^3$, S is a σ_2-regular surface. Then*

$$\begin{aligned}\mu_1\left(U^j_{2\sigma_2}-\Gamma^j_{11\sigma_2}\right) &= \mu_2\left(U^j_{3\sigma_2}-\Gamma^j_{12\sigma_2}\right)\\ \mu_1\left(U^j_{4\sigma_2}-\Gamma^j_{12\sigma_2}\right) &= \mu_2\left(U^j_{5\sigma_2}-\Gamma^j_{22\sigma_2}\right), \quad j\in\{1,2\},\end{aligned} \tag{5.19}$$

and

$$\begin{aligned}\mu_1\left(U^3_{2\sigma_2}-A_{\sigma_2}\right) &= \mu_2\left(U^3_{2\sigma_2}-B_{\sigma_2}\right)\\ \mu_1\left(U^3_{4\sigma_2}-B_{\sigma_2}\right) &= \mu_2\left(U^3_{5\sigma_2}-C_{\sigma_2}\right).\end{aligned} \tag{5.20}$$

Definition 5.15 The equations (5.19) will be called the $\sigma_2\mu_1\mu_2$-Gauss equations.

Definition 5.16 The equations (5.20) will be called the $\sigma_2\mu_1\mu_2$-Godazzi-Mainardi equations.

Proof. We have

$$\begin{aligned}f_{t_1}^{\Delta_1^2\sigma_2} &= U^1_{2\sigma_2}f_{t_1}^{\Delta_1\sigma_2}+U^2_{2\sigma_2}f_{t_2}^{\Delta_2}+U^3_{2\sigma_2}N_{\sigma_2}\\ &= f_{t_1}^{\Delta_1^2}+\mu_2 f_{t_1t_2}^{\Delta_1^2\Delta_2}\\ &= \Gamma^1_{11\sigma_2}f_{t_1}^{\Delta_1\sigma_2}+\Gamma^2_{11\sigma_2}f_{t_2}^{\Delta_2}+A_{\sigma_2}N_{\sigma_2}+\mu_2 f_{t_1t_2}^{\Delta_1^2\Delta_2},\end{aligned}$$

whereupon

$$\mu_1\left(U^1_{2\sigma_2}-\Gamma^1_{11\sigma_2}\right)f_{t_1}^{\Delta_1\sigma_2}+\mu_1\left(U^2_{2\sigma_2}-\Gamma^2_{11\sigma_2}\right)f_{t_2}^{\Delta_2}+\mu_1\left(U^3_{2\sigma_2}-A_{\sigma_2}\right)N_{\sigma_2}=\mu_1\mu_2 f_{t_1t_2}^{\Delta_1^2\Delta_2}. \tag{5.21}$$

Also,

$$\begin{aligned}f_{t_1t_2}^{\Delta_1\Delta_2\sigma_1} &= U^1_{3\sigma_2}f_{t_1}^{\Delta_1\sigma_2}+U^2_{3\sigma_2}f_{t_2}^{\Delta_2}+U^3_{3\sigma_2}N_{\sigma_2}\\ &= f_{t_1t_2}^{\Delta_1\Delta_2}+\mu_1 f_{t_1}^{\Delta_1^2\Delta_2}\\ &= \Gamma^1_{12\sigma_2}f_{t_1}^{\Delta_1\sigma_2}+\Gamma^2_{12\sigma_2}f_{t_2}^{\Delta_2}+B_{\sigma_2}N_{\sigma_2}+f_{t_1t_2}^{\Delta_1^2\Delta_2},\end{aligned}$$

from where

$$\mu_2\left(U^1_{3\sigma_2}-\Gamma^1_{12\sigma_2}\right)+\mu_2\left(U^2_{3\sigma_2}-\Gamma^2_{12\sigma_2}\right)+\mu_2\left(U^3_{3\sigma_2}-B_{\sigma_2}\right)=\mu_1\mu_2 f_{t_1t_2}^{\Delta_1^2\Delta_2}.$$

Now, we equate the last equation and equation (5.21) and we get the first and second equations of $\sigma_2\mu_1\mu_2$-Gauss equations and the first equation of the $\sigma_2\mu_1\mu_2$-Godazzi-Mainardi equations. Next,

$$\begin{aligned}f_{t_1t_2}^{\Delta_1\Delta_2\sigma_2} &= U^1_{4\sigma_2}f_{t_1}^{\Delta_1\sigma_2}+U^2_{4\sigma_2}f_{t_2}^{\Delta_2}+U^3_{4\sigma_2}N_{\sigma_2}\\ &= f_{t_1t_2}^{\Delta_1\Delta_2}+\mu_2 f_{t_1t_2}^{\Delta_1\Delta_2^2}\\ &= \Gamma^1_{12\sigma_2}f_{t_1}^{\Delta_1\sigma_2}+\Gamma^2_{12\sigma_2}f_{t_2}^{\Delta_2}+B_{\sigma_2}N_{\sigma_2}+\mu_2 f_{t_1t_2}^{\Delta_1\Delta_2^2},\end{aligned}$$

whereupon

$$\mu_1\left(U^1_{4\sigma_2}-\Gamma^1_{12\sigma_2}\right)f_{t_1}^{\Delta_1\sigma_2}+\mu_1\left(U^2_{4\sigma_2}-\Gamma^2_{12\sigma_2}\right)f_{t_2}^{\Delta_2}+\mu_1\left(U^3_{4\sigma_2}-B_{\sigma_2}\right)N_{\sigma_2}=\mu_1\mu_2 f_{t_1t_2}^{\Delta_1\Delta_2^2}. \tag{5.22}$$

Moreover,

$$\begin{aligned}f_{t_2}^{\Delta_2^2\sigma_1}&=U^1_{5\sigma_2}f_{t_1}^{\Delta_1\sigma_2}+U^2_{5\sigma_2}f_{t_2}^{\Delta_2}+U^3_{5\sigma_2}N_{\sigma_2}\\&=f_{t_2}^{\Delta_2^2}+\mu_1 f_{t_1t_2}^{\Delta_1\Delta_2^2}\\&=\Gamma^1_{22\sigma_2}f_{t_1}^{\Delta_1\sigma_2}+\Gamma^2_{22\sigma_2}f_{t_2}^{\Delta_2}+C_{\sigma_2}N_{\sigma_2}+\mu_1 f_{t_1t_2}^{\Delta_1\Delta_2^2}.\end{aligned}$$

Hence,

$$\mu_2\left(U^1_{5\sigma_2}-\Gamma^1_{22\sigma_2}\right)f_{t_1}^{\Delta_1\sigma_2}+\mu_2\left(U^2_{5\sigma_2}-\Gamma^2_{22\sigma_2}\right)f_{t_2}^{\Delta_2}+\mu_2\left(U^3_{5\sigma_2}-C_{\sigma_2}\right)N_{\sigma_2}=\mu_1\mu_2 f_{t_1t_2}^{\Delta_1\Delta_2^2}.$$

By the last equation and (5.22), we find the third and fourth equations of the $\sigma_2\mu_1\mu_2$-Gauss equations and the second equation of the $\sigma_2\mu_1\mu_2$-Godazzi-Mainardi equations. This completes the proof.

5.4 Darboux Frames

Let (I,g) be a regular curve which is parameterized by arc length and whose support lies on the surface S, and $M=g(t_0)$, $t_0\in I$, be a point on the curve, and $e_1=g^{\Delta}(t_0)$, that is, e_1 is the unit tangent vector to g at M, and $\{e_1,e_2,e_3\}$ is the Frenet frame.

1. Assume that S is a σ_1-regular surface and N_{σ_1} is the σ_1-normal to S at M. Set

$$P_{\sigma_1}=N_{\sigma_1}\times e_1.$$

Definition 5.17 The frame $\{M,e_1,N_{\sigma_1},P_{\sigma_1}\}$ is called a σ_1-Darboux frame.

Denote

$$\theta_{\sigma_1}=\angle(N_{\sigma_1},e_2).$$

Then

$$\begin{aligned}e_2&=\cos\angle(P_{\sigma_1},e_2)P_{\sigma_1}+\sin\angle(P_{\sigma_1},e_2)N_{\sigma_1}\\e_3&=\cos\angle(P_{\sigma_1},e_3)P_{\sigma_1}+\sin\angle(P_{\sigma_1},e_3)N_{\sigma_1}.\end{aligned}$$

We have that P_{σ_1} lies in the normal plane of the curve at M.

Definition 5.18 The vector P_{σ_1} is called the unit σ_1-tangential normal vector of the curve.

Since

$$\angle(P_{\sigma_1}, e_2) = \frac{\pi}{2} - \theta_{\sigma_1} \quad \text{and} \quad \angle(P_{\sigma_1}, e_3) = \pi - \theta_{\sigma_1},$$

we get

$$e_2 = \sin\theta_{\sigma_1} P_{\sigma_1} + \cos\theta_{\sigma_1} N_{\sigma_1}$$

$$e_3 = -\cos\theta_{\sigma_1} P_{\sigma_1} + \sin\theta_{\sigma_1} N_{\sigma_1}.$$

By the last system, we find

$$P_{\sigma_1} = \sin\theta_{\sigma_1} e_2 - \cos\theta_{\sigma_1} e_3$$

$$N_{\sigma_1} = \cos\theta_{\sigma_1} e_2 + \sin\theta_{\sigma_1} e_3.$$

The derivatives of the σ_1-Darboux frame can be represented in the following way:

$$\begin{aligned} e_{1t}^{\Delta} &= c_{11\sigma_1} e_1 + c_{12\sigma_1} N_{\sigma_1} + c_{13\sigma_1} P_{\sigma_1} \\ P_{\sigma_1 t}^{\Delta} &= c_{21\sigma_1} e_1 + c_{22\sigma_1} N_{\sigma_1} + c_{23\sigma_1} P_{\sigma_1} \\ N_{\sigma_1 t}^{\Delta} &= c_{31\sigma_1} e_1 + c_{32\sigma_1} N_{\sigma_1} + c_{33\sigma_1} P_{\sigma_1}. \end{aligned} \tag{5.23}$$

Definition 5.19 The equations (5.23) will be called the σ_1-Darboux formulae.

2. Assume that S is a σ_2-regular surface and N_{σ_2} is the σ_2-normal to S at M. Set

$$P_{\sigma_2} = N_{\sigma_2} \times e_1.$$

Definition 5.20 The frame $\{M, e_1, N_{\sigma_2}, P_{\sigma_2}\}$ is called the σ_2-Darboux frame.

Definition 5.21 The vector P_{σ_2} is called the unit the σ_2-tangential normal vector of the curve.

Denote

$$\theta_{\sigma_2} = \angle(N_{\sigma_2}, e_2).$$

Then

$$e_2 = \sin\theta_{\sigma_2} P_{\sigma_2} + \cos\theta_{\sigma_2} N_{\sigma_2}$$

$$e_3 = -\cos\theta_{\sigma_2} P_{\sigma_2} + \sin\theta_{\sigma_2} N_{\sigma_2}$$

and

$$P_{\sigma_2} = \sin\theta_{\sigma_2} e_2 - \cos\theta_{\sigma_2} e_3$$

$$N_{\sigma_2} = \cos\theta_{\sigma_2} e_2 + \sin\theta_{\sigma_2} e_3.$$

The derivatives of the σ_2-Darboux frame can be represented in the following way.

$$\begin{aligned} e_{1t}^{\Delta} &= c_{11\sigma_2} e_1 + c_{12\sigma_2} N_{\sigma_2} + c_{13\sigma_2} P_{\sigma_2} \\ P_{\sigma_2 t}^{\Delta} &= c_{21\sigma_2} e_1 + c_{22\sigma_2} N_{\sigma_2} + c_{23\sigma_2} P_{\sigma_2} \\ N_{\sigma_2 t}^{\Delta} &= c_{31\sigma_2} e_1 + c_{32\sigma_2} N_{\sigma_2} + c_{33\sigma_2} P_{\sigma_2}. \end{aligned} \tag{5.24}$$

Definition 5.22 The equations (5.24) will be called the σ_2-Darboux formulae.

3. Assume that S is a $\sigma_1^2\sigma_2$-regular surface and $N_{\sigma_1^2\sigma_2}$ is the $\sigma_1^2\sigma_2$-normal to S at M. Set

$$P_{\sigma_1^2\sigma_2} = N_{\sigma_1^2\sigma_2} \times e_1.$$

Definition 5.23 The frame $\{M, e_1, N_{\sigma_1^2\sigma_2}, P_{\sigma_1^2\sigma_2}\}$ is called the $\sigma_1^2\sigma_2$-Darboux frame.

Definition 5.24 The vector $P_{\sigma_1^2\sigma_2}$ is called the unit $\sigma_1^2\sigma_2$-tangential normal vector of the curve.

Denote

$$\theta_{\sigma_1^2\sigma_2} = \angle(N_{\sigma_1^2\sigma_2}, e_2).$$

Then

$$\begin{aligned} e_2 &= \sin\theta_{\sigma_1^2\sigma_2} P_{\sigma_1^2\sigma_2} + \cos\theta_{\sigma_1^2\sigma_2} N_{\sigma_1^2\sigma_2} \\ e_3 &= -\cos\theta_{\sigma_1^2\sigma_2} P_{\sigma_1^2\sigma_2} + \sin\theta_{\sigma_1^2\sigma_2} N_{\sigma_1^2\sigma_2} \end{aligned}$$

and

$$\begin{aligned} P_{\sigma_1^2\sigma_2} &= \sin\theta_{\sigma_1^2\sigma_2} e_2 - \cos\theta_{\sigma_1^2\sigma_2} e_3 \\ N_{\sigma_1^2\sigma_2} &= \cos\theta_{\sigma_1^2\sigma_2} e_2 + \sin\theta_{\sigma_1^2\sigma_2} e_3. \end{aligned}$$

The derivatives of the $\sigma_1^2\sigma_2$-Darboux frame can be represented in the following way.

$$\begin{aligned} e_{1t}^{\Delta} &= c_{11\sigma_1^2\sigma_2} e_1 + c_{12\sigma_1^2\sigma_2} N_{\sigma_1^2\sigma_2} + c_{13\sigma_1^2\sigma_2} P_{\sigma_1^2\sigma_2} \\ P_{\sigma_1^2\sigma_2 t}^{\Delta} &= c_{21\sigma_1^2\sigma_2} e_1 + c_{22\sigma_1^2\sigma_2} N_{\sigma_1^2\sigma_2} + c_{23\sigma_1^2\sigma_2} P_{\sigma_1^2\sigma_2} \\ N_{\sigma_1^2\sigma_2 t}^{\Delta} &= c_{31\sigma_1^2\sigma_2} e_1 + c_{32\sigma_1^2\sigma_2} N_{\sigma_1^2\sigma_2} + c_{33\sigma_1^2\sigma_2} P_{\sigma_1^2\sigma_2}. \end{aligned} \tag{5.25}$$

Definition 5.25 The equations (5.25) will be called the $\sigma_1^2\sigma_2$-Darboux formulae.

4. Assume that S is a $\sigma_1\sigma_2^2$-regular surface and $N_{\sigma_1\sigma_2^2}$ is the $\sigma_1\sigma_2^2$-normal to S at M. Set

$$P_{\sigma_1\sigma_2^2} = N_{\sigma_1\sigma_2^2} \times e_1.$$

Definition 5.26 The frame $\{M, e_1, N_{\sigma_1\sigma_2^2}, P_{\sigma_1\sigma_2^2}\}$ is called the $\sigma_1\sigma_2^2$-Darboux frame.

Definition 5.27 The vector $P_{\sigma_1\sigma_2^2}$ is called the unit $\sigma_1\sigma_2^2$-tangential normal vector of the curve.

Denote

$$\theta_{\sigma_1\sigma_2^2} = \angle(N_{\sigma_1\sigma_2^2}, e_2).$$

Then

$$e_2 = \sin\theta_{\sigma_1\sigma_2^2} P_{\sigma_1\sigma_2^2} + \cos\theta_{\sigma_1\sigma_2^2} N_{\sigma_1\sigma_2^2}$$
$$e_3 = -\cos\theta_{\sigma_1\sigma_2^2} P_{\sigma_1\sigma_2^2} + \sin\theta_{\sigma_1\sigma_2^2} N_{\sigma_1\sigma_2^2}$$

and

$$P_{\sigma_1\sigma_2^2} = \sin\theta_{\sigma_1\sigma_2^2} e_2 - \cos\theta_{\sigma_1\sigma_2^2} e_3$$
$$N_{\sigma_1\sigma_2^2} = \cos\theta_{\sigma_1\sigma_2^2} e_2 + \sin\theta_{\sigma_1\sigma_2^2} e_3.$$

The derivatives of the $\sigma_1\sigma_2^2$-Darboux frame can be represented in the following way.

$$\begin{aligned} e_{1t}^{\Delta} &= c_{11\sigma_1\sigma_2^2} e_1 + c_{12\sigma_1\sigma_2^2} N_{\sigma_1\sigma_2^2} + c_{13\sigma_1\sigma_2^2} P_{\sigma_1\sigma_2^2} \\ P_{\sigma_1\sigma_2^2 t}^{\Delta} &= c_{21\sigma_1\sigma_2^2} e_1 + c_{22\sigma_1\sigma_2^2} N_{\sigma_1\sigma_2^2} + c_{23\sigma_1\sigma_2^2} P_{\sigma_1\sigma_2^2} \\ N_{\sigma_1\sigma_2^2 t}^{\Delta} &= c_{31\sigma_1\sigma_2^2} e_1 + c_{32\sigma_1\sigma_2^2} N_{\sigma_1\sigma_2^2} + c_{33\sigma_1\sigma_2^2} P_{\sigma_1\sigma_2^2}. \end{aligned} \tag{5.26}$$

Definition 5.28 The equations (5.26) will be called the $\sigma_1\sigma_2^2$-Darboux formulae.

5.5 The Geodesic Torsions

Suppose that a_{ij}, $i,j \in \{1,2,3\}$, are the coefficients in the Frenet formulae for the curve g.

1. Let S be a σ_1-regular surface. We have

$$P_{\sigma_1} = \sin\theta_{\sigma_1} e_2 - \cos\theta_{\sigma_1} e_3.$$

Applying the Pötzsche chain rule, we find

$$\begin{aligned} P_{\sigma_1 t}^{\Delta} &= \left(\int_0^1 \cos(\theta_{\sigma_1} + h\mu\theta_{\sigma_1}^{\Delta})dh\right)\theta_{\sigma_1}^{\Delta} e_2 + \sin\theta_{\sigma_1}^{\sigma} e_2^{\Delta} \\ &\quad + \left(\int_0^1 \sin(\theta_{\sigma_1} + h\mu\theta_{\sigma_1}^{\Delta})dh\right)\theta_{\sigma_1}^{\Delta} e_3 - \cos\theta_{\sigma_1}^{\sigma} e_3^{\Delta} \\ &= \left(\int_0^1 \cos(\theta_{\sigma_1} + h\mu\theta_{\sigma_1}^{\Delta})dh\right)\theta_{\sigma_1}^{\Delta} e_2 \\ &\quad + \left(\int_0^1 \sin(\theta_{\sigma_1} + h\mu\theta_{\sigma_1}^{\Delta})dh\right)\theta_{\sigma_1}^{\Delta} e_3 \\ &\quad + \sin\theta_{\sigma_1}^{\sigma}(a_{21}e_1 + a_{22}e_2 + a_{23}e_3) \\ &\quad - \cos\theta_{\sigma_1}^{\sigma}(a_{31}e_1 + a_{32}e_2 + a_{33}e_3) \end{aligned}$$

$$
\begin{aligned}
&= \left(a_{21}\sin\theta^{\sigma}_{\sigma_1} - a_{31}\cos\theta^{\sigma}_{\sigma_1}\right)e_1 \\
&\quad + \left(\left(\int_0^1 \cos(\theta_{\sigma_1} + h\mu\theta^{\Delta}_{\sigma_1})dh\right)\theta^{\Delta}_{\sigma_1} + \sin\theta^{\sigma}_{\sigma_1}a_{22} - \cos\theta^{\sigma}_{\sigma_1}a_{32}\right)e_2 \\
&\quad + \left(\left(\int_0^1 \sin(\theta_{\sigma_1} + h\mu\theta^{\Delta}_{\sigma_1})dh\right)\theta^{\Delta}_{\sigma_1} + \sin\theta^{\sigma}_{\sigma_1}a_{23} - \cos\theta^{\sigma}_{\sigma_1}a_{33}\right)e_3 \\
&= (a_{21}\sin\theta_{\sigma_1} - a_{31}\cos\theta_{\sigma_1})e_1 \\
&\quad + \left(\left(\int_0^1 \cos(\theta_{\sigma_1} + h\mu\theta^{\Delta}_{\sigma_1})dh\right)\theta^{\Delta}_{\sigma_1} + \sin\theta^{\sigma}_{\sigma_1}a_{22} - \cos\theta^{\sigma}_{\sigma_1}a_{32}\right) \\
&\quad (\sin\theta_{\sigma_1}P_{\sigma_1} + \cos\theta_{\sigma_1}N_{\sigma_1}) \\
&\quad + \left(\left(\int_0^1 \sin(\theta_{\sigma_1} + h\mu\theta^{\Delta}_{\sigma_1})dh\right)\theta^{\Delta}_{\sigma_1} + \sin\theta^{\sigma}_{\sigma_1}a_{23} - \cos\theta^{\sigma}_{\sigma_1}a_{33}\right) \\
&\quad (-\cos\theta_{\sigma_1}P_{\sigma_1} + \sin\theta_{\sigma_1}N_{\sigma_1}) \\
&= \left(a_{21}\sin\theta^{\sigma}_{\sigma_1} - a_{31}\cos\theta^{\sigma}_{\sigma_1}\right)e_1 \\
&\quad + \left(\left(\int_0^1 \cos(\theta_{\sigma_1} + h\mu\theta^{\Delta}_{\sigma_1})dh\right)\theta^{\Delta}_{\sigma_1}\sin\theta_{\sigma_1}\right. \\
&\quad + \left(\sin\theta^{\sigma}_{\sigma_1}a_{22} - \cos\theta^{\sigma}_{\sigma_1}a_{32}\right)\sin\theta_{\sigma_1} \\
&\quad - \left(\sin\theta^{\sigma}_{\sigma_1}a_{23} - \cos\theta^{\sigma}_{\sigma_1}a_{33}\right)\cos\theta_{\sigma_1} \\
&\quad \left. - \left(\int_0^1 \sin(\theta_{\sigma_1} + h\mu\theta^{\Delta}_{\sigma_1})dh\right)\theta^{\Delta}_{\sigma_1}\cos\theta_{\sigma_1}\right)P_{\sigma_1} \\
&\quad + \left(\left(\int_0^1 \cos(\theta_{\sigma_1} + h\mu\theta^{\Delta}_{\sigma_1})dh\right)\theta^{\Delta}_{\sigma_1}\cos\theta_{\sigma_1}\right. \\
&\quad + \left(\sin\theta^{\sigma}_{\sigma_1}a_{22} - \cos\theta^{\sigma}_{\sigma_1}a_{32}\right)\cos\theta_{\sigma_1} \\
&\quad + \left(\sin\theta^{\sigma}_{\sigma_1}a_{23} - \cos\theta^{\sigma}_{\sigma_1}a_{33}\right)\sin\theta_{\sigma_1} \\
&\quad \left. + \left(\int_0^1 \sin(\theta_{\sigma_1} + h\mu\theta^{\Delta}_{\sigma_1})dh\right)\theta^{\Delta}_{\sigma_1}\sin\theta_{\sigma_1}\right)N_{\sigma_1}.
\end{aligned}
$$

Definition 5.29 The function

$$
\begin{aligned}
d_{\sigma_1} = &\left(\left(\int_0^1 \cos(\theta_{\sigma_1} + h\mu\theta^{\Delta}_{\sigma_1})dh\right)\theta^{\Delta}_{\sigma_1}\cos\theta^{\sigma}_{\sigma_1}\right. \\
&+ \left(\sin\theta^{\sigma}_{\sigma_1}a_{22} - \cos\theta^{\sigma}_{\sigma_1}a_{32}\right)\cos\theta_{\sigma_1} \\
&+ \left(\sin\theta^{\sigma}_{\sigma_1}a_{23} - \cos\theta^{\sigma}_{\sigma_1}a_{33}\right)\sin\theta_{\sigma_1} \\
&\left. + \left(\int_0^1 \sin(\theta_{\sigma_1} + h\mu\theta^{\Delta}_{\sigma_1})dh\right)\theta^{\Delta}_{\sigma_1}\sin\theta^{\sigma}_{\sigma_1}\right)
\end{aligned}
$$

will be called the geodesic σ_1-torsion.

2. Let S be a σ_2-regular surface.

Definition 5.30 The function

$$d_{\sigma_2} = \Big(\Big(\int_0^1 \cos(\theta_{\sigma_2} + h\mu\theta_{\sigma_2}^{\Delta}) dh \Big) \theta_{\sigma_2}^{\Delta} \cos\theta_{\sigma_2}^{\sigma} + \big(\sin\theta_{\sigma_2}^{\sigma} a_{22} - \cos\theta_{\sigma_2}^{\sigma} a_{32} \big) \cos\theta_{\sigma_2} + \big(\sin\theta_{\sigma_2}^{\sigma} a_{23} - \cos\theta_{\sigma_2}^{\sigma} a_{33} \big) \sin\theta_{\sigma_2} + \Big(\int_0^1 \sin(\theta_{\sigma_2} + h\mu\theta_{\sigma_2}^{\Delta}) dh \Big) \theta_{\sigma_2}^{\Delta} \sin\theta_{\sigma_2}^{\sigma} \Big)$$

will be called the geodesic σ_2-torsion.

3. Let S be a $\sigma_1^2\sigma_2$-regular surface.

Definition 5.31 The function

$$d_{\sigma_1^2\sigma_2} = \Big(\Big(\int_0^1 \cos(\theta_{\sigma_1^2\sigma_2} + h\mu\theta_{\sigma_1^2\sigma_2}^{\Delta}) dh \Big) \theta_{\sigma_1^2\sigma_2}^{\Delta} \cos\theta_{\sigma_1^2\sigma_2}^{\sigma} + \Big(\sin\theta_{\sigma_1^2\sigma_2}^{\sigma} a_{22} - \cos\theta_{\sigma_1^2\sigma_2}^{\sigma} a_{32} \Big) \cos\theta_{\sigma_1^2\sigma_2} + \Big(\sin\theta_{\sigma_1^2\sigma_2}^{\sigma} a_{23} - \cos\theta_{\sigma_1^2\sigma_2}^{\sigma} a_{33} \Big) \sin\theta_{\sigma_1^2\sigma_2} + \Big(\int_0^1 \sin(\theta_{\sigma_1^2\sigma_2} + h\mu\theta_{\sigma_1^2\sigma_2}^{\Delta}) dh \Big) \theta_{\sigma_1^2\sigma_2}^{\Delta} \sin\theta_{\sigma_1^2\sigma_2}^{\sigma} \Big)$$

will be called the geodesic $\sigma_1^2\sigma_2$-torsion.

4. Let S be a $\sigma_1\sigma_2^2$-regular surface.

Definition 5.32 The function

$$d_{\sigma_1\sigma_2^2} = \Big(\Big(\int_0^1 \cos(\theta_{\sigma_1\sigma_2^2} + h\mu\theta_{\sigma_1\sigma_2^2}^{\Delta}) dh \Big) \theta_{\sigma_1\sigma_2^2}^{\Delta} \cos\theta_{\sigma_1\sigma_2^2}^{\sigma} + \big(\sin\theta_{\sigma_1}^{\sigma} a_{22} - \cos\theta_{\sigma_1}^{\sigma} a_{32} \big) \cos\theta_{\sigma_1\sigma_2^2} + \big(\sin\theta_{\sigma_1}^{\sigma} a_{23} - \cos\theta_{\sigma_1}^{\sigma} a_{33} \big) \sin\theta_{\sigma_1\sigma_2^2} + \Big(\int_0^1 \sin(\theta_{\sigma_1\sigma_2^2} + h\mu\theta_{\sigma_1\sigma_2^2}^{\Delta}) dh \Big) \theta_{\sigma_1\sigma_2^2}^{\Delta} \sin\theta_{\sigma_1\sigma_2^2}^{\sigma} \Big)$$

will be called the geodesic $\sigma_1\sigma_2^2$-torsion.

5.6 The Geodesic Curvatures. The Geodesic Lines

1. Let S be a σ_1-regular surface.

Definition 5.33 The geodesic σ_1-curvature is defined as follows.

$$k_{g\sigma_1} = e_1^{\Delta} \cdot P_{\sigma_1}.$$

We have

$$\begin{aligned} k_{g\sigma_1} &= g^{\Delta^2} \cdot P_{\sigma_1} \\ &= g^{\Delta^2} \cdot (N_{\sigma_1} \times e_1) \\ &= g^{\Delta^2} \cdot (N_{\sigma_1} \times g^{\Delta}) \\ &= \left(g^{\Delta}, g^{\Delta^2}, N_{\sigma_1}\right). \end{aligned}$$

Definition 5.34 The curve g is called a σ_1-geodesic line if its geodesic σ_1-curvature vanishes.

2. Let S be a σ_2-regular surface.

Definition 5.35 The geodesic σ_2-curvature is defined as follows.

$$k_{g\sigma_2} = e_1^{\Delta} \cdot P_{\sigma_2}.$$

We have

$$k_{g\sigma_2} = \left(g^{\Delta}, g^{\Delta^2}, N_{\sigma_2}\right).$$

Definition 5.36 The curve g is called a σ_2-geodesic line if its geodesic σ_2-curvature vanishes.

3. Let S be a $\sigma_1^2\sigma_2$-regular surface.

Definition 5.37 The geodesic $\sigma_1^2\sigma_2$-curvature is defined as follows.

$$k_{g\sigma_1^2\sigma_2} = e_1^{\Delta} \cdot P_{\sigma_1^2\sigma_2}.$$

We have

$$k_{g\sigma_1^2\sigma_2} = \left(g^{\Delta}, g^{\Delta^2}, N_{\sigma_1^2\sigma_2}\right).$$

Definition 5.38 The curve g is called $\sigma_1^2\sigma_2$-geodesic line if its geodesic $\sigma_1^2\sigma_2$-curvature vanishes.

4. Let S is a $\sigma_1\sigma_2^2$-regular surface.

Definition 5.39 The geodesic $\sigma_1\sigma_2^2$-curvature is defined as follows.

$$k_{g\sigma_1\sigma_2^2} = e_1^{\Delta} \cdot P_{\sigma_1\sigma_2^2}.$$

We have

$$k_{g\sigma_1\sigma_2^2} = \left(g^{\Delta}, g^{\Delta^2}, N_{\sigma_1\sigma_2^2}\right).$$

Definition 5.40 The curve g is called $\sigma_1\sigma_2^2$-geodesic line if its geodesic $\sigma_1\sigma_2^2$-curvature vanishes.

5.7 Advanced Practical Problems

Problem 5.1 Let S be a σ_1-regular surface. Find in the terms of the σ_1-Christoffel coefficients

1. $f_{t_1}^{\Delta_1\sigma_2} + \Gamma^1_{22\sigma_1} f_{t_2}^{\Delta_2\sigma_1}$,
2. $f_{t_2}^{\Delta_2\sigma_1} - 3\Gamma^1_{11\sigma_1} f_{t_1}^{\Delta_1} + f_{t_1}^{\Delta_1\sigma_2}$,
3. $7f_{t_1}^{\Delta_1\sigma_1} - 4\Gamma^1_{11\sigma_1} f_{t_1}^{\Delta_1} + f_{t_2}^{\Delta_2\sigma_2}$.

Problem 5.2 Let S be a σ_2-regular surface. Find expressions for

1. $f_{t_1}^{\Delta_1\sigma_2} - 2\Gamma^1_{11\sigma_2} f_{t_2}^{\Delta_2\sigma_2} + \Gamma^2_{22\sigma_2} f_{t_1}^{\Delta_1}$,
2. $f_{t_1}^{\Delta_1\sigma_1} - \Gamma^2_{11\sigma_2} f_{t_2}^{\Delta_2} + \Gamma^1_{11\sigma_2} f_{t_2}^{\Delta_2\sigma_2}$,
3. $f_{t_1}^{\Delta_1} + \Gamma^1_{12\sigma_2} f_{t_2}^{\Delta_2\sigma_1} + \Gamma^1_{22\sigma_2} f_{t_2}^{\Delta_2}$.

in the terms of the σ_2-Christoffel coefficients.

Problem 5.3 Suppose S is a σ_1-regular surface, σ_1 is Δ_1-differentiable and σ_2 is Δ_2-differentiable. Find

1. $4f_{t_1}^{\Delta_1\sigma_1\sigma_2} + 2f_{t_1}^{\Delta_1\sigma_1^2\sigma_2} - f_{t_1}^{\Delta_1\sigma_1\sigma_2^2}$,
2. $3f_{t_2}^{\Delta_1\sigma_1\sigma_2^2} + f_{t_1}^{\Delta_1\sigma_1^2\sigma_2} + 6f_{t_1}^{\Delta_1\sigma_1\sigma_2^2}$,
3. $2f_{t_2}^{\Delta_2\sigma_1^2\sigma_2} + f_{t_1}^{\Delta_1\sigma_1\sigma_2} - 5f_{t_1}^{\Delta_1\sigma_1^2\sigma_2} + f_{t_1}^{\Delta_1\sigma_1\sigma_2^2}$.

Problem 5.4 Suppose S is a σ_2-regular surface, σ_1 is Δ_1-differentiable and σ_2 is Δ_2-differentiable. Find

1. $\Gamma^2_{22\sigma_1\sigma_2} f_{t_2}^{\Delta_1\sigma_1^2\sigma_2} + 3f_{t_2}^{\Delta_2\sigma_1^2\sigma_2} + 4f_{t_2}^{\Delta_2\sigma_1\sigma_2^2}$,
2. $f_{t_2}^{\Delta_2\sigma_1\sigma_2} - 3f_{t_2}^{\Delta_1\sigma_1^2\sigma_2} + 4f_{t_2}^{\Delta_2\sigma_1^2\sigma_2}$,
3. $2f_{t_2}^{\Delta_2\sigma_1\sigma_2} + f_{t_2}^{\Delta_2\sigma_1^2\sigma_2} + 2f_{t_2}^{\Delta_2\sigma_1\sigma_2^2}$.

Problem 5.5 Suppose S is a $\sigma_1^2\sigma_2$-regular surface, σ_1 is Δ_1-differentiable and σ_2 is Δ_2-differentiable. Find

1. $2W^2_{2\sigma_1^2\sigma_2} f_{t_2}^{\Delta_2\sigma_1^2\sigma_2} + N^{\sigma_1}_{\sigma_1^2\sigma_2} - 4N^{\sigma_2}_{\sigma_1^2\sigma_2}$,
2. $3W^{1\sigma_2}_{1\sigma_1^2\sigma_2} f_{t_1}^{\Delta_1\sigma_1\sigma_2} + N^{\sigma_1}_{\sigma_1^2\sigma_2}$,
3. $W^1_{2\sigma_1^2\sigma_2} f_{t_1}^{\Delta_1\sigma_1\sigma_2} + N^{\sigma_1}_{\sigma_1^2\sigma_2} + 4N^{\sigma_2}_{\sigma_1^2\sigma_2}$.

Problem 5.6 Suppose S is a $\sigma_1\sigma_2^2$-regular surface, σ_1 is Δ_1-differentiable, σ_2 is Δ_2-differentiable. Find

1. $2N_{\sigma_1\sigma_2^2} - 4N^{\sigma_1}_{\sigma_1\sigma_2^2} + 3N^{\sigma_2}_{\sigma_1\sigma_2^2}$,
2. $N_{\sigma_1\sigma_2^2} - 3f_{t_1}^{\Delta_1\sigma_1\sigma_2^2} + 4f_{t_2}^{\Delta_2\sigma_1\sigma_2} + 5N^{\sigma_1}_{\sigma_1\sigma_2^2}$,
3. $N^{\sigma_2}_{\sigma_1\sigma_2^2} + 2W^1_{2\sigma_1\sigma_2^2}f_{t_2}^{\Delta_2\sigma_1\sigma_2} + f_{t_1}^{\Delta_1\sigma_1\sigma_2^2}$.

5.8 Notes and References

In this chapter we introduce the main differentiation rules using the first and second fundamental forms of a surface. The Christoffel and Weingarten coefficients are defined and the Gauss and Godazzi-Mainardi equations are deduced, using them. Darboux frames, geodesic torsions, geodesic curvatures and geodesic lines are also defined in this chapter.

6

Minimal Surfaces

Let $\mathbb{T}_1$ and $\mathbb{T}_2$ be time scales with forward jump operators and delta differentiation operators σ_1, σ_2 and Δ_1, Δ_2, respectively. With $\mathscr{C}_{rd}$ we denote the set of functions $f(x,y)$ on $\mathbb{T}_1 \times \mathbb{T}_2$ with the following properties.

1. f is rd-continuous in x for fixed y.
2. f is rd-continuous in y for fixed x.
3. If $(x_0,y_0) \in \mathbb{T}_1 \times \mathbb{T}_2$ with x_0 right-dense or maximal and y_0 right-dense or maximal, then f is continuous at (x_0,y_0).
4. If x_0 and y_0 are both left-sided, then the limit $f(x,y)$ exists(finite) as (x,y) approaches (x_0,y_0) along any path in

$$\{(x,y) \in \mathbb{T}_1 \times \mathbb{T}_2 : x < x_0, \quad y < y_0\}.$$

By $\mathscr{C}_{rd}^{(1)}$ we denote the set of all continuous functions for which both the Δ_1-partial derivative and the Δ_2-partial derivative exist and are of the class C_{rd}.

6.1 Statement of the Variational Problem

Let $E \subset \mathbb{T}_1 \times \mathbb{T}_2$ be a set of type ω and let Γ be its positively oriented fence. Suppose that a function

$$L(x,y,u,p,q), \quad (x,y) \in E \bigcup \Gamma \quad and \quad (u,p,q) \in \mathbb{R}^3,$$

is given and it is continuous together with its partial delta derivatives of the first and second order with respect to x, y and partial usual derivatives of the first and second order with respect to u, p, q. Consider the functional

$$\mathscr{L}(u) = \int\int_E L(x,y,u(\sigma_1(x),\sigma_2(y)),u^{\Delta_1}(x,\sigma_2(y)),u^{\Delta_2}(\sigma_1(x),y))\Delta_1 x \Delta_2 y \qquad (6.1)$$

whose domain of definition $D(\mathscr{L})$ consists of functions $u \in \mathscr{C}_{rd}^{(1)}(E \bigcup \Gamma)$ satisfying the "boundary conditions"

$$u = g(x,y) \quad on \quad \Gamma, \qquad (6.2)$$

where g is a fixed function defined and continuous on Γ.

DOI: 10.1201/9781003205265-6

Definition 6.1 We call functions $u \in D(\mathscr{L})$ admissible.

Definition 6.2 The functions $\eta \in \mathscr{C}_{rd}^{(1)}(E \bigcup \Gamma)$ and $\eta = 0$ on Γ, are called admissible variations.

If $f \in \mathscr{C}_{rd}^{(1)}(E \bigcup \Gamma)$, we define the norm

$$\|f\| = \sup_{(x,y)\in E\bigcup\Gamma} |f(x,y)| + \sup_{(x,y)\in E} \left|f^{\Delta_1}(x,\sigma_2(y))\right| + \sup_{(x,y)\in E} \left|f^{\Delta_2}(\sigma_1(x),y)\right|.$$

Definition 6.3 A function $\hat{u} \in D(\mathscr{L})$ is called a weak local minimum of $\mathscr{L}$ provided there exists a $\delta > 0$ such that

$$\mathscr{L}(\hat{u}) \leq \mathscr{L}(u)$$

for all $u \in D(\mathscr{L})$ with

$$\|u - \hat{u}\| < \delta.$$

If

$$\mathscr{L}(\hat{u}) < \mathscr{L}(u)$$

for all such $u \neq \hat{u}$, then $\hat{u}$ is said to be proper weak local minimum.

6.2 First and Second Variation

For a fixed element $u \in D(\mathscr{L})$ and a fixed admissible variation η, we define $\Phi : \mathbb{R} \to \mathbb{R}$ as follows.

$$\Phi(\epsilon) = \mathscr{L}(u + \epsilon\eta).$$

Definition 6.4 The first and second variation $\mathscr{L}$ at the point u are defined by

$$\mathscr{L}_1(u,\eta) = \Phi'(0) \quad \textit{and} \quad \mathscr{L}_2(u,\eta) = \Phi''(0),$$

respectively.

Theorem 6.1 **(Necessary Condition)** *If $\hat{u} \in D(\mathscr{L})$ is a local minimum of $\mathscr{L}$, then*

$$\mathscr{L}_1(u,\eta) = 0 \quad \textit{and} \quad \mathscr{L}_2(u,\eta) \geq 0$$

for all admissible variations η.

Proof. Assume that $\mathscr{L}$ has a local minimum at $\hat{u} \in D(\mathscr{L})$. Let η be an arbitrary admissible variation. Then

$$\Phi'(0) = \mathscr{L}_1(\hat{u},\eta) \quad \textit{and} \quad \Phi''(0) = \mathscr{L}_2(\hat{u},\eta).$$

By Taylor's formula, we get

$$\Phi(\epsilon) = \Phi(0) + \Phi'(0)\epsilon + \frac{1}{2!}\Phi''(\alpha)\epsilon^2,$$

where $|\alpha| \in (0, |\epsilon|)$.

If $|\epsilon|$ is sufficiently small, then

$$\|u + \epsilon\eta - u\| = |\epsilon|\,\|\eta\|$$

will be as small as we please. Hence, from the definition of a local minimum, we obtain

$$\mathscr{L}(u + \epsilon\eta) \geq \mathscr{L}(u),$$

that is,

$$\Phi(\epsilon) \geq \Phi(0).$$

Therefore Φ has a local minimum for $\epsilon = 0$. From here,

$$\Phi'(0) = 0,$$

or, equivalently,

$$\mathscr{L}_1(u, \eta) = 0.$$

Since $\Phi'(0) = 0$, we have

$$\Phi(\epsilon) - \Phi(0) = \frac{1}{2}\Phi''(\alpha)\epsilon^2.$$

Therefore $\Phi''(\alpha) \geq 0$ for all ϵ whose absolute values are sufficiently small. Letting $\epsilon \to 0$ and using the fact that $\alpha \to 0$, as $\epsilon \to 0$, and Φ'' is continuous, we get

$$\Phi''(0) \geq 0,$$

or, equivalently,

$$\mathscr{L}_2(u, \eta) \geq 0.$$

This completes the proof.

Theorem 6.2 (Sufficient Condition) *Let $u \in D(\mathscr{L})$ be such that*

$$\mathscr{L}_1(u, \eta) = 0$$

for all admissible variations η. If $\mathscr{L}_2(u, \eta) \leq 0$ for all $u \in D(\mathscr{L})$ and all admissible variations η, then $\mathscr{L}$ has an absolute minimum at the point u. If $\mathscr{L}_2(u, \eta) \leq 0$ for all u in some neighbourhood of the point u and all admissible variations η, then the functional $\mathscr{L}$ has a local maximum at u.

Proof. For the function Φ we have

$$\Phi(1) = \Phi(0) + \Phi'(0) + \frac{1}{2!}\Phi''(\alpha), \quad \alpha \in (0, 1). \tag{6.3}$$

Note that

$$\begin{aligned}
\Phi(1) &= \mathscr{L}(\varkappa+\eta), \quad \Phi(0) = \mathscr{L}(\varkappa),\\
\Phi'(0) &= \mathscr{L}_1(\varkappa,\eta)\\
&= 0,\\
\Phi''(\alpha) &= \left(\frac{d^2}{d\epsilon^2}\mathscr{L}(\varkappa+\epsilon\eta)\right)\Big|_{\epsilon=\alpha}\\
&= \left(\frac{d^2}{d\beta^2}\mathscr{L}(\varkappa+\alpha\eta+\beta\eta)\right)\Big|_{\beta=0}\\
&= \mathscr{L}_2(\varkappa+\alpha\eta,\eta).
\end{aligned}$$

Hence and (6.3), we obtain

$$\mathscr{L}(\varkappa+\eta) = \mathscr{L}(\varkappa) + \frac{1}{2!}\mathscr{L}_2(\varkappa+\alpha\eta,\eta)$$

for all admissible variations η, where $\alpha \in (0,1)$. Suppose that $\mathscr{L}_2(u,\eta) \geq 0$ for all $u \in D(\mathscr{L})$ and all admissible variations η. If $u \in D(\mathscr{L})$, then putting

$$\eta = u - \varkappa,$$

we get

$$\mathscr{L}(u) \geq \mathscr{L}(\varkappa).$$

Then $\mathscr{L}$ has an absolute minimum at the point $\varkappa$. Now we suppose that $\mathscr{L}_2(u,\eta) \leq 0$ for all u in some neighbourhood of the point $\varkappa$ and all admissible variations η. There exists $r > 0$ such that for $u \in D(\mathscr{L})$ and

$$\|u - \varkappa\| < r,$$

we have $\mathscr{L}_2(u,\eta) \leq 0$ for all admissible variations η. We take such an element u and we put $\eta = u - \varkappa$. Then

$$\mathscr{L}(u) = \mathscr{L}(\varkappa) + \frac{1}{2}\mathscr{L}_2(\varkappa+\alpha\eta,\eta).$$

Note that

$$\begin{aligned}
\|\varkappa+\alpha\eta-\varkappa\| &= \|\alpha\eta\|\\
&= |\alpha|\|\eta\|\\
&\leq \|\eta\|\\
&= \|u-\varkappa\|\\
&< r.
\end{aligned}$$

Hence,

$$\mathscr{L}_2(\varkappa+\alpha\eta,\eta) \leq 0,$$

and, then

$$\mathscr{L}(u) \leq \mathscr{L}(\tilde{u}).$$

This completes the proof.

By Theorem 6.1 and Theorem 6.2, it follows that

$$\begin{aligned}\mathscr{L}_1(u,\eta) = \int\int_E \Big(& L_u(x,y,u(\sigma_1(x),\sigma_2(y)),u^{\Delta_1}(x,\sigma_2(y)),u^{\Delta_2}(\sigma_1(x),y))\eta(\sigma_1(x),\sigma_2(y)) \\ & + L_p(x,y,u(\sigma_1(x),\sigma_2(y)),u^{\Delta_1}(x,\sigma_2(y)),u^{\Delta_2}(\sigma_1(x),y))\eta^{\Delta_1}(x,\sigma_2(y)) \\ & + L_q(x,y,u(\sigma_1(x),\sigma_2(y)),u^{\Delta_1}(x,\sigma_2(y)),u^{\Delta_2}(\sigma_1(x),y))\eta^{\Delta_2}(\sigma_1(x),y) \Big) \Delta_1 x \Delta_2 y,\end{aligned} \tag{6.4}$$

and

$$\begin{aligned}\mathscr{L}_2(u,\eta) = \int\int_E \Big(& L_{uu}(x,y,u(\sigma_1(x),\sigma_2(y)),u^{\Delta_1}(x,\sigma_2(y)),u^{\Delta_2}(\sigma_1(x),y))\left(\eta(\sigma_1(x),\sigma_2(y))\right)^2 \\ & + L_{pp}(x,y,u(\sigma_1(x),\sigma_2(y)),u^{\Delta_1}(x,\sigma_2(y)),u^{\Delta_2}(\sigma_1(x),y))\left(\eta^{\Delta_1}(x,\sigma_2(y))\right)^2 \\ & + L_{qq}(x,y,u(\sigma_1(x),\sigma_2(y)),u^{\Delta_1}(x,\sigma_2(y)),u^{\Delta_2}(\sigma_1(x),y))\left(\eta^{\Delta_2}(\sigma_1(x),y)\right)^2 \\ & + 2L_{up}(x,y,u(\sigma_1(x),\sigma_2(y)),u^{\Delta_1}(x,\sigma_2(y)),u^{\Delta_2}(\sigma_1(x),y))\eta(\sigma_1(x),\sigma_2(y)) \\ & \quad \eta^{\Delta_1}(x,\sigma_2(y)) \\ & + 2L_{uq}(x,y,u(\sigma_1(x),\sigma_2(y)),u^{\Delta_1}(x,\sigma_2(y)),u^{\Delta_2}(\sigma_1(x),y))\eta(\sigma_1(x),\sigma_2(y)) \\ & \quad \eta^{\Delta_2}(\sigma_1(x),y) \\ & + 2L_{pq}(x,y,u(\sigma_1(x),\sigma_2(y)),u^{\Delta_1}(x,\sigma_2(y)),u^{\Delta_2}(\sigma_1(x),y))\eta^{\Delta_1}(x,\sigma_2(y)) \\ & \quad \eta^{\Delta_2}(\sigma_1(x),y) \Big) \Delta_1 x \Delta_2 y.\end{aligned} \tag{6.5}$$

Example 6.1 Let

$$\begin{aligned} & L(x,y,u(\sigma_1(x),\sigma_2(y)),u^{\Delta_1}(x,\sigma_2(y)),u^{\Delta_2}(\sigma_1(x),y)) \\ & = x+y+u(\sigma_1(x),\sigma_2(y)) + \left(u^{\Delta_1}(x,\sigma_2(y))\right)^2 \\ & \quad + \left(u^{\Delta_2}(\sigma_1(x),y)\right)^3.\end{aligned}$$

Here

$$L(x,y,u,p,q) = x+y+u+p^2+q^3.$$

Then

$$L_u(x,y,u,p,q) = 1,$$

$$L_p(x,y,u,p,q) = 2p,$$
$$L_q(x,y,u,p,q) = 3q^2,$$
$$L_{uu}(x,y,u,p,q) = 0,$$
$$L_{pp}(x,y,u,p,q) = 2,$$
$$L_{qq}(x,y,u,p,q) = 6q,$$
$$L_{uq}(x,y,u,p,q) = 0,$$
$$L_{up}(x,y,u,p,q) = 0,$$
$$L_{pq}(x,y,u,p,q) = 0.$$

Therefore the equations (6.4) and (**??**) take the form

$$\mathscr{L}_1(u,\eta) = \int\int_E \Bigg(\eta(\sigma_1(x),\sigma_2(y)) + 2u^{\Delta_1}(x,\sigma_2(y))\eta^{\Delta_1}(x,\sigma_2(y))$$
$$+3\left(u^{\Delta_1}(\sigma_1(x),y)\right)^2 \eta^{\Delta_2}(\sigma_1(x),y) \Bigg) \Delta_1 x \Delta_2 y,$$

$$\mathscr{L}_2(u,\eta) = \int\int_E \left(2\left(\eta^{\Delta_1}(x,\sigma_2(y))\right)^2 + 6u^{\Delta_2}(\sigma_1(x),y)\left(\eta^{\Delta_2}(\sigma_1(x),y)\right)^2\right) \Delta_1 x \Delta_2 y.$$

Exercise 6.1 Write the equations (6.4) and (**??**) for the following functionals.

1. $L(x,y,u,p,q) = xy + u^2 + upq,$
2. $L(x,y,u,p,q) = xyupq,$
3. $L(x,y,u,p,q) = x^2 + u^2p^2 + u^2q^2,$
4. $L(x,y,u,p,q) = x^2 + yupq,$
5. $L(x,y,u,p,q) = (x - yu)^2 + (y + pq)^2.$

6.3 Euler's Condition

Let E be an ω-type subset of $\mathbb{T}_1 \times \mathbb{T}_2$ and Γ be the positively oriented fence of E. We set

$$E^\sigma = \left\{(x,y) \in E : (\sigma_1(x),\sigma_2(y)) \in E\right\}.$$

Lemma 6.1 (Dubois-Reymond's Lemma) *If $M(x,y)$ is continuous on $E\bigcup\Gamma$ with*

$$\int\int_E M(x,y)\eta(\sigma_1(x),\sigma_2(y))\Delta_1 x \Delta_2 y = 0$$

for every admissible variation η, *then*

$$M(x,y) = 0 \quad \textit{for} \quad \textit{all} \quad (x,y) \in E^{\sigma}.$$

Proof. Assume the contrary. Without loss of generality, we suppose that $(x_0,y_0) \in E^{\sigma}$ is such that $M(x_0,y_0) > 0$. The continuity of $M(x,y)$ ensures that $M(x,y)$ is positive in a rectangle

$$\Omega = [x_0,x_1) \times [y_0,y_1) \subset E$$

for some points $x_1 \in \mathbb{T}_1$, $y_1 \in \mathbb{T}_2$ such that

$$\sigma_1(x_0) \leq x_1 \quad \textit{and} \quad \sigma_2(y_0) \leq y_1.$$

We set

$$\eta(x,y) = \begin{cases} (x-x_0)^2(x-\sigma_1(x_1))^2(y-y_0)(y-\sigma_2(y_1))^2 & \textit{for} \quad (x,y) \in \Omega \\ 0 \quad \textit{for} \quad (x,y) \in E\backslash\Omega. \end{cases}$$

We have that $\eta \in \mathscr{C}^{(1)}_{rd}(E\bigcup\Gamma)$, $\eta\Big|_{\Gamma} = 0$, that is, η is an admissible variation. We have that

$$\int\int_E M(x,y)\eta(\sigma_1(x),\sigma_2(y))\Delta_1 x\Delta_2 y = \int\int_\Omega M(x,y)\eta(\sigma_1(x),\sigma_2(y))\Delta_1 x\Delta_2 y$$
$$> 0,$$

which is a contradiction. This completes the proof.

Theorem 6.3 (Euler's Necessary Condition) *Suppose that an admissible function* u *provides a local minimum for* $\mathscr{L}$ *and the function* u *has continuous partial delta derivatives of the second order. Then* u *satisfies the Euler-Lagrange equation*

$$\begin{aligned} 0 &= L_u(x,y,u(\sigma_1(x),\sigma_2(y)),u^{\Delta_1}(x,\sigma_2(y)),u^{\Delta_2}(\sigma_1(x),y)) \\ &\quad -L_p^{\Delta_1}(x,y,u(\sigma_1(x),\sigma_2(y)),u^{\Delta_1}(x,\sigma_2(y)),u^{\Delta_2}(\sigma_1(x),y)) \\ &\quad -L_q^{\Delta_2}(x,y,u(\sigma_1(x),\sigma_2(y)),u^{\Delta_1}(x,\sigma_2(y)),u^{\Delta_2}(\sigma_1(x),y)) \end{aligned} \tag{6.6}$$

for $(x,y) \in E^{\sigma}$.

Proof. Since u is a local minimum for $\mathscr{L}$, by Theorem 6.1, it follows that

$$\mathscr{L}_1(u,\eta) = 0$$

for all admissible variations η. Hence and by (6.4), applying integration by parts and Green's formula, we get

$$0 = \mathscr{L}_1(u,\eta)$$
$$= \int\int_E \Big(L_u(x,y,u(\sigma_1(x),\sigma_2(y)),u^{\Delta_1}(x,\sigma_2(y)),u^{\Delta_2}(\sigma_1(x),y))$$

$$\times\eta(\sigma_1(x),\sigma_2(y))\Delta_1 x\Delta_2 y$$

$$+L_p(x,y,u(\sigma_1(x),\sigma_2(y)),u^{\Delta_1}(x,\sigma_2(y)),u^{\Delta_2}(\sigma_1(x),y))$$

$$\times\eta^{\Delta_1}(x,\sigma_2(y))$$

$$+L_q(x,y,u(\sigma_1(x),\sigma_2(y)),u^{\Delta_1}(x,\sigma_2(y)),u^{\Delta_2}(\sigma_1(x),y))\eta^{\Delta_2}(\sigma_1(x),y)\Bigg)\Delta_1 x\Delta_2 y$$

$$=\int\int_E L_u(x,y,u(\sigma_1(x),\sigma_2(y)),u^{\Delta_1}(x,\sigma_2(y)),u^{\Delta_2}(\sigma_1(x),y))$$

$$\times\eta(\sigma_1(x),\sigma_2(y))\Delta_1 x\Delta_2 y$$

$$+\int\int_E\Bigg(L_p(x,y,u(\sigma_1(x),\sigma_2(y)),u^{\Delta_1}(x,\sigma_2(y)),u^{\Delta_2}(\sigma_1(x),y))$$

$$\times\eta^{\Delta_1}(x,\sigma_2(y))$$

$$+L_q(x,y,u(\sigma_1(x),\sigma_2(y)),u^{\Delta_1}(x,\sigma_2(y)),u^{\Delta_2}(\sigma_1(x),y))$$

$$\times\eta^{\Delta_2}(\sigma_1(x),y)\Bigg)\Delta_1 x\Delta_2 y$$

$$=\int\int_E L_u(x,y,u(\sigma_1(x),\sigma_2(y)),u^{\Delta_1}(x,\sigma_2(y)),u^{\Delta_2}(\sigma_1(x),y))$$

$$\times\eta(\sigma_1(x),\sigma_2(y))\Delta_1 x\Delta_2 y$$

$$+\int\int_E\Bigg(\frac{\partial}{\Delta_1 x}\Bigg(L_p(x,y,u(\sigma_1(x),\sigma_2(y)),u^{\Delta_1}(x,\sigma_2(y)),u^{\Delta_2}(\sigma_1(x),y))$$

$$\times\eta(x,\sigma_2(y))\Bigg)$$

$$+\frac{\partial}{\Delta_2 y}\Big(L_q(x,y,u(\sigma_1(x),\sigma_2(y)),u^{\Delta_1}(x,\sigma_2(y)),u^{\Delta_2}(\sigma_1(x),y))$$

$$\times\eta(\sigma_1(x),y)\Big)\Bigg)\Delta_1 x\Delta_2 y$$

$$-\int\int_E\Bigg(L_p^{\Delta_1}(x,y,u(\sigma_1(x),\sigma_2(y)),u^{\Delta_1}(x,\sigma_2(y)),u^{\Delta_2}(\sigma_1(x),y))$$

$$+L_q^{\Delta_2}(x,y,u(\sigma_1(x),\sigma_2(y)),u^{\Delta_1}(x,\sigma_2(y)),u^{\Delta_2}(\sigma_1(x),y))\Bigg)$$

$$\times\eta(\sigma_1(x),\sigma_2(y))\Delta_1 x\Delta_2 y$$

$$
\begin{aligned}
&= \iint_E L_u(x,y,u(\sigma_1(x),\sigma_2(y)),u^{\Delta_1}(x,\sigma_2(y)),u^{\Delta_2}(\sigma_1(x),y)) \\
&\quad \times \eta(\sigma_1(x),\sigma_2(y))\Delta_1 x\Delta_2 y \\
&\quad - \iint_E \Big(L_p^{\Delta_1}(x,y,u(\sigma_1(x),\sigma_2(y)),u^{\Delta_1}(x,\sigma_2(y)),u^{\Delta_2}(\sigma_1(x),y)) \\
&\quad + L_q^{\Delta_2}(x,y,u(\sigma_1(x),\sigma_2(y)),u^{\Delta_1}(x,\sigma_2(y)),u^{\Delta_2}(\sigma_1(x),y)) \Big) \\
&\quad \times \eta(\sigma_1(x),\sigma_2(y))\Delta_1 x\Delta_2 y \\
&\quad + \int_\Gamma \Big(L_p^{\Delta_1}(x,y,u(\sigma_1(x),\sigma_2(y)),u^{\Delta_1}(x,\sigma_2(y)),u^{\Delta_2}(\sigma_1(x),y)) \\
&\quad \times \eta(x,\sigma_2(y))\Delta_2 y \\
&\quad + L_q^{\Delta_2}(x,y,u(\sigma_1(x),\sigma_2(y)),u^{\Delta_1}(x,\sigma_2(y)),u^{\Delta_2}(\sigma_1(x),y)) \\
&\quad \times \eta(\sigma_1(x),y)\Delta_1 x \Big) \\
&= \iint_E L_u(x,y,u(\sigma_1(x),\sigma_2(y)),u^{\Delta_1}(x,\sigma_2(y)),u^{\Delta_2}(\sigma_1(x),y)) \\
&\quad \times \eta(\sigma_1(x),\sigma_2(y))\Delta_1 x\Delta_2 y \\
&\quad - \iint_E \Big(L_p^{\Delta_1}(x,y,u(\sigma_1(x),\sigma_2(y)),u^{\Delta_1}(x,\sigma_2(y)),u^{\Delta_2}(\sigma_1(x),y)) \\
&\quad + L_q^{\Delta_2}(x,y,u(\sigma_1(x),\sigma_2(y)),u^{\Delta_1}(x,\sigma_2(y)),u^{\Delta_2}(\sigma_1(x),y)) \Big) \\
&\quad \times \eta(\sigma_1(x),\sigma_2(y))\Delta_1 x\Delta_2 y.
\end{aligned}
$$

From here and from Lemma 6.1, we get (6.6). This completes the proof.

Example 6.2 Let $\mathbb{T}_1 = 2^{\mathbb{N}_0}$ and $\mathbb{T}_2 = 3^{\mathbb{N}_0}$. Consider the variational problem

$$\mathscr{L}(u) = \iint_E x^2y^3u(2x,3y)u^{\Delta_1}(x,3y)u^{\Delta_2}(2x,y)\Delta_1 x\Delta_2 y \longrightarrow \min,$$

where

$$E = \{(x,y) \in \mathbb{T}_1 \times \mathbb{T}_2 : 1 \le x \le 8, \quad 1 \le y \le 27\}.$$

Here

$$L(x,y,u,p,q) = x^2y^3upq,$$

$$\sigma_1(x) = 2x, \quad x \in \mathbb{T}_1,$$

$$\sigma_2(y) = 3y, \quad y \in \mathbb{T}_2.$$

Then

$$\begin{aligned}
L_u(x,y,u,p,q) &= x^2y^3pq, \\
L_p(x,y,u,p,q) &= x^2y^3uq, \\
L_q(x,y,u,p,q) &= x^2y^3up, \\
L_p^{\Delta_1}(x,y,u,p,q) &= (\sigma_1(x)+x)y^3uq \\
&= (2x+x)y^3uq \\
&= 3xy^3uq, \\
L_q^{\Delta_2}(x,y,u,p,q) &= x^2\left((\sigma_2(y))^2 + y\sigma_2(y) + y^2\right)up \\
&= x^2\left((3y)^2 + y(3y) + y^2\right)up \\
&= x^2(9y^2+3y^2+y^2)up \\
&= 13x^2y^2up.
\end{aligned}$$

The Euler-Lagrange equation takes the form

$$\begin{aligned}
&x^2y^3u^{\Delta_1}(x,3y)u^{\Delta_2}(2x,y) - 3xy^3u(2x,3y)u^{\Delta_2}(2x,y) \\
&- 13x^2y^2u(2x,3y)u^{\Delta_1}(x,3y) = 0, \quad (x,y) \in E.
\end{aligned}$$

Exercise 6.2 Write the Euler-Lagrange equation for the variational problem

$$\mathscr{L}(y) = \int\int_E \left((x^2+y^2)\,u(\sigma_1(x),\sigma_2(y)) + \left(u^{\Delta_1}(x,\sigma_2(y))\right)^2 u^{\Delta_2}(\sigma_1(x),y)\right)\Delta_1x\Delta_2y \longrightarrow \min,$$

in the cases

1. $\mathbb{T}_1 = \mathbb{Z}$, $\mathbb{T}_2 = 2\mathbb{Z}$,
2. $\mathbb{T}_1 = \mathbb{T}_2 = 3\mathbb{Z}$,
3. $\mathbb{T}_1 = 2^{\mathbb{N}_0}$, $\mathbb{T}_2 = 3\mathbb{Z}$,
4. $\mathbb{T}_1 = \mathbb{N}_0$, $\mathbb{T}_2 = 3^{\mathbb{N}_0}$,
5. $\mathbb{T}_1 = \mathbb{T}_2 = \mathbb{N}_0^3$.

6.4 Minimal Surfaces

Suppose that $\mathbb{T}$, $\mathbb{T}_1$, $\mathbb{T}_2$, $\mathbb{T}_3$, $\mathbb{T}_{(1)}$, $\mathbb{T}_{(2)}$, $\mathbb{T}_{(3)}$ are time scales with forward jump operators and delta differentiation operators σ, σ_1, σ_2, σ_3, $\sigma_{(1)}$, $\sigma_{(2)}$, $\sigma_{(3)}$ and Δ, Δ_1, Δ_2, Δ_3, $\Delta_{(1)}$, $\Delta_{(2)}$ and $\Delta_{(3)}$, respectively. Let $I \subseteq \mathbb{T}$, $U, U_1, W_1 \subseteq \mathbb{T}_1 \times \mathbb{T}_2$, $W_2 \subseteq \mathbb{T}_{(1)} \times \mathbb{T}_{(2)} \times \mathbb{T}_{(3)}$ and $V \subseteq \mathbb{T}_1 \times \mathbb{T}_2 \times \mathbb{T}_3$.

Let S be a σ_1- or σ_2- or $\sigma_e^2\sigma_2$- or $\sigma_1\sigma_2^2$-oriented surface with a local σ_1- or σ_2- or $\sigma_1^2\sigma_2$- or $\sigma_1\sigma_2^2$- parametrization (U,f).

Definition 6.5 Let S be a σ_1-regular surface, σ_1 be Δ_1-differentiable on $\mathbb{T}_1$ and $f_{t_1}^{\Delta_1}$ and $f_{t_2}^{\Delta_2\sigma_1}$ be σ_1-completely delta differentiable on U. Then S is said to be a $\sigma_1\sigma_1\sigma_1$-minimal surface and its $\sigma_1\sigma_1\sigma_1$-mean curvature vanishes.

Theorem 6.4 *Let S be a σ_1-regular surface, σ_1 be Δ_1-differentiable on $\mathbb{T}_1$ and $f_{t_1}^{\Delta_1}$ and $f_{t_2}^{\Delta_2\sigma_1}$ be σ_1-completely delta differentiable on U. Then S is a $\sigma_1\sigma_1\sigma_1$-minimal surface if and only if*

$$\left(f_{t_1}^{\Delta_1^2}+f_{t_2}^{\Delta_2^2\sigma_1^2}\right)\cdot N_{\sigma_1}^{\sigma_1\sigma_2}=0.$$

Definition 6.6 Let S be a σ_1-regular surface, σ_1 be Δ_1-differentiable on $\mathbb{T}_1$, $f_{t_1}^{\Delta_1}$ be σ_1-completely delta differentiable and $f_{t_2}^{\Delta_2\sigma_1}$ be σ_2-completely delta differentiable on U. Then S is said to be a $\sigma_1\sigma_1\sigma_2$-minimal surface if and only if its $\sigma_1\sigma_1\sigma_2$-mean curvature vanishes.

Theorem 6.5 *Let S be a σ_1-regular surface, σ_1 be Δ_1-differentiable on $\mathbb{T}_1$, $f_{t_1}^{\Delta_1}$ be σ_1-completely delta differentiable and $f_{t_2}^{\Delta_2\sigma_1}$ be σ_2-completely delta differentiable on U. Then S is a $\sigma_1\sigma_1\sigma_2$-minimal surface if and only if*

$$\left(f_{t_1}^{\Delta_1^2}+f_{t_2}^{\Delta_2^2\sigma_1}\right)\cdot N_{\sigma_1}^{\sigma_1\sigma_2}=0.$$

Definition 6.7 Let S be a σ_1-regular surface, σ_1 be Δ_1-differentiable on $\mathbb{T}_1$, $f_{t_1}^{\Delta_1}$ be σ_2-completely delta differentiable and $f_{t_2}^{\Delta_2\sigma_1}$ be σ_1-completely delta differentiable on U. Then S is said to be a $\sigma_1\sigma_2\sigma_1$-minimal surface if its $\sigma_1\sigma_2\sigma_1$-mean curvature vanishes.

Theorem 6.6 *Let S be a σ_1-regular surface, σ_1 be Δ_1-differentiable on $\mathbb{T}_1$, $f_{t_1}^{\Delta_1}$ be σ_2-completely delta differentiable and $f_{t_2}^{\Delta_2\sigma_1}$ be σ_1-completely delta differentiable on U. Then S is a $\sigma_1\sigma_2\sigma_1$-minimal surface if and only if*

$$\left(f_{t_1}^{\Delta_1^2\sigma_2}+f_{t_2}^{\Delta_2^2\sigma_1^2}\right)\cdot N_{\sigma_1}^{\sigma_1\sigma_2}=0.$$

Definition 6.8 Let S be a σ_1-regular surface, σ_1 be Δ_1-differentiable on $\mathbb{T}_1$ and $f_{t_1}^{\Delta_1}$ and $f_{t_2}^{\Delta_2\sigma_1}$ be σ_2-completely delta differentiable on U. Then S is said to be a $\sigma_1\sigma_2\sigma_2$-minimal surface if its $\sigma_1\sigma_2\sigma_2$-mean curvature vanishes.

Theorem 6.7 *Let S be a σ_1-regular surface, σ_1 be Δ_1-differentiable on $\mathbb{T}_1$ and $f_{t_1}^{\Delta_1}$ and $f_{t_2}^{\Delta_2\sigma_1}$ be σ_2-completely delta differentiable on U. Then S is a $\sigma_1\sigma_2\sigma_2$-minimal surface if and only if*

$$\left(f_{t_1}^{\Delta_1^2\sigma_2}+f_{t_2}^{\Delta_2^2\sigma_1}\right)\cdot N_{\sigma_1}^{\sigma_1\sigma_2}=0.$$

Definition 6.9 Let S be a σ_2-regular surface, σ_2 be Δ_2-differentiable on $\mathbb{T}_2$ and $f_{t_1}^{\Delta_1\sigma_2}$ and $f_{t_2}^{\Delta_2}$ be σ_1-completely delta differentiable on U. Then S is said to be a $\sigma_2\sigma_1\sigma_1$-minimal surface if its $\sigma_2\sigma_1\sigma_1$-mean curvature vanishes.

Theorem 6.8 *Let S be a σ_2-regular surface, σ_2 be Δ_2-differentiable on $\mathbb{T}_2$ and $f_{t_1}^{\Delta_1\sigma_2}$ and $f_{t_2}^{\Delta_2}$ be σ_1-completely delta differentiable on U. Then S is a $\sigma_2\sigma_1\sigma_1$-minimal surface if and only if*

$$\left(f_{t_1}^{\Delta_1^2} + f_{t_2}^{\Delta_2^2\sigma_1}\right) \cdot N_{\sigma_2}^{\sigma_1\sigma_2} = 0.$$

Definition 6.10 Let S_2 be a σ_2-regular surface, σ_2 be Δ_2-differentiable on $\mathbb{T}_1$, $f_{t_1}^{\Delta_1\sigma_2}$ be σ_1-completely delta differentiable and $f_{t_2}^{\Delta_2}$ be σ_2-completely delta differentiable on U. Then S is said to be a $\sigma_2\sigma_1\sigma_2$-minimal surface if its $\sigma_2\sigma_1\sigma_2$-mean curvature vanishes.

Theorem 6.9 *Let S_2 be a σ_2-regular surface, σ_2 be Δ_2-differentiable on $\mathbb{T}_1$, $f_{t_1}^{\Delta_1\sigma_2}$ be σ_1-completely delta differentiable and $f_{t_2}^{\Delta_2}$ be σ_2-completely delta differentiable on U. Then S is a $\sigma_2\sigma_1\sigma_2$-minimal surface if and only if*

$$\left(f_{t_1}^{\Delta_1^2} + f_{t_2}^{\Delta_2^2}\right) \cdot N_{\sigma_2}^{\sigma_1\sigma_2} = 0.$$

Definition 6.11 Let S be a σ_2-regular surface, σ_2 be Δ_2-differentiable on $\mathbb{T}_2$, $f_{t_1}^{\Delta_1\sigma_2}$ be σ_2-completely delta differentiable and $f_{t_2}^{\Delta_2}$ be σ_1-completely delta differentiable on U. Then S is said to be a $\sigma_2\sigma_2\sigma_1$-minimal surface if its $\sigma_2\sigma_2\sigma_1$-mean curvature vanishes.

Theorem 6.10 *Let S be a σ_2-regular surface, σ_2 be Δ_2-differentiable on $\mathbb{T}_2$, $f_{t_1}^{\Delta_1\sigma_2}$ be σ_2-completely delta differentiable and $f_{t_2}^{\Delta_2}$ be σ_1-completely delta differentiable on U. Then S is a $\sigma_2\sigma_2\sigma_1$-minimal surface if and only if*

$$\left(f_{t_1}^{\Delta_1^2\sigma_2^2} + f_{t_2}^{\Delta_2^2\sigma_1}\right) \cdot N_{\sigma_2}^{\sigma_1\sigma_2} = 0.$$

Definition 6.12 Let S be a σ_2-regular surface, σ_2 be Δ_2-differentiable on $\mathbb{T}_1$ be $f_{t_1}^{\Delta_1\sigma_2}$ and $f_{t_2}^{\Delta_2}$ be σ_2-completely delta differentiable on U. Then S is said to be a $\sigma_1^2\sigma_2\sigma_1\sigma_1$-minimal surface if its $\sigma_2\sigma_2\sigma_2$-mean curvature vanishes.

Theorem 6.11 *Let S be a σ_2-regular surface, σ_2 be Δ_2-differentiable on $\mathbb{T}_1$ and $f_{t_1}^{\Delta_1\sigma_2}$ and $f_{t_2}^{\Delta_2}$ be σ_2-completely delta differentiable on U. Then S is a $\sigma_2\sigma_2\sigma_2$-minimal surface if and only if*

$$\left(f_{t_1}^{\Delta_1^2\sigma_2^2} + f_{t_2}^{\Delta_2^2}\right) \cdot N_{\sigma_2}^{\sigma_1\sigma_2} = 0.$$

Definition 6.13 Let S be a $\sigma_1^2\sigma_2$-regular surface, σ_1 be Δ_1-differentiable on $\mathbb{T}_1$ and $f_{t_1}^{\Delta_1}$ and $f_{t_2}^{\Delta_2\sigma_1}$ be σ_1-completely delta differentiable on U. Then S is said to be a $\sigma_1^2\sigma_2\sigma_1\sigma_1$-minimal surface if its $\sigma_1^2\sigma_2\sigma_1\sigma_1$-mean curvature vanishes.

Theorem 6.12 *Let S be a $\sigma_1^2\sigma_2$-regular surface, σ_1 be Δ_1-differentiable on $\mathbb{T}_1$ and $f_{t_1}^{\Delta_1}$ and $f_{t_2}^{\Delta_2\sigma_1}$ be σ_1-completely delta differentiable on U. Then S is a $\sigma_1^2\sigma_2\sigma_1\sigma_1$-minimal surface if and only if*

$$\left(f_{t_1}^{\Delta_1^2} + f_{t_2}^{\Delta_2^2\sigma_1^2}\right) \cdot N_{\sigma_1\sigma_2^2}^{\sigma_1\sigma_2} = 0.$$

Definition 6.14 Let S be a $\sigma_1^2\sigma_2$-regular surface, σ_1 be Δ_1-differentiable on $\mathbb{T}_1$, $f_{t_1}^{\Delta_1}$ be σ_1-completely delta differentiable and $f_{t_2}^{\Delta_2\sigma_1}$ be σ_2-completely delta differentiable on U. Then S is said to be a $\sigma_1^2\sigma_2\sigma_1\sigma_2$-minimal surface if its $\sigma_1^2\sigma_2\sigma_1\sigma_2$-mean curvature vanishes.

Theorem 6.13 *Let S be a $\sigma_1^2\sigma_2$-regular surface, σ_1 be Δ_1-differentiable on $\mathbb{T}_1$, $f_{t_1}^{\Delta_1}$ be σ_1-completely delta differentiable and $f_{t_2}^{\Delta_2\sigma_1}$ be σ_2-completely delta differentiable on U. Then S is a $\sigma_1^2\sigma_2\sigma_1\sigma_2$-minimal surface if and only if*

$$\left(f_{t_1}^{\Delta_1^2} + f_{t_2}^{\Delta_2^2\sigma_1}\right) \cdot N_{\sigma_1\sigma_2^2}^{\sigma_1\sigma_2} = 0.$$

Definition 6.15 *Let S be a $\sigma_1^2\sigma_2$-regular surface, σ_1 be Δ_1-differentiable on $\mathbb{T}_1$, $f_{t_1}^{\Delta_1}$ be σ_2-completely delta differentiable and $f_{t_2}^{\Delta_2\sigma_1}$ be σ_1-completely delta differentiable on U. Then S is said to be a $\sigma_1^2\sigma_2\sigma_2\sigma_1$-minimal surface if its $\sigma_1^2\sigma_2\sigma_2\sigma_1$-mean curvature vanishes.*

Theorem 6.14 *Let S be a $\sigma_1^2\sigma_2$-regular surface, σ_1 be Δ_1-differentiable on $\mathbb{T}_1$, $f_{t_1}^{\Delta_1}$ be σ_2-completely delta differentiable and $f_{t_2}^{\Delta_2\sigma_1}$ be σ_1-completely delta differentiable on U. Then S is a $\sigma_1^2\sigma_2\sigma_2\sigma_1$-minimal surface if and only if*

$$\left(f_{t_1}^{\Delta_1^2\sigma_2} + f_{t_2}^{\Delta_2^2\sigma_1^2}\right) \cdot N_{\sigma_1\sigma_2^2}^{\sigma_1\sigma_2} = 0.$$

Definition 6.16 Let S be a $\sigma_1^2\sigma_2$-regular surface, σ_1 be Δ_1-differentiable on $\mathbb{T}_1$ and $f_{t_1}^{\Delta_1}$ and $f_{t_2}^{\Delta_2\sigma_1}$ be σ_2-completely delta differentiable on U. Then S is said to be a $\sigma_1^2\sigma_2\sigma_2\sigma_2$-minimal surface if its $\sigma_1^2\sigma_2\sigma_2\sigma_2$-mean curvature vanishes.

Theorem 6.15 *Let S be a $\sigma_1^2\sigma_2$-regular surface, σ_1 be Δ_1-differentiable on $\mathbb{T}_1$ and $f_{t_1}^{\Delta_1}$ and $f_{t_2}^{\Delta_2\sigma_1}$ be σ_2-completely delta differentiable on U. Then S is a $\sigma_1^2\sigma_2\sigma_2\sigma_2$-minimal surface if and only if*

$$\left(f_{t_1}^{\Delta_1^2\sigma_2} + f_{t_2}^{\Delta_2^2\sigma_1}\right) \cdot N_{\sigma_1^2\sigma_2}^{\sigma_1\sigma_2} = 0.$$

Definition 6.17 Let S be a $\sigma_1\sigma_2^2$-regular surface, σ_2 be Δ_2-differentiable on $\mathbb{T}_2$ And $f_{t_1}^{\Delta_1\sigma_2}$ and $f_{t_2}^{\Delta_2}$ be σ_1-completely delta differentiable on U. Then S is said to be a $\sigma_1\sigma_2^2\sigma_1\sigma_1$-minimal surface if its $\sigma_1\sigma_2^2\sigma_1\sigma_1$-mean curvature vanishes.

Theorem 6.16 *Let S be a $\sigma_1\sigma_2^2$-regular surface, σ_2 be Δ_2-differentiable on $\mathbb{T}_2$ and $f_{t_1}^{\Delta_1\sigma_2}$ and $f_{t_2}^{\Delta_2}$ be σ_1-completely delta differentiable on U. Then S is a $\sigma_1\sigma_2^2\sigma_1\sigma_1$-minimal surface if and only if*

$$\left(f_{t_1}^{\Delta_1^2} + f_{t_2}^{\Delta_2^2\sigma_1}\right) \cdot N_{\sigma_1\sigma_2^2}^{\sigma_1\sigma_2} = 0.$$

Definition 6.18 Let S be a $\sigma_1\sigma_2^2$-regular surface, σ_2 be Δ_2-differentiable on $\mathbb{T}_2$ and $f_{t_1}^{\Delta_1\sigma_2}$ be σ_1-completely delta differentiable and be $f_{t_2}^{\Delta_2}$ is σ_2-completely delta differentiable on U. Then S is said to be a $\sigma_1\sigma_2^2\sigma_1\sigma_2$-minimal surface if its $\sigma_1\sigma_2^2\sigma_1\sigma_2$-mean curvature vanishes.

Theorem 6.17 *Let S be a $\sigma_1\sigma_2^2$-regular surface, σ_2 be Δ_2-differentiable on $\mathbb{T}_2$, $f_{t_1}^{\Delta_1\sigma_2}$ be σ_1-completely delta differentiable and $f_{t_2}^{\Delta_2}$ be σ_2-completely delta differentiable on U. Then S is a $\sigma_1\sigma_2^2\sigma_1\sigma_2$-minimal surface if and only if*

$$\left(f_{t_1}^{\Delta_1^2}+f_{t_2}^{\Delta_2^2}\right)\cdot N_{\sigma_1\sigma_2^2}^{\sigma_1\sigma_2}=0.$$

Definition 6.19 Let S be a $\sigma_1\sigma_2^2$-regular surface, σ_2 be Δ_2-differentiable on $\mathbb{T}_2$, $f_{t_1}^{\Delta_1\sigma_2}$ be σ_2-completely delta differentiable and $f_{t_2}^{\Delta_2}$ is σ_1-completely delta differentiable on U. Then S is said to be a $\sigma_1\sigma_2^2\sigma_2\sigma_1$-minimal surface if its $\sigma_1\sigma_2^2\sigma_2\sigma_1$-mean curvature vanishes.

Theorem 6.18 *Let S be a $\sigma_1\sigma_2^2$-regular surface, σ_2 beΔ_2-differentiable on $\mathbb{T}_2$, $f_{t_1}^{\Delta_1\sigma_2}$ is be σ_2-completely delta differentiable and $f_{t_2}^{\Delta_2}$ be σ_1-completely delta differentiable on U. Then S is a $\sigma_1\sigma_2^2\sigma_2\sigma_1$-minimal surface if and only if*

$$\left(f_{t_1}^{\Delta_1^2\sigma_2^2}+f_{t_2}^{\Delta_2^2\sigma_1}\right)\cdot N_{\sigma_1\sigma_2^2}^{\sigma_1\sigma_2}=0.$$

Definition 6.20 Let S be a $\sigma_1\sigma_2^2$-regular surface, σ_2 be Δ_2-differentiable on $\mathbb{T}_2$ and $f_{t_1}^{\Delta_1\sigma_2}$ and $f_{t_2}^{\Delta_2}$ be σ_2-completely delta differentiable on U. Then S is said to be a $\sigma_1\sigma_2^2\sigma_2\sigma_2$-minimal surface if its $\sigma_1\sigma_2^2\sigma_2\sigma_2$-mean curvature vanishes.

Theorem 6.19 *Let S be a $\sigma_1\sigma_2^2$-regular surface, σ_2 be Δ_2-differentiable on $\mathbb{T}_2$ and $f_{t_1}^{\Delta_1\sigma_2}$ be $f_{t_2}^{\Delta_2}$ be σ_2-completely delta differentiable on U. Then S is a $\sigma_1\sigma_2^2\sigma_2\sigma_2$-minimal surface if and only if*

$$\left(f_{t_1}^{\Delta_1^2\sigma_2^2}+f_{t_2}^{\Delta_2^2}\right)\cdot N_{\sigma_1\sigma_2^2}^{\sigma_1\sigma_2}=0.$$

6.5 Advanced Practical Problems

Problem 6.1 Write the equations (6.4) and (**??**) for the following functionals.

1. $L(x,y,u,p,q)=x+p+q+u^2-(y+2p)^2$,
2. $L(x,y,u,p,q)=(x^2+p)^3+yuq$,
3. $L(x,y,u,p,q)=(x-p)^2+(y+q)^3$,
4. $L(x,y,u,p,q)=(x+y+u+p+q)^2$,
5. $L(x,y,u,p,q)=(x-u-p)^2-q^4$.

Problem 6.2 Write the Euler-Lagrange equation for the variational problem

$$\mathscr{L}(u)=\int\int_E\left((x^3-3xy+y^2+y^4)\,(u(\sigma_1(x),\sigma_2(y)))^2\right)\Delta_1x\Delta_2y\longrightarrow\min,$$

in the cases

1. $\mathbb{T}_1 = \mathbb{Z}$, $\mathbb{T}_2 = 2\mathbb{Z}$,
2. $\mathbb{T}_1 = \mathbb{T}_2 = 3\mathbb{Z}$,
3. $\mathbb{T}_1 = 2^{\mathbb{N}_0}$, $\mathbb{T}_2 = 3\mathbb{Z}$,
4. $\mathbb{T}_1 = \mathbb{N}_0$, $\mathbb{T}_2 = 3^{\mathbb{N}_0}$,
5. $\mathbb{T}_1 = \mathbb{T}_2 = \mathbb{N}_0^3$.

6.6 Notes and References

In this chapter we introduce the first and second variations. They are given necessary and sufficient conditions for the extremum of a functional. The Euler equations are deduced. Also in this chapter the minimal sufraces are defined and some of their properties are deduced.

7

The Delta Nature Connection

7.1 The Directional Derivative

Let $\mathbb{T}$ be a time scale with forward jump operator σ and delta differentiation operator Δ. Also let, $x^0 \in \mathbb{T}$, $v = (v_1, v_2, \ldots, v_n) \in \mathbb{R}^n$ be a unit vector and $(t^0, t_1^0, t_2^0, \ldots, t_n^0) \in \mathbb{R}^{n+1}$ be a fixed point. Set

$$\mathbb{T}_j = \{t_j - t_j^0 + (\xi - x^0)v_j : \xi \in \mathbb{T}\}, \quad j \in \{1, \ldots, n\}.$$

Note that $\mathbb{T}_j$, $j \in \{1, \ldots, n\}$, are time scales. Let σ_j and Δ_j be the forward jump operators and delta differentiation operators, respectively, of $\mathbb{T}_j$, $j \in \{1, \ldots, n\}$. Denote $\Lambda^n = \mathbb{T}_1 \times \mathbb{T}_2, \ldots, \times \mathbb{T}_n$.

Definition 7.1 Let $f : \Lambda^n \to \mathbb{R}$ be a given function. Also let,

$$F(\xi) = f\left(t^0 + (\xi - x^0)v\right), \quad \xi \in \mathbb{T}.$$

The directional delta derivative of the function f at the point x^0 in the direction of the vector v is defined as the number

$$F^{\Delta}(x^0)$$

and we will denote

$$\frac{\partial f}{\Delta v}(t^0) = F^{\Delta}(x^0).$$

Remark 7.1 If f is σ_j-completely delta differentiable for some $j \in \{1, \ldots, n\}$, we get

$$\begin{aligned}
\frac{\partial f}{\Delta v}(t^0) &= f_{t_j}^{\Delta_j}(t^0)v_j \\
&\quad + f_{t_{j-1}}^{\Delta_{j-1}}\left(t_1^0, \ldots, t_{j-1}^0, \sigma_j(t_j^0), t_{j+1}^0, \ldots, t_n^0\right)v_{j-1} \\
&\quad + \cdots \\
&\quad + f_{t_1}^{\Delta_1}\left(t_1^0, \sigma_2(t_2^0), \ldots, \sigma_{j-1}(t_{j-1}^0), \sigma_j(t_j^0), t_{j+1}^0, \ldots, t_n^0\right)v_1 \\
&\quad + f_{t_{j+1}}^{\Delta_{j+1}}\left(\sigma_1(t_1^0), \sigma_2(t_2^0), \ldots, \sigma_j(t_j^0), t_{j+1}^0, \ldots, t_n^0\right)v_{j+1} \\
&\quad + \cdots \\
&\quad + f_{t_n}^{\Delta_n}\left(\sigma_1(t_1^0), \ldots, \sigma_{n-1}(t_{n-1}^0), t_n^0\right)v_n.
\end{aligned}$$

DOI: 10.1201/9781003205265-7

Example 7.1 Let $\mathbb{T} = \mathbb{Z}$, $x^0 = 2$, $v = \left(-\frac{1}{\sqrt{2}}, \frac{1}{\sqrt{2}}\right) \in \mathbb{R}^2$. We note that v is a unit vector in $\mathbb{R}^2$. Take $t^0 = (1,1)$. Set

$$\begin{aligned}
\mathbb{T}_1 &= \left\{ t_1 = 1 + (\xi - 2)\left(-\frac{1}{\sqrt{2}}\right) : \xi \in \mathbb{T} \right\} \\
&= \left\{ t_1 = -\frac{1}{\sqrt{2}}\xi + 1 + \sqrt{2} : \xi \in \mathbb{T} \right\} \\
&= \frac{1}{\sqrt{2}}\mathbb{Z} + 1 + \sqrt{2}, \\
\mathbb{T}_2 &= \left\{ t_2 = 1 + (\xi - 2)\frac{1}{\sqrt{2}} : \xi \in \mathbb{T} \right\} \\
&= \left\{ t_2 = \frac{1}{\sqrt{2}}\xi - \sqrt{2} + 1 : \xi \in \mathbb{T} \right\} \\
&= \frac{1}{\sqrt{2}}\mathbb{Z} + 1 - \sqrt{2}.
\end{aligned}$$

We have that $\mathbb{T}_1$ and $\mathbb{T}_2$ are time scales and

$$\sigma_1(t_1) = t_1 + \frac{1}{\sqrt{2}}, \quad t_1 \in \mathbb{T}_1,$$

$$\sigma_2(t_2) = t_2 + \frac{1}{\sqrt{2}}, \quad t_2 \in \mathbb{T}_2.$$

Consider the function $f : \mathbb{T}_1 \times \mathbb{T}_2 \to \mathbb{R}$ given by

$$f(t_1, t_2) = t_1^2 + t_1 t_2 + t_2^2, \quad (t_1, t_2) \in \mathbb{T}_1 \times \mathbb{T}_2.$$

We have that f is σ_1- and σ_2-completely delta differentiable on $\mathbb{T}_1 \times \mathbb{T}_2$ and

$$\begin{aligned}
f_{t_1}^{\Delta_1}(t_1, t_2) &= \sigma_1(t_1) + t_1 + t_2 \\
&= t_1 + \frac{1}{\sqrt{2}} + t_1 + t_2 \\
&= 2t_1 + t_2 + \frac{1}{\sqrt{2}}, \\
f_{t_2}^{\Delta_2}(t_1, t_2) &= t_1 + \sigma_2(t_2) + t_2 \\
&= t_1 + t_2 + \frac{1}{\sqrt{2}} + t_2 \\
&= t_1 + 2t_2 + \frac{1}{\sqrt{2}}, \\
f_{t_2}^{\Delta_2}(\sigma_1(t_1), t_2) &= f_{t_2}^{\Delta_2}\left(t_1 + \frac{1}{\sqrt{2}}, t_2\right)
\end{aligned}$$

$$
\begin{aligned}
&= t_1 + \frac{1}{\sqrt{2}} + 2t_2 + \frac{1}{\sqrt{2}} \\
&= t_1 + 2t_2 + \frac{2}{\sqrt{2}} \\
&= t_1 + 2t_2 + \sqrt{2}, \quad (t_1, t_2) \in \mathbb{T}_1 \times \mathbb{T}_2.
\end{aligned}
$$

Hence,

$$
\begin{aligned}
f_{t_1}^{\Delta_1}(1,1) &= 2 + 1 + \frac{1}{\sqrt{2}} \\
&= 3 + \frac{1}{\sqrt{2}}, \\
f_{t_2}^{\Delta_2}(\sigma_1(1),1) &= 1 + 2 + \sqrt{2} \\
&= 3 + \sqrt{2}
\end{aligned}
$$

and

$$
\begin{aligned}
\frac{\partial f}{\Delta v}(2) &= f_{t_1}^{\Delta_1}(1,1)v_1 + f_{t_2}^{\Delta_2}(\sigma_1(1),1)v_2 \\
&= \left(3 + \frac{1}{\sqrt{2}}\right)\left(-\frac{1}{\sqrt{2}}\right) + (3+\sqrt{2})\left(\frac{1}{\sqrt{2}}\right) \\
&= -\frac{3}{\sqrt{2}} - \frac{1}{2} + \frac{3}{\sqrt{2}} + 1 \\
&= \frac{1}{2}.
\end{aligned}
$$

Exercise 7.1 Let $\mathbb{T} = 3\mathbb{Z}$, $x^0 = 3$, $v = \left(-\frac{1}{\sqrt{3}}, \sqrt{\frac{2}{3}}\right) \in \mathbb{R}^2$.

1. Find $\mathbb{T}_1$ and $\mathbb{T}_2$.
2. Let $f(t_1,t_2) = t_1^3 + t_1^2 t_2 + t_2^2$, $(t_1,t_2) \in \mathbb{T}_1 \times \mathbb{T}_2$. Find
$$\frac{\partial f}{\Delta v}(t^0).$$

Theorem 7.1 *Let $f, g : \Lambda^n \to \mathbb{R}$ be σ_j-completely delta differentiable for some $j \in \{1, \ldots, n\}$. Then*

$$\frac{\partial(af+bg)}{\Delta v}(t^0) = a\frac{\partial f}{\Delta v}(t^0) + b\frac{\partial g}{\Delta v}(t^0)$$

for any $a, b \in \mathbb{R}$.

Proof. We have

$$\frac{\partial(af+bg)}{\Delta v}(t^0) = (af+bg)_{t_j}^{\Delta_j}(t^0)v_j$$

$$
\begin{aligned}
&+(af+bg)_{t_{j-1}}^{\Delta_{j-1}}\left(t_1^0,\ldots,t_{j-1}^0,\sigma_j(t_j^0),t_{j+1}^0,\ldots,t_n^0\right)v_{j-1}\\
&+\cdots\\
&+(af+bg)_{t_1}^{\Delta_1}\left(t_1^0,\sigma_2(t_2^0),\ldots,\sigma_{j-1}(t_{j-1}^0),\sigma_j(t_j^0),t_{j+1}^0,\ldots,t_n^0\right)v_1\\
&+(af+bg)_{t_{j+1}}^{\Delta_{j+1}}\left(\sigma_1(t_1^0),\sigma_2(t_2^0),\ldots,\sigma_j(t_j^0),t_{j+1}^0,\ldots,t_n^0\right)v_{j+1}\\
&+\cdots\\
&+(af+bg)_{t_n}^{\Delta_n}\left(\sigma_1(t_1^0),\ldots,\sigma_{n-1}(t_{n-1}^0),t_n^0\right)v_n\\
=\ & af_{t_j}^{\Delta_j}(t^0)v_j\\
&+af_{t_{j-1}}^{\Delta_{j-1}}\left(t_1^0,\ldots,t_{j-1}^0,\sigma_j(t_j^0),t_{j+1}^0,\ldots,t_n^0\right)v_{j-1}\\
&+\cdots\\
&+af_{t_1}^{\Delta_1}\left(t_1^0,\sigma_2(t_2^0),\ldots,\sigma_{j-1}(t_{j-1}^0),\sigma_j(t_j^0),t_{j+1}^0,\ldots,t_n^0\right)v_1\\
&+af_{t_{j+1}}^{\Delta_{j+1}}\left(\sigma_1(t_1^0),\sigma_2(t_2^0),\ldots,\sigma_j(t_j^0),t_{j+1}^0,\ldots,t_n^0\right)v_{j+1}\\
&+\cdots\\
&+af_{t_n}^{\Delta_n}\left(\sigma_1(t_1^0),\ldots,\sigma_{n-1}(t_{n-1}^0),t_n^0\right)v_n\\
&+bg_{t_j}^{\Delta_j}(t^0)v_j\\
&+bg_{t_{j-1}}^{\Delta_{j-1}}\left(t_1^0,\ldots,t_{j-1}^0,\sigma_j(t_j^0),t_{j+1}^0,\ldots,t_n^0\right)v_{j-1}\\
&+\cdots\\
&+bg_{t_1}^{\Delta_1}\left(t_1^0,\sigma_2(t_2^0),\ldots,\sigma_{j-1}(t_{j-1}^0),\sigma_j(t_j^0),t_{j+1}^0,\ldots,t_n^0\right)v_1\\
&+bg_{t_{j+1}}^{\Delta_{j+1}}\left(\sigma_1(t_1^0),\sigma_2(t_2^0),\ldots,\sigma_j(t_j^0),t_{j+1}^0,\ldots,t_n^0\right)v_{j+1}\\
&+\cdots\\
&+bg_{t_n}^{\Delta_n}\left(\sigma_1(t_1^0),\ldots,\sigma_{n-1}(t_{n-1}^0),t_n^0\right)v_n\\
=\ & a\frac{\partial f}{\Delta v}(t^0)+b\frac{\partial g}{\Delta v}(t^0).
\end{aligned}
$$

This completes the proof.

Theorem 7.2 *Let $n=2, f,g:\Lambda^2\to\mathbb{R}$ be σ_1-completely delta differentiable. Then*

$$
\begin{aligned}
\frac{\partial(fg)}{\Delta v}(t_1^0,t_2^0) &= f(t_1^0,t_2^0)\frac{\partial g}{\Delta v}(t_1^0,t_2^0)+g(\sigma_1(t_1^0),t_2^0)\frac{\partial f}{\Delta v}(t_1^0,t_2^0)\\
&\quad+\left(\mu_1(t_1^0)f_{t_1}^{\Delta_1}(t_1^0,t_2^0)+\mu_2(t_2^0)f_{t_2}^{\Delta_2}(\sigma_1(t_1^0),t_2^0)\right)g_{t_2}^{\Delta_2}(\sigma_1(t_1^0),t_2^0)v_2\\
&= f(t_1^0,t_2^0)\frac{\partial g}{\Delta v}(t_1^0,t_2^0)+g(\sigma_1(t_1^0),\sigma_2(t_2^0))\frac{\partial f}{\Delta v}(t_1^0,t_2^0)\\
&\quad+\left(\mu_1(t_1^0)v_2-\mu_2(t_2^0)v_1\right)f_{t_1}^{\Delta_1}(t_1^0,t_2^0)g_{t_2}^{\Delta_2}(\sigma_1(t_1^0),t_2^0)
\end{aligned}
$$

$$
\begin{aligned}
&= f(\sigma_1(t_1^0), t_2^0)\frac{\partial g}{\Delta v}(t_1^0, t_2^0) + g(\sigma_1(t_1^0), t_2^0)\frac{\partial f}{\Delta v}(t_1^0, t_2^0) \\
&\quad - \mu_1(t_1^0) f_{t_1}^{\Delta_1}(t_1^0, t_2^0) g_{t_1}^{\Delta_1}(t_1^0, t_2^0) v_1 \\
&\quad + \mu_2(t_2^0) f_{t_2}^{\Delta_2}(\sigma_1(t_1^0), t_2^0) g_{t_2}^{\Delta_2}(\sigma_1(t_1^0), t_2^0) v_2 \\
&= f(t_1^0, t_2^0)\frac{\partial g}{\Delta v}(t_1^0, t_2^0) + g(\sigma_1(t_1^0), t_2^0)\frac{\partial f}{\Delta v}(t_1^0, t_2^0) \\
&\quad + \mu_1(t_1^0) f_{t_1}^{\Delta_1}(t_1^0, t_2^0) g_{t_2}^{\Delta_2}(\sigma_1(t_1^0), t_2^0) v_2 \\
&\quad + \mu_2(t_2^0) f_{t_2}^{\Delta_2}(\sigma_1(t_1^0), t_2^0) g_{t_2}^{\Delta_2}(\sigma_1(t_1^0), t_2^0) v_2 \\
&= f(\sigma_1(t_1^0), t_2^0)\frac{\partial g}{\Delta v}(t_1^0, t_2^0) + g(t_1^0, t_2^0)\frac{\partial f}{\Delta v}(t_1^0, t_2^0) \\
&\quad + \left(\mu_2(t_2^0) g_{t_2}^{\Delta_2}(\sigma_1(t_1^0), t_2^0) + \mu_1(t_1^0) g_{t_1}^{\Delta_1}(t_1^0, t_2^0)\right) f_{t_2}^{\Delta_2}(\sigma_1(t_1^0), t_2^0) v_2 \\
&= f(t_1^0, t_2^0)\frac{\partial g}{\Delta v}(t_1^0, t_2^0) + g(\sigma_1(t_1^0), t_2^0)\frac{\partial f}{\Delta v}(t_1^0, t_2^0) \\
&\quad + \left(\mu_1(t_1^0) f_{t_1}^{\Delta_1}(t_1^0, t_2^0) + \mu_2(t_2^0) f_{t_2}^{\Delta_2}(\sigma_1(t_1^0), t_2^0)\right) g_{t_2}^{\Delta_2}(\sigma_1(t_1^0), t_2^0) v_2 \\
&= f(\sigma_1(t_1^0), \sigma_2(t_2^0))\frac{\partial g}{\Delta v}(t_1^0, t_2^0) + g(\sigma_1(t_1^0), t_2^0)\frac{\partial f}{\Delta v}(t_1^0, t_2^0) \\
&\quad - \left(\mu_1(t_1^0) f_{t_1}^{\Delta_1}(t_1^0, t_2^0) + \mu_2(t_2^0) f_{t_2}^{\Delta_2}(\sigma_1(t_1^0), t_2^0)\right) g_{t_1}^{\Delta_1}(t_1^0, t_2^0) v_1 \\
&= f(\sigma_1(t_1^0), t_2^0)\frac{\partial g}{\Delta v}(t_1^0, t_2^0) + g(t_1^0, t_2^0)\frac{\partial f}{\Delta v}(t_1^0, t_2^0) \\
&\quad + f_{t_2}^{\Delta_2}(\sigma_1(t_1^0), t_2^0)\left(\mu_2(t_2^0) g_{t_2}^{\Delta_2}(\sigma_1(t_1^0), t_2^0) + \mu_1(t_1^0) g_{t_1}^{\Delta_1}(t_1^0, t_2^0)\right) v_2 \\
&= f(\sigma_1(t_1^0), t_2^0)\frac{\partial g}{\partial v}(t_1^0, t_2^0) + g(\sigma_1(t_1^0), t_2^0)\frac{\partial f}{\Delta v}(t_1^0, t_2^0) \\
&\quad + \mu_2(t_2^0) f_{t_2}^{\Delta_2}(\sigma_1(t_1^0), t_2^0) g_{t_2}^{\Delta_2}(\sigma_1(t_1^0), t_2^0) v_2 \\
&\quad - \mu_1(t_1^0) g_{t_1}^{\Delta_1}(t_1^0, t_2^0) f_{t_1}^{\Delta_1}(t_1^0, t_2^0) v_1 \\
&= f(\sigma_1(t_1^0), \sigma_2(t_2^0))\frac{\partial g}{\Delta \nu}(t_1^0, t_2^0) + g(t_1^0, t_2^0)\frac{\partial f}{\Delta v}(t_1^0, t_2^0) \\
&\quad + \left(\mu_1(t_1^0) v_2 - \mu_2(t_2^0) v_1\right) g_{t_1}^{\Delta_1}(t_1^0, t_2^0) f_{t_2}^{\Delta_2}(\sigma_1(t_1^0), t_2^0) \\
&= f(\sigma_1(t_1^0), \sigma_2(t_2^0))\frac{\partial g}{\Delta v}(t_1^0, t_2^0) + g(\sigma_1(t_1^0), t_2^0)\frac{\partial f}{\Delta v}(t_1^0, t_2^0) \\
&\quad - \left(\mu_1(t_1^0) f_{t_1}^{\Delta_1}(t_1^0, t_2^0) + \mu_2(t_2^0) f_{t_2}^{\Delta_2}(\sigma_1(t_1^0), t_2^0)\right) g_{t_1}^{\Delta_1}(t_1^0, t_2^0) v_1.
\end{aligned}
$$

Proof. We have that $fg : \Lambda^2 \to \mathbb{R}$ is σ_1-completely delta differentiable and

$$
\begin{aligned}
\frac{\partial (fg)}{\Delta v}(t_1^0, t_2^0) &= (fg)_{t_1}^{\Delta_1}(t_1^0, t_2^0) v_1 + (fg)_{t_2}^{\Delta_2}(\sigma_1(t_1^0), t_2^0) v_2 \\
&= f(t_1^0, t_2^0) g_{t_1}^{\Delta_1}(t_1^0, t_2^0) v_1 + g(\sigma_1(t_1^0), t_2^0) f_{t_1}^{\Delta_1}(t_1^0, t_2^0) v_1
\end{aligned}
$$

$$
\begin{aligned}
&+f(\sigma_1(t_1^0),t_2^0)g_{t_2}^{\Delta_2}(\sigma_1(t_1^0),t_2^0)v_2+g(\sigma_1(t_1^0),\sigma_2(t_2^0))f_{t_2}^{\Delta_2}(\sigma_1(t_1^0),t_2^0)v_2\\
&=\left(f(t_1^0,t_2^0)g_{t_1}^{\Delta_1}(t_1^0,t_2^0)v_1+f(t_1^0,t_2^0)g_{t_2}^{\Delta_2}(\sigma_1(t_1^0),t_2^0)v_2\right)\\
&\quad+\left(f(\sigma_1(t_1^0),t_2^0)-f(t_1^0,t_2^0)\right)g_{t_2}^{\Delta_2}(\sigma_1(t_1^0),t_2^0)v_2\\
&\quad+\left(g(\sigma_1(t_1^0),t_2^0)f_{t_1}^{\Delta_1}(t_1^0,t_2^0)v_1+g(\sigma_1(t_1^0),t_2^0)f_{t_2}^{\Delta_2}(\sigma_1(t_1^0),t_2^0)v_2\right)\\
&\quad+\left(g(\sigma_1(t_1^0),\sigma_2(t_2^0))-g(\sigma_1(t_1^0),t_2^0)\right)f_{t_2}^{\Delta_2}(\sigma_1(t_1^0),t_2^0)v_2\\
&=f(t_1^0,t_2^0)\frac{\partial g}{\Delta v}(t_1^0,t_2^0)+\mu_1(t_1^0)f_{t_1}^{\Delta_1}(t_1^0,t_2^0)g_{t_2}^{\Delta_2}(\sigma_1(t_1^0),t_2^0)v_2\\
&\quad+g(\sigma_1(t_1^0),t_2^0)\frac{\partial f}{\Delta v}(t_1^0,t_2^0)+\mu_2(t_2^0)g_{t_2}^{\Delta_2}(\sigma_1(t_1^0),t_2^0)f_{t_2}^{\Delta_2}(\sigma_1(t_1^0),t_2^0)v_2\\
&=f(t_1^0,t_2^0)\frac{\partial g}{\Delta v}(t_1^0,t_2^0)+g(\sigma_1(t_1^0),t_2^0)\frac{\partial f}{\Delta v}(t_1^0,t_2^0)\\
&\quad+\left(\mu_1(t_1^0)f_{t_1}^{\Delta_1}(t_1^0,t_2^0)+\mu_2(t_2^0)f_{t_2}^{\Delta_2}(\sigma_1(t_1^0),t_2^0)\right)g_{t_2}^{\Delta_2}(\sigma_1(t_1^0),t_2^0)v_2\\
&=f(t_1^0,t_2^0)\left(g_{t_1}^{\Delta_1}(t_1^0,t_2^0)v_1+g_{t_2}^{\Delta_2}(\sigma_1(t_1^0),f_2^0)v_2\right)\\
&\quad+g(\sigma_1(t_1^0),\sigma_2(t_2^0))\left(f_{t_1}^{\Delta_1}(t_1^0,t_2^0)v_1+f_{t_2}^{\Delta_2}(\sigma_1(t_1^0),t_2^0)v_2\right)\\
&\quad+\left(f(\sigma_1(t_1^0),t_2^0)-f(t_1^0,t_2^0)\right)g_{t_2}^{\Delta_2}(\sigma_1(t_1^0),t_2^0)v_2\\
&\quad-\left(g(\sigma_1(t_1^0),\sigma_2(t_2^0))-g(\sigma_1(t_1^0),t_2^0)\right)f_{t_1}^{\Delta_1}(t_1^0,t_2^0)v_1\\
&=f(t_1^0,t_2^0)\frac{\partial g}{\Delta v}(t_1^0,t_2^0)+g(\sigma_1(t_1^0),\sigma_2(t_2^0))\frac{\partial f}{\Delta v}(t_1^0,t_2^0)\\
&\quad+\mu_1(t_1)f_{t_1}^{\Delta_1}(t_1^0,t_2^0)g_{t_2}^{\Delta_2}(\sigma_1(t_1^0),t_2^0)v_2\\
&\quad-\mu_2(t_2)g_{t_2}^{\Delta_2}(\sigma_1(t_1^0),t_2^0)f_{t_1}^{\Delta_1}(t_1^0,t_2^0)v_1\\
&=f(t_1^0,t_2^0)\frac{\partial g}{\Delta v}(t_1^0,t_2^0)+g(\sigma_1(t_1^0),\sigma_2(t_2^0))\frac{\partial f}{\Delta v}(t_1^0,t_2^0)\\
&\quad+\left(\mu_1(t_1^0)v_2-\mu_2(t_2^0)v_1\right)f_{t_1}^{\Delta_1}(t_1^0,t_2^0)g_{t_2}^{\Delta_2}(\sigma_1(t_1^0),t_2^0)\\
&=f(\sigma_1(t_1^0),t_2^0)\left(g_{t_1}^{\Delta_1}(t_1^0,t_2^0)v_1+g_{t_2}^{\Delta_2}(\sigma_1(t_1^0),t_2^0)v_2\right)\\
&\quad+g(\sigma_1(t_1^0),t_2^0)\left(f_{t_1}^{\Delta_1}(t_1^0,t_2^0)v_1+f_{t_2}^{\Delta_2}(\sigma_1(t_1^0),t_2^0)v_2\right)\\
&\quad-\left(f(\sigma_1(t_1^0),t_2^0)-f(t_1^0,t_2^0)\right)g_{t_1}^{\Delta_1}(t_1^0,t_2^0)v_1\\
&\quad+\left(g(\sigma_1(t_1^0),\sigma_2(t_2^0))-g(\sigma_1(t_1^0),t_2^0)\right)f_{t_2}^{\Delta_2}(\sigma_1(t_1^0),t_2^0)v_2\\
&=f(\sigma_1(t_1^0),t_2^0)\frac{\partial g}{\Delta v}(t_1^0,t_2^0)+g(\sigma_1(t_1^0),t_2^0)\frac{\partial f}{\Delta v}(t_1^0,t_2^0)\\
&\quad-\mu_1(t_1^0)f_{t_1}^{\Delta_1}(t_1^0,t_2^0)g_{t_1}^{\Delta_1}(t_1^0,t_2^0)v_1\\
&\quad+\mu_2(t_2^0)g_{t_2}^{\Delta_2}(\sigma_1(t_1^0),t_2^0)f_{t_2}^{\Delta_2}(\sigma_1(t_1^0),t_2^0)v_2\\
&=f(t_1^0,t_2^0)\left(g_{t_1}^{\Delta_1}(t_1^0,t_2^0)v_1+g_{t_2}^{\Delta_2}(\sigma_1(t_1^0),t_2^0)v_2\right)
\end{aligned}
$$

$$
\begin{aligned}
&+g(\sigma_1(t_1^0),t_2^0)\left(f_{t_1}^{\Delta_1}(t_1^0,t_2^0)v_1+f_{t_2}^{\Delta_2}(\sigma_1(t_1^0),t_2^0)v_2\right)\\
&+\left(f(\sigma_1(t_1^0),t_2^0)-f(t_1^0,t_2^0)\right)g_{t_2}^{\Delta_2}(\sigma_1(t_1^0),t_2^0)v_2\\
&+\left(g(\sigma_1(t_1^0),\sigma_2(t_2^0))-g(\sigma_1(t_1^0),t_2^0)\right)f_{t_2}^{\Delta_2}(\sigma_1(t_1^0),t_2^0)v_2\\
=&f(t_1^0,t_2^0)\frac{\partial g}{\Delta v}(t_1^0,t_2^0)+g(\sigma_1(t_1^0),t_2^0)\frac{\partial f}{\Delta v}(t_1^0,t_2^0)\\
&+\mu_1(t_1^0)f_{t_1}^{\Delta_1}(t_1^0,t_2^0)g_{t_2}^{\Delta_2}(\sigma_1(t_1^0),t_2^0)v_2\\
&+\mu_2(t_2^0)g_{t_2}^{\Delta_2}(\sigma_1(t_1^0),t_2^0)f_{t_2}^{\Delta_2}(\sigma_1(t_1^0),t_2^0)v_2
\end{aligned}
$$

and

$$
\begin{aligned}
\frac{\partial(fg)}{\Delta v}(t_1^0,t_2^0)=&f(\sigma_1(t_1^0),t_2^0)g_{t_1}^{\Delta_1}(t_1^0,t_2^0)v_1+g(t_1^0,t_2^0)f_{t_1}^{\Delta_1}(t_1^0,t_2^0)v_1\\
&+f(\sigma_1(t_1^0),t_2^0)g_{t_2}^{\Delta_2}(\sigma_1(t_1^0),t_2^0)v_2+g(\sigma_1(t_1^0),\sigma_2(t_2^0))f_{t_2}^{\Delta_2}(\sigma_1(t_1^0),t_2^0)v_2\\
=&f(\sigma_1(t_1^0),t_2^0)\left(g_{t_1}^{\Delta_1}(t_1^0,t_2^0)v_1+g_{t_2}^{\Delta_2}(\sigma_1(t_1^0),t_2^0)v_2\right)\\
&+g(t_1^0,t_2^0)\left(f_{t_1}^{\Delta_1}(t_1^0,t_2^0)v_1+f_{t_2}^{\Delta_2}(\sigma_1(t_1^0),t_2^0)v_2\right)\\
&+\left(g(\sigma_1(t_1^0),\sigma_2(t_2^0))-g(t_1^0,t_2^0)\right)f_{t_2}^{\Delta_2}(\sigma_1(t_1^0),t_2^0)v_2\\
=&f(\sigma_1(t_1^0),t_2^0)\frac{\partial g}{\Delta v}(t_1^0,t_2^0)+g(t_1^0,t_2^0)\frac{\partial f}{\Delta v}(t_1^0,t_2^0)\\
&+\left(g(\sigma_1(t_1^0),\sigma_2(t_2^0))-g(\sigma_1(t_1^0),t_2^0)+g(\sigma_1(t_1^0),t_2^0)-g(t_1^0,t_2^0)\right)f_{t_2}^{\Delta_2}(\sigma_1(t_1^0),t_2^0)v_2\\
=&f(\sigma_1(t_1^0),t_2^0)\frac{\partial g}{\Delta v}(t_1^0,t_2^0)+g(t_1^0,t_2^0)\frac{\partial f}{\Delta v}(t_1^0,t_2^0)\\
&+\left(\mu_2(t_2^0)g_{t_2}^{\Delta_2}(\sigma_1(t_1^0),t_2^0)+\mu_1(t_1^0)g_{t_1}^{\Delta_1}(t_1^0,t_2^0)\right)f_{t_2}^{\Delta_2}(\sigma_1(t_1^0),t_2^0)v_2,
\end{aligned}
$$

and

$$
\begin{aligned}
\frac{\partial(fg)}{\Delta v}(t_1^0,t_2^0)=&f(t_1^0,t_2^0)g_{t_1}^{\Delta_1}(t_1^0,t_2^0)v_1+g(\sigma_1(t_1^0),t_2^0)f_{t_1}^{\Delta_1}(t_1^0,t_2^0)v_1\\
&+f(\sigma_1(t_1^0),\sigma_2(t_2^0))g_{t_2}^{\Delta_2}(\sigma_1(t_1^0),t_2^0)v_2+g(\sigma_1(t_1^0),t_2^0)f_{t_2}^{\Delta_2}(\sigma_1(t_1^0),t_2^0)v_2\\
=&f(t_1^0,t_2^0)\left(g_{t_1}^{\Delta_1}(t_1^0,t_2^0)v_1+g_{t_2}^{\Delta_2}(\sigma_1(t_1^0),t_2^0)v_2\right)\\
&+\left(f(\sigma_1(t_1^0),\sigma_2(t_2^0))-f(t_1^0,t_2^0)\right)g_{t_2}^{\Delta_2}(\sigma_1(t_1^0),t_2^0)v_2\\
&+g(\sigma_1(t_1^0),t_2^0)\left(f_{t_1}^{\Delta_1}(t_1^0,t_2^0)v_1+f_{t_2}^{\Delta_2}(\sigma_1(t_1^0),t_2^0)v_2\right)\\
=&f(t_1^0,t_2^0)\frac{\partial g}{\Delta v}(t_1^0,t_2^0)+g(\sigma_1(t_1^0),t_2^0)\frac{\partial f}{\Delta v}(t_1^0,t_2^0)\\
&+\left(f(\sigma_1(t_1^0),\sigma_2(t_2^0))-f(\sigma_1(t_1^0),t_2^0)+f(\sigma_1(t_1^0),t_2^0)-f(t_1^0,t_2^0)\right)g_{t_2}^{\Delta_2}(\sigma_1(t_1^0),t_2^0)v_2\\
=&f(t_1^0,t_2^0)\frac{\partial g}{\Delta v}(t_1^0,t_2^0)+g(\sigma_1(t_1^0),t_2^0)\frac{\partial f}{\Delta v}(t_1^0,t_2^0)\\
&+\left(\mu_1(t_1^0)f_{t_1}^{\Delta_1}(t_1^0,t_2^0)+\mu_2(t_2^0)f_{t_2}^{\Delta_2}(\sigma_1(t_1^0),t_2^0)\right)g_{t_2}^{\Delta_2}(\sigma_1(t_1^0),t_2^0)v_2
\end{aligned}
$$

$$
\begin{aligned}
&= f(\sigma_1(t_1^0), \sigma_2(t_2^0)) \left(g_{t_1}^{\Delta_1}(t_1^0, t_2^0) v_1 + g_{t_2}^{\Delta_2}(\sigma_1(t_1^0), t_2^0) v_2 \right) \\
&\quad + g(\sigma_1(t_1^0), t_2^0) \left(f_{t_1}^{\Delta_1}(t_1^0, t_2^0) v_1 + f_{t_2}^{\Delta_2}(\sigma_1(t_1^0), t_2^0) v_2 \right) \\
&\quad - \left(f(\sigma_1(t_1^0), \sigma_2(t_2^0)) - f(t_1^0, t_2^0) \right) g_{t_1}^{\Delta_1}(t_1^0, t_2^0) v_1 \\
&= f(\sigma_1(t_1^0), \sigma_2(t_2^0)) \frac{\partial g}{\Delta v}(t_1^0, t_2^0) + g(\sigma_1(t_1^0), t_2^0) \frac{\partial f}{\Delta v}(t_1^0, t_2^0) \\
&\quad - \left(f(\sigma_1(t_1^0), \sigma_2(t_2^0)) - f(\sigma_1(t_1^0), t_2^0) + f(\sigma_1(t_1^0), t_2^0) - f(t_1^0, t_2^0) \right) g_{t_1}^{\Delta_1}(t_1^0, t_2^0) v_1 \\
&= f(\sigma_1(t_1^0), \sigma_2(t_2^0)) \frac{\partial g}{\Delta v}(t_1^0, t_2^0) + g(\sigma_1(t_1^0), t_2^0) \frac{\partial f}{\Delta v}(t_1^0, t_2^0) \\
&\quad - \left(\mu_1(t_1^0) f_{t_1}^{\Delta_1}(t_1^0, t_2^0) + \mu_2(t_2^0) f_{t_2}^{\Delta_2}(\sigma_1(t_1^0), t_2^0) \right) g_{t_1}^{\Delta_1}(t_1^0, t_2^0) v_1,
\end{aligned}
$$

and

$$
\begin{aligned}
\frac{\partial (fg)}{\Delta v}(t_1^0, t_2^0) &= f(\sigma_1(t_1^0), t_2^0) g_{t_1}^{\Delta_1}(t_1^0, t_2^0) v_1 + g(t_1^0, t_2^0) f_{t_1}^{\Delta_1}(t_1^0, t_2^0) v_1 \\
&\quad + f(\sigma_1(t_1^0), \sigma_2(t_2^0)) g_{t_2}^{\Delta_2}(\sigma_1(t_1^0), t_2^0) v_2 + g(\sigma_1(t_1^0), t_2^0) f_{t_2}^{\Delta_2}(\sigma_1(t_1^0), t_2^0) v_2 \\
&= f(\sigma_1(t_1^0), t_2^0) \left(g_{t_1}^{\Delta_1}(t_1^0, t_2^0) v_1 + g_{t_2}^{\Delta_2}(\sigma_1(t_1^0), t_2^0) v_2 \right) \\
&\quad + \left(f(\sigma_1(t_1^0), \sigma_2(t_2^0) - f(\sigma_1(t_1^0), t_2^0) \right) g_{t_2}^{\Delta_2}(\sigma_1(t_1^0), t_2^0) v_2 \\
&\quad + g(t_1^0, t_2^0) \left(f_{t_1}^{\Delta_1}(t_1^0, t_2^0) v_1 + f_{t_2}^{\Delta_2}(\sigma_1(t_1^0), t_2^0) v_2 \right) \\
&\quad + \left(g(\sigma_1(t_1^0), t_2^0) - g(t_1^0, t_2^0) \right) f_{t_2}^{\Delta_2}(\sigma_1(t_1^0), t_2^0) v_2 \\
&= f(\sigma_1(t_1^0), t_2^0) \frac{\partial g}{\Delta v}(t_1^0, t_2^0) + g(t_1^0, t_2^0) \frac{\partial f}{\Delta v}(t_1^0, t_2^0) \\
&\quad + f_{t_2}^{\Delta_2}(\sigma_1(t_1^0), t_2^0) \left(\mu_2(t_2^0) g_{t_2}^{\Delta_2}(\sigma_1(t_1^0), t_2^0) + \mu_1(t_1^0) g_{t_1}^{\Delta_1}(t_1^0, t_2^0) \right) v_2 \\
&= f(\sigma_1(t_1^0), t_2^0) \left(g_{t_1}^{\Delta_1}(t_1^0, t_2^0) v_1 + g_{t_2}^{\Delta_2}(\sigma_1(t_1^0), t_2^0) v_2 \right) \\
&\quad + \left(f(\sigma_1(t_1^0), \sigma_2(t_2^0)) - f(\sigma_1(t_1^0), t_2^0) \right) g_{t_2}^{\Delta_2}(\sigma_1(t_1^0), t_2^0) v_2 \\
&\quad + g(\sigma_1(t_1^0), t_2^0) \left(f_{t_1}^{\Delta_1}(t_1^0, t_2^0) v_1 + f_{t_2}^{\Delta_2}(\sigma_1(t_1^0), t_2^0) v_2 \right) \\
&\quad - \left(g(\sigma_1(t_1^0), t_2^0) - g(t_1^0, t_2^0) \right) f_{t_1}^{\Delta_1}(t_1^0, t_2^0) v_1 \\
&= f(\sigma_1(t_1^0), t_2^0) \frac{\partial g}{\partial v}(t_1^0, t_2^0) + g(\sigma_1(t_1^0), t_2^0) \frac{\partial f}{\Delta v}(t_1^0, t_2^0) \\
&\quad + \mu_2(t_2^0) f_{t_2}^{\Delta_2}(\sigma_1(t_1^0), t_2^0) g_{t_2}^{\Delta_2}(\sigma_1(t_1^0), t_2^0) v_2 \\
&\quad - \mu_1(t_1^0) g_{t_1}^{\Delta_1}(t_1^0, t_2^0) f_{t_1}^{\Delta_1}(t_1^0, t_2^0) v_1 \\
&= f(\sigma_1(t_1^0), \sigma_2(t_2^0)) \left(g_{t_1}^{\Delta_1}(t_1^0, t_2^0) v_1 + g_{t_2}^{\Delta_2}(\sigma_1(t_1^0), t_2^0) v_2 \right) \\
&\quad - \left(f(\sigma_1(t_1^0), \sigma_2(t_2^0)) - f(\sigma_1(t_1^0), t_2^0) \right) g_{t_1}^{\Delta_1}(t_1^0, t_2^0) v_1 \\
&\quad + g(t_1^0, t_2^0) \left(f_{t_1}^{\Delta_1}(t_1^0, t_2^0) v_1 + f_{t_2}^{\Delta_2}(\sigma_1(t_1^0), t_2^0) v_2 \right) \\
&\quad + \left(g(\sigma_1(t_1^0), t_2^0) - g(t_1^0, t_2^0) \right) f_{t_2}^{\Delta_2}(\sigma_1(t_1^0), t_2^0) v_2
\end{aligned}
$$

$$
\begin{aligned}
&= f(\sigma_1(t_1^0), \sigma_2(t_2^0)) \frac{\partial g}{\Delta v}(t_1^0, t_2^0) + g(t_1^0, t_2^0) \frac{\partial f}{\Delta v}(t_1^0, t_2^0) \\
&\quad - \mu_2(t_2^0) f_{t_2}^{\Delta_2}(\sigma_1(t_1^0), t_2^0) g_{t_1}^{\Delta_1}(t_1^0, t_2^0) v_1 \\
&\quad + \mu_1(t_1^0) g_{t_1}^{\Delta_1}(t_1^0, t_2^0) f_{t_2}^{\Delta_2}(\sigma_1(t_1^0), t_2^0) v_2 \\
&= f(\sigma_1(t_1^0), \sigma_2(t_2^0)) \frac{\partial g}{\delta v}(t_1^0, t_2^0) + g(t_1^0, t_2^0) \frac{\partial f}{\Delta v}(t_1^0, t_2^0) \\
&\quad + \left(\mu_1(t_1^0) v_2 - \mu_2(t_2^0) v_1\right) g_{t_1}^{\Delta_1}(t_1^0, t_2^0) f_{t_2}^{\Delta_2}(\sigma_1(t_1^0), t_2^0) \\
&= f(\sigma_1(t_1^0), \sigma_2(t_2^0)) \left(g_{t_1}^{\Delta_1}(t_1^0, t_2^0) v_1 + g_{t_2}^{\Delta_2}(\sigma_1(t_1^0), t_2^0) v_2\right) \\
&\quad - \left(f(\sigma_1(t_1^0), \sigma_2(t_2^0)) - f(\sigma_1(t_1^0), t_2^0)\right) g_{t_1}^{\Delta_1}(t_1^0, t_2^0) v_1 \\
&\quad + g(\sigma_1(t_1^0), t_2^0) \left(f_{t_1}^{\Delta_1}(t_1^0, t_2^0) v_1 + f_{t_2}^{\Delta_2}(\sigma_1(t_1^0), t_2^0) v_2\right) \\
&\quad - \left(g(\sigma_1(t_1^0), t_2^0) - g(t_1^0, t_2^0)\right) f_{t_1}^{\Delta_1}(t_1^0, t_2^0) v_1 \\
&= f(\sigma_1(t_1^0), \sigma_2(t_2^0)) \frac{\partial g}{\Delta v}(t_1^0, t_2^0) + g(\sigma_1(t_1^0), t_2^0) \frac{\partial f}{\Delta v}(t_1^0, t_2^0) \\
&\quad - \mu_2(t_2^0) f_{t_2}^{\Delta_2}(\sigma_1(t_1^0), t_2^0) g_{t_1}^{\Delta_1}(t_1^0, t_2^0) v_1 \\
&\quad - \mu_1(t_1^0) g_{t_1}^{\Delta_1}(t_1^0, t_2^0) f_{t_1}^{\Delta_1}(t_1^0, t_2^0) v_1 \\
&= f(\sigma_1(t_1^0), \sigma_2(t_2^0)) \frac{\partial g}{\Delta v}(t_1^0, t_2^0) + g(\sigma_1(t_1^0), t_2^0) \frac{\partial f}{\Delta v}(t_1^0, t_2^0) \\
&\quad - \left(\mu_1(t_1^0) f_{t_1}^{\Delta_1}(t_1^0, t_2^0) + \mu_2(t_2^0) f_{t_2}^{\Delta_2}(\sigma_1(t_1^0), t_2^0)\right) g_{t_1}^{\Delta_1}(t_1^0, t_2^0) v_1.
\end{aligned}
$$

This completes the proof.

Theorem 7.3 *Let $f, g : \Lambda^2 \to \mathbb{R}$ be σ_2-completely delta differentiable. Then*

$$
\begin{aligned}
\frac{\partial (fg)}{\Delta v} &= f(t_1^0, t_2^0) \frac{\partial g}{\Delta v}(t_1^0, t_2^0) + g_1(t_1^0, \sigma_2(t_2^0)) \frac{\partial f}{\Delta v}(t_1^0, t_2^0) \\
&\quad + \mu_2(t_2^0) f_{t_2}^{\Delta_2}(t_1^0, t_2^0) g_{t_1}^{\Delta_1}(t_1^0, \sigma_2(t_2^0)) v_1 \\
&\quad + \mu_1(t_1^0) f_{t_1}^{\Delta_1}(t_1^0, \sigma_2(t_2^0)) g_{t_1}^{\Delta_1}(t_1^0, \sigma_2(t_2^0)) v_1 \\
&= f(t_1^0, \sigma_2(t_2^0)) \frac{\partial g}{\Delta v}(t_1^0, t_2^0) + g(t_1^0, \sigma_2(t_2^0)) \frac{\partial f}{\Delta v}(t_1^0, t_2^0) \\
&\quad - \mu_2(t_2^0) f_{t_2}^{\Delta_2}(t_1^0, t_2^0) g_{t_2}^{\Delta_2}(t_1^0, t_2^0) v_2 \\
&\quad + \mu_1(t_2^0) g_{t_1}^{\Delta_1}(t_1^0, \sigma_2(t_2^0)) f_{t_1}^{\Delta_1}(t_1^0, \sigma_2(t_2^0)) v_1 \\
&= f(t_1^0, \sigma_2(t_2^0)) \frac{\partial g}{\Delta v}(t_1^0, t_2^0) + g(\sigma_1(t_1^0), \sigma_2(t_2^0)) \frac{\partial f}{\Delta v}(t_1^0, t_2^0) \\
&\quad - \mu_2(t_2^0) f_{t_2}^{\Delta_2}(t_1^0, t_2^0) g_{t_2}^{\Delta_2}(t_1^0, t_2^0) v_2 \\
&\quad - \mu_1(t_1^0) g_{t_1}^{\Delta_1}(t_1^0, \sigma_2(t_2^0)) f_{t_2}^{\Delta_2}(t_1^0, t_2^0)
\end{aligned}
$$

$$
\begin{aligned}
&= f(\sigma_1(t_1^0), \sigma_2(t_2^0)) \frac{\partial g}{\Delta v}(t_1^0, t_2^0) + g(t_1^0, \sigma_2(t_2^0)) \frac{\partial f}{\Delta v}(t_1^0, t_2^0) \\
&\quad - \left(\mu_2(t_2^0) f_{t_2}^{\Delta_2}(\sigma_1(t_1^0), t_2^0) + \mu_1(t_1^0) f_{t_1}^{\Delta_1}(t_1^0, t_2^0)\right) g_{t_2}^{\Delta_2}(t_1^0, t_2^0) v_2 \\
&= f(t_1^0, t_2^0) \frac{\partial g}{\Delta v}(t_1^0, t_2^0) + g(t_1^0, \sigma_2(t_2^0)) \frac{\partial f}{\Delta v}(t_1^0, t_2^0) \\
&\quad + \left(\mu_2(t_2^0) f_{t_2}^{\Delta_2}(\sigma_1(t_1^0), t_2^0) + \mu_1(t_1^0) f_{t_1}^{\Delta_1}(t_1^0, t_2^0)\right) g_{t_1}^{\Delta_1}(t_1^0, \sigma_2(t_2^0)) v_1 \\
&= f(\sigma_1(t_1^0), \sigma_2(t_2^0)) \frac{\partial g}{\Delta v}(t_1^0, t_2^0) + g(t_1^0, \sigma_2(t_2^0)) \frac{\partial f}{\Delta v}(t_1^0, t_2^0) \\
&\quad - \mu_1(t_1^0) f_{t_1}^{\Delta_1}(t_1^0, \sigma_2(t_2^0)) g_{t_2}^{\Delta_2}(t_1^0, t_2^0) v_2 \\
&\quad - \mu_2(t_2^0) g_{t_2}^{\Delta_2}(t_1^0, t_2^0) f_{t_2}^{\Delta_2}(t_1^0, t_2^0) v_2 \\
&= f(\sigma_1(t_1^0), \sigma_2(t_2^0)) \frac{\partial g}{\Delta v}(t_1^0, t_2^0) + g(t_1^0, t_2^0) \frac{\partial f}{\Delta v}(t_1^0, t_2^0) \\
&\quad - \mu_1(t_1^0) f_{t_1}^{\Delta_1}(t_1^0, \sigma_2(t_2^0)) g_{t_2}^{\Delta_2}(t_1^0, t_2^0) v_2 \\
&\quad + \mu_2(t_2^0) f_{t_1}^{\Delta_1}(t_1^0, \sigma_2(t_2^0)) g_{t_2}^{\Delta_2}(t_1^0, t_2^0) v_1 \\
&= f(t_1^0, \sigma_2(t_2^0)) \frac{\partial g}{\Delta v}(t_1^0, t_2^0) + g(t_1^0, t_2^0) \frac{\partial f}{\Delta v}(t_1^0, t_2^0) \\
&\quad + \mu_1(t_1^0) f_{t_1}^{\Delta_1}(t_1^0, \sigma_2(t_2^0)) g_{t_1}^{\Delta_1}(t_1^0, \sigma_2(t_2^0)) v_1 \\
&\quad + \mu_2(t_2^0) g_{t_2}^{\Delta_2}(t_1^0, t_2^0) f_{t_1}^{\Delta_1}(t_1^0, \sigma_2(t_2^0)) v_1.
\end{aligned}
$$

Proof. We have fg is σ_2-completely delta differentiable and

$$
\begin{aligned}
\frac{\partial(fg)}{\Delta v} &= (fg)_{t_1}^{\Delta_1}(t_1^0, \sigma_2(t_2^0)) v_1 + (fg)_{t_2}^{\Delta_2}(t_1^0, t_2^0) v_2 \\
&= f(t_1^0, \sigma_2(t_2^0)) g_{t_1}^{\Delta_1}(t_1^0, \sigma_2(t_2^0)) v_1 + g(\sigma_1(t_1^0), \sigma_2(t_2^0)) f_{t_1}^{\Delta_1}(t_1^0, \sigma_2(t_2^0)) v_1 \\
&\quad + f(t_1^0, t_2^0) g_{t_2}^{\Delta_2}(t_1^0, t_2^0) v_2 + g(t_1^0, \sigma_2(t_2^0)) f_{t_2}^{\Delta_2}(t_1^0, t_2^0) v_2 \\
&= f(t_1^0, t_2^0) \left(g_{t_1}^{\Delta_1}(t_1^0, \sigma_2(t_2^0)) v_1 + g_{t_2}^{\Delta_2}(t_1^0, t_2^0) v_2\right) \\
&\quad + \left(f(t_1^0, \sigma_2(t_2^0)) - f(t_1^0, t_2^0)\right) g_{t_1}^{\Delta_1}(t_1^0, \sigma_2(t_2^0)) v_1 \\
&\quad + g(t_1^0, \sigma_2(t_2^0)) \left(f_{t_1}^{\Delta_1}(t_1^0, \sigma_2(t_2^0)) v_1 + f_{t_2}^{\Delta_2}(t_1^0, t_2^0) v_2\right) \\
&\quad + \left(g(\sigma_1(t_1^0), \sigma_2(t_2^0)) - g(t_1^0, \sigma_2(t_2^0))\right) f_{t_1}^{\Delta_1}(t_1^0, \sigma_2(t_2^0)) v_1 \\
&= f(t_1^0, t_2^0) \frac{\partial g}{\Delta v}(t_1^0, t_2^0) + g_1(t_1^0, \sigma_2(t_2^0)) \frac{\partial f}{\Delta v}(t_1^0, t_2^0) \\
&\quad + \mu_2(t_2^0) f_{t_2}^{\Delta_2}(t_1^0, t_2^0) g_{t_1}^{\Delta_1}(t_1^0, \sigma_2(t_2^0)) v_1 \\
&\quad + \mu_1(t_1^0) f_{t_1}^{\Delta_1}(t_1^0, \sigma_2(t_2^0)) g_{t_1}^{\Delta_1}(t_1^0, \sigma_2(t_2^0)) v_1 \\
&= f(t_1^0, \sigma_2(t_2^0)) \left(g_{t_1}^{\Delta_1}(t_1^0, \sigma_2(t_2^0)) v_1 + g_{t_2}^{\Delta_2}(t_1^0, t_2^0) v_2\right) \\
&\quad - \left(f(t_1^0, \sigma_2(t_2^0)) - f(t_1^0, t_2^0)\right) g_{t_2}^{\Delta_2}(t_1^0, t_2^0) v_2
\end{aligned}
$$

$$
\begin{aligned}
&+g(t_1^0,\sigma_2(t_2^0))\left(f_{t_1}^{\Delta_1}(t_1^0,\sigma_2(t_2^0))v_1+f_{t_2}^{\Delta_2}(t_1^0,t_2^0)v_2\right)\\
&+\left(g(\sigma_1(t_1^0),\sigma_2(t_2^0))-g(t_1^0,\sigma_2(t_2^0))\right)f_{t_1}^{\Delta_1}(t_1^0,\sigma_2(t_2^0))v_1\\
=\;&f(t_1^0,\sigma_2(t_2^0))\frac{\partial g}{\Delta v}(t_1^0,t_2^0)+g(t_1^0,\sigma_2(t_2^0))\frac{\partial f}{\Delta v}(t_1^0,t_2^0)\\
&-\mu_2(t_2^0)f_{t_2}^{\Delta_2}(t_1^0,t_2^0)g_{t_2}^{\Delta_2}(t_1^0,t_2^0)v_2\\
&+\mu_1(t_2^0)g_{t_1}^{\Delta_1}(t_1^0,\sigma_2(t_2^0))f_{t_1}^{\Delta_1}(t_1^0,\sigma_2(t_2^0))v_1\\
=\;&f(t_1^0,\sigma_2(t_2^0))\left(g_{t_1}^{\Delta_1}(t_1^0,\sigma_2(t_2^0))v_1+g_{t_2}^{\Delta_2}(t_1^0,t_2^0)v_2\right)\\
&-\left(f(t_1^0,\sigma_2(t_2^0))-f(t_1^0,t_2^0)\right)g_{t_2}^{\Delta_2}(t_1^0,t_2^0)v_2\\
&+g(\sigma_1(t_1^0),\sigma_2(t_2^0))\left(f_{t_1}^{\Delta_1}(t_1^0,\sigma_2(t_2^0))v_1+f_{t_2}^{\Delta_2}(t_1^0,t_2^0)v_2\right)\\
&-\left(g(\sigma_1(t_1^0),\sigma_2(t_2^0))-g(t_1^0,\sigma_2(t_2^0))\right)f_{t_2}^{\Delta_2}(t_1^0,t_2^0)\\
=\;&f(t_1^0,\sigma_2(t_2^0))\frac{\partial g}{\Delta v}(t_1^0,t_2^0)+g(\sigma_1(t_1^0),\sigma_2(t_2^0))\frac{\partial f}{\Delta v}(t_1^0,t_2^0)\\
&-\mu_2(t_2^0)f_{t_2}^{\Delta_2}(t_1^0,t_2^0)g_{t_2}^{\Delta_2}(t_1^0,t_2^0)v_2\\
&-\mu_1(t_1^0)g_{t_1}^{\Delta_1}(t_1^0,\sigma_2(t_2^0))f_{t_2}^{\Delta_2}(t_1^0,t_2^0)
\end{aligned}
$$

and

$$
\begin{aligned}
\frac{\partial(fg)}{\Delta v}(t_1^0,t_2^0)=\;&(fg)_{t_1}^{\Delta_1}(t_1^0,\sigma_2(t_2^0))v_1+(fg)_{t_2}^{\Delta_2}(t_1^0,t_2^0)v_2\\
=\;&f(\sigma_1(t_1^0),\sigma_2(t_2^0))g_{t_1}^{\Delta_1}(t_1^0,\sigma_2(t_2^0))v_1+g(t_1^0,\sigma_2(t_2^0))f_{t_1}^{\Delta_1}(t_1^0,\sigma_2(t_2^0))v_1\\
&+f(t_1^0,t_2^0)g_{t_2}^{\Delta_2}(t_1^0,t_2^0)v_2+g(t_1^0,\sigma_2(t_2^0))f_{t_2}^{\Delta_2}(t_1^0,t_2^0)v_2\\
=\;&f(\sigma_1(t_1^0),\sigma_2(t_2^0))\left(g_{t_1}^{\Delta_1}(t_1^0,\sigma_2(t_2^0))v_1+g_{t_2}^{\Delta_2}(t_1^0,t_2^0)v_2\right)\\
&+g(t_1^0,\sigma_2(t_2^0))\left(f_{t_1}^{\Delta_1}(t_1^0,\sigma_2(t_2^0))v_1+f_{t_2}^{\Delta_2}(t_1^0,t_2^0)v_2\right)\\
&-\left(f(\sigma_1(t_1^0),\sigma_2(t_2^0))-f(t_1^0,t_2^0)\right)g_{t_2}^{\Delta_2}(t_1^0,t_2^0)v_2\\
=\;&f(\sigma_1(t_1^0),\sigma_2(t_2^0))\frac{\partial g}{\Delta v}(t_1^0,t_2^0)+g(t_1^0,\sigma_2(t_2^0))\frac{\partial f}{\Delta v}(t_1^0,t_2^0)\\
&-\left(f(\sigma_1(t_1^0),\sigma_2(t_2^0))-f(\sigma_1(t_1^0),t_2^0)+f(\sigma_1(t_1^0),t_2^0)-f(t_1^0,t_2^0)\right)g_{t_2}^{\Delta_2}(t_1^0,t_2^0)v_2\\
=\;&f(\sigma_1(t_1^0),\sigma_2(t_2^0))\frac{\partial g}{\Delta v}(t_1^0,t_2^0)+g(t_1^0,\sigma_2(t_2^0))\frac{\partial f}{\Delta v}(t_1^0,t_2^0)\\
&-\left(\mu_2(t_2^0)f_{t_2}^{\Delta_2}(\sigma_1(t_1^0),t_2^0)+\mu_1(t_1^0)f_{t_1}^{\Delta_1}(t_1^0,t_2^0)\right)g_{t_2}^{\Delta_2}(t_1^0,t_2^0)v_2\\
=\;&f(t_1^0,t_2^0)\left(g_{t_1}^{\Delta_1}(t_1^0,\sigma_2(t_2^0))v_1+g_{t_2}^{\Delta_2}(t_1^0,t_2^0)v_2\right)\\
&+g(t_1^0,\sigma_2(t_2^0))\left(f_{t_1}^{\Delta_1}(t_1^0,\sigma_2(t_2^0))v_1+f_{t_2}^{\Delta_2}(t_1^0,t_2^0)v_2\right)\\
&+\left(f(\sigma_1(t_1^0),\sigma_2(t_2^0))-f(t_1^0,t_2^0)\right)g_{t_1}^{\Delta_1}(t_1^0,\sigma_2(t_2^0))v_1
\end{aligned}
$$

$$
\begin{aligned}
&= f(t_1^0,t_2^0)\frac{\partial g}{\Delta v}(t_1^0,t_2^0) + g(t_1^0,\sigma_2(t_2^0))\frac{\partial f}{\Delta v}(t_1^0,t_2^0)\\
&\quad + \left(f(\sigma_1(t_1^0),\sigma_2(t_2^0)) - f(\sigma_1(t_1^0),t_2^0) + f(\sigma_1(t_1^0),t_2^0) - f(t_1^0,t_2^0)\right) g_{t_1}^{\Delta_1}(t_1^0,\sigma_2(t_2^0))v_1\\
&= f(t_1^0,t_2^0)\frac{\partial g}{\Delta v}(t_1^0,t_2^0) + g(t_1^0,\sigma_2(t_2^0))\frac{\partial f}{\Delta v}(t_1^0,t_2^0)\\
&\quad + \left(\mu_2(t_2^0)f_{t_2}^{\Delta_2}(\sigma_1(t_1^0),t_2^0) + \mu_1(t_1^0)f_{t_1}^{\Delta_1}(t_1^0,t_2^0)\right) g_{t_1}^{\Delta_1}(t_1^0,\sigma_2(t_2^0))v_1,
\end{aligned}
$$

and

$$
\begin{aligned}
\frac{\partial(fg)}{\Delta v}(t_1^0,t_2^0) &= (fg)_{t_1}^{\Delta_1}(t_1^0,\sigma_2(t_2^0))v_1 + (fg)_{t_2}^{\Delta_2}(t_1^0,t_2^0)v_2\\
&= f(\sigma_1(t_1^0),\sigma_2(t_2^0))g_{t_1}^{\Delta_1}(t_1^0,\sigma_2(t_2^0))v_1 + g(t_1^0,\sigma_2(t_2^0))f_{t_1}^{\Delta_1}(t_1^0,\sigma_2(t_2^0))v_1\\
&\quad + f(t_1^0,\sigma_2(t_2^0))g_{t_2}^{\Delta_2}(t_1^0,t_2^0)v_2 + g(t_1^0,t_2^0)f_{t_2}^{\Delta_2}(t_1^0,t_2^0)v_2\\
&= f(\sigma_1(t_1^0),\sigma_2(t_2^0))\left(g_{t_1}^{\Delta_1}(t_1^0,\sigma_2(t_2^0))v_1 + g_{t_2}^{\Delta_2}(t_1^0,t_2^0)v_2\right)\\
&\quad + g(t_1^0,\sigma_2(t_2^0))\left(f_{t_1}^{\Delta_1}(t_1^0,\sigma_2(t_2^0))v_1 + f_{t_2}^{\Delta_2}(t_1^0,t_2^0)v_2\right)\\
&\quad - \left(f(\sigma_1(t_1^0),\sigma_2(t_2^0)) - f(t_1^0,\sigma_2(t_2^0))\right) g_{t_2}^{\Delta_2}(t_1^0,t_2^0)v_2\\
&\quad - \left(g(t_1^0,\sigma_2(t_2^0)) - g(t_1^0,t_2^0)\right) f_{t_2}^{\Delta_2}(t_1^0,t_2^0)v_2\\
&= f(\sigma_1(t_1^0),\sigma_2(t_2^0))\frac{\partial g}{\Delta v}(t_1^0,t_2^0) + g(t_1^0,\sigma_2(t_2^0))\frac{\partial f}{\Delta v}(t_1^0,t_2^0)\\
&\quad - \mu_1(t_1^0)f_{t_1}^{\Delta_1}(t_1^0,\sigma_2(t_2^0))g_{t_2}^{\Delta_2}(t_1^0,t_2^0)v_2\\
&\quad - \mu_2(t_2^0)g_{t_2}^{\Delta_2}(t_1^0,t_2^0)f_{t_2}^{\Delta_2}(t_1^0,t_2^0)v_2\\
&= f(\sigma_1(t_1^0),\sigma_2(t_2^0))\left(g_{t_1}^{\Delta_1}(t_1^0,\sigma_2(t_2^0))v_1 + g_{t_2}^{\Delta_2}(t_1^0,t_2^0)v_2\right)\\
&\quad + g(t_1^0,t_2^0)\left(f_{t_1}^{\Delta_1}(t_1^0,\sigma_2(t_2^0))v_1 + f_{t_2}^{\Delta_2}(t_1^0,t_2^0)v_2\right)\\
&\quad - \left(f(\sigma_1(t_1^0),\sigma_2(t_2^0)) - f(t_1^0,\sigma_2(t_2^0))\right) g_{t_2}^{\Delta_2}(t_1^0,t_2^0)v_2\\
&\quad + \left(g(t_1^0,\sigma_2(t_2^0)) - g(t_1^0,t_2^0)\right) f_{t_1}^{\Delta_1}(t_1^0,\sigma_2(t_2^0))v_1\\
&= f(\sigma_1(t_1^0),\sigma_2(t_2^0))\frac{\partial g}{\Delta v}(t_1^0,t_2^0) + g(t_1^0,t_2^0)\frac{\partial f}{\Delta v}(t_1^0,t_2^0)\\
&\quad - \mu_1(t_1^0)f_{t_1}^{\Delta_1}(t_1^0,\sigma_2(t_2^0))g_{t_2}^{\Delta_2}(t_1^0,t_2^0)v_2\\
&\quad + \mu_2(t_2^0)f_{t_1}^{\Delta_1}(t_1^0,\sigma_2(t_2^0))g_{t_2}^{\Delta_2}(t_1^0,t_2^0)v_1\\
&= f(t_1^0,\sigma_2(t_2^0))\left(g_{t_1}^{\Delta_1}(t_1^0,\sigma_2(t_2^0))v_1 + g_{t_2}^{\Delta_2}(t_1^0,t_2^0)v_2\right)\\
&\quad + g(t_1^0,t_2^0)\left(f_{t_1}^{\Delta_1}(t_1^0,\sigma_2(t_2^0))v_1 + f_{t_2}^{\Delta_2}(t_1^0,t_2^0)v_2\right)\\
&\quad + \left(f(\sigma_1(t_1^0),\sigma_2(t_2^0)) - f(t_1^0,\sigma_2(t_2^0))\right) g_{t_1}^{\Delta_1}(t_1^0,\sigma_2(t_2^0))v_1\\
&\quad + \left(g(t_1^0,\sigma_2(t_2^0)) - g(t_1^0,t_2^0)\right) f_{t_1}^{\Delta_1}(t_1^0,\sigma_2(t_2^0))v_1
\end{aligned}
$$

$$
\begin{aligned}
&= f(t_1^0, \sigma_2(t_2^0)) \frac{\partial g}{\Delta v}(t_1^0, t_2^0) + g(t_1^0, t_2^0) \frac{\partial f}{\Delta v}(t_1^0, t_2^0) \\
&+ \mu_1(t_1^0) f_{t_1}^{\Delta_1}(t_1^0, \sigma_2(t_2^0)) g_{t_1}^{\Delta_1}(t_1^0, \sigma_2(t_2^0)) v_1 \\
&+ \mu_2(t_2^0) g_{t_2}^{\Delta_2}(t_1^0, t_2^0) f_{t_1}^{\Delta_1}(t_1^0, \sigma_2(t_2^0)) v_1.
\end{aligned}
$$

This completes the proof.

Theorem 7.4 *Let $f : \Lambda^2 \to \mathbb{R}$ be σ_1-completely delta differentiable and*

$$
f(t_1^0, t_2^0), \quad f(\sigma_1(t_1^0), t_2^0), \quad f(\sigma_1(t_1^0), \sigma_2(t_2^0)) \neq 0.
$$

Then

$$
\begin{aligned}
\frac{\partial}{\Delta v}\left(\frac{1}{f}\right) &= -\frac{1}{f(t_1^0, t_2^0) f(\sigma_1(t_1^0), t_2^0) f(\sigma_1(t_1^0), \sigma_2(t_2^0))} \Bigg(f(\sigma_1(t_1^0), \sigma_2(t_2^0)) \frac{\partial f}{\Delta v}(t_1^0, t_2^0) \\
&- \left(\mu_1(t_1^0) f_{t_1}^{\Delta_1}(t_1^0, t_2^0) + \mu_2(t_2^0) f_{t_2}^{\Delta_2}(\sigma_1(t_1^0), t_2^0) f_{t_2}^{\Delta_2}(\sigma_1(t_1^0), t_2^0) v_2\right) \\
&= -\frac{1}{f(t_1^0, t_2^0) f(\sigma_1(t_1^0), t_2^0) f(\sigma_1(t_1^0), \sigma_2(t_2^0))} \Bigg(f(t_1^0, t_2^0) \frac{\partial f}{\Delta v}(t_1^0, ty_2^0) \\
&+ \left(\mu_1(t_1^0) f_{t_1}^{\Delta_1}(t_1^0, t_2^0) + \mu_2(t_2^0) f_{t_2}^{\Delta_2}(\sigma_1(t_1^0), t_2^0)\right) f_{t_1}^{\Delta_1}(t_1^0, t_2^0) v_1 \Bigg).
\end{aligned}
$$

Proof. By the definition of the directional derivative, we obtain

$$
\begin{aligned}
\frac{\partial}{\Delta v}\left(\frac{1}{f}\right) &= \frac{\partial}{\Delta_1 t_1}\left(\frac{1}{f}\right)(t_1^0, t_2^0) v_1 + \frac{\partial}{\Delta_2 t_2}\left(\frac{1}{f}\right)(\sigma_1(t_1^0), t_2^0) v_2 \\
&= -\left(\frac{f_{t_1}^{\Delta_1}(t_1^0, t_2^0)}{f(t_1^0, t_2^0) f(\sigma_1(t_1^0), t_2^0)} v_1 + \frac{f_{t_2}^{\Delta_2}(\sigma_1(t_1^0), t_2^0)}{f(\sigma_1(t_1^0), t_2^0) f(\sigma_1(t_1^0), \sigma_2(t_2^0))} v_2 \right) \\
&= -\frac{1}{f(t_1^0, t_2^0) f(\sigma_1(t_1^0), t_2^0) f(\sigma_1(t_1^0), \sigma_2(t_2^0))} \Bigg(f(\sigma_1(t_1^0), \sigma_2(t_2^0)) f_{t_1}^{\Delta_1}(t_1^0, t_2^0) v_1 \\
&+ f(t_1^0, t_2^0) f_{t_2}^{\Delta_2}(\sigma_1(t_1^0), t_2^0) v_2 \Bigg) \\
&= -\frac{1}{f(t_1^0, t_2^0) f(\sigma_1(t_1^0), t_2^0) f(\sigma_1(t_1^0), \sigma_2(t_2^0))} \Bigg(f(\sigma_1(t_1^0), \sigma_2(t_2^0)) \Big(f_{t_1}^{\Delta_1}(t_1^0, t_2^0) v_1 \\
&+ f_{t_2}^{\Delta_2}(\sigma_1(t_1^0), t_2^0) v_2 \Big) + \left(f(t_1^0, t_2^0) - f(\sigma_1(t_1^0), \sigma_2(t_2^0)) \right) f_{t_2}^{\Delta_2}(\sigma_1(t_1^0), t_2^0) v_2 \Bigg) \\
&= -\frac{1}{f(t_1^0, t_2^0) f(\sigma_1(t_1^0), t_2^0) f(\sigma_1(t_1^0), \sigma_2(t_2^0))} \Bigg(f(\sigma_1(t_1^0), \sigma_2(t_2^0)) \frac{\partial f}{\Delta v}(t_1^0, t_2^0) \\
&+ \left(-f(\sigma_1(t_1^0), t_2^0) + f(t_1^0, t_2^0) + f(\sigma_1(t_1^0), t_2^0) - f(\sigma_1(t_1^0), \sigma_2(t_2^0)) \right) f_{t_2}^{\Delta_2}(\sigma_1(t_1^0), t_2^0) v_2 \Bigg) \\
&= -\frac{1}{f(t_1^0, t_2^0) f(\sigma_1(t_1^0), t_2^0) f(\sigma_1(t_1^0), \sigma_2(t_2^0))} \Bigg(f(\sigma_1(t_1^0), \sigma_2(t_2^0)) \frac{\partial f}{\Delta v}(t_1^0, t_2^0)
\end{aligned}
$$

$$+\left(-\mu_1(t_1^0)f_{t_1}^{\Delta_1}(t_1^0,t_2^0)-\mu_2(t_2^0)f_{t_2}^{\Delta_2}(\sigma_1(t_1^0),t_2^0)\right)f_{t_2}^{\Delta_2}(\sigma_1(t_1^0),t_2^0)v_2)$$

$$=-\frac{1}{f(t_1^0,t_2^0)f(\sigma_1(t_1^0),t_2^0)f(\sigma_1(t_1^0),\sigma_2(t_2^0))}\left(f(t_1^0,t_2^0)\left(f_{t_1}^{\Delta_1}(t_1^0,t_2^0)v_1+f_{t_2}^{\Delta_2}(\sigma_1(t_1^0),t_2^0)v_2\right)\right.$$

$$\left.+\left(f(\sigma_1(t_1^0),\sigma_2(t_2^0))-f(t_1^0,t_2^0)\right)f_{t_1}^{\Delta_1}(t_1^0,t_2^0)v_1\right)$$

$$=-\frac{1}{f(t_1^0,t_2^0)f(\sigma_1(t_1^0),t_2^0)f(\sigma_1(t_1^0),\sigma_2(t_2^0))}\left(f(t_1^0,t_2^0)\frac{\partial f}{\Delta v}(t_1^0,t_2^0)\right.$$

$$\left.+\left(f(\sigma_1(t_1^0),\sigma_2(t_2^0))-f(\sigma_1(t_1^0),t_2^0)+f(\sigma_1(t_1^0),t_2^0)-f(t_1^0,t_2^0)\right)f_{t_1}^{\Delta_1}(t_1^0,t_2^0)v_1\right)$$

$$=-\frac{1}{f(t_1^0,t_2^0)f(\sigma_1(t_1^0),t_2^0)f(\sigma_1(t_1^0),\sigma_2(t_2^0))}\left(f(t_1^0,t_2^0)\frac{\partial f}{\Delta v}(t_1^0,t_2^0)\right.$$

$$\left.+\left(\mu_1(t_1^0)f_{t_1}^{\Delta_1}(t_1^0,t_2^0)+\mu_2(t_2^0)f_{t_2}^{\Delta_2}(\sigma_1(t_1^0),t_2^0)\right)f_{t_1}^{\Delta_1}(t_1^0,t_2^0)v_1\right).$$

This completes the proof.

Theorem 7.5 *Let f be σ_2-completely delta differentiable and*

$$f(t_1^0,t_2^0),\quad f(\sigma_1(t_1^0),\sigma_2(t_2^0)),\quad f(t_1^0,\sigma_2(t_2^0))\neq 0.$$

Then

$$\frac{\partial}{\Delta v}\left(\frac{1}{f}\right)(t_1^0,t_2^0)=-\frac{1}{f(t_1^0,t_2^0)f(t_1^0,\sigma_2(t_2^0))f(\sigma_1(t_1^0),\sigma_2(t_2^0))}\left(f(t_1^0,\sigma_2(t_2^0))\frac{\partial f}{\Delta v}(t_1^0,t_2^0)\right.$$

$$+\left(\mu_2(t_2^0)f_{t_2}^{\Delta_2}(\sigma_1(t_1^0),t_2^0)\right.$$

$$\left.\left.+\mu_1(t_1^0)f_{t_1}^{\Delta_1}(t_1^0,\sigma_2(t_2^0))\right)f_{t_2}^{\Delta_2}(t_1^0,t_2^0)v_2\right)$$

$$=-\frac{1}{f(t_1^0,t_2^0)f(t_1^0,\sigma_2(t_2^0))f(\sigma_1(t_1^0),\sigma_2(t_2^0))}\left(f(\sigma_1(t_1^0),\sigma_2(t_2^0))\frac{\partial f}{\Delta v}(t_1^0,t_2^0)\right.$$

$$\left.-\left(\mu_1(t_1^0)f_{t_1}^{\Delta_1}(t_1^0,t_2^0)+\mu_2(t_2^0)f_{t_2}^{\Delta_2}(\sigma_1(t_1^0),t_2^0)\right)f_{t_1}^{\Delta_1}(t_1^0,\sigma_2(t_2^0))v_1\right).$$

Proof. By the definition of the directional derivative, we find

$$\frac{\partial}{\Delta v}\left(\frac{1}{f}\right)(t_1^0,t_2^0)=\frac{\partial}{\Delta_1 t_1}\left(\frac{1}{f}\right)(t_1^0,\sigma_2(t_2^0))v_1+\frac{\partial}{\Delta_2 t_2}\left(\frac{1}{f}\right)(t_1^0,t_2^0)$$

$$=-\left(\frac{f_{t_1}^{\Delta_1}(t_1^0,\sigma_2(t_2^0))}{f(t_1^0,\sigma_2(t_2^0))f(\sigma_1(t_1^0),\sigma_2(t_2^0))}v_1+\frac{f_{t_2}^{\Delta_2}(t_1^0,t_2^0)}{f(t_1^0,t_2^0)f(t_1^0,\sigma_2(t_2^0))}v_2\right)$$

$$=-\frac{1}{f(t_1^0,t_2^0)f(t_1^0,\sigma_2(t_2^0))f(\sigma_1(t_1^0),\sigma_2(t_2^0))}\left(f(t_1^0,t_2^0)f_{t_1}^{\Delta_1}(t_1^0,\sigma_2(t_2^0))v_1\right.$$

$$+f(\sigma_1(t_1^0),\sigma_2(t_2^0))f_{t_2}^{\Delta_2}(t_1^0,t_2^0)v_2\Big)$$

$$= -\frac{1}{f(t_1^0,t_2^0)f(t_1^0,\sigma_2(t_2^0))f(\sigma_1(t_1^0),\sigma_2(t_2^0))}\Big(f(t_1^0,t_2^0)\Big(f_{t_1}^{\Delta_1}(t_1^0,\sigma_2(t_2^0))v_1$$

$$+f_{t_2}^{\Delta_2}(t_1^0,t_2^0)v_2\Big) + \big(f(\sigma_1(t_1^0),\sigma_2(t_2^0)) - f(t_1^0,t_2^0)\big)f_{t_2}^{\Delta_2}(t_1^0,t_2^0)v_2\Big)$$

$$= -\frac{1}{f(t_1^0,t_2^0)f(t_1^0,\sigma_2(t_2^0))f(\sigma_1(t_1^0),\sigma_2(t_2^0))}\Big(f(t_1^0,t_2^0)\frac{\partial f}{\Delta v}(t_1^0,t_2^0)$$

$$+\Big(\mu_2(t_2^0)f_{t_2}^{\Delta_2}(\sigma_1(t_1^0),t_2^0) + \mu_1(t_1^0)f_{t_1}^{\Delta_1}(t_1^0,t_2^0)\Big)f_{t_2}^{\Delta_2}(t_1^0,t_2^0)v_2\Big)$$

$$= -\frac{1}{f(t_1^0,t_2^0)f(t_1^0,\sigma_2(t_2^0))f(\sigma_1(t_1^0),\sigma_2(t_2^0))}\Big(f(\sigma_1(t_1^0),\sigma_2(t_2^0))\Big(f_{t_1}^{\Delta_1}(t_1^0,\sigma_2(t_2^0))v_1$$

$$+f_{t_2}^{\Delta_2}(t_1^0,t_2^0)v_1\Big) + \big(f(t_1^0,t_2^0) - f(\sigma_1(t_1^0),\sigma_2(t_2^0))\big)f_{t_1}^{\Delta_1}(t_1^0,\sigma_2(t_2^0))v_1\Big)$$

$$= -\frac{1}{f(t_1^0,t_2^0)f(t_1^0,\sigma_2(t_2^0))f(\sigma_1(t_1^0),\sigma_2(t_2^0))}\Big(f(\sigma_1(t_1^0),\sigma_2(t_2^0))\frac{\partial f}{\Delta v}(t_1^0,t_2^0)$$

$$-\big(\mu_1(t_1^0)f_{t_1}^{\Delta_1}(t_1^0,t_2^0) + \mu_2(t_2^0)f_{t_2}^{\Delta_2}(\sigma_1(t_1^0),t_2^0)\big)f_{t_1}^{\Delta_1}(t_1^0,\sigma_2(t_2^0))v_1\Big).$$

This completes the proof.

Remark 7.2 Note that Theorem 7.2 - Theorem 7.5 can be generalized for arbitrary $n \in \mathbb{N}$, $n \geq 2$.

Exercise 7.2 Let $f,g : \Lambda^2 \to \mathbb{R}$ be σ_1-completely delta differentiable and

$$g(t_1^0,t_2^0), \quad g(\sigma_1(t_1^0),t_2^0), \quad g(t_1^0,\sigma_2(t_2^0)) \neq 0.$$

Find expressions for

$$\frac{\partial}{\Delta v}\left(\frac{f}{g}\right)(t_1^0,t_2^0).$$

Hint. Use $\frac{f}{g} = f \cdot \frac{1}{g}$. Then apply Theorem 7.2 and Theorem 7.4.

Exercise 7.3 Let $f,g : \Lambda^2 \to \mathbb{R}$ be σ_2-completely delta differentiable and

$$g(t_1^0,t_2^0), \quad g(\sigma_1(t_1^0),\sigma_2(t_2^0)), \quad g(t_1^0,\sigma_2(t_2^0)) \neq 0.$$

Find expressions for

$$\frac{\partial}{\Delta v}\left(\frac{f}{g}\right)(t_1^0,t_2^0).$$

Hint. Use $\frac{f}{g} = f \cdot \frac{1}{g}$. Then apply Theorem 7.3 and Theorem 7.5.

7.2 Tangent Spaces. Vector Fields

Definition 7.2 We call Λ^n an n-dimensional time scale Euclidean space, shortly n-dimensional Euclidean space.

Let $p \in \Lambda^n$. A tangent vector v_p to Λ^n consists of two points of Λ^n: its vector part v and its point of applications p.

Definition 7.3 The set $T_p(\Lambda^n)$ consisting of all tangent vectors that have p as point of application is called tangent space of Λ^n at p.

Definition 7.4 A vector field W on Λ^n is a function that assigns to each point p of Λ^n a tangent vector v_p to Λ^n at p. Generally, a vector field W is denoted by

$$W = \sum_{j=1}^{n} w_j(t) \frac{\partial}{\Delta_j t_j}, \quad t \in \Lambda^n,$$

where $w_j : \Lambda^n \to \mathbb{R}$ and have partial delta derivatives on Λ^n. The set

$$\left\{ \frac{\partial}{\Delta_1 t_1}, \frac{\partial}{\Delta_2 t_2}, \ldots, \frac{\partial}{\Delta_n t_n} \right\}$$

is called the basis of $T_p(\Lambda^n)$. If the functions $w_j, j \in \{1, \ldots, n\}$, are σ_k-completely delta differentiable for some $k \in \{1, \ldots, n\}$, we say that the vector field W is σ_k-completely delta differentiable. With $\chi_k(\Lambda^n)$ we will denote the set of σ_k-completely delta differentiable vector fields on Λ^n. Define, for $k \in \{1, \ldots, n\}$,

$$W^{\sigma_k} = \sum_{j=1}^{n} w_j^{\sigma_k} \frac{\partial}{\Delta t_j}.$$

Definition 7.5 If we have two vector fields

$$V_1 = \sum_{j=1}^{n} V_{1j} \frac{\partial}{\Delta t_j}, \quad V_2 = \sum_{j=1}^{n} V_{2j} \frac{\partial}{\Delta t_j},$$

where $V_{ij} : \Lambda^n \to \mathbb{R}, i \in \{1,2\}, j \in \{1, \ldots, n\}$, are given functions. Then, for $a, b \in \mathbb{R}$, $f : \Lambda^n \to \mathbb{R}$, define

$$aV_1 + bV_2 = \sum_{j=1}^{n} (aV_{1j} + bV_{2j}) \frac{\partial}{\Delta t_j}$$

and

$$fV_1 = \sum_{j=1}^{n} (fV_{1j}) \frac{\partial}{\Delta t_j}.$$

Example 7.2 Let $n = 3$ and

$$V_1 = (t_1 + t_2)\frac{\partial}{\Delta t_1} + t_2^2\frac{\partial}{\Delta t_2} + t_1 t_2 t_3\frac{\partial}{\Delta t_3},$$

$$V_2 = t_2\frac{\partial}{\Delta t_1} + t_1 t_2 t_2\frac{\partial}{\Delta t_3}, \quad (t_1, t_2, t_3) \in \Lambda^3.$$

Then

$$V_1 - V_2 = t_1\frac{\partial}{\Delta t_1} + t_2^2\frac{\partial}{\Delta t_2}, \quad (t_1, t_2, t_3) \in \Lambda^3.$$

Exercise 7.4 Let $n = 3$ and

$$V_1 = \left(t_1^2 - t_2^2\right)\frac{\partial}{\Delta t_1} + \left(t_1^2 + t_2^2\right)\frac{\partial}{\Delta t_2} + t_2 t_3\frac{\partial}{\Delta t_3},$$

$$V_2 = t_2^2\frac{\partial}{\Delta t_1} - \left(t_1 + 3t_2^2\right)\frac{\partial}{\Delta t_2} - 3t_2 t_3\frac{\partial}{\Delta t_3},$$

$$V_3 = t_2^2\frac{\partial}{\Delta t_1} - t_1^2\frac{\partial}{\Delta t_2} + t_1 t_2\frac{\partial}{\Delta t_3}, \quad (t_1, t_2, t_3) \in \Lambda^3.$$

Find

1. $2V_1 - 3V_2$,
2. $V_1 + V_2 + V_3$,
3. $t_1 V_1 + t_2 V_2 + t_3 V_3$.

7.3 Delta Covariant Differentiation. The Delta Nature Connection

In this section, we will define the covariant derivative and we will prove some of its properties. Assume that $t_0 \in \mathbb{T}$.

Definition 7.6 Let W be a vector field on Λ^n and $v_p \in V_p(\Lambda^n)$. Assume that

$$Y(t) = W(p + (t - t_0)v), \quad t \in \mathbb{T}.$$

The derivative $Y^{\Delta}(t_0)$ is called a delta covariant derivative of the vector field W in the direction of the tangent vector v_p and it is denoted by $\Delta_{v_p} W$. Note that $\Delta_{v_p} W$ is also a tangent vector at the point p.

Theorem 7.6 *Let $w_j : \Lambda^n \to \mathbb{R}$ be differentiable, $j \in \{1, \ldots, n\}$. Let also, $v_p \in V_p(\Lambda^n)$ and*

$$W = \sum_{j=1}^{n} w_j \frac{\partial}{\Delta t_j}.$$

Then

$$\Delta_{v_p} W = \sum_{j=1}^{n} \frac{\partial w_j}{\Delta v_p} \frac{\partial}{\Delta t_j}(p).$$

Proof. Let

$$Y(t) = W(p + (t - t_0)v), \quad t \in \mathbb{T}.$$

Then

$$Y(t) = \sum_{j=1}^{n} w_j(p + (t - t_0)v) \frac{\partial}{\Delta t_j}, \quad t \in \mathbb{T}.$$

Thus,

$$\Delta_{v_p} W = \sum_{j=1}^{n} \frac{\partial w_j}{\Delta v_p}(p) \frac{\partial}{\Delta t_j}(p).$$

This completes the proof.

Definition 7.7 Let U and W be vector fields on Λ^n. The delta covariant derivative of W with respect to U is defined by

$$\Delta_U W(p) = \Delta_{U(p)} W.$$

Definition 7.8 The mapping $(U, W) \to \Delta_U W$ will be called the delta nature connection. If

$$W = \sum_{j=1}^{n} w_j \frac{\partial}{\Delta t_j}$$

is given, then we can define the delta nature connection as follows

$$\Delta_U W = \sum_{j=1}^{n} \frac{\partial w_j}{\Delta U} \frac{\partial w_j}{\Delta t_j}.$$

Remark 7.3 If we take the vector field U as a delta tangent vector f^{Δ} to the curve f on the time scale, the delta nature connection will be

$$\Delta_{f^{\Delta}(t)} f = \sum_{j=1}^{n} \frac{\partial f_j}{\Delta_{f^{\Delta}(t)}} \frac{\partial}{\Delta t_j}.$$

Theorem 7.7 *Let U, W and Z be vector fields. Then*

$$\Delta_U(aW + bZ) = a\Delta_U W + b\Delta_U Z$$

for any $a, b \in \mathbb{R}$.

Proof. Let

$$W=\sum_{j=1}^{n} w_j \frac{\partial}{\Delta t_j}, \quad Z=\sum_{j=1}^{n} z_j \frac{\partial}{\Delta t_j}.$$

Then

$$aW+bZ=\sum_{j=1}^{n}(aw_j+bz_j)\frac{\partial}{\Delta t_j}.$$

Hence,

$$\begin{aligned}\Delta_U(aW+bZ) &= \sum_{j=1}^{n} \frac{\partial(aw_j+bz_j)}{\Delta U}\frac{\partial}{\Delta t_j} \\ &= \sum_{j=1}^{n}\left(a\frac{\partial w_j}{\Delta U}+b\frac{\partial z_j}{\Delta U}\right)\frac{\partial}{\Delta t_j} \\ &= a\sum_{j=1}^{n}\frac{\partial w_j}{\Delta U}\frac{\partial}{\Delta t_j}+b\sum_{j=1}^{n}\frac{\partial z_j}{\Delta U}\frac{\partial}{\Delta t_j} \\ &= a\Delta_U W+b\Delta_U Z.\end{aligned}$$

This completes the proof.

Theorem 7.8 *Let $n=2$, V and G be vector fields and*

$$G=\sum_{j=1}^{2} g_j \frac{\partial}{\Delta t_j},$$

f, g_j, $j\in\{1,2\}$ and be σ_1-completely delta differentiable on Λ^2. Then

$$\Delta_V(fG)=f\Delta_V G+\frac{\partial f}{\Delta V}G^{\sigma_1}+\left(\mu_1 f_{t_1}^{\Delta_1}+\mu_2 f_{t_2}^{\Delta_2\sigma_1}\right)V_2\sum_{j=1}^{2} g_{jt_2}^{\Delta_2\sigma_1}\frac{\partial}{\Delta t_j}.$$

Proof. We have

$$fG=\sum_{j=1}^{2} fg_j\frac{\partial}{\Delta t_j}$$

and

$$\Delta_V(fG)=\sum_{j=1}^{2}\frac{\partial(fg_j)}{\Delta V}\frac{\partial}{\Delta t_j}.$$

Now, we use the first equation of Theorem 7.2 and we get

$$\Delta_V(fG)=\sum_{j=1}^{2}\left(f\frac{\partial g_j}{\Delta V}+g_j^{\sigma_1}\frac{\partial f}{\Delta V}+\left(\mu_1 f_{t_1}^{\Delta_1}+\mu_2 f_{t_2}^{\Delta_2\sigma_1}\right)g_{jt_2}^{\Delta_2\sigma_1}V_2\right)\frac{\partial}{\Delta t_j}$$

$$=f\sum_{j=1}^{2}\frac{\partial g_j}{\Delta V}\frac{\partial}{\Delta t_j}+\frac{\partial f}{\Delta V}\sum_{j=1}^{2}g_j^{\sigma_1}\frac{\partial}{\Delta t_j}+\left(\mu_1 f_{t_1}^{\Delta_1}+\mu_2 f_{t_2}^{\Delta_2\sigma_1}\right)V_2\sum_{j=1}^{2}g_{jt_2}^{\Delta_2\sigma_1}\frac{\partial}{\Delta t_j}$$

$$=f\Delta_V G+\frac{\partial f}{\Delta V}G^{\sigma_1}+\left(\mu_1 f_{t_1}^{\Delta_1}+\mu_2 f_{t_2}^{\Delta_2\sigma_1}\right)V_2\sum_{j=1}^{2}g_{jt_2}^{\Delta_2\sigma_1}\frac{\partial}{\Delta t_j}.$$

This completes the proof.

Remark 7.4 Using the other equations of Theorem 7.2, we can obtain the other variants of Theorem 7.8.

Theorem 7.9 *Let $n=2$, V and G be vector fields,*

$$G=\sum_{j=1}^{2}g_j\frac{\partial}{\Delta t_j},$$

f, g_j, $j\in\{1,2\}$, be σ_2-completely delta differentiable. Then

$$\Delta_V(fG)=f\Delta_V G+\frac{\partial f}{\Delta V}G^{\sigma_2}+\mu_2 f_{t_2}^{\Delta_2}V_1\sum_{j=1}^{2}g_{jt_1}^{\Delta_1\sigma_2}\frac{\partial}{\Delta t_j}+\mu_1 f_{t_1}^{\Delta_1\sigma_2}V_1\sum_{j=1}^{2}g_{jt_1}^{\Delta_1\sigma_2}\frac{\partial}{\Delta t_j}.$$

Proof. By the first equation of Theorem 7.3, we find

$$\Delta_V(fG)=\sum_{j=1}^{2}\frac{\partial(fg_j)}{\Delta V}\frac{\partial}{\Delta t_j}$$

$$=\sum_{j=1}^{2}\left(f\frac{\partial g_j}{\Delta V}+g_j^{\sigma_2}\frac{\partial f}{\Delta V}+\mu_2 f_{t_2}^{\Delta_2}g_{jt_1}^{\Delta_1\sigma_2}V_1+\mu_1 f_{t_1}^{\Delta_1\sigma_2}g_{jt_1}^{\Delta_1\sigma_2}V_1\right)\frac{\partial}{\Delta t_j}$$

$$=f\sum_{j=1}^{2}\frac{\partial g_j}{\Delta V}\frac{\partial}{\Delta t_j}+\frac{\partial f}{\Delta V}\sum_{j=1}^{2}g_j^{\sigma_2}\frac{\partial}{\Delta t_j}+\mu_2 f_{t_2}^{\Delta_2}V_1\sum_{j=1}^{2}g_{jt_1}^{\Delta_1\sigma_2}\frac{\partial}{\Delta t_j}$$

$$+\mu_1 f_{t_1}^{\Delta_1\sigma_2}V_1\sum_{j=1}^{2}g_{jt_1}^{\Delta_1\sigma_2}\frac{\partial}{\Delta t_j}$$

$$=f\Delta_V G+\frac{\partial f}{\Delta V}G^{\sigma_2}+\mu_2 f_{t_2}^{\Delta_2}V_1\sum_{j=1}^{2}g_{jt_1}^{\Delta_1\sigma_2}\frac{\partial}{\Delta t_j}$$

$$+\mu_1 f_{t_1}^{\Delta_1\sigma_2}V_1\sum_{j=1}^{2}g_{jt_1}^{\Delta_1\sigma_2}\frac{\partial}{\Delta t_j}.$$

This completes the proof.

Remark 7.5 Using the other equations of Theorem 7.3, one can deduce the other variants of Theorem 7.9.

Remark 7.6 Theorem 7.8 and Theorem 7.9 can be deduced for arbitrary $n \in \mathbb{N}$.

7.4 The Lie Brackets

Suppose that V_1, V_2 and V_3 are vector fields such that

$$V_1 = \sum_{j=1}^{n} V_{1j} \frac{\partial}{\Delta t_j},$$

$$V_2 = \sum_{j=1}^{n} V_{2j} \frac{\partial}{\Delta t_j},$$

$$V_3 = \sum_{j=1}^{n} V_{3j} \frac{\partial}{\Delta t_j},$$

where $V_{ij}, i \in \{1,2,3\}, j \in \{1,\ldots,n\}$, are delta differentiable on Λ^n.

Definition 7.9 For a function $f : \Lambda^n \to \mathbb{R}$ that is delta differentiable, define $V_1 [V_2]_\Delta (f)$ in the following manner

$$V_1 [V_2]_\Delta (f) = \sum_{j=1}^{n} V_{1j} \frac{\partial}{\Delta t_j} \left(\sum_{k=1}^{n} V_{2k} \frac{\partial f}{\Delta t_k} \right).$$

Example 7.3 Let $n = 2$, $\mathbb{T}_1 = 2^{\mathbb{N}_0}$, $\mathbb{T}_2 = 3^{\mathbb{N}_0}$ and

$$V_1 = \left(t_1^2 + t_2^2\right) \frac{\partial}{\Delta t_1} + t_1 t_2 \frac{\partial}{\Delta t_2},$$

$$V_2 = t_1 t_2 \frac{\partial}{\Delta t_1} + \left(t_1^2 - t_2^2\right) \frac{\partial}{\Delta t_2},$$

$$f(t_1,t_2) = t_1 + t_2, \quad (t_1,t_2) \in \Lambda^2.$$

Here

$$\sigma_1(t_1) = 2t_1, \quad t_1 \in \mathbb{T}_1,$$

$$\sigma_2(t_2) = 3t_2, \quad t_2 \in \mathbb{T}_2,$$

$$V_{11}(t_1,t_2) = t_1^2 + t_2^2,$$

$$V_{12}(t_1,t_2) = t_1 t_2,$$

$$V_{21}(t_1,t_2) = t_1 t_2,$$

$$V_{22}(t_1,t_2) = t_1^2 - t_2^2, \quad (t_1,t_2) \in \Lambda^2.$$

Then

$$\begin{aligned}
(V_{11})_{t_1}^{\Delta_1}(t_1,t_2) &= \sigma_1(t_1) + t_1 \\
&= 2t_1 + t_1 \\
&= 3t_1, \\
(V_{11})_{t_2}^{\Delta_2}(t_1,t_2) &= \sigma_2(t_2) + t_2 \\
&= 3t_2 + t_2 \\
&= 4t_2, \\
(V_{12})_{t_1}^{\Delta_1}(t_1,t_2) &= t_2, \\
(V_{12})_{t_2}^{\Delta_2}(t_1,t_2) &= t_1, \\
(V_{21})_{t_1}^{\Delta_1}(t_1,t_2) &= t_2, \\
(V_{22})_{t_2}^{\Delta_2}(t_1,t_2) &= t_1, \\
(V_{22})_{t_1}^{\Delta_1}(t_1,t_2) &= \sigma_1(t_1) + t_1 \\
&= 2t_1 + t_1 \\
&= 3t_1, \\
(V_{22})_{t_2}^{\Delta_2}(t_1,t_2) &= -\sigma_2(t_2) - t_2 \\
&= -3t_2 - t_2 \\
&= -4t_2, \\
f_{t_1}^{\Delta_1}(t_1,t_2) &= 1, \\
f_{t_2}^{\Delta_2}(t_1,t_2) &= 1, \\
f_{t_1}^{\Delta_1^2}(t_1,t_2) &= 0, \\
f_{t_1t_2}^{\Delta_1\Delta_2}(t_1,t_2) &= 0, \\
f_{t_2}^{\Delta_2^2}(t_1,t_2) &= 0, \quad (t_1,t_2) \in \Lambda^2,
\end{aligned}$$

and

$$\begin{aligned}
V_1[V_2]_\Delta(f) = V_{11} &\left((V_{21})_{t_1}^{\Delta_1}(t_1,t_2)\frac{\partial f}{\Delta t_1} + (V_{22})_{t_1}^{\Delta_1}(t_1,t_2)\frac{\partial f}{\Delta t_2} \right) \\
+V_{12} &\left((V_{21})_{t_2}^{\Delta_2}(t_1,t_2)\frac{\partial f}{\Delta t_1} + (V_{22})_{t_2}^{\Delta_2}(t_1,t_2)\frac{\partial f}{\Delta t_2} \right) \\
+V_{11} &\left(V_{21}^{\sigma_1}(t_1,t_2)\frac{\partial^2 f}{\Delta t_1^2} + V_{22}^{\sigma_1}(t_1,t_2)\frac{\partial^2 f}{\Delta t_1 \Delta t_2} \right) \\
+V_{12} &\left(V_{21}^{\sigma_2}(t_1,t_2)\frac{\partial^2 f}{\Delta t_1 \Delta t_2} + V_{22}^{\sigma_2}(t_1,t_2)\frac{\partial^2 f}{\Delta t_2^2} \right)
\end{aligned}$$

$$
\begin{aligned}
&= \left(t_1^2+t_2^2\right)\left(t_2\frac{\partial f}{\Delta t_1}+3t_1\frac{\partial f}{\Delta t_2}\right)+t_1t_2\left(t_1\frac{\partial f}{\Delta t_1}-4t_2\frac{\partial f}{\Delta t_2}\right)\\
&= \left(t_1^2t_2+t_2^3+t_1^2t_2\right)\frac{\partial f}{\Delta t_1}+\left(3t_1^3+3t_1t_2^2-4t_1t_2^2\right)\frac{\partial f}{\Delta t_2}\\
&= \left(2t_1^2t_2+t_2^3\right)\frac{\partial f}{\Delta t_1}+\left(3t_1^3-t_1t_2^2\right)\frac{\partial f}{\Delta t_2}\\
&= 2t_1^2t_2+t_2^3+3t_1^3-t_1t_2^2, \quad (t_1,t_2)\in\Lambda^2,
\end{aligned}
$$

and

$$
\begin{aligned}
V_2[V_1]_\Delta(f) &= V_{21}\left((V_{11})_{t_1}^{\Delta_1}(t_1,t_2)\frac{\partial f}{\Delta t_1}+(V_{12})_{t_1}^{\Delta_1}(t_1,t_2)\frac{\partial f}{\Delta t_2}\right)\\
&\quad +V_{22}\left((V_{11})_{t_2}^{\Delta_2}(t_1,t_2)\frac{\partial f}{\Delta t_1}+(V_{12})_{t_2}^{\Delta_2}(t_1,t_2)\frac{\partial f}{\Delta t_2}\right)\\
&\quad +V_{21}\left(V_{11}^{\sigma_1}(t_1,t_2)\frac{\partial^2 f}{\Delta t_1^2}+V_{12}^{\sigma_1}(t_1,t_2)\frac{\partial^2 f}{\Delta t_1\Delta t_2}\right)\\
&\quad +V_{22}\left(V_{11}^{\sigma_2}(t_1,t_2)\frac{\partial^2 f}{\Delta t_1\Delta t_2}+V_{12}^{\sigma_2}(t_1,t_2)\frac{\partial^2 f}{\Delta t_2^2}\right)\\
&= t_1t_2\left(3t_1\frac{\partial f}{\Delta t_1}+t_2\frac{\partial f}{\Delta t_2}\right)+\left(t_1^2-t_2^2\right)\left(4t_2\frac{\partial f}{\Delta t_1}+t_1\frac{\partial f}{\Delta t_2}\right)\\
&= \left(3t_1^2t_2+4t_1^2t_2-4t_2^3\right)\frac{\partial f}{\Delta t_1}+\left(t_1t_2^2+t_1^3-t_1t_2^2\right)\frac{\partial f}{\Delta t_2}\\
&= \left(7t_1^2t_2-4t_2^3\right)\frac{\partial f}{\Delta t_1}+t_1^3\frac{\partial f}{\Delta t_2}\\
&= 7t_1^2t_2-4t_2^3+t_1^3, \quad (t_1,t_2)\in\Lambda^2.
\end{aligned}
$$

Exercise 7.5 Let $n=2$, $\mathbb{T}_1=\mathbb{Z}$, $\mathbb{T}_2=3\mathbb{Z}$ and

$$
\begin{aligned}
V_1 &= \left(t_1^2-t_1t_2+t_2^3\right)\frac{\partial}{\Delta t_1}+\left(t_1^2+t_2+\frac{t_1+t_2}{1+t_1^2}\right)\frac{\partial}{\Delta t_2},\\
V_2 &= (t_1-t_2)\frac{\partial}{\Delta t_1}+t_1^3t_2^3\frac{\partial}{\Delta t_2},\\
V_3 &= t_1t_2^2\frac{\partial}{\delta t_1}+t_1\frac{\partial}{\Delta t_2},\\
f(t_1,t_2) &= t_1^2+3t_1t_2+t_2^3, \quad (t_1,t_2)\in\Lambda^3.
\end{aligned}
$$

Find

1. $V_1[V_2]_\Delta(f)$,
2. $V_1[V_3]_\Delta(f)-(V_1+V_2)[V_3]_\Delta(f)$,
3. $(2v_1+4V_2)[V_3]_\Delta(f)$.

Remark 7.7 By Example 7.3, it follows that in the general case, we have

$$V_1[V_2]_\Delta \neq V_2[V_1]_\Delta.$$

Theorem 7.10 *Let V_j, $j \in \{1,\ldots,4\}$, be vector fields such that*

$$V_j = \sum_{k=1}^{n} V_{jk} \frac{\partial}{\Delta t_k}, \quad j \in \{1,\ldots,4\}.$$

Then, for any $a,b,c,d \in \mathbb{R}$, we have

$$\begin{aligned}(aV_1 + bV_2)[cV_3 + dV_4]_\Delta &= ac(V_1[V_3]_\Delta) + ad(V_1[V_4]_\Delta)\\ &\quad + bc(V_2[V_3]_\Delta) + bd(V_2[V_4]_\Delta).\end{aligned}$$

Proof. We have

$$\begin{aligned}aV_1 &= a\left(\sum_{k=1}^{n} V_{1k}\frac{\partial}{\Delta t_k}\right)\\ &= \sum_{k=1}^{n}(aV_{1k})\frac{\partial}{\Delta t_k},\\ bV_2 &= b\left(\sum_{k=1}^{n} V_{2k}\frac{\partial}{\Delta t_k}\right)\\ &= \sum_{k=1}^{n}(bV_{2k})\frac{\partial}{\Delta t_k},\\ cV_3 &= c\left(\sum_{k=1}^{n} V_{3k}\frac{\partial}{\Delta t_k}\right)\\ &= \sum_{k=1}^{n}(cV_{3k})\frac{\partial}{\Delta t_k},\\ dV_4 &= d\left(\sum_{k=1}^{n} V_{4k}\frac{\partial}{\Delta t_k}\right)\\ &= \sum_{k=1}^{n}(dV_{4k})\frac{\partial}{\Delta t_k}\end{aligned}$$

and

$$\begin{aligned}aV_1 + bV_2 &= \sum_{k=1}^{n}(aV_{1k})\frac{\partial}{\Delta t_k} + \sum_{k=1}^{n}(bV_{2k})\frac{\partial}{\Delta t_k}\\ &= \sum_{k=1}^{n}(aV_{1k} + bV_{2k})\frac{\partial}{\Delta t_k},\end{aligned}$$

$$\begin{aligned} cV_3 + dV_4 &= \sum_{k=1}^{n} (cV_{3k}) \frac{\partial}{\Delta t_k} + \sum_{k=1}^{n} (dV_{4k}) \frac{\partial}{\Delta t_k} \\ &= \sum_{k=1}^{n} (cV_{3k} + dV_{4k}) \frac{\partial}{\Delta t_k}. \end{aligned}$$

Hence,

$$\begin{aligned} (aV_1 + bV_2)[cV_3 + dV_4]_\Delta &= \sum_{j=1}^{n} \left(aV_{1j} + bV_{2j}\right) \frac{\partial}{\Delta t_j} \left(\sum_{k=1}^{n} (cV_{3k} + dV_{4k}) \frac{\partial}{\Delta t_k} \right) \\ &= \sum_{j=1}^{n} \left(aV_{1j} + bV_{2j}\right) \left(c \frac{\partial}{\Delta t_j} \left(\sum_{k=1}^{n} V_{3k} \frac{\partial}{\Delta t_k} \right) \right. \\ &\quad \left. + b \frac{\partial}{\Delta t_j} \left(\sum_{k=1}^{n} V_{4k} \frac{\partial}{\Delta t_k} \right) \right) \\ &= \left(a \sum_{j=1}^{n} V_{1j} + b \sum_{j=1}^{n} V_{2j} \right) \left(c \frac{\partial}{\Delta t_j} \left(\sum_{k=1}^{n} V_{3k} \frac{\partial}{\Delta t_k} \right) \right. \\ &\quad \left. + d \frac{\partial}{\Delta t_j} \left(\sum_{k=1}^{n} V_{4k} \frac{\partial}{\Delta t_k} \right) \right) \\ &= a \sum_{j=1}^{n} V_{1j} \left(c \frac{\partial}{\Delta t_j} \left(\sum_{k=1}^{n} V_{3k} \frac{\partial}{\Delta t_k} \right) + d \frac{\partial}{\Delta t_j} \left(\sum_{k=1}^{n} V_{4k} \frac{\partial}{\Delta t_k} \right) \right) \\ &\quad + b \sum_{j=1}^{n} V_{2j} \left(c \frac{\partial}{\Delta t_j} \left(\sum_{k=1}^{n} V_{3k} \frac{\partial}{\Delta t_k} \right) + d \frac{\partial}{\Delta t_j} \left(\sum_{k=1}^{n} V_{4k} \frac{\partial}{\Delta t_k} \right) \right) \\ &= ac \sum_{j=1}^{n} V_{1j} \frac{\partial}{\Delta t_j} \left(\sum_{k=1}^{n} V_{3k} \frac{\partial}{\Delta t_k} \right) + ad \sum_{j=1}^{n} V_{1j} \frac{\partial}{\Delta t_j} \left(\sum_{k=1}^{n} V_{4k} \frac{\partial}{\Delta t_k} \right) \\ &\quad + bc \sum_{j=1}^{n} V_{2j} \frac{\partial}{\Delta t_j} \left(\sum_{k=1}^{n} V_{3k} \frac{\partial}{\Delta t_k} \right) + bd \sum_{j=1}^{n} V_{2j} \frac{\partial}{\Delta t_j} \left(\sum_{k=1}^{n} V_{4k} \frac{\partial}{\Delta t_k} \right) \\ &= ac(V_1[V_3]_\Delta) + ad(V_1[V_4]_\Delta) + bc(V_2[V_3]_\Delta) + bd(V_2[V_4]_\Delta). \end{aligned}$$

This completes the proof.

Theorem 7.11 *Let V_j, $j \in \{1,2\}$, be vector fields such that*

$$V_j = \sum_{k=1}^{n} V_{jk} \frac{\partial}{\Delta t_k}, \quad j \in \{1,2\},$$

where V_{jk}, $j \in \{1,2\}$, $k \in \{1,\ldots,n\}$, are delta differentiable and $f : \Lambda^n \to \mathbb{R}$ is delta differentiable. Then

$$V_1\left[fV_2\right]_\Delta = \sum_{j=1}^{n} V_{1j}\left(\sum_{k=1}^{n} f^{\sigma_j}\frac{\partial}{\Delta t_j}\left(V_{2k}\frac{\partial}{\Delta t_k}\right)\right)$$
$$+\sum_{j=1}^{n} V_{1j}\left(\sum_{k=1}^{n}\frac{\partial}{\Delta t_j}fV_{2k}\frac{\partial}{\Delta t_k}\right).$$

Proof. We have

$$\begin{aligned}V_1\left[fV_2\right]_\Delta &= \sum_{j=1}^{n} V_{1j}\frac{\partial}{\Delta t_j}\left(\sum_{k=1}^{n} fV_{2k}\frac{\partial}{\Delta t_k}\right)\\ &= \sum_{j=1}^{n} V_{1j}\left(\sum_{k=1}^{n} f^{\sigma_j}\frac{\partial}{\Delta t_j}\left(V_{2k}\frac{\partial}{\Delta t_k}\right)\right)\\ &\quad+\sum_{j=1}^{n} V_{1j}\left(\sum_{k=1}^{n}\frac{\partial}{\Delta t_j}fV_{2k}\frac{\partial}{\Delta t_k}\right).\end{aligned}$$

This completes the proof.

Definition 7.10 The equation

$$[V_1, V_2]_\Delta = V_1\,[V_2]_\Delta - V_2\,[V_1]_\Delta$$

will be called the delta Lie brackets on the time scale, shortly Lie brackets.

Example 7.4 Let n, $\mathbb{T}_1$, $\mathbb{T}_2$, V_1, V_2 and f be as in Example 7.3. Then

$$\begin{aligned}[V_1, V_2]_\Delta(f) &= V_1\,[V_2]_\Delta(f) - V_2\,[V_1]_\Delta(f)\\ &= \left(2t_1^2t_2 + t_2^3\right)\frac{\partial f}{\Delta t_1} + \left(3t_1^3 - t_1t_2^2\right)\frac{\partial f}{\Delta t_2}\\ &\quad - \left(7t_1^2t_2 - 4t_2^3\right)\frac{\partial f}{\Delta t_1} - t_1^3\frac{\partial f}{\Delta t_2}\\ &= \left(5t_2^3 - 5t_1^2t_2\right)\frac{\partial f}{\Delta t_1} + \left(2t_1^3 - t_1t_2^2\right)\frac{\partial f}{\Delta t_2}\\ &= 2t_1^3 + 5t_2^3 - 5t_1^2t_2 - t_1t_2^2, \quad (t_1,t_2) \in \Lambda^2.\end{aligned}$$

Exercise 7.6 Let $n = 2$, $\mathbb{T}_1 = 2^{\mathbb{N}_0}$, $\mathbb{T}_2 = 3\mathbb{Z}$ and

$$V_1 = (t_1 + t_2)\frac{\partial}{\Delta t_1} + t_1^2t_2\frac{\partial}{\Delta t_2},$$

$$V_2 = (t_1^2 - t_2)\frac{\partial}{\Delta t_1} + t_1t_2^3\frac{\partial}{\Delta t_2},$$

$$f(t_1,t_2) = t_1(t_1 + t_2^2), \quad (t_1,t_2) \in \Lambda^2.$$

Find

$$[V_1,V_2]_\Delta (f).$$

By the definition, it follows that

$$[V_1,V_1]_\Delta = 0.$$

Below, we will deduce some of the properties of the Lie brackets.

Theorem 7.12 *Let V_j, $j \in \{1,2\}$, be vector fields such that*

$$V_j = \sum_{k=1}^{n} V_{jk} \frac{\partial}{\Delta t_k}, \quad j \in \{1,2\},$$

where $V_{jk} : \Lambda^n \to \mathbb{R}$ are delta differentiable, $j \in \{1,2\}$, $k \in \{1,\dots,n\}$. Then

$$[V_1,V_2]_\Delta = -[V_2,V_1]_\Delta.$$

Proof. We have

$$\begin{aligned}
[V_1,V_2]_\Delta &= V_1[V_2]_\Delta - V_2[V_1]_\Delta \\
&= -(V_2[V_1]_\Delta - V_1[V_2]_\Delta) \\
&= -[V_2,V_1]_\Delta.
\end{aligned}$$

This completes the proof.

Theorem 7.13 *Let V_j, $j \in \{1,2,3\}$, be vector fields such that*

$$V_j = \sum_{k=1}^{n} V_{jk} \frac{\partial}{\Delta t_k}, \quad j \in \{1,2\},$$

where $V_{jk} : \Lambda^n \to \mathbb{R}$ are delta differentiable, $j \in \{1,2,3\}$, $k \in \{1,\dots,n\}$. Then

$$[aV_1 + bV_2, V_3]_\Delta = a[V_1,V_3]_\Delta + b[V_2,V_3]_\Delta$$

for any $a,b \in \mathbb{R}$.

Proof. We apply Theorem 7.10 and we find

$$\begin{aligned}
[aV_1 + bV_2, V_3]_\Delta &= (aV_1 + bV_2)[V_3]_\Delta - V_3[aV_1 + bV_2]_\Delta \\
&= a(V_1[V_3]_\Delta) + b(V_2[V_3]_\Delta) - a(V_3[V_1]_\Delta) - b(V_3[V_2]_\Delta) \\
&= a(V_1[V_3]_\Delta - V_3[V_1]_\Delta) + b(V_2[V_3]_\Delta - V_3[V_2]_\Delta) \\
&= a[V_1,V_3]_\Delta + b[V_2,V_3]_\Delta.
\end{aligned}$$

This completes the proof.

Theorem 7.14 *Let V_j, $j \in \{1,2,3\}$, be vector fields such that*

$$V_j = \sum_{k=1}^{n} V_{jk} \frac{\partial}{\Delta t_k}, \quad j \in \{1,2\},$$

where $V_{jk} : \Lambda^n \to \mathbb{R}$ are delta differentiable, $j \in \{1,2,3\}$, $k \in \{1,\ldots,n\}$. Then

$$[V_1,[V_2,V_3]_\Delta]_\Delta + [V_2,[V_3,V_1]_\Delta]_\Delta + [V_3,[V_1,V_2]_\Delta]_\Delta = 0.$$

Proof. We have

$$\begin{aligned}
[V_1,[V_2,V_3]_\Delta]_\Delta &= V_1([V_2,V_3]_\Delta)_\Delta - ([V_2,V_3]_\Delta)[V_1]_\Delta \\
&= V_1(V_2[V_3]_\Delta)_\Delta - V_1(V_3[V_2]_\Delta)_\Delta \\
&\quad -(V_2[V_3]_\Delta - V_3[V_2]_\Delta)[V_1]_\Delta \\
&= V_1(V_2[V_3]_\Delta)_\Delta - V_1(V_3[V_2]_\Delta)_\Delta \\
&\quad -(V_2[V_3]_\Delta)[V_1]_\Delta + (V_3[V_2]_\Delta)[V_1]_\Delta \\
&= V_1(V_2[V_3]_\Delta)_\Delta - V_1(V_3[V_2]_\Delta)_\Delta \\
&\quad -V_2(V_3[V_1]_\Delta)_\Delta + V_3(V_2[V_1]_\Delta)_\Delta
\end{aligned}$$

and

$$\begin{aligned}
[V_2,[V_3,V_1]_\Delta]_\Delta &= V_2[V_3,V_1]_\Delta - ([V_3,V_1]_\Delta)[V_2]_\Delta \\
&= V_2(V_3[V_1]_\Delta - V_1[V_3]_\Delta)_\Delta \\
&\quad -(V_3[V_1]_\Delta - V_1[V_3]_\Delta)[V_2]_\Delta \\
&= V_2(V_3[V_1]_\Delta)_\Delta - V_2(V_1[V_3]_\Delta)_\Delta \\
&\quad -V_3(V_1[V_2]_\Delta)_\Delta + V_1(V_3[V_2]_\Delta)_\Delta,
\end{aligned}$$

and

$$\begin{aligned}
[V_3,[V_1,V_2]_\Delta]_\Delta &= V_3([V_1,V_2]_\Delta)_\Delta - ([V_1,V_2]_\Delta)[V_3]_\Delta \\
&= V_3(V_1[V_2]_\Delta - V_2[V_1]_\Delta)_\Delta \\
&\quad -(V_1[V_2]_\Delta - V_2[V_1]_\Delta)[V_3]_\Delta \\
&= V_3(V_1[V_2]_\Delta - V_2[V_1]_\Delta)_\Delta \\
&\quad -V_1(V_2[V_3]_\Delta)_\Delta + V_2(V_1[V_3]_\Delta)_\Delta,
\end{aligned}$$

whereupon we get the desired result.

7.5 The Algebra of Dynamic Forms

In this section we introduce dynamic forms as formal objects and how to do algebra with them. For convenience, in this section we consider the case

$n = 3$. Let $\mathbb{T}_1$, $\mathbb{T}_2$ and $\mathbb{T}_3$ be time scales with forward jump operators and delta differentiation operators σ_1, σ_2, σ_3 and Δ_1, Δ_2, Δ_3, respectively. Set $\Lambda^3 = \mathbb{T}_1 \times \mathbb{T}_2 \times \mathbb{T}_3$.

Definition 7.11 1. A 0-dynamic form ϕ is a function $f : \Lambda^3 \to \mathbb{R}$, that is, $\phi = f$.

2. A 1-dynamic form is

$$\phi = f\Delta_1 t_1 + g\Delta_2 t_2 + h\Delta_3 t_3,$$

where $f, g, h : \Lambda^3 \to \mathbb{R}$.

3. A 2-dynamic form is

$$\phi = f\Delta_1 t_1 \Delta_2 t_2 + g\Delta_2 t_2 \Delta_3 t_3 + h\Delta_1 t_1 \Delta_3 t_3,$$

where $f, g, h : \Lambda^3 \to \mathbb{R}$.

4. A 3-dynamic form is

$$\phi = f\Delta_1 t_1 \Delta_2 t_2 \Delta_3 t_3,$$

where $f : \Lambda^3 \to \mathbb{R}$.

Example 7.5 The dynamic form

$$\phi = (t_1 + t_2)\Delta_1 t_1 + t_2 \Delta_3 t_3, \quad (t_1, t_2, t_3) \in \Lambda^3,$$

is a 1-dynamic form.

Example 7.6 The dynamic form

$$\phi = t_1 t_2 t_3 \Delta_1 t_1 \Delta_2 t_2 \Delta_3 t_3, \quad (t_1, t_2, t_3) \in \Lambda^3,$$

is a 3-dynamic form.

So far, $\Delta_1 t_1$, $\Delta_2 t_2$ and $\Delta_3 t_3$ are just symbols. Note that every term in a k-dynamic form, $k \in \{1,2,3\}$, has k-occurrences of $\Delta_1 t_1$, $\Delta_2 t_2$ and $\Delta_3 t_3$.

Definition 7.12 Let

$$\phi_1 = f_1 \Delta_1 t_1 + g_1 \Delta_2 t_2 + h_1 \Delta_3 t_3,$$
$$\phi_2 = f_2 \Delta_1 t_1 + g_2 \Delta_2 t_2 + h_2 \Delta_3 t_3,$$

where $f_j, g_j, h_j : \Lambda^3 \to \mathbb{R}$, $j \in \{1,2\}$, are given functions. Then, we define addition and substraction of ϕ_1 and ϕ_2 in the following manner

$$\phi_1 + \phi_2 = (f_1 + f_2)\Delta_1 t_1 + (g_1 + g_2)\Delta_2 t_2 + (h_1 + h_2)\Delta_3 t_3$$

and

$$\phi_1 - \phi_2 = (f_1 - f_2)\Delta_1 t_1 + (g_1 - g_2)\Delta_2 t_2 + (h_1 - h_2)\Delta_3 t_3,$$

respectively.

Definition 7.13 Let

$$\phi_1 = f_1 \Delta_1 t_1 \Delta_2 t_2 + g_1 \Delta_2 t_2 \Delta_3 t_3 + h_1 \Delta_1 t_1 \Delta_3 t_3,$$
$$\phi_2 = f_2 \Delta_1 t_1 \Delta_2 t_2 + g_2 \Delta_2 t_2 \Delta_3 t_3 + h_2 \Delta_1 t_1 \Delta_3 t_3,$$

where $f_j, g_j, h_j : \Lambda^3 \to \mathbb{R}$, $j \in \{1,2\}$, are given functions. Then, we define addition and substraction of ϕ_1 and ϕ_2 in the following manner

$$\phi_1 + \phi_2 = (f_1 + f_2)\Delta_1 t_1 \Delta_2 t_2 + (g_1 + g_2)\Delta_2 t_2 \Delta_3 t_3 + (h_1 + h_2)\Delta_1 t_1 \Delta_3 t_3$$

and

$$\phi_1 - \phi_2 = (f_1 - f_2)\Delta_1 t_1 \Delta_2 t_2 + (g_1 - g_2)\Delta_2 t_2 \Delta_3 t_3 + (h_1 - h_2)\Delta_1 t_1 \Delta_3 t_3,$$

respectively.

Definition 7.14 Let

$$\phi_1 = f \Delta_1 t_1 \Delta_2 t_2 \Delta_3 t_3,$$
$$\phi_2 = g \Delta_1 t_1 \Delta_1 t_2 \Delta_3 t_3,$$

where $f, g : \Lambda^3 \to \mathbb{R}$ are given functions. Then, we define the addition and substraction of ϕ_1 and ϕ_2 in the following manner

$$\phi_1 + \phi_2 = (f + g)\Delta_1 t_1 \Delta_2 t_2 \Delta_3 t_3$$

and

$$\phi_1 - \phi_2 = (f - g)\Delta_1 t_1 \Delta_2 t_2 \Delta_3 t_3.$$

Remark 7.8 Note that we can add and subtract two k-dynamic forms, not a k-dynamic form and an l-dynamic form, $k \neq l$, $k, l \in \{0,1,2,3\}$.

Example 7.7 Let

$$\phi_1 = (t_1 + t_2)\Delta_1 t_1 + t_1^2 \Delta_2 t_2 + (t_1 - t_2)\Delta_3 t_3,$$
$$\phi_2 = (3t_1 - t_2)\Delta_1 t_1 + t_1^2 \Delta_2 t_2 + (t_1 + t_2)\Delta_3 t_3, \quad (t_1, t_2, t_3) \in \Lambda^3.$$

Then

$$\phi_1 + \phi_2 = 4t_1 \Delta_1 t_1 + 2t_1^2 \Delta_2 t_2 + 2t_1 \Delta_3 t_3, \quad (t_1, t_2, t_3) \in \Lambda^3,$$

and

$$\phi_1 - \phi_2 = 2(-t_1 + t_2)\Delta_1 t_1 - 2t_2 \Delta_3 t_3, \quad (t_1, t_2, t_3) \in \Lambda^3.$$

Example 7.8 Let

$$\phi_1 = 4t_1 t_2 t_3 \Delta_1 t_1 \Delta_2 t_2 \Delta_3 t_3,$$
$$\phi_2 = (t_1 t_2 t_3 + t_1)\Delta_1 t_1 \Delta_2 t_2 \Delta_3 t_3, \quad (t_1, t_2, t_3) \in \Lambda^3.$$

Then

$$\phi_1+\phi_2=(5t_1t_2t_3+t_1)\Delta_1t_1\Delta_2t_2\Delta_3t_3, \quad (t_1,t_2,t_3)\in\Lambda^3,$$

and

$$\phi_1-\phi_2=(3t_1t_2t_3-t_1)\Delta_1t_1\Delta_2t_2\Delta_3t_3, \quad (t_1,t_2,t_3)\in\Lambda^3.$$

Exercise 7.7 Let

$$\phi_1=t_1t_2\Delta_1t_1\Delta_2t_2+t_3^2\Delta_2t_2\Delta_3t_3-t_1t_2t_3\Delta_1t_1\Delta_3t_3,$$
$$\phi_2=3t_1t_2\Delta_1t_1\Delta_2t_2-t_3^2\Delta_2t_2\Delta_3t_3+t_1^2\Delta_2t_2\Delta_3t_3,$$

$(t_1,t_2,t_3)\in\Lambda^3$. Find

$$\phi_1+\phi_2 \quad \text{and} \quad \phi_1-\phi_3.$$

Definition 7.15 Let $f:\Lambda^3\to\mathbb{R}$ is a given function and

$$\phi_1=f_1\Delta_1t_1+f_2\Delta_2t_2+f_3\Delta t_3,$$
$$\phi_2=g_1\Delta_1t_1\Delta_2t_2+g_2\Delta_1t_1\Delta_3t_3+g_3\Delta_2t_2\Delta_3t_3,$$
$$\phi_3=g\Delta_1t_1\Delta_2t_2\Delta_3t_3,$$

where $g,f_j,g_j:\Lambda^3\to\mathbb{R}$, $j\in\{1,2,3\}$, are given functions.Then, we define

$$f\phi_1=(ff_1)\Delta_1t_1+(ff_2)\Delta_2t_2+(ff_3)\Delta t_3,$$
$$f\phi_2=(fg_1)\Delta_1t_1\Delta_2t_2+(fg_2)\Delta_1t_1\Delta_3t_3+(fg_3)\Delta_2t_2\Delta_3t_3,$$
$$f\phi_3=(fg)\Delta_1t_1\Delta_2t_2\Delta_3t_3.$$

Example 7.9 Let

$$f(t_1,t_2,t_3)=t_1-t_2+t_3, \quad (t_1,t_2,t_3)\in\Lambda^3,$$

and

$$\phi=t_1\Delta_1t_1+t_2\Delta_2t_2+t_3\Delta_3t_3, \quad (t_1,t_2,t_3)\in\Lambda^3.$$

Then

$$f\phi=(t_1^2-t_1t_2+t_1t_3)\Delta_1t_1+(t_1t_2-t_2^2+t_2t_3)\Delta_2t_2+(t_1t_3-t_2t_3+t_3^2)\Delta_3t_3, \quad (t_1,t_2,t_3)\in\Lambda^3.$$

Exercise 7.8 Let

$$\phi_1=(t_1+2t_2)\Delta_1t_1-t_3\Delta_2t_2-(t_1+t_2)\Delta_3t_3,$$
$$\phi_2=(t_1-t_2)\Delta_1t_1+t_2\Delta_2t_2+t_3\Delta_3t_3.$$

Find

1. $\phi_1+2\phi_2$.
2. $t_1\phi_1-t_2\phi_2$.

We introduce the following rules.

1. $\Delta_i t_i \Delta_i t_i = 0$, $i \in \{1,2,3\}$.
2. $\Delta_i t_i \Delta_j t_j = -\Delta_j t_j \Delta_i t_i$, $i,j \in \{1,2,3\}$.

Example 7.10 Let

$$\begin{aligned}\phi_1 &= t_2 \Delta_1 t_1 - t_1 \Delta_2 t_2 + t_3 \Delta_3 t_3,\\ \phi_2 &= t_1^2 \Delta_1 t_1 + t_2^2 \Delta_2 t_2 + t_3 \Delta_3 t_3, \quad (t_1,t_2,t_3) \in \Lambda^3.\end{aligned}$$

Then

$$\begin{aligned}\phi_1\phi_2 &= (t_2 \Delta_1 t_1 - t_1 \Delta_2 t_2 + t_3 \Delta_3 t_3)\left(t_1^2 \Delta_1 t_1 + t_2^2 \Delta_2 t_2 + t_3 \Delta_3 t_3\right)\\ &= t_1^2 t_2 \Delta_1 t_1 \Delta_1 t_1 + t_2^3 \Delta_1 t_1 \Delta_2 t_2 + t_2 t_3 \Delta_1 t_1 \Delta_3 t_3\\ &\quad - t_1^3 \Delta_2 t_2 \Delta_1 t_1 - t_1 t_2^2 \Delta_2 t_2 \Delta_2 t_2 - t_1 t_3 \Delta_2 t_2 \Delta_3 t_3\\ &\quad + t_1^2 t_3 \Delta_3 t_3 \Delta_1 t_1 + t_2^2 t_3 \Delta_3 t_3 \Delta_2 t_2 + t_3^2 \Delta_3 t_3 \Delta_3 t_3\\ &= t_2^3 \Delta_1 t_1 \Delta_2 t_2 + t_2 t_3 \Delta_1 t_1 \Delta_3 t_3 - t_1^3 \Delta_2 t_2 \Delta_1 t_1 - t_1 t_3 \Delta_2 t_2 \Delta_3 t_3\\ &\quad + t_1^2 t_3 \Delta_3 t_3 \Delta_1 t_1 + t_2^2 t_3 \Delta_3 t_3 \Delta_2 t_2\\ &= (t_1^3 + t_2^3) \Delta_1 t_1 \Delta_2 t_2 + (t_2 - t_1^2) t_3 \Delta_1 t_1 \Delta_3 t_3 - (t_1 + t_2^2) t_3 \Delta_2 t_2 \Delta_3 t_3,\end{aligned}$$

$(t_1,t_2,t_3) \in \Lambda^3$.

Example 7.11 Let

$$\begin{aligned}\phi_1 &= t_1 \Delta_1 t_1 + t_2 \Delta_3 t_3,\\ \phi_2 &= (t_1 + t_2 + t_3) \Delta_1 t_1 \Delta_2 t_2 + \Delta_1 t_1 \Delta_3 t_3, \quad (t_1,t_2,t_3) \in \Lambda^3.\end{aligned}$$

Then

$$\begin{aligned}\phi_1\phi_2 &= (t_1 \Delta_1 t_1 + t_2 \Delta_3 t_3)((t_1 + t_2 + t_3) \Delta_1 t_1 \Delta_2 t_2 + \Delta_1 t_1 \Delta_3 t_3)\\ &= (t_1^2 + t_1 t_2 + t_1 t_3) \Delta_1 t_1 \Delta_1 t_1 \Delta_2 t_2 + t_1 \Delta_1 t_1 \Delta_1 t_1 \Delta_3 t_3\\ &\quad + (t_1 t_2 + t_2^2 + t_2 t_3) \Delta_3 t_3 \Delta_1 t_1 \Delta_2 t_2 + t_2 \Delta_3 t_3 \Delta_1 t_1 \Delta_3 t_3\\ &= (t_1 t_2 + t_2^2 + t_2 t_3) \Delta_1 t_1 \Delta_2 t_2 \Delta_3 t_3, \quad (t_1,t_2,t_3) \in \Lambda^3.\end{aligned}$$

Exercise 7.9 Let

$$\begin{aligned}\phi_1 &= \frac{1+t_1}{1+t_2^2} \Delta_1 t_1 + t_1 t_2 \Delta_2 t_2 + t_3^3 \Delta_3 t_3,\\ \phi_2 &= (t_1 - t_3) \Delta_1 t_1 + t_2^2 \Delta_2 t_2 + t_3 \Delta_3 t_3,\\ \phi_3 &= \Delta_1 t_1 \Delta_2 t_2 + 3 \Delta_1 t_1 \Delta_3 t_3 + t_1 \Delta_2 t_2 \Delta_3 t_3, \quad (t_1,t_2,t_3) \in \Lambda^3.\end{aligned}$$

Find

1. $\phi_1 - 3\phi_2$,
2. $t_1\phi_1 + t_2\phi_2$,

3. $\phi_1\phi_2 + \phi_3$,
4. $\phi_1\phi_3$.

Theorem 7.15 *Let ϕ be an 1-dynamic form or a 3-dynamic form. Then $\phi^2 = 0$.*

Proof. 1. Let ϕ be an 1-dynamic form and

$$\phi = f\Delta_1 t_1 + g\Delta_2 t_2 + h\Delta_3 t_3,$$

where $f, g, h : \Lambda^3 \to \mathbb{R}$. Then

$$\begin{aligned}\phi^2 &= (f\Delta_1 t_1 + g\Delta_2 t_2 + h\Delta_3 t_3)(f\Delta_1 t_1 + g\Delta_2 t_2 + h\Delta_3 t_3)\\ &= f^2\Delta_1 t_1\Delta_1 t_1 + (fg)\Delta_1 t_1\Delta_2 t_2 + (fh)\Delta_1 t_1\Delta_3 t_3\\ &\quad + (fg)\Delta_2 t_2\Delta_1 t_1 + g^2\Delta_2 t_2\Delta_2 t_2 + (gh)\Delta_2 t_2\Delta_3 t_3\\ &\quad + (hf)\Delta_3 t_3\Delta_1 t_1 + (hg)\Delta_3 t_3\Delta_2 t_2 + h^2\Delta_3 t_3\Delta_3 t_3\\ &= 0.\end{aligned}$$

2. Let ϕ be a 3-dynamic form and

$$\phi = f\Delta_1 t_1\Delta_2 t_2\Delta_3 t_3,$$

where $f : \Lambda^3 \to \mathbb{R}$ is a given function. Then

$$\begin{aligned}\phi^2 &= (f\Delta_1 t_1\Delta_2 t_2\Delta_3 t_3)(f\Delta_1 t_1\Delta_2 t_2\Delta_3 t_3)\\ &= 0.\end{aligned}$$

This completes the proof.

7.6 Exterior Differentiations

In this section we introduce the conception of exterior differentiation. For convenience, consider the case $n = 3$.

Definition 7.16 Let $f : \Lambda^3 \to \mathbb{R}$ be σ_1-completely delta differentiable. Then:

1. we define the $\sigma_2\sigma_3$-exterior differentiation $\Delta_{\sigma_2\sigma_3} f$ in the following manner

$$\Delta_{\sigma_2\sigma_3} f = f_{t_1}^{\Delta_1}\Delta_1 t_1 + f_{t_2}^{\Delta_2\sigma_1}\Delta_2 t_2 + f_{t_3}^{\Delta_3\sigma_1\sigma_2}\Delta_3 t_3.$$

2. we define the $\sigma_3\sigma_2$-differentiation $\Delta_{\sigma_3\sigma_2} f$ in the following manner

$$\Delta_{\sigma_3\sigma_2} f = f_{t_1}^{\Delta_1}\Delta_1 t_1 + f_{t_2}^{\Delta_2\sigma_1\sigma_3}\Delta_2 t_2 + f_{t_3}^{\Delta_3\sigma_1}\Delta_3 t_3.$$

Example 7.12 Let $\mathbb{T}_1 = \mathbb{Z}$, $\mathbb{T}_2 = 2^{\mathbb{N}_0}$, $\mathbb{T}_3 = 3^{\mathbb{N}_0}$ and

$$f(t_1,t_2,t_3) = t_1^2 + \frac{t_3}{1+t_2}, \quad (t_1,t_2,t_3) \in \Lambda^3.$$

Then

$$\sigma_1(t_1) = t_1 + 1, \quad t_1 \in \mathbb{T}_1,$$
$$\sigma_2(t_2) = 2t_2, \quad t_2 \in \mathbb{T}_2,$$
$$\sigma_3(t_3) = 3t_3, \quad t_3 \in \mathbb{T}_3,$$

and

$$\begin{aligned} f_{t_1}^{\Delta_1}(t_1,t_2,t_3) &= \sigma_1(t_1) + t_1 \\ &= t_1 + 1 + t_1 \\ &= 2t_1 + 1, \\ f_{t_2}^{\Delta_2}(t_1,t_2,t_3) &= -\frac{t_3}{(1+t_2)(1+\sigma_2(t_2))} \\ &= -\frac{t_3}{(1+t_2)(1+2t_2)}, \\ f_{t_2}^{\Delta_2\sigma_1}(t_1,t_2,t_3) &= -\frac{t_3}{(1+t_2)(1+2t_2)}, \\ f_{t_2}^{\Delta_2\sigma_1\sigma_3}(t_1,t_2,t_3) &= -\frac{\sigma_3(t_3)}{(1+t_2)(1+2t_2)} \\ &= -\frac{3t_3}{(1+t_2)(1+2t_2)}, \\ f_{t_3}^{\Delta_3}(t_1,t_2,t_3) &= \frac{1}{1+t_2}, \\ f_{t_3}^{\Delta_3\sigma_1}(t_1,t_2,t_3) &= \frac{1}{1+t_2}, \\ f_{t_3}^{\Delta_3\sigma_1\sigma_2}(t_1,t_2,t_3) &= \frac{1}{1+\sigma_2(t_2)} \\ &= \frac{1}{1+2t_2}, \quad (t_1,t_2,t_3) \in \Lambda^3. \end{aligned}$$

Therefore

$$\begin{aligned} \Delta_{\sigma_2\sigma_3} f &= f_{t_1}^{\Delta_1}(t_1,t_2,t_3)\Delta_1 t_1 + f_{t_2}^{\Delta_2\sigma_1}(t_1,t_2,t_3)\Delta_2 t_2 + f_{t_3}^{\Delta_3\sigma_1\sigma_2}(t_1,t_2,t_3)\Delta_3 t_3 \\ &= (2t_1+1)\Delta_1 t_1 - \frac{t_3}{(1+t_2)(1+2t_2)}\Delta_2 t_2 + \frac{1}{1+2t_2}\Delta_3 t_3, \quad (t_1,t_2,t_3) \in \Lambda^3, \end{aligned}$$

and

$$\Delta_{\sigma_3\sigma_2} f = f_{t_1}^{\Delta_1}(t_1,t_2,t_3)\Delta_1 t_1 + f_{t_2}^{\Delta_2\sigma_1\sigma_3}(t_1,t_2,t_3)\Delta_2 t_2 + f_{t_3}^{\Delta_3\sigma_1}(t_1,t_2,t_3)\Delta_3 t_3$$

$$= (2t_1+1)\Delta_1 t_1 - \frac{3t_3}{(1+t_2)(1+2t_2)}\Delta_2 t_2 + \frac{1}{1+t_2}\Delta_3 t_3, \quad (t_1,t_2,t_3) \in \Lambda^3.$$

Exercise 7.10 Let $\mathbb{T}_1 = 2^{\mathbb{N}_0}$, $\mathbb{T}_2 = 3\mathbb{Z}$, $\mathbb{T}_3 = 4^{\mathbb{N}_0}$ and

$$f(t_1,t_2,t_3) = e_1(t_1,1)t_2 + \frac{t_3+2t_2}{1+t_1^2} + \sin_4(t_3,1), \quad (t_1,t_2,t_3) \in \Lambda^3.$$

Find

$$\Delta_{\sigma_2\sigma_3} f \quad \text{and} \quad \Delta_{\sigma_3\sigma_2} f.$$

Definition 7.17 Let $f : \Lambda^3 \to \mathbb{R}$ be σ_2-completely delta differentiable. Then:

1. we define $\sigma_1\sigma_3$- exterior differentiation $\Delta_{\sigma_1\sigma_3} f$ in the following manner

$$\Delta_{\sigma_1\sigma_3} f = f_{t_1}^{\Delta_1\sigma_2}(t_1,t_2,t_3)\Delta_1 t_1 + f_{t_2}^{\Delta_2}(t_1,t_2,t_3)\Delta_2 t_2 + f_{t_3}^{\Delta_3\sigma_1\sigma_2}(t_1,t_2,t_3)\Delta_3 t_3,$$

$(t_1,t_2,t_3) \in \Lambda^3$.

2. we define $\sigma_3\sigma_1$-exterior differentiation $\Delta_{\sigma_3\sigma_1} f$ in the following manner

$$\Delta_{\sigma_3\sigma_1} f = f_{t_1}^{\Delta_1\sigma_2\sigma_3}(t_1,t_2,t_3)\Delta_1 t_1 + f_{t_2}^{\Delta_2}(t_1,t_2,t_3)\Delta_2 t_2 + f_{t_3}^{\Delta_3\sigma_2}(t_1,t_2,t_3)\Delta_3 t_3,$$

$(t_1,t_2,t_3) \in \Lambda^3$.

Example 7.13 Let $\mathbb{T}_1$, $\mathbb{T}_2$, $\mathbb{T}_3$ and f be as in Example 7.12. Then

$$\begin{aligned} f_{t_1}^{\Delta_1\sigma_2}(t_1,t_2,t_3) &= f_{t_1}^{\Delta_1\sigma_2\sigma_3}(t_1,t_2,t_3) \\ &= 2t_1+1, \\ f_{t_3}^{\Delta_3\sigma_2}(t_1,t_2,t_3) &= f_{t_3}^{\Delta_3\sigma_1\sigma_2}(t_1,t_2,t_3) \\ &= \frac{1}{1+2t_2}, \quad (t_1,t_2,t_3) \in \Lambda^3. \end{aligned}$$

Hence,

$$\begin{aligned} \Delta_{\sigma_1\sigma_3} f &= \Delta_{\sigma_3\sigma_1} f \\ &= (2t_1+1)\Delta_1 t_1 - \frac{t_3}{(1+t_2)(1+2t_2)}\Delta_2 t_2 + \frac{1}{1+2t_2}\Delta_3 t_3, \quad (t_1,t_2,t_3) \in \Lambda^3. \end{aligned}$$

Exercise 7.11 Let $\mathbb{T}_1 = 4\mathbb{Z}$, $\mathbb{T}_2 = 3^{\mathbb{N}_0}$, $\mathbb{T}_3 = 2^{\mathbb{N}_0}$ and

$$f(t_1,t_2,t_3) = \frac{1+t_1}{1+t_1+t_2+t_3}, \quad (t_1,t_2,t_3) \in \Lambda^3.$$

Find

$$\Delta_{\sigma_1\sigma_3} f \quad \text{and} \quad \Delta_{\sigma_3\sigma_1} f.$$

Definition 7.18 Let $f : \Lambda^3 \to \mathbb{R}$ be σ_3-completely delta differentiable. Then:

1. we define the $\sigma_1\sigma_2$-exterior differentiation $\Delta_{\sigma_1\sigma_2}f$ in the following manner

$$\Delta_{\sigma_1\sigma_2}f = f_{t_1}^{\Delta_1\sigma_3}(t_1,t_2,t_3)\Delta_1 t_1 + f_{t_2}^{\Delta_2\sigma_1\sigma_3}(t_1,t_2,t_3)\Delta_2 t_2 + f_{t_3}^{\Delta_3}(t_1,t_2,t_3)\Delta_3 t_3,$$

$(t_1,t_2,t_3) \in \Lambda^3$.

2. we define the $\sigma_2\sigma_1$-exterior differentiation $\Delta_{\sigma_2\sigma_1}f$ in the following manner

$$\Delta_{\sigma_2\sigma_1}f = f_{t_1}^{\Delta_1\sigma_2\sigma_3}(t_1,t_2,t_3)\Delta_1 t_1 + f_{t_2}^{\Delta_2\sigma_3}(t_1,t_2,t_3)\Delta_2 t_2 + f_{t_3}^{\Delta_3}(t_1,t_2,t_3)\Delta_3 t_3,$$

$(t_1,t_2,t_3) \in \Lambda^3$.

Example 7.14 Let $\mathbb{T}_1$, $\mathbb{T}_2$, $\mathbb{T}_3$ and f be as in Example 7.12. Then

$$\begin{aligned}
f_{t_1}^{\Delta_1\sigma_3}(t_1,t_2,t_3) &= f_{t_1}^{\Delta_1\sigma_2\sigma_3}(t_1,t_2,t_3)\\
&= 2t_1+1,\\
f_{t_2}^{\Delta_2\sigma_1\sigma_3}(t_1,t_2,t_3) &= f_{t_2}^{\Delta_2\sigma_3}(t_1,t_2,t_3)\\
&= -\frac{\sigma_3(t_3)}{(1+t_2)(1+2t_2)}\\
&= -\frac{3t_3}{(1+t_2)(1+2t_2)}, \quad (t_1,t_2,t_3) \in \Lambda^3.
\end{aligned}$$

Hence,

$$\begin{aligned}
\Delta_{\sigma_1\sigma_2}f &= \Delta_{\sigma_2\sigma_1}f\\
&= (2t_1+1)\Delta_1 t_1 - \frac{3t_3}{(1+t_2)(1+2t_2)}\Delta_2 t_2 + \frac{1}{1+t_2}\Delta_3 t_3,
\end{aligned}$$

$(t_1,t_2,t_3) \in \Lambda^3$.

Exercise 7.12 Let $\mathbb{T}_1 = 2\mathbb{Z}$, $\mathbb{T}_2 = 3^{\mathbb{N}_0}$, $\mathbb{T}_3 = 2^{\mathbb{N}_0}$ and

$$f(t_1,t_2,t_3) = \frac{1+t_1+t_3}{1+t_2^2+t_3^2}, \quad (t_1,t_2,t_3) \in \Lambda^3.$$

Find

$$\Delta_{\sigma_1\sigma_2}f \quad \text{and} \quad \Delta_{\sigma_2\sigma_1}f.$$

Definition 7.19 Let ϕ be a k-dynamic form. The $\Delta_{\sigma_i\sigma_j}$ exterior differentiation of ϕ is the $(k+1)$-dynamic form obtained from ϕ by applying $\Delta_{\sigma_i\sigma_j}$, $i,j \in \{1,2,3\}$, to each of the functions included in ϕ.

Remark 7.9 Let

$$\phi = f\Delta_1 t_1 \Delta_2 t_2 \Delta_3 t_3,$$

where $f : \Lambda^3 \to \mathbb{R}$ is delta differentiable. Then

$$\Delta_{\sigma_i\sigma_j}\phi = 0, \quad i,j \in \{1,2,3\}.$$

Example 7.15 Let $\mathbb{T}_1 = \mathbb{T}_2 = \mathbb{T}_3 = \mathbb{Z}$. Consider the 1-dynamic form

$$\phi = (t_1 + t_2 + t_3)\Delta_1 t_1 + t_1^2 \Delta_2 t_2 + t_1 t_2^2 \Delta_3 t_3, \quad (t_1, t_2, t_3) \in \Lambda^3.$$

Let

$$\begin{aligned} \phi_1(t_1, t_2, t_3) &= t_1 + t_2 + t_3, \\ \phi_2(t_1, t_2, t_3) &= t_1^2, \\ \phi_3(t_1, t_2, t_3) &= t_1 t_2^2, \quad (t_1, t_2, t_3) \in \Lambda^3. \end{aligned}$$

Then

$$\begin{aligned} \phi_{1t_1}^{\Delta_1}(t_1, t_2, t_3) &= \phi_{1t_2}^{\Delta_2}(t_1, t_2, t_3) \\ &= \phi_{1t_3}^{\Delta_3}(t_1, t_2, t_3) \\ &= 1, \quad (t_1, t_2, t_3) \in \Lambda^3, \end{aligned}$$

and

$$\Delta_{\sigma_2\sigma_3}\phi_1 = \Delta_1 t_1 + \Delta_2 t_2 + \Delta_3 t_3, \quad (t_1, t_2, t_3) \in \Lambda^3.$$

Next,

$$\begin{aligned} \phi_{2t_1}^{\Delta_1}(t_1, t_2, t_3) &= \sigma_1(t_1) + t_1 \\ &= t_1 + 1 + t_1 \\ &= 2t_1 + 1, \quad (t_1, t_2, t_3) \in \Lambda^3, \end{aligned}$$

and

$$\Delta_{\sigma_2\sigma_3}\phi_2 = (2t_1 + 1)\Delta_1 t_1, \quad (t_1, t_2, t_3) \in \Lambda^3.$$

Moreover,

$$\begin{aligned} \phi_{3t_1}^{\Delta_1}(t_1, t_2, t_3) &= t_2^2, \\ \phi_{3t_2}^{\Delta_2}(t_1, t_2, t_3) &= (\sigma_2(t_2) + t_2)t_1 \\ &= (t_2 + 1 + t_2)t_1 \\ &= (2t_2 + 1)t_1, \\ \phi_{3t_2}^{\Delta_2\sigma_1}(t_1, t_2, t_3) &= (2t_2 + 1)\sigma_1(t_1) \\ &= (t_1 + 1)(2t_2 + 1), \quad (t_1, t_2, t_3) \in \Lambda^3, \end{aligned}$$

and

$$\Delta_{\sigma_2\sigma_3}\phi_3 = t_2^2 \Delta_1 t_1 + (t_1 + 1)(2t_2 + 1)\Delta_2 t_2, \quad (t_1, t_2, t_3) \in \Lambda^3.$$

Consequently

$$\begin{aligned} \Delta_{\sigma_2\sigma_3}\phi &= (\Delta_{\sigma_2\sigma_3}\phi_1)\,\Delta_1 t_1 + (\Delta_{\sigma_2\sigma_3}\phi_2)\Delta_2 t_2 + (\Delta_{\sigma_2\sigma_3}\phi_3)\,\Delta_3 t_3 \\ &= (\Delta_1 t_1 + \Delta_2 t_2 + \Delta_3 t_3)\,\Delta_1 t_1 + ((2t_1 + 1)\Delta_1 t_1)\Delta_2 t_2 \\ &\quad + \left(t_2^2 \Delta_1 t_1 + (t_1 + 1)(2t_2 + 1)\Delta_2 t_2\right)\Delta_3 t_3 \end{aligned}$$

$$\begin{aligned}
&= \Delta_2 t_2 \Delta_1 t_1 + \Delta_3 t_3 \Delta_1 t_1 + (2t_1+1)\Delta_1 t_1 \Delta_2 t_2 \\
&\quad + t_2^2 \Delta_1 t_1 \Delta_3 t_3 + (t_1+1)(2t_2+1)\Delta_2 t_2 \Delta_3 t_3 \\
&= -\Delta_1 t_1 \Delta_2 t_2 - \Delta_1 t_1 \Delta_3 t_3 + (2t_1+1)\Delta_1 t_1 \Delta_2 t_2 \\
&\quad + t_2^2 \Delta_1 t_1 \Delta_3 t_3 + (t_1+1)(2t_2+1)\Delta_2 t_2 \Delta_3 t_3 \\
&= 2t_1 \Delta_1 t_1 \Delta_2 t_2 + (t_2^2-1)\Delta_1 t_1 \Delta_3 t_3 + (t_1+1)(2t_2+1)\Delta_2 t_2 \Delta_3 t_3,
\end{aligned}$$

$(t_1,t_2,t_3) \in \Lambda^3$.

Example 7.16 Let

$$\begin{aligned}
\phi_1 &= f \Delta_1 t_1 + g \Delta_2 t_2 + h \Delta_3 t_3, \\
\phi_2 &= f \Delta_1 t_1 \Delta_2 t_2 + g \Delta_1 t_1 \Delta_3 t_3 + h \Delta_2 t_2 \Delta_3 t_3,
\end{aligned}$$

where $f,g,h : \Lambda^3 \to \mathbb{R}$ are σ_1-completely delta differentiable. We will find representations for $\Delta_{\sigma_2\sigma_3}\phi_1$ and $\Delta_{\sigma_2\sigma_3}\phi_2$. We have

$$\begin{aligned}
\Delta_{\sigma_2\sigma_3}\phi_1 &= (\Delta_{\sigma_2\sigma_3} f)\,\Delta_1 t_1 + (\Delta_{\sigma_2\sigma_3} g)\,\Delta_2 t_2 + (\Delta_{\sigma_2\sigma_3} h)\,\Delta_3 t_3 \\
&= \left(f_{t_1}^{\Delta_1}\,\Delta_1 t_1 + f_{t_2}^{\Delta_2\sigma_1}\,\Delta_2 t_2 + f_{t_3}^{\Delta_3\sigma_1\sigma_2}\,\Delta_3 t_3\right)\Delta_1 t_1 \\
&\quad + \left(g_{t_1}^{\Delta_1}\,\Delta_1 t_1 + g_{t_2}^{\Delta_2\sigma_1}\,\Delta_2 t_2 + g_{t_3}^{\Delta_3\sigma_1\sigma_2}\,\Delta_3 t_3\right)\Delta_2 t_2 \\
&\quad + \left(h_{t_1}^{\Delta_1}\,\Delta_1 t_1 + h_{t_2}^{\Delta_2\sigma_1}\,\Delta_2 t_2 + h_{t_3}^{\Delta_3\sigma_1\sigma_2}\,\Delta_3 t_3\right)\Delta_3 t_3 \\
&= -f_{t_2}^{\Delta_2\sigma_1}\,\Delta_1 t_1 \Delta_2 t_2 - f_{t_3}^{\Delta_3\sigma_1\sigma_2}\,\Delta_1 t_1 \Delta_3 t_3 \\
&\quad + g_{t_1}^{\Delta_1}\,\Delta_1 t_1 \Delta_2 t_2 - g_{t_3}^{\Delta_3\sigma_1\sigma_2}\,\Delta_2 t_2 \Delta_3 t_3 \\
&\quad + h_{t_1}^{\Delta_1}\,\Delta_1 t_1 \Delta_3 t_3 + h_{t_2}^{\Delta_2\sigma_1}\,\Delta_2 t_2 \Delta_3 t_3 \\
&= \left(g_{t_1}^{\Delta_1} - f_{t_2}^{\Delta_2\sigma_1}\right)\Delta_1 t_1 \Delta_2 t_2 + \left(h_{t_1}^{\Delta_1} - f_{t_3}^{\Delta_3\sigma_1\sigma_2}\right)\Delta_1 t_1 \Delta_3 t_3 \\
&\quad + \left(h_{t_2}^{\Delta_2\sigma_1} - g_{t_3}^{\Delta_3\sigma_1\sigma_2}\right)\Delta_2 t_2 \Delta_3 t_3
\end{aligned}$$

and

$$\begin{aligned}
\Delta_{\sigma_2\sigma_3}\phi_2 &= (\Delta_{\sigma_2\sigma_3} f)\,\Delta_1 t_1 \Delta_2 t_2 + (\Delta_{\sigma_2\sigma_3} g)\,\Delta_1 t_1 \Delta_3 t_3 + (\Delta_{\sigma_2\sigma_3} h)\,\Delta_2 t_2 \Delta_3 t_3 \\
&= \left(f_{t_1}^{\Delta_1}\,\Delta_1 t_1 + f_{t_2}^{\Delta_2\sigma_1}\,\Delta_2 t_2 + f_{t_3}^{\Delta_3\sigma_1\sigma_2}\,\Delta_3 t_3\right)\Delta_1 t_1 \Delta_2 t_2 \\
&\quad + \left(g_{t_1}^{\Delta_1}\,\Delta_1 t_1 + g_{t_2}^{\Delta_2\sigma_1}\,\Delta_2 t_2 + g_{t_3}^{\Delta_3\sigma_1\sigma_2}\,\Delta_3 t_3\right)\Delta_1 t_1 \Delta_3 t_3 \\
&\quad + \left(h_{t_1}^{\Delta_1}\,\Delta_1 t_1 + h_{t_2}^{\Delta_2\sigma_1}\,\Delta_2 t_2 + h_{t_3}^{\Delta_3\sigma_1\sigma_2}\,\Delta_3 t_3\right)\Delta_2 t_2 \Delta_3 t_3 \\
&= f_{t_3}^{\Delta_3\sigma_1\sigma_2}\,\Delta_1 t_1 \Delta_2 t_2 \Delta_3 t_3 - g_{t_2}^{\Delta_2\sigma_1}\,\Delta_1 t_1 \Delta_2 t_2 \Delta_3 t_3 \\
&\quad + h_{t_1}^{\Delta_1}\,\Delta_1 t_1 \Delta_2 t_2 \Delta_3 t_3 \\
&= \left(f_{t_3}^{\Delta_3\sigma_1\sigma_2} - g_{t_2}^{\Delta_2\sigma_1} + h_{t_1}^{\Delta_1}\right)\Delta_1 t_1 \Delta_2 t_2 \Delta_3 t_3.
\end{aligned}$$

If in addition, $f, g : \Lambda^3 \to \mathbb{R}$ are twice σ_1-completely delta differentiable, σ_1 is Δ_1-differentiable and σ_2 is Δ_2-differentiable, then

$$\begin{aligned}\Delta_{\sigma_2\sigma_3}(\Delta_{\sigma_2\sigma_3}\phi_1) &= \Bigg(\left(g_{t_1}^{\Delta_1} - f_{t_2}^{\Delta_2\sigma_1}\right)_{t_3}^{\Delta_3\sigma_1\sigma_2} - \left(h_{t_1}^{\Delta_1} - f_{t_3}^{\Delta_3\sigma_1\sigma_2}\right)_{t_2}^{\Delta_2\sigma_1} \\ &\quad + \left(h_{t_2}^{\Delta_2\sigma_1} - g_{t_3}^{\Delta_3\sigma_1\sigma_2}\right)_{t_1}^{\Delta_1} \Bigg) \Delta_1 t_1 \Delta_2 t_2 \Delta_3 t_3 \\ &= \Bigg(g_{t_1t_3}^{\Delta_1\Delta_3\sigma_1\sigma_2} - f_{t_2t_3}^{\Delta_2\Delta_3\sigma_1^2\sigma_2} - h_{t_1t_2}^{\Delta_1\Delta_2\sigma_1} + \left(1+\mu_2^{\Delta_2}\right) f_{t_2t_3}^{\Delta_2\Delta_3\sigma_1^2\sigma_2} \\ &\quad + \left(1+\mu_1^{\Delta_1}\right) h_{t_1t_2}^{\Delta_1\Delta_2\sigma_1} - \left(1+\mu_1^{\Delta_1}\right) g_{t_1t_3}^{\Delta_1\Delta_3\sigma_1\sigma_2} \Bigg) \Delta_1 t_1 \Delta_2 t_2 \Delta_3 t_3 \\ &= \left(\mu_1^{\Delta_1} \left(h_{t_1t_2}^{\Delta_1\Delta_2\sigma_1} - g_{t_1t_3}^{\Delta_1\Delta_3\sigma_1\sigma_2}\right) + \mu_2^{\Delta_2} f_{t_2t_3}^{\Delta_2\Delta_3\sigma_1^2\sigma_2} \right) \Delta_1 t_1 \Delta_2 t_2 \Delta_3 t_3.\end{aligned}$$

Remark 7.10 By the last example, it follows that in the general case

$$\Delta_{\sigma_2\sigma_3}(\Delta_{\sigma_2\sigma_3}\phi_1) \neq 0.$$

Exercise 7.13 Let $\mathbb{T}_1 = \mathbb{Z}$, $\mathbb{T}_2 = 2^{\mathbb{N}_0}$, $\mathbb{T}_3 = 3^{\mathbb{N}_0}$ and

$$\phi = \left(t_2^2 + t_3^2\right) \Delta_1 t_1 \Delta_2 t_2 + (t_2 t_3) \Delta_1 t_1 \Delta_3 t_3 + \left(t_1^2 + t_2^2 + t_3^2\right) \Delta_2 t_2 \Delta_3 t_3,$$

$(t_1, t_2, t_3) \in \Lambda^3$. Find

1. $\Delta_{\sigma_1\sigma_3}\phi$,
2. $\Delta_{\sigma_2\sigma_3}\phi$,
3. $\Delta_{\sigma_1\sigma_2}\phi + \Delta_{\sigma_1\sigma_3}\phi$.

Definition 7.20 If ϕ is a dynamic form with the property

$$\Delta_{\sigma_i\sigma_j}\phi = 0, \quad i, j \in \{1, 2, 3\},$$

then ϕ is called the $\sigma_i\sigma_j$-closed dynamic form.

Example 7.17 Let

$$\phi = t_1 \Delta_1 t_1 \Delta_2 t_2 + t_3 \Delta_1 t_1 \Delta_3 t_3 + t_2 \Delta_2 t_2 \Delta_3 t_3, \quad (t_1, t_2, t_3) \in \Lambda^3.$$

Then

$$\begin{aligned}\Delta_{\sigma_1\sigma_j} &= \Delta_1 t_1 \Delta_1 t_1 \Delta_2 t_2 + \Delta_3 t_3 \Delta_1 t_1 \Delta_3 t_3 + \Delta_2 t_2 \Delta_2 t_2 \Delta_3 t_3 \\ &= 0, \quad (t_1, t_2, t_3) \in \Lambda^3.\end{aligned}$$

Thus, ϕ is a σ_{ij}-exact 2-dynamic form.

Definition 7.21 If ϕ is a dynamic form with the property

$$\phi = \Delta_{\sigma_i\sigma_j}\psi$$

for some dynamic form ψ, or $\phi = 0$, then ϕ is called a $\sigma_i\sigma_j$-exact dynamic form.

Example 7.18

$$\phi = 2t_1\Delta_1 t_1 \Delta_2 t_2 + (t_2^2 - 1)\Delta_1 t_1 \Delta_3 t_3 + (t_1+1)(2t_2+1)\Delta_2 t_2 \Delta_3 t_3,$$
$$\psi = (t_1+t_2+t_3)\Delta_1 t_1 + t_1^2 \Delta_2 t_2 + t_1 t_2^2 \Delta_3 t_3, \quad (t_1,t_2,t_3) \in \Lambda^3.$$

By Example 7.15, it follows that

$$\Delta_{\sigma_2\sigma_3}\psi = \phi.$$

Therefore ϕ is a $\sigma_2\sigma_3$-exact form.

Remark 7.11 By Example 7.16, it follows that not any $\sigma_i\sigma_j$-exact dynamic form is a σ_{ij}-closed form, $i,j \in \{1,2,3\}$.

Exercise 7.14 Let ϕ_1 and ϕ_2 be two k-dynamic forms. Prove that

$$\Delta_{\sigma_i\sigma_j}(a\phi_1 + b\phi_2) = a\Delta_{\sigma_i\sigma_j}\phi_1 + b\Delta_{\sigma_i\sigma_j}\phi_2.$$

7.7 Advanced Practical Problems

Problem 7.1 Let $\mathbb{T} = 4\mathbb{Z}$, $x^0 = 1$, $t^0(-2,1)$, $v = \left(\frac{1}{2}, \frac{\sqrt{3}}{2}\right) \in \mathbb{R}^2$.

1. Find $\mathbb{T}_1$ and $\mathbb{T}_2$.
2. Let $f(t_1,t_2) = \frac{t_1+t_2}{1+t_2^2}$, $(t_1,t_2) \in \mathbb{T}_1 \times \mathbb{T}_2$. Find

$$\frac{\partial f}{\Delta v}(t^0).$$

Problem 7.2 Let $n = 4$, $\mathbb{T}_1 = \mathbb{T}_2 = \mathbb{Z}$, $\mathbb{T}_3 = 4^{\mathbb{N}_0}$, $\mathbb{T}_4 = 2\mathbb{Z}$ and

$$V_1 = (t_1+t_2+t_3+t_4)\frac{\partial}{\Delta t_1} + \left(t_1^2 - t_2^3 + t_3 t_4\right)\frac{\partial}{\Delta t_2} + t_4\frac{\partial}{\Delta t_4}$$

$$V_2 = \left(t_1^2 - t_2 t_3^2 + t_4^3\right)\frac{\partial}{\Delta t_1} + t_1 t_2 \frac{\partial}{\Delta t_3},$$

$$V_3 = (1+t_1)\frac{\partial}{\Delta t_1} + t_1 t_2 \frac{\partial}{\Delta t_3}.$$

Find

1. $2V_1 - 3V_2 + 4V_3$,
2. $t_1 V_2 - t_2 V_3$,
3. $3t_1 V_1 + t_2 V_2 - 7V_3$.

Problem 7.3 Let $n = 4$ and

$$V_1 = (t_1 + t_2)\frac{\partial}{\Delta t_1} + t_1 t_2 t_4 \frac{\partial}{\Delta t_3} + (t_1 + t_4)^2 \frac{\partial}{\Delta t_4},$$

$$V_2 = t_1 \frac{\partial}{\Delta t_1} + t_2 \frac{\partial}{\Delta t_2} + t_3 \frac{\partial}{\Delta t_3} + t_4 \frac{\partial}{\Delta t_4},$$

$$V_3 = t_1^2 \frac{\partial}{\Delta t_1} + t_2 \frac{\partial}{\Delta t_3} + t_1^4 t_2^2 \frac{\partial}{\Delta t_4}.$$

Find

1. $V_1 - 3V_3$,
2. $V_1 + t_1 V_2$,
3. $t_1 V_1 - t_2 V_2 + t_3 V_3$.

Problem 7.4 Let $n = 2$, $\mathbb{T}_1 = 2\mathbb{Z}$, $\mathbb{T}_2 = 3\mathbb{Z}$ and

$$V_1 = t_1^2 t_2 \frac{\partial}{\Delta t_1} + t_1 t_2 \frac{\partial}{\Delta t_2},$$

$$V_2 = (t_1 + t_2)^2 \frac{\partial}{\Delta t_1} + t_1^2 t_2^2 \frac{\partial}{\Delta t_2},$$

$$V_3 = (t_1 + t_2 + t_1^2)\frac{\partial}{\Delta t_1} + (3t_1^2 - 2t_2^3)\frac{\partial}{\Delta t_2}.$$

Find

1. $V_1 [V_3]_\Delta$,
2. $[V_1, V_2]_\Delta$,
3. $2[V_1, V_2]_\Delta - 3[V_2, V_3]_\Delta$.

Problem 7.5 Let

$$\phi_1 = (1 + t_1 + t_2 + t_3^2)\Delta_1 t_1 + t_2^3 \Delta_2 t_2 + t_3 \Delta_3 t_3,$$

$$\phi_2 = \frac{1 + t_2}{1 + t_3 + t_3^2}\Delta_1 t_1 + t_3^4 \Delta_3 t_3,$$

$$\phi_3 = (t_1 - t_2)\Delta_1 t_1 \Delta_2 t_2 + t_2 \Delta_1 t_1 \Delta_3 t_3 + t_1^2 \Delta_2 t_2 \Delta_3 t_3, \quad (t_1, t_2, t_3) \in \Lambda^3.$$

Find

1. $2\phi_1 + 4\phi_2$,
2. $t_2\phi_1 - t_3\phi_2$,
3. $2\phi_1\phi_2 + 4\phi_3$,
4. $\phi_1\phi_3$.

Problem 7.6 Let $\mathbb{T}_1 = \mathbb{Z}$, $\mathbb{T}_2 = 2^{\mathbb{N}_0}$, $\mathbb{T}_3 = 4^{\mathbb{N}_0}$ and

$$f(t_1,t_2,t_3) = e_1(t_2,1) + \frac{t_1+t_2}{1+t_1^2+t_3^2}\sin_1(t_3,1),$$

$(t_1,t_2,t_3) \in \Lambda^3$. Find

1. $\Delta_{\sigma_1\sigma_2}f$, $\Delta_{\sigma_1\sigma_3}f$ and $\Delta_{\sigma_2\sigma_3}f$,
2. $2\Delta_{\sigma_1\sigma_2}f - \Delta_{\sigma_1\sigma_3}f$,
3. $(t_1+t_2)\Delta_{\sigma_2\sigma_3}f + 3\Delta_{\sigma_1\sigma_3}f$.

Problem 7.7 Let $\mathbb{T}_1 = 2\mathbb{Z}$, $\mathbb{T}_2 = 3^{\mathbb{N}_0}$, $\mathbb{T}_3 = 2^{\mathbb{N}_0}$ and

$$\phi = \frac{t_1+1}{1+3t_2}\Delta_1 t_1 + \frac{1+t_2}{1+t_2^2+t_3^2}\Delta_2 t_2 + \frac{2+t_1}{1+t_2+t_3}\Delta_3 t_3,$$

$(t_1,t_2,t_3) \in \Lambda^3$. Find

1. $\Delta_{\sigma_2\sigma_3}\phi$,
2. $\Delta_{\sigma_1\sigma_2}\phi$,
3. $3t_1\Delta_{\sigma_1\sigma_2}\phi + t_2\Delta_{\sigma_2\sigma_3}\phi$.

7.8 Notes and References

In this chapter we define directional derivative and deduce some of its properties. Using the directional derivative we introduce the tangent space on time scales and covariant differentiation on time scales and the delta nature connection on time scales is defined. The Lie bracket is defined, and some of its properties are given. We give a definition for k-dynamic forms on time scales and the $\Delta_{\sigma_{ij}}$-exterior differentiation of k-dynamic forms. Some of the results in this chapter can be found in [1], [6], [7] and [8].

Appendix A

Implicit Function Theorem

Let $\mathbb{T}$, $\mathbb{T}_1$ be time scales with forward jump operators and delta differentiation operators σ, σ_1 and Δ, Δ_1, respectively. Also, let $a, b \in \mathbb{T}$, $a < b$, and $(x_0, y_0) \in [a,b] \times \mathbb{R}$. Take $c \in \mathbb{R}$, $c > 0$, and

$$D = \{(x,y) \in \mathbb{T} \times \mathbb{R} : a \le x \le b, \quad y_0 - c \le y \le y_0 + c\}.$$

We have that $(x_0, y_0) \in D$.

Theorem A.1 *Let $f : D \to \mathbb{R}$ be a continuous function that satisfies the Lipschitz condition with a constant $L \in (0,1)$ with respect to its second argument. Also, let*

$$f(x_0, y_0) = 0. \tag{A.1}$$

Then there exists a $\delta > 0$ for which there exists a unique function $\phi \in \mathscr{C}([x_0 - \delta, x_0 + \delta] \cap [a,b])$

such that

$$\phi(x) = y_0 + f(x, \phi(x)), \quad x \in [x_0 - \delta, x_0 + \delta] \cap [a,b], \quad \textit{and} \quad y_0 = \phi(x_0).$$

Proof. Let $q > 0$ be chosen so that

$$q < (1 - L)c.$$

Since $f(\cdot, y_0)$ is a continuous function, by (A.1), it follows that there is a $\delta > 0$ such that

$$|f(x, y_0) - f(x_0, y_0)| < q, \quad x \in [x_0 - \delta, x_0 + \delta] \cap [a,b].$$

Define

$$X = \{g \in \mathscr{C}([x_0 - \delta, x_0 + \delta] \cap [a,b]) : |g(x) - y_0| \le c, \quad x \in [x_0 - \delta, x_0 + \delta] \cap [a,b]\}$$

and in X define the metric

$$d(y,z) = \max_{[x_0 - \delta, x_0 + \delta] \cap [a,b]} |y(x) - z(x)|, \quad y, z \in X.$$

Note that X is a complete metric space with respect to the metric d. For $y \in X$, define the operator

$$Fy(x) = y_0 + f(x, y(x)), \quad x \in [x_0 - \delta, x_0 + \delta] \cap [a,b]. \tag{A.2}$$

We have that $Fy \in \mathscr{C}([x_0 - \delta, x_0 + \delta] \cap [a,b])$ for any $y \in X$ and

$$|Fy(x) - y_0| = |f(x, y(x))|$$

$$\begin{aligned}
&\leq |f(x,y(x)) - f(x,y_0)| + |f(x,y_0)| \\
&\leq L|y(x) - y_0| + (1-L)c \\
&\leq Lc + (1-L)c \\
&= c, \quad x \in [x_0 - \delta, x_0 + \delta] \cap [a,b],
\end{aligned}$$

for any $y \in X$.

Therefore $F : X \to X$.

Next, take $y_1, y_2 \in X$.

Then

$$\begin{aligned}
Fy_1(x) &= y_0 + f(x, y_1(x)), \\
Fy_2(x) &= y_0 + f(x, y_2(x)), \quad x \in [x_0 - \delta, x_0 + \delta] \cap [a,b],
\end{aligned}$$

and

$$\begin{aligned}
|Fy_1(x) - Fy_2(x)| &= |y_0 + f(x, y_1(x)) - y_0 - f(x, y_2(x))| \\
&= |f(x, y_1(x)) - f(x, y_2(x))| \\
&\leq L|y_1(x) - y_2(x)| \\
&\leq Ld(y_1, y_2), \quad x \in [x_0 - \delta, x_0 + \delta] \cap [a,b].
\end{aligned}$$

Therefore

$$d(Fy_1, Fy_2) \leq Ld(y_1, y_2)$$

and $F : X \to X$ is a contraction. Therefore there exists a unique $\phi \in X$ so that

$$\phi(x) = y_0 + f(x, \phi(x)), \quad x \in [x_0 - \delta, x_0 + \delta] \cap [a,b].$$

Now, construct the sequence

$$\begin{aligned}
&y_0(x) = y_0, \\
&y_n(x) = y_0 + f(x, y_{n-1}(x)), \quad x \in [x_0 - \delta, x_0 + \delta] \cap [a,b], \quad n \in \mathbb{N}.
\end{aligned}$$

We have

$$\begin{aligned}
|y_1(x) - y_0| &= |f(x, y_0)| \\
&\leq |f(x, y_0) - f(x_0, y_0)| + |f(x_0, y_0)| \\
&\leq (1-L)c \\
&< c, \\
|y_2(x) - y_1(x)| &= |f(x, y_1(x)) - f(x, y_0)| \\
&\leq L|y_1(x) - y_0| \\
&\leq Lc,
\end{aligned}$$

$$\begin{aligned}|y_3(x) - y_2(x)| &= |f(x, y_2(x)) - f(x, y_1(x))| \\ &\le L|y_2(x) - y_1(x)| \\ &\le L^2 c, \quad x \in [x_0 - \delta, x_0 + \delta] \cap [a, b].\end{aligned}$$

Assume that

$$|y_n(x) - y_{n-1}(x)| \le L^{n-1} c, \quad x \in [x_0 - \delta, x_0 + \delta] \cap [a, b],$$

for some $n \in \mathbb{N}$. Then

$$\begin{aligned}|y_{n+1}(x) - y_n(x)| &= |f(x, y_n(x)) - f(x, y_{n-1}(x))| \\ &\le L|y_n(x) - y_{n-1}(x)| \\ &\le L^n c, \quad x \in [x_0 - \delta, x_0 + \delta] \cap [a, b].\end{aligned}$$

Therefore

$$\begin{aligned}\sum_{n=0}^{\infty} |y_{n+1}(x) - y_n(x)| &\le c \sum_{n=0}^{\infty} L^n \\ &= \frac{c}{1-L}, \quad x \in [x_0 - \delta, x_0 + \delta] \cap [a, b].\end{aligned}$$

Consequently

$$y_n \to \phi, \quad \text{as} \quad n \to \infty,$$

uniformly on $[x_0 - \delta, x_0 + \delta] \cap [a, b]$. Note that

$$\begin{aligned}y_1(x_0) &= y_0 + f(x_0, y_0) \\ &= y_0, \\ y_2(x_0) &= y_0 + f(x_0, y_1(x_0)) \\ &= y_0 + f(x_0, y_0) \\ &= y_0.\end{aligned}$$

Assume that

$$y_n(x_0) = y_0$$

for some $n \in \mathbb{N}$. Then

$$\begin{aligned}y_{n+1}(x_0) &= y_0 + f(x_0, y_n(x_0)) \\ &= y_0 + f(x_0, y_0) \\ &= y_0.\end{aligned}$$

Therefore

$$\begin{aligned}y_0 &= \lim_{n \to \infty} y_n(x_0) \\ &= \phi(x_0).\end{aligned}$$

This completes the proof.

Theorem A.2 *Let $f : D \to \mathbb{R}$ be a continuous function and the classical derivative f'_y exists and it is continuous on D. Let also,*

$$f(x_0, y_0) = 0,$$
$$f'_y(x_0, y_0) \neq 0.$$

Then there exists a $\delta > 0$ for which there exists a unique function $\phi \in \mathscr{C}([x_0 - \delta, x_0 + \delta] \cap [a, b])$ such that

$$f(x, \phi(x)) = 0, \quad x \in [x_0 - \delta, x_0 + \delta] \cap [a, b], \quad \textit{and} \quad y_0 = \phi(x_0).$$

Proof. Let

$$\lambda = -\frac{1}{f'_y(x_0, y_0)}$$

and

$$g(x, y) = y - y_0 + \lambda f(x, y), \quad (x, y) \in D.$$

Then

$$g'_y(x, y) = 1 + \lambda f'_y(x, y), \quad (x, y) \in D,$$

and

$$\begin{aligned} g'_y(x_0, y_0) &= 1 + \lambda f'_y(x_0, y_0) \\ &= 1 - \frac{f'_y(x_0, y_0)}{f'_y(x_0, y_0)} \\ &= 0. \end{aligned}$$

Take $k \in (0, 1)$. Then, using that f'_y is continuous on D, there exists a $\delta_1 > 0$ such that

$$D_1 = \{(x, y) \in [a, b] \times \mathbb{R} : x_0 - \delta_1 \leq x \leq x_0 + \delta_1, \quad y_0 - \delta_1 \leq y \leq y_0 + \delta_1\} \subset D$$

and

$$|g'_y(x, y)| \leq k, \quad (x, y) \in D_1.$$

Let

$$D_2 = \left\{(x, y) \in D_1 : x_0 - \frac{\delta_1}{2} \leq x \leq x_0 + \frac{\delta_1}{2}, \quad y_0 - \frac{\delta_1}{2} \leq y \leq y_0 + \frac{\delta_1}{2}\right\}.$$

Applying the classical mean value theorem, we get

$$\begin{aligned} |g(x, y_1) - g(x, y_2)| &= |g'_y(x, \xi)||y_1 - y_2| \\ &\leq k|y_1 - y_2|, \quad (x, y_1), (x, y_2) \in D_2, \end{aligned}$$

where ξ is between y_1 and y_2. Note that

$$\begin{aligned} g(x_0, y_0) &= y_0 - y_0 + \lambda f(x_0, y_0) \\ &= 0. \end{aligned}$$

Therefore g satisfies all conditions of Theorem A.1. Then there exists a $\delta > 0$ for which there exists a unique continuous function

$$\phi \in \mathscr{C}([x_0 - \delta, x_0 + \delta] \cap [a,b])$$

and

$$\phi(x) = y_0 + g(x, \phi(x)), \quad x \in [x_0 - \delta, x_0 + \delta] \cap [a,b], \tag{A.3}$$

and

$$y_0 = \phi(x_0).$$

By (A.3), we get

$$\begin{aligned}\phi(x) &= y_0 + \phi(x) - y_0 + \lambda f(x, \phi(x)) \\ &= \phi(x) + \lambda f(x, \phi(x)), \quad x \in [x_0 - \delta, x_0 + \delta] \cap [a,b],\end{aligned}$$

whereupon

$$f(x, \phi(x)) = 0, \quad x \in [x_0 - \delta, x_0 + \delta] \cap [a,b].$$

This completes the proof.

Remark A.1 Suppose that all conditions of Theorem A.2 hold. If ϕ is delta differentiable on $[x_0 - \delta, x_0 + \delta] \cap [a,b]$, then by the equation

$$f(x, \phi(x)) = 0, \quad x \in [x_0 - \delta, x_0 + \delta] \cap [a,b],$$

we get

$$\begin{aligned} 0 &= f_x^{\Delta}(x, \phi(x)) \\ &\quad + \left(\int_0^1 f_y'(\sigma(x), \phi(x) + h\mu(x)\phi^{\Delta}(x)) dh \right) \phi^{\Delta}(x), \end{aligned} \tag{A.4}$$

$x \in [x_0 - \delta, x_0 + \delta] \cap [a,b]$.

Let $y_0, c_1, c_2 \in \mathbb{T}_1$, $c_1 \leq y_0 \leq c_2$,

$$\tilde{D} = \{(x,y) \in \mathbb{T} \times \mathbb{T}_1 : a \leq x \leq b, \quad c_1 \leq y \leq c_2\}.$$

Then $(x_0, y_0) \in \tilde{D}$. As we have proved Theorem A.1, one can prove the following theorem.

Theorem A.3 *Let $f : \tilde{D} \to \mathbb{R}$ be a continuous function that satisfies the Lipschitz condition with a constant $L \in (0,1)$ with respect to its second argument. Also, let*

$$f(x_0, y_0) = 0.$$

Then there exists a $\delta > 0$ for which there exists a unique function $\phi \in \mathscr{C}([x_0 - \delta, x_0 + \delta] \cap [a,b])$ such that

$$\phi(x) = y_0 + f(x, \phi(x)), \quad x \in [x_0 - \delta, x_0 + \delta] \cap [a,b], \quad \textit{and} \quad y_0 = \phi(x_0).$$

Theorem A.4 *Let $f : \widetilde{D} \to \mathbb{R}$ be a continuous function and the Δ_1 partial derivative $f_y^{\Delta_1}$ exists and it is continuous on $\widetilde{D}$. Let also,*

$$f(x_0, y_0) = 0,$$
$$f_y^{\Delta_1}(x_0, y_0) \neq 0.$$

Then there exists a $\delta > 0$ for which there exists a unique function $\phi \in \mathscr{C}([x_0 - \delta, x_0 + \delta] \cap [a,b])$ such that

$$f(x, \phi(x)) = 0, \quad x \in [x_0 - \delta, x_0 + \delta] \cap [a,b], \quad \text{and} \quad y_0 = \phi(x_0).$$

Proof. The proof repeats the main steps of the proof of Theorem A.2. Let

$$\lambda = -\frac{1}{f_y^{\Delta_1}(x_0, y_0)}$$

and

$$g(x,y) = y - y_0 + \lambda f(x,y), \quad (x,y) \in \widetilde{D}.$$

Then

$$g_y^{\Delta_1}(x,y) = 1 + \lambda f_y^{\Delta_1}(x,y), \quad (x,y) \in \widetilde{D},$$

and

$$g_y^{\Delta_1}(x_0, y_0) = 0.$$

Take $k \in (0,1)$. Then, using that $f_y^{\Delta_1}$ is continuous on D, there exists a $\delta_1 > 0$ such that

$$\widetilde{D}_1 = \{(x,y) \in [a,b] \times [c_1, c_2] : x_0 - \delta_1 \le x \le x_0 + \delta_1, \quad y_0 - \delta_1 \le y \le y_0 + \delta_1\}$$

and

$$|g_y^{\Delta_1}(x,y)| \le k, \quad (x,y) \in \widetilde{D}_1.$$

Let

$$\widetilde{D}_2 = \left\{(x,y) \in \widetilde{D}_1 : x_0 - \frac{\delta_1}{2} \le x \le x_0 + \frac{\delta_1}{2}, \quad y_0 - \frac{\delta_1}{2} \le y \le y_0 + \frac{\delta_1}{2}\right\}.$$

Applying the time scales mean value theorem, we get

$$\begin{aligned}|g(x,y_1) - g(x,y_2)| &\le \sup_{\xi \in [y_1, y_2]} |g_y^{\Delta_1}(x,\xi)||y_1 - y_2| \\ &\le k|y_1 - y_2|, \quad (x,y_1),(x,y_2) \in \widetilde{D}_2.\end{aligned}$$

Also,

$$g(x_0, y_0) = 0.$$

Thus, g satisfies all conditions of Theorem A.3. Then there exists a $\delta > 0$ for which there exists a unique continuous function $\phi \in \mathscr{C}([x_0 - \delta, x_0 + \delta] \cap [a,b])$ and

$$\phi(x) = y_0 + g(x, \phi(x)), \quad x \in [x_0 - \delta, x_0 + \delta] \cap [a,b], \tag{A.5}$$

and

$$y_0 = \phi(x_0).$$

By (A.5), we get

$$f(x,\phi(x)) = 0, \quad x \in [x_0 - \delta, x_0 + \delta] \cap [a,b].$$

This completes the proof.

Remark A.2 Suppose that all conditions of Theorem A.4 hold and ϕ is Δ-differentiable.

1. If the classical partial derivative f_y' exists, then (A.4) holds.
2. Let

$$\phi(\mathbb{T}) = \mathbb{T}_1 \quad \text{and} \quad \phi(\sigma(x)) = \sigma_1(\phi(x)), \quad x \in \mathbb{T}.$$

   1. If f is σ-completely delta differentiable, then

$$0 = f_x^{\Delta}(x,\phi(x)) + f_y^{\Delta_1}(\sigma(x),\phi(x))\phi^{\Delta}(x).$$

   2. If f is σ_1-completely delta differentiable, then

$$\begin{aligned} 0 &= f_x^{\Delta}(x,\sigma_1(\phi(x))) + f_y^{\Delta_1}(x,\phi(x))\phi^{\Delta}(x) \\ &= f_x^{\Delta}(x,\phi(\sigma(x))) + f_y^{\Delta_1}(x,\phi(x))\phi^{\Delta}(x). \end{aligned}$$

References

1. N. Aktan, M. Sariskaya, K. İlarslan and İ. Günaltili. Nabla 1-forms on n-dimensional time scales, *Int. J. Open Problems Comput. Math.*, Vol. 5, 3, 2012, pp. 96–110.
2. S. Atmaca and Ö. Akgüller. Surfaces on time scales and their metric properties, *Advances in Difference Equations*, 2013.
3. S. Atmaca. Normal and osculating planes for Δ-regular curves, *Abstract and Applied Analysis*, Vol. 2010, Article ID 923916, 8 pages.
4. G. Guseinov and E. Özylmaz. Tangent lines of generalized regular curves parameterized by time scales, *Turk. J. Math.*, Vol. 25, 2001, pp. 553–562.
5. S. Hilger. *Ein Maßkettenkalkül mit Anwendung auf Zentrumsmannigfaltigkeiten*. PhD thesis, Universität Würzburg, 1988.
6. H. Kusak and A. Galiskan. The delta nature connection on time scales, *J. Math. Anal. Appl.*, Vol. 375, 2011, pp. 323–330.
7. H. Kusak and A. Galiskan. The Lie brackets on time scales, *Abstract and Applied Analysis*, Vol. 2012, Article ID 303706, 12 pages.
8. E. Özylmaz. Tangent space and derivative mapping, *International J. Math. Combin.*, Vol. 2, 2009, pp. 01–10.
9. H. Samanci. The matrix representation of the delta shape operator on time scales, *Advances in Difference Equations*, 2016, pp. 1–14.
10. E. Yilmaz, M. Aydin and T. Gulsen. A certain class of surfaces on product time scales with interpretations from economics, Filomat, 2018, pp. 5297–5306.

Index

0-Level Set, 74
$A_{a\sigma_2}$, 116
$P_{\sigma_1\sigma_2^2}$, 303
$P_{\sigma_1^2\sigma_2}$, 303
P_{σ_1}, 301
P_{σ_2}, 302
$T_{a\sigma_1\sigma_2^2}S$, 79
$T_{a\sigma_1^2\sigma_2}S$, 78
$T_{a\sigma_1}S$, 78
$T_{a\sigma_2}S$, 78
$\mathscr{A}_{a\sigma_1}$, 106
$\phi^1_{\sigma_1\sigma_1\sigma_1}$, 158, 161
$\phi^1_{\sigma_1\sigma_2\sigma_1}$, 163
$\phi^1_{\sigma_1\sigma_2\sigma_2}$, 165
$\phi^1_{\sigma_1\sigma_2^2\sigma_1\sigma_1}$, 181
$\phi^1_{\sigma_1\sigma_2^2\sigma_1\sigma_2}$, 181
$\phi^1_{\sigma_1\sigma_2^2\sigma_2\sigma_1}$, 182
$\phi^1_{\sigma_1\sigma_2^2\sigma_2\sigma_2}$, 182
$\phi^1_{\sigma_1^2\sigma_2\sigma_1\sigma_1}$, 177, 178
$\phi^1_{\sigma_1^2\sigma_2\sigma_2\sigma_1}$, 179
$\phi^1_{\sigma_1^2\sigma_2\sigma_2\sigma_2}$, 179
$\phi^1_{\sigma_2\sigma_1\sigma_1}$, 167
$\phi^1_{\sigma_2\sigma_1\sigma_2}$, 169
$\phi^1_{\sigma_2\sigma_2\sigma_1}$, 171
$\phi^1_{\sigma_2\sigma_2\sigma_2}$, 174
σ_1-, σ_2-, $\sigma_1^2\sigma_2$-, $\sigma_1\sigma_2^2$-Elliptic Points, 187
σ_1-, σ_2-, $\sigma_1^2\sigma_2$-, $\sigma_1\sigma_2^2$-Flat, 187
σ_1-, σ_2-, $\sigma_1^2\sigma_2$-, $\sigma_1\sigma_2^2$-Hyperbolic Points, 187
σ_1-, σ_2-, $\sigma_1^2\sigma_2$-, $\sigma_1\sigma_2^2$-Parabolic Points, 187
σ_1-Angle, 139
σ_1-Area of a Surface, 142
σ_1-Asymptotic Direction, 183
σ_1-Asymptotic Line, 183
σ_1-Christoffel Coefficients, 257
σ_1-Compatible Parameterization, 94
σ_1-Curve, 183
σ_1-Darboux Formulae, 302
σ_1-Differential of j-th kind, 102
σ_1-Gauss Equations, 290
σ_1-Godazzi-Mainardi Equations, 290
σ_1-Length of a Curve on a Surface, 133
σ_1-Normal Curvature, 158
σ_1-Normal to a Surface, 84
σ_1-Orientable Surface, 94
σ_1-Oriented Surface, 94
σ_1-Regular Parameterized Surface, 69
σ_1-Spherical Map, 105
σ_1-Tangent Plane, 79
σ_1-Tangent Vector Space, 78
σ_1-Tangential Normal Vector, 301
σ_1-Weingarten Coefficients, 265
$\sigma_1\mu_1\mu_1$-Gauss Equations, 295
$\sigma_1\mu_1\mu_1$-Godazzi-Mainardi Equations, 295
$\sigma_1\sigma_1\sigma_1$-Curvature Line, 187
$\sigma_1\sigma_1\sigma_1$-Gaussian Curvature, 189
$\sigma_1\sigma_1\sigma_1$-Mean Curvature, 189
$\sigma_1\sigma_1\sigma_1$-Minimal Surface, 321
$\sigma_1\sigma_1\sigma_1$-Principal Curvature, 187
$\sigma_1\sigma_1\sigma_1$-Principal Direction, 187
$\sigma_1\sigma_1\sigma_1$-Principal Line, 187
$\sigma_1\sigma_1\sigma_1$-Total Curvature, 189
$\sigma_1\sigma_1\sigma_2$-Curvature Line, 189
$\sigma_1\sigma_1\sigma_2$-Gaussian Curvature, 191
$\sigma_1\sigma_1\sigma_2$-Mean Curvature, 191
$\sigma_1\sigma_1\sigma_2$-Minimal Surface, 321
$\sigma_1\sigma_1\sigma_2$-Principal Curvature, 189
$\sigma_1\sigma_1\sigma_2$-Principal Direction, 189
$\sigma_1\sigma_1\sigma_2$-Principal Line, 189
$\sigma_1\sigma_1\sigma_2$-Total Curvature, 191
$\sigma_1\sigma_2$-Exterior Differentiation, 362
$\sigma_1\sigma_2\sigma_1$-Curvature Line, 191
$\sigma_1\sigma_2\sigma_1$-Gaussian Curvature, 192
$\sigma_1\sigma_2\sigma_1$-Mean Curvature, 192
$\sigma_1\sigma_2\sigma_1$-Principal Curvature, 191
$\sigma_1\sigma_2\sigma_1$-Principal Direction, 191
$\sigma_1\sigma_2\sigma_1$-Principal Line, 191
$\sigma_1\sigma_2\sigma_1$-Total Curvature, 192
$\sigma_1\sigma_2\sigma_2$-Curvature Line, 192
$\sigma_1\sigma_2\sigma_2$-Gaussian Curvature, 193
$\sigma_1\sigma_2\sigma_2$-Mean Curvature, 193

$\sigma_1\sigma_2\sigma_2$-Minimal Surface, 321
$\sigma_1\sigma_2\sigma_2$-Principal Curvature, 192
$\sigma_1\sigma_2\sigma_2$-Principal Direction, 192
$\sigma_1\sigma_2\sigma_2$-Principal Line, 192
$\sigma_1\sigma_2\sigma_2$-Total Curvature, 193
$\sigma_1\sigma_2\sigma_{(1)}\sigma_{(2)}\sigma_{(3)}$-Map, 66
$\sigma_1\sigma_2^2$-Angle, 142
$\sigma_1\sigma_2^2$-Area of a Surface, 147
$\sigma_1\sigma_2^2$-Asymptotic Direction, 186
$\sigma_1\sigma_2^2$-Asymptotic Line, 186
$\sigma_1\sigma_2^2$-Christoffel Coefficients, 264
$\sigma_1\sigma_2^2$-Curve, 186
$\sigma_1\sigma_2^2$-Darboux Formulae, 304
$\sigma_1\sigma_2^2$-Length of a Curve on a Surface, 138
$\sigma_1\sigma_2^2$-Normal Curvature, 180
$\sigma_1\sigma_2^2$-Normal to a Surface, 84
$\sigma_1\sigma_2^2$-Orientable Surface, 94
$\sigma_1\sigma_2^2$-Oriented Surface, 94
$\sigma_1\sigma_2^2$-Regular Parameterized Surface, 72
$\sigma_1\sigma_2^2$-Shape Operator, 121
$\sigma_1\sigma_2^2$-Spherical Map, 106
$\sigma_1\sigma_2^2$-Tangent Plane, 79
$\sigma_1\sigma_2^2$-Tangent Vector Space, 79
$\sigma_1\sigma_2^2$-Tangential Normal Vector, 303
$\sigma_1\sigma_2^2$-Weingarten Coefficients, 288
$\sigma_1\sigma_2^2\sigma_1\sigma_1$-Curvature Line, 204
$\sigma_1\sigma_2^2\sigma_1\sigma_1$-Gaussian Curvature, 205
$\sigma_1\sigma_2^2\sigma_1\sigma_1$-Mean Curvature, 205
$\sigma_1\sigma_2^2\sigma_1\sigma_1$-Minimal Surface, 323
$\sigma_1\sigma_2^2\sigma_1\sigma_1$-Principal Curvature, 204
$\sigma_1\sigma_2^2\sigma_1\sigma_1$-Principal Direction, 204
$\sigma_1\sigma_2^2\sigma_1\sigma_1$-Principal Line, 204
$\sigma_1\sigma_2^2\sigma_1\sigma_1$-Total Curvature, 205
$\sigma_1\sigma_2^2\sigma_1\sigma_2$-Curvature Line, 205
$\sigma_1\sigma_2^2\sigma_1\sigma_2$-Gaussian Curvature, 206
$\sigma_1\sigma_2^2\sigma_1\sigma_2$-Mean Curvature, 206
$\sigma_1\sigma_2^2\sigma_1\sigma_2$-Minimal Surface, 323
$\sigma_1\sigma_2^2\sigma_1\sigma_2$-Principal Curvature, 205
$\sigma_1\sigma_2^2\sigma_1\sigma_2$-Principal Direction, 205
$\sigma_1\sigma_2^2\sigma_1\sigma_2$-Principal Line, 205
$\sigma_1\sigma_2^2\sigma_1\sigma_2$-Total Curvature, 206
$\sigma_1\sigma_2^2\sigma_2\sigma_1$-Curvature Line, 207
$\sigma_1\sigma_2^2\sigma_2\sigma_1$-Gaussian Curvature, 208
$\sigma_1\sigma_2^2\sigma_2\sigma_1$-Mean Curvature, 208
$\sigma_1\sigma_2^2\sigma_2\sigma_1$-Minimal Surface, 324
$\sigma_1\sigma_2^2\sigma_2\sigma_1$-Principal Curvature, 207
$\sigma_1\sigma_2^2\sigma_2\sigma_1$-Principal Direction, 207
$\sigma_1\sigma_2^2\sigma_2\sigma_1$-Principal Line, 207
$\sigma_1\sigma_2^2\sigma_2\sigma_1$-Total Curvature, 208
$\sigma_1\sigma_2^2\sigma_2\sigma_2$-Curvature Line, 208
$\sigma_1\sigma_2^2\sigma_2\sigma_2$-Gaussian Curvature, 209
$\sigma_1\sigma_2^2\sigma_2\sigma_2$-Mean Curvature, 209
$\sigma_1\sigma_2^2\sigma_2\sigma_2$-Minimal Surface, 324
$\sigma_1\sigma_2^2\sigma_2\sigma_2$-Principal Curvature, 208
$\sigma_1\sigma_2^2\sigma_2\sigma_2$-Principal Direction, 208
$\sigma_1\sigma_2^2\sigma_2\sigma_2$-Principal Line, 208
$\sigma_1\sigma_2^2\sigma_2\sigma_2$-Total Curvature, 209
$\sigma_1\sigma_3$-Exterior Differentiation, 361
$\sigma_1^2\sigma_2$-Angle, 141
$\sigma_1^2\sigma_2$-Area of a Surface, 147
$\sigma_1^2\sigma_2$-Asymptotic Direction, 185
$\sigma_1^2\sigma_2$-Asymptotic Line, 185
$\sigma_1^2\sigma_2$-Christoffel Coefficients, 262
$\sigma_1^2\sigma_2$-Curve, 185
$\sigma_1^2\sigma_2$-Darboux Formulae, 303
$\sigma_1^2\sigma_2$-Length of a Curve on a Surface, 138
$\sigma_1^2\sigma_2$-Normal Curvature, 177
$\sigma_1^2\sigma_2$-Normal to a Surface, 84
$\sigma_1^2\sigma_2$-Orientable Surface, 94
$\sigma_1^2\sigma_2$-Oriented Surface, 94
$\sigma_1^2\sigma_2$-Regular Parameterized Surface, 72
$\sigma_1^2\sigma_2$-Shape Operator, 118
$\sigma_1^2\sigma_2$-Spherical Map, 106
$\sigma_1^2\sigma_2$-Tangent Plane, 79
$\sigma_1^2\sigma_2$-Tangent Vector Space, 78
$\sigma_1^2\sigma_2$-Tangential Normal Vector, 303
$\sigma_1^2\sigma_2$-Weingarten Coefficients, 288
$\sigma_1^2\sigma_2\sigma_1\sigma_1$-Curvature Line, 199
$\sigma_1^2\sigma_2\sigma_1\sigma_1$-Gaussian Curvature, 200
$\sigma_1^2\sigma_2\sigma_1\sigma_1$-Mean Curvature, 200
$\sigma_1^2\sigma_2\sigma_1\sigma_1$-Minimal Surface, 322
$\sigma_1^2\sigma_2\sigma_1\sigma_1$-Principal Curvature, 199
$\sigma_1^2\sigma_2\sigma_1\sigma_1$-Principal Direction, 199
$\sigma_1^2\sigma_2\sigma_1\sigma_1$-Principal Line, 199
$\sigma_1^2\sigma_2\sigma_1\sigma_1$-Total Curvature, 200
$\sigma_1^2\sigma_2\sigma_1\sigma_2$-Curvature Line, 200
$\sigma_1^2\sigma_2\sigma_1\sigma_2$-Gaussian Curvature, 201
$\sigma_1^2\sigma_2\sigma_1\sigma_2$-Mean Curvature, 201
$\sigma_1^2\sigma_2\sigma_1\sigma_2$-Minimal Surface, 323
$\sigma_1^2\sigma_2\sigma_1\sigma_2$-Principal Curvature, 200
$\sigma_1^2\sigma_2\sigma_1\sigma_2$-Principal Direction, 200
$\sigma_1^2\sigma_2\sigma_1\sigma_2$-Principal Line, 200
$\sigma_1^2\sigma_2\sigma_1\sigma_2$-Total Curvature, 201
$\sigma_1^2\sigma_2\sigma_2\sigma_1$-Curvature Line, 201
$\sigma_1^2\sigma_2\sigma_2\sigma_1$-Gaussian Curvature, 203
$\sigma_1^2\sigma_2\sigma_2\sigma_1$-Mean Curvature, 203
$\sigma_1^2\sigma_2\sigma_2\sigma_1$-Minimal Surface, 323
$\sigma_1^2\sigma_2\sigma_2\sigma_1$-Principal Curvature, 201

$\sigma_1^2\sigma_2\sigma_2\sigma_1$-Principal Direction, 201
$\sigma_1^2\sigma_2\sigma_2\sigma_1$-Principal Line, 201
$\sigma_1^2\sigma_2\sigma_2\sigma_1$-Total Curvature, 203
$\sigma_1^2\sigma_2\sigma_2\sigma_2$-Curvature Line, 203
$\sigma_1^2\sigma_2\sigma_2\sigma_2$-Gaussian Curvature, 204
$\sigma_1^2\sigma_2\sigma_2\sigma_2$-Mean Curvature, 204
$\sigma_1^2\sigma_2\sigma_2\sigma_2$-Minimal Surface, 323
$\sigma_1^2\sigma_2\sigma_2\sigma_2$-Principal Curvature, 203
$\sigma_1^2\sigma_2\sigma_2\sigma_2$-Principal Direction, 203
$\sigma_1^2\sigma_2\sigma_2\sigma_2$-Principal Line, 203
$\sigma_1^2\sigma_2\sigma_2\sigma_2$-Total Curvature, 204
σ_2-Angle, 140
σ_2-Area of a Surface, 144
σ_2-Asymptotic Direction, 185
σ_2-Asymptotic Line, 185
σ_2-Christoffel Coefficients, 260
σ_2-Compatible Parameterization, 94
σ_2-Curve, 185
σ_2-Darboux Formulae, 302
σ_2-Differential of j-th kind, 102
σ_2-Gauss Equations, 296
σ_2-Godazzi-Mainardi Equations, 297
σ_2-Length of a Curve on a Surface, 133
σ_2-Normal Curvature, 167
σ_2-Normal to a Surface, 84
σ_2-Orientable Surface, 94
σ_2-Oriented Surface, 94
σ_2-Regular Parameterized Surface, 70
σ_2-Regular Surface, 73
σ_2-Simple Surface, 73
σ_2-Spherical Map, 105
σ_2-Tangent Plane, 79
σ_2-Tangent Vector Space, 78
σ_2-Tangential Normal Vector, 302
σ_2-Weingarten Coefficients, 276
$\sigma_2\mu_1\mu_2$-Gauss Equations, 300
$\sigma_2\mu_1\mu_2$-Godazzi-Mainardi Equations, 300
$\sigma_2\sigma_1$-Exterior Differentiation, 362
$\sigma_2\sigma_1\sigma_1$-Curvature Line, 193
$\sigma_2\sigma_1\sigma_1$-Gaussian Curvature, 194
$\sigma_2\sigma_1\sigma_1$-Mean Curvature, 194
$\sigma_2\sigma_1\sigma_1$-Minimal Surface, 321
$\sigma_2\sigma_1\sigma_1$-Principal Curvature, 193
$\sigma_2\sigma_1\sigma_1$-Principal Direction, 193
$\sigma_2\sigma_1\sigma_1$-Principal Line, 193
$\sigma_2\sigma_1\sigma_1$-Total Curvature, 194
$\sigma_2\sigma_1\sigma_2$-Curvature Line, 195
$\sigma_2\sigma_1\sigma_2$-Gaussian Curvature, 196
$\sigma_2\sigma_1\sigma_2$-Mean Curvature, 196
$\sigma_2\sigma_1\sigma_2$-Minimal Surface, 322
$\sigma_2\sigma_1\sigma_2$-Principal Curvature, 195
$\sigma_2\sigma_1\sigma_2$-Principal Direction, 195
$\sigma_2\sigma_1\sigma_2$-Principal Line, 195
$\sigma_2\sigma_1\sigma_2$-Total Curvature, 196
$\sigma_2\sigma_2\sigma_1$-Curvature Line, 196
$\sigma_2\sigma_2\sigma_1$-Gaussian Curvature, 197
$\sigma_2\sigma_2\sigma_1$-Mean Curvature, 197
$\sigma_2\sigma_2\sigma_1$-Minimal Surface, 322
$\sigma_2\sigma_2\sigma_1$-Principal Curvature, 196
$\sigma_2\sigma_2\sigma_1$-Principal Direction, 196
$\sigma_2\sigma_2\sigma_1$-Principal Line, 196
$\sigma_2\sigma_2\sigma_1$-Total Curvature, 197
$\sigma_2\sigma_2\sigma_2$-Curvature Line, 197
$\sigma_2\sigma_2\sigma_2$-Gaussian Curvature, 198
$\sigma_2\sigma_2\sigma_2$-Mean Curvature, 198
$\sigma_2\sigma_2\sigma_2$-Minimal Surface, 322
$\sigma_2\sigma_2\sigma_2$-Principal Curvature, 197
$\sigma_2\sigma_2\sigma_2$-Principal Direction, 197
$\sigma_2\sigma_2\sigma_2$-Principal Line, 197
$\sigma_2\sigma_2\sigma_2$-Total Curvature, 198
$\sigma_2\sigma_3$-Exterior Differentiation, 359
σ_3-Differential of j-th kind, 102
$\sigma_3\sigma_1$-Exterior Differentiation, 361
$\sigma_3\sigma_2$-Exterior Differention, 359
$\sigma_i\sigma_j$-Closed Dynamic Form, 365
$\sigma_i\sigma_j$-Exact Dynamic Form, 365
σ_k-Completely Delta Differentiable Vector Field, 342
$\sigma_{(1)}\sigma_{(2)}\sigma_{(3)}\sigma_1\sigma_2$-Map, 66
$\sigma_{(k)}$-Differential of j-th Kind, 105
$d_{\sigma_1\sigma_2^2}$, 306
d_{σ_1}, 305
d_{σ_2}, 305
e_1-Evolute, 46
e_1-Focal Curve, 46
e_1-Osculating Circle, 46
e_2-Evolute, 45
e_2-Focal Curve, 45
e_2-Osculating Circle, 45
k-Dynamic Form, 355
$k_{g\sigma_1\sigma_2^2}$, 307
$k_{g\sigma_1^2\sigma_2}$, 307
$k_{g\sigma_1}$, 306
$k_{g\sigma_2}$, 306
$k_{n\sigma_1\sigma_2^2}$, 180
$k_{n\sigma_1^2\sigma_2}$, 177

$k_{n\sigma_1}$, 158
$k_{n\sigma_2}$, 167
(ALP), 9

σ_1-Shape Operator, 106
σ_2-Shape Operator, 116

Addition of k-Dynamic Forms, 356
Admissible Functions, 312
Admissible Parametrizing Four, 3
Admissible Variation, 312
Arc Length Parameter, 6
Arc Length Property, 9

Basis of $\sigma_1\sigma_1\sigma_1$-Principal Directions, 188
Basis of $\sigma_1\sigma_1\sigma_2$-Principal Directions, 190
Basis of $\sigma_1\sigma_2\sigma_1$-Principal Directions, 191
Basis of $\sigma_1\sigma_2\sigma_2$-Principal Directions, 193
Basis of $\sigma_1\sigma_2^2\sigma_1\sigma_1$-Principal Directions, 205
Basis of $\sigma_1\sigma_2^2\sigma_1\sigma_2$-Principal Directions, 206
Basis of $\sigma_1\sigma_2^2\sigma_2\sigma_1$-Principal Directions, 207
Basis of $\sigma_1\sigma_2^2\sigma_2\sigma_2$-Principal Directions, 208
Basis of $\sigma_1^2\sigma_2\sigma_1\sigma_1$-Principal Directions, 199
Basis of $\sigma_1^2\sigma_2\sigma_1\sigma_2$-Principal Directions, 200
Basis of $\sigma_1^2\sigma_2\sigma_2\sigma_1$-Principal Directions, 202
Basis of $\sigma_1^2\sigma_2\sigma_2\sigma_2$-Principal Directions, 203
Basis of $\sigma_2\sigma_1\sigma_1$-Principal Directions, 194
Basis of $\sigma_2\sigma_1\sigma_2$-Principal Directions, 195
Basis of $\sigma_2\sigma_2\sigma_1$-Principal Directions, 196
Basis of $\sigma_2\sigma_2\sigma_2$-Principal Directions, 198
Basis of Tangent Space, 342
Bertrand Curves, 51
Bertrand Mates, 51
Binormal of a Space Curve, 43
Biregular Curve, 35

Chart, 77
Compatible Local Parameterization, 49
Compatible Parameterization, 94
Conjugate Bertrand Curves, 51
Curvature, 41
Curvature Vector, 41
Curve, 1
Curvilinear Coordinate System, 77

Delta Covariant Derivative, 343, 344
Diffeomorphism, 66, 99
Differentiable Map on a Surface, 97
Directional Delta Derivative, 327
Dubois-Reymond Lemma, 317

Equivalent Curves, 3
Euler Formula, 188, 190, 191, 193–195, 197–199, 201–203, 205–207, 209
Euler Necessary Condition, 317
Euler-Lagrange Equation, 317
Evolute, 48
Explicit Equation of a Plane Curve, 16
Exporession of a Map in Curvilinear Coordinates, 97

First Fundamental Form of a Surface, 123, 128–130
First Variation, 312
Focal Curve, 48
Frenet n-Frame, 42
Frenet Curve, 13

General Helix, 49
Geodesic σ_1-Torsion, 305
Geodesic $\sigma_1\sigma_2^2$-Torsion, 306
Geodesic σ_2-Torsion, 305
Geodesic σ_1-Curvature, 306
Geodesic $\sigma_1\sigma_2^2$-Curvature, 307
Geodesic $\sigma_1^2\sigma_2$-Curvature, 307
Geodesic σ_2-Curvature, 306
Global σ_2-Parameterization of a Surface, 73
Global Parameterization of a Surface, 73

Homeomorphism, 66
Homogeneous Part of a Rigid Motion, 55

Joachimstahl Theorem, 210, 212, 213

Lie Brackets, 352
Local σ_2-Parameterization of a Surface, 73
Local Parameterization of a Surface, 73

Meusnier Theorem, 158, 161, 163, 165, 167, 172, 174, 178, 179, 181–183

Nonparametric Form of a Plane Curve, 16
Normal Line, 23
Normal Plane, 23

Opposite σ_1-Orientation, 94
Opposite $\sigma_1\sigma_2^2$-Orientation, 94
Opposite $\sigma_1^2\sigma_2$-Orientation, 94
Opposite σ_2-Orientation, 94
Orientation of a Regular Curve, 49
Oriented Regular Curve, 49
Osculating Circle, 48
Osculating Plane, 38

Parametric Curve, 1
Parametric Equations of a Surface, 73
Parametric Representation of a Plane Curve, 15
Parameterization, 1
Principal Normal of a Space Curve, 43
Proper Weak Local Minimum, 312

Rectifying Plane, 43
Regular Parameterized Curve, 1
Regular Surface, 73
Rigid Motion, 55

Second σ_1-Fundamental Form of a Surface, 151
Second $\sigma_1\sigma_2^2$-Fundamental Form of a Surface, 156
Second $\sigma_1^2\sigma_2$-Fundamental Form of a Surface, 154
Second σ_2-Fundamental Form of a Surface, 153
Second Variation, 312
Simple Surface, 73
Smooth Map, 65, 99
Smooth Map on a Surface, 97
Substraction of k-Dynamic Forms, 356
Support of σ_1-Regular Parameterized Surface, 72
Support of σ_2-Regular Parameterized Surface, 72

Tangent Line, 22
Tangent Space, 342
Total Length, 6
Two-Dimensional Torus, 75

Vector Field, 342

Weak Local Minimum, 312

www.ingramcontent.com/pod-product-compliance
Lightning Source LLC
Chambersburg PA
CBHW060754220326
41598CB00022B/2429

* 9 7 8 1 0 3 2 0 7 0 8 0 3 *